0		000	301	477	602	699	778	845	903	954
1	000	041	079	114	146	176	204	230	255	279
2	301	322	342	362	380	398	415	431	447	462
3	477	491	505	519	531	544	556	568	580	591
4	602	613	623	633	643	653	663	672	681	690
5	699	708	716	724	732	740	748	756	763	771
6	778	785	792	799	806	813	820	826	833	839
7	845	851	857	863	869	875	881	886	892	898
8	903	908	914	919	924	929	934	940	944	949
9	954	959	964	968	973	978	982	987	991	996
10	000	004	009	013	017	021	025	029	033	037
11	041	045	049	053	057	061	064	068	072	076
12	079	083	086	090	093	097	100	104	107	111
13	114	117	121	124	127	130	134	137	140	143
14	146	149	152	155	158	161	164	167	170	173
15	176	179	182	185	188	190	193	196	199	201
16	204	207	210	212	215	217	220	223	225	228
17	230	233	236	238	241	243	246	248	250	253
18	255	258	260	262	265	267	270	272	274	276
19	279	281	283	286	288	290	292	294	297	299
20	301	303	305	308	310	312	314	316	318	320
21	322	324	326	328	330	332	334	336	338	340
22	342	344	346	348	350	352	354	356	358	360
23	362	364	365	367	369	371	373	375	377	378
24	380	382	384	386	387	389	391	393	394	396
25	398	400	401	403	405	407	408	410	412	413
26	415	417	418	420	422	423	425	427	428	430
27	431	433	435	436	438	439	441	442	444	446
28	447	449	450	452	453	455	456	458	459	461
29	462	464	465	467	468	470	471	473	474	476
30	477	479	480	481	483	484	486	487	489	490
31	491	493	494	496	497	498	500	501	502	504
32	505	507	508	509	511	512	513	515	516	517
33	519	520	521	522	524	525	526	528	529	530
34	531	533	534	535	537	538	539	540	542	543
35	544	545	547	548	549	550	551	553	554	555
36	556	558	559	560	561	562	563	565	566	567
37	568	569	571	572	573	574	575	576	577	579
38	580	581	582	583	584	585	587	588	589	590
39	591	592	593	594	596	597	598	599	600	601
40	602	603	604	605	606	607	609	610	611	612
41	613	614	615	616	617	618	619	620	621	622
42	623	624	625	626	627	628	629	630	631	632
43	633	634	635	636	637	638	639	640	641	642
44	643	644	645	646	647	648	649	650	651	652
45	653	654	655	656	657	658	659	660	661	662
46	663	664	665	666	667	667	668	669	670	671
47	672	673	674	675	676	677	678	679	679	680
48	681	682	683	684	685	686	687	688	688	689
49	690	691	692	693	694	695	695	696	697	698

50	699	700	701	702	702	703	704	705	706	70
51	708	708	709	710	711	712	713	713	714	715
52	716	[illegible]	[illegible]	[illegible]	[illegible]	[illegible]	[illegible]	[illegible]	723	723
53	[illegible]	[illegible]	[illegible]	[illegible]	[illegible]	[illegible]	[illegible]	[illegible]	731	732
54	[illegible]	[illegible]	[illegible]	[illegible]	[illegible]	[illegible]	[illegible]	[illegible]	739	740
55	[illegible]	[illegible]	[illegible]	[illegible]	744	744	745	746	747	747
56	748	749	750	751	751	752	753	754	754	755
57	756	757	757	758	759	760	760	761	762	763
58	763	764	765	766	766	767	768	769	769	770
59	771	772	772	773	774	775	775	776	777	777
60	778	779	780	780	781	782	782	783	784	785
61	785	786	787	787	788	789	790	790	791	792
62	792	793	794	794	795	796	797	797	798	799
63	799	800	801	801	802	803	803	804	805	806
64	806	807	808	808	809	810	810	811	812	812
65	813	814	814	815	816	816	817	818	818	819
66	820	820	821	822	822	823	823	824	825	825
67	826	827	827	828	829	829	830	831	831	832
68	833	833	834	834	835	836	836	837	838	838
69	839	839	840	841	841	842	843	843	844	844
70	845	846	846	847	848	848	849	849	850	851
71	851	852	852	853	854	854	855	856	856	857
72	857	858	859	859	860	860	861	862	862	863
73	863	864	865	865	866	866	867	867	868	869
74	869	870	870	871	872	872	873	873	874	874
75	875	876	876	877	877	878	879	879	880	880
76	881	881	882	883	883	884	884	885	885	886
77	886	887	888	888	889	889	890	890	891	892
78	892	893	893	894	894	895	895	896	897	897
79	898	898	899	899	900	900	901	901	902	903
80	903	904	904	905	905	906	906	907	907	908
81	908	909	910	910	911	911	912	912	913	913
82	914	914	915	915	916	916	917	918	918	919
83	919	920	920	921	921	922	922	923	923	924
84	924	925	925	926	926	927	927	928	928	929
85	929	930	930	931	931	932	932	933	933	934
86	934	935	936	936	937	937	938	938	939	939
87	940	940	941	941	942	942	942	943	943	944
88	944	945	945	946	946	947	947	948	948	949
89	949	950	950	951	951	952	952	953	953	954
90	954	955	955	956	956	957	957	958	958	959
91	959	960	960	960	961	961	962	962	963	963
92	964	964	965	965	966	966	967	967	968	968
93	968	969	969	970	970	971	971	972	972	973
94	973	974	974	975	975	975	976	976	977	977
95	978	978	979	979	980	980	980	981	981	982
96	982	983	983	984	984	985	985	985	986	986
97	987	987	988	988	989	989	989	990	990	99
98	991	992	992	993	993	993	994	994	995	99
99	996	996	997	997	997	998	998	999	999	000

BASIC MATHEMATICS FOR ELECTRONICS

fourth edition

NELSON M. COOKE
Late President
Cooke Engineering Company

HERBERT F. R. ADAMS
Chief Instructor
Electronics Division
British Columbia Vocational School—Burnaby

McGraw-Hill Book Company/Gregg Division
New York St. Louis Dallas San Francisco Auckland Düsseldorf
Johannesburg Kuala Lumpur London Mexico Montreal New Delhi
Panama Paris São Paulo Singapore Sydney Tokyo Toronto

Library of Congress Cataloging in Publication Data

Cooke, Nelson Magor.
Basic mathematics for electronics.

First ed. published in 1942 under title: Mathematics for electricians and radiomen.
Includes index.
1. Electric engineering—Mathematics. 2. Electronics—Mathematics. I. Adams, Herbert F. R., joint author. II. Title.
TK153.C63 1976 510'.2'46213 75-37798
ISBN 0-07-012512-0

BASIC MATHEMATICS FOR ELECTRONICS

5 6 7 8 9 0 V H V H 8 5 4 3 2 1 0

The editors for this book were Gerald O. Stoner and Alice V. Manning,
the designer was Charles A. Carson,
and the production supervisors were Patricia Ollague and Iris Levy.
It was set in Cairo by York Graphic Services, Inc.
It was printed and bound by Von Hoffman Press, Inc.

contents

Preface v

CHAPTERS

1 Introduction 1
2 Algebra—General Numbers 6
3 Algebra—Addition and Subtraction 13
4 Algebra—Multiplication and Division 24
5 Equations 39
6 Powers of 10 53
7 Units and Dimensions 67
8 Ohm's Law—Series Circuits 87
9 Resistance—Wire Sizes 107
10 Special Products and Factoring 117
11 Algebraic Fractions 135
12 Fractional Equations 153
13 Ohm's Law—Parallel Circuits 171
14 Meter Circuits 181
15 Divider Circuits and Wheatstone Bridges 192
16 Graphs 202
17 Simultaneous Equations 220
18 Determinants 234
19 Batteries 247
20 Exponents and Radicals 254
21 Quadratic Equations 271
22 Network Simplification 292
23 Angles 312
24 Trigonometric Functions 322
25 Trigonometric Tables 337
26 Solution of Right Triangles 350
27 Trigonometric Identities and Equations 360
28 Elementary Plane Vectors 369
29 Periodic Functions 380
30 Alternating Currents—Fundamental Ideas 392
31 Phasor Algebra 404

32 Alternating Currents—Series Circuits 412
33 Alternating Currents—Parallel Circuits 439
34 Logarithms 459
35 Applications of Logarithms 492
36 Number Systems for Computers 515
37 Boolean Algebra 525

TABLES

1 Mathematical Symbols 538
2 Letter Symbols 538
3 Abbreviations 539
4 Greek Alphabet 541
5 Copper Wire Tables 542
6 Conversion Factors 544
7 Common Logarithms 548
8 Common Antilogarithms 550
9 Natural Sines 552
10 Natural Cosines 554
11 Natural Tangents 556
12 Natural Logarithms 558
13 Exponential Functions 560
14 Decimal Multipliers 562
15 Rounded Values of Preferred Numbers 563
16 Degrees to Radians 564
17 Radians to Degrees 566
Answers to Odd-Numbered Problems 568

Index 611

preface

This is the fourth edition of the textbook originally entitled *Mathematics for Electricians and Radiomen* by the late Nelson M. Cooke. The text of the third edition was well settled, and the decisions as to the deletions from and additions to the second edition were all resolved prior to the unexpected and untimely death of Mr. Cooke. This new metric edition is his continuing monument, and we hope that its usefulness to the electronics technicians and technologists of the metric age will be equal to the value of the previous editions.

This book is designed to be used by students in the field of electronics both in schools and in private study. It should be used in conjunction with theoretical and practical studies in electronics. The various chapters which represent applications of the mathematical developments will fit various courses of study and may be stressed, omitted, repeated, or adjusted to accommodate the requirements of the course being followed by individuals or classes.

The material is offered in "block" form: algebra, trigonometry, logarithms, and computer mathematics, but teachers will find that the studies in one main subject may be interrupted to fit suitably in another, or the topics may be interleaved, so that, after the initial chapters, studies in algebra and trigonometry may proceed together. Everything possible has been done to promote individual flexibility. Some sections dealing with practical applications may be delayed until the appropriate theory or laboratory work has been covered.

As a result of extensive correspondence with teachers and students using the first, second, and third editions of the book, and on the basis of a nationwide survey conducted by the publisher, the consensus indicates it desirable to continue the deletion of the chapters on arithmetic review in order to make room for more pressing subjects, such as determinants, number systems, and Boolean algebra. We assume that a reasonable background of high school mathematics is part of the preparation of most students reaching the level of this book and that such a background will be brought to this study by most students interested in electronics.

For students who have been away from formal schooling for some years, *Arithmetic Review for Electronics* provides instruction on the deleted arithmetic topics in an inexpensive form. For students who do not need such an intensive review, the fourth edition of *Basic Mathematics for Electronics* will introduce a new supplement: Adams: *Study Guide to accompany Basic Mathematics for Electronics*, Fourth edition. This study guide provides general educational information on electronic calculators; review information on arithmetic skills; and step-by-step review, problem, and self-test material for each chapter of the text.

To achieve a greater concentration on low-power electronics circuitry, some sections dealing with electric distribution circuits, motors, and generators have been deleted from this edition, since electronics technicians who find it necessary to deal with these special "power" subjects will be able to quickly pick them up from power-oriented textbooks.

This new metric edition has been developed to aid and encourage students who have been educated to date in the "English" or "traditional" *foot-pound* units system.

We have found it desirable to deviate from the "pure" metric notation in one regard: In electronics, the *symbol* for current is I, the *abbreviation* for the unit of measure, the ampere, is A. The metric purist has learned to refer to A as the *symbol* for ampere. We hope the deviation will be understood and accepted by the users of the book as being more meaningful in the context of circuit analysis.

Teachers and students alike must learn to think and solve problems in the new metric language. For that essential reason, very little is offered in this first metric edition by way of conversions from one system to the other. Fortunately, the electrical units of our studies in electronics *are* the SI metric units. We have only to learn to think in the metric units of length and weight (mass) to become proficient in metric calculations. One area of study where conversions will continue to be important is in the adjustment of older handbook formulas which use the traditional system of units—a number of exercises involving such conversions is offered in this edition.

Many people have shared in the effort to make this a useful and valuable text: teachers, resident and home-study students, and practicing engineers and technicians. All of them have our gratitude and the satisfaction of knowing that they have contributed to the improvement of the original text. Students at the British Columbia Vocational School—Burnaby helped to polish the wording of the problems and to check the accuracy of the answers. The helpful comments to the third edition of a McGraw-Hill author, Russell Heiserman, also were greatly appreciated.

In addition, many friends and colleagues have advised and encouraged me in this revision. Special thanks must be given to Fred Bailey and Peter Dell, British Columbia Vocational School—Burnaby, and to Reg Ridsdale, Head of the Department of Electricity and Electronics, British Columbia Institute of Technology.

As usual, comments and criticisms are always welcome. The reviser has always been critical of textbook errors, but he has learned how difficult it is to avoid them. It is requested that comments be addressed to him in care of the publisher.

HERBERT F. R. ADAMS

1

introduction

In the legions of textbooks on the subject of mathematics, all the basic principles contained here have been expounded in admirable fashion. However, students of electricity, radio, and electronics have need for a course in mathematics that is directly concerned with application to electric and electronic circuits. This book is intended to provide those students with a sound mathematical background as well as further their understanding of basic circuitry.

1-1 MATHEMATICS—A LANGUAGE

The study of mathematics may be likened to the study of a language. In fact, mathematics is a language, the language of number and size. Just as the rules of grammar must be studied in order to master English, so must certain concepts, definitions, rules, terms, and words be learned in the pursuit of mathematical knowledge. These form the vocabulary or structure of the language. The more a language is studied and used, the greater becomes the vocabulary; the more mathematics is studied and applied, the greater its usefulness becomes.

There is one marked difference, however, between the study of a language and the study of mathematics. A language is based on words, phrases, expressions, and usages that have been brought together through the ages in more or less haphazard fashion according to the customs of the times. Mathematics is built upon the firm foundation of sound logic and orderly reasoning and progresses smoothly, step by step, from the simplest numerical processes to the most complicated and advanced applications, each step along the way resting squarely upon those which have been taken before. This makes mathematics the fascinating subject that it is.

1-2 MATHEMATICS—A TOOL

As the builder works with a square and compasses, so the engineer employs mathematics. A thorough grounding in this subject is essential to proficiency in any of the numerous branches of engineering. In no other branch is this more apparent than in the study of electrical and electronic subjects, for most of our basic ideas of electrical phenomena are based upon mathematical reasoning and stated in mathematical terms. This is a fortunate circumstance, for it enables us to build a structure of electrical knowledge with precision, assembling and expressing the components in clear and concise mathematical terms and arranging the whole in logical order. Without mathematical assistance, technicians must be content with the long and painful process of accumulating bits of information, details of experience, etc., and they may never achieve a thorough understanding of the field in which they live and work.

1-3 MATHEMATICS—A TEACHER

In addition to laying a foundation for technical knowledge and assisting in the practical application of knowledge already possessed, mathematics offers unlimited advantages in respect to mental training. The solution of a problem, no matter how simple, demands logical thinking for it to be possible to state the facts of the problem in mathematical terms and then proceed with the solution. Continued study in this orderly manner will increase your mental capacity and enable you to solve more difficult problems, understand more complicated engineering principles, and cope more successfully with the everyday problems of life.

1-4 METHODS OF STUDY

Before beginning detailed study of this text, you should carefully analyze it, in its entirety, in order to form a mental outline of its content, scope, and arrangement. You should make another preliminary survey of each individual chapter before attempting detailed study of the subject matter. After the detailed study, you should work problems until all principles are fixed firmly in your mind before proceeding to new material.

In working problems, the same general procedure is recommended. First, analyze a problem in order to determine the best method of solution. Then state the problem in mathematical terms by utilizing the principles that are applicable. If you make but little progress, it is probable you have not completely mastered the principles explained in the text, and a review is in order.

The authors are firm believers in the use of a workbook, preferably in the form of a loose-leaf notebook, which contains all the problems you have worked, together with the numerous notes made while studying the text. Such a book is an invaluable aid for purposes of review. The habit of jotting down notes during reading or studying should be cultivated. Such notes in your own words will provide a better understanding of a concept.

1-5 RATE OF PROGRESS

Home-study students should guard against too rapid progress. There is a tendency, especially in studying a chapter whose contents are familiar or easy to comprehend, to hurry on to the next chapter. Hasty reading may cause the loss of the meaning that a particular section or paragraph is intended to convey. Proficiency in mathematics depends upon thorough understanding of each step as it is encountered so that it can be used to master the one which follows.

1-6 IMPORTANCE OF PROBLEMS

Full advantage should be taken of the many problems distributed throughout the text. There is no approach to a full and complete understanding of any branch of mathematics other than the solution of numerous problems. Application of what has been learned from the text to practical problems in which you are primarily interested will not only help with the subject matter of the problem but also serve the purpose of fixing in mind the mathematical principles involved.

In general, the arrangement of problems is such that the most difficult appear at the end of each group. It is apparent that the working of the simpler problems first will tend to make the more difficult ones easier to solve. The home-study student is, therefore, urged to work all problems in the order given. At times, this may appear to be useless, and you may have the desire to proceed to more interesting things, but time spent in working problems will amply repay you in giving you a depth of understanding to be obtained in no other manner. This does not mean that progress should cease if a particular problem appears to be impossible to solve. Return to such problems when your mind is fresh, or mark them for solution during a review period.

1-7 ILLUSTRATIVE EXAMPLES

Each of the illustrative examples in this book is intended to make clear some important principle or method of solution. The subject matter of these examples will be more thoroughly assimilated if, after careful analysis of the problem set forth, you make an independent solution and compare the method and results with the illustrative example.

1-8 REVIEW

Too much stress cannot be placed upon the necessity for frequent and thorough review. Points that have been missed in the original study of the text will often stand out clearly upon careful review. A review of each chapter before proceeding with the next is recommended.

1-9 SECTION REFERENCES

Throughout this book you will be referred to earlier sections for review or to bring to attention similar material pertaining to the subject under discussion. For the purpose of ready reference and convenience, at the bottom of each right-hand page, two sets of numbers are listed: the first number refers to the *first* text or problem section on the left-hand page, and the second number refers to the *last* text or problem section on the right-hand page. Thus, wherever you open the book, these numbers show the section (or sections) covered on the pages in view. For example, Sec. 4-10 is easily found on page 33 by leafing through the book while noting the inclusive numbers.

1-10 ABBREVIATIONS

Every profession, every technology has its own jargon—the particular words and phrases which describe the phenomena with which it deals. Electronics is particularly noteworthy in this respect, with inductance, capacitance, resistance, impedance, and frequency leading a host of others. Each phenomenon must be measured and described in understandable units so that other workers in the field will be able to understand exactly what is involved. After establishing such a vocabulary and list of units, the next logical development is a system of abbreviations—shorthand symbols which everyone will recognize as standing for the units and dimensions of the technology. For many years there was no single agreed-upon list of electronics abbreviations, and most of us had to be able to recognize several variations as acceptable abbreviations of the same term. For instance, A, a, amp, Amp, amps, and Amps were all used to represent *amperes*, depending upon the teacher, the author, and the publisher involved.

Even today, the exhaustive list of standard abbreviations recommended by the Institute of Electrical and Electronics Engineers is not wholly acceptable to all branches of the industry, and local variations and established forms continue to be used. Some publishers are still reluctant to use the single-letter abbreviations for fear of introducing ambiguities. Some of us were reluctant to adopt Hertz (Hz) in place of cycles per second (c/s or c/sec or cps), not so much that we did not honor Hertz as that the name does not make obvious the "per time" relationship involved in frequency. However, the name is now being used widely, and Hz is used in this edition as a reflection of what you may expect when you step out of school and into industry.

One drawback to all this is that although we have, at the publisher's request, attempted to be uniform in the matter of abbreviations, you will nevertheless meet, and must be able to deal with, several variations for many years to come. However, in an attempt to keep before you the dimensional aspect of units, we have used rev/min rather than rpm, Ω/m rather than Ω/M, and so on. You should study the tables of symbols and abbreviations carefully and repeatedly so that you achieve an early and complete mastery of abbreviations. It has been our aim to supply the full name the first time the term is used, follow it immediately with the abbreviation in parentheses, and then use the abbreviation at every opportunity thereafter.

1-11 SIGNIFICANT FIGURES

The resistors, capacitors, and other devices used in electronic circuitry are often manufactured to convenient tolerances: 5%, 10%, and 20% being the most common. Accordingly, it is meaningless to calculate a resistance value to many decimal places, or to many "significant figures," when the circuit is to be constructed with a standard off-the-shelf resistor made to, say, ±10% accuracy. (Obviously a shunt to be made by hand may well be accurate to $\frac{1}{2}$%, and then this argument would not apply.)

A ten-inch slide rule can be relied upon to give a satisfactory answer (three significant figures) to most of the problems at the level of study in this text. Answers computed by logarithms or by long multiplication or division will disagree with slide rule answers and with each other if they are taken to enough decimal places.

There are occasions, of course, when three significant figures may not be sufficient: accountants and auditors will want your financial calculations to be correct to the nearest cent, even when thousands of dollars are involved; the FCC will not be satisfied with a carrier frequency correct to only three significant figures; logarithms and trigonometric functions are given to four places, or five, or ten, and the answers achieved will reflect the accuracy of the tables used; angles greater than 90° must be converted into equivalents less than 90° for purposes of calculations, and should not be rounded off prior to conversion. All the answers in this text reflect these notions, and you are accordingly encouraged to start using a good slide rule early in your career. (See Chap. 6 before purchasing a slide rule.)

1-12 METRICATION

Since the publication of the third edition of this book, both Canada and the United States have taken lengthy strides toward the adoption of the metric system of units. The metric system has undergone many refinements during its lifetime, from CGS through MKS to MKSA. Now, in the most logical refinement of all, the International System of Metric Units, known universally as SI (for Système International des Unités), is being phased into the North American measurement scene. Since the SI metric units of electricity are those with which most of the users of this book are already familiar (amperes, volts, ohms, hertz, etc.), it only remains for us to learn to think in terms of meters for length and kilograms for weight (mass) to become proficient in the basic electronics requirements of metric units.

The SI units which you must know in order to study this textbook are introduced in Chap. 7, which has been completely rewritten for this fourth edition.

2

algebra—general numbers

In general, arithmetic consists of the operations of addition, subtraction, multiplication, and division of a type of numbers represented by the digits 0, 1, 2, 3, . . . , 9. By using the above operations or combinations of them, we are able to solve many problems. However, a knowledge of mathematics limited to arithmetic is inadequate and a severe handicap to anyone interested in acquiring an understanding of electric circuits. Proficiency in performing even the most simple operations of algebra enables you to solve problems and determine relations that would be impossible with arithmetic alone.

2-1 THE GENERAL NUMBER

Algebra may be thought of as a continuation of arithmetic in which letters and symbols are used to represent definite quantities whose actual values may or may not be known. For example, in electrical and radio texts, it is customary to represent currents by the letters I or i; voltages by E, e, V, or v; resistances by R or r; etc. The base of a triangle is often represented by b, and the altitude may be specified as a. Such letters or symbols used for representing quantities in a general way are known as *general numbers* or *literal numbers*.

The importance of the general-number idea cannot be overemphasized. Although it is possible to express the various laws and facts concerning electricity in English, they are more concisely and compactly expressed in mathematical form in terms of general numbers. As an example, Ohm's law states, in part, that the current in a certain part of a circuit is proportional to the potential difference (voltage) across that part of the circuit and inversely proportional to the resistance of that part. This same statement, in mathematical terms, says

$$I = \frac{E}{R}$$

where I represents the current, E is the potential difference, and R is the resistance. Such an expression is known as a *formula*.

Although expressing various laws and relationships of science as formulas gives us a more compact form of notation, that is not the real value of the formula. As you attain proficiency in algebra, the value of general formulas will become more apparent. Our studies of algebra will consist mainly in learning how to add, subtract, multiply, divide, and solve general algebraic expressions, or formulas, in order to attain a better understanding of the fundamentals of electricity and related fields.

2-2 SIGNS OF OPERATION

In algebra the signs of operation $+$, $-$, $\times$, and $\div$ have the same meanings as in arithmetic. The sign $\times$ is generally omitted between literal numbers. For example, $I \times R$ is written IR and means that I is to be multiplied by R. Similarly, $2\pi fL$ means 2 times π times f times L. Sometimes the symbol $\cdot$ is used to denote multiplication. Thus $I \times R$, $I \cdot R$, and IR all mean I times R.

2-3 THE ORDER OF SIGNS OF OPERATION

In performing a series of different operations, we will follow convention and perform the multiplications first, next the divisions, and then the additions and subtractions. Thus,

$$\begin{aligned} 16 \div 4 + 8 + 4 \times 5 - 3 &= 4 + 8 + 20 - 3 \\ &= 29 \end{aligned}$$

2-4 ALGEBRAIC EXPRESSIONS

An *algebraic expression* is one that expresses or represents a number by the signs and symbols of algebra. A *numerical algebraic expression* is one consisting entirely of signs and numerals. A *literal algebraic expression* is one containing general numbers or letters. An example of a numerical algebraic expression is $8 - (6 + 2)$, and I^2R is a literal algebraic expression.

2-5 THE PRODUCT

As in arithmetic, a *product* is the result obtained by multiplying two or more numbers. Thus, 12 is the product of 6×2.

2-6 THE FACTOR

If two or more numbers are multiplied together, each of them or the product of any combination of them is called a *factor* of the product. For example, in the product $2xy$, 2, x, y, $2x$, $2y$, and xy are all factors of $2xy$.

2-7 COEFFICIENTS

Any factor of a product is known as the *coefficient* of the product of the remaining factors. In the foregoing example, 2 is the coefficient of xy, x is the coefficient of $2y$, y is the coefficient of $2x$, etc. It is common practice to speak of the numerical part of an expression as the *coefficient* or as the *numerical coefficient*. If an expression contains no numerical coefficient, 1 is understood to be the numerical coefficient. Thus, $1abc$ is the same as abc.

2-8 PRIMES AND SUBSCRIPTS

When, for example, two resistances are being compared in a formula or it is desirable to make a distinction between them, the resistances may be represented by R_1 and R_2 or R_a and R_b. The small numbers or letters written at the right of and below the R's are called *subscripts*. They are generally used to denote different values of the same units. R_1 and R_2 are read "R sub one" and "R sub two" or simply "R one" and "R two."

Care must be used in distinguishing between subscripts and exponents. Thus E^2 is an indicated operation that means $E \cdot E$, whereas E_2 is used to distinguish one quantity from another of the same kind.

Primes and *seconds*, instead of subscripts, are often used to denote quantities. Thus one current might be denoted by I' and another by I''. The first is read "I prime" and the latter is read "I second." I' resembles I^1 (I to the first power), but in general this causes little confusion.

2-9 EVALUATION

To *evaluate* an algebraic expression is to find its numerical value. In Sec. 2-1, it was stated that in algebra certain signs and symbols are used to represent definite quantities. Also, in Sec. 2-4, an algebraic expression was defined as one that represents a number by the signs and symbols of algebra. We can find the numerical, or definite, value of an algebraic expression only when we know the values of the letters in the expression.

example 1 Find the value of $2ir$ if $i = 5$ and $r = 11$.

solution $2ir = 2 \times 5 \times 11 = 110$

example 2 Evaluate the expression $23E - 3ir$ if $E = 10$, $i = 3$, and $r = 22$.

solution

$$\begin{aligned} 23E - 3ir &= 23 \times 10 - 3 \times 3 \times 22 \\ &= 230 - 198 = 32 \end{aligned}$$

example 3 Find the value of $\frac{E}{R} - 3I$ if $E = 230$, $R = 5$, and $I = 8$.

solution

$$\frac{E}{R} - 3I = \frac{230}{5} - 3 \times 8 = 46 - 24 = 22$$

PROBLEMS 2-1

note The accuracy of answers to numerical computations is, in general, that obtained with a ten-inch slide rule.

1. (*a*) What does the expression $(25)(R)$ mean?
   (*b*) What is the meaning of $6 \cdot r$?
   (*c*) What does $0.25I$ mean?
2. What is the value of:
   (*a*) $5i$ when $i = 7$ amperes (A)?
   (*b*) $4Z$ when $Z = 16$ ohms (Ω)?
   (*c*) $16V$ when $V = 110$ volts (V)?
3. One electrolytic capacitor costs \$2.75.
   (*a*) What will one gross of capacitors cost?
   (*b*) What will n capacitors cost?
4. One dozen resistors cost a total of \$2.04.
   (*a*) What is the cost of each resistor?
   (*b*) What is the cost of p resistors?
5. The current in a certain circuit is $25I$ A. What is the current if it is reduced to one-half its original value?
6. There are three resistances, of which the second is twice the first and one-sixth the third. If R represents the first resistance, what expressions describe the other two?
7. There are four capacitances, of which the second is two-thirds the first, the third is six times the second, and the fourth is twelve times the third. If C represents the first capacitance, in picofarads (pF), what expressions describe the other three?

8. If $P = 3$, $X = 5$, and $\psi = 12$, evaluate:

(a) $P + \psi$ (b) $\psi + X - P$ (c) $\dfrac{\psi}{X}$ (d) $\dfrac{X - P}{\psi}$ (e) $\dfrac{P + \psi}{X}$

9. Write the expression which will represent each of the following:
(a) A resistance which is $R\ \Omega$ greater than $16\ \Omega$.
(b) A voltage which is 220 V more than e V.
(c) A current which is I A less than i A.

10. A circuit has a resistance of $125\ \Omega$. Express a resistance which is $R\ \Omega$ less than six times this resistance.

11. An inductance L_1 exceeds another inductance L_2 by 125 millihenrys (mH). Express the inductance L_2 in terms of L_1.

12. When two capacitors C_1 and C_2 are connected in series, the resultant capacitance C_s of the combination is expressed by the formula

$$C_s = \frac{C_1 C_2}{C_1 + C_2}$$

What is the resultant capacitance if:
(a) 5 pF is connected in series with 15 pF?
(b) 150 pF is connected in series with 475 pF?

13. The current in any part of a circuit is given by $I = \dfrac{E}{R}$, in which I is the current in amperes through that part, E is the electromotive force (emf) in volts across that part, and R is the resistance in ohms of that part. What will be the current through a circuit with:
(a) An emf of 220 V and a resistance of $5\ \Omega$?
(b) An emf of 50 V and a resistance of $200\ \Omega$?

14. The time interval between the transmission of a radar pulse and the reception of its echo off a target is $t = \dfrac{2R}{c}$ seconds (s), where t is the time interval in seconds, R is the range in kilometers (km), and c is the speed of light, at which radio waves travel. [c = 300 000 kilometers per second (km/s).] What is the time between the transmission of a pulse and the reception of its echo from a target at a distance of 124 km?

15. The relation $t = \dfrac{2R}{c}$ in Prob. 14 is applicable to the transmission of sound in air and water. Owing to slower speeds of transmission, R is usually expressed in meters (m), and c is expressed in meters per second (m/s). (In air, $c \simeq 335$ m/s, and in salt water, $c \simeq 1460$ m/s. The sign $\simeq$ means "is approximately equal to.")
(a) What is the time between the transmission of a short pulse of sound through air and the reception of its echo at a distance of 500 m?

(*b*) What time will elapse if the sound pulse is transmitted through seawater at the same distance?

16. The relationship between the wavelength λ of a wave, the frequency f in hertz (Hz, or cycles per second), and the speed c at which the wave is propagated is $\lambda = \frac{c}{f}$. If λ is expressed in meters (m), then c must be expressed in meters per second (m/s); that is, λ and c must be expressed in the same units of length.

(*a*) What is the wavelength in kilometers of a radio wave having a frequency of 980 kilohertz (kHz) (980 kHz = 980 000 Hz)?

(*b*) What is the wavelength in meters of a radio wave having a frequency of 121.5 megahertz (MHz) (121.5 MHz = 121 500 000 Hz; c = 300 000 km/s = 300 000 000 m/s)?

17. The distance between a dipole antenna and its reflector is usually one-fifth of a wavelength. What will be this spacing for a signal at 205 MHz in meters?

2-10 EXPONENTS

To express "x is to be taken as a factor four times," we could write $xxxx$, but the general agreement is to write x^4 instead.

An *exponent*, or *power*, is a number written at the right of and above a second number to indicate how many times the second number is to be taken as a factor. The number to be multiplied by itself is called the *base*.

Thus, I^2 is read "I square" or "I second power" and means that I is to be taken twice as a factor; e^3 is read "e cube" or "e third power" and means that e is to be taken as a factor three times. Likewise, 5^4 is read "5 fourth power" and means that 5 is to be taken as a factor four times; thus,

$$5^4 = 5 \times 5 \times 5 \times 5 = 625$$

When no exponent, or power, is indicated, the exponent is understood to be 1. Thus, x is the same as x^1.

2-11 THE RADICAL SIGN

The radical sign $\sqrt{\ }$ has the same meaning in algebra as in arithmetic; $\sqrt{e}$ means the square root of e, $\sqrt[3]{x}$ means the cube root of x, $\sqrt[4]{i}$ means the fourth root of i, etc. The small number in the angle of a radical sign, like the 4 in $\sqrt[4]{i}$, is known as the *index* of the root.

2-12 TERMS

A *term* is an expression containing literal and/or numerical parts which are not separated by plus or minus signs. Terms may be parts of larger expressions in which the terms are separated by plus or minus signs. $3E^2$, IR, and $-2e$ are all terms of the expression $3E^2 + IR - 2e$.

Although the value of a term depends upon the values of the literal factors of the term, it is customary to refer to a term whose sign is plus as a *positive term*. Likewise, we refer to a term whose sign is minus as a *negative term*.

Terms having the same literal parts are called *like terms* or *similar terms*. $2a^2bx$, $-a^2bx$, $18a^2bx$, and $-4a^2bx$ are like terms.

Terms that are not alike in their literal parts are called *unlike terms* or *dissimilar terms*. $5xy$, $6ac$, $9I^2R$, and EI are *unlike terms*.

An algebraic expression consisting of but one term is known as a *monomial*.

A *polynomial*, or *multinomial*, is an algebraic expression consisting of two or more terms.

A *binomial* is a polynomial of two terms. $e + ir$, $a - 2b$, and $2x^2y + xyz^2$ are binomials.

A *trinomial* is a polynomial consisting of three terms. $2a + 3b - c$, $IR + 3e - E^2$, and $8ab^3c + 3d + 2xy$ are trinomials.

PROBLEMS 2-2

1. If $a = 3$, $b = 6$, and $c = 2$, evaluate the following:

(a) $2abc$ (b) $5a^2b + 3c$ (c) $a^2b^2c^2$

(d) $12ac^2 - 2b^2$ (e) $\sqrt{4a^2b^2}$ (f) $5\sqrt{9b^2c^2} + 3a^2$

2. If $E = 110$, $I = 6$, and $R = 25$, evaluate the following:

(a) $5EI$ (b) EI^2R (c) $I^2R + \dfrac{12E^2}{R}$

(d) $\dfrac{25I^3R^2}{6IR} - \sqrt{\dfrac{100E^2}{R}}$ (e) $\dfrac{36E^2IR}{I^3R} - 3R^2$

3. State which of the following are monomials, binomials, and trinomials:

(a) $\dfrac{E}{I}$ (b) I^2R (c) $2\pi fL$

(d) $a + jb$ (e) $\Phi + \theta + 90°$ (f) $E_s - E_g$

(g) $I + \sqrt{\dfrac{P}{R}} + \dfrac{E}{R}$ (h) $a^2 + 2ab + b^2$ (i) $\dfrac{5(\mu E_g)^2}{16r_p}$

(j) $\dfrac{1}{R_1} + \dfrac{1}{R_2} + \dfrac{1}{R_3}$

4. In Probs. 1, 2, and 3, state which expressions are polynomials.

5. Write the following statements in algebraic symbols:

(a) I is equal to E divided by R.

(b) E is equal to I times R.

(c) P is equal to R times the square of I.

(d) R_1 is equal to the sum of R_2 and R_3.

(e) K is equal to M divided by the square root of the product of L_1 and L_2.

(f) R_p is equal to the product of R_1 and R_2 divided by their sum.

(g) The meter multiplier N is equal to the meter resistance R_m divided by the shunt resistance R_s all plus 1.

6. The approximate inductance of a single-layer air-core coil, such as used in the tuning circuits of radio receivers, can be calculated by the formula

$$L = \frac{2.54r^2n^2}{9r + 10l} \qquad \text{microhenrys } (\mu\text{H})$$

where L = inductance, μH
r = radius of winding, centimeters (cm)
n = number of turns of wire in winding
l = length of coil, cm

What is the inductance of a coil that is 3 cm in diameter and 10 cm long and has 150 turns of wire?

7. The winding in Prob. 6 is removed from the coil form, and smaller wire is substituted, so that, in the same length of coil, the number of turns is tripled. What is the inductance?

8. The power in any part of an electric circuit is given by the formula

$$P = I^2R \qquad \text{watts (W)}$$

where P = power, W
I = current, A
R = resistance, Ω

Find the power expended when:
(*a*) The current is 0.25 A and the resistance is 10 000 Ω.
(*b*) The current is 30 A and the resistance is 0.5 Ω.

9. In Prob. 8, if the resistance is kept constant, what happens to the power if the current is (*a*) doubled, (*b*) tripled, (*c*) halved?

10. The power in any part of an electric circuit is also given by the formula

$$P = \frac{E^2}{R} \qquad \text{W}$$

where P = power expended, W
E = electromotive force, V
R = resistance, Ω

What happens to the power if:
(*a*) The voltage is doubled and the resistance is unchanged?
(*b*) The voltage is halved and the resistance is unchanged?
(*c*) The resistance is doubled and the voltage is unchanged?
(*d*) The resistance is halved and the voltage is unchanged?

3

algebra—addition and subtraction

The problems of arithmetic deal with positive numbers only. A *positive number* may be defined as any number greater than zero. Accepting this definition, we know that when such numbers are added, multiplied, and divided, the results are always positive. Such is the case in subtraction if a number is subtracted from a larger one. However, if we attempt to subtract a number from a smaller one, arithmetic furnishes us with neither a rule for carrying out this operation nor a meaning for the result.

3-1 NEGATIVE NUMBERS

Limiting our knowledge of mathematics to positive numbers would place us under a severe handicap, for there are many instances when it becomes necessary to deal with numbers that are called negative. Often, a negative number is defined as a number less than zero. Numerous examples of the uses of negative numbers could be cited. For example, zero degrees on the Celsius thermometer has been chosen as the temperature of melting ice—commonly referred to as freezing temperature. Now, everyone knows that in some climates it gets much colder than "freezing." Such temperatures are referred to as so many "degrees below zero." How shall we state, in the language of mathematics, a temperature of "10 degrees below zero"? Ten degrees above zero would be written $10°$. Because $0°$ is the reference point, it is logical to assume that $10°$ below zero would be written as $-10°$, which, for our purposes, makes it a negative number.

Therefore, we see that a definition making a negative number less than zero is not completely correct. A negative number is some quantity away from a reference point in one direction (the defined negative direction), whereas the same positive quantity is simply the same quantity in the opposite direction (the defined positive direction).

Negative numbers are prefixed with the minus sign. Thus, negative 2 is written -2, negative $3ac$ is written $-3ac$, etc. If no sign precedes it, a number is assumed to be positive.

3-2 PRACTICAL NEED FOR NEGATIVE NUMBERS

The need for negative numbers often arises in the consideration of voltages or currents in electric and electronic circuits. It is com-

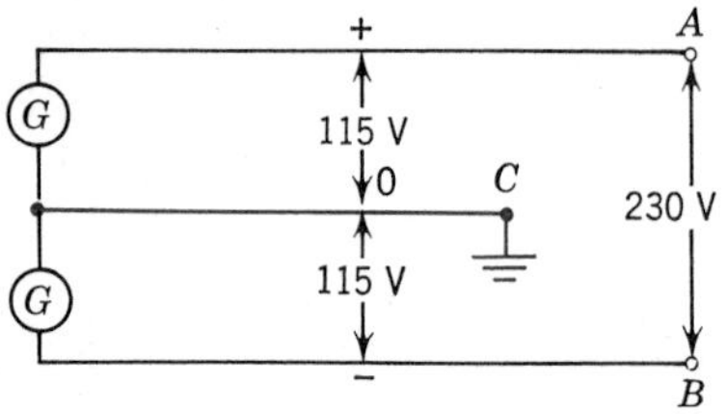

Fig. 3-1 Two 115-V Generators Connected in Series with Neutral Wire Grounded

mon practice to select the ground, or earth, as a point of zero potential. This does not mean, however, that there can be no potentials below ground, or zero, potential. Consider the case of the three wire feeders connected as shown in Fig. 3-1.

The generators G, which maintain a voltage of 115 V each, are connected in series so that their voltages add to give a voltage of 230 V across points A and B, and the neutral wire is grounded at C. Since C is at ground, or zero, potential, point A is 115 V positive with respect to C and point B is 115 V negative with respect to C. Therefore, the voltage at A with respect to ground, or zero, potential could be denoted as 115 V and the voltage at B with respect to ground could be denoted as -115 V.

Similar conditions exist in vacuum-tube circuits, as illustrated by the schematic circuit diagram of a type 6C5 vacuum tube in Fig. 3-2. The plate current indicated by the arrow flows through the cathode resistor R and creates a difference of potential of 8 V across R, so that point A is $+8$ V with respect to ground. Since the grid G is connected directly to ground, the grid is -8 V with respect to the cathode K.

Fig. 3-2 The Grid G Is Negative with Respect to Cathode K

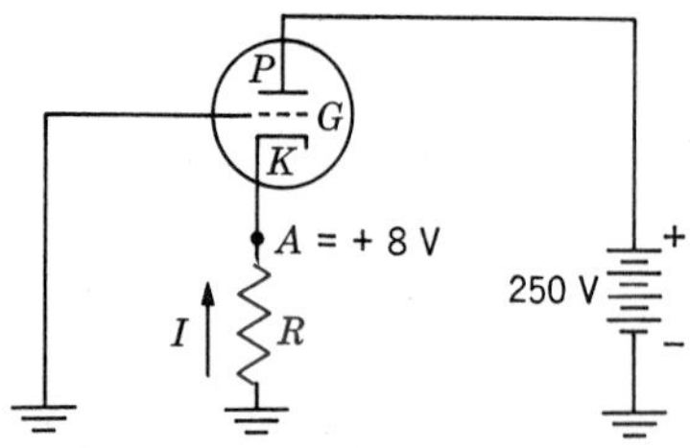

3-3 THE MATHEMATICAL NEED FOR NEGATIVE NUMBERS

From a purely mathematical viewpoint the need for negative numbers can be seen from the following succession of subtractions in which we subtract successively larger numbers from 5:

$$\begin{array}{ccccc} 5 & 5 & 5 & 5 & 5 \\ 0 & 1 & 2 & 3 & 4 \\ \hline 5 & 4 & 3 & 2 & 1 \end{array}$$

$$\begin{array}{cccc} 5 & 5 & 5 & 5 \\ 5 & 6 & 7 & 8 \\ \hline 0 & -1 & -2 & -3 \end{array}$$

The above subtractions result in the remainders becoming less until zero is reached. When the remainder becomes less than zero, the fact is indicated by placing the negative sign before the remainder. This is one reason for defining a negative number as a number less than zero. Mathematically, the definition is correct if we consider only the signs that precede the numbers.

You must not lose sight of the fact, however, that as far as magnitude, or size, is concerned, a negative number may represent a larger absolute value than some positive number. *The positive and negative signs simply denote reference from zero.* For example, if some point in an electric circuit is 1000 V negative with respect to ground, you can say so by writing -1000 V. But if you make good contact with your body between that point and ground, your chances of being electrocuted are just as good as if that point were positive 1000 V with respect to ground—and you wrote it $+1000$ V! In this

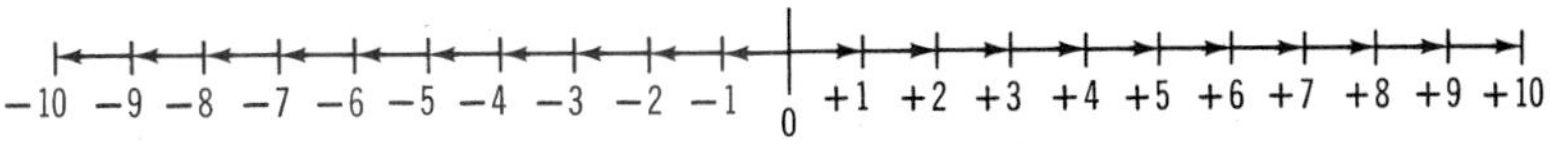

Fig. 3-3 Graphical Representation of Numbers from −10 to +10

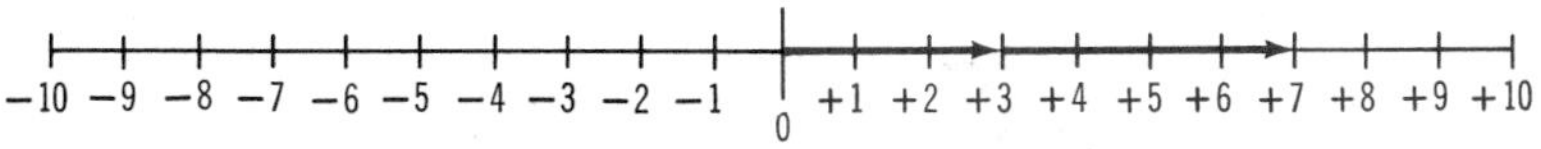

Fig. 3-4 Graphical Addition of 3 and 4 to Obtain 7

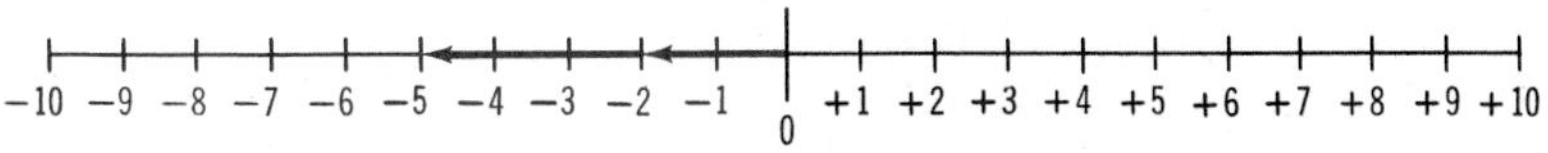

Fig. 3-5 Illustrating the Addition of −2 and −3. The Result Is −5

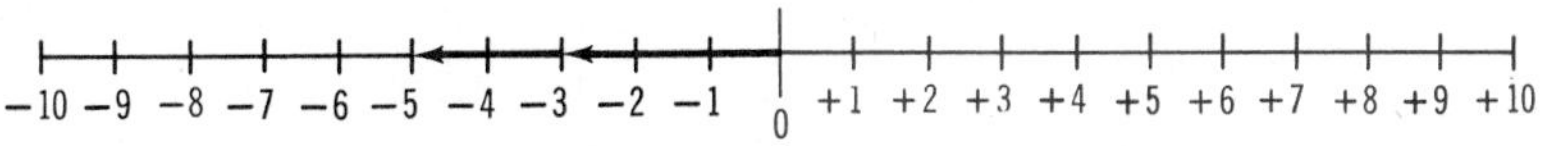

Fig. 3-6 Adding −3 and −2 Is the Same as Adding −2 and −3; Each Result Is −5

case, *how much* is far more important than a matter of sign preceding the number. Similarly, −1000 V is greater than +500 V, but of different polarity. −\$10 000 is greater than +\$6000, except that it is owed, rather than owned.

3-4 THE ABSOLUTE VALUE OF A NUMBER

The numerical, or absolute, value of a number is the value of the number without regard to sign. Thus, the absolute values of numbers such as −1, +4, −6, and +3 are 1, 4, 6, and 3, respectively. Note that different numbers, such as −9 and +9, may have the same absolute value. To specify the absolute value of a number, such as Z, we write $|Z|$. This is often referred to as "the modulus of Z," or simply, "mod Z."

3-5 ADDITION OF POSITIVE AND NEGATIVE NUMBERS

Positive and negative numbers can be represented graphically as in Fig. 3-3. Positive numbers are shown as being directed toward the right of zero, which is the reference point, whereas negative numbers are directed toward the left.

Such a scale of numbers can be used to illustrate both addition and subtraction as performed in arithmetic. Thus, in adding 3 to 4, we can begin at 3 and count 4 units to the right to obtain the sum 7. Or, because these are positive numbers directed toward the right, we could draw them to scale, place them end to end, and measure their total length to obtain a length of 7 units in the positive direction. This is illustrated in Fig. 3-4.

In like manner, −2 and −3 can be added to obtain −5 as shown in Fig. 3-5.

Note that adding −3 and −2 is the same as adding −2 and −3 as in the foregoing example. The sum −5 is obtained, as shown in Fig. 3-6.

Suppose we want to add +6 and −10. We could accomplish this on the scale by first counting 6 units to the right and from *that* point counting 10 units to the left. In so doing,

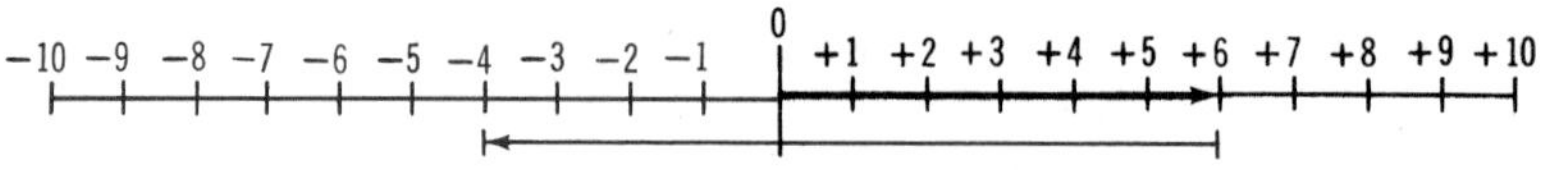

Fig. 3-7 Graphical Addition of +6 and −10

we would end up at −4, which is the sum of +6 and −10. Similarly, we could have started by first counting 10 units to the left, from zero, and from that point counting 6 units to the right for the +6. Again we would have arrived at −4.

Adding +6 and −10 can be accomplished graphically as in Fig. 3-7. The +6 is drawn to scale, and then the tail of the −10 is joined with the head of +6. The head of the −10 is then on −4. As would be expected, the same result is obtained by first drawing in the −10 and then the +6.

The following examples can be checked graphically in order to verify their correctness:

+8	+9	+6	−5	−7	−17
+4	−3	−9	+2	+9	−14
+12	+6	−3	−3	+2	−31

Consideration of the above examples enables us to establish the following rule:

Rule

1. To add two or more numbers with like signs, find the sum of their absolute values and prefix this sum with the common sign.
2. To add a positive number to a negative number, find the difference of their absolute values and prefix to the result the sign of the number that has the greater absolute value.

When three or more algebraic numbers that differ in signs are to be added, find the sum of the positive numbers and then the sum of the negative numbers. Add these sums algebraically, and use Rule 2 to obtain the total algebraic sum.

The *algebraic sum* of two or more numbers is the result obtained by adding them according to the preceding rules. Hereafter, the word "add" will mean "find the algebraic sum."

PROBLEMS 3-1

Add:

1. 28, 43 **2.** 36, −18 **3.** −82, 36 **4.** −18, −47 **5.** 124, −96

6. 165, −572 **7.** −286, −795 **8.** 0.0007, −0.0052 **9.** 175.03, −2.75, 36.28 **10.** −97.63, 5.74, −26.32

11. $7\frac{1}{2}$, $-3\frac{1}{2}$ **12.** $-6\frac{1}{4}$, $2\frac{1}{8}$ **13.** $-3\frac{1}{32}$, $-7\frac{3}{16}$ **14.** $-\frac{5}{8}$, $3\frac{1}{4}$ **15.** $\frac{1}{3}$, $-\frac{1}{5}$

3-6 THE SUBTRACTION OF POSITIVE AND NEGATIVE NUMBERS

We may think of subtraction as the process of determining what number must be added to a given number in order to produce another given number. Thus, when we subtract 5 from 9 and get 4, we have found that 4 must be added to 5 in order to obtain 9. From this it is seen that subtraction is the inverse of addition.

example 1 $(+5) - (+2) = ?$

solution In this example the question is asked, "What number added to $+2$ will give $+5$?" Using the scale of Fig. 3-8, start at $+2$ and count to the right (positive direction), until you reach $+5$. This requires three units. Therefore, the difference is $+3$, or $(+5) - (+2) = +3$.

example 2 $(+5) - (-2) = ?$

solution In this example the question is asked, "What number added to -2 will give $+5$?" Using the scale, start at -2 and count the number of units to $+5$. This requires seven units, and because it was necessary to count in the positive direction, the difference is $+7$, or $(+5) - (-2) = +7$.

example 3 $(-5) - (+2) = ?$

solution In this example the question is, "What number added to $+2$ will give -5?" Again using the scale, we start at $+2$ and count the number of units to -5. This requires seven units, but because it was necessary to count in the negative direction, the difference is -7, or $(-5) - (+2) = -7$.

example 4 $(-5) - (-2) = ?$

solution Here the question is, "What number added to -2 will give -5?" Using the scale, we start at -2 and count the number of units to -5. This requires three units in the negative direction. Hence, $(-5) - (-2) = -3$.

Summing up Examples 1 to 4, we have the following subtractions:

$$\begin{array}{cccc} +5 & +5 & -5 & -5 \\ \underline{+2} & \underline{-2} & \underline{+2} & \underline{-2} \\ +3 & +7 & -7 & -3 \end{array}$$

A study of the foregoing subtractions illustrates the following principles:

1. Subtracting a positive number is equivalent to adding a negative number of the same absolute value.
2. Subtracting a negative number is equivalent to adding a positive number of the same absolute value.

These principles can be used for the purpose of establishing the following rule:

Rule

To subtract one number from another, change the sign of the subtrahend and add algebraically.

Fig. 3-8 Scale for Graphical Subtraction of Positive and Negative Numbers

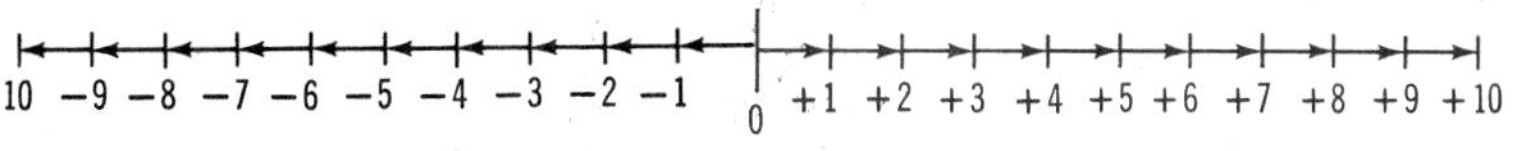

As in arithmetic, the number to be subtracted is called the *subtrahend*. The number from which the subtrahend is subtracted is called the *minuend*. The result is called the *remainder* or *difference*.

$$\begin{array}{lcr} \text{Minuend} & = & -642 \\ \text{Subtrahend} & = & 403 \\ \hline \text{Remainder} & = & -1045 \end{array}$$

PROBLEMS 3-2

Subtract the second line from the first:

1. $\begin{array}{r} 87 \\ 26 \\ \hline \end{array}$ **2.** $\begin{array}{r} 25 \\ -96 \\ \hline \end{array}$ **3.** $\begin{array}{r} -362 \\ -575 \\ \hline \end{array}$ **4.** $\begin{array}{r} -125 \\ 252 \\ \hline \end{array}$ **5.** $\begin{array}{r} 596 \\ -398 \\ \hline \end{array}$

6. $\begin{array}{r} 0.009\,25 \\ 0.072\,54 \\ \hline \end{array}$ **7.** $\begin{array}{r} -3.08 \\ -6.92 \\ \hline \end{array}$ **8.** $\begin{array}{r} 5\frac{2}{3} \\ -2\frac{3}{4} \\ \hline \end{array}$ **9.** $\begin{array}{r} -12\frac{7}{16} \\ -2\frac{3}{8} \\ \hline \end{array}$ **10.** $\begin{array}{r} -\frac{1}{2} \\ \frac{5}{8} \\ \hline \end{array}$

11. How many degrees must the temperature rise to change from (*a*) $+6°$ to $+73°$, (*b*) $-12°$ to $+14°$, and (c) $-273°$ to $-114°$?

12. How many degrees must the temperature fall to change from (*a*) $+212°$ to $+32°$, (*b*) $+55°$ to $-16°$, and (c) $-6°$ to $-42°$?

13. What amount of money is required to change an account from a debit of \$124.50 to a credit of \$240.30?

14. A certain point in a circuit is 570 V negative with respect to ground. Another point in the same circuit is 115 V positive with respect to ground. What is the potential difference between the two points?

15. In Fig. 3-2 what is the potential difference between the plate P and the cathode K?

3-7 ADDITION AND SUBTRACTION OF LIKE TERMS

In arithmetic, it is never possible to add unlike quantities. For example, we should not add inches and gallons and expect to obtain a sensible answer. Neither should we attempt to add volts and amperes, kilohertz and microfarads, ohms and watts, etc. So it goes on through algebra—we can never add quantities unless they are expressed in the same units.

The addition of two like terms such as $6EI + 12EI = 18EI$ can be checked by substituting numbers for the literal factors. Thus, if $E = 1$ and $I = 2$,

$$\begin{array}{rcrcrcr} 6EI & = & 6 \times 1 \times 2 & = & 6 \times 2 & = & 12 \\ 12EI & = & 12 \times 1 \times 2 & = & 12 \times 2 & = & 24 \\ \hline 18EI & = & 18 \times 1 \times 2 & = & 18 \times 2 & = & 36 \end{array}$$

From the foregoing, it is apparent that like terms may be added or subtracted by adding or subtracting their coefficients.

The addition or subtraction of unlike terms cannot be carried out but can only be indicated, because the unlike literal factors may stand for entirely different quantities.

example 5 Addition of like terms:

$$\begin{array}{r} -3i^2r \\ 8i^2r \\ \hline 5i^2r \end{array} \qquad \begin{array}{r} -16IR \\ 14IR \\ -3IR \\ \hline -5IR \end{array} \qquad \begin{array}{r} 13jIX \\ -20jIX \\ -32jIX \\ \hline -39jIX \end{array}$$

example 6 Subtraction of like terms:

$$\begin{array}{r} -8e_1 \\ 3e_1 \\ \hline -11e_1 \end{array} \qquad \begin{array}{r} 6iZ \\ -13iZ \\ \hline 19iZ \end{array} \qquad \begin{array}{r} -28L^2R \\ -29L^2R \\ \hline L^2R \end{array}$$

example 7 Addition of unlike terms:

$$\begin{array}{l} 3e \\ -3IX \\ 4E \\ \hline 3e - 3IX + 4E \end{array} \qquad \begin{array}{l} -3r \\ 4R \\ -16R_t \\ \hline 4R - 3r - 16R_t \end{array}$$

$$\begin{array}{l} 3EI \\ 10I^2R \\ -46W \\ \hline 3EI + 10I^2R - 46W \end{array}$$

3-8 ADDITION AND SUBTRACTION OF POLYNOMIALS

Polynomials are added or subtracted by arranging like terms in the same column and then combining terms in each column, as with monomials.

example 8 Addition of polynomials

$$\begin{array}{l} -3ab + 6cd + x^2y \\ 14ab \qquad\quad - 5x^2y \\ ab - 3cd \\ \hline 12ab + 3cd - 4x^2y \end{array} \qquad \begin{array}{r} 6E + 3RI - 8IZ \\ RI - 2IZ \\ -7E \qquad\quad + 3IZ \\ \hline -E + 4RI - 7IZ \end{array}$$

example 9 Subtraction of polynomials:

$$\begin{array}{l} 3mn + 16pq - xy^2 \\ -9mn \qquad\quad + 7xy^2 \\ \hline 12mn + 16pq - 8xy^2 \end{array} \qquad \begin{array}{l} 11R + 4x \\ 15R \qquad - 18Z \\ \hline -4R + 4x + 18Z \end{array}$$

PROBLEMS 3-3

Add:

1. $2i, 6i, -5i, 8i$
2. $5i^2r, 10i^2r, -26i^2r, 3i^2r$
3. $27IZ, 165IZ, -64IZ, -32IZ, 16IZ$
4. $65IR, -8.7IR, IR, -16.6IR, 15.2IR$
5. $3i + 16I, -8i - 12I$
6. $8jX, 26jX, -30jX, 18R, -5jX, 12R$
7. $25IR + 3E, -4IR - 2E, -18IR + 12E$
8. $12\,\Omega, 2\omega, -16\omega, 4\,\Omega$
9. $25\phi + 41\theta, 36\theta - 82\phi, -53\phi + 51\theta$
10. $5L, 4R, -27L, -5Z, 36L, 7R - 2Z$
11. $$\begin{array}{l} 6i^2r + 8W - 6ei + 32w \\ -3i^2r + 3W + 8ei + 18w \\ 24i^2r - W - 5ei - w \\ \hline \end{array}$$
12. $$\begin{array}{l} 25IX - 16IZ + 3IR \\ 14IZ + 2IX - IR \\ 8IR + 4IX - 3IZ \\ \hline \end{array}$$
13. $$\begin{array}{l} 1.65eI + 3.07W - 1.46I^2r \\ 0.025W - 1.11eI - 0.85I^2r \\ 3.06I^2r + 0.92eI + 0.725W \\ \hline \end{array}$$

14. $2.15ei + 1.64\frac{e^2}{r} - 3.82i^2r$, $0.57\frac{e^2}{r} + 1.94i^2r$

15. $\frac{1}{4}\pi ft$, $-3\pi Z$, $-\frac{2}{3}\pi ft$, $\frac{3}{16}\pi ft$, $\frac{7}{8}\pi Z$

16. To $47IR + 3IZ$ add $-15IR - 4IZ$.

17. From $25\phi + 3\theta$ subtract $15\phi - 7\theta$.

18. From $17.2\omega L + 5X_C - 13.2Z$ subtract $4.5\omega L - 3.2X_C + 5.6Z$.

19. From the sum of $26.2\frac{E^2}{R} + 14.6EI - 3I^2R$ and $6.2I^2R - 3.8EI + 19.6\frac{E^2}{R}$ subtract $27.2EI - 2.6I^2R - 1.8\frac{E^2}{R}$.

20. Subtract $9.5X_C + \frac{3.26}{\omega C}$ from the sum of $-8.7X_C + \frac{2.46}{\omega C}$ and $-4.6X_C - \frac{1.98}{\omega C}$.

21. Take $1.25IR + 0.64IX - 2.81IZ$ from $-0.06IR + 0.23IX + 1.09IZ$.

22. How much more than $5E_g - 2iR$ is $3E_g + 6iR$?

23. What must be added to $3\psi + 2.8\lambda$ to obtain $9.64\psi - 4.3\lambda$?

24. What must be subtracted from $16.2\gamma - 3.3\alpha + 2.8\beta$ to obtain $8.1\alpha + 1.7\gamma - 2.6\beta$?

3-9 SIGNS OF GROUPING

Often it is necessary to express or group together quantities that are to be affected by the same operation. Also, it is desirable to be able to represent that two or more terms are to be considered as one quantity.

In order to meet the above requirements, signs of grouping have been adopted. These signs are the *parentheses* ( ), the *brackets* [ ], the *braces* { }, and the *vinculum* ——. The first three are placed around the terms to be grouped, as $(E - IR)$, $[a + 3b]$, and $\{x^2 + 4y\}$. All have the same meaning: that the enclosed terms are to be considered as one quantity.

Thus, $16 - (12 - 5)$ means that the quantity $(12 - 5)$ is to be subtracted from 16. That is, 5 is to be subtracted from 12, and then the remainder 7 is to be subtracted from 16 to give a final remainder of 9. In like manner, $E - (IR + e)$ means that the sum of $(IR + e)$ is to be subtracted from E.

Carefully note that the sign preceding a sign of grouping, as the minus sign between E and $(IR + e)$ above, is a sign of *operation* and does not denote that $(IR + e)$ is a negative quantity.

The vinculum is used mainly with radical signs and fractions, as

$$\sqrt{7245} \qquad \text{and} \qquad \frac{a+b}{x-y}$$

In the latter case the vinculum denotes the division of $a + b$ by $x - y$, in addition to grouping the terms in the numerator and denominator. When studying later chapters, you will avoid many mistakes by remembering that *the vinculum is a sign of grouping.*

In working problems involving signs of grouping, the operations within the signs of grouping should be performed first.

example 10 $a + (b + c) = ?$

solution This means, "What result will be obtained when the sum of $b + c$ is added to a?" Because both b and c are denoted as positive, it follows that we can write

$$a + (b + c) = a + b + c$$

because it makes no difference in which order we add.

example 11 $a + (b - c) = ?$

solution This means, "What result will be obtained when the difference of $b - c$ is added to a?" Again, because it makes no difference in which order we add, we can write

$$a + (b - c) = a + b - c$$

example 12 $a - (b + c) = ?$

solution Here the sum of $b + c$ is to be subtracted from a. This is the same as if we first subtract b from a and from this remainder subtract c. Therefore,

$$a - (b + c) = a - b - c$$

or, because this is subtraction, we could change the signs and add algebraically, remembering that b and c are denoted as positive, as shown below:

$$\begin{array}{r} a \\ b + c \\ \hline a - b - c \end{array}$$

example 13 $a - (-b - c) = ?$

solution This means that the quantity $-b - c$ is to be subtracted from a. Performing this subtraction, we obtain

$$\begin{array}{r} a \\ -b - c \\ \hline a + b + c \end{array}$$

Therefore,

$$a - (-b - c) = a + b + c$$

A study of Examples 10 to 13 enables us to state the following:

Rules

1. Parentheses or other signs of grouping preceded by a plus sign can be removed without any other change.
2. To remove parentheses or other signs of grouping preceded by a minus sign, change the sign of every term within the sign of grouping.

Although not apparent in the examples, another rule can be added as follows:

3. If parentheses or other signs of grouping occur one within another, remove the inner grouping first.

examples

$$(x + y) + (2x - 3y)$$
$$= x + y + 2x - 3y = 3x - 2y$$

$$3a - (2b + c) - a$$
$$= 3a - 2b - c - a = 2a - 2b - c$$

$$10x - (-3x - 4y) + 2y$$
$$= 10x + 3x + 4y + 2y = 13x + 6y$$

$$x - [2x + 3y - (3x - y) - 4x]$$
$$= x - [2x + 3y - 3x + y - 4x]$$
$$= x - 2x - 3y + 3x - y + 4x$$
$$= 6x - 4y$$

PROBLEMS 3-4

Simplify by removing the signs of grouping and combining the similar terms:

1. $(x - 3y - 4) - (x + 4y - 7)$
2. $(5\lambda + 3\theta) - (-4\lambda + 5\theta + 6)$
3. $(4R + 5Z + 6X) - (9X - 6R + 5Z - 3)$
4. $6I^2R + [-5EI + (-I^2R - 3EI) - 7EI] + 5$
5. $8\frac{E^2}{R} - \left[-6I^2R - \left(5\frac{E^2}{R} - \overline{6I^2R + 3\frac{E^2}{R} + 3EI}\right)\right]$
6. $X_L - \{3L - [2R - (X_L + 5L)]\}$
7. $5\alpha - \{\alpha + \beta - [\gamma + \alpha + \beta - (\alpha + \beta + \gamma) - 3\alpha] - 3\beta\}$
8. $-\{-\theta - [\phi + \omega - 2\phi - (\omega + \phi) - \phi] - 2\theta - 3\omega\}$
9. $4a - [-5a - (-6b + 3c) - (8a - \overline{4b - 3c})]$
10. $5.4R - 2.6Z - \overline{1.5IX - 7R} - [4.6Z - (3X_C - 5.7IX) - 4.32R + 27]$

3-10 INSERTING SIGNS OF GROUPING

To enclose terms within signs of grouping preceded by a plus sign, rewrite the terms without changing their signs.

example 14 $a + b - c + d = a + (b - c + d)$

To enclose terms within signs of grouping preceded by a minus sign, rewrite the terms and change the signs of the terms enclosed.

example 15 $a + b - c + d = a + b - (c - d)$

No difficulty need be encountered when inserting signs of grouping because, by removing the signs of grouping from the result, the original expression should be obtained.

example 16 $x - 3y + z = x - (3y - z)$
$= x - 3y + z$

PROBLEMS 3-5

1. Enclose the last three terms of each of the following expressions in parentheses preceded by a plus sign:
   (a) $3X + X_C - X_L + Z$ (b) $\alpha + 6\beta - 3\phi + \lambda$
   (c) $5W + 6I^2R - 3EI + 7I^2Z$ (d) $\frac{E^2}{R} - 3I^2R + 7I^2Z - 4EI$
   (e) $8\lambda + 3\mu - 7\theta - 3\phi + 6\alpha$
2. Enclose the last three terms of each of the following expressions in parentheses preceded by a minus sign:
   (a) $8EI + 5I^2R - 6W + 4\frac{E^2}{R}$ (b) $5a + 3b - 6c + 4d + 5e$
   (c) $8\omega + 13\phi - 3\lambda + 2.7r$ (d) $4.6^\circ - 3\theta + 3.8^r - 0.52\phi$
   (e) $\frac{E^2}{R} + W + I^2Z - 6EI$

3. Write the amount by which N is less than $(X^2 + R^2)$.
4. The sum of two currents is 526 milliamperes (mA). The larger of the two currents is i mA. What is the smaller?
5. The difference between two voltages is 16.8 V. The smaller voltage is e V. What is the greater?
6. Write the amount by which X_L exceeds $\frac{1}{2\pi fC}$.
7. What is the larger part of Z if $\sqrt{r^2 + x^2}$ is the smaller part?
8. Write the amount by which E is greater than $e - IR$.
9. Write the amount by which P exceeds $I^2R + \frac{E^2}{R}$.
10. The difference between two numbers is 19.6. If the larger number is β, what is the smaller?
11. Write the smaller part of X_C if $\frac{1}{2\pi fC_1}$ is the larger part.
12. The difference between two numbers is X^2 and the larger of the two is Z^2. Write the relationship which describes R^2, which is the smaller of the two.

---

4 algebra—multiplication and division

Multiplication is often defined as the *process of repeated addition.* Thus, 2×3 may be thought of as adding 2 three times, or $2 + 2 + 2 = 6$.

Considering multiplication as a shortened form of addition is not satisfactory, however, when the multiplier is a fraction. For example, it would not be sensible to say that $5 \times \frac{2}{7}$ was adding 5 two-sevenths of a time. This problem could be rewritten as $\frac{2}{7} \times 5$, which would be the same as adding $\frac{2}{7}$ five times. But this is only a temporary help, for if two fractions are to be multiplied together, as $\frac{3}{4} \times \frac{5}{6}$, the original definition of multiplication will not apply. However, the definition has been extended to include such cases, and the product of $5 \times \frac{2}{7}$ is taken to mean 5 multiplied by 2 and this product divided by 7; that is, by $5 \times \frac{2}{7}$ is meant $\frac{5 \times 2}{7}$. Also,

$$\frac{3}{4} \times \frac{5}{6} = \frac{3 \times 5}{4 \times 6} = \frac{15}{24}$$

4-1 MULTIPLICATION OF POSITIVE AND NEGATIVE NUMBERS

Because we are now dealing with both positive and negative numbers, it becomes necessary to determine what sign the product will have when combinations of these numbers are multiplied.

When only two numbers are to be multiplied, there can be but four possible combinations of signs, as follows:

(1) $(+2) \times (+3) = ?$
(2) $(-2) \times (+3) = ?$
(3) $(+2) \times (-3) = ?$
(4) $(-2) \times (-3) = ?$

Combination (1) means that $+2$ is to be added three times:

$$(+2) + (+2) + (+2) = +6$$

or

$$(+2) \times (+3) = +6$$

In the same manner, combination (2) means that -2 is to be added three times:

$$(-2) + (-2) + (-2) = -6$$

or

$$(-2) \times (+3) = -6$$

Combination (3) means that $+2$ is to be subtracted three times:

$$-(+2) - (+2) - (+2) = -6$$

or

$$(+2) \times (-3) = -6$$

Note that this is the same as subtracting 6 once, -6 being thus obtained.

Combination (4) means that -2 is to be subtracted three times:

$$-(-2) - (-2) - (-2) = +6$$

or

$$(-2) \times (-3) = +6$$

This may be considered to be the same as subtracting -6 once, and because subtracting -6 once is the same as adding $+6$, we obtain $+6$ as above.

From the foregoing we have these rules:

Rules

1. The product of two numbers having like signs is positive.
2. The product of two numbers having unlike signs is negative.
3. If more than two factors are multiplied, Rules 1 and 2 are to be used successively.
4. The product of an even number of negative factors is positive. The product of an odd number of negative factors is negative.

These rules can be summarized in general terms as follows:

- Rule 1 $(+a)(+b) = +ab$
- Rule 1 $(-a)(-b) = +ab$
- Rule 2 $(+a)(-b) = -ab$
- Rule 2 $(-a)(+b) = -ab$
- Rule 3 $(-a)(+b)(-c) = +abc$
- Rule 4 $(-a)(-b)(-c)(-d) = +abcd$
- Rule 4 $(-a)(-b)(-c) = -abc$

PROBLEMS 4-1

Find the products of the following factors:

1. 3, 4

2. 6, -5

3. -9.1, -1.5

4. -1.7, 6.5, -7.3

5. $\frac{3}{16}$, $-\frac{5}{8}$, $\frac{1}{2}$

6. $\frac{2}{3}$, $-\frac{3}{4}$, $-\frac{7}{8}$

7. -0.025, -0.0005, -2.5, -0.03

8. 3000, -0.06, 250, -0.002

9. $-e$, $-i$, t

10. q, $-r$, $-s$, t

11. $2\pi f$, L_1, L_2

12. θ^2, ϕ^2, λ^2

13. $\frac{1}{2}$, $\frac{1}{\pi}$, $\frac{1}{f}$, $\frac{1}{C_p}$

14. $\frac{1}{a}$, $-\frac{1}{b}$, $-\frac{1}{c}$, $-\frac{1}{d}$

15. ψ, $\frac{1}{\theta}$, $-\frac{1}{\phi}$, μ

4-2 GRAPHICAL REPRESENTATION

Our system of representing numbers is a graphical one, as previously illustrated in Fig. 3-3. It might be well at this time to consider certain facts regarding multiplication.

When a number is multiplied by any other number except 1, we can think of the operation as having changed the absolute value of the multiplicand. Thus, 3 in $\times$ 4 becomes 12 in, 6 A $\times$ 3 becomes 18 A, etc. Such multiplications could be represented graphically by simply extending the multiplicand the proper amount, as shown in Fig. 4-1.

The multiplication of a negative number by a positive number is shown in Fig. 4-2.

From these examples, it is evident that a positive multiplier simply changes the absolute value, or magnitude, of the number being multiplied. What happens if the multiplier is negative? As an example, consider $2 \times (-3) = -6$. How will this be represented graphically?

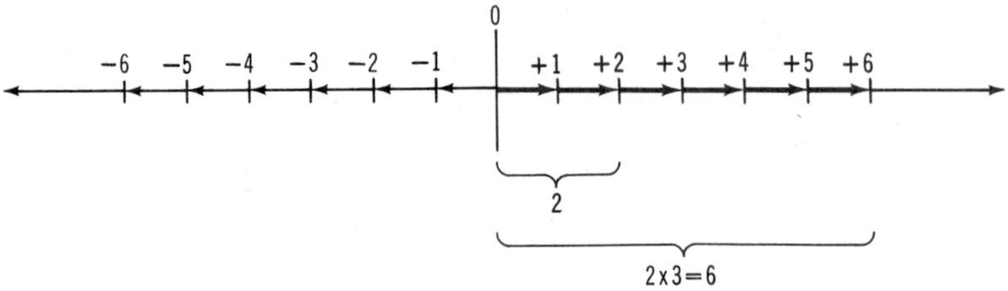

Fig. 4-1 Representation of the Multiplication $2 \times 3 = 6$

Now, $2 \times (-3) = -6$ is the same as

$$2 \times (+3) \times (-1) = -6$$

Therefore, let us first multiply 2×3 to obtain $+6$ and represent it as shown in Fig. 4-1. We must multiply by -1 to complete the problem and in so doing should obtain -6, but -6 must be represented as a number six units in length and directed toward the left, as illustrated in Fig. 4-2. We therefore agree that multiplication by -1 causes counterclockwise rotation of a number in a direction that will be exactly opposite from its original direction. This is illustrated in Fig. 4-3.

If both multiplicand and multiplier are negative, as

$$(-2) \times (-3) = +6$$

the representation is as illustrated in Fig. 4-4. Again,

$$(-2) \times (-3) = +6$$

is the same as

$$(-2) \times (+3) \times (-1) = +6$$

The product has an absolute value of 6, and at the same time there has been rotation to $+6$ because of multiplication by -1.

The foregoing representations are also applicable to division, since the law of signs is the same as in multiplication.

The important thing to bear in mind is that multiplication or division by -1 causes counterclockwise rotation of a number to a direction exactly opposite the original direction. The number -1, when used as a multiplier or divisor, should be considered as an *operator* for the purpose of rotation. It is important that you clearly understand this concept, for you will encounter it later on.

4-3 LAW OF EXPONENTS IN MULTIPLICATION

As explained in Sec. 2-10, an exponent indicates how many times a number is to be

Fig. 4-2 Representation of the Multiplication $-2 \times 3 = -6$

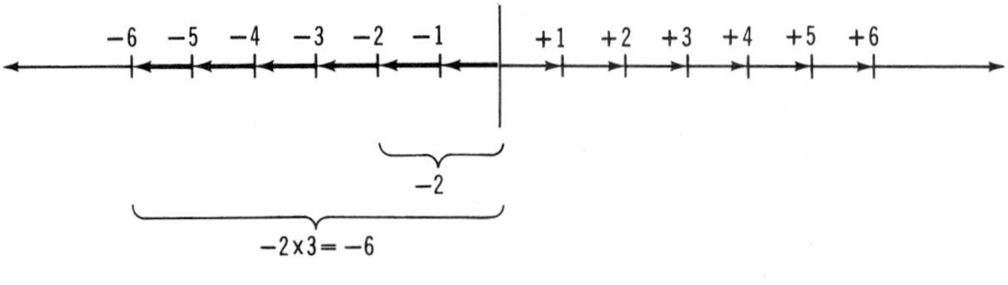

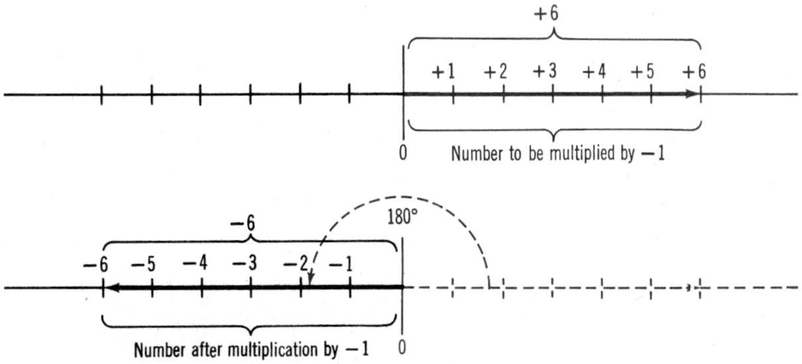

Fig. 4-3 Multiplication by −1 Rotates Multiplicand Counterclockwise through 180°

taken as a factor. Thus $x^4 = x \cdot x \cdot x \cdot x$, $a^3 = a \cdot a \cdot a$, etc.

Because $x^4 = x \cdot x \cdot x \cdot x$
and $x^3 = x \cdot x \cdot x$
then $x^4 \cdot x^3 = x \cdot x \cdot x \cdot x \cdot x \cdot x \cdot x = x^7$
or $x^4 \cdot x^3 = x^{4+3} = x^7$

Thus, we have the rule:

Rule
To find the product of two or more powers having the same base, add the exponents.

examples

$$a^3 \cdot a^2 = a^{3+2} = a^5$$
$$x^4 \cdot x^4 = x^{4+4} = x^8$$
$$6^2 \cdot 6^3 \cdot 6^5 = 6^{2+3+5} = 6^{10}$$
$$a^2 \cdot b^3 \cdot b^3 \cdot a^5 = a^{2+5} \cdot b^{3+3} = a^7 b^6$$
$$e \cdot e^3 = e^{1+3} = e^4$$
$$3^2 \cdot 3^4 = 3^{2+4} = 3^6$$
$$e^a \cdot e^b = e^{a+b}$$

From the foregoing examples, it is seen that the law of exponents can be expressed in the well-known general form

$$a^m \cdot a^n = a^{m+n}$$

where $a \neq 0$ and m and n are literal numbers and may represent any number of factors.

Fig. 4-4 Illustration of −6 Rotated Counterclockwise through 180° to Become +6 Due to Multiplication by −1

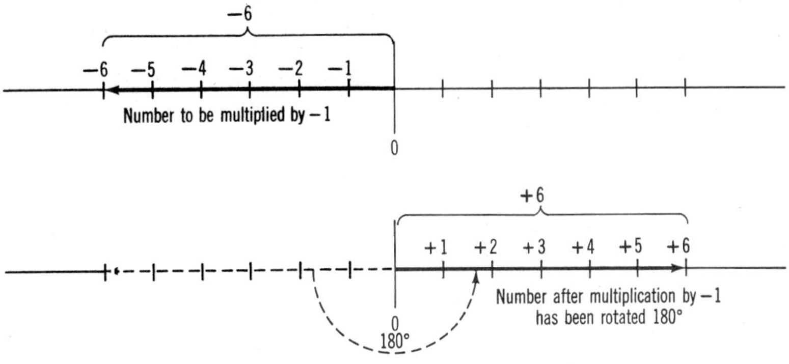

4-4 MULTIPLICATION OF MONOMIALS

Rules

1. Find the product of the numerical coefficients and give it the proper sign, plus or minus, according to the rules for multiplication (Sec. 4-1).

2. Multiply this numerical product by the product of the literal factors. Use the law of exponents as applicable.

example 1 Multiply $3a^2b$ by $4ab^3$.

solution

$$(3a^2b)(4ab^3) = +(3 \cdot 4) \cdot a^{2+1} \cdot b^{1+3} = 12a^3b^4$$

example 2 Multiply $-6x^3y^2$ by $3xy^2$.

solution

$$(-6x^3y^2)(3xy^2) = -(6 \cdot 3) \cdot x^{3+1} \cdot y^{2+2} = -18x^4y^4$$

example 3 Multiply $-5e^2x^4y$ by $-3e^2x^2p$.

solution

$$(-5e^2x^4y)(-3e^2x^2p) = +(5 \cdot 3)e^{2+2} \cdot p \cdot x^{4+2} \cdot y = 15e^4px^6y$$

PROBLEMS 4-2

Multiply:

1. $x^3 \cdot x^2$
2. $-b^3 \cdot b^5$
3. $e^2 \cdot e^3 \cdot -e^5$
4. $-\lambda \cdot \lambda^2 \cdot -\theta^3$
5. $(2m^2)(3m^2)$
6. $(6\alpha)(-3\beta^3)$
7. $(4x)(5m^3)(-3x^2m)$
8. $(-5\mu)^2$
9. $(am^n)(bm^p)$
10. $(13b^x)(-2b^{a+y})$
11. $(2p)^3$
12. $(-5\lambda^2)^3$
13. $(-3a^2b^3cd^2)(-2abc^2d^5)$
14. $(\frac{1}{4}a^3)(-\frac{2}{3}ab^2)$
15. $(\frac{3}{16}X_L)(\frac{2}{3}M)(-2\pi)$
16. $(14a^2b^3cd)(-\frac{2}{7}ab^2de)$
17. $(0.5e^2i)(3i^2r)(-0.05ei)(w)$
18. $(\frac{5}{16}\theta\phi)(-\frac{3}{25}\mu\theta)(-\frac{24}{27}\theta^2\omega)$
19. $(a^3)^2$
20. $(3p^q)^r$

4-5 MULTIPLICATION OF POLYNOMIALS BY MONOMIALS

Another method of graphically representing the product of two numbers is as shown in Fig. 4-5. The product $5 \times 6 = 30$ is shown as a rectangle whose sides are 5 and 6 units in length; therefore, the rectangle contains 30 square units.

Fig. 4-5 Graphical Representation of the Multiplication $5 \times 6 = 30$

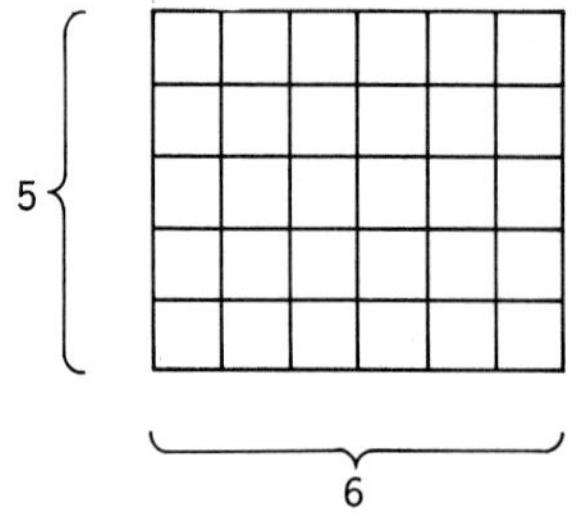

Similarly, the product of $5(6 + 9)$ can be represented as illustrated in Fig. 4-6.

Thus, $5(6 + 9)$
$= 5 \times 15$
$= 75$

Also, $5(6 + 9)$
$= (5 \times 6) + (5 \times 9)$
$= 30 + 45$
$= 75$

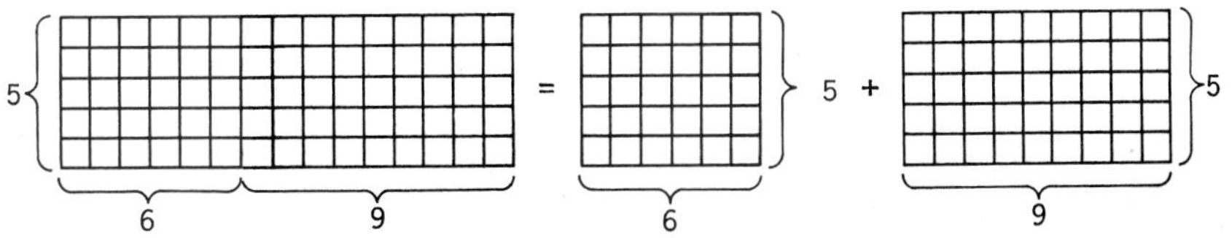

Fig. 4-6 Graphical Representation of the Multiplication $5(6 + 9) = 5 \times 6 + 5 \times 9 = 75$

In like manner the product

$$a(c + d) = ac + ad$$

can be illustrated as in Fig. 4-7.

From the foregoing, you can show that

$$3(4 + 2) = 3 \times 4 + 3 \times 2 = 12 + 6 = 18$$
$$4(5 + 3 + 4) = 4 \times 5 + 4 \times 3 + 4 \times 4$$
$$= 20 + 12 + 16 = 48$$
$$x(y + z) = xy + xz$$
$$p(q + r + s) = pq + pr + ps$$

Note that, in all cases, each term of the polynomial (the terms enclosed in parentheses) is multiplied by the monomial. From these examples, we develop the following rule:

Rule

To multiply a polynomial by a monomial, multiply each term of the polynomial by the monomial and write in succession the resulting terms with their proper signs.

example 4 $3x(3x^2y - 4xy^2 + 6y^3) = ?$

solution

Multiplicand $= 3x^2y - 4xy^2 + 6y^3$
Multiplier $= 3x$
Product $= 9x^3y - 12x^2y^2 + 18xy^3$

example 5
$-2ac(-10a^3 + 4a^2b - 5ab^2c + 7bc^2) = ?$

solution

Multiplicand $= -10a^3 + 4a^2b - 5ab^2c + 7bc^2$
Multiplier $= -2ac$
Product $= 20a^4c - 8a^3bc + 10a^2b^2c^2 - 14abc^3$

example 6 Simplify $5(2e - 3) - 3(e + 4)$.

solution First multiply $5(2e - 3)$ and $3(e + 4)$, and then subtract the second result from the first, thus:

$$5(2e - 3) - 3(e + 4) = (10e - 15) - (3e + 12)$$
$$= 10e - 15 - 3e - 12$$
$$= 7e - 27$$

Fig. 4-7 Illustration of the Product $a(c + d) = ac + ad$

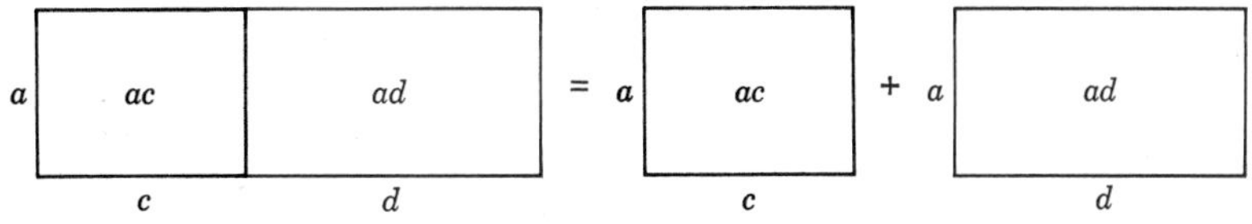

PROBLEMS 4-3

Multiply:

1. $3a + 5b$ by 6
2. $2a + 3$ by $3a$
3. $2R_1 + 4R_2$ by $2I^2$
4. $5.8a - jb$ by b^2

SECTIONS 4-4 TO 4-5

5. $\lambda^2 + 2\theta - 3\mu$ by 4.7ϕ

6. $2\alpha^3 - 3\alpha^2 + 4\alpha$ by -5α

7. $4\alpha^3\beta + 3\alpha^2\beta^2 - 5\alpha\beta^3$ by $0.5\alpha\beta$

8. $2\theta^2\phi - 5\alpha\theta^2 - 4\alpha\beta + 3$ by $3\alpha\phi$

9. $-5a^2r_1 - 2ar_1^2 + 6r_1^3$ by $-3ar_2$

10. $3\omega^2L_1^2 - 5\omega^2M + 7\omega^2L_2^2$ by $-2\omega L_1L_2$

11. $\frac{1}{2}I^2R - \frac{1}{4}I^2R^2 - \frac{1}{3}iZ$ by $\frac{2}{3}iIZ$

12. $8ab + 4ab^2 + 4$ by $-\frac{1}{4}ab^2$

13. $\dfrac{I^2R}{4} - \dfrac{i^2r}{2} + \dfrac{P}{6}$ by $12IP$

14. $5\mu^2k^2 + 3\eta k - 2\mu\eta^2$ by $-3\theta\omega$

15. $0.025E^3Z^2 + 0.05EZ^4 - 1.67Z^5$ by $6.28IZ$

Simplify:

16. $3ars(-4ar^2 + 2rs - 6as^2)$

17. $3(6\phi - 5\theta) - 3(\phi + 2\theta)$

18. $\mu(\alpha - j\beta) + \mu(\alpha + j\beta)$

19. $\theta(\theta^2 + \phi) - \phi(\theta + \phi^2)$

20. $3Z(2I^2 - i^2) - Z(6I^2 - 5i^2)$

21. $0.5\omega(6\pi + 5\eta\omega - \pi\omega^2) - 3\pi(0.7\omega - \eta\pi + 2\omega^3)$

22. $\frac{1}{2}\gamma\beta(4\gamma^2\beta - 2\gamma\beta^2 - 10\gamma^3 + 5\beta^3)$

23. $8\lambda\left(\dfrac{E^2}{3} + \dfrac{Ee}{2} - \dfrac{e^2}{16}\right)$

24. $5\theta(2\theta^2 + 3\theta\phi - 6\phi^2) - 3(6\theta^3 - 2\theta^2\phi - 7\theta\phi^2)$

25. $0.25I\left(\dfrac{R}{5} - \dfrac{R_1}{10} + \dfrac{3R_2}{5}\right) + 1.5(0.05IR - 0.375IR_2)$

26. $\frac{1}{2}\lambda\mu(6\lambda^2\mu - 5\lambda\mu^2 + 12\lambda - 4\mu)$

27. $4\theta(3\theta^2 + 2\theta\phi - 3\phi^2) - 2\theta(6\theta^2 + 4\theta\phi - 6\phi^2)$

28. $5\gamma^2\left(\dfrac{\gamma\lambda}{3} - \dfrac{\beta\lambda^2}{5} - \dfrac{\theta\beta^2}{10}\right) + 6\beta^2\left(\dfrac{\beta\lambda}{2} + \dfrac{\gamma\lambda}{5} - \dfrac{\gamma^2\theta}{12}\right)$

29. $6\left(\dfrac{s}{3} - \dfrac{s}{2} + \dfrac{2s}{3}\right) + 8\left(\dfrac{s}{4} - \dfrac{s}{2} + \dfrac{3s}{4}\right)$

30. $0.5I^2(2R_1 + 3R_2 - 5R_3) - 0.8I^2(-0.5R_1 - 2R_2 - 0.25R_3)$

4-6 MULTIPLICATION OF A POLYNOMIAL BY A POLYNOMIAL

It is apparent that

$$(3 + 4)(6 - 3) = 7 \times 3 = 21$$

The above multiplication can also be accomplished in the following manner:

$$\begin{aligned}(3 + 4)(6 - 3) &= 3(6 - 3) + 4(6 - 3)\\ &= (18 - 9) + (24 - 12)\\ &= 9 + 12\\ &= 21\end{aligned}$$

Similarly,

$$\begin{aligned}(2a - 3b)(a + 5b)&\\ &= 2a(a + 5b) - 3b(a + 5b)\\ &= (2a^2 + 10ab) - (3ab + 15b^2)\\ &= 2a^2 + 10ab - 3ab - 15b^2\\ &= 2a^2 + 7ab - 15b^2\end{aligned}$$

From the foregoing, we have the following:

Rule

To multiply polynomials, multiply every term of the multiplicand by each term of the multiplier and add the partial products.

example 7 Multiply $2i - 3$ by $i + 2$.

solution

$$\begin{array}{lrl} \text{Multiplicand} & = & 2i - 3 \\ \text{Multiplier} & = & \underline{i + 2} \\ i \text{ times } (2i - 3) & = & 2i^2 - 3i \\ 2 \text{ times } (2i - 3) & = & \underline{4i - 6} \\ \text{Adding, product} & = & 2i^2 + i - 6 \end{array}$$

example 8 Multiply $a^2 - 3ab + 2b^2$ by $2a^2 - 3b^2$.

solution

$$\begin{array}{lrl} \text{Multiplicand} & = & a^2 - 3ab + 2b^2 \\ \text{Multiplier} & = & \underline{2a^2 - 3b^2} \\ 2a^2 \text{ times} & & \\ (a^2 - 3ab + 2b^2) & = & 2a^4 - 6a^3b + 4a^2b^2 \\ -3b^2 \text{ times} & & \\ (a^2 - 3ab + 2b^2) & = & \underline{ - 3a^2b^2 + 9ab^3 - 6b^4} \\ \text{Adding, product} & = & 2a^4 - 6a^3b + a^2b^2 + 9ab^3 - 6b^4 \end{array}$$

Products obtained by multiplication can be tested by substituting any convenient numerical values for the literal numbers. It is not good practice to substitute the numbers 1 and 2. If there are exponents, then the use of 1 will not be a proof of correct work, for 1 to any power is still 1. Similarly, if addition should be involved, the use of 2 could give an incorrect indication, because $2 + 2 = 4$, and $2 \times 2 = 4$.

example 9 Multiply $a^2 - 4ab - b^2$ by $a + b$, and test by letting $a = 3$ and $b = 4$.

solution

$$\begin{array}{ll} a^2 - 4ab - b^2 & = 9 - 48 - 16 = -55 \\ \underline{a + b} & = 3 + 4 = \underline{7} \\ a^3 - 4a^2b - ab^2 & -385 \\ \underline{\phantom{a^3 - {}} a^2b - 4ab^2 - b^3} & \\ a^3 - 3a^2b - 5ab^2 - b^3 & = 27 - 108 - 240 - 64 \\ & = -385 \end{array}$$

PROBLEMS 4-4

Multiply

1. $\alpha + 1$ by $\alpha + 1$	**2.** $\alpha + 1$ by $\alpha - 1$	**3.** $\alpha - 1$ by $\alpha - 1$
4. $\beta + 2$ by $\beta + 2$	**5.** $\beta + 3$ by $\beta - 3$	**6.** $\beta - 3$ by $\beta - 3$
7. $p + 3$ by $p + 5$	**8.** $X_C - 6$ by $X_C - 4$	**9.** $r - 11$ by $r + 3$
10. $j + 2$ by $j - 2$		

note Parentheses or other signs of grouping are often used to indicate a product. Thus, $(ir + e)(2ir - 3e)$ means $ir + e$ multiplied by $2ir - 3e$. Perform the indicated multiplications:

11. $(m + 4)(m + 2)$	**12.** $(4C + L)(3C + L)$
13. $(\alpha + 7\beta)(3\alpha - 6\beta)$	**14.** $(ax + bx)(cx + dx)$
15. $(2\theta + \lambda)(3\theta - 5\lambda)$	**16.** $(17EI - 2I^2R)(2EI - 6I^2R)$
17. $(3m + 2n)(2m - 3n)$	**18.** $(1.5\psi + 0.5\phi)(2\psi + 1.75\phi)$
19. $(R - 3Z)(5R - 2Z)$	**20.** $(\frac{1}{3}m - \frac{1}{2}q)(\frac{3}{4}m + \frac{5}{6}q)$
21. $(2a^2 + 5a - 1)(3a + 1)$	**22.** $(3\theta^2 - 4\theta - 7)(\theta + 3)$
23. $(R + r)(2R^2 - 4Rr + 2r^2)$	**24.** $(x + y)(x + y)(x + y)$
25. $(a + b)(a - b)(a - b)$	**26.** $(p - q)(p - q)(p - q)$
27. $(\theta - \phi)(\theta + \phi)(\theta - \phi)$	**28.** $(IR + P)(I^2R^2 - 2IRP + P^2)$
29. $(a^2 + 2ab + b^2)(a + b)$	**30.** $(x + 1)^2$

31. $(x + y)^2$ **32.** $(x - y)^2$
33. $(M - N)^2$ **34.** $(2\theta\phi + \psi + 1)^2$
35. $(2\alpha + 2w)^3$
36. $(3\alpha + 7)(\alpha - 5) + (2\alpha - 3)(4\alpha - 1)$
37. $2(3I^2R + 1)(4I^2R - 5) - 4(2I^2R - 2)(I^2R + 3)$
38. $4(3\theta - 2\phi + \lambda)(2\theta + 2\phi - \lambda) - 6(\theta + 2\phi + \lambda)(2\theta - \phi - 2\lambda)$
39. $3a(2a + b - 1)^2 - 2a(a + 2b + 1)^2$
40. $3\theta^2(5\omega - \lambda + \theta)^2 - \theta^2(\omega + 7\lambda - 2\theta)^2$

4-7 DIVISION

The division of algebraic expressions requires the development of certain rules and new methods in connection with operations involving exponents. However, if you have mastered the processes of the preceding sections, algebraic division will be an easy subject.

For the purpose of review the following definitions are given:

1. The *dividend* is a number, or quantity, that is to be divided.
2. The *divisor* is a number by which a number, or quantity, is to be divided.
3. The *quotient* is the result obtained by division. That is,

$$\frac{\text{Dividend}}{\text{Divisor}} = \text{quotient}$$

4-8 DIVISION OF POSITIVE AND NEGATIVE NUMBERS

Because division is the inverse of multiplication, the methods of the latter will serve as an aid in developing methods for division. For example,

because $6 \times 4 = 24$
then $24 \div 6 = 4$
and $24 \div 4 = 6$

These relations can be used in applying the rules for multiplication to division.

All the possible cases can be represented as follows:

$$(+24) \div (+6) = ?$$
$$(-24) \div (+6) = ?$$
$$(+24) \div (-6) = ?$$
$$(-24) \div (-6) = ?$$

Because division is the inverse of multiplication, we apply the rules for multiplication of positive and negative numbers and obtain the following:

$(+24) \div (+6) = +4$ because $(+4) \times (+6) = +24$
$(-24) \div (+6) = -4$ because $(-4) \times (+6) = -24$
$(+24) \div (-6) = -4$ because $(-4) \times (-6) = +24$
$(-24) \div (-6) = +4$ because $(+4) \times (-6) = -24$

Therefore, we have the following:

Rule

To divide positive and negative numbers,
1. If dividend and divisor have like signs, the quotient is positive.
2. If dividend and divisor have unlike signs, the quotient is negative.

PROBLEMS 4-5
Divide the first number by the second in Probs. 1 to 10:

1. 25, 5 **2.** −16, 4 **3.** −30, −6 **4.** −6.4, −800

5. $-\frac{2}{3}, \frac{1}{2}$ 6. $\frac{21}{64}, \frac{7}{16}$ 7. $2\pi fC, -1$ 8. $R, E - e$
9. $E \times 10^8, L_v$ 10. $\omega L, Q$

Supply the missing divisors:

11. $\frac{-24}{?} = 4$ 12. $\frac{16}{?} = -2$ 13. $\frac{75}{?} = -\frac{1}{3}$ 14. $-\frac{27}{?} = -\frac{1}{3}$

15. $\frac{-\frac{15}{16}}{?} = \frac{3}{8}$

4-9 THE LAW OF EXPONENTS IN DIVISION

By previous definition of an exponent (Sec. 2-10),

$$x^6 = x \cdot x \cdot x \cdot x \cdot x \cdot x$$

and $x^3 = x \cdot x \cdot x$

Then $x^6 \div x^3 = \frac{x^6}{x^3} = \frac{\not{x} \cdot \not{x} \cdot \not{x} \cdot x \cdot x \cdot x}{\not{x} \cdot \not{x} \cdot \not{x}} = x^3$

This result is obtained by canceling common factors in numerator and denominator. The above could be expressed as

$$x^6 \div x^3 = \frac{x^6}{x^3} = x^{6-3} = x^3$$

In like manner,

$$\frac{a^7}{a^3} = a^{7-3} = a^4$$

From the foregoing, it is seen that the law of exponents can be expressed in the general form

$$a^m \div a^n = \frac{a^m}{a^n} = a^{m-n}$$

where $a \neq 0$ and m and n are general numbers.

4-10 THE ZERO EXPONENT

Any number, except zero, divided by itself results in a quotient of 1. Thus,

$$\frac{6}{6} = 1$$

Also, $\frac{a^3}{a^3} = 1$

Therefore, $\frac{a^3}{a^3} = a^{3-3} = a^0 = 1$

Then, in the general form, $\frac{a^m}{a^n} = a^{m-n}$

If $m = n$
then $m - n = 0$
and $\frac{a^m}{a^n} = a^{m-n} = a^0 = 1$

The foregoing leads to the definition that

Any base, except zero, affected by zero exponent is equal to 1.

Thus, a^0, x^0, y^0, 3^0, 4^0, etc., all equal 1.

4-11 THE NEGATIVE EXPONENT

If the law of exponents in division is to apply to all cases, it must apply when n is greater than m. Thus,

$$\frac{a^2}{a^5} = \frac{\not{a} \cdot \not{a}}{\not{a} \cdot \not{a} \cdot a \cdot a \cdot a} = \frac{1}{a^3}$$

or $$\frac{a^2}{a^5} = a^{2-5} = a^{-3}$$

Therefore, $$a^{-3} = \frac{1}{a^3}$$

Also, $$a^{-n} = \frac{1}{a^n}$$

This leads to the definition that

Any base affected by a negative exponent is the same as 1 divided by that same base but affected by a positive exponent of the same absolute value as the negative exponent.

examples

$$x^{-4} = \frac{1}{x^4}$$

$$2^{-2} = \frac{1}{2^2} = \frac{1}{4}$$

$$3^{-3} = \frac{1}{3^3} = \frac{1}{27}$$

$$\frac{4^3}{4^5} = \frac{\not{4} \times \not{4} \times \not{4}}{\not{4} \times \not{4} \times \not{4} \times 4 \times 4}$$

$$= \frac{1}{4 \times 4} = \frac{1}{4^2} = 4^{-2}$$

or $$\frac{4^3}{4^5} = 4^{3-5} = 4^{-2}$$

It follows, from the consideration of negative exponents, that

Any *factor* of an algebraic term may be transferred from numerator to denominator, or vice versa, by changing the sign of the exponent of the *factor*.

example 10

$$3a^2x^3 = \frac{3a^2}{x^{-3}} = \frac{3}{a^{-2}x^{-3}} = \frac{3x^3}{a^{-2}}$$

4-12 DIVISION OF ONE MONOMIAL BY ANOTHER

Rule

To divide one monomial by another,

1. Find the quotient of the absolute values of the numerical coefficients and affix the proper sign according to the rules for division of positive and negative numbers (Sec. 4-8).
2. Determine the literal coefficients with their proper exponents, and write them after the numerical coefficient found in 1 above.

example 11 Divide $-12a^3x^4y$ by $4a^2x^2y$.

solution $$\frac{-12a^3x^4y}{4a^2x^2y} = -3ax^2$$

example 12 Divide $-7a^2b^4c$ by $-14ab^2c^3$. Express the quotient with positive exponents.

solution $$\frac{-7a^2b^4c}{-14ab^2c^3} = \frac{ab^2}{2c^2}$$

example 13 Divide $15a^{-2}b^2c^3d^{-4}$ by $-5a^2bc^{-1}d^{-2}$. Express the quotient with positive exponents.

solution $$\frac{15a^{-2}b^2c^3d^{-4}}{-5a^2bc^{-1}d^{-2}} = -\frac{3bc^4}{a^4d^2}$$

Division can be checked by substituting convenient numerical values for the literal factors or by multiplying the divisor by the quotient, the product of which should result in the dividend.

PROBLEMS 4-6

Divide:

1. $20x^4y^6$ by $5x^2y^2$
2. $-32x^8y^4z^6$ by $-8x^4yz^5$
3. $18\theta^4\phi^3\psi^5$ by $-9\theta^3\phi\psi^2$
4. $-96\omega^4L^8M^2$ by $-24\omega^2L^6M^2$
5. $108X_C{}^4Z^5$ by $-81X_C{}^3Z^3$
6. $-10^3a^{12}b^9c^{14}d^7$ by $10a^5b^5c^4d^4$
7. $-33\eta^6\lambda^4\pi^{10}$ by $-11\eta^2\lambda\pi^9$
8. $13k^4\Delta^3\varepsilon^5$ into $-39k^8\Delta^4\varepsilon^5$
9. $-\frac{7}{16}m^4n^5p^2$ into $-\frac{21}{64}m^5n^7p^3$
10. $-\frac{5}{8}x^4y^{12}z^8$ into $-\frac{25}{48}x^3y^{14}z^6$
11. $\dfrac{18c^9d^2e^3}{-2c^8d^2e^3}$
12. $\dfrac{33\theta^3\phi^2\alpha}{3\theta^6\phi\alpha^2}$
13. $\dfrac{108\lambda^5\psi^6Q^2}{-27\lambda^2\psi^2Q^2}$
14. $\dfrac{35i^4r^5p^3w^5}{-0.7ir^5p^4w^3}$
15. $\dfrac{-9a^{-3}b^4c^{-4}d^2}{-27a^{-2}b^{-3}c^2d^{-2}}$
16. $\dfrac{-21rs^2t^{-4}u^6}{-63r^2s^{-2}t^{-3}u^2}$
17. $\dfrac{13\phi^2\theta^{-6}\psi\Omega^{-1}}{-52\phi^{-4}\theta^6\psi^2\Omega^2}$
18. $\dfrac{\frac{7}{16}x^3y^4\alpha^{-3}}{\frac{3}{8}x^{-2}y^3\alpha^{-1}}$
19. $\dfrac{360\alpha^4\beta^3\gamma^{-7}}{0.004\alpha^2\beta^{-2}\gamma^{-5}}$
20. $\dfrac{-0.000\,256I^4R^3Z}{0.016I^{-2}R^{-1}Z^3}$

4-13 DIVISION OF A POLYNOMIAL BY A MONOMIAL

Because $\quad 2 \times 8 = 16$

then $\quad \frac{16}{2} = 8$

Also, because

$$3(a + 4) = 3a + 12$$

then

$$\frac{3a + 12}{3} = a + 4$$

Similarly, because

$$3I(2R + 3r) = 6IR + 9Ir$$

then

$$\frac{6IR + 9Ir}{3I} = 2R + 3r$$

From the foregoing we have the following:

Rule

To divide a polynomial by a monomial,

1. Divide each term of the dividend by the divisor.
2. Unite the results with the proper signs obtained by the division.

example 14

Divide $8a^2b^3c - 12a^3b^2c^2 + 4a^2b^2c$ by $4a^2b^2c$.

solution

$$\frac{8a^2b^3c - 12a^3b^2c^2 + 4a^2b^2c}{4a^2b^2c} = 2b - 3ac + 1$$

example 15

Divide $-27x^3y^2z^5 + 3x^4y^2z^4 - 9x^4y^3z^5$ by $-3x^3y^2z^4$.

solution

$$\frac{-27x^3y^2z^5 + 3x^4y^2z^4 - 9x^4y^3z^5}{-3x^3y^2z^4}$$

$$= 9z - x + 3xyz$$

PROBLEMS 4-7

Divide:

1. $8x + 10y$ by 2
2. $12\theta - 6\phi$ by 3

3. $108\alpha^2 - 81\beta^2$ by 9

4. $16\phi^6 - 8\phi^4 + 24\phi^2$ by $4\phi^2$

5. $24R_1 + 48R_2 - 32R_3$ by 8

6. $X_C^{\,6} - 12X_C^{\,4} - 18X_C^{\,2}$ by $3X_C$

7. $0.025\mu^4\pi^2 + 50\mu^2\pi^4$ by $5\mu\pi^3$

8. $8.1\alpha^3\beta^2\gamma + 7.2\alpha^2\beta\gamma^3 - 3.6\alpha\beta\gamma$ by $0.09\alpha^2\beta^2\gamma^2$

9. $\frac{1}{2}m^2$ into $\frac{3}{20}m^5 - \frac{7}{10}m^3 - \frac{3}{5}m$

10. $-\frac{3}{4}I^2R$ into $\frac{5}{16}I^4R^2 - \frac{3}{8}I^2R + \frac{3}{10}I^{-2}R^{-1} + \frac{3}{4}I^{-4}R^{-2}$

11. $\dfrac{102xyz + 170x^2yz^2 - 85x^3yz^3 - 51x^5y^5z}{17xyz}$

12. $R^2(I + i) + r^2(I + i)$ by $I + i$

13. $8(\theta + \phi)^2 - 16(\theta + \phi)^4 + 12(\theta + \phi)^6$ by $4(\theta + \phi)$

14. $\lambda(\alpha^2 + \beta^2)^2 - \pi(\alpha^2 + \beta^2)^2$ by $-(\alpha^2 + \beta^2)^2$

15. $8\pi(EI + P)^4 - 32\pi(EI + P)^2 + 96\pi(EI + P)$ by $16\pi(EI + P)^2$

16. $\dfrac{6I^2(R + r)(R - r) + 10I^4(R + r)^2(R - r)^2 - 12I^8(R + r)^4(R - r)^4}{-2I(R + r)(R - r)}$

17. $\dfrac{5I\left(\omega L - \dfrac{1}{\omega C}\right) - 10I^3\left(\omega L - \dfrac{1}{\omega C}\right)^3 - 25I^5\left(\omega L - \dfrac{1}{\omega C}\right)^5}{5I^2\left(\omega L - \dfrac{1}{\omega C}\right)^2}$

18. $\dfrac{(2\beta + 7)(\beta + 1) - (3\beta + 2)(\beta + 1)^2}{\beta + 1}$

19. $\dfrac{-36\omega(\theta + \phi)(\theta - \phi) + 12\omega(\theta + \phi)^2(\theta - \phi)^2 - 24\omega(\theta + \phi)^3(\theta - \phi)^3}{4\omega(\theta + \phi)^2}$

20. $\dfrac{54E^2(R + R_1)(r + r_1) + 36E^4(R + R_1)^2(r + r_1)^2 - 108E^6(R + R_1)^3(r + r_1)^3}{9E(R + R_1)(r + r_1)}$

4-14 DIVISION OF ONE POLYNOMIAL BY ANOTHER

Rule

To divide one polynomial by another,

1. Arrange the dividend and divisor in ascending or descending powers of some common literal factor.
2. Divide the first term of the dividend by the first term of the divisor, and write the result as the first term of the quotient.
3. Multiply the entire divisor by the first term of the quotient, write the product under the proper terms of the dividend, and subtract it from the dividend.
4. Consider the remainder a new dividend, and repeat 1, 2, and 3 until there is no remainder or until there is a remainder that cannot be divided by the divisor.

example 16 Divide $x^2 + 5x + 6$ by $x + 2$.

solution Write the divisor and dividend in the usual positions for long division and eliminate the terms of the dividend, one by one:

$$\begin{array}{r} x + 3 \\ x + 2\overline{)\,x^2 + 5x + 6} \\ \underline{x^2 + 2x\phantom{{}+6}} \\ 3x + 6 \\ \underline{3x + 6} \end{array}$$

x, the first term of the divisor divides into x^2, the first term of the dividend, x times. Therefore, x is written as the first term of the quotient. The product of the first term of the quotient and the divisor $x^2 + 2x$ is then written under like terms in the dividend and subtracted. The first term of the remainder then serves as a new dividend, and the process of division is continued.

This result can be checked by multiplying the divisor by the quotient.

$$\begin{array}{rl} \text{Divisor} = & x + 2 \\ \text{Quotient} = & \underline{x + 3} \\ & x^2 + 2x \\ & \underline{\qquad 3x + 6} \\ \text{Dividend} = & x^2 + 5x + 6 \end{array}$$

example 17

Divide $a^2b^2 + a^4 + b^4$ by $-ab + b^2 + a^2$.

solution First arrange the dividend and divisor according to step 1 of the rule. Because there are no a^3b or ab^3 terms, allowance is made by supplying 0 terms. Thus,

$$\begin{array}{r|l} & a^2 + ab + b^2 \\ \hline a^2 - ab + b^2 & a^4 + 0 + a^2b^2 + 0 + b^4 \\ & \underline{a^4 - a^3b + a^2b^2} \\ & a^3b \\ & \underline{a^3b - a^2b^2 + ab^3} \\ & a^2b^2 - ab^3 + b^4 \\ & \underline{a^2b^2 - ab^3 + b^4} \end{array}$$

example 18

Divide $4 + x^4 + 3x^2$ by $x^2 - 2$.

solution

$$\begin{array}{r|l} & x^2 + 5 \\ \hline x^2 - 2 & x^4 + 3x^2 + 4 \\ & \underline{x^4 - 2x^2} \\ & 5x^2 + 4 \\ & \underline{5x^2 - 10} \\ & 14 = \text{remainder} \end{array}$$

This result is written

$$x^2 + 5 + \frac{14}{x^2 - 2}$$

which is as it would be written in an arithmetical division that did not divide out evenly.

PROBLEMS 4-8

Divide:

1. $x^2 + 2x + 1$ by $x + 1$

2. $9p^2 + 9p - 40$ by $3p - 5$

3. $\theta^2 + 7\theta + 12$ by $\theta + 4$

4. $12\omega^2 + 29\omega + 14$ by $4\omega + 7$

5. $6E^2 - 22E + 12$ by $3E - 2$

6. $6\phi^2 - 13\phi\theta + 6\theta^2$ by $2\phi - 3\theta$

7. $3R^3 + 9R^2 - 7R - 4RZ - 12Z - 21$ by $R + 3$

8. $\phi^3 + 3\phi^2\omega + 3\phi\omega^2 + \omega^3$ by $\phi + \omega$

9. $K^3 + 6K^2 + 7K - 8$ by $K - 1$

10. $12\lambda^2 - 36\phi^2 - 11\lambda\phi$ by $4\lambda - 9\phi$

11. $E^2 - e^2$ by $E - e$

12. $E^3 - e^3$ by $E - e$

13. $E^4 - e^4$ by $E - e$

14. $E^3 + I^3R^3$ by $E + IR$

15. $E^4 - I^4R^4$ by $E^2 - I^2R^2$

16. $\theta^5 + \phi^5$ by $\theta + \phi$

17. $X^6 - Y^6$ by $X + Y$

18. $X^6 + Y^6$ by $X^2 + Y^2$
19. $\theta^3 + 3\theta^2\phi + 3\theta\phi^2 + \phi^3$ by $\theta + \phi$
20. $L_1{}^4 - L_2{}^4$ by $L_1 + L_2$
21. $6R_2{}^3 - R_2{}^2 - 14R_2 + 3$ by $3R_2{}^2 + 4R_2 - 1$
22. $1 + 2m^4 + 4m^2 - m^3 + 7m$ by $3 + m^2 - m$
23. $30E^4 + 3 - 82E^2 - 5E + 11E^3$ by $3E^2 - 4 + 2E$
24. $\frac{1}{8}\theta^3 - \frac{9}{4}\theta^2\phi + \frac{27}{2}\theta\phi^2 - 27\phi^3$ by $\frac{1}{2}\theta - 3\phi$
25. $6R^2 - \frac{5}{6}R - \frac{1}{6}$ by $2R - \frac{1}{2}$
26. $n^3 - \frac{9}{5}n^2 - \frac{9}{25}n - \frac{27}{125}$ by $n - \frac{3}{5}$
27. $36x^2 + \frac{1}{9}y^2 + \frac{1}{4} - 4xy - 6x + \frac{1}{3}y$ by $6x - \frac{1}{3}y - \frac{1}{2}$
28. $\frac{1}{27}K^3 - \frac{1}{12}K^2 + \frac{1}{16}K - \frac{1}{64}$ by $\frac{1}{3}K - \frac{1}{4}$
29. $\frac{3}{2}L_1{}^2 - L_1 - \frac{8}{3}$ into $\frac{9}{16}L_1{}^4 - \frac{3}{4}L_1{}^3 - \frac{7}{4}L_1{}^2 + \frac{4}{3}L_1 + \frac{16}{9}$
30. $R_1{}^7 + \left(\frac{E}{I}\right)^7$ by $R_1 + \frac{E}{I}$

5

equations

In the preceding chapters, considerable time has been spent in the study of the fundamental operations of algebra. These fundamentals will be of little value unless they can be put to practical use in the solution of problems. This is accomplished by use of the equation, the most valuable tool in mathematics.

5-1 DEFINITIONS

An *equation* is a mathematical statement that two numbers, or quantities, are equal. The *equality sign* (=) is used to separate the two equal quantities. The terms to the left of the equality sign are known as the *left member* of the equation, and the terms to the right are known as the *right member* of the equation. For example, in the equation

$$3E + 4 = 2E + 6$$

$3E + 4$ is the left member and is equal to $2E + 6$, which is the right member.

An *identical equation*, or *identity*, is an equation whose members are equal for all values of the literal numbers contained in the equation. The equation

$$4I(r + R) = 4Ir + 4IR$$

is an identity because if $I = 2$, $r = 3$, and $R = 1$, then

$$4I(r + R) = 4 \cdot 2(3 + 1) = 32$$

Also, $$4Ir + 4IR = 4 \cdot 2 \cdot 3 + 4 \cdot 2 \cdot 1 = 24 + 8 = 32$$

Any other values of I, r, and R substituted in the equation will produce equal numerical results in the two members of the equation.

An equation is said to be *satisfied* if, when numerical values are substituted for the literal numbers, the equation becomes an identity. Thus, the equation

$$ir - iR = 3r - 3R$$

is satisfied by $i = 3$, because when we substitute this value in the equation, we obtain

$$3r - 3R = 3r - 3R$$

which is an identity.

A *conditional equation* is one consisting of one or more literal numbers that is not satisfied by all values of the literal numbers. Thus, the equation

$$e + 3 = 7$$

is not satisfied by any value of e except $e = 4$.

To *solve* an equation is to find the value or values of the unknown number that will satisfy the equation. This value is called the *root* of the equation. Thus, if

$$i + 6 = 14$$

the equation becomes an identity only when i is 8, and therefore 8 is the root of the equation.

5-2 AXIOMS

An *axiom* is a truth, or fact, that is self-evident and needs no formal proof. The various methods of solving equations are derived from the following axioms:

1. If equal numbers are added to equal numbers, the sums are equal.

example 1 If $x = x$,

then $x + 2 = x + 2$
because, if $x = 4$,
$4 + 2 = 4 + 2$
or $6 = 6$

Therefore, *the same number can be added to both members of an equation without destroying the equality.*

2. If equal numbers are subtracted from equal numbers, the remainders are equal.

example 2 If $x = x$,

then $x - 2 = x - 2$
because, if $x = 4$,
$4 - 2 = 4 - 2$
or $2 = 2$

Therefore, *the same number can be subtracted from both members of an equation without destroying the equality.*

3. If equal numbers are multiplied by equal numbers, their products are equal.

example 3 If $x = x$,

then $3x = 3x$
because, if $x = 4$,
$3 \cdot 4 = 3 \cdot 4$
or $12 = 12$

Therefore, *both members of an equation can be multiplied by the same number without destroying the equality.*

4. If equal numbers are divided by equal numbers, their quotients are equal.

example 4 If $x = x$,

then $\frac{x}{2} = \frac{x}{2}$

because, if $x = 4$,

$\frac{4}{2} = \frac{4}{2}$

or $2 = 2$

Therefore, *both members of an equation can be divided by the same number without destroying the equality, except that division by zero is not allowed.*

5. Numbers that are equal to the same number or equal numbers are equal to each other.

example 5 If $a = x$ and $b = x$,

then $a = b$
because, if $x = 4$,
$a = 4$
and $b = 4$

Therefore, *an equal quantity can be substituted for any term of an equation without destroying the equality.*

6. Like powers of equal numbers are equal.

example 6 If $x = x$,

then $x^3 = x^3$
because, if $x = 4$,
$4^3 = 4^3$
or $64 = 64$

Therefore, *both members of an equation can be raised to the same power without destroying the equality.*

7. Like roots of equal numbers are equal.

example 7 If $x = x$,

then $\sqrt{x} = \sqrt{x}$
because, if $x = 4$,
$\sqrt{4} = \sqrt{4}$
or $2 = 2$

Therefore, *like roots of both members of an equation can be extracted without destroying the equality.*

5-3 NOTATION

In order to shorten the *explanations* of the solutions of various equations, we shall employ the letters **A, S, M,** and **D** for "add," "subtract," "multiply," and "divide," respectively. Thus,

- **A:** 6 will mean "add 6 to both members of the equation."
- **S:** $-6x$ will mean "subtract $-6x$ from both members of the equation."
- **M:** $-3a$ will mean "multiply both members of the equation by $-3a$."
- **D:** 2 will mean "divide both members of the equation by 2."

5-4 THE SOLUTION OF EQUATIONS

A considerable amount of time and drill must be spent in order to become proficient in the solution of equations. It is in this branch of mathematics that you will find you must be familiar with the more elementary parts of algebra.

Some of the methods used in the solutions are very easy—so easy, in fact, that there is a tendency to employ them mechanically. This is all very well, but you should not become so mechanical that you forget the reason for performing certain operations.

We shall begin the solution of equations with very easy cases and attempt to build up general methods of procedure for all equations as we proceed to the more difficult problems.

If you are studying equations for the first time, you are urged to study the following examples carefully until you thoroughly understand the methods and the reasons behind them.

example 8 Find the value of x if $x - 3 = 2$.

solution In this equation, it is seen by inspection that x must be equal to 5. However, to make the solution by the methods of algebra, proceed as follows:

Given $x - 3 = 2$
A: 3, $x = 2 + 3$ (Axiom 1)
Collecting terms, $x = 5$

example 9 Solve for e if $e + 4 = 12$.

solution

Given $e + 4 = 12$

S: 4,	$e = 12 - 4$	(Axiom 2)
Collecting terms,	$e = 8$	

example 10 Solve for i if $3i + 5 = 20$.

solution

Given	$3i + 5 = 20$	
S: 5,	$3i = 20 - 5$	(Axiom 2)
Collecting terms,	$3i = 15$	
D: 3,	$i = 5$	(Axiom 4)

example 11

Solve for r if $40r - 10 = 15r + 90$.

solution

Given	$40r - 10 = 15r + 90$	
S: $15r$,	$40r - 10 - 15r = 90$	(Axiom 2)
A: 10,	$40r - 15r = 90 + 10$	(Axiom 1)
Collecting terms,	$25r = 100$	
D: 25	$r = 4$	(Axiom 4)

From the foregoing examples, it will be noted that adding or subtracting a term from both members of an equation is equivalent to *transposing* that number from one member to the other and changing its sign. This fact leads to the following rule:

Rule
A *term* can be transposed from one member of an equation to the other provided that its sign is changed.

By transposing all terms containing the unknown to the left member and all others to the right member, by collecting terms and dividing both members by the numerical coefficient of the unknown, the equation has been solved for the value of the unknown.

5-5 CANCELING TERMS IN AN EQUATION

example 12 Solve for x if $x + y = z + y$.

solution

Given	$x + y = z + y$	
S: y,	$x = z$	(Axiom 2)

The term y in both members of the given equation does not appear in the next equation as the result of subtraction. The result is the same as if the term were dropped from both members. This fact leads to the following rule:

Rule
If the same *term* preceded by the same sign occurs in both members of an equation, it can be canceled.

5-6 CHANGING SIGNS IN AN EQUATION

example 13 Solve for x if $8 - x = 3$.

solution

Given	$8 - x = 3$	
S: 8,	$-x = 3 - 8$	(Axiom 2)
M: -1,	$x = -3 + 8$	(Axiom 3)
Collecting terms,	$x = 5$	

Note that multiplication by -1 has the effect of changing the signs of all terms. This gives the following rule:

Rule
The signs of all the *terms* of an equation can be changed without destroying the equality.

Although the foregoing rules involving mechanical methods are valuable, you should not lose sight of the fact that they are all derived from fundamentals, or axioms, as outlined in Sec. 5-2.

5-7 CHECKING THE SOLUTION

If there is any doubt that the value of the unknown is correct, the solution can be checked by substituting the value of the unknown in the original equation. If the two members reduce to an identity, the value of the unknown is correct.

example 14
Solve and test $3i + 14 + 2i = i + 26$.

solution

Given $\quad 3i + 14 + 2i = i + 26$

Transposing,

$$3i + 2i - i = 26 - 14$$

Collecting terms, $\quad 4i = 12$

D: 4, $\quad i = 3$

Test by substituting $i = 3$ in given equation.

check

$$(3 \cdot 3) + 14 + (2 \cdot 3) = 3 + 26$$
$$9 + 14 + 6 = 3 + 26$$
$$29 = 29$$

PROBLEMS 5-1
Solve for the unknown in the following equations:

1. $3x - 6 = 6$
2. $4\theta - 1 = 3\theta + 3$
3. $k - 10 = 5 + 4k$
4. $l - 9l = -6l - 2$
5. $6p + 3 - 2p = 27$
6. $16 - 9\mu = 5\mu - 12$
7. $11\pi - 22 = 4\pi + 13$
8. $5M + 2 = 3 + 4M$
9. $21 - 15IR = -8IR - 7$
10. $27Q + 22 = 30 + 17Q$
11. $8\alpha - 5(4\alpha + 3) = -3 - 4(2\alpha - 7)$
12. $3(\lambda - 2) - 10(\lambda - 6) = 5$
13. $4 + 3(E - 7) = 16 + 2(5E + 1)$
14. $4(K - 5) - 3(K - 2) = 2(K - 1)$
15. $0 = 18 - 4Q + 27 + 9Q - 3 + 16Q$
16. $25R_1 - 19 - [3 - (4R_1 - 15)] - 3R_1 + (6R_1 + 21) = 0$
17. $19 - 5I(4I + 1) = 40 - 10I(2I - 1)$
18. $(\phi + 5)(\phi - 4) + 4\phi^2 = (5\phi + 3)(\phi - 4) + 2(\phi - 4) + 64$
19. $6(\beta - 1)(\beta - 2) - 4(\beta + 2)(\beta + 1) = 2(\beta + 1)(\beta - 1) - 24$
20. $18 - 3Z(2Z + 1) - [3 - 2(Z + 2)(Z - 3)] = 18 - 6Z - 4(Z - 5)(Z + 2)$

5-8 FORMING AND SOLVING EQUATIONS

As previously stated, we are continually trying to express certain laws and relations in the language of mathematics.

examples

Words	*Algebraic Symbols*
The sum of the voltages E and e	$E + e$
The difference between resistances R and R_1	$R - R_1$
The excess of current I_1 over current I_2	$I_1 - I_2$
The number of centimeters in l meters	$100l$
The number of cents in d dollars	$100d$
The voltage E is equal to the product of the current I and the resistance R	$E = IR$

The solution of most problems consists in writing an equation that connects various observed data with known facts. This, then, is nothing more than translating from ordinary English, or words, into the symbolic language of mathematics. In relatively simple problems the translation can be made directly, almost word by word, into algebraic symbols (Examples 15 and 16).

It is almost impossible to lay down a set of rules for the solution of general problems, for they could not be made applicable to all cases. However, no rules will be needed if you thoroughly understand what is to be translated into the language of mathematics from the wording or facts of the problem at hand. The following outline will serve as a guide:

1. Read the problem carefully so that you understand every fact in it and recognize the relationships between the facts.
2. Determine what is to be found (the unknown quantity), and denote it by some letter. If there are two unknowns, try to represent one of them in terms of the other. If there are more than two unknowns, try to represent all but one of them in terms of that one.
3. Find two expressions which, according to the facts of the problem, represent the same quantity, and set them equal to each other. You can then solve the resulting equation for the unknown.

example 15 Five times a certain voltage diminished by 3,

$$5 \quad \times \qquad E \qquad - \qquad 3$$

gives the same result as the voltage increased by 125.

$$= \qquad E \qquad + \qquad 125$$

That is, $5E - 3 = E + 125$

or $E = 32\text{ V}$

example 16 What number increased by 42 is equal to 110?

$$x \qquad + \qquad 42 \qquad = \qquad 110$$

That is, $x + 42 = 110$

or $x = 68$ ***check*** $68 + 42 = 110$

PROBLEMS 5-2

1. The sum of two voltages is E V. One voltage is 75 V. What is the other?
2. The difference between two resistances is 10.5 Ω. One resistance is R Ω. What is the other?
3. How great a distance d will you travel in t hours (h) at r kilometers per hour (km/h)?
4. What is the fraction f whose numerator n is 3 less than its denominator?
5. An electric timer has a guarantee of y years. We have been using it for t years. For how many years longer will the guarantee apply?
6. An oscilloscope is guaranteed for q years, and it has been in service for m months. How much longer is it covered by the guarantee?
7. At what speed must a missile be traveling to cover Z km in t min?
8. From what number must 8 be subtracted in order that the remainder may be 27?
9. If a certain voltage is doubled and the result is diminished by 15, the remainder is 205 V. What is the voltage?
10. The volume of a parts container is v cubic centimeters (cm^3). Express the height in centimeters if the width is w cm and the length is l cm.
11. Write algebraically that the current is equal to the voltage divided by the resistance.
12. Write algebraically that the power dissipated by a resistor is equal to the square of the current multiplied by the resistance.
13. A stock room is twice as long as it is wide, and its perimeter is 36 m. Find its length and width.
14. A multimeter and an oscilloscope together cost \$574. The oscilloscope costs \$356 more than the meter. Find the cost of (*a*) the oscilloscope and (*b*) the multimeter.
15. Find the three sides of a triangle whose perimeter is 23.5 m if one side is 6.5 m shorter than the second side, and one-half the third side.
16. The sum of the three angles in any triangle is 180°. The smallest angle in a given triangle is one-half the second angle and 52° smaller than the largest angle. How many degrees does each angle contain?
17. Write algebraically that the square on the hypotenuse h of a right triangle is equal to the sum of the squares on the other two sides, which are identified as a and b.
18. The sum of two consecutive numbers is 31. What are the numbers?
19. The sum of three consecutive numbers is 192. What are the numbers?
20. Write algebraically that the product of the impressed emf E and the resultant current I in a circuit is equivalent to the square of the emf divided by R, the resistance in the circuit.

5-9 LITERAL EQUATIONS—FORMULAS

A *formula* is a rule, or law, generally pertaining to some scientific relationship expressed as an equation by means of letters, symbols, and constant terms.

example 17 The area A of a rectangle is equal to the product of its base b and its altitude h. This statement written as a formula is

$$A = bh$$

example 18 The power P expended in an electric circuit is equal to its current I squared times the resistance R of the circuit. Stated as a formula

$$P = I^2R$$

The ability to handle formulas is of the utmost importance. The usual formula expresses one quantity in terms of other quantities, and it is often desirable to solve for *any* quantity contained in a formula. This is readily accomplished by using the knowledge gained in solving equations.

example 19 The voltage E across a part of a circuit is given by the current I through that part of the circuit times the resistance R of that part. That is,

$$E = IR$$

Suppose E and I are given but it is desired to find R.

Given	$E = IR$	
D: I,	$\frac{E}{I} = R$	(Axiom 4)
or	$R = \frac{E}{I}$	

Similarly, if we wanted to solve for I,

Given	$E = IR$	
D: R,	$\frac{E}{R} = I$	(Axiom 4)
or	$I = \frac{E}{R}$	

example 20 Solve for I if $e = E - IR$.

solution

Given	$e = E - IR$	
Transposing,	$IR = E - e$	
D: R,	$I = \frac{E - e}{R}$	(Axiom 4)

example 21 Solve for C if $X_C = \frac{1}{2\pi fC}$.

solution

Given	$X_C = \frac{1}{2\pi fC}$	
D: X_C,	$1 = \frac{1}{2\pi fCX_C}$	(Axiom 4)
M: C,	$C = \frac{1}{2\pi fX_C}$	(Axiom 3)

It will be noted from the foregoing examples that if the numerator of a member of an equation contains but one term, any *factor* of that term may be transferred to the denominator of the other member as a *factor*. In like manner if the denominator of a member of an equation contains but one term, any *factor* of that term may be transferred to the numerator of the other member as a *factor*. These mechanical transformations simply make use of Axioms 3 and 4, and you should not lose sight of the real reasons behind them.

PROBLEMS 5-3

Given:	Solve for:	Given:	Solve for:
1. $Q = CV$	C and V	**2.** $I = \frac{E}{Z}$	E and Z
3. $R^2 = Z^2 - X^2$	Z^2 and X^2	**4.** $R = \frac{P}{I^2}$	P and I^2
5. $L = \frac{Rm}{K}$	R, K, and m	**6.** $R_2 = R_t - R_1 - R_3$	R_t, R_1, and R_3
7. $f = \frac{v}{\lambda}$	λ and v	**8.** $C = 2\pi r$	r and π
9. $R = \frac{\omega L}{Q}$	L, Q, and ω	**10.** $L = \frac{X_L}{2\pi f}$	X_L and f
11. $C = \frac{1}{2\pi f X_C}$	X_C and f	**12.** $S = 2\pi rh$	r and h
13. $H = \frac{\phi}{A}$	ϕ and A	**14.** $N_s = \frac{E_s N_p}{E_p}$	N_p, E_s, and E_p
15. $B = \frac{E10^8}{Lv}$	E, L, and v	**16.** $T = ph + 2A$	h and A
17. $E_s I_s = E_p I_p$	E_s	**18.** $L = \frac{F}{Hi}$	F, H, and i
19. $R = \frac{E - e}{I}$	I, E, and e	**20.** $\mu = g_m r_p$	g_m
21. $t = \frac{\theta}{\omega}$	θ and ω	**22.** $h = \frac{V^2}{2g}$	g and V^2
23. $V_0 = 2V - V_t$	V and V_t	**24.** $n = \frac{\omega}{2\pi}$	ω
25. $A = \frac{4}{3}\pi r^3$	r^3	**26.** $\mu = \frac{B^2 Al}{8\omega}$	l, ω, and A
27. $C = \frac{F(R - r)}{Z_t}$	Z_t, F, R, and r	**28.** $r = \frac{F}{4\pi^2 n^2 m}$	m and F
29. $R_L = \frac{E_b - e_b}{i}$	E_b, e_b, and i	**30.** $t = \frac{T(C - F)}{C}$	T and F
31. $R = \frac{\rho l}{d^2}$	l, ρ, and d^2	**32.** $PF = \frac{R}{X}$	R and X
33. $C = \frac{0.0884KA(n - 1)}{d}$	A and n	**34.** $M = k\sqrt{L_1 L_2}$	k

35. $Z_r = \dfrac{L}{RC}$ L, C, and R

36. $F = \dfrac{eI}{2kT_g}$ T_g

37. $\omega = \dfrac{\eta\beta}{\gamma\alpha}$ β

38. $\dfrac{P_{so}}{P_{no}} = \dfrac{P_{si}}{P_{ni}}$ P_{no}

39. $\rho = \dfrac{Qe}{hv}$ Q

40. $E_b = \dfrac{V_B + V_{pt}}{W}$ V_{pt}

41. $V_2 = (1 - \omega^2 LC_2)V_3$ C_2

42. $4a = \dfrac{h + 2b}{v}$ b

43. $Q = I_p p + I_n n$ I_n

44. $G_o = G + \dfrac{g_m}{1 + n}$ g_m

45. $\omega_{01} = \dfrac{1}{C(R_1 + R_2)}$ R_1

note When solving numerical problems which involve the solution of formulas, always solve the formula algebraically for the wanted factor before substituting the numerical values. This procedure permits you to check your work more easily, because the letters retain their identity through the various algebraic procedures, whereas once numbers are added, multiplied, etc., their identity is lost, and your audit becomes more difficult.

46. The power P in any part of an electric circuit is given by $P = \dfrac{E^2}{R}$ W, in which E is the emf applied to that part of the circuit and R is the resistance of that part. What is the resistance of a circuit in which 1.21 W is expended at an emf of 110 V?

47. The voltage drop E across any part of a circuit can be computed by the formula $E = IZ$ V, where I is the current in amperes through that part of the circuit and Z is the impedance in ohms of that part. What is the impedance of a circuit in which a voltage drop of 460 V is produced by a current flow of 0.115 A?

48. To find the frequency f of an alternator in hertz (Hz), that is, cycles per second, the number of pairs of poles P is multiplied by the speed of the armature S in revolutions per second (rev/s), $f = PS$. A tachometer connected to the rotor of a 60-Hz alternator reads 3600 revolutions per minute (rev/min). How many poles has the alternator?

49. For radio waves, the relationship between frequency f in megahertz (MHz) and wavelength λ in meters is expressed by the formula $f = \dfrac{3 \times 10^2}{\lambda}$ MHz. What is the wavelength of a radio wave at 60 MHz?

50. The length of a broadband dipole L_{fD} used for television reception can be computed by the formula $L_{fD} = \dfrac{141.3}{f}$ m, where f is the frequency in mega-

hertz. The folded dipole in Fig. 5-1 is 0.7976 m. For what frequency was it constructed?

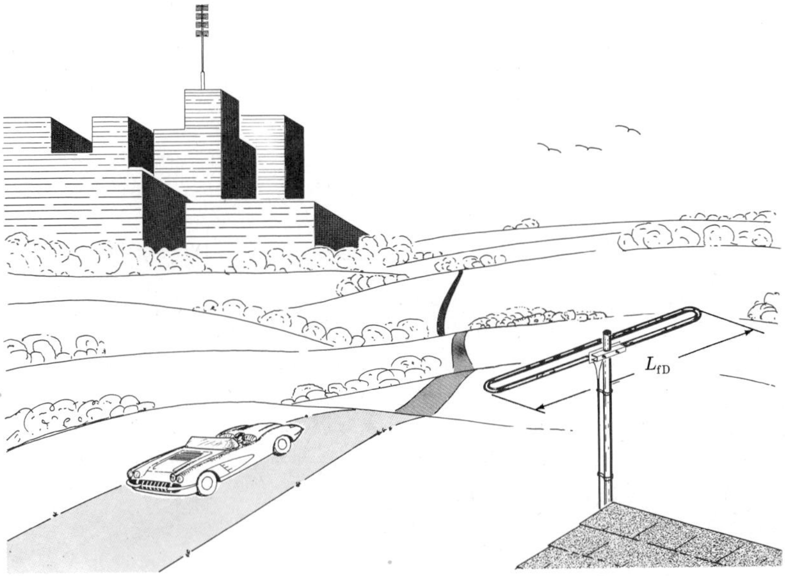

Fig. 5-1 Folded Dipole of Prob. 50

5-10 RATIO AND PROPORTION

Because proportions are special forms of equations, it is expedient to look now at the twin subjects ratio and proportion.

A ratio is a comparison of two things expressed in one of two ways: first, the "old-fashioned" method, $a:b$, pronounced "a is to b"; and second, as found in newer books, $\frac{a}{b}$. If the ratio of x to y is 1 to 4, or $\frac{1}{4}$, then x is one-quarter of y. Alternatively, y is four times as great as x.

example 22 Write the ratio of 25 cents (¢) to \$3.00.

solution 25¢ to \$3.00 may be written simply as 25¢:\$3.00, but this does not tell us much. It is more helpful to convert both quantities to the same units:

$$\frac{25¢}{\$3.00} = \frac{25¢}{300¢} = \frac{1}{12}$$

Note that the two parts being compared are given the same units, in this case cents, so that when the simplification is performed, not only the numbers but also the units are canceled. Thus a *true ratio* is a "pure," or dimensionless, number. Notice also that a ratio may be an integer; that is, a fraction whose denominator is 1.

PROBLEMS 5-4

Write as a fraction the ratio of:

1. 3 cm to 12 cm

2. 12 square meters (m^2) to 18 m^2

3. 15 000 Ω to 12 000 Ω
4. \$5.00 to 25¢
5. Write two different sets of numbers in the ratio 2:3.
6. Write two different sets of numbers in the ratio 0.125:1.
7. A recipe for ceramic insulators calls for 8 parts of type *A* clay to 24 parts of type *B*. What is the ratio of type *A* to type *B*?
8. In Prob. 7, what is the ratio of the weight of type *A* to the weight of the total mixture?
9. The mechanical advantage (MA) of any machine is the ratio of load moved to effort applied. What is the MA of a system in which 24 kilograms (kg) of effort just starts motion of a 768-kg load?
10. In a certain alloy, 55% of the material is copper and 22% is zinc. What is the ratio of zinc to copper?

Just as ratios compare two things, so proportions are equalities of pairs of ratios.

When we draw a map to scale, the proportions on the map should equal those on the ground. If the scale is 1 cm to 10 km, then a trip which is 3 cm on the map must be 30 km on the ground. The proportion here is $\frac{1\text{ cm}}{3\text{ cm}} = \frac{10\text{ km}}{30\text{ km}}$ and, since the units cancel, our true proportion is an equality of two pure numbers. We could also write this proportion as $\frac{1\text{ cm}}{10\text{ km}} = \frac{3\text{ cm}}{30\text{ km}}$. Note that it is essential that the units on one side of the proportion be equal to those on the other side. This provides one good way of checking your work. If you perform a wrong operation such as multiplying instead of dividing, you will find that your units will reveal an error. The solution may read "cm/km = cm · km" and such an imbalance of units is a sure indication that you have made an error.

The usual purpose of proportions is to solve one part when the other three parts are known.

example 23 Given the proportion $\frac{18}{a} = \frac{6}{5}$, solve for a.

solution Obeying the usual rules of equations,

$$18 \times 5 = 6 \times a$$
$$a = 15$$

In the older form of writing ratios and proportions, $\frac{a}{b} = \frac{c}{d}$ would be written $a:b::c:d$, and pronounced "a is to b as c is to d." The elements on the outsides of the proportion were called the "extremes," and those in the middle the "means." Based on these definitions, you can prove the old law of proportions:

Rule

In a proportion, the product of the means equals the product of the extremes.

PROBLEMS 5-5

Find the missing term in each of the following proportions:

1. $\frac{5}{8} = \frac{?}{16}$
2. $\frac{3}{7} = \frac{?}{56}$
3. $\frac{?}{15} = \frac{40}{3}$
4. $\frac{80}{?} = \frac{60}{12}$

5. $\frac{X}{90} = \frac{60}{360}$ 6. $\frac{5}{i} = \frac{0.2}{36}$ 7. $\frac{6}{IR} = \frac{9}{12}$ 8. $\frac{0.6}{1.2} = \frac{0.4}{d}$

9. $\frac{0.007}{0.200} = \frac{Q}{0.04}$ 10. $\frac{16}{Z} = \frac{Z}{4}$

5-11 VARIATION AND PROPORTIONALITY

Often, in the study of electronics, you will hear such expressions as "the current is proportional to the voltage and inversely proportional to the resistance" and "the force is jointly proportional to the charges and inversely proportional to the square of the distance between them."

Sometimes an equivalent expression is used: "the current varies directly as the voltage," etc.

Two forms may be used to express mathematically the words "the current varies as the voltage." The first uses the symbol of proportionality: $I \propto E$. The second substitutes for the symbol $\propto$ the equivalent "$= k$," where k is the "konstant" of proportionality: $I = kE$. Other symbols such as b, c, n, etc., also are used as constants.

Similarly, the expression "the current is inversely proportional to the resistance" may be written $I \propto \frac{1}{R}$, or $I = k\frac{1}{R}$, or simply, $I = \frac{k}{R}$.

"Jointly proportional" means "proportional to the product," so that "the force is jointly proportional to the masses" may be written $F \propto m_1 m_2$ or $F = km_1 m_2$.

Often, past experience, tables, measurements, as well as calculations, may reveal the value of the constant of proportionality. For example, we know that the circumference of a circle is proportional to its radius. We may write this $C \propto R$, or $C = kR$. However, from previous knowledge, we can replace the general constant k by the known constant of proportionality 2π, and we can write $C = 2\pi R$.

example 24 If a varies directly as ρ and if $a = 8$ when $\rho = 4$, what will be the value of a when $\rho = 7$?

solution $a \propto \rho = k\rho$. We know that $8 = k4$, from which $k = 2$. Substitute this value of k into the second condition:

$$a = k \times 7 = 2 \times 7 = 14$$

PROBLEMS 5-6

Write the following expressions in "proportionality" form and in "equation" form:

1. The distance D varies directly as the rate R.
2. The cost C varies directly as the weight W.
3. The capacitance C varies directly as the area A.
4. The reactance X_L varies jointly with the frequency f and the inductance L.
5. The capacitive reactance X_C varies inversely as the capacitance C.
6. The resistance varies directly as the length l and inversely as the cross-sectional area A.

7. The period T of vibration of a reed is directly proportional to the square root of the length l.
8. The volume of a sphere V is proportional to the cube of its radius r.
9. The volume of a gas V varies inversely as the pressure P.
10. The ratio of the similar areas A_1 and A_2 is proportional to the square of the ratio of corresponding lengths l_1 and l_2.
11. The illumination L of an object varies inversely as the square of the distance d from the source of light.
12. If the current I varies directly as the voltage E and if $I = 0.5$ A when $E = 30$ V, what will be the value of I when $E = 75$ V?
13. In a certain varistor the current is proportional to the square of the voltage. If $I = 0.006$ A when $E = 110$ V, what voltage will produce a current of 1.5 A?
14. The resistance R of a wire varies directly as the wire length l and inversely as the square of the wire diameter d. If $R = 3.277\ \Omega$ when $l = 1$ km, and $d = 2.588$ mm, what will be the resistance of a 500-m length of wire 1.45 mm in diameter?
15. The load that a beam of given thickness can carry safely is directly proportional to its width and inversely proportional to its length. If a beam 10 m long and 50 mm wide can support 9000 kg, what load could be supported by a beam of identical thickness 25 m long and 75 mm wide?

6

powers of ten

If you have not yet learned to operate a slide rule, now is a good time to begin. The methods explained in this chapter not only will allow you to make accurate computations with cumbersome numbers but will be of considerable assistance in obtaining correct answers from your slide rule calculations.

The slide rule is an instrument, or tool, designed for the purpose of saving time and labor in calculating. Every technical person should be proficient in the operation of some type of slide rule. The solution of every practical problem, when a concrete answer is desired, eventually reduces to an arithmetical computation. Valuable time is wasted in performing a series of multiplications, divisions, square roots, etc., with a pencil and paper when there is available an instrument that will do the work satisfactorily in a fraction of the time and with a fairly high degree of accuracy. Very few people enjoy performing numerical computations simply for the joy of "figuring." The practical person wants concrete answers; therefore, such a person should use whatever tools or devices are available to help arrive at those answers with a minimum expenditure of time and effort.

6-1 TYPES OF SLIDE RULE

A complete description of various slide rules or of a particular type of rule is not within the scope of this book. Briefly, the slide rule is a mechanical equivalent of a table of logarithms. In the modern sense the slide rule is a mechanical analog computer consisting of a number of scales so graduated and arranged that multiplication, division, raising to powers, extracting roots, and many other operations can be performed with facility.

Types of slide rules range from inexpensive beginner's slide rules to those comparable to calculating machines. Most of them are designed for use in general mathematical operations; some are designed especially for use in specific professions or trades.

No attempt is made here to advise you as to just what type of rule is best suited to your use. If you are attending a technical school, your instructors are qualified to advise as to the type of rule they believe best. If you are professionally employed, your technical associates will be able to assist in your selection of a rule.

Among the many types developed, the Cooke Radio Slide Rule, developed by the

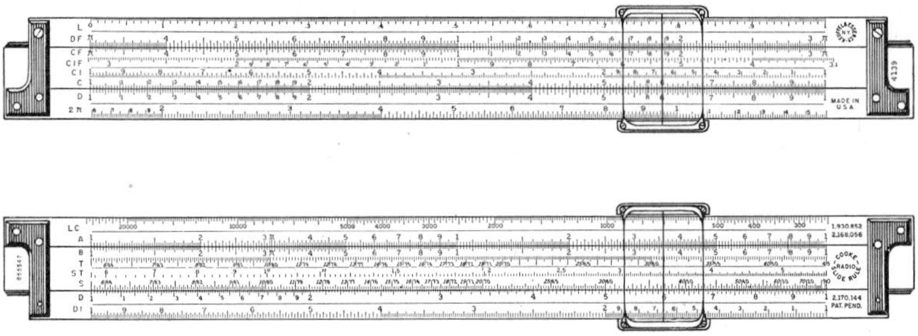

Fig. 6-1 Front and Back Views of Cooke Radio Slide Rule (*By Courtesy of Keuffel & Esser Company, Morristown, New Jersey*)

late Nelson M. Cooke, (Fig. 6-1) has met with moderate success. This rule employs a minimum number of scales but at the same time allows almost as wide a mathematical scope as may be desired. The scales have been designed and arranged for the express purpose of completing the more common electronics and electrical problems in a simple and straightforward manner.

More recently, the Keuffel and Esser Company has produced the Decilon Slide Rule (Fig. 6-2), and this has been generally accepted by electronics technicians and engineers because of its more generalized scale selection.

Instruction books are furnished with all slide rules; thus, the beginner needs no instructor but merely a reasonable amount of practice in order to become proficient in using the rule.

It is therefore strongly recommended that, if you do not have a slide rule, you acquire one and learn to use it while studying this text. You will save many hours that otherwise would be devoted to figuring with a pencil and can be well spent in the study of mathematics or other essential subjects, to say nothing of lightening otherwise tedious computations.

6-2 POCKET CALCULATORS

As this edition of *Basic Mathematics for Electronics* is being rewritten, integrated circuit technology has enabled manufacturers to produce pocket calculators for quite reasonable prices. Of course, the cheaper calculators are designed to perform only basic arithmetic, with perhaps square roots as an extra option. Naturally, those equipped to deal with the additional problems which people in the various fields of electronics have to solve (powers and roots, calculations involving π, common and natural logarithms, and the direct and inverse trigonometric

Fig. 6-2 Front View of Decilon Slide Rule (*Courtesy of Keuffel & Esser Company, Morristown, New Jersey*)

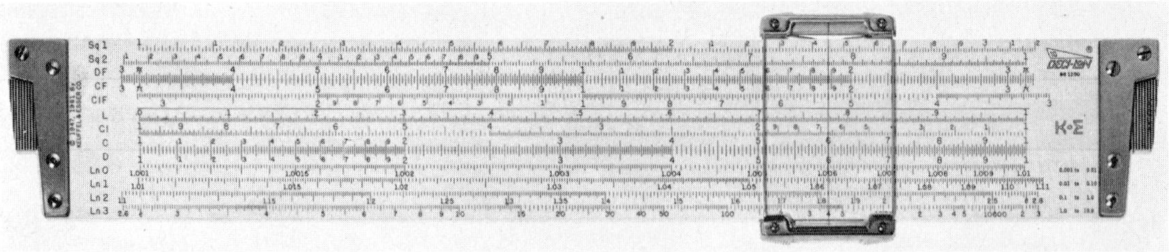

(a)

(b)

Fig. 6-3 Two Popular Models of Pocket Calculators from the Hewlett-Packard Line. (a) The HP-21, Simplest of the Series, with All the Functions Required for the Study of the Material in This Book. (b) The HP-55, Which Provides All the Features of the HP-21, Plus Addressable Storage and the Ability to be Programmed for the Solution of Very Complicated Calculations. (*Courtesy of Hewlett-Packard, Palo Alto, California*)

functions, together with several stages of storage) are considerably more expensive.

Therefore, while many users of this book will replace their slide rules with calculators, many others will continue to use slide rules extensively, and the original notes about slide rule operations are retained in this edition.

Users of pocket calculators should not neglect the importance of the *ideas* of powers of 10, even if they do not feel the urgency to use the techniques extensively. Most pocket calculators give only 6 to 10 places, and this is insufficient when dealing with the wide range of units—picowatts to megohms—with which we are involved.

6-3 ACCURACY OF SLIDE RULES

From an electronics or electrical viewpoint, except possibly where extremely accurate measurements are needed, the accuracy of a slide rule leaves nothing to be desired. Its accuracy is nearly proportional to the length of scales used. The scales of a ten-inch rule give results accurate to within 1 part in 1000, or one-tenth of 1%.

When practical electronic or electric circuits are taken into consideration, slide rule computations are more accurate than the circuit components involved. For example, the tolerances of resistors, inductors, and capacitors used in the usual radio and television receivers do well to average ±10%. Also, the average switchboard meter is seldom correct to within 3% throughout its calibration. Suppose we go into a store to buy a 10% tolerance, 10 000-Ω resistor and ask the salesperson to check the resistance on an ohmmeter. If the resistance measures anywhere between 9000 and 11 000 Ω, which is within the $\pm 10\%$ tolerance, we should be satisfied. However, if the ohmmeter has an accuracy within $\pm 2\%$, the salesperson is to be congratulated on having a good meter. Because, in all probability, this person does not know just how accurate the meter is, we leave the store *hoping* we have a resistor somewhere near the correct value. Actually, such a resistor would be entirely satisfactory for ordinary requirements, as we shall see later.

Other circuit components, except those used in the laboratory, vary in much the same manner, and when temperature, humidity, and other variations are taken into consideration, the results obtained with the slide rule more than meet all practical needs.

From the foregoing, it might appear that mathematical accuracy in the calculation of electric circuits is unnecessary. Far from it—the laws of electricity follow concise mathematical concepts, and we *can* construct circuit components and measuring equipment that are very precise. However, mainly for economic reasons, it is neither practical nor necessary to maintain such a high degree of accuracy in average circuits.

The important point is that we must first know how accurate our available circuit components and measuring equipment are and then depend upon this accuracy to a reasonable extent. Some students thoughtlessly make computations of quantities that have been found by measurements, instrument readings, etc., and carry the operations to several unnecessary decimal places. Moreover, this computation consumes a considerable amount of time; and worse still, the results often give a false impression of accuracy. In this connection, it is safe to assume that the constants of any electronic or electric circuit components or the calibration of meters, excluding precision measuring equipments, are generally not correct beyond three significant figures.

6-4 SIGNIFICANT FIGURES

In mathematics, a number is generally considered as being exact. For example, 220 would mean 220.0000, etc., for as many added zeros as desired. However, a meter reading, for example, is always an *approximation*. We might read 220 V on a certain switchboard type of voltmeter, but a precision instrument might show that voltage to be 220.3 V, and a series of precise measurements might show the voltage to be 220.36 V. It should be noted that the position of the decimal point does not determine the accuracy of a number. For example, 115 V, 0.115 kV, and 115 000 mV are of identical value and equally accurate.

Any number representing a measurement, or the amount of some quantity, expresses the accuracy of the measurement. The figures required are known as *significant figures*.

The *significant figures* of any number are the figures 1, 2, 3, 4, . . . , 9, in addition to such ciphers, or zeros, as may occur between them or as may have been retained in properly rounding them off.

examples

0.002 36	is correct to *three* significant figures.
3.141 59	is correct to *six* significant figures.
980 000.0	is correct to *seven* significant figures.
24.	is correct to *two* significant figures.
24.0	is correct to *three* significant figures.
0.025 00	is correct to *four* significant figures.

PROBLEMS 6-1

To how many significant figures have the following numbers been expressed?

1. 2.718 28	**2.** 0.000 003 14	**3.** 300 000	**4.** 23.0055
5. 1.00	**6.** 1	**7.** 0.000 01	**8.** 6.28
9. 0.000 025 38	**10.** 2726.375		

6-5 ROUNDED NUMBERS

A number is *rounded off* by dropping one or more figures at its right. If the last figure dropped is 6 or more, we increase the last figure retained by 1. Thus 3867 would be rounded off to 3870, 3900, or 4000. If the last figure dropped is 4 or less, we leave the last figure retained as it is. Thus 5134 would be rounded off to 5130, 5100, or 5000. If the last figure dropped is 5, add 1 if it will make the last figure retained *even*; otherwise do not. Thus, $55.7\not{5} = 55.8$, but $67.6\not{5} = 67.6$.

6-6 DECIMALS

Two important considerations arise in making computations involving decimals:

1. A slide rule gives only the significant figures of the result of a mathematical operation. For example, suppose that we have performed some operation on the slide rule and read as the result the significant figures 432. Now the slide rule does not indicate whether this answer is 0.0432, 0.432, 4.32, 4320, 43 200, etc. Therefore, it becomes necessary for the slide rule operator to fix the decimal point; that is, the operator must first determine the *approximate* answer in order to use the more accurate figures taken from the slide rule scales.
2. Unfortunately, electrical engineers and particularly electronics engineers are required to handle cumbersome numbers ranging from extremely small fractions of electrical units to very large numbers, as represented by radio frequencies. The fact that these wide limits of numbers are encountered in the same problem does not simplify matters. This situation is becoming more complicated owing to the trend to the higher radio frequencies with attendant smaller fractions of units represented by circuit components.

For these reasons, in using a slide rule, the decimal point *cannot* be fixed "by inspection" except in the simpler problems. Accordingly, many beginners interested in using the slide rule for solving electronics and electrical problems have become discouraged by the difficulty of placing the decimal points due to the above-mentioned wide range of numbers encountered in the average problem.

The problem of properly placing the decimal point and thus reducing unnecessary work presents little difficulty to the person who has a working knowledge of the powers of 10.

6-7 POWERS OF 10

The powers of 10 are sometimes termed the "engineer's shorthand." A thorough knowledge of the powers of 10 and the ability to apply the theory of exponents will greatly assist in determining an approximation. If a slide rule is used with the powers of 10, the average problem reduces to the usual slide rule operations plus simple mental arithmetic.

Table 6-1

Number	Power of 10	Expressed in English
0.000 001 =	10^{-6} =	ten to the negative *sixth* power
0.000 01 =	10^{-5} =	ten to the negative *fifth* power
0.000 1 =	10^{-4} =	ten to the negative *fourth* power
0.001 =	10^{-3} =	ten to the negative *third* power
0.01 =	10^{-2} =	ten to the negative *second* power
0.1 =	10^{-1} =	ten to the negative *first* power
1 =	10^{0} =	ten to the *zero* power
10 =	10^{1} =	ten to the *first* power
100 =	10^{2} =	ten to the *second* power
1 000 =	10^{3} =	ten to the *third* power
10 000 =	10^{4} =	ten to the *fourth* power
100 000 =	10^{5} =	ten to the *fifth* power
1 000 000 =	10^{6} =	ten to the *sixth* power

If a slide rule is not used for computation, the powers of 10 enable one to work all problems by using convenient whole numbers. Either offers a convenient method for obtaining a final answer with the decimal point in its proper place.

Some of the multiples of 10 may be represented as shown in Table 6-1. From the table it is seen that any decimal may be written as a whole number times some negative power of 10. This may be expressed by the following:

Rule

To express a decimal as a whole number times a power of 10, move the decimal point to the right and count the number of places to the original point. The number of places counted is the proper negative power of 10.

examples

$$0.006\,87 = 6.87 \times 10^{-3}$$
$$0.000\,048\,2 = 4.82 \times 10^{-5}$$
$$0.346 = 34.6 \times 10^{-2}$$
$$0.086\,43 = 86.43 \times 10^{-3}$$

Also, it is seen that any large number can be expressed as some smaller number times the proper power of 10. This can be expressed by the following rule:

Rule

To express a large number as a smaller number times a power of 10, move the decimal point to the left and count the number of places to the original decimal point. The number of places counted will give the proper positive power of 10.

examples

$$435 = 4.35 \times 10^{2}$$
$$964\,000 = 96.4 \times 10^{4}$$
$$6835.2 = 6.8352 \times 10^{3}$$
$$5723 = 5.723 \times 10^{3}$$

PROBLEMS 6-2

Express the following numbers to three significant figures and write them as numbers between 1 and 10 times the proper power of 10:

1. 643 000
2. 13.6
3. 6534
4. 0.0963
5. 0.000 000 009 435
6. 8 743 000
7. 0.367
8. 59 235
9. 250×10^{-3}
10. $0.000\,086 \times 10^{6}$
11. $0.000\,399 \times 10^{8}$
12. $0.000\,399\,5 \times 10^{8}$
13. 259×10^{-4}
14. 0.031 415 9
15. 276 492.536 24
16. $1\,254\,325 \times 10^{-12}$
17. 0.000 000 107 52
18. $0.000\,008\,145\,73 \times 10^{12}$
19. 3.000 725
20. $0.000\,055\,55 \times 10^{-3}$

6-8 MULTIPLICATION WITH POWERS OF 10

In Sec. 4-3 the law of exponents in multiplication was expressed in the general form

$$a^m \cdot a^n = a^{m+n} \qquad (\text{where } a \neq 0)$$

This law is directly applicable to the powers of 10.

example 1 Multiply 1000 by 100 000.

solution

$$1000 = 10^3$$

and

$$100\,000 = 10^5$$

then

$$\begin{aligned} 1000 \times 100\,000 &= 10^3 \times 10^5 \\ &= 10^{3+5} \\ &= 10^8 \end{aligned}$$

example 2 Multiply 0.000 001 by 0.001.

solution

$$0.000\,001 = 10^{-6}$$

and

$$0.001 = 10^{-3}$$

then

$$\begin{aligned} 0.000\,001 \times 0.001 &= 10^{-6} \times 10^{-3} \\ &= 10^{-6+(-3)} \\ &= 10^{-6-3} = 10^{-9} \end{aligned}$$

example 3 Multiply 23 000 by 7000.

solution

$$23\,000 = 2.3 \times 10^4$$

and

$$7000 = 7 \times 10^3$$

then

$$\begin{aligned} 23\,000 \times 7000 &= 2.3 \times 10^4 \times 7 \times 10^3 \\ &= 2.3 \times 7 \times 10^7 \\ &= 16.1 \times 10^7, \text{ or } 161\,000\,000 \end{aligned}$$

example 4 Multiply 0.000 037 by 600.

solution

$$\begin{aligned} 0.000\,037 \times 600 &= 3.7 \times 10^{-5} \times 6 \times 10^2 \\ &= 3.7 \times 6 \times 10^{-3} \\ &= 22.2 \times 10^{-3}, \text{ or } 0.0222 \end{aligned}$$

example 5
Multiply 72 000 × 0.000 025 × 4600.

solution

$$\begin{aligned} &72\,000 \times 0.000\,025 \times 4600 \\ &\quad = 7.2 \times 10^4 \times 2.5 \times 10^{-5} \times 4.6 \times 10^3 \\ &\quad = 7.2 \times 2.5 \times 4.6 \times 10^2 \\ &\quad = 82.8 \times 10^2, \text{ or } 8280 \end{aligned}$$

You will find that by expressing all numbers as numbers between 1 and 10 times the proper power of 10, the determination of the proper place for the decimal point will become a matter of inspection.

PROBLEMS 6-3

Multiply the following. Although all factors are not expressed to three significant figures, express answers to three significant figures as numbers between 1 and 10 times the proper power of 10.

1. $10\,000 \times 0.01 \times 0.0001$ **2.** $0.000\,01 \times 10^5 \times 100$
3. 0.0004×980
4. $0.000\,25 \times 16 \times 10^{-4} \times 20 \times 10^5$
5. $0.000\,008\,4 \times 0.005 \times 0.000\,17$

6. $35\,000\,000 \times 680 \times 10^{-9} \times 5.5 \times 10^{-5}$
7. $9.34 \times 10^{12} \times 628\,000 \times 0.000\,053 \times 10^{-3}$
8. $500 \times 10^{-6} \times 782 \times 10^{4} \times 0.000\,037 \times 10^{-8}$
9. $5\,960\,000 \times 0.000\,888 \times 604 \times 10^{-5}$
10. $2.846 \times 10^{3} \times 0.009\,438 \times 10^{6} \times 0.6848 \times 10^{4}$

The alternating-current inductive reactance of a circuit or an inductor is given by

$$X_L = 2\pi fL \qquad \Omega$$

where X_L = inductive reactance, Ω
f = frequency of alternating current, Hz
L = inductance of circuit, or inductor, henrys (H)

Compute the inductive reactance when:

11. $f = 60$ Hz, $L = 0.015$ H
12. $f = 1000$ Hz, $L = 0.015$ H
13. $f = 1\,000\,000$ Hz, $L = 0.015$ H
14. $f = 60$ Hz, $L = 1.5$ H
15. $f = 10\,000$ Hz, $L = 0.000\,003\,5$ H

6-9 DIVISION WITH POWERS OF 10

The law of exponents in division (Secs. 4-9 to 4-11) can be summed up in the following general form:

$$\frac{a^m}{a^n} = a^{m-n} \qquad \text{(where } a \neq 0\text{)}$$

example 6

$$\frac{10^5}{10^3} = 10^{5-3} = 10^2$$

or

$$\frac{10^5}{10^3} = 10^5 \times 10^{-3} = 10^2$$

example 7

$$\frac{72\,000}{0.0008} = \frac{72 \times 10^3}{8 \times 10^{-4}}$$

$$= \frac{72}{8} \times 10^{3+4}$$

$$= 9 \times 10^7$$

or

$$\frac{72\,000}{0.0008} = \frac{72 \times 10^3}{8 \times 10^{-4}}$$

$$\frac{72\,000}{0.0008} = \frac{72}{8} \times 10^3 \times 10^4$$

$$= 9 \times 10^7$$

example 8

$$\frac{169 \times 10^5}{13 \times 10^5} = \frac{169}{13} \times 10^{5-5}$$

$$= 13 \times 10^0$$

$$= 13 \times 1 = 13$$

or

$$\frac{169 \times 10^5}{13 \times 10^5} = 13$$

It is apparent that powers of 10 which are factors that have the same exponents in numerator and denominator can be canceled. Also, you will note that powers of 10 which are factors can be transferred at will from denominator to numerator, or vice versa, if the sign of the exponent is changed when the transfer is made (Sec. 4-11).

6-10 APPROXIMATIONS

Multiplying 37 by 26 is very close to multiplying 40 by 25. The approximation 1000 is "within the order" of the actual product, 962. Usually, approximations which are within reason may be arrived at, and they serve as a guide to what the actual answer should be.

Such approximations should be made quickly before making slide rule or other exact calculations. The "order" of the calculated answer should be of the "order" of the approximation. If you expect an answer of the order of 1000 and you actually come up with 940 or 1050, the answer is probably correct. If, however, you arrive at an answer of 9.62, you should suspect that you have lost a factor of 10^2 somewhere, and you should check out your calculations. Although approximations will not guarantee the correctness of the calculated answer, they will reveal possible errors.

6-11 COMBINED MULTIPLICATION AND DIVISION

Combined multiplication and division is most conveniently accomplished by alternately multiplying and dividing until the problem is completed.

example 9 Simplify

$$\frac{0.000\,644 \times 96\,000 \times 3300}{161\,000 \times 0.000\,001\,20}$$

solution First convert all numbers in the problem to numbers between 1 and 10 times their proper power of 10, thus:

$$\frac{6.44 \times 10^{-4} \times 9.6 \times 10^{4} \times 3.3 \times 10^{3}}{1.61 \times 10^{5} \times 1.2 \times 10^{-6}} = \frac{6.44 \times 9.6 \times 3.3 \times 10^{4}}{1.61 \times 1.2}$$

The problem as now written consists of multiplication and division of simple numbers. The answer approximates to

$$\frac{6 \times 10 \times 3 \times 10^{4}}{2 \times 1} = 90 \times 10^{4}$$

If the remainder of the problem is computed by slide rule, then the answer 1056 from the slide rule can easily be adjusted to read 105.6×10^4, or 1.056×10^6. If the problem is solved without the aid of a slide rule, there are no small decimals and no cumbersome large numbers to handle.

Instead of first finding the product of the numerator and dividing it by the product of the denominator, it is best to divide and multiply alternately. Thus, we divide 6.44 by 1.61 to obtain 4. Then we multiply this 4 by 9.6 to obtain 38.4. We then divide 38.4 by 1.2, which results in a quotient of 32. Finally, we multiply 32 by 3.3, which results in a product of 105.6. Because we still have a factor of 10^4, the answer is 105.6×10^4. If we desire to express the answer in powers of 10, we would write it 1.056×10^6, but written out, without the power of 10, it would be 1 056 000.

The method of alternately dividing and multiplying offers the slide rule operator the advantage of working the problem straight through without the necessity of jotting down the product of the factors of the numerator before proceeding to find the product of the denominator factors.

6-12 RECIPROCALS

In radio and electrical problems, many formulas are used that involve reciprocals, such as

$$\frac{1}{R_t} = \frac{1}{R_1} + \frac{1}{R_2}$$

$$X_C = \frac{1}{2\pi fC}$$

$$f = \frac{1}{2\pi\sqrt{LC}}$$

The *reciprocal* of a number is 1 divided by that number. Such problems present no difficulty if the powers of 10 are used properly.

example 10

Simplify $\dfrac{1}{40\,000 \times 0.000\,25 \times 125 \times 10^{-6}}$.

solution First convert all numbers in the denominator to numbers between 1 and 10 times their proper power of 10, thus:

$$\frac{1}{4 \times 10^4 \times 2.5 \times 10^{-4} \times 1.25 \times 10^{-4}} = \frac{10^4}{4 \times 2.5 \times 1.25}$$

Multiplying the factors of the denominator results in

$$\frac{10^4}{12.5}$$

Instead of writing out the numerator as 10 000 and then dividing by 12.5, we could write the numerator as two factors in order better to divide mentally. That is, we can write the problem as

$$\frac{10^2 \times 10^2}{12.5} \text{ or } \frac{100}{12.5} \times 10^2 = 8 \times 10^2$$

This method is of particular advantage to the slide rule operator because of the ease of estimating the final result.

If the final result is a decimal, rewriting the numerator into two factors allows fixing the decimal point with the least effort.

example 11

Simplify $\dfrac{1}{625 \times 10^4 \times 2000 \times 64\,000}$.

solution First convert all numbers in the denominator to numbers between 1 and 10 times their proper power of 10, thus:

$$\frac{1}{6.25 \times 10^6 \times 2 \times 10^3 \times 6.4 \times 10^4} = \frac{10^{-13}}{6.25 \times 2 \times 6.4}$$

Multiplying the factors in the denominator results in

$$\frac{10^{-13}}{80}$$

Instead of writing out the numerator as 0.000 000 000 000 1 and dividing it by 80, we write the numerator as two factors in order better to divide mentally:

$$\frac{10^2 \times 10^{-15}}{80} \text{ or } \frac{100}{80} \times 10^{-15} = 1.25 \times 10^{-15}$$

If the value of the denominator product were over 100 and less than 1000, we would break up the numerator so that one of the factors would be 10^3 or 1000, and so on. This method will always result in a final quotient of a number between 1 and 10 times the proper power of 10.

PROBLEMS 6-4

Perform the indicated operations. Round off the figures in the results, if necessary, and express answers to three significant figures as a number between 1 and 10 times the proper power of 10:

1. $\dfrac{0.000\,25}{500}$

2. $\dfrac{10}{0.000\,125 \times 80\,000}$

3. $\dfrac{0.6043}{5763}$

4. $\dfrac{420 \times 0.036}{0.0090}$

5. $\dfrac{0.256 \times 338 \times 10^{-9}}{865\,000}$

6. $\dfrac{1}{6.28 \times 452\,000 \times 0.000\,155}$

7. $\dfrac{2804 \times 74.23}{0.000\,900\,6 \times 0.008\,040}$

8. $\dfrac{1000}{248\,000 \times 5630 \times 10^{-3} \times 0.000\,090\,3 \times 10^{2}}$

9. $\dfrac{1 \times 10^{6}}{6.28 \times 10^{3} \times 2500 \times 10^{3} \times 0.25 \times 10^{-6}}$

10. $\dfrac{1}{6.28 \times 400 \times 10^{6} \times 50 \times 10^{-12}}$

11. $\dfrac{150 \times 216 \times 1.78}{4.77 \times 10^{2} \times 1.23 \times 6.03 \times 10^{4}}$

12. $\dfrac{65.3 \times 10^{-6} \times 504 \times 10^{6} \times 12\,700}{312 \times 10^{6} \times 0.007 \times 6.82}$

The alternating-current capacitive reactance of a circuit, or capacitor, is given by the formula

$$X_C = \frac{1}{2\pi f C} \qquad \Omega$$

where X_C = capacitive reactance, Ω
f = frequency of the alternating current, Hz
C = capacitance of the circuit, or capacitor, farads (F)

Compute the capacitive reactances when

13. $f = 60$ Hz, $C = 0.000\,004$ F
14. $f = 28\,000\,000$ Hz, $C = 0.000\,000\,000\,025$ F
15. $f = 225\,000\,000\,000\,000$ Hz, $C = 0.000\,000\,000\,563$ F

6-13 THE POWER OF A POWER

It becomes necessary, in order to work a variety of problems utilizing the powers of 10, to consider a few new definitions concerning the laws of exponents before we study them in algebra. This, however, should present no difficulty.

In finding the power of a power the exponents are multiplied. That is, in general,

$$(a^m)^n = a^{mn} \qquad (\text{where } a \neq 0)$$

example 12

$$100^3 = 100 \times 100 \times 100 = 1\,000\,000 = 10^6$$

or

$$100^3 = 10^2 \times 10^2 \times 10^2 = 10^6$$

then

$$100^3 = (10^2)^3 = 10^{2\times3} = 10^6$$

Numbers can be factored when raised to a power in order to reduce the labor in obtaining the correct number of significant figures, or properly fixing the decimal point.

example 13

$$19\,000^3 = (1.9 \times 10^4)^3$$
$$= 1.9^3 \times 10^{4\times3} = 6.859 \times 10^{12}$$

example 14

$$0.000\,007\,5^2 = (7.5 \times 10^{-6})^2$$
$$= 7.5^2 \times 10^{(-6)\times2}$$
$$= 56.25 \times 10^{-12}$$
$$= 5.625 \times 10^{-11}$$

In Example 13, 19 000 was factored into 1.9×10^4 in order to allow an easy mental check. Because 1.9 is nearly 2 and $2^3 = 8$, it is apparent that the result of cubing 1.9 must be 6.859, not 0.6859 or 68.59.

In Example 14, the 0.000 007 5 was factored for the same reason. We know that $7^2 = 49$; therefore the result of squaring 7.5 must be 56.25, not 0.5625 or 5.625.

6-14 THE POWER OF A PRODUCT

The power of a product is the same as the product of the powers of the factors. That is, in general,

$$(abc)^m = a^m b^m c^m$$

example 15

$$(10^5 \times 10^3)^3 = 10^{5\times3} \times 10^{3\times3}$$
$$= 10^{15} \times 10^9 = 10^{24}$$

or

$$(10^5 \times 10^3)^3 = (10^8)^3 = 10^{8\times3} = 10^{24}$$

6-15 THE POWER OF A FRACTION

The power of a fraction equals the power of the numerator divided by the power of the denominator. That is,

$$\left(\frac{a}{b}\right)^m = \frac{a^m}{b^m}$$

example 16 $\left(\frac{10^5}{10^3}\right)^2 = \frac{10^{5\times2}}{10^{3\times2}} = \frac{10^{10}}{10^6} = 10^4$

The above can be solved by first clearing the exponents inside the parentheses and then raising to the required power. Thus,

$$\left(\frac{10^5}{10^3}\right)^2 = (10^{5-3})^2 = (10^2)^2 = 10^4$$

6-16 THE ROOT OF A POWER

The root of a power in exponents is given by

$$\sqrt[n]{a^m} = a^{m\div n} \qquad (\text{where } a \text{ and } n \neq 0)$$

example 17 $\sqrt{25 \times 10^8} = \sqrt{25} \times \sqrt{10^8}$
$= 5 \times 10^{8\div2} = 5 \times 10^4$

example 18 $\sqrt[3]{125 \times 10^6} = \sqrt[3]{125} \times \sqrt[3]{10^6}$
$= 5 \times 10^{6\div3} = 5 \times 10^2$

In the general case when m is evenly divisible by n, the process of extracting roots is comparatively simple. When m is not evenly divisible by n, the result obtained by extracting the root is a fractional power.

example 19 $\sqrt{10^5} = 10^{5\div2} = 10^{\frac{5}{2}}$, or $10^{2.5}$

Such fractional exponents are encountered in various phases of engineering mathematics and are conveniently solved by the use of logarithms. However, in using the powers of 10, the fractional exponent is cumbersome for obtaining a final answer. It becomes necessary, therefore, to devise some means of extracting a root whereby an integer can be obtained as an exponent in the final result. The means found is to express the number, the root of which is desired, as some number times a power of 10 that is evenly divisible by the index of the required root. As an example, suppose it is desired to extract the square root of 400 000. Though it is true that

$$\sqrt{400\,000} = \sqrt{4 \times 10^5}$$
$$= \sqrt{4} \times \sqrt{10^5}$$
$$= 2 \times 10^{2.5}$$

we have a fractional exponent that is not readily reduced to actual figures. However, if we express the number differently, we obtain an integer as an exponent. Thus,

$$\sqrt{400\,000} = \sqrt{40 \times 10^4}$$
$$= \sqrt{40} \times \sqrt{10^4}$$
$$= 6.32 \times 10^2$$

It will be noted that there are a number of ways of expressing the above square root, such as

$$\sqrt{400\,000} = \sqrt{0.4 \times 10^6}$$

or $$\sqrt{4000 \times 10^2}$$

or $$\sqrt{0.004 \times 10^8}$$

All are equally correct, but you should try to write the problem in a form that will allow a rough mental approximation in order that the decimal may be properly placed with respect to the significant figures.

PROBLEMS 6-5

Perform the indicated operations. When answers do not come out in round numbers, express them to three significant figures.

1. $(10^3)^4$ **2.** $(10^{-4})^3$ **3.** $(10^2 \times 10^3)^4$

4. $(4 \times 10^{-4})^2$ **5.** $(5 \times 10^3)^4$ **6.** $(3 \times 10^{-2})^3$

7. $(2 \times 10^4 \times 8 \times 10^{-5})^2$ **8.** $\left(\frac{32 \times 10^3}{8 \times 10^4}\right)^2$ **9.** $\sqrt{0.0625 \times 0.0004}$

10. $\sqrt{0.000\,36 \times 0.009}$ **11.** $\sqrt{36 \times 10^2 \times 25 \times 10^{-2}}$

12. $\sqrt[3]{27 \times 10^{-3} \times 8 \times 10^{12}}$ **13.** $\frac{1}{6.28\sqrt{250 \times 10^{-3} \times 10^{-9}}}$

14. $\left(\frac{63 \times 10^6 \times 460 \times 10^{-12}}{5.1 \times 10^{-6}}\right)^2$

The resonant frequency of a circuit is given by the formula

$$f = \frac{1}{2\pi\sqrt{LC}} \qquad \text{Hz}$$

where f = resonant frequency, Hz
L = inductance of circuit, H
C = capacitance of circuit, F

Compute the resonant frequencies when:

15. $L = 0.000\,045$ H, $C = 0.000\,000\,000\,250$ F
16. $L = 0.000\,018$ H, $C = 100 \times 10^{-12}$ F
17. $L = 8 \times 10^{-6}$ H, $C = 56.3 \times 10^{-12}$ F

18. $L = 0.000\,23$ H, $C = 0.000\,000\,000\,5$ F
19. $L = 70.4 \times 10^{-6}$ H, $C = 250 \times 10^{-12}$ F
20. $L = 40$ H, $C = 7 \times 10^{-6}$ F

6-17 ADDITION AND SUBTRACTION WITH POWERS OF 10

Sometimes it becomes necessary, when making calculations, to perform additions and subtractions with powers of 10. These present no difficulties if you remember that you are dealing with the addition and subtraction of terms as described in Sec. 3-7. For example, you would not write $3x^2 + 5x^3 = 8x^5$, because $3x^2$ and $5x^3$ are unlike quantities. Similarly, you would not write

$$3 \times 10^2 + 5 \times 10^3 = 8 \times 10^5$$

because 3×10^2 and 5×10^3 are also unlike quantities.

The foregoing addition of

$$3 \times 10^2 + 5 \times 10^3$$

can be performed by either of two methods. You can convert the numbers so that no powers of 10 are involved and write $300 + 5000 = 5300$. Also, you can rewrite the terms to be added so that like powers of 10 are added, such as

$$3 \times 10^2 + 50 \times 10^2 = 53 \times 10^2$$

or

$$0.3 \times 10^3 + 5 \times 10^3 = 5.3 \times 10^3$$

This is the same as adding like terms.

example 20 Add 8.3×10^4 and 3.6×10^2.

solution

$$\begin{aligned} 8.3 \times 10^4 &= 83\,000 = 830 \times 10^2 \\ 3.6 \times 10^2 &= \underline{\quad 360} = \underline{\quad 3.6 \times 10^2} \\ &\ 83\,360 = 833.6 \times 10^2 \\ & = 8.336 \times 10^4 \end{aligned}$$

PROBLEMS 6-6

Perform the indicated operations. Express all answers (*a*) in ordinary form and (*b*) to three significant figures as numbers between 1 and 10 times the proper power of 10.

1. $3 \times 10^3 + 1 \times 10^2$
2. $25 \times 10^6 + 3.4 \times 10^3$
3. $1.73 \times 10^{12} + 2.46 \times 10^{12}$
4. $2 \times 10^3 + 4 \times 10^{-1}$
5. $6.28 \times 10^6 - 159 \times 10^{-3}$

7

units and dimensions

As previously stated, the solution of every practical problem, when a concrete answer is desired, eventually reduces to an arithmetical computation; that is, the answer reduces to some *number*. In order for this answer, or number, to have a concrete meaning, it must be expressed in some *unit*. For example, if you were told that the resistance of a circuit is 16, the information would have no meaning unless you knew to what unit the 16 referred.

From the foregoing it is apparent that the expression for the magnitude of any physical quantity must consist of two parts. The first part, which is a number, specifies "how much"; the second part specifies the unit of measurement, or "what," as, for example, in 16 Ω, 20 A, or 100 m, the Ω, A, or m.

It is necessary, therefore, before beginning the study of circuits, to define a few of the more common electrical and dimensional units used in electrical and electronics engineering.

7-1 SYSTEMS OF MEASUREMENT

Over the years, the systems by which we have made measurements have changed considerably. We do not often now deal with grains of corn or the length of a man's forearm. Occasionally the civil engineer surveying an antenna site will talk about "chains" when we would have said "hundreds of feet" or "meters." We in electronics are primarily concerned with three specific fields of measurement: distance-mass, time, and electric charge. The electrical quantities are all fundamentally related to these measurements, as you will discover if you study higher mathematics.

Generally speaking, North Americans have used two main systems for measuring some quantities, whereas the units of other quantities are the same in both systems. One of these systems is the so-called English FPS (foot-pound-second) system, which was widely, almost exclusively, used by engineers in English-speaking countries until very recently. The other is the metric MKS (meter-kilogram-second) system, which has grown in importance over the last century. The modern refinement of the MKS system is called SI, for Système International des Unités. It is used by well over 90% of the world's population, and has become even more widely used in the last few years with the conversion of Great Britain, India, Australia, Canada, and, as metric legislation is passed, a growing number of industries in the United States to the SI metric system. You should note that in both the English system and the SI metric system several of the units are the same: seconds, volts, ohms, and amperes, especially.[1]

7-2 THE ENGLISH SYSTEM

The English (or traditional, or foot-pound) system, developed over many centuries, contains many quite arbitrary relationships between the units, and no systematic correlations. It is being superseded by the very logical SI metric units (Sec. 7-3). The small list given here shows only a very few of the many conversions which have been developed over the years.

12 inches (in) = 1 foot (ft)
3 feet (ft) = 1 yard (yd)
5280 feet (ft) = 1 statute mile (mi)
16 ounces (oz) = 1 pound (lb)
2000 pounds (lb) = 1 ton

7-3 THE SI METRIC SYSTEM

The metric system is a relatively newer, more orderly system related originally to the measurement of the earth itself. It uses decimal relationships throughout, rather than the arbitrary, hard-to-memorize conversions of the English system. The metric system started out as the centimeter-gram-second (CGS) system. Later, the CGS units were modified; the basic defined units were the meter, the kilogram, and the second, and the name of the system was changed to MKS. Later still, the ampere was added in order to elevate electrical units to the "physical" ones, and the name changed again to MKSA.

The most recent development is known as SI, for Système International des Unités. It is the result of many years of concentrated international cooperation and agreement, and has been published by the General Conference on Weights and Measures and the International Standards Organization. This system defines seven base measuring units, but goes much farther in including other specific details. Altogether, SI involves:

- Seven base measuring units
- An added collection of related units which can be defined in terms of the seven base units
- An orderly use of decimal calculations, with powers of 10 notation, and special word prefixes representing numbers
- An international system of abbreviations and symbols
- A comprehensive international system of standards

Some of the SI units have little direct meaning to us in electronics and are listed below only for the sake of providing you with a complete list. Others are daily necessities, and you will find these used repeatedly throughout this book.

7-4 THE SEVEN BASE SI UNITS

Length/meter The meter was originally defined to be 1×10^{-7} of the length of the line of longitude passing through Paris from the equator to the North Pole. The present-day definition is that the standard meter is the length of 1 650 763.73 wavelengths in a vacuum of the radiation corresponding to the

[1] A quite thorough introduction to the SI metric system was written by Mr. Adams and published by McGraw-Hill in a revised edition in June 1974. It is entitled *SI Metric Units: An Introduction*. Students and teachers with no background in the metric system at all may find it useful.

unperturbed transition between the energy levels $2p_{10}$ and $5d_5$ of the krypton 86 atom. This orange-red line has a wavelength of 6057.802×10^{-10} m.

1 meter (m) = 39.370 079 inches (in)

Mass/kilogram The kilogram is simply defined to be the mass of a special cylinder of platinum-iridium alloy which is in the safekeeping of the International Bureau of Weights and Measures. This cylinder is called the *International Prototype Kilogram*.

1 kilogram (kg) = 2.204 622 6 pounds (lb)

Time/second The second is specifically defined as the duration of 9 192 631 870 periods of the radiation corresponding to the transition between the two hyperfine levels of the ground state of the atom of cesium 133. There are other special definitions of time based on the sun, stars, and moon, but the definition above is one which can be duplicated in laboratories of Bureaus of Standards anywhere.

Electric current/ampere That intensity of electric current known as an ampere is defined as the constant current which, if maintained in two straight parallel conductors of infinite length and of negligible cross section in a vacuum exactly 1 meter (m) apart, will produce a force between these conductors of 2×10^{-7} newton (N) per meter length of wire. The symbol for current is I, and the abbreviation for amperes is A.

Temperature/kelvin The standard SI unit of temperature is the kelvin. The freezing point of pure water is 273.15 K (*not* °K). Ordinary temperature readings will be made on the Celsius scale, on which the freezing point of pure water is 0°C and the boiling point is 100°C (= 373.15 K).

Luminous intensity/candela The standard SI unit of luminous intensity is the candela (cd). This is the amount of luminosity which will produce a luminous flux of 1 lumen (l) within a solid angle of 1 steradian (sr). (We will not concern ourselves with illumination in this book.)

Molecular substance/mole The mole is the standard SI unit which gives the gram molecular weight of a substance. (We will not concern ourselves with this more or less pure science unit in this book.)

Angles/radians and steradians In addition to these seven base units, there are two supplementary units for the measurement of angles: the radian for the measurement of plane angles, and the steradian for the measurement of solid angles. We will study plane angles in Chap. 23.

7-5 THE ADDITIONAL DEFINED SI UNITS

In addition to the base or standard SI units, there are sixteen other units which are used so often that they have been given special names. Many of these are important to us in electronics, and these are listed first in the descriptions which follow. Again, all sixteen units are listed for the benefit of users of this book who are studying beyond the limitations of the book.

Electric charge/coulomb The coulomb is defined as the ampere-second. A reverse definition is that one ampere (A) is the current intensity when one coulomb (C) flows in a circuit for one second (s). The coulomb is also defined as $6.241\,96 \times 10^{18}$ electronic charges.

Electric potential/volt The volt is the practical unit of electromotive force (emf), or potential difference. It is defined as the watt per ampere. (See watt below.) A more common understanding of the volt is that it is the potential difference which will drive a current of one ampere (A) through a resistance of one ohm (Ω). The symbols for voltage are E, e, V, and v, and the abbreviation for volt is V.

Electric resistance/ohm The ohm is the practical unit of resistance. It is defined as the volt per ampere; that is, the ohm is the amount of resistance which limits the current flow to one ampere (A) when the applied electromotive force (potential) is one volt (V). The symbol for resistance is R, and the abbreviation for ohms is Ω.

Electric conductance/siemens The siemens is the practical unit of conductance. The siemens is the reciprocal of the ohm since the conductance is the reciprocal of the resistance. The relationship between ohms and siemens is given by $G = \frac{1}{R}$ siemens.

If resistance is thought of as representing the difficulty with which an electric current is forced to flow through a circuit, conductivity may be thought of as the ease with which a current will pass through the same circuit.

A conductance of one siemens will permit a current flow of one ampere (A) under an electrical pressure of one volt (V). The siemens is a new unit honoring a pioneer in electricity. Formerly, the unit of conductance was called the mho. The symbol for conductance is G, and the abbreviation for siemens is S.

Electric capacitance/farad The farad is the unit of capacitance. It is the ampere-second per volt. A circuit, or capacitor, is said to have a capacitance of one farad when a change of one volt per second across it produces a current of one ampere. The symbol for capacitance is C, and the abbreviation for farad is F. Capacitance will be further discussed in Chap. 32.

Electric inductance/henry The henry is the unit of inductance. It is the volt-second per ampere. A circuit, or inductor, is said to have a self-inductance of one henry when a counterelectromotive force of one volt is generated within it by a rate of change of current of one ampere per second. The symbol for inductance is L, and the abbreviation for henry is H. Inductance will be further discussed in Chap. 32.

Frequency/hertz The SI unit of frequency is the hertz, which was formerly called the cycle per second. Since cycle is not a unit as such, it is sufficient to describe the hertz as the reciprocal of time.

Magnetic flux/weber Magnetic flux is fully described as the volt-second.

Magnetic flux density/tesla Tesla is the special name given to the "density" relationship of webers per square meter.

Luminous flux/lumen This SI unit relates the amount of radiant energy in terms of candelas of luminous intensity multiplied by the solid angle in steradians from which the radiant flux "flows."

Illumination/lux This unit describes the lumens per square meter relationship of luminous flux.

Energy/joule The SI unit for energy of all forms—mechanical work, electric energy, heat quantity, etc.—is the joule. The joule may be expressed in terms of newtons of force multiplied by the distance in meters through which the force moves in the direction of its application.

Force/newton The newton is the SI unit describing joules per meter.

Pressure/pascal The pascal is defined as the relationship newtons per square meter. It is a very small unit of measurement.

Power/watt The watt is the SI unit for power of all forms—electric, mechanical, and so on. It is defined as the energy in joules expended per unit of time in seconds. The symbol for power is *P*, and the abbreviation for watt is W.

In direct-current circuits the power in watts is the product of the voltage and the current, or

$$P = EI \qquad \text{W}$$

The watthour is the unit of electric energy, and its abbreviation is Wh. It is the amount of energy delivered by a power of one watt over a period of one hour.

Customary temperature/degree Celsius Ordinary (nonscientific) temperature measurements will be made on the Celsius scale which relates to the Kelvin scale: $°C = K - 273.15$.

7-6 SOME SI METRIC INTERRELATIONSHIPS

The fundamental relationships between metric units are decimal relationships. The following equations show some of the simpler multiples and submultiples. You will be involved in many such conversions as you continue your studies in electronics.

$$1 \text{ millimeter (mm)} = \frac{1}{1000} \text{ meter} = 10^{-3} \text{ m}$$

$$1 \text{ centimeter (cm)} = \frac{1}{100} \text{ meter} = 10^{-2} \text{ m}$$

$$1 \text{ kilometer (km)} = 1000 \text{ meters} = 10^{3} \text{ m}$$

$$1 \text{ gram (g)} = \frac{1}{1000} \text{ kilogram (kg)}$$
$$= 10^{-3} \text{ kg}$$

7-7 RELATIONS BETWEEN THE SYSTEMS

Since the metric system is based on a decimal plan and the English system is not, there is no one numerical factor or constant which can be used for the conversion of one system to the other. Although Table 6 in the Appendix contains some conversion factors, a few approximate equivalents are given for your convenience:

1 inch (in) = 2.540 centimeters (cm)
1 foot (ft) = 0.3048 meter (m)
1 meter (m) = 39.37 inches (in)
1 mile (mi) = 1.609 kilometers (km)
1 kilometer (km) = 0.6214 mile (mi)
1 kilogram (kg) = 2.205 pounds (lb)
1 pound (lb) = 0.4536 kilogram (kg)

If you are unfamiliar with the metric system, try to visualize these relationships for future convenience. What is the weight in kilograms of a loaf of bread in your community? What is the distance in kilometers from your home to your work? What is your height in centimeters?

The units of time (seconds) and of electricity are identical in the two systems, and we will now deal with them in more detail.

Since the SI units will become increasingly important as Canada and the United States convert to the metric system, you should make the habit of thinking in metric units. Do not keep translating metric quantities into the old English units.

7-8 FREQUENCY

A current which reverses itself at intervals is called an *alternating current.* When this current rises from zero value to maximum value and returns to zero and then increases to maximum value in the opposite direction and again returns to zero, it is said to have completed *one cycle.* The number of times this cycle is repeated in one second is known as the *frequency* of the alternating current. Thus, the average house current is 60 cycles per second (cps). The frequency of radio waves may be as high as several hundred million cycles per second. Note that frequency involves our other main unit, time, by measuring the number of events per second. In both the English and SI systems,

60 seconds (s) = 1 minute (min)
60 minutes (min) = 1 hour (h)
24 hours (h) = 1 day (d)

The International Electrotechnical Commission (IEC), the International Organization for Standardization (ISO), and the Conférence Générale des Poids et Mesures (CGPM) have adopted the name *hertz* (Hz) as the unit of frequency.

1 hertz = 1 cycle per second

7-9 RANGES OF UNITS

As stated in Sec. 6-6, the fields of communication and electrical engineering embrace extremely wide ranges in values of the foregoing units. For example, at the input of a radio receiver, we deal in millionths of a volt, whereas the output circuit of a transmitter may develop hundreds of thousands of volts. An electric clock might consume a fraction of a watt, whereas the powerhouse furnishing this power probably has a capability of millions of watts.

Furthermore, two of these units, the henry and the farad, are very large units, especially the latter. The average radio receiver employs inductances ranging from a few millionths of a henry, as represented by tuning inductance, to several henrys for power filters. The farad is so large that even the largest capacitors are rated in millionths of a farad. Smaller capacitors used in radio circuits are often rated in terms of so many millionths of one-millionth of a farad.

The use of some power of 10 is very convenient in converting to larger multiples or

Table 7-1 DECIMAL MULTIPLIERS

Number		*Power of 10*		*Expressed in English*		*Prefix*	*Abbreviation*
0.000 000 000 000 000 001	=	10^{-18}	=	ten to the negative *eighteenth* power	=	atto	a
0.000 000 000 000 001	=	10^{-15}	=	ten to the negative *fifteenth* power	=	femto	f
0.000 000 000 001	=	10^{-12}	=	ten to the negative *twelfth* power	=	pico	p
0.000 000 001	=	10^{-9}	=	ten to the negative *ninth* power	=	nano	n
0.000 001	=	10^{-6}	=	ten to the negative *sixth* power	=	micro	μ
0.001	=	10^{-3}	=	ten to the negative *third* power	=	milli	m
1	=	10^{0}	=	ten to the *zero* power	=	unit	
1 000	=	10^{3}	=	ten to the *third* power	=	kilo	k
1 000 000	=	10^{6}	=	ten to the *sixth* power	=	mega	M
1 000 000 000	=	10^{9}	=	ten to the *ninth* power	=	giga	G
1 000 000 000 000	=	10^{12}	=	ten to the *twelfth* power	=	tera	T
1 000 000 000 000 000	=	10^{15}	=	ten to the *fifteenth* power	=	peta	P
1 000 000 000 000 000 000	=	10^{18}	=	ten to the *eighteenth* power	=	exa	E

Table 7-2 DENIGRATED POWERS OF TEN

Number	Power of 10	Expressed in English	Prefix	Abbreviation
100	$= 10^2$	= ten squared	= hecto	h
10	$= 10^1$	= ten	= deca	da
0.1	$= 10^{-1}$	= ten to the negative *first* power	= deci	d
0.01	$= 10^{-2}$	= ten to the negative *second* power	= centi	c

smaller fractions of the basic units, called *practical units.*

7-10 DECIMAL MULTIPLIERS

Some of the more common multipliers and their unit names are explained below, and all of them are shown in Table 7-1.

MILLIUNITS The milliunit is one-thousandth of a unit. Thus, it takes 1000 millivolts to equal 1 volt, 500 milliamperes to equal 0.5 ampere, etc. This unit is commonly used with volts, amperes, henrys, and watts. It is abbreviated m. Thus, 10 mH = 10 millihenrys.* Mathematically, milli = 10^{-3}. 1 mW = 10^{-3} W.

MICROUNITS The microunit is one-millionth of a unit. That is, it takes 1 000 000 microamperes to make 1 ampere, 2 000 000 microfarads to equal to 2 farads, etc. This unit, abbreviated μ (greek letter mu), is commonly used with volts, amperes, ohms, siemens, henrys, and farads. Thus

$$5\ \mu\text{F} = 5 \text{ microfarads}$$

Mathematically, micro = 10^{-6}. 1 μs = 10^{-6} s.

PICOUNITS The picounit, formerly called the micromicrounit, is one-millionth of one-millionth of a unit. That is, 1 farad is equivalent to 1 000 000 000 000, or 10^{12}, picofarads. This unit is seldom used for anything other than farads. It is represented by p. Thus, 250 pF = 250 picofarads. Mathematically, pico = 10^{-12}. Older texts use the micromicrounit, abbreviated $\mu\mu$. Thus,

$$2\ \mu\mu\text{F} = 2 \text{ micromicrofarads} = 2 \text{ pF}$$

KILOUNITS The kilounit is one thousand basic units. Thus, 1 kilovolt is equivalent to 1000 volts. This unit is commonly used with cycles, volts, amperes, ohms, watts, and volt-amperes. It is abbreviated k. Thus, 35 kW means 35 kilowatts; 2000 hertz (Hz) = 2 kilohertz (kHz). Mathematically, kilo = 10^3.

MEGAUNITS The megaunit is one million basic units. Thus, 1 megohm is equal to 1 000 000 ohms. This unit is used mainly with ohms and hertz. It is abbreviated M. Thus, 3 MHz = 3 megahertz. Mathematically, mega = 10^6.

7-11 PREFERRED DECIMAL MULTIPLIERS

Table 7-1 gives the prefix names and abbreviations for the *third powers of 10.* These are the preferred powers, and, therefore, the preferred prefixes. Almost every calculation in electronics will result in a quantity involving one of the third powers of 10 prefixes: *milli*watts, *micro*amperes, *mega*hertz.

In a few cases, prefixes are also used for other powers of 10. Table 7-2 lists these denigrated, or nonstandard, powers of 10

*See Table 3 in the Appendix for abbreviations.

In order to use the preferred third powers of 10, quantities will normally be expressed as numbers between 0.1 and 1000.

In Examples 1 to 3, express numbers using preferred third powers of 10 prefixes.

example 1 0.01 A

solution $0.01\ A = 1 \times 10^{-2}\ A$

Rewriting to a third power,

$$1 \times 10^{-2} = 10 \times 10^{-3}$$

Therefore,

$$0.01\ A = 10 \times 10^{-3}\ A = 10\ mA$$

example 2 1320 kHz

solution

$$\begin{aligned} 1320\ kHz &= 1.32 \times 10^3\ kHz \\ 1320\ kHz &= 1.32 \times 10^3 \times 10^3\ Hz \\ &= 1.32 \times 10^6\ Hz \\ &= 1.32\ MHz \end{aligned}$$

example 3 0.872 H

solution This is a perfectly good number and need not be changed. However, some people may prefer to rewrite it as 872 mH, which is equally good.

7-12 DECIMAL CONVERSION FACTORS

Often it becomes necessary to convert microamperes to milliamperes, gigahertz to kilohertz, megawatts to watts, and so on. The more common conversions in simplified form are listed in Table 7-3.

example 4 Convert 8 μF to farads.

solution $8\ \mu F = 8 \times 10^{-6}\ F$

Table 7-3 CONVERSION FACTORS

Multiply	*By*	*To Obtain*
Picounits	10^{-6}	Microunits
Picounits	10^{-9}	Milliunits
Picounits	10^{-12}	Units
Microunits	10^{6}	Picounits
Microunits	10^{-3}	Milliunits
Microunits	10^{-6}	Units
Milliunits	10^{9}	Picounits
Milliunits	10^{3}	Microunits
Milliunits	10^{-3}	Units
Units	10^{12}	Picounits
Units	10^{6}	Microunits
Units	10^{3}	Milliunits
Units	10^{-3}	Kilounits
Units	10^{-6}	Megaunits
Kilounits	10^{3}	Units
Kilounits	10^{-3}	Megaunits
Megaunits	10^{6}	Units
Megaunits	10^{3}	Kilounits

example 5 Convert 250 mA to amperes.

solution

$$\begin{aligned} 250\ mA &= 250 \times 10^{-3}\ A \\ &= 2.50 \times 10^{-1}\ A \\ \text{or} \quad &= 0.250\ A \end{aligned}$$

example 6 Convert 1500 W to kilowatts.

solution

$$\begin{aligned} 1500\ W &= 1500 \times 10^{-3}\ kW \\ \text{or} \quad &= 1.5\ kW \end{aligned}$$

example 7 Convert 200 000 Ω to megohms.

solution

$$200\,000\ \Omega = 200\,000 \times 10^{-6}\ M\Omega = 0.2\ M\Omega$$

example 8 Convert 2500 kHz to megahertz.

solution

$2500\ \text{kHz} = 2500 \times 10^{-3}\ \text{MHz} = 2.500\ \text{MHz}$

example 9 Convert 0.000 450 S to microsiemens.

solution

$0.000\,450\ \text{S} = 0.000\,450 \times 10^{6}\ \mu\text{S}$

or $= 450\ \mu\text{S}$

example 10 Convert 5 μs to seconds.

solution $5\ \mu\text{s} = 5 \times 10^{-6}\ \text{s}$

PROBLEMS 7-1

Express answers to three significant figures as numbers between 1 and 10 times the proper power of 10:

1. 4300 V	= (a) ____ mV	= (b) ____ μV	= (c) ____ kV
2. 6.85 A	= (a) ____ mA	= (b) ____ μA	
3. 1.35 V	= (a) ____ kV	= (b) ____ μV	= (c) ____ mV
4. 125 mA	= (a) ____ μA	= (b) ____ A	
5. 3300 Ω	= (a) ____ kΩ	= (b) ____ MΩ	= (c) ____ S
6. 50 μF	= (a) ____ F	= (b) ____ pF	
7. 20000 pF	= (a) ____ F	= (b) ____ μF	
8. 16.5 mH	= (a) ____ H	= (b) ____ μH	
9. 347 W	= (a) ____ kW	= (b) ____ mW	= (c) ____ μW
10. 25.3 s	= (a) ____ ms	= (b) ____ μs	
11. 1320 kHz	= (a) ____ MHz	= (b) ____ Hz	
12. 47 kΩ	= (a) ____ Ω	= (b) ____ MΩ	= (c) ____ S
13. 400 mW	= (a) ____ W	= (b) ____ kW	
14. 220 μH	= (a) ____ mH	= (b) ____ H	
15. 15 kHz	= (a) ____ MHz	= (b) ____ Hz	
16. 8 μs	= (a) ____ ms	= (b) ____ s	= (c) ____ ns
17. 0.055 A	= (a) ____ μA	= (b) ____ mA	
18. 325 kV	= (a) ____ V	= (b) ____ MV	
19. 2.7 MΩ	= (a) ____ Ω	= (b) ____ kΩ	
20. 3.7 kWh	= (a) ____ Wh	= (b) ____ mWh	
21. 3350 mH	= (a) ____ μH	= (b) ____ H	
22. 506 MHz	= (a) ____ kHz	= (b) ____ Hz	
23. 0.000 50 μF	= (a) ____ pF	= (b) ____ F	
24. 1500 ms	= (a) ____ μs	= (b) ____ s	= (c) ____ ns
25. 2.5 S	= (a) ____ μS	= (b) ____ Ω	
26. 5000 μS	= (a) ____ S	= (b) ____ Ω	
27. 2350 μA	= (a) ____ mA	= (b) ____ A	
28. 0.15 kV	= (a) ____ V	= (b) ____ mV	
29. 150 MW	= (a) ____ W	= (b) ____ kW	
30. 980 000 Hz	= (a) ____ kHz	= (b) ____ MHz	

7-13 INTERSYSTEM CONVERSIONS

In the early sections of this chapter we briefly reviewed the two systems with which we most often deal, and we listed some common conversion factors. Some books of tables give hundreds of such interrelationships, and you will meet them as you continue your studies.

You must realize that, without the units, your calculations are incomplete. When measurements are added, subtracted, multiplied, or divided, then the units pertaining to those measurements must also take part in the calculations.

example 11 Add 6 V and 12 V.

solution $6\text{ V} + 12\text{ V} = 18\text{ V}$

example 12 Add 3 m and 75 cm.

solution In the metric system, the values of the prefixes represent decimal multipliers:

$$75\text{ cm} = \frac{75}{100}\text{ m} = 0.75\text{ m}$$

(*a*) Adding:

$$\begin{array}{rcl} 3\text{ m} & = & 3\text{ m} \\ +75\text{ cm} & & +0.75\text{ m} \\ \hline & = & 3.75\text{ m} \end{array}$$

(*b*) Alternatively, 3 m = 300 cm

Adding:

$$\begin{array}{rcl} 3\text{ m} & = & 300\text{ cm} \\ +75\text{ cm} & & +\ 75\text{ cm} \\ \hline & = & 375\text{ cm} \end{array}$$

Either answer is correct, but one form may be more acceptable than the other under some circumstances. The same person might properly describe the height of a child as 112 cm and later refer to a folded dipole antenna as 1.12 m long.

example 13 What is the speed of an object that traverses 30 m in 2 s?

solution Speed is given in units of distance per unit of time. In this case, the speed is

$$\frac{30\text{ m}}{2\text{ s}} = 15\,\frac{\text{m}}{\text{s}} \qquad \text{(usually written m/s*)}$$

example 14 What is the area of a room 12 m long and 18 m wide?

solution Areas are given in square measure:

$$(12\text{ m})(18\text{ m}) = 216\text{ square meters (m}^2\text{)}$$

example 15

$$3\,\Omega + 6\,\Omega = 9\,\Omega$$
$$230\text{ V} - 115\text{ V} = 115\text{ V}$$

example 16

$$2\text{ m} \times 4\text{ m} = 2 \times 4 \times \text{m} \times \text{m} = 8\text{ m}^{2\dagger}$$
$$0.7\text{ m} \times 1.6\text{ m} = 0.7 \times 1.6 \times \text{m} \times \text{m} = 1.12\text{ m}^2$$
$$20\text{ cm} \times 1.2\text{ m} = 0.2\text{ m} \times 1.2\text{ m} = 0.24\text{ m}^2$$
$$3\text{ m} \times 5\text{ m} \times 2\text{ m} = 3 \times 5 \times 2 \times \text{m} \times \text{m} \times \text{m} = 30\text{ m}^3$$
$$6\text{ m} \times 10\text{ m} = 6 \times 10\text{ meters} \times \text{meters} = 60\text{ meters}^2 = 60\text{ m}^2$$

When a ratio between identical units is expressed, such as $\frac{60\text{ m}}{12\text{ m}}$, the units cancel, and the result of the division is a pure number with no dimension.

example 17 $\frac{60\text{ m}}{12\text{ m}} = \frac{60\,\cancel{\text{m}}}{12\,\cancel{\text{m}}} = 5$

*m/s (*a shilling fraction*) has exactly the same meaning as $\frac{\text{m}}{\text{s}}$ (*a built-up fraction*); the *only* difference is in the manner of printing.

†It is generally preferred that areas be written in the form 8 m² rather than 8 sq m.

When quantities having different units are multiplied or divided, the result must express the operation.

example 18 $4\text{ m} \times 5\text{ kg} = 4 \times 5 \times \text{m} \times \text{kg}$
$$= 20\text{ kg}\cdot\text{m}$$

example 19 $$\frac{30\text{ m}}{10\text{ s}} = \frac{30}{10}\,\frac{\text{m}}{\text{s}} = 3\,\frac{\text{m}}{\text{s}}$$
$$= 3\text{ m/s} = 3\text{ m}\cdot\text{s}^{-1}$$

example 20 $$\frac{45\,\Omega}{15\text{ m}} = \frac{45}{15}\,\frac{\Omega}{\text{m}} = 3\,\Omega/\text{m}$$

In Example 19 above, note that m/s is read as "meters per second," and in Example 20, Ω/m is read as "ohms per meter." Per means *divided by*.

Thus some of the equivalent lengths stated in Sec. 7-7 can be expressed as follows:

- There are 2.540 cm/in.
- There is 0.3048 m/ft.
- There are 1.609 km/mi.
- There are 39.37 in/m.
- There is 0.6214 mi/km.

Using relations in forms so that units are treated mathematically as literal factors facilitates conversions and assures that results will be obtained with correct units.

example 21 Convert 3 in to centimeters.

solution

$$3\text{ in} \times 2.54\,\frac{\text{cm}}{\text{in}} = 3 \times 2.54 \cdot \cancel{\text{in}} \cdot \frac{\text{cm}}{\cancel{\text{in}}}$$
$$= 7.62\text{ cm}$$

example 22 How many meters are there in 236 ft?

solution

$$236\text{ ft} \times 0.3048\,\frac{\text{m}}{\text{ft}} = 236 \times 0.3048 \cdot \cancel{\text{ft}} \cdot \frac{\text{m}}{\cancel{\text{ft}}}$$
$$= 71.93\text{ m}$$

example 23 A certain resistance wire has a resistance of 3 Ω/m. What is the resistance of 6 m of this wire?

solution

$$3\,\frac{\Omega}{\text{m}} \times 6\text{ m} = 3 \times 6 \cdot \frac{\Omega}{\cancel{\text{m}}} \cdot \cancel{\text{m}} = 18\,\Omega$$

example 24 Convert 1500 kHz to hertz.

solution There are 10^3 Hz per kilohertz, that is, $10^3\,\dfrac{\text{Hz}}{\text{kHz}}$. Then

$$1500\text{ kHz} = 1500\,\frac{\text{kcycles}}{\text{s}} \times 10^3\,\frac{\text{cycles}}{\text{kcycle}}$$
$$= 1500 \times 10^3\,\frac{\cancel{\text{kcycles}}}{\text{s}} \cdot \frac{\text{cycles}}{\cancel{\text{kcycles}}}$$
$$= 1.5 \times 10^6\text{ cycles/s}$$
$$= 1.5 \times 10^6\text{ Hz}$$

example 25 The wavelength λ of a radio wave in meters, the frequency *f* of the wave in hertz, and the velocity of propagation c in meters per second are related to one another by the formula

$$\lambda = \frac{c}{f}$$

or $$\lambda = \frac{3 \times 10^8}{f} \qquad \text{m}$$

Derive a formula for wavelength expressed in feet.

solution Since there are 3.28 ft/m, this factor must be applied to express λ in feet. Thus,

$$\lambda = \frac{3 \times 10^8}{f} \text{ m} \times 3.28 \frac{\text{ft}}{\text{m}}$$

$$= \frac{3 \times 3.28 \times 10^8}{f} \cdot \text{m} \cdot \frac{\text{ft}}{\text{m}}$$

$$= \frac{9.84 \times 10^8}{f} \text{ ft*}$$

*Some users of this book will be interested in the 1972 report issued by the National Bureau of Standards in Boulder, Colorado, which gives the value of c as 299 792.4562 km/s ± 1.1 m/s. 3×10^8 is a sufficiently accurate approximation for the correct solution to all the problems in this book, and, indeed, for most of the problems ever to be solved by the majority of electronics technicians anywhere.

example 26 By using the formula $\lambda = (3 \times 10^8)/f$ m, derive a formula for wavelength in meters when the frequency is expressed in megahertz.

solution In the above formula f is expressed in hertz and it is desired to express the frequency in megahertz. Since

$$\text{MHz} = \text{Hz} \times 10^6,$$

this is substituted for f in the formula. Thus,

$$\lambda = \frac{3 \times 10^8}{f \times 10^6} \text{ m} = \frac{3 \times 10^2}{f} \text{ m} = \frac{300}{f} \text{ m}$$

PROBLEMS 7-2

1. 9 ft = (*a*) ____ in = (*b*) ____ cm = (*c*) ____ mm
2. 3500 mm = (*a*) ____ km = (*b*) ____ ft = (*c*) ____ yd
3. 2.05 m = (*a*) ____ in = (*b*) ____ cm = (*c*) ____ yd
4. 15 840 ft = (*a*) ____ km = (*b*) ____ mi = (*c*) ____ cm
5. 5064 yd = (*a*) ____ mi = (*b*) ____ m = (*c*) ____ km
6. An automobile is traveling at a rate of 90 mi/h. What is its speed in meters per second?
7. The radius of No. 14 wire is 32-thousandths of an inch. What is its diameter in millimeters?
8. Radio waves are often referred to by wavelength instead of frequency. The wavelength of waves at a frequency of 3000 MHz is 10 cm. What is that wavelength in millimeters?
9. A power transmission line 120 km long was found to have a total inductance of 0.4488 H. What is the inductance per kilometer?
10. The capacitance of a power line was measured at 4.98 nF/km. What is the capacitance per meter?
11. A transmission line 250 m long was found to have an attenuation loss of 0.15 decibel (dB). What is the attenuation in decibels per hundred meters?
12. A twisted-pair transmission line 200 m long has a loss of 42 dB. What is the loss in decibels per centimeter?
13. The high-frequency resistance of No. 10 copper wire was measured some years ago, using a 6-ft length of wire. At 100 MHz, the resistance was found to be 0.588 Ω. What is the resistance in ohms per centimeter at the same frequency?
14. The speed of free electrons in random motion is approximately 100 000 m/s. What is this speed in miles per hour?
15. The speed of electrons "drifting" in an electric current flow is about 0.2 cm/s. What is this speed in inches per minute?

7-14 PRACTICAL CONSIDERATIONS

In Secs. 6-6 and 7-9 and in several instances through the use of examples and problems, attempts have been made to emphasize the fact that extremely wide ranges in values of units are encountered in electrical and electronics computations. This has been done in order to impress you with the necessity of exercising care in making computations if you are to obtain accurate results. For example, in computing inductive reactances, the frequency may be in megahertz and the inductance in microhenrys. In radar and other applications we are concerned with the velocity of propagation of radio waves (3×10^8 m/s) and with time intervals in microseconds. This is equally true in television reception, particularly as it relates to the production of duplicate images, usually called *ghosts*. As an example, Fig. 7-1 illustrates how a television receiver can receive a picture signal from a transmitting station by different paths. The direct wave is received from the transmitter along one path, while the other signal arrives at the receiving antenna via a path 1 km longer than the direct path as a result of being reflected. Because the velocity of radio waves is 3×10^8 m/s, the reflected signal arrives at the receiver $1/(3 \times 10^5)$ s, or about 3.3 μs later than the signal received via the direct path between transmitter and receiver. Since the electron scanning beam scans one horizontal line in approximately 55 μs, on a picture 50 cm wide the beam will scan about 1 cm in 1.1 μs. Therefore, the reflected signal arriving 3.3 μs late will produce a second

Fig. 7-1 Antenna Receiving Picture Signal via Two Paths

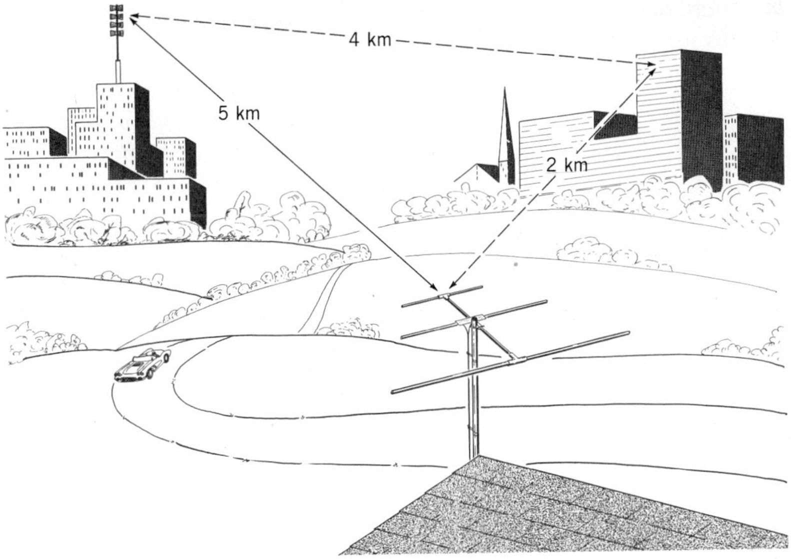

Fig. 7-2 Television Ghost (*Courtesy of Radio Corporation of America*)

picture 3 cm to the right in the direction of scanning as shown in Fig. 7-2. This duplicate image produced by the reflected wave is called a *ghost*.

7-15 SIGNIFICANT FIGURES

The subjects of accuracy and significant figures were discussed in Secs. 6-3 and 6-4. Now that we have some idea of the various units used in electrical and radio problems, two questions arise:

1. To how many significant figures should an answer be expressed?
2. How can we definitely show that an answer is correct to just so many significant figures?

The answer to the first question is comparatively easy. No answer can be more accurate than the figures, or data, used in the problem. As stated in Sec. 6-3, it is safe to assume that the values of the average circuit components and calibrations of meters that we use in our everyday work are not known beyond three significant figures. Therefore, in the future we will round off long answers and express them to three significant figures. The exception will be when it is necessary to carry figures out in order to demonstrate some fact or law, carefully.

The second question brings up some interesting points. As an example, suppose we have a resistance of 500 000 Ω and we want to write this value so that it will be apparent to anyone that the figure 500 000 is correct to three significant figures. We can do so by writing

$$500 \times 10^3\ \Omega \quad \text{or} \quad 500\ \text{k}\Omega$$
$$50.0 \times 10^4\ \Omega$$
$$5.00 \times 10^5\ \Omega \quad \text{or} \quad 0.500\ \text{M}\Omega,\ \text{etc.}$$

Any one of these expressions definitely shows that the resistance is correct to three significant figures. Similarly, suppose we had measured the capacitance of a capacitor to be 3500 pF. How can we specify that the figure 3500 is correct to three significant figures? Again we can do so by writing

$$350 \times 10\ \text{pF}$$
$$35.0 \times 10^2\ \text{pF}$$
$$3.50 \times 10^3\ \text{pF} \quad \text{or} \quad 3.50\ \text{nF, etc.}$$

As in the preceding example, there are definitely three figures in the first factor that show the degree of accuracy.

7-16 CALCULATIONS WITH UNITS

In Sec. 7-14 we emphasized the necessity of keeping track of the units involved when performing calculations. The necessity becomes even more apparent when decimal multipliers of basic units are involved, or when you are unsure how to proceed with a solution involving units of different measurements such as decibels and meters, ohms and meters, and hours and kilometers.

As long as your calculations are made in basic units, which are directly related, you will have no difficulty. For example, you know that

$$\text{Ohms} = \frac{\text{volts}}{\text{amperes}}$$

and

$$\text{Ohms} \neq \frac{\text{volts}}{\text{milliamperes}}$$

The milliamperes must be converted to amperes in order to keep the basic relationship in units. Therefore,

$$\text{Ohms} = \text{the number of } \frac{\text{volts}}{\text{milliamperes} \times 10^{-3}}$$

Of course, you could make up your own formulas for special cases and write, for example,

$$\text{Ohms} = \text{the number of } \frac{\text{volts} \times 10^{3}}{\text{milliamperes}}$$

but the task would be endless. Some frequently used formulas are derived for convenience, and you will derive some of them in Problems 7-3. However, when performing calculations you will never go wrong if you first convert to basic units.

example 27 The voltage across a circuit is 250 V, and the current is 5 mA. What is the resistance of the circuit?

solution Since $\text{ohms} = \dfrac{\text{volts}}{\text{amperes}}$, it is necessary to convert the current of 5 mA into amperes before calculating:

$$R = \frac{E}{I} = \frac{250}{5 \times 10^{-3}}$$
$$= 50 \times 10^{3}\ \Omega$$
$$= 50\ \text{k}\Omega$$

example 28 A current of 150μA flows through a resistance of 30 kΩ. What is the voltage across the resistance?

solution Since the current is in microunits and the resistance is in kilounits, both must be converted into basic units (amperes and ohms) before calculating:

$$\text{Volts} = \text{amperes} \times \text{ohms}$$

or

$$E = I \times R$$
$$= (150 \times 10^{-6})(30 \times 10^{3})$$
$$= 4.5\ \text{V}$$

You will encounter cases in which you may be unsure how to proceed, particularly when you deal with units of differing measurements, such as Ω/m, μF/km, m/s, kg/m², and dB/100 m. Keeping track of your units and handling them as literal numbers will ensure a correct numerical answer expressed in the proper units.

example 29 How long will it take to travel 225 km at an average speed of 45 km/h?

solution Here we have kilometers and kilometers per hour, and we know the answer must be expressed in hours. Also, we know that

$$\text{Distance} = \text{speed} \times \text{time}$$

or

$$\text{Time} = \frac{\text{distance}}{\text{speed}}$$

That is

$$\text{h} = \frac{\text{km}}{\frac{\text{km}}{\text{h}}}$$
$$= \cancel{\text{km}} \cdot \frac{\text{h}}{\cancel{\text{km}}} = \text{h}$$

Knowing that the answer will be expressed in the proper unit, we can complete the calculation:

$$\text{Time} = \frac{225\ \text{km}}{45\ \frac{\text{km}}{\text{h}}}$$
$$= \frac{225}{45}\ \cancel{\text{km}} \cdot \frac{\text{h}}{\cancel{\text{km}}}$$
$$= 5\ \text{h}$$

example 30 A 3-km roll of No. 10 wire is measured and found to have a resistance of 9.81 Ω. What is the resistance of 100 m of this wire?

solution The resistance must be expressed in ohms. Since the measurement was $\frac{9.81\ \Omega}{3000\ \text{m}}$,

$$\frac{9.81}{3000}\frac{\Omega}{\text{m}} = 3.27 \times 10^{-3}\frac{\Omega}{\text{m}}$$

Then the resistance of 100 m of this wire is

$$3.27 \times 10^{-3}\frac{\Omega}{\cancel{\text{m}}} \times 100\,\cancel{\text{m}} = 0.327\ \Omega$$

This could be written as 0.327 Ω/100 m.

In the problems which follow, you will be asked to make conversions to accommodate readings in units which do not exactly fit the formulas relating the dimensions, as in Example 27, in which 5 mA had to be converted into amperes before proceeding. You will also be asked to convert the basic or classic formulas to adjust for units other than the basic ones. When both of these conversions are asked for in a single problem, follow this rule:

Rule

Adjust the units in which the measurements were made so that they will agree with the units for which the formula was developed. Then convert to other units as required.

PROBLEMS 7-3

1. The capacitive reactance of a circuit, or a capacitor, is given by the formula

$$X_C = \frac{1}{2\pi fC} \qquad \Omega$$

where X_C = capacitive reactance, Ω
f = frequency, Hz
C = capacitance of circuit, or capacitor, F

Show that
$$X_C = \frac{159 \times 10^3}{fC} \qquad \Omega$$

when f = frequency, MHz
C = capacitance, pF

2. Referring to Prob. 1, what is the capacitive reactance of a capacitor of 0.000 50 μF at a frequency of 4000 MHz?

3. The inductive reactance of a circuit, or an inductor, is given by the formula

$$X_L = 2\pi fL \qquad \Omega$$

where X_L = inductive reactance, Ω
f = frequency, Hz
L = inductance of circuit, or inductor, H

Derive a formula for X_L
when f = frequency, MHz
L = inductance, μH

4. Referring to Prob. 3, an amplifier coil has an inductance of 27 μH. What is its inductive reactance at 6 MHz?

5. The resonant frequency of any circuit is given by the formula

$$f = \frac{1}{2\pi\sqrt{LC}} \qquad \text{Hz}$$

where f = frequency, Hz
L = inductance of circuit, H
C = capacitance of circuit, F
Derive a formula expressing f in megahertz
when L = inductance, μH
C = capacitance, pF

6. Referring to Prob. 5, what is the resonant frequency of a circuit with an inductance of 0.25 μH and a capacitance of 16 pF?

7. In copper conductors used in transmission lines, the depth of penetration of high-frequency currents is given by the formula

$$\delta = \frac{2.61 \times 10^{-3}}{\sqrt{f}} \qquad \text{in}$$

where f = frequency, Hz
Derive a formula for current penetration depth in centimeters when f is the frequency in megahertz.

8. Referring to Prob. 7, to what depth in millimeters will a current at 3.75 GHz penetrate a copper conductor?

9. The high-frequency resistance of a round copper wire or of round copper tubing was found in an old handbook to be

$$R_{ac} = 9.98 \times 10^{-4} \frac{\sqrt{f}}{d} \qquad \Omega/\text{ft}$$

where R_{ac} = high-frequency resistance, Ω/ft
f = frequency, MHz
d = outside diameter of conductor, in
Derive a formula for R_{ac} in ohms per centimeter when f is given in megahertz and d is given in centimeters.

10. Referring to Prob. 9, No. 36 wire has a diameter of 0.127 mm. What is the resistance per centimeter of the wire at a frequency of 85 MHz?

11. Use the formula in Example 25 to show that $\lambda = \frac{3 \times 10^4}{f}$ cm when f is given in megahertz.

12. Use the formula in Example 25 to derive a formula for wavelength (λ) in centimeters when f is given in kilohertz.

13. The midfrequency of television channel 4 is 69 MHz. Using the formula derived in Prob. 12, what is the length of one wavelength in centimeters?

14. The great majority of television receiving antennas consist of various combinations of dipoles. A dipole antenna is one that is approximately one-half wavelength long (0.5λ), as illustrated in Fig. 7-3. The actual length is slightly

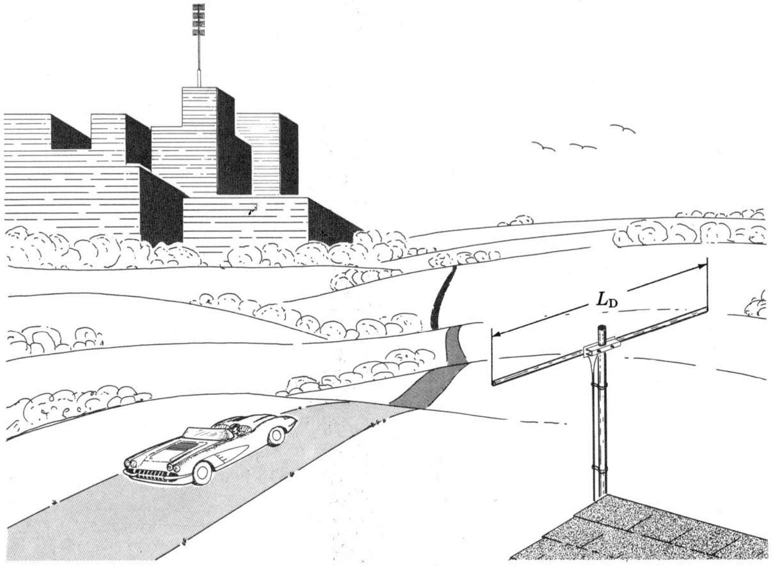

Fig. 7-3 Dipole Antenna

less than a half wave owing to "end effect" caused by the capacitance of the antenna, and it has been determined that dipoles used for television reception should be approximately 6% shorter than one-half wavelength. Use the formula derived in Prob. 12 to derive a formula for the length of a dipole antenna in centimeters when the frequency is in megahertz.

15. The midfrequency of television channel 13 is 213 MHz. Using the formula derived in Prob. 14, what length would you make a receiving antenna for this channel?

16. If a wire approximately one-half wavelength long is placed behind a dipole antenna, the wire acts as a reflector and increases the directivity of the antenna. This results in the reception of stronger signals when the dipole and the reflector are pointed at the transmitting station as illustrated in Fig. 7-4. For best results, the reflector should be 5% longer than the dipole. Referring to the formula for the length of a dipole derived in Prob. 14, derive a formula for the length of a reflector in centimeters when f is in megahertz.

17. The distance between a dipole and its reflector should be approximately one-fifth of one wavelength (0.2λ) as shown in Fig. 7-4. Referring to previously derived formulas, compute the following dimensions for the midfrequency of television channel 10, which is 195 MHz: (*a*) length of dipole, (*b*) length of reflector, and (*c*) spacing between dipole and reflector.

18. The directivity of a dipole-reflector combination, as shown in Fig. 7-4, can be increased by the addition of a conductor in front of the dipole as illustrated in Fig. 7-5. This wire or tube, which is known as a director, is

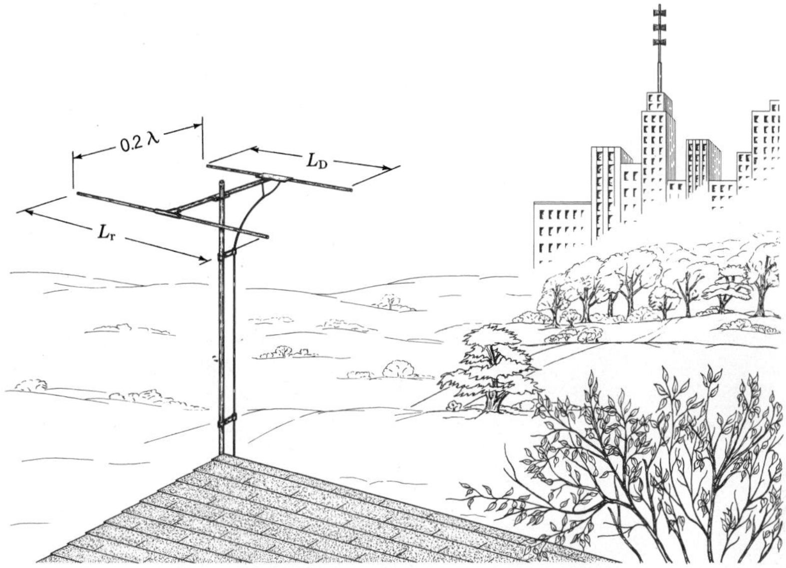

Fig. 7-4 Dipole Antenna with Reflector

usually placed one-tenth wavelength (0.1λ) from the dipole, and it should be about 5% shorter than the dipole. Derive a formula for the length of a director in centimeters when f is in megahertz.

19. Referring to Fig. 7-5, compute the following dimensions for the midfrequency

Fig. 7-5 Dipole Antenna with Reflector and Director

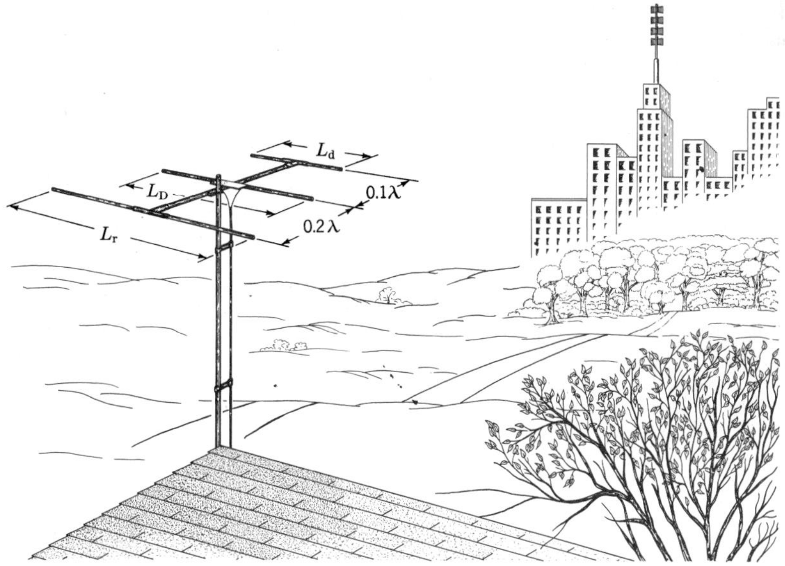

of television channel 10, which is 195 MHz: (*a*) length of dipole, (*b*) length of reflector, (*c*) length of director, (*d*) spacing between dipole and reflector, and (*e*) spacing between dipole and director.

20. Ohm's law may be stated in the form $E = IR$, where E is measured in volts, I in amperes, and R in ohms. What voltage will appear across a resistor measuring 680 MΩ when a current of 0.250 μA flows through it?

8

ohm's law—series circuits

Ohm's law for the electric circuit is the foundation of electric circuit analysis and is therefore of fundamental importance. The various relations of Ohm's law are easily learned and readily applied to practical circuits. A thorough knowledge of these relations and their applications is essential for understanding electric circuits.

This chapter is concerned with the study of Ohm's law in dc series circuits as applied to *parts* of a circuit. For this reason, the internal resistance of a source of voltage, such as a generator or a battery, and the resistance of the wires connecting the parts of a circuit are not discussed in this chapter.

8-1 THE ELECTRIC CIRCUIT

An electric circuit consists of a source of voltage connected by conductors to the apparatus that is to use the electric energy.

An electric current will flow between two points in a conductor when a difference of potential exists across those points. The most generally accepted concept of an electric current is that it consists of a motion, or flow, of electrons from the negative toward a more positive point in a circuit. The force that causes the motion of electrons is called an *electromotive force*, a *potential difference*, or a *voltage*, and the opposition to the motion is called *resistance*.

The basic theories of electrical phenomena and the methods of producing currents are not within the scope of this book. You will find them adequately treated in the great majority of textbooks on the subject.

8-2 OHM'S LAW

Ohm's law for the electric circuit, reduced to plain terms, states the relation that exists among voltage, current, and resistance. One way of stating this relation is as follows: The voltage across any *part* of a circuit is proportional to the product of the current through that *part* of the circuit and the resistance of that *part* of the circuit. Stated as a formula the foregoing is expressed as

$$E = IR \qquad \text{V} \tag{1}$$

where E = voltage, or potential difference, V
I = current, A
R = resistance, Ω

If any two factors are known, the third can be found by solving Eq. (1). Thus,

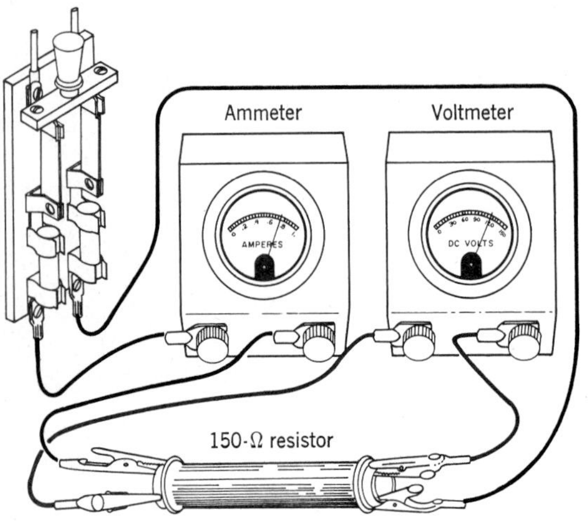

Fig. 8-1 Sketch of the Circuit of Example 1 Showing How the Parts Are Connected to Form the Circuit

$$I = \frac{E}{R} \quad \text{A} \tag{2}$$

and

$$R = \frac{E}{I} \quad \Omega \tag{3}$$

8-3 METHODS OF SOLUTION

The general outline for working problems given in Sec. 5-8 is applicable to the solution of circuit problems. In addition, a neat, simplified diagram of the circuit should be drawn for each problem. The diagram should be labeled with all the known values of the circuit such as voltage, current, and resistances. In this manner the circuit and problem can be visualized and understood. Solving a problem by making purely mechanical substitutions in the proper formulas is not conducive to gaining a complete understanding of any problem.

example 1 How much current will flow through a resistance of 150 Ω if the applied voltage across the resistance is 117 V?

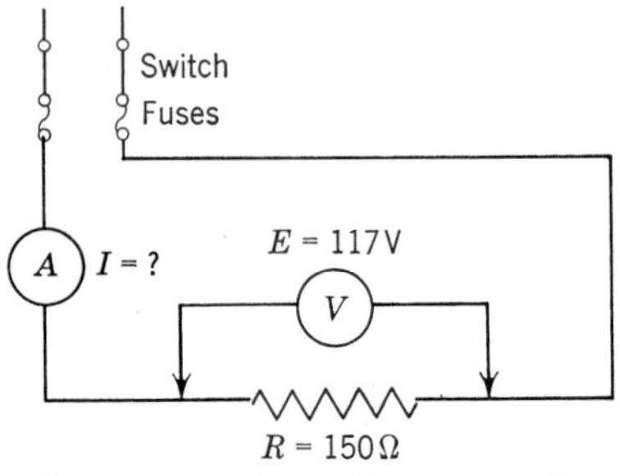

Fig. 8-2 Schematic Circuit Diagram of Example 1

solution The circuit is represented in Figs. 8-1 and 8-2.

Given $E = 117$ V $\quad R = 150\ \Omega$

$I = ?$

$$I = \frac{E}{R} = \frac{117}{150} = 0.780 \text{ A}$$

example 2 A voltmeter connected across a resistance reads 22 V, and an ammeter connected in series with the resistance reads 2.60 A. What is the value of the resistance?

solution The circuit is represented in Fig. 8-3.

Given $E = 220$ V $\quad I = 2.60$ A

$R = ?$

$$R = \frac{E}{I} = \frac{220}{2.60} = 84.6\ \Omega$$

Fig. 8-3 Circuit of Example 2

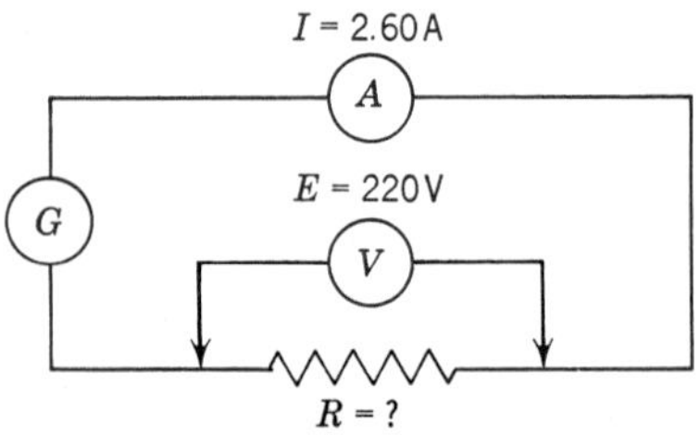

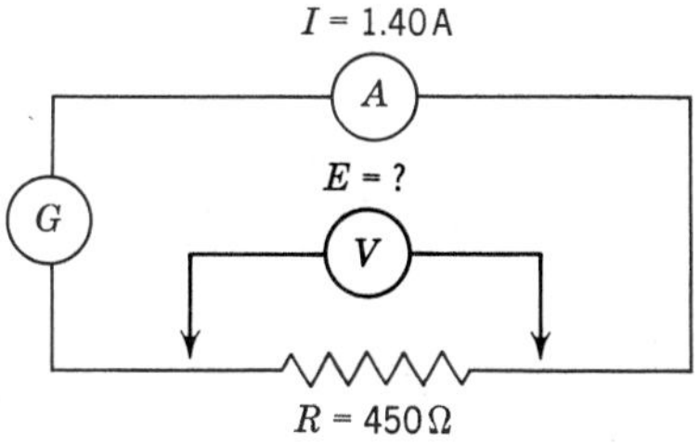

Fig. 8-4 Circuit of Example 3

example 3 A current of 1.40 A flows through a resistance of 450 Ω. What should the reading be of a voltmeter when it is connected across the resistance?

solution The diagram of the circuit is shown in Fig. 8-4.

Given $I = 1.40\ \text{A}$ $\quad R = 450\ \Omega$

$E = ?$

$E = IR = 1.40 \times 450 = 630\ \text{V}$

example 4 A measurement shows a potential difference of 63.0 μV across a resistance of 300 Ω. How much current is flowing through the resistance?

solution The circuit is represented in Fig. 8-5.

Fig. 8-5 Circuit of Example 4

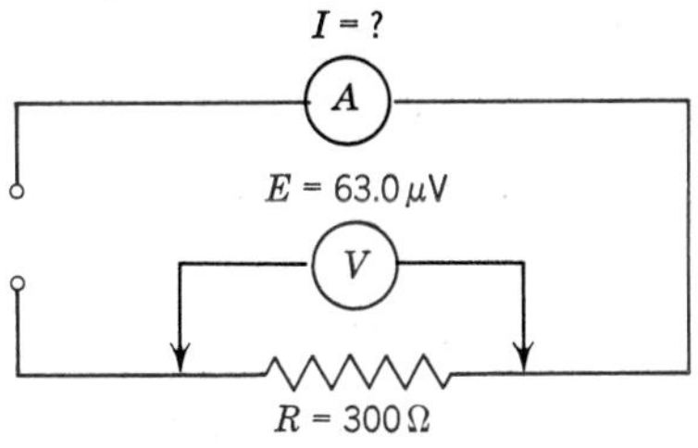

Given

$E = 63.0\ \mu\text{V} = 6.3 \times 10^{-5}\ \text{V} \qquad R = 300\ \Omega$

$I = ?$

$$I = \frac{E}{R} = \frac{6.3 \times 10^{-5}}{300} = \frac{6.3 \times 10^{-7}}{3.00}$$

$= 2.1 \times 10^{-7}\ \text{A}$

or $I = 0.21\ \mu\text{A}$

example 5 A current of 8.60 mA flows through a resistance of 500 Ω. What voltage exists across the resistance?

solution The circuit is represented in Fig. 8-6.

Given

$I = 8.60\ \text{mA} = 8.60 \times 10^{-3}\ \text{A} \qquad R = 500\ \Omega$

$E = ?$

$E = IR = 8.60 \times 10^{-3} \times 500$

$= 8.60 \times 10^{-3} \times 5 \times 10^{2}$

$= 8.60 \times 5 \times 10^{-1} = 4.30\ \text{V}$

Carefully note, as illustrated in Examples 4 and 5, that the equations expressing Ohm's law are in units, that is, volts, amperes, and ohms.

Fig. 8-6 Circuit of Example 5

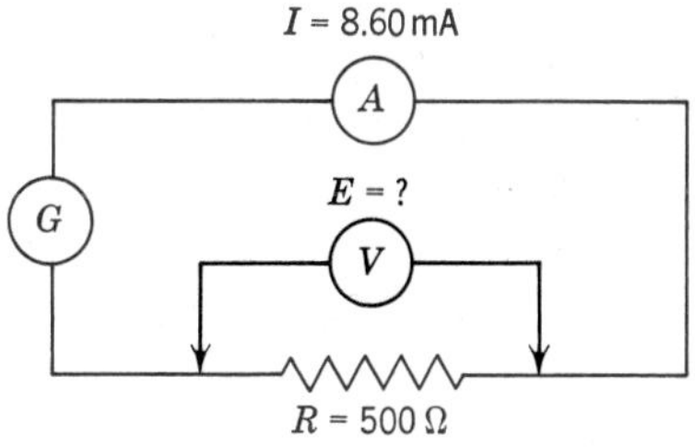

PROBLEMS 8-1

1. How much current will flow through a resistance of 50.0 Ω if a potential of 220 V is applied across it?
2. A certain soldering iron draws 1.35 A from a 120-V line. What is the resistance of the heating unit of the soldering iron?

3. What current will flow when an emf of 440 V is impressed across a 71.0-Ω resistor?
4. A milliammeter connected in series with a 10-kΩ resistor reads 8.0 mA. What is the voltage across the resistor?
5. A microvoltmeter connected across a 500-Ω resistor reads 40 μV. What current is flowing through the resistor?
6. What voltage is required to cause a current flow of 6.2 mA through a resistance of 7.1 kΩ?
7. A certain milliammeter, with a scale of 0 to 1.0 mA, has a resistance of 32 Ω. If this milliammeter is connected directly across a 120-V line, how much current will flow through the meter? What conclusion do you draw?
8. The current flowing through a 3.3-kΩ resistor is 4.3 mA. What should a voltmeter read when it is connected across the resistor?
9. The cold resistance of a carbon filament lamp is 210 Ω, and the hot resistance is 189 Ω. What is the current flow (*a*) the instant the lamp is switched across a 120-V line and (*b*) when constant operating temperature is reached?
10. A type SN954 half-wave rectifier tube filament draws a current of 450 mA at its rated voltage of 6.3 V. What is the resistance of the filament when the tube is in operation?

8-4 POWER

In specifying the rating of electrical equipment, it is customary to state not only the voltage at which the equipment was designed to operate but also the rate at which the equipment produces or consumes electric energy.

The rate of producing or consuming energy is called *power*, and electric energy is measured in watts or kilowatts. Thus, your study lamp may be rated 100 W at 120 V; a generator may be rated 2000 kW at 440 V; etc.

Electric motors were formerly rated in terms of the mechanical energy output, measured in *horsepower*, which they could develop. The conversion from electric energy to this older unit of mechanical energy is given by the relation

$$746 \text{ W} = 1 \text{ horsepower (hp)}$$

With the advent of metrication, the SI unit of power will be used more and more, and the *watt*, or *joule per second*, will replace the horsepower rating.

$$1 \text{ watt (W)} = 1 \text{ J/s}$$

Because users of this book will undoubtedly be called upon to handle older motors rated in horsepower, a number of problems involving this older unit have been retained in this edition.

8-5 THE WATT

Energy is expended at a rate of one wattsecond (Ws or W · s)* (joule, J) every second when one volt causes a current of one ampere to flow. In this case, we say that the

* The use of the center dot in the abbreviations for wattsecond and watthour is preferred in general physics relationships. However, it is customarily omitted in electricity and electronics usage.

power represented when one volt causes one ampere to flow is one watt. This relation is expressed as

$$P = EI \qquad \text{W} \tag{4}$$

This is a useful equation when the voltage and current are known.

Because, by Ohm's law, $E = IR$, this value of E can be substituted in Eq. (4). Thus,

$$P = (IR)I$$

or

$$P = I^2R \qquad \text{W} \tag{5}$$

This is a useful equation when the current and resistance are known.

By substituting the value of I of Eq. (2) in Eq. (4),

$$P = E\frac{E}{R}$$

or

$$P = \frac{E^2}{R} \qquad \text{W} \tag{6}$$

This is a useful equation when the voltage and resistance are known.

WATTHOURS–KILOWATTHOURS The consumer of electric energy pays for the amount of energy used by his electrical equipment. This is measured by instruments known as *watthour* or *kilowatthour meters*. These meters record the amount of energy taken by the consumer.

Electric energy is sold at so much per kilowatthour (kWh). One watthour of energy is consumed when one watt of power continues in action for one hour. Similarly, 1 kWh is consumed when the power is 1000 W and the action continues for 1 h or when a 100-W rate persists for 10 h, etc. Thus, the amount of energy consumed is the product of the power and the time. Perhaps in time kilowatthour meters will be replaced by megajoule meters.

8-6 LOSSES

The study of the various forms in which energy may occur and the transformation of one kind of energy into another has led to the important principle known as the principle of the *conservation of energy*. Briefly, this states that energy can never be created or destroyed. It can be transformed from one form to another, but the total amount remains unchanged. Thus, an electric motor converts electric energy into mechanical energy, the incandescent lamp changes electric energy into heat energy, the loudspeaker converts electric energy into sound energy, the generator converts mechanical energy into electric energy, etc. In each instance the transformation from one type of energy to another is not accomplished with 100% efficiency because some energy is converted into heat and does no useful work as far as that particular conversion is concerned.

Resistance in a circuit may serve a number of useful purposes, but unless it has been specifically designed for heating or dissipation purposes, the energy transformed in the resistance generally serves no useful purpose.

8-7 EFFICIENCY

Because all electrical equipment contains resistance, some heat always develops when current flows. Unless the equipment is to be used for producing heat, the heat due to the resistance of the equipment represents wasted energy. No electrical equipment or other machine is capable of converting energy received into useful work without some loss.

The power that is furnished a machine is called its *input*, and the power received from a machine is called its *output*. The efficiency of a machine is equal to the ratio of the output to the input. That is,

$$\text{Efficiency} = \frac{\text{output}}{\text{input}} \qquad (7)$$

It is evident that the efficiency, as given in Eq. (7), is always a decimal, that is, a number less than 1. Naturally, in Eq. (7), the output and input must be expressed in the same units. Hence, if the output is expressed in kilowatts, then the input must be expressed in kilowatts; if the output is expressed in horsepower, then the input must be expressed in horsepower; etc.

example 6 A voltage of 110 V across a resistor causes a current of 5 A to flow through the resistor. How much power is expended in the resistor?

solution 1 The circuit is represented in Fig. 8-7.

Given $E = 110\text{ V}$ $I = 5\text{ A}$
$P = ?$

Using Eq. (4),

$$P = EI = 110 \times 5 = 550\text{ W}$$

solution 2 Find the value of the resistance and use it to solve for *P*. Thus, using Eq. (3),

$$R = \frac{E}{I} = \frac{110}{5} = 22\ \Omega$$

Using Eq. (5),

$$P = I^2R = 5^2 \times 22 = 5 \times 5 \times 22 = 550\text{ W}$$

Fig. 8-7 Circuit of Example 6

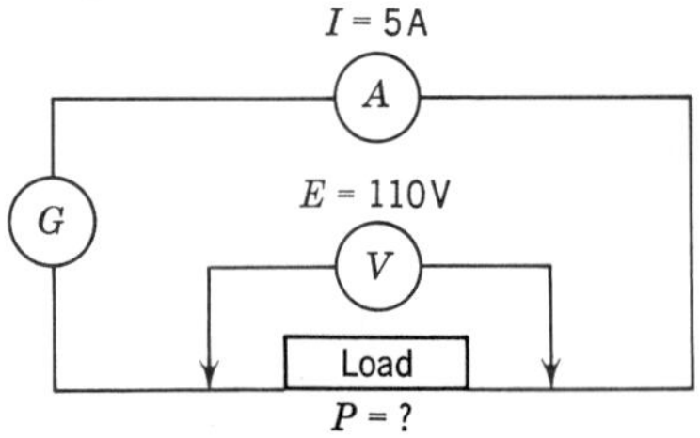

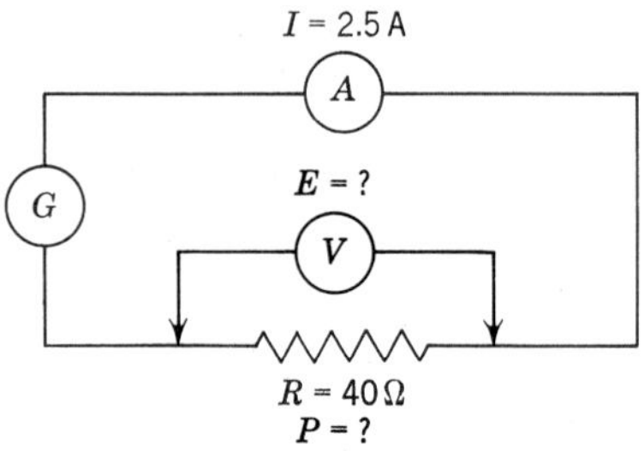

Fig. 8-8 Circuit of Example 7

solution 3 Using Eq. (6),

$$P = \frac{E^2}{R} = \frac{110^2}{22} = \frac{110 \times 110}{22} = 550\text{ W}$$

Solving a problem by two methods serves as an excellent check on the results, for there is little chance of making the same error twice, as happens too often when the same method of solution is repeated.

example 7 A current of 2.5 A flows through a resistance of 40 Ω.
(*a*) How much power is expended in the resistor?
(*b*) What is the potential difference across the resistor?

solution 1 The circuit is represented in Fig. 8-8.

Given $I = 2.5\text{ A}$ $R = 40\ \Omega$
$P = ?$ $E = ?$

(*a*) $P = I^2R = 2.5^2 \times 40 = 2.5 \times 2.5 \times 40 = 250\text{ W}$

(*b*) $E = IR = 2.5 \times 40 = 100\text{ V}$

solution 2 (*a*) Find *E*, as above, and use it to solve for *P*. Thus,

$$P = \frac{E^2}{R} = \frac{100^2}{40} = \frac{100 \times 100}{40} = 250\text{ W}$$

or $P = EI = 100 \times 2.5 = 250\text{ W}$

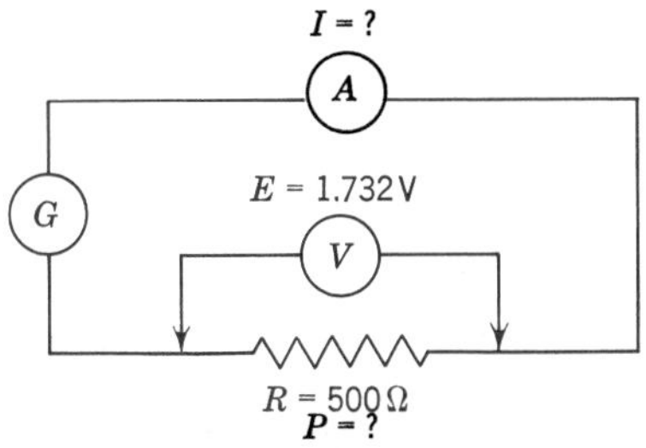

Fig. 8-9 Circuit of Example 8

example 8 A voltage of 1.732 V is applied across a 500-Ω resistor.

(a) How much power is expended in the resistor?

(b) How much current flows through the resistor?

solution A diagram of the circuit is shown in Fig. 8-9.

Given $E = 1.732\text{ V}$ $R = 500\ \Omega$ $P = ?$ $I = ?$

(a) $$P = \frac{E^2}{R} = \frac{1.732^2}{500} = \frac{1.732^2}{5 \times 10^2}$$

$$= \frac{1.732^2}{5} \times 10^{-2} = 0.006\text{ W}$$

or $P = 6\text{ mW}$

(b) $$I = \frac{E}{R} = \frac{1.732}{500} = \frac{1.732}{5} \times 10^{-2}$$

$$= 0.346 \times 10^{-2}\text{ A}$$

or $I = 3.46\text{ mA}$

Check the foregoing solution for power by using an alternative method.

example 9

(a) What is the hot resistance of a 100-W 110-V lamp?

(b) How much current does the lamp take?

(c) At 4¢/kWh, how much does it cost to operate this lamp for 24 h?

solution 1 The circuit is represented in Fig. 8-10.

Given $P = 100\text{ W}$ $E = 110\text{ V}$

(a) Because the power and voltage are known and the resistance is unknown, an equation that contains these three must be used. Thus,

$$P = \frac{E^2}{R}$$

hence, $$R = \frac{E^2}{P} = \frac{110^2}{100} = 121\ \Omega$$

(b) $$I = \frac{E}{R} = \frac{110}{121} = 0.909\text{ A}$$

(c) If the lamp is lighted for 24 h, it will consume

$$100 \times 24 = 2400\text{ Wh} = 2.40\text{ kWh}$$

At 4¢/kWh the cost would be

$$2.4 \times 4 = 9.6¢$$

solution 2 The current may be found first by making use of the relation

$$P = EI$$

which results in $I = \dfrac{P}{E} = \dfrac{100}{110} = 0.909\text{ A}$

The resistance can now be determined by

Fig. 8-10 Circuit of Example 9

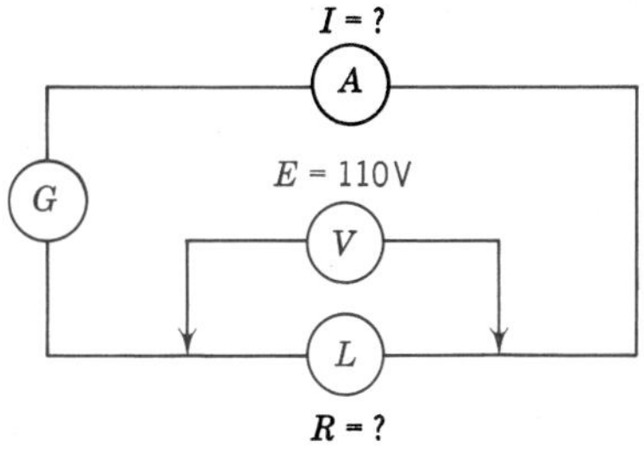

$$R = \frac{E}{I} = \frac{110}{0.909} = 121\ \Omega$$

The solution can be checked by

$$P = I^2R = 0.909^2 \times 121 = 100\text{ W}$$

which is the power rating of the lamp as given in the example. The cost is computed as before.

example 10 A motor delivering 6.50 mechanical horsepower is drawing 26.5 A from a 220-V line.
(a) How much electric power is the motor taking from the line?
(b) What is the efficiency of the motor?
(c) If power costs 3¢/kWh, how much does it cost to run the motor for 8 h?

solution A diagram of the circuit is shown in Fig. 8-11.

Given $E = 220\text{ V}$ $I = 26.5\text{ A}$

and mechanical horsepower

Fig. 8-11 Circuit of Example 10

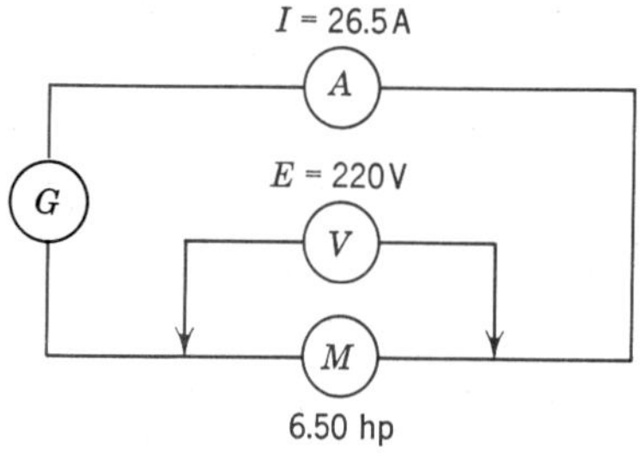

$$P = 6.5\text{ hp} = 6.5 \times 746$$
$$= 4850\text{ W} = 4.85\text{ kW}$$

(a) The power taken by the motor is

$$P = EI = 220 \times 26.5 = 5830\text{ W}$$
$$= 5.83\text{ kW}$$

$$(b)\ \text{Efficiency} = \frac{\text{output}}{\text{input}} = \frac{4.85}{5.83} = 0.832$$
$$= 83.2\%$$

(c) Because the motor consumes 5.83 kW, in 8 hr it would take

$$5.83 \times 8 = 46.6\text{ kWh}$$

At 3¢/kWh, the cost would be

$$46.6 \times 0.03 = \$1.40$$

note The cost was computed in two steps for the purpose of illustrating the solution. When you have become familiar with the method, the cost should be computed in one step. Thus,

$$\text{Cost} = 5.83 \times 8 \times 0.03 = \$1.40$$

From the foregoing examples, it will be noted that computations involving power consist mainly in the applications of Ohm's law. Little trouble will be encountered if each problem is given careful thought and the systematic procedure previously outlined is followed in finding the solution.

PROBLEMS 8-2

1. 7.5 hp = (a) ____W = (b) ____kW
2. 29.84 kW = (a) ____W = (b) ____hp
3. What current is drawn by a 100-W soldering iron that is connected to a 120-V line?
4. How much power is expended in a 120-Ω resistor through which a current of 15 A flows?

5. What is the electric horsepower of a generator which delivers a current of 50.9 A at 220 V?
6. A voltmeter connected across a 2.2-kΩ resistor reads 120 V. How much power is being expended in the resistor?
7. A diesel engine is rated at 1500 hp. What is its electrical rating in kilowatts?
8. An ammeter is connected in the circuit of a 440-V motor. When the motor is running, the ammeter reads 2.27 A. How much power is being absorbed from the line?
9. The resistance of a certain ammeter is 0.012 Ω. Determine the power expended in the meter when it reads 3 A.
10. The resistance of a certain voltmeter is 300 kΩ. Determine the power expended in the voltmeter when it is connected across a 220-V line.
11. A type 6F6 vacuum tube, used in the output stage of a radio receiver, has a cathode-biasing resistor of 470 Ω. A voltmeter connected across this resistor reads 16.5 V.
    (*a*) How much power must the resistor be able to radiate continuously while in operation?
    (*b*) What is the current flow through the resistor?
12. A type 6C5 vacuum tube is operating with a cathode-biasing resistor of 1 kΩ through which flows a current of 8 mA.
    (*a*) How much power is being expended in the resistor?
    (*b*) What is the voltage across the resistor?
13. An emf of 90 μV is applied across a 390-Ω resistor.
    (*a*) How much power is expended in the resistor?
    (*b*) How much current will flow through the resistor?
14. A 1-kΩ resistor in the emitter circuit of a 2N1414 transistor produces a voltage drop of 6 V between collector and emitter.
    (*a*) What is the emitter current?
    (*b*) What is the power loss in this bias resistor?
15. A radar antenna motor is delivering 10 hp. A kilowattmeter that measures the power taken by the motor reads 8.24 kW.
    (*a*) What is the efficiency of the motor?
    (*b*) At 2.5¢/kWh, how much would it cost to run the motor continuously for 5 days?
16. A 440-V 10-hp forced-draft fan motor has an efficiency of 80%.
    (*a*) How many kilowatts does it consume?
    (*b*) How much current does it draw from the line?
    (c) At 2.5¢/kWh, how much would it cost to run this motor continuously for 1 week?
17. A generator which is 80% efficient delivers 50 A at 220 V. What must be the output of the diesel engine which drives the generator?
18. 23.9 kW is required to operate a 25-hp forced-draft fan motor.
    (*a*) What is its efficiency?
    (*b*) How much power is lost in the motor?

19. A generator delivers 80 A at 220 V with an efficiency of 88%. How much power is lost in the generator?

20. A 230-V $7\frac{1}{2}$-hp motor, which has an efficiency of 85%, is driving a radio transmitter 2-kV generator which has an efficiency of 80%. The motor is running fully loaded.

(a) How much power does the motor take from the line?
(b) How much current does the motor draw?
(c) How much power will the generator deliver?
(d) How much current will the generator deliver?
(e) What is the overall efficiency; that is, what is the efficiency from motor input to generator output?

8-8 RESISTANCES IN SERIES

So far, our studies of the electric circuit have taken into consideration but one electric component in the circuit, excluding the source of voltage. This is all very well for the purpose of becoming familiar with simple Ohm's law for power relations. However, practical circuits consist of more than one piece of equipment as far as circuit computations are concerned.

In a *series circuit* the various components comprising the circuit are so connected that the current, starting from the voltage source, must flow through each circuit component, in turn, before returning to the other side of the source.

There are three important facts concerning series circuits that must be borne in mind in order to understand thoroughly the action of such circuits and to facilitate their solution. In a series circuit:

1. The total voltage is equal to the sum of the voltages across the different parts of the circuit.
2. The current in any part of the circuit is the same.
3. The total resistance of the circuit is equal to the sum of the resistances of the different parts.

Point 1 is practically self-evident. If the sum of all the potential differences (voltage drops) around the circuit were not equal to the applied voltage, there would be some voltage left over which would cause an increase in current. This increase in current would continue until it caused enough voltage drop across some resistance just to balance the applied voltage. Hence,

$$E_t = E_1 + E_2 + E_3 + \cdots \qquad (8)$$

Point 2 is evident, for the circuit components are so connected that the current must flow through each part in turn and there are no other paths back to the source.

To some, point 3 might not be self-evident. However, because it is agreed that the current I in Figs. 8-12 and 8-13 flows through all resistors, Eq. (8) can be used to demonstrate the truth of point 3. Thus, by dividing each member of Eq. (8) by I, we have

$$\frac{E_t}{I} = \frac{E_1 + E_2 + E_3}{I} + \cdots$$

or

$$\frac{E_t}{I} = \frac{E_1}{I} + \frac{E_2}{I} + \frac{E_3}{I} + \cdots$$

and by substituting R for $\frac{E}{I}$, we have

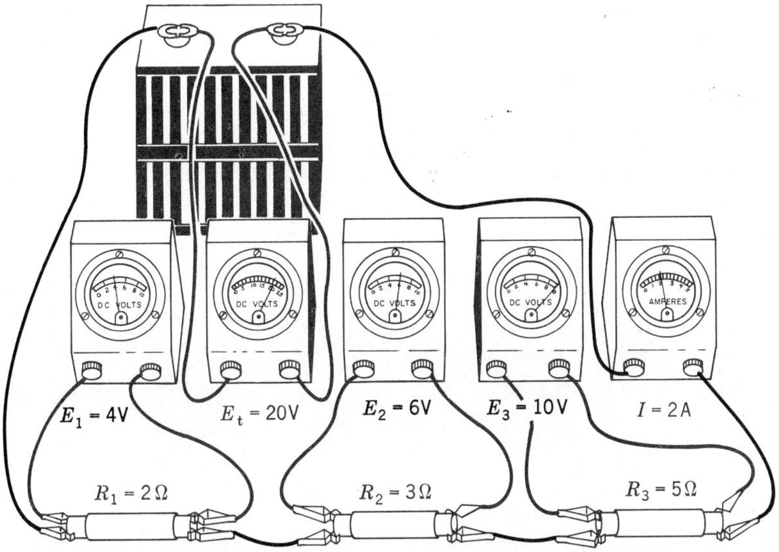

Fig. 8-12 Three Resistors Connected in Series with a Voltmeter Connected across Each Resistor. The Sum of the Voltages across the Resistors Is Equal to the Battery Voltage

$$R_t = R_1 + R_2 + R_3 + \cdots \qquad (9)$$

note E_t and R_t are used to denote total voltage and total resistance respectively.

example 11 Three resistors $R_1 = 30\ \Omega$, $R_2 = 160\ \Omega$, and $R_3 = 40\ \Omega$ are connected in series across a generator. A voltmeter connected across R_2 reads 80 V. What is the voltage of the generator?

solution Figure 8-14 is a diagram of the circuit.

$$I = \frac{E_2}{R_2} = \frac{80}{160} = 0.5\ \text{A}$$

$$R_t = R_1 + R_2 + R_3$$
$$= 30 + 160 + 40 = 230\ \Omega$$
$$E_t = IR_t = 0.5 \times 230 = 115\ \text{V}$$

example 12 A 300-Ω relay must be operated from a 120-V line. How much resist-

Fig. 8-13 Schematic Diagram of the Circuit Represented in Fig. 8-12

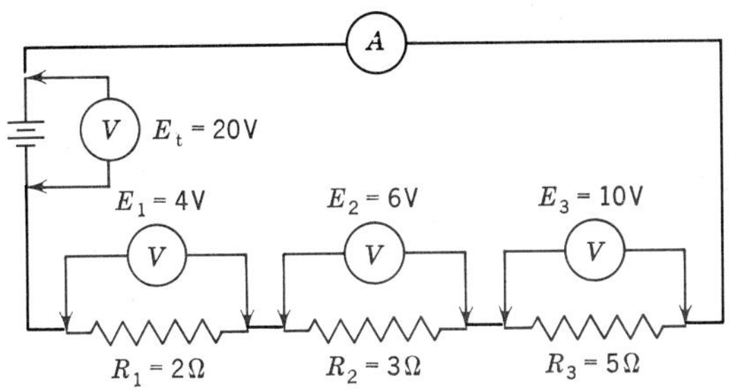

Fig. 8-14 Circuit of Example 11

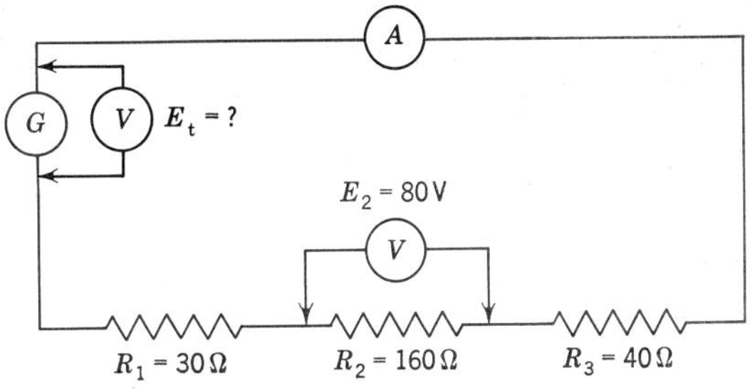

ance must be added in series with the relay coil to limit the current through it to 250 mA?

solution 1 The circuit is represented in Fig. 8-15. For a current of 250 mA to flow in a 120-V circuit, the total resistance must be

$$R_t = \frac{E}{I} = \frac{120}{0.250} = 480\ \Omega$$

Because the relay coil has a resistance of 300 Ω, the resistance to be added is

$$R_x = R_t - R_c = 480 - 300 = 180\ \Omega$$

solution 2 For 0.250 A to flow through the relay coil, the voltage across the coil must be

$$E_c = IR_c = 0.250 \times 300 = 75\ \text{V}$$

Because the line voltage is 120 V, the voltage across the added resistance must be

$$E_x = E - E_c = 120 - 75 = 45\ \text{V}$$

Then the value of resistance to be added is

$$R_x = \frac{E_x}{I} = \frac{45}{0.250} = 180\ \Omega$$

Fig. 8-15 Circuit of Example 12

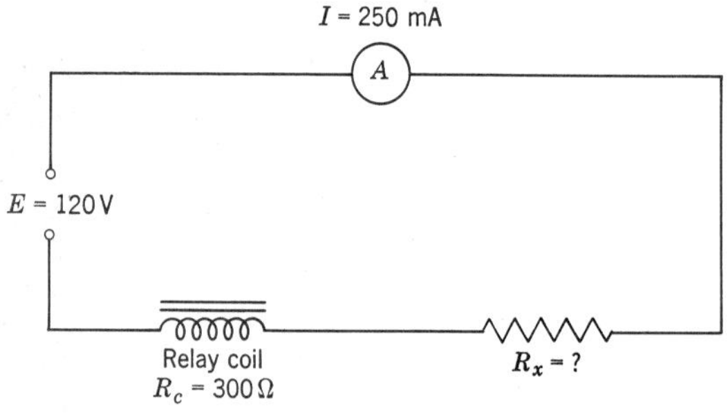

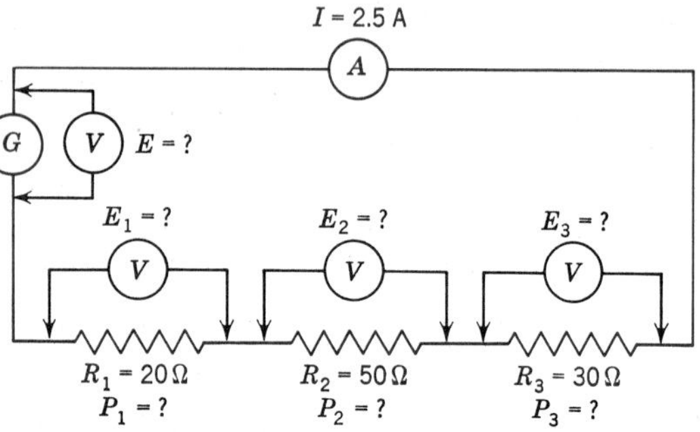

Fig. 8-16 Circuit of Example 13

example 13 Three resistors $R_1 = 20\ \Omega$, $R_2 = 50\ \Omega$, and $R_3 = 30\ \Omega$ are connected in series across a generator. The current through the circuit is 2.5 A.

(*a*) What is the generator voltage?

(*b*) What is the voltage across each resistor?

(*c*) How much power is expended in each resistor?

(*d*) What is the total power expended?

solution The circuit is represented in Fig. 8-16.

(*a*)
$$R_t = R_1 + R_2 + R_3 = 20 + 50 + 30 = 100\ \Omega$$
$$E = IR_t = 2.5 \times 100 = 250\ \text{V}$$

(*b*)
$$E_1 = IR_1 = 2.5 \times 20 = 50\ \text{V}$$
$$E_2 = IR_2 = 2.5 \times 50 = 125\ \text{V}$$
$$E_3 = IR_3 = 2.5 \times 30 = 75\ \text{V}$$

check

$$E = E_1 + E_2 + E_3 = 50 + 125 + 75 = 250\ \text{V}$$

(*c*) Power in R_1,

$$P_1 = E_1 I = 50 \times 2.5 = 125\ \text{W}$$

check

$$P_1 = I^2R_1 = 2.5^2 \times 20 = 125\ \text{W}$$

Power in R_2,

$$P_2 = E_2 I = 125 \times 2.5 = 312.5 \text{ W}$$

check

$$P_2 = I^2 R_2 = 2.5^2 \times 50 = 312.5 \text{ W}$$

Power in R_3,

$$P_3 = E_3 I = 75 \times 2.5 = 187.5 \text{ W}$$

check $P_3 = I^2 R_3 = 2.5^2 \times 30 = 187.5 \text{ W}$

(*d*) Total power,

$$P_t = P_1 + P_2 + P_3$$
$$= 125 + 312.5 + 187.5 = 625 \text{ W}$$

check

$$P_t = I^2 R_t = 2.5^2 \times 100 = 625 \text{ W}$$

or $$P_t = \frac{E^2}{R_t} = \frac{250^2}{100} = 625 \text{ W}$$

PROBLEMS 8-3

1. Three resistors, $R_1 = 330\ \Omega$, $R_2 = 680\ \Omega$, and $R_3 = 570\ \Omega$, are connected in series across 110 V.
(*a*) How much current flows in the circuit?
(*b*) What is the voltage drop across R_2?
(*c*) How much power is expended in R_1?

2. Three resistors, $R_1 = 2.2\ \text{k}\Omega$, $R_2 = 5.7\ \text{k}\Omega$, and $R_3 = 1.5\ \text{k}\Omega$, are connected in series across 450 V.
(*a*) How much current flows through the circuit?
(*b*) What is the voltage drop across each resistor?

3. A 115-V soldering iron which is rated at 100 W is to be used on a 220-V line.
(*a*) How much resistance must be connected in series with the iron to limit the current to rated value?
(*b*) If a standard resistor of 150 Ω is used in place of this calculated value, what minimum power rating must be specified for this resistor?
(*c*) If the standard resistor of (*b*) is used, what actual power will be delivered to the soldering iron?

4. Four identical 100-W lamps are connected in series across a 440-V line. The hot resistance of each lamp is 121 Ω.
(*a*) What is the current through the lamps?
(*b*) What is the voltage drop across each lamp?
(*c*) What is the power dissipated by each lamp?

5. Three identical lamps are connected in series across a 440-V line. If the current through the lamps is 820 mA, what is the hot resistance of each lamp?

6. Three resistors, R_1, R_2, and R_3, are connected in series across a 470-V power supply. A voltmeter connected across R_1 reads 76 V. When connected across R_2, the voltmeter reads 51 V. R_3 is 150 kΩ.
(*a*) What is the current flowing through the circuit?
(*b*) What is the value of R_1?

(c) What is the value of R_2?
(d) What is the wattage dissipated by each resistor?

7. Three resistors of 12, 18, and 47 Ω are connected in series across a 12-V source. If the current through the circuit is 153 mA, what is the resistance of the connecting wires and connections?

8. A certain broadcast tuner has been designed to use one each of the following tubes: 12BE6, 12BA6, 12AT6, and 35W4. The first three tubes require 12.6 V each for heaters (filaments), and the 35W4 requires 35 V. Since all the heaters are designed for 150 mA, they can be operated in series. What value of series resistance R_s is required for operation from a 115-V line?

9. Three resistors, $R_1 = 1.2\ \Omega$, R_2, and R_3, are connected in series across a 125-V generator, which delivers a current of 27.8 A. The voltage drop across R_3 is 50 V.
(a) What is the value of R_3?
(b) What is the value of R_2?
(c) How much power is expended in the circuit?

10. Four resistors, $R_1 = 820\ \Omega$, $R_2 = 270\ \Omega$, $R_3 = 1.5\ \text{k}\Omega$, and $R_4 = 390\ \Omega$, are connected in series across a generator. The voltage appearing across R_3 is 504 V.
(a) What is the generator voltage?
(b) What is the power being dissipated by each resistor?

Fig. 8-17 Evolution of Vacuum Tubes: (a) T-9 Octal Base, (b) Glass Miniature, (c) Glass Subminiature, and (d) Ceramic Microminiature (Shown Approximately Three-Fourths of Actual Size) (*Courtesy of General Electric Company*)

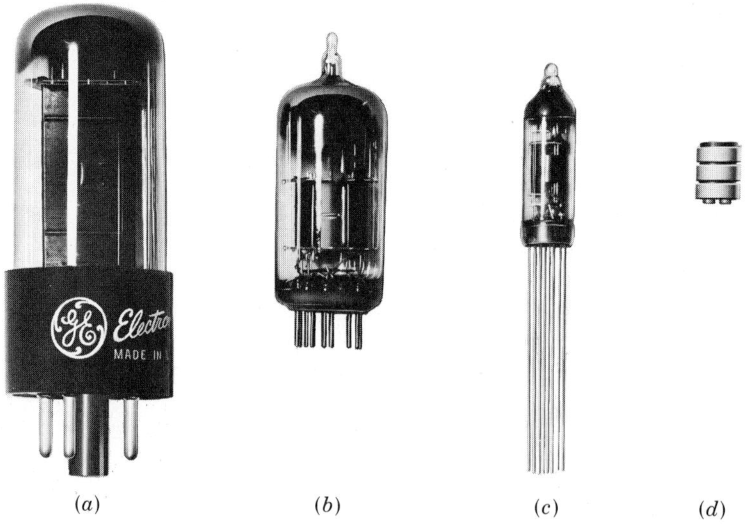

Fig. 8-18 Cutaway of Type GL-5751 Vacuum Tube, Shown Approximately $2\frac{1}{2}$ Times Actual Size (*Courtesy of General Electric Company*)

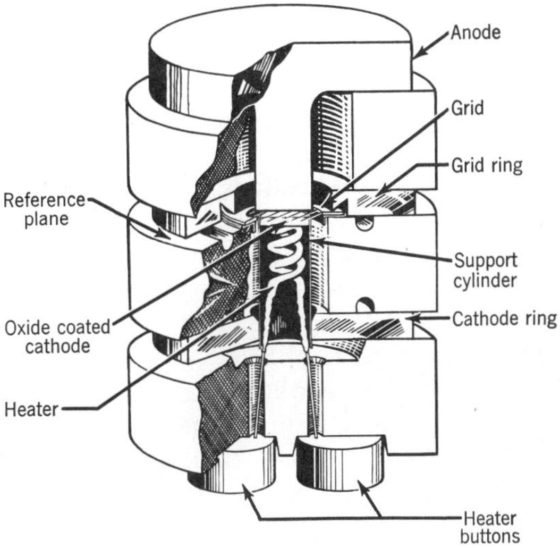

Fig. 8-19 Basic Physical Construction of Type 6BY4 Vacuum Tube Illustrated in Fig. 8-17d (*Courtesy of General Electric Company*)

8-9 BIAS RESISTORS—TUBES

The great majority of vacuum-tube applications require that the control grid G of the tube be maintained at a negative potential with respect to the cathode K. There are several methods of accomplishing this, and they largely depend upon the use of the tube and the circuit with which it is used. However, the most common source of bias voltage is a resistance R_k inserted in the cathode circuit, where the cathode current I_k must flow through it. The voltage drop across this resistance is employed as a bias voltage as illustrated in Fig. 8-20, which illustrates schematically a type 6C5 triode operating with a bias voltage of $E_c = -8$ V. Since the plate supply voltage maintains the plate P positive with respect to the cathode K, electrons flow from cathode to plate, and these constitute the plate current I_b.

As far as the dc circuit, and therefore the bias voltage is concerned, Fig. 8-20 can be reduced to the equivalent series circuit of Fig. 8-21 wherein the signal voltage source E_s has been eliminated and the equivalent

Fig. 8-20 Grid G is Biased −8 V with Respect to Cathode K

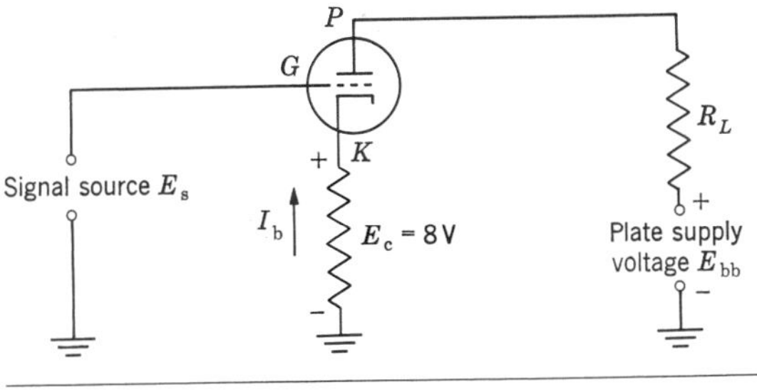

Fig. 8-21 Equivalent Circuit of Fig. 8-20

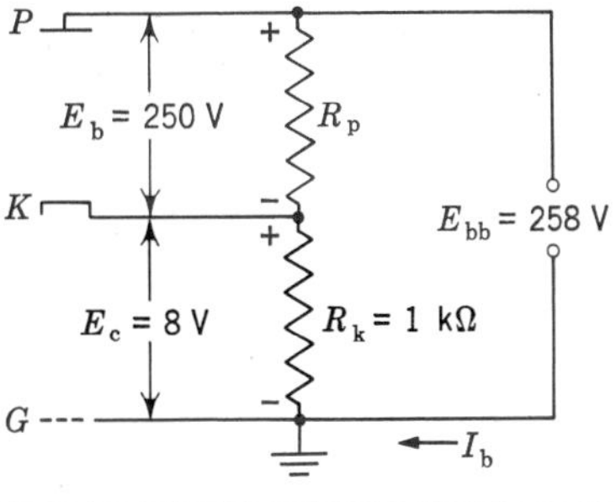

plate resistance R_p has been substituted for the cathode-to-plate electron circuit. For the purpose of illustration, the plate load resistance R_L has been eliminated from this particular example, but this elimination is not possible in all applications, as will be shown later. In this circuit the tube is operating with a plate supply voltage of $E_{bb} = 258$ V, a plate voltage with respect to cathode of $E_p = 250$ V, and a grid bias voltage with respect to cathode of $E_c = -8$ V.

Starting at the negative source of the plate supply voltage E_{bb}, the plate current I_p of 8 mA flows through the 1-kΩ cathode-biasing resistor R_k, which results in a voltage of 8 V across this resistor. The polarity is such that the cathode is 8 V positive with respect to the negative source of plate voltage, ground potential, and the grid, since all are connected together. This is the same as saying that the grid is 8 V negative with respect to the cathode. The remaining 250 V exists between plate P and cathode K, with the plate 250 V positive with respect to cathode.

example 14 The type 6A3 triode power amplifier tube, when operating as a class A amplifier, has a plate current of 60 mA when the plate voltage is 250 V and the grid bias E_c is −45 V.

(a) What value of cathode-biasing resistor R_k is necessary?

(b) How much power is consumed in the biasing resistor?

(c) Disregarding plate load resistance R_L what is the value of the plate voltage supply E_{bb}?

(d) How much power P_b is taken from the plate voltage supply?

solution The circuit is shown schematically in Fig. 8-22.

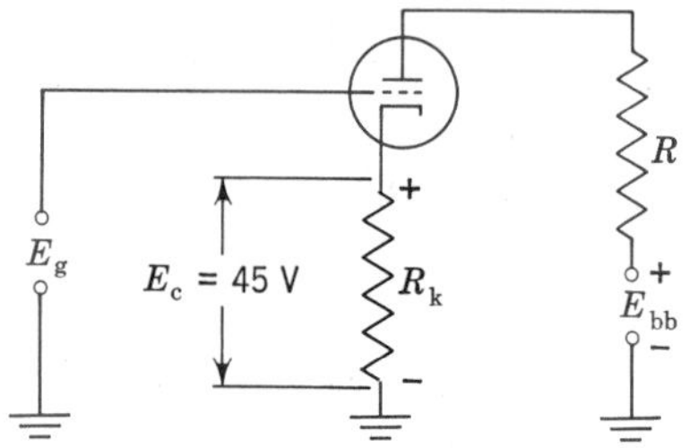

Fig. 8-22 Circuit of Example 14

(a) $R_k = \dfrac{E_c}{I_b} = \dfrac{45}{0.060} = \dfrac{45}{6 \times 10^{-2}}$

$= \dfrac{45}{6} \times 10^2 = 750\ \Omega$

(b) $P_k = I_b{}^2 R_k = (6 \times 10^{-2})^2 \times 750 = 2.7$ W

check

$$P = \frac{E_c{}^2}{R_k} = \frac{45^2}{750} = 2.7 \text{ W}$$

(c) $E_{bb} = E_b + E_c = 250 + 45 = 295$ V

(d) $P_b = E_{bb} I_b = 17.7$ W

example 15 If the tube of Example 14 is to work into a dc load resistance of $R_L = 2.5$ kΩ, what plate supply voltage E_{bb} will be required?

solution The circuit is illustrated in Fig. 8-23. The voltage across the load resistance R_L is

$$E_L = I_b R_L = 0.060 \times 2500 = 150 \text{ V}$$

In order to maintain the original tube oper-

Fig. 8-23 Circuit of Example 15

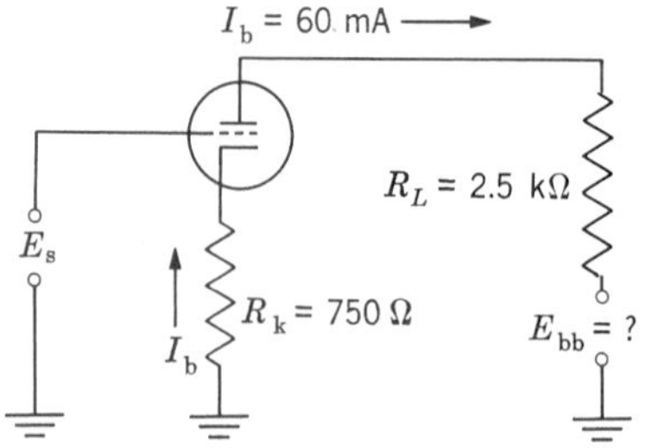

ating voltages, the supply voltage must be increased by 150 V. That is,

$$E_{bb} = 295 + 150 = 445 \text{ V}$$

or

$$E_{bb} = E_b + E_c + E_L = 250 + 45 + 150 = 445 \text{ V}$$

example 16 The type 6SK7 pentode has the following characteristics: Plate voltage $E_b = 250$ V, grid bias $E_c = -3$ V, plate current $I_b = 9.2$ mA, screen voltage $E_{sg} = 100$ V, screen current $I_{sg} = 2.4$ mA. Disregarding the plate load resistance R_L,

(*a*) What value of cathode-biasing resistor is necessary?

(*b*) What value of series screen grid resistor R_{sg} is needed if the screen grid voltage is to be supplied from the positive side of the plate voltage supply?

solution The circuit is illustrated in Fig. 8-24. The control grid, which is nearest the cathode, is to be 3 V negative with respect to cathode. The suppressor grid, which is nearest the plate, is connected directly to the cathode to suppress secondary emission. The screen grid, which is between control grid and suppressor grid, is to be operated at 100 V positive with respect to the cathode.

(*a*) Since the plate current I_p and the screen current I_{sg} both flow through the cathode, the cathode current from the supply is

$$I_k = I_b + I_{sg} = 9.2 + 2.4 = 11.6 \text{ mA}$$

$$\text{then } R_k = \frac{E_c}{I_k} = \frac{3}{11.6 \times 10^{-3}} = \frac{30}{11.6} \times 10^2 = 259\ \Omega$$

(*b*) The series screen grid dropping resistor must reduce the plate voltage of 250 to 100 V on the screen. Therefore, the voltage drop across this resistor must be

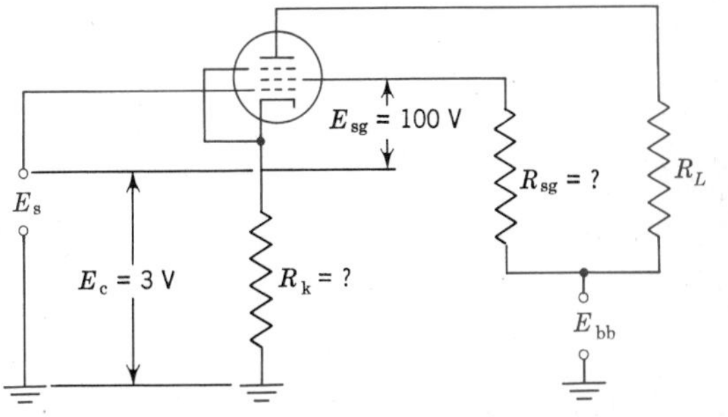

Fig. 8-24 Circuit of Example 16

$$E = E_b - E_{sg} = 250 - 100 = 150 \text{ V}$$

$$\text{then } R_{sg} = \frac{E}{I_{sg}} = \frac{150}{2.4 \times 10^{-3}} = \frac{15}{2.4} \times 10^4 = 62.5 \text{ k}\Omega$$

8-10 BIAS RESISTORS—TRANSISTORS

Proper operation of a transistor circuit requires that the emitter-base junction of the transistor be forward-biased and that the collector-base junction be reverse-biased, as shown in Fig. 8-25.

Sometimes the use of two different batteries is avoided by utilizing bias resistors, as in tube circuits. In addition, resistor values are chosen to limit current flows to acceptable

Fig. 8-25 *NPN* Transistor Biased for Proper Operation. The *N*-Type Emitter is Forward-Biased for Low Effective Resistance, and the *N*-Type Collector is Reverse-Biased for High Effective Resistance.

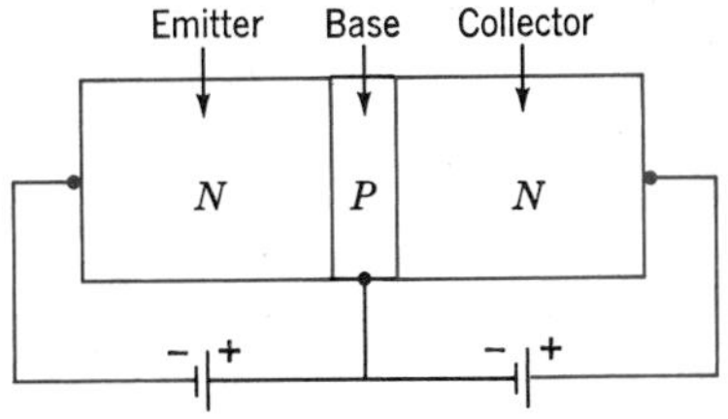

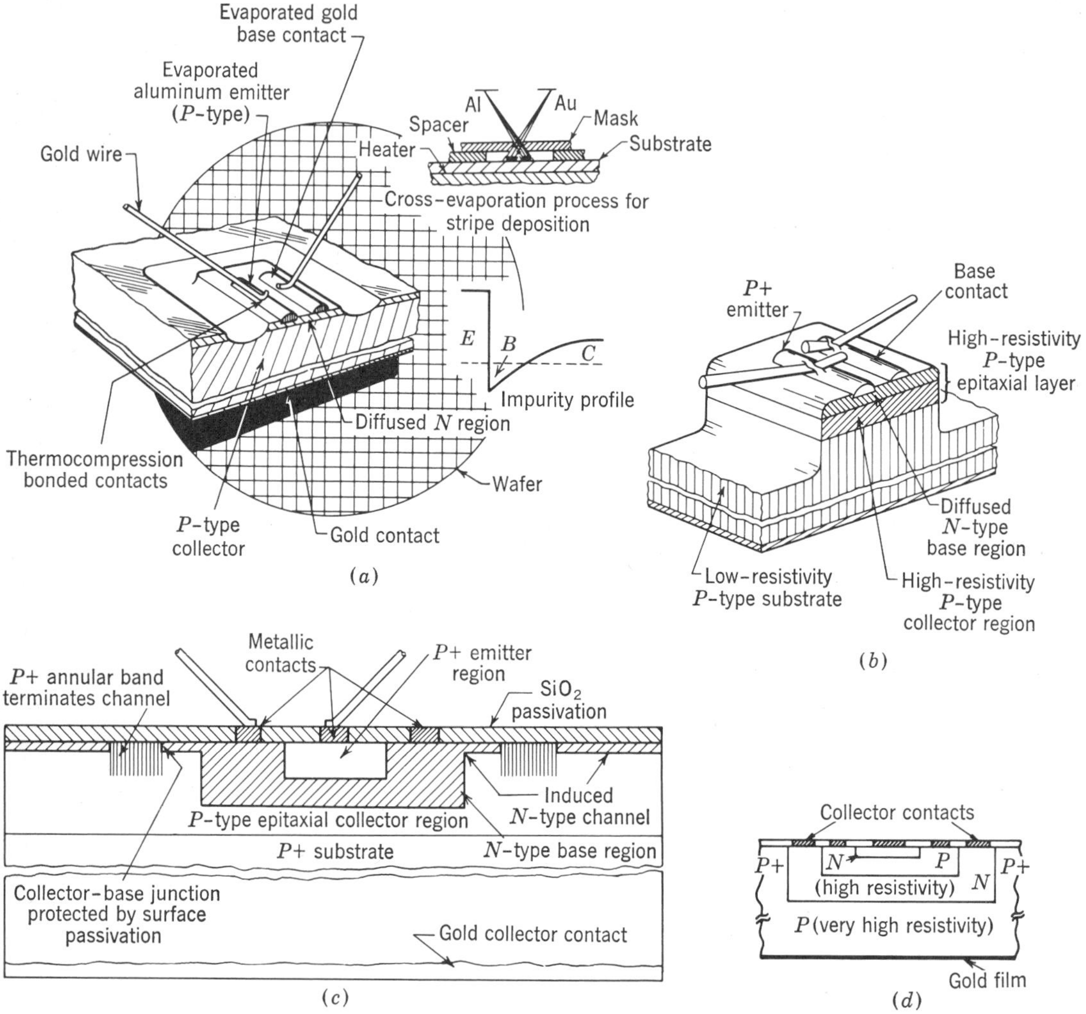

Fig. 8-26 Evolution of Transistors (*Courtesy of Lothar Stern, "Fundamentals of Integrated Circuits," Hayden Book Co., Inc., 1968*)
(*a*) Diffused Base Mesa
(*b*) Epitaxial Mesa
(*c*) Annular
(*d*) Basic Integrated

levels. Figure 8-27 shows a simple circuit in which transistor Q_1 is supplied by a single battery E_B. The resistor in the base circuit R_B is chosen to regulate the base-emitter current I_B, and the output signal is taken across the load resistor R_L as the collector current I_C flows through it.

example 17 In Fig. 8-27, assuming that the voltage drop across the emitter-base junc-

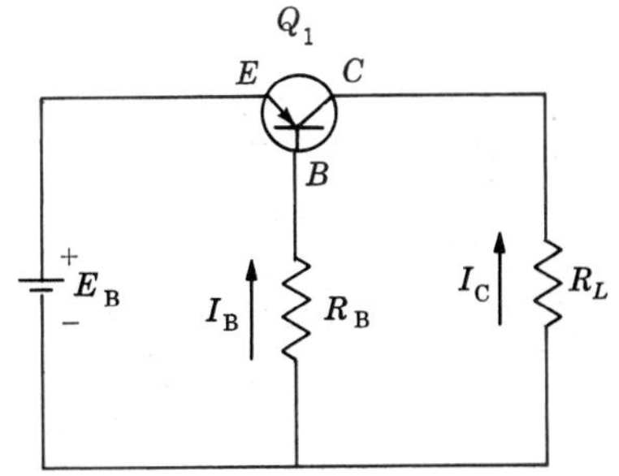

Fig. 8-27 Simple Single-Battery Transistor Biasing Circuit for *PNP* Transistor Q_1

tion is negligible, what must be the value of R_B if the base current must be limited to 80 μA? $E_B = 6$ V.

solution

$$R_B = \frac{E_B}{I_B} = \frac{6}{80 \times 10^{-6}} = 75\text{ k}\Omega$$

When two batteries are used, as in Fig. 8-28, an analysis based upon constant-emitter-current bias reveals that

$$R_E = \frac{E_E}{I_E}$$

$$I_C = \alpha I_E + I_{CO}$$

$$I_B = (1 - \alpha)I_E - I_{CO}$$

where I_{CO} = the very small leakage current in the collector circuit at room temperature

α = the current amplification factor under certain circuit arrangements; its value is usually slightly less than 1

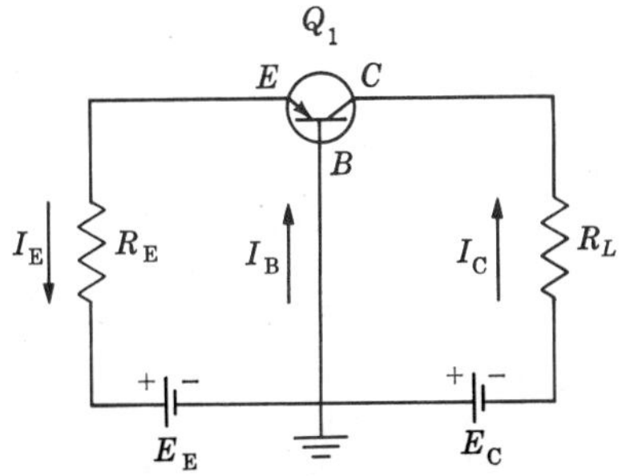

Fig. 8-28 *PNP* Transistor Q_1 Biased by Means of Two Batteries, E_e and E_c

example 18 In Fig. 8-28, the applied emf $E_E = 12$ V and the specifications for transistor Q_1 indicate that the emitter current I_E should be limited to 10 mA. What value of resistor R_E should be chosen?

solution $R_E = \frac{E_E}{I_E} = \frac{12}{10 \times 10^{-3}} = 1.2\text{ k}\Omega$

example 19 For the circuit of Fig. 8-28, $E_E = 12$ V, $I_E = 8$ mA, $\alpha = 0.95$, and $I_{CO} = 50$ μA. Find (*a*) R_E, (*b*) I_C, and (*c*) I_B.

solution

(*a*) $R_E = \frac{E_E}{I_E} = \frac{12}{0.008} = 1.5\text{ k}\Omega$

(*b*) $I_C = \alpha I_E + I_{CO}$
$= (0.95)(0.008) + 0.000\,050$
$= 7.65$ mA

(*c*) $I_B = (1 - 0.95)(0.008) - 0.000\,050$
$= 350$ μA

PROBLEMS 8-4

1. The type 6A5G triode power amplifier, when operating as a class A amplifier with a plate voltage of 300 V, draws 11 mA of plate current when the grid bias is −10.5 V.

(*a*) What is the value of the cathode bias resistor?

(b) Disregarding plate load resistance, what is the plate supply voltage?

2. The type 6AF5G triode, when operating as a class A amplifier with a plate voltage of 180 V, draws 7 mA of plate current when the grid bias is -18 V.
   (a) What is the value of the cathode bias resistor?
   (b) How much power is expended in the bias resistor?
   (c) Disregarding plate load resistance, what is the plate supply voltage?
3. The type 12E5GT triode, when operating as a class A amplifier with a plate voltage of 250 V, draws 50 mA of plate current when the grid bias is -10.5 V. The plate load resistance is 1 kΩ.
   (a) What is the value of the cathode bias resistor?
   (b) How much power is expended in the bias resistor?
   (c) What is the plate supply voltage?
   (d) How much power is taken from the plate supply?
4. The type 14V7 high-frequency pentode, when operating as a class A amplifier with a plate voltage of 300 V and a screen voltage of 150 V, draws 9.6 mA of plate current and 3.9 mA of screen current when the grid bias is -2 V.
   (a) What is the value of the cathode bias resistor?
   (b) What is the value of the screen dropping resistor?
5. The type 6M7G pentode, when operating as a class A amplifier with a plate voltage of 250 V and a screen voltage of 125 V draws 10.5 mA of plate current and 2.8 mA of screen current when the grid bias is -2.5 V.
   (a) What is the value of the cathode bias resistor?
   (b) How much power is expended in the bias resistor?
   (c) What is the value of the screen dropping resistor?
   (d) How much power is expended in the screen dropping resistor?
   (e) Disregarding load resistance, what is the plate supply voltage?
   (f) How much power is taken from the plate supply?
6. In Fig. 8-27, assuming that the voltage drop across the emitter-base junction is negligible, what must be the value of R_B if the base current must be limited to 90 μA? $E_B = 6$ V.
7. It is desired to operate a transistor in grounded-base connection (Fig. 8-28) with a fixed bias of 6 V. The maximum current in the base circuit is 100 μA.
   (a) What is the value of the resistor which will provide this voltage?
   (b) What is the power which this resistor must radiate?
8. In the circuit of Fig. 8-28, emf $E_E = 6$ V, and the emitter current I_E should be limited to 8 mA. What value of resistor R_E should be chosen?
9. In the circuit of Fig. 8-28, what value should R_E be if $E_E = 30$ V and I_E must be kept to 12 mA or less?
10. In the circuit of Fig. 8-28, $E_E = 12$ V and $E_C = 15$ V. $I_E = 10$ mA, $\alpha = 0.98$, and $I_{CO} = 75$ μA. Find (a) R_E, (b) I_C, and (c) I_B.

9

resistance—wire sizes

The effects of resistance in series circuits were discussed in the preceding chapter. However, in order to prevent confusion while the more simple relations of Ohm's law were being discussed, the nature of resistance and the resistance of wires used for connecting sources of voltage with their respective loads were not mentioned.

In the consideration of practical circuits two important features must be taken into account: the resistance of the wires between the source of power and the electronic equipment that is to be furnished with power and the current-carrying capacity of these wires for a given temperature rise.

9-1 RESISTANCE

There is a wide variation in the ease (conductance) of current flow through different materials. No material is a perfect conductor, and the amount of opposition (resistance) to current flow within it is governed by the specific resistance of the material, its length, cross-sectional area, and temperature. Thus, for the same material and cross-sectional area, a long conductor will have a greater resistance than a shorter one. That is, *the resistance of a conductor of uniform cross-sectional area is directly proportional to its length*. This is conveniently expressed as

$$\frac{R_1}{R_2} = \frac{L_1}{L_2} \tag{1}$$

where R_1 and R_2 are the resistances of conductors with lengths L_1 and L_2, respectively.

example 1 The resistance of No. 8 copper wire is 2.06 Ω/km. What is the resistance of 175 m of the wire?

solution Given $R_1 = 2.06\ \Omega$, $L_1 = 1000$ m, and $L_2 = 175$ m, $R_2 = ?$ Solving Eq. (1) for R_2, we have

$$R_2 = \frac{R_1 L_2}{L_1} \frac{\Omega \cdot \text{m}}{\text{m}} = \frac{2.06 \times 175}{1000} \frac{\Omega \cdot \cancel{\text{m}}}{\cancel{\text{m}}} = 0.3605\ \Omega$$

For the same material and length, one conductor will have more resistance than another with a larger cross-sectional area. That is, *the resistance of a conductor is inversely proportional to its cross-sectional area*. Expressed as an equation,

$$\frac{R_1}{R_2} = \frac{A_2}{A_1} \tag{2}$$

where R_1 and R_2 are the resistances of conductors with cross-sectional areas A_1 and A_2, respectively.

Because most wires are drawn round, Eq. (2) can be rearranged into a more convenient form. For example, let A_1 and A_2 represent the cross-sectional areas of two equal lengths of round wires with diameters d_1 and d_2, respectively. Because the area A of a circle of a diameter d is given by

$$A = \frac{\pi d^2}{4}$$

then

$$A_1 = \frac{\pi d_1^{\,2}}{4} \quad \text{and} \quad A_2 = \frac{\pi d_2^{\,2}}{4}$$

Substituting in Eq. (2)

$$\frac{R_1}{R_2} = \frac{\dfrac{\pi d_2^{\,2}}{4}}{\dfrac{\pi d_1^{\,2}}{4}}$$

or

$$\frac{R_1}{R_2} = \frac{d_2^{\,2}}{d_1^{\,2}} \tag{3}$$

Hence, the resistance of a round conductor varies inversely as the square of its diameter.

example 2 A rectangular conductor with a cross-sectional area of 1.04 square millimeters (mm²) has a resistance of 0.075 Ω. What would be its resistance if its cross-sectional area were 2.08 mm²?

solution Given $R_1 = 0.075\ \Omega$, $A_1 = 1.04\ \text{mm}^2$, and $A_2 = 2.08\ \text{mm}^2$, $R_2 = ?$ Solving Eq. (2) for R_2,

$$R_2 = \frac{R_1 A_1}{A_2}\ \frac{\Omega \cdot \text{mm}^2}{\text{mm}^2}$$

$$= \frac{0.075 \times 1.04}{2.08}\ \frac{\Omega \cdot \cancel{\text{mm}^2}}{\cancel{\text{mm}^2}} = 0.0375\ \Omega$$

example 3 A round conductor with a diameter of 0.25 mm has a resistance of 8 Ω. What would be its resistance if its diameter were 0.5 mm?

solution Given $d_1 = 0.25$ mm, $R_1 = 8\ \Omega$, and $d_2 = 0.5$ mm, $R_2 = ?$ Solving Eq. (3) for R_2,

$$R_2 = \frac{R_1 d_1^{\,2}}{d_2^{\,2}}\ \frac{\Omega \cdot \text{mm}^2}{\text{mm}^2} = \frac{8 \times 0.25^2}{0.5^2}\ \frac{\Omega \cdot \cancel{\text{mm}^2}}{\cancel{\text{mm}^2}} = 2\ \Omega$$

Hence, if the diameter is doubled, the cross-sectional area is increased four times and the resistance is reduced to one-quarter of its original value.

PROBLEMS 9-1

1. Number 14 copper wire has a resistance of 8.28 Ω/km.
   (a) What is the resistance of 500 m of this wire?
   (b) What is the resistance of 20 m of this wire?
2. Number 30 copper wire has a resistance of 340 Ω/km.
   (a) What is the resistance of 800 m of this wire?
   (b) What is the resistance of 1 m?
3. Using the information of Prob. 2, what is the resistance of a coil that has a mean diameter of 40 mm and is wound with 6280 turns of No. 30 copper wire?
4. The resistance of a 1-km run of No. 10 copper wire telephone line is measured and found to be 3.277 Ω.
   (a) What is the resistance per meter?

(b) What is the resistance of a 720-m line?
(c) What is the resistance of 20 m?

5. The telephone line of Prob. 4 is replaced with No. 8 copper wire, which has a resistance of 2.061 Ω/km. What is the resistance of the 720-m section?
6. A length of square conductor that is 0.5 cm on a side has a resistance of 0.0756 Ω. What will be the resistance of a similar length of 1.5 cm square conductor?
7. One kilometer of No. 6 wire, which has a diameter of 4.115 mm, has a resistance of 1.297 Ω. What is the resistance of 1 km of No. 2 wire whose diameter is 6.543 mm?
8. The resistance of 30 m of a specially drawn wire is found to be 32.1 Ω. A coil wound with identical wire has a measured resistance of 702 Ω. What is the length of wire in the coil?
9. It is desired to wind a milliammeter shunt having a resistance of 4.62 Ω, and No. 40 enameled copper wire with a resistance of 3.54 kΩ/km is available. What length of wire is required?
10. It is desired to wind a microammeter shunt having a resistance of 0.280 Ω, and No. 36 enameled copper wire with a resistance of 1.36 kΩ/km is available. What length of wire is required?

9-2 MICROHM-METER

Equations (1) and (2) from Sec. 9-1 can be combined to form the compound proportion

$$\frac{R_1}{R_2} = \frac{L_1}{L_2} \cdot \frac{A_2}{A_1}$$

Such ratios are extremely helpful in solving problems when sufficient information is available. However, as a statement regarding a single conductor, we fall back on the simple proportionality (Sec. 5-11)

$$R \propto \frac{L}{A}$$

When this proportionality is written as an equation with a constant of proportionality, it becomes

$$R = \rho \frac{L}{A} \qquad \Omega \qquad (4)$$

where R = resistance of wire
ρ = specific resistance of material of which wire is made
L = length of wire
A = cross-sectional area of wire

From an algebraic rearrangement of Eq. (4), you can see that the units of ρ, the specific resistance, must relate to the units of L and A:

$$\rho = \frac{RA}{L}$$

In the older, or traditional, system of units L was measured in feet and A was measured in circular mils (the circular-mil area was defined to be equal to the square of the diameter when the diameter was given in mils, or thousandths of an inch). This resulted in ρ being expressed in Ω-cmils/ft, commonly pronounced Ω/cmil-ft. (See Secs. 9-2 and 9-3 of the third edition of this book if you require

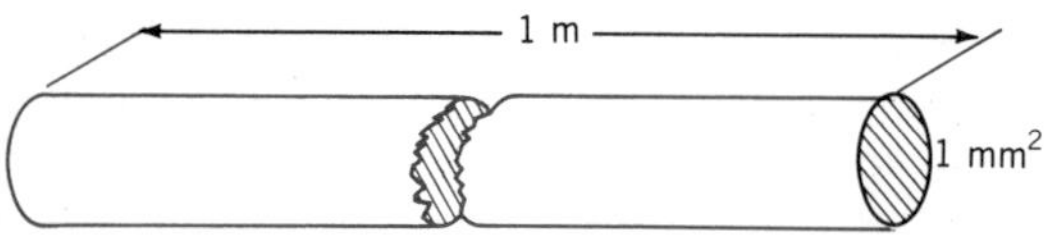

Fig. 9-1 Representation of 1 Microhm-meter Conductor

more information about this superseded set of units.)

In the SI metric system, the unit of length is the meter, and the unit of area is the square meter. Using these units, ρ would be expressed as

$$\rho = \frac{RA}{L} = \frac{\Omega \cdot m^2}{m} = \Omega \cdot m$$

Sometimes you will see tables of specific resistance giving values of ρ in ohm-meters. However, since more realistic sizes of wire will be given in square millimeters (see Appendix, Table 5), more often than not you will see practical values of ρ as:

$$\rho = \frac{RA}{l} = \frac{\Omega \cdot mm^2}{m} = \frac{\Omega \cdot m^2 \times 10^{-6}}{m}$$
$$= \Omega \cdot m \times 10^{-6} = \text{microhm-meters}$$

Table 9-1 gives specific resistances of common conductive materials in microhm-meters.

Table 9-1 SPECIFIC RESISTANCE AT 20°C

Material	$\mu\Omega \cdot m$
Aluminum	0.028 24
Brass	0.070 0
Constantan	0.490 0
Copper, hard drawn	0.017 71
Gold	0.024 4
Iron	0.100 0
Lead	0.220 0
Mercury	0.957 8
Nickel	0.078 0
Silver	0.015 9
Tin	0.115 0
Zinc	0.058 0

Equation (4) and its definition block may now be adjusted to read

$$R = \rho \frac{L}{A} \quad \Omega \tag{5}$$

where R = resistance of wire, Ω
ρ = specific resistance of wire, $\mu\Omega \cdot m$
L = length of wire, m
A = cross-sectional area of wire, mm^2

example 4 What is the resistance at 20°C of a copper wire 250 m long and 1.63 mm in diameter?

solution Given $L = 250$ m, $d = 1.63$ mm ($r = 0.815$ mm), and, from Table 9-1, $\rho = 0.017\,71\ \mu\Omega \cdot m$, $R = ?$ Substituting in Eq. (4),

$$R = \frac{0.017\,71 \times 250}{\pi(0.815)^2} = 2.122\ \Omega$$

example 5 The resistance of a conductor 1 km long and 2.05 mm in diameter is found to be 4.82 Ω at 20°C. What is the specific resistance of the wire?

solution Given $L = 1000$ m, $d = 2.05$ mm ($r = 1.025$ mm), and $R = 4.82\ \Omega$, $\rho = ?$ Solving Eq. (5) for ρ,

$$\rho = \frac{RA}{L} = \frac{4.82\,\pi(1.025)^2}{10^3} \quad \mu\Omega \cdot m$$
$$= 15.91 \times 10^{-3}$$
$$= 0.015\,91\ \mu\Omega \cdot m$$

example 6 A roll of copper wire is found to have a resistance of 2.54 Ω at 20°C. The diameter of the wire is measured as 1.63 mm. How long is the wire?

solution Solving Eq. (5) for L,

$$L = \frac{RA}{\rho} \qquad \text{m}$$

Substituting known values,

$$L = \frac{2.54\pi(0.815)^2}{0.017\,71} \text{ m}$$
$$= 299 \text{ m}$$

PROBLEMS 9-2

note In the following problems, consider that all the wire temperatures are 20°C.

1. What is the resistance of a copper wire 75 m long and 0.361 mm in diameter?
2. With reference to Prob. 1, what is the resistance of an otherwise identical wire of aluminum?
3. With reference to Prob. 1, what is the resistance of an otherwise identical wire of iron?
4. What is the resistance of 200 m of copper wire with a diameter of 0.254 mm?
5. A special alloy wire 10 m long and 0.079 mm in diameter has a resistance of 78 Ω. What is the specific resistance of the alloy?
6. A constantan wire that has a specific resistance of 0.49 μΩ · m has a diameter of 0.511 mm and a length of 1.1 m. What is its resistance?
7. How many kilometers of copper wire 3.264 mm in diameter will it take to make 5.00 Ω of resistance?
8. What is the resistance of the wire in Prob. 7 in ohms per kilometer?
9. A coil of copper wire has a resistance of 2.38 Ω. If the diameter of the wire is 2.05 mm, find the length of the wire.
10. What is the resistance of 2 km of the wire in Prob. 9?

9-3 TEMPERATURE EFFECTS

In the preceding section the specific resistance of certain materials was given at a temperature of 20°C. The reason for stating the temperature is that the resistance of all pure metals increases with a rise in temperature. The results of experiments show that over ordinary temperature ranges this variation in resistance is directly proportional to the temperature. Hence, for each degree rise in temperature above some reference value, each ohm of resistance is increased by a constant amount α, called the *temperature coefficient of resistance*. The relation between temperature and resistance can be expressed by the equation

$$R_t = R_0(1 + \alpha t) \qquad \Omega \tag{6}$$

where R_t = resistance at a temperature of t°C*

R_0 = resistance at 0°C

α = temperature coefficient of resistance at 0°C

The temperature coefficient for copper is 0.004 27. That is, if a copper wire has a resistance of 1 Ω at 0°C, it will have a resistance of 1 + 0.004 27 = 1.004 27 Ω at 1°C. The value of the temperature coefficient for copper is essentially the same as that for most of the unalloyed metals, such as gold, silver, aluminum, and lead.

* °C stands for degrees Celsius.

The value of α varies with the temperature at which it is determined: 0.004 27 is only valid when the reference temperature is 0°C. If 20°C is taken as the reference temperature, the value of α changes to 0.003 93. Thus it is important to always relate to the reference temperature for which the value of α is specified. If the resistance is known at a certain temperature, and you are required to determine the resistance corresponding to some other temperature, you *must* first calculate, as an intermediate step, the resistance for the reference temperature.

example 7 The resistance of a coil of copper wire is 34 Ω at 15°C. What is the resistance at 70°C?

solution 1 Using $\alpha = 0.004\,27$, and relating to 0°C, $R_t = 34\ \Omega$ and $t = 15$ (15°C − 0°C).

$$R_t = R_0(1 + \alpha t)$$

Solving for R_0,

$$R_0 = \frac{R_t}{1 + \alpha t} = \frac{34}{1 + (0.004\,27)(15)} = \frac{34}{1.064\,05} = 31.95\ \Omega$$

Using R_0, solve for R_{70}:

$$R_{70} = 31.95(1 + 0.004\,27 \times 70) = 31.95(1.2989) = 41.5\ \Omega$$

solution 2 Using $\alpha = 0.003\,93$, and relating to 20°C, $R_t = 34\ \Omega$ and

$$t = -5\ (15°\text{C} - 20°\text{C})$$

Solving for R_{20},

$$R_{20} = \frac{34}{1 + (0.003\,93)(-5)} = \frac{34}{1 - 0.019\,65} = \frac{34}{0.980\,35} = 34.68\ \Omega$$

Using R_{20}, solve for R_{70}:

$$R_{70} = 34.68(1 + 0.003\,93 \times 50) = 34.68(1.1965) = 41.5\ \Omega$$

A more convenient relation is derived by assuming that the proportionality between resistance and temperature extends linearly to the point where copper has a resistance of 0 Ω at a temperature of −234.5°C. This results in the ratio

$$\frac{R_2}{R_1} = \frac{234.5 + t_2}{234.5 + t_1} \qquad (7)$$

where R_1 = resistance of copper in ohms at a temperature of t_1°C
R_2 = resistance of copper in ohms at a temperature of t_2°C

example 8 The resistance of a coil of copper wire is 34 Ω at 15°C. What is its resistance at 70°C?

solution Given $R_1 = 34\ \Omega$, $t_1 = 15$°C, and $t_2 = 70$°C. $R_2 = ?$ Solving Eq. (7) for R_2,

$$R_2 = \frac{234.5 + t_2}{234.5 + t_1} R_1$$

Substituting the known values,

$$R_2 = \frac{234.5 + 70}{234.5 + 15} \times 34 = 41.5\ \Omega$$

The specifications for electric machines generally include a provision that the temperature of the coils, etc., when the machines are operating under a specified load for a specified time, must not rise more than a certain number of degrees. Temperature rise can be computed by measuring the resistance of the coils at room temperature and again at the end of the test.

example 9 The field coils of a shunt motor have a resistance of 90 Ω at 20°C. After the motor was run for 3 h, the resistance of the field coils was 146 Ω. What was the temperature of the coils?

solution Given $R_1 = 90\ \Omega$, $t_1 = 20°C$, $R_2 = 146\ \Omega$. $t_2 = ?$ Solving Eq. (7) for t_2,

$$t_2 = \frac{234.5 + t_1}{R_1} R_2 - 234.5$$

Substituting the known values,

$$t_2 = \frac{234.5 + 20}{90} \times 146 - 234.5$$
$$= 413 - 234.5 = 178.5°$$

The actual temperature rise is

$$t_2 - t_1 = 178.5° - 20° = 158.5°$$

PROBLEMS 9-3

1. The resistance of a coil of copper wire at 40°C is 5.38 Ω. What will its resistance be at 0°C?
2. If the resistance of a copper coil is 3.07 Ω at 0°C, what will it be at 20°C?
3. The dc resistance of an inductor is 19.5 Ω at 80°C. What will be the resistance when the inductor is operated at an ambient temperature of 20°C?
4. The resistance of the primary winding of a transformer was 2.95 Ω at 20°C. After operating for 3 h, the resistance increased to 3.28 Ω. What was the final operating temperature?
5. The specifications for a high-power transformer included a provision that it was to operate continuously under full load with the winding temperature not to exceed 55°C. The resistance of the primary coil was measured before the transformer was put on test, at 22°C, and found to be 52.7 Ω. After a day's test at rated load, the resistance was again measured and it was found to be 60.0 Ω. Did the transformer meet the specifications?

9-4 WIRE MEASURE

Wire sizes are designated by numbers in a system known as the American wire gage (formerly Brown and Sharpe gage). These gage numbers, ranging from 0000, the largest size, to 40, the smallest size, are based on a constant ratio between successive gage numbers. The wire sizes and other pertinent data are listed in Table 5 in the Appendix.

Inspection of the wire table will reveal the progression formed by the wire sizes. As the sizes become smaller, every third gage number results in one-half the area and, therefore, double the resistance.

example 10 Number 10 wire has an area of 5.261 mm^2 and a resistance of 3.277 Ω/km. Three sizes smaller, No. 13 wire has an area of 2.63 mm^2 (almost exactly one-half of 5.261), and a resistance of 6.56 Ω (almost double 3.277). Similarly, half the resistance of No. 10 wire is provided by No. 7:

$1.634\ \Omega$/km, with an area of $10.55\ mm^2$ (double 5.261).

9-5 FACTORS GOVERNING WIRE SIZE IN PRACTICE

From an electrical viewpoint, three factors govern the selection of the size of wire to be used for transmitting current:

1. The safe current-carrying capacity of the wire
2. The power lost in the wire
3. The allowable voltage variation, or the voltage drop, in the wire

It must be remembered that the length of wire, for the purpose of computing wire resistance and its effects, is always twice the distance from the source of power to the load (outgoing and return leads).

example 11 A motor receives its power through No. 4 wire from a generator located at a distance of 1 km. The voltage across the motor is 220 V, and the current taken by the motor is 19.8 A. What is the terminal voltage of the generator?

solution The circuit is represented in Fig. 9-2. Note that it consists of a simple series circuit which can be simplified to that of Fig. 9-3. The resistance of the 1 km of No. 4 wire from the generator to the motor is represented by R_o; reference to Table 5 shows it to be $0.8152\ \Omega$. Similarly, the resistance from the motor back to the generator, which is represented by R_r, is also $0.8152\ \Omega$. The voltage drop in *each* wire is

Fig. 9-2 Generator *G* Supplying Power to Motor *M* at a Distance of 1 km

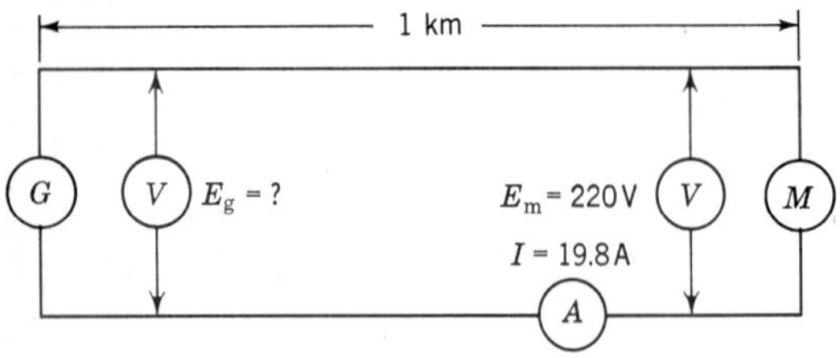

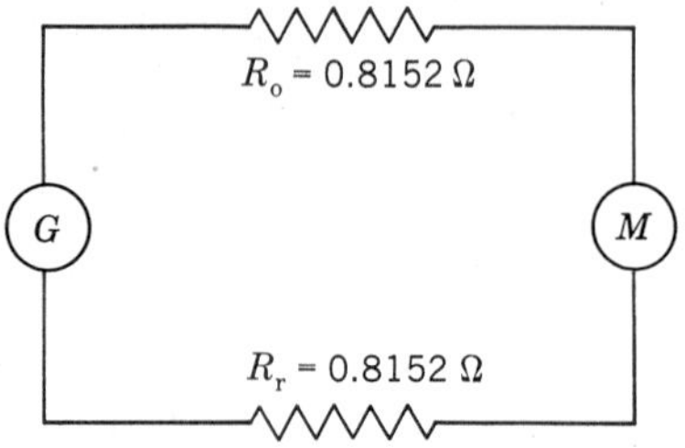

Fig. 9-3 Simplified Form of Circuit Shown in Fig. 9-2

$$E = IR_o = IR_r = 19.8 \times 0.8152 = 16.14\ V$$

Since the applied voltage must equal the sum of all the voltage drops around the circuit (Sec. 8-8), the terminal voltage of the generator is

$$E_g = 220 + 16.14 + 16.14 = 252.28 \text{ or } 252\ V$$

Since the resistance out R_o is equal to the return resistance R_r, the foregoing solution is simplified by taking twice the actual wire distance for the length of wire that constitutes the resistance of the feeders. Therefore, the length of No. 4 wire between generator and motor is 2 km, which results in a line resistance R_L of

$$2 \times 0.8152 = 1.6304\ \Omega$$

The circuit can be further simplified as shown in Fig. 9-4. Thus, the generator terminal voltage is

$$\begin{aligned} E_g &= 220 + IR_L \\ &= 220 + (19.8 \times 1.6304) = 252\ V \end{aligned}$$

The power lost in the line is

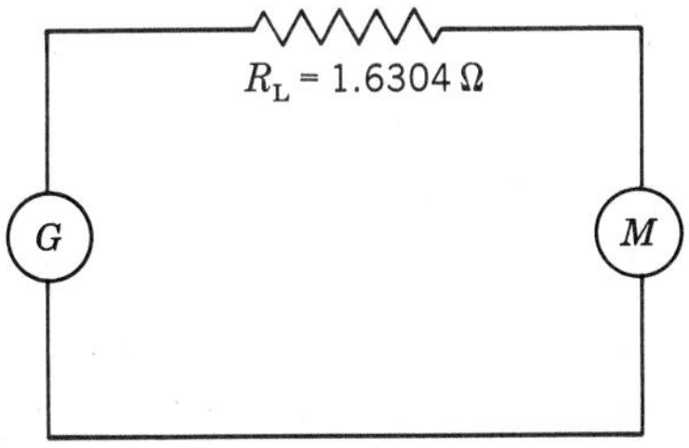

Fig. 9-4 Equivalent Circuit of Circuits Shown in Figs. 9-2 and 9-3

$$P_L = I^2 R_L$$
$$= 19.8^2 \times 1.6304 = 639 \text{ W}$$

The power taken by the motor is

$$P_M = E_M I$$
$$= 220 \times 19.8$$
$$= 4356 \text{ W} = 4.356 \text{ kW}$$

The power delivered by the generator is

$$P_G = P_L + P_M$$
$$= 639 + 4356 = 4995 \text{ W}$$

Efficiency of transmission

$$= \frac{\text{power delivered to load}}{\text{power delivered by generator}}$$
$$= \frac{4356}{4995} = 0.872 = 87.2\% \qquad (8)$$

The efficiency of transmission is obtainable in terms of the generator terminal voltage E_G and the voltage across the load E_L. Because

$$\text{Power delivered to load} = E_L I$$

and

$$\text{Power delivered by generator} = E_G I$$

substituting in Eq. (8) gives us

$$\text{Efficiency of transmission} = \frac{E_L I}{E_G I} = \frac{E_L}{E_G} \qquad (9)$$

and substituting the voltages in Eq. (9) gives us

$$\text{Efficiency of transmission} = \frac{220}{252}$$
$$= 0.873 = 87.3\%$$

PROBLEMS 9-4

note All wires in the following problems are of copper with characteristics as listed in Table 5 of the Appendix.

1. (*a*) What is the resistance of 2500 m of No. 00 wire?
(*b*) What is its weight?

2. (*a*) What is the resistance of 1.8 m of No. 8 wire?
(*b*) What is its weight?

3. (*a*) What is the length of a 120-kg coil of No. 12 wire?
(*b*) What is its resistance?

4. (*a*) What is the length of a 100-kg coil of No. 16 wire?
(*b*) What is its resistance?

5. A telephone cable consisting of several pairs of No. 19 wire connects two cities 40 km apart. If a pair is short-circuited at one end, what will be the resistance of the loop thus formed?

6. A relay is to be wound with 1500 turns of No. 22 wire. The average diameter of a turn is 46 mm.
   (*a*) What will be the resistance?
   (*b*) What will be the weight of the coil?

7. Fifteen kilowatts of power is to be transmitted 200 m from a generator that maintains a constant terminal voltage of 240 V. If not over 5% line drop is allowed, what size wire must be used?

8. A generator with a constant brush potential of 230 V is feeding a motor 100 m away. The feeders are No. 6 wire, and the motor current is 27.7 A.
   (*a*) What would a voltmeter read if connected across the motor brushes?
   (*b*) What is the efficiency of transmission?

9. A motor requiring 34 A at 230 V is located 125 m from a generator that maintains a constant terminal voltage of 240 V.
   (*a*) What size wire must be used between generator and motor in order to supply the motor with rated current and voltage?
   (*b*) What will be the efficiency of transmission?

10. A 25-hp 230-V motor is to be installed 120 m from a generator that maintains a constant potential of 240 V.
   (*a*) If the motor is 84% efficient, what size wire should be used between motor and generator?
   (*b*) If the wire specified in (*a*) is used, what will be the motor voltage under rated load condition?

10

special products and factoring

In the study of arithmetic, it is necessary to memorize the multiplication tables as an aid to rapid computation. Similarly, in the study of algebra, certain forms of expressions occur so frequently that it is essential to be able to multiply, divide, or factor them by inspection.

10-1 FACTORING

To *factor* an algebraic expression means to find two or more expressions that when multiplied will result in the original expression.

example 1 $2 \times 3 \times 4 = 24$. Thus, 2, 3, and 4 are some of the factors of 24.

example 2 $b(x + y) = bx + by$. b and $(x + y)$ are the factors of $bx + by$.

example 3 $(x + 4)(x - 3) = x^2 + x - 12$. The quantities $(x + 4)$ and $(x - 3)$ are the factors of $x^2 + x - 12$.

10-2 PRIME NUMBERS

A number that has no factor other than itself and unity is known as a *prime number*. Thus, 3, 5, 13, x, and $(a + b)$ are prime numbers.

10-3 SQUARE OF A MONOMIAL

At this point you should review the law of exponents for multiplication in Sec. 4-3.

example 4 $(2ab^2)^2 = (2ab^2)(2ab^2) = 4a^2b^4$

example 5

$$(-3x^2y^3)^2 = (-3x^2y^3)(-3x^2y^3) = 9x^4y^6$$

By application of the rules for the multiplication of numbers having like signs and the law of exponents, we have the following rule:

Rule

To square a monomial, square the numerical coefficient, multiply this product by the literal factors of the monomial, and multiply the exponent of each letter by 2.

10-4 CUBE OF A MONOMIAL

example 6

$$(3a^2b)^3 = (3a^2b)(3a^2b)(3a^2b) = 27a^6b^3$$

example 7

$$(-2xy^3)^3 = (-2xy^3)(-2xy^3)(-2xy^3)$$
$$= -8x^3y^9$$

Note that the cube of a *positive* number is always *positive* and that the cube of a *negative* number is always *negative*. Again, by application of the rules for the multiplication of positive and negative numbers and the law of exponents, we have the following rule:

Rule

To cube a monomial, cube the numerical coefficient, multiply this product by the literal factors of the monomial, multiply the exponent of each letter by 3, and affix the same sign as the monomial.

PROBLEMS 10-1

Find the values of the following indicated powers:

1. $(xy)^2$ **2.** $(\theta\lambda)^3$ **3.** $(ei^2Z)^3$ **4.** $\pi\left(\frac{D}{2}\right)^2$

5. $(-4\pi\phi)^2$ **6.** $(3\alpha^2\omega^3)^2$ **7.** $(-2IR)^3$ **8.** $\left(3\frac{e}{i}\right)^2$

9. $2\pi(X_L)^2$ **10.** $\left(-3\frac{ir}{e}\right)^3$ **11.** $-\left(\frac{1}{2\pi fC}\right)^2$ **12.** $-(-13x^3y)^2$

13. $\left(-\frac{5P^2}{EI}\right)^3$ **14.** $\frac{(E_sN_p)^2}{E_p^{\ 3}}$ **15.** $-\left(\frac{V^2}{2g}\right)^3$ **16.** $\left(\frac{120f}{N}\right)^2$

17. $\left(\frac{B^2Al}{8\omega}\right)^3$ **18.** $-(2\pi fL)^3$ **19.** $-(\frac{4}{3}\pi R^3)^2$ **20.** $(\frac{5}{8}u^2v^3wx^4y^5)^3$

21. $\left(\frac{x^4y^6}{p^5}\right)^3$

10-5 SQUARE ROOT OF A MONOMIAL

The *square root* of an expression is one of its equal factors.

example 8 $\sqrt{3}$ is a number such that

$$\sqrt{3}\cdot\sqrt{3} = 3$$

example 9 $\sqrt{n}$ is a number such that

$$\sqrt{n}\cdot\sqrt{n} = n$$

Because $(+2)(+2) = +4$
and $(-2)(-2) = +4$

it is apparent that 4 has two square roots, +2 and −2. Similarly, 16 has two square roots, +4 and −4.

In general, every number has two square roots equal in magnitude, one positive and one negative. The positive root is known as the *principal root;* if no sign precedes the radical, the positive root is understood. Thus, in practical numerical computations, the following is understood:

$$\sqrt{4} = +2$$
and
$$-\sqrt{4} = -2$$

In dealing with literal numbers, the values

of the various factors often are unknown. Therefore, when we extract a square root, we affix the double sign $\pm$ to denote "plus or minus."

example 10 Since $a^4 \cdot a^4 = a^8$ and $(-a^4)(-a^4) = a^8$,

then $\sqrt{a^8} = \pm a^4$

example 11 Since $x^2y^3 \cdot x^2y^3 = x^4y^6$ and $(-x^2y^3)(-x^2y^3) = x^4y^6$,

then $\sqrt{x^4y^6} = \pm x^2y^3$

From the foregoing examples, we formulate the following:

Rule

To extract the square root of a monomial, extract the square root of the numerical coefficient, divide the exponents of the letters by 2, and affix the $\pm$ sign.

example 12 $\sqrt{4a^4b^2} = \pm 2a^2b$

example 13 $\sqrt{\frac{1}{9}x^2y^6z^4} = \pm\frac{1}{3}xy^3z^2$

note A perfect monomial square is one that is positive and has a perfect square numerical coefficient and has only even numbers as exponents.

10-6 CUBE ROOT OF A MONOMIAL

The *cube root* of a monomial is one of its three equal factors.

Because	$(+2)(+2)(+2) = 8$
then	$\sqrt[3]{8} = 2$
Similarly,	$(-2)(-2)(-2) = -8$
and	$\sqrt[3]{-8} = -2$

From this it is evident that the cube root of a monomial has the same sign as the monomial itself.

Because	$x^2y^3 \cdot x^2y^3 \cdot x^2y^3 = x^6y^9$
then	$\sqrt[3]{x^6y^9} = x^2y^3$

The above results can be stated as follows:

Rule

To extract the cube root of a monomial, extract the cube root of the numerical coefficient, divide the exponents of the letters by 3, and affix the same sign as the monomial.

example 14 $\sqrt[3]{8x^6y^3z^{12}} = 2x^2yz^4$

example 15 $\sqrt[3]{-27a^3b^9c^6} = -3ab^3c^2$

note A perfect cube monomial has a positive or negative perfect cube numerical coefficient and exponents that are exactly divisible by 3.

PROBLEMS 10-2

Find the value of the following:

1. $\sqrt{a^2}$
2. $\sqrt{\omega^4}$
3. $\sqrt{9i^2}$
4. $\sqrt{6^2}$
5. $\sqrt{(-\omega)^2}$
6. $\sqrt{100m^2n^{12}}$
7. $\sqrt{25\lambda^4\Omega^6}$
8. $5\sqrt{64\phi^4}$
9. $\sqrt[3]{27x^6}$
10. $\sqrt[3]{-64\theta^3}$
11. $\sqrt[3]{(-2)^6}$
12. $\sqrt{4\pi^2f^2L^2 \times 10^2}$
13. $\sqrt{169m^4n^2p^6}$
14. $\sqrt[5]{32\lambda^5\psi^{10}}$
15. $\sqrt[3]{27\theta^6\phi^{12}\omega^3}$
16. $\sqrt{121x^{10}y^{12}z^6}$
17. $\sqrt{\dfrac{256\pi^2r^2x^4}{289z^6\phi^4}}$
18. $\sqrt{\dfrac{25m^4n^2p^8}{64a^4b^2c^6}}$

19. $-\sqrt{\dfrac{625r^6s^4t^8}{16x^6z^{10}}}$ 20. $\sqrt[3]{\dfrac{-8\pi^3X_L{}^3}{27Z^6X_C{}^{12}}}$ 21. $-\sqrt[3]{\dfrac{-64a^3\omega^6}{125x^6z^{12}}}$

22. $\sqrt{\dfrac{196h^2n^4p^6}{121a^2b^4c^2}}$ 23. $\sqrt{\dfrac{25v^2t^2}{256a^8b^2x^2}}$ 24. $\sqrt[3]{-\dfrac{1}{64}a^3b^{12}c^{15}}$

10-7 POLYNOMIALS WITH A COMMON MONOMIAL FACTOR

Type: $a(b + c + d) = ab + ac + ad$

Rule

To factor polynomials whose terms contain a common monomial factor:

1. Determine by inspection the greatest common factor of its terms.
2. Divide the polynomial by this factor.
3. Write the quotient in parentheses preceded by the monomial factor.

example 16 Factor $3x^2 - 9xy^2$.

solution The common monomial factor of both terms is $3x$.

$$\therefore 3x^2 - 9xy^2 = 3x(x - 3y^2)$$

example 17
Factor $2a - 6a^2b + 4ax - 10ay^3$.

solution Each term contains the factor $2a$.

$$\therefore 2a - 6a^2b + 4ax - 10ay^3 = 2a(1 - 3ab + 2x - 5y^3)$$

example 18
Factor $14x^2yz^3 - 7xy^2z^2 + 35xz^5$.

solution Each term contains the factor $7xz^2$.

$$\therefore 14x^2yz^3 - 7xy^2z^2 + 35xz^5 = 7xz^2(2xyz - y^2 + 5z^3)$$

PROBLEMS 10-3
Factor:

1. $2a + 6$
2. $\dfrac{1}{3}m + \dfrac{q}{3}$
3. $3\theta + \theta\phi + 4\theta\omega$
4. $\frac{1}{2}ab^3 - \frac{1}{6}a^2b^2 + \frac{1}{8}a^3b$
5. $20ir - 10iz$
6. $48\theta^3\phi - 144\theta^2\phi^2 + 108\theta\phi^3$
7. $\dfrac{a^2y^2}{9} + \dfrac{a^3y}{3} - \dfrac{ay^3}{12}$
8. $4\omega^4X_L{}^4 - 12\omega X_L + 28\omega^2X_L{}^2$
9. $2a^3b^2c + 8a^2bc^3 + 12a^2b^2c^2$
10. $\frac{1}{4}I^2R^2Z^2 - \frac{1}{12}IRZ^2 + \frac{1}{16}I^2RZ$
11. $36\alpha^4\beta^3\omega^2 - 72\alpha^2\beta^2\omega^5 + 180\alpha^2\beta^5\omega^2$
12. $54\theta^3\lambda^2\phi + 81\theta^2\lambda\phi^3 - 108\theta\lambda^3\phi^2$
13. $\frac{1}{64}I^2i^3 + \frac{1}{76}Ii^4 - \frac{1}{48}I^3i^2$
14. $\frac{1}{36}X_L{}^3X_C{}^2 - \frac{1}{18}X_L{}^2X_C{}^3 + \frac{1}{72}X_L{}^2X_C{}^2$
15. $720\eta^4\theta^2\phi^3\omega + 1080\eta^2\theta^4\phi\omega^2 + 600\eta^3\theta^3\phi\omega^2 - 480\eta\theta^6\phi\omega$

10-8 SQUARE OF A BINOMIAL

Type: $(a + b)^2 = a^2 + 2ab + b^2$

The multiplication

$$\begin{array}{r} a + b \\ a + b \\ \hline a^2 + ab \qquad \\ + ab + b^2 \\ \hline a^2 + 2ab + b^2 \end{array}$$

results in the formula

$$(a + b)^2 = a^2 + 2ab + b^2$$

which can be expressed by the following rule:

Rule

To square the sum of two terms, square the first term, add twice the product of the two terms, and add the square of the second term.

example 19 Square $2b + 4cd$.

solution

$$\begin{aligned}(2b + 4cd)^2 &= (2b)^2 + 2(2b)(4cd) + (4cd)^2 \\ &= 4b^2 + 16bcd + 16c^2d^2\end{aligned}$$

example 20 Let x and y be represented by lengths. Then

$$(x + y)^2 = x^2 + 2xy + y^2$$

can be illustrated graphically as shown in Fig. 10-1.

The multiplication

$$\begin{array}{r} a - b \\ a - b \\ \hline a^2 - ab \qquad \\ - ab + b^2 \\ \hline a^2 - 2ab + b^2 \end{array}$$

results in the formula

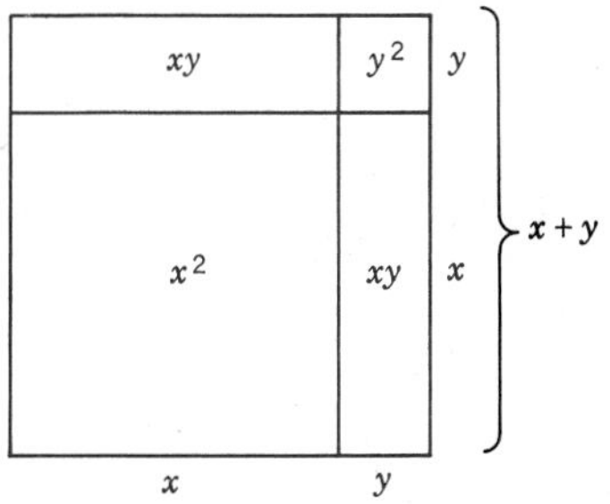

Fig. 10-1 Graphical Illustration of $(x + y)^2 = x^2 + 2xy + y^2$

$$(a - b)^2 = a^2 - 2ab + b^2$$

which can be expressed as follows:

Rule

To square the difference of two terms, square the first term, subtract twice the product of the two terms, and add the square of the second term.

example 21 Square $3a^2 - 5xy$.

solution

$$\begin{aligned}(3a^2 - 5xy)^2 &= (3a^2)^2 - 2(3a^2)(5xy) + (5xy)^2 \\ &= 9a^4 - 30a^2xy + 25x^2y^2\end{aligned}$$

example 22 Let x and y be represented by lengths. Then

$$(x - y)^2 = x^2 - 2xy + y^2$$

can be illustrated graphically as shown in Fig. 10-2. x^2 is the large square. The figure

Fig. 10-2 Graphical Illustration of $(x - y)^2 = x^2 - 2xy + y^2$

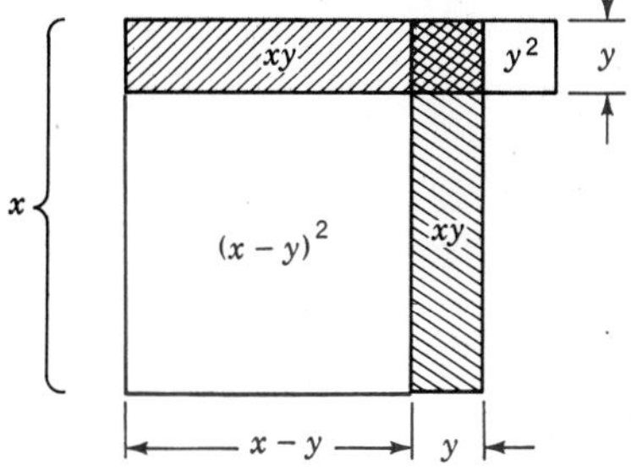

shows that the two rectangles taken from x^2 leave $(x - y)^2$. Since an amount y^2 is a part of one xy that has been subtracted from x^2 and is outside x^2, we must add it. Hence, we obtain

$$(x - y)^2 = x^2 - 2xy + y^2$$

Mentally, practice squaring sums and differences of binomials by following the foregoing rules. Proficiency in these and later methods will greatly reduce the labor in performing multiplications.

PROBLEMS 10-4

Mentally, square the following:

1. $\theta + 3$	**2.** $a + 6$	**3.** $m - R$	**4.** $I - 5$
5. $\alpha + 16$	**6.** $p - 4$	**7.** $3X - R$	**8.** $2r + 3R$
9. $F - f$	**10.** $2\alpha - 3\beta$	**11.** $5\theta + 4\phi$	**12.** $2\lambda - 5\mu$
13. $9r_1 - 3r_2$	**14.** $m^2 + 6$	**15.** $1 + X_L^2$	**16.** $2\theta^2 - 13\phi$
17. $6v^2 - 2t^3$	**18.** $20 + 2$	**19.** $30 - 3$	**20.** $30 + 5$
21. $6\pi R^2 - 2\pi r^2$	**22.** $2\pi f L_1 - Z$	**23.** $1.5\theta^2 - 0.5\alpha$	**24.** $\frac{1}{2}R_1 + \frac{1}{4}R_2$
25. $\frac{3}{4}X^2 - \frac{1}{2}Z^2$	**26.** $\frac{1}{3}\phi^3\lambda + \frac{1}{2}\alpha^2$	**27.** $6\phi^2\omega - \frac{1}{4}\lambda^2$	

Expand:

28. $(a + 5)^2$	**29.** $(x + \frac{1}{2})^2$	**30.** $(\alpha + \frac{1}{3})^2$	**31.** $(\frac{1}{2} - E)^2$
32. $(\mu - \frac{1}{12})^2$	**33.** $(1 + e^3)^2$	**34.** $(X_1^2 + \frac{2}{3})^2$	**35.** $(L^2 - \frac{7}{8}P)^2$
36. $\left(\frac{X}{2} + Y\right)^2$	**37.** $\left(\frac{b}{3} + \frac{m}{2}\right)^2$	**38.** $(3 + 2ab)^2$	**39.** $(R_1 - \frac{5}{8}R_2)^2$

40. $(2\phi + \frac{3}{4}\theta^2)^2$

41. Develop a graphical illustration of $(x + y)(x - y)$.

10-9 SQUARE ROOT OF A TRINOMIAL

In the preceding section, it was shown that

$$(a + b)^2 = a^2 + 2ab + b^2$$

and

$$(a - b)^2 = a^2 - 2ab + b^2$$

From these and other binomials that have been squared, it is evident that a trinomial is a perfect square if

1. Two terms are squares of monomials and positive.
2. The other term is twice the product of these monomials and has affixed either a plus or a minus sign.

example 23 $x^2 + 2xy + y^2$ is a perfect trinomial square because x^2 and y^2 are the squares of the monomials x and y, respectively, and $2xy$ is twice the product of the monomials. Therefore,

$$x^2 + 2xy + y^2 = (x + y)^2$$

example 24 $4a^2 - 12ab + 9b^2$ is a perfect trinomial square because $4a^2$ and $9b^2$ are the squares of $2a$ and $3b$, respectively, and the other term is $-2(2a)(3b)$. Therefore,

$$4a^2 - 12ab + 9b^2 = (2a - 3b)^2$$

Rule

To extract the square root of a perfect trinomial square, extract the square root of the two perfect square monomials and connect them with the sign of the remaining term.

example 25 Supply the missing term in $x^4 + ? + 16$ so that the three terms will form a perfect trinomial square.

solution The missing term is twice the product of the monomials whose squares result in the two known terms; that is, $2(x^2)(4) = 8x^2$. Hence,

$$x^4 + 8x^2 + 16 = (x^2 + 4)^2$$

example 26 Supply the missing term in $25a^2 + 30ab + ?$ so that the three terms will form a perfect trinomial square.

solution The square root of the first term is $5a$. The missing term is the square of some number N such that $2(5a)(N) = 30ab$. Then by multiplying, we obtain $10aN = 30ab$, or $N = 3b$. Therefore,

$$25a^2 + 30ab + 9b^2 = (5a + 3b)^2$$

PROBLEMS 10-5

Supply the missing terms so that the three terms form perfect trinomial squares:

1. $e^2 + ? + 9$
2. $I^2 + ? + 4$
3. $\lambda^2 - ? + 4$
4. $F^2 - ? + f^2$
5. $25x^2 - ? + y^2$
6. $25X_C^2 - ? + 4$
7. $49\omega^2 + ? + \pi^2$
8. $100L_1^2 + ? + 16M^2$
9. $4m^2 + ? + 9p^2$
10. $r^2 - 2rs + ?$
11. $E^2 + 2EI + ?$
12. $? - 90xy + 25y^2$
13. $? + 80pq + 100q^2$
14. $Z^2 + 12XZ + ?$
15. $\frac{1}{9}\theta^2\phi^2 - ? + \frac{1}{4}\omega^2$
16. $\frac{1}{16}\eta^4 - \frac{1}{4}\eta^2\theta + ?$
17. $? - \dfrac{\pi\phi}{3} + \dfrac{1}{4}\phi^2$
18. $\dfrac{R_1^2}{64} - \dfrac{1}{24}XR_1 + ?$

Extract the square roots of the following:

19. $M^2 + 2M + 1$
20. $a^2 - 10ab + 25b^2$
21. $16q_1^2 + 8q_1q_2 + q_2^2$
22. $E^2 + 12EI + 36I^2$
23. $9\alpha^4\beta^2 + 54\alpha^2\beta\gamma + 81\gamma^2$
24. $64\omega^2\lambda^2 + 16\omega\lambda\Omega^2 + \Omega^4$
25. $\frac{9}{25}\pi^2R^4 + \frac{4}{5}\pi R^2 + \frac{4}{9}$
26. $9R_1^2 + \frac{4}{25}r^2 - \frac{12}{5}R_1r$
27. $\dfrac{10\phi\lambda}{21} + \dfrac{25\phi^2}{36} + \dfrac{4\lambda^2}{49}$
28. $-\dfrac{4Z^4M^2}{27} + \dfrac{4Z^8}{81} + \dfrac{M^4}{9}$

10-10 PRIME FACTORS OF AN EXPRESSION

In factoring a number, all its prime factors should be obtained. After an expression is factored once, it may be possible to factor it again.

example 27 Find the prime factors of $12i^2r + 12iIr + 3I^2r$.

solution

$$\begin{aligned} 12i^2r + 12iIr + 3I^2r &= 3r(4i^2 + 4iI + I^2) \\ &= 3r(2i + I)(2i + I) \\ &= 3r(2i + I)^2 \end{aligned}$$

PROBLEMS 10-6

Find the prime factors of the following:

1. $3ac + 6bc$

2. $15qrx + 35rtx$

3. $2\lambda\theta^2 + 4\theta\lambda\phi + 2\phi^2\lambda$

4. $5E^2i^2 - 10EIi^2R + 5I^2i^2R^2$

5. $24\alpha^4 + 120\alpha^3\beta + 150\alpha^2\beta^2$

6. $\dfrac{2E^4}{3r} + \dfrac{4E^2e^2}{r} + \dfrac{6e^4}{r}$

7. $\dfrac{20f_o\omega_1^2}{\omega} - \dfrac{40f_o\omega_1\omega_2}{\omega} + \dfrac{20f_o\omega_2^2}{\omega}$

8. $\dfrac{3X_L^2}{4X_C} + \dfrac{3KMX_L}{4X_C} + \dfrac{3K^2M^2}{16X_C}$

9. $\dfrac{5r\lambda^2}{16e} - \dfrac{5f^2r\lambda}{2e} + \dfrac{5f^4r}{e}$

10. $\dfrac{200Ff^2}{C} + \dfrac{480Ffx}{C} + \dfrac{288Fx^2}{C}$

10-11 PRODUCT OF THE SUM AND DIFFERENCE OF TWO NUMBERS

Type: $(a + b)(a - b) = a^2 - b^2$

The multiplication of the sum and difference of two general numbers, such as

$$\begin{array}{r} a + b \\ a - b \\ \hline a^2 + ab \qquad \\ - ab - b^2 \\ \hline a^2 \qquad\quad - b^2 \end{array}$$

results in the formula

$$(a + b)(a - b) = a^2 - b^2$$

which can be expressed by the following:

Rule

The product of the sum and difference of two numbers is equal to the difference of their squares.

example 28

$$(3x + 4y)(3x - 4y) = 9x^2 - 16y^2$$

example 29

$$(6ab^2 + 7c^3d)(6ab^2 - 7c^3d) = 36a^2b^4 - 49c^6d^2$$

PROBLEMS 10-7

Multiply by inspection:

1. $(\theta + 2)(\theta - 2)$

2. $(\phi - 4)(\phi + 4)$

3. $(I + i)(I - i)$

4. $(I + Z)(I - Z)$

5. $(3Q - 2L)(3Q + 2L)$

6. $(2\pi R_1 + 2\pi R_2)(2\pi R_1 - 2\pi R_2)$

7. $(\frac{2}{3}EI + P)(\frac{2}{3}EI - P)$

8. $(\frac{3}{4}\omega^2 - \frac{2}{5}\lambda)(\frac{3}{4}\omega^2 + \frac{2}{5}\lambda)$

9. $\left(\frac{2E^2}{R} + \frac{3I^2R}{P}\right)\left(\frac{2E^2}{R} - \frac{3I^2R}{P}\right)$ 10. $\left(\frac{\theta^2}{2\phi} + \frac{3\alpha}{2\beta}\right)\left(\frac{\theta^2}{2\phi} - \frac{3\alpha}{2\beta}\right)$

10-12 FACTORING THE DIFFERENCE OF TWO SQUARES

Rule

To factor the difference of two squares, extract the square root of the two squares, add the roots for one factor, and subtract the second root from the first for the other factor.

example 30 $x^2 - y^2 = (x + y)(x - y)$

example 31

$$9a^2c^4 - 36d^6 = (3ac^2 + 6d^3)(3ac^2 - 6d^3)$$

PROBLEMS 10-8

Factor:

1. $a^2 - b^2$ 2. $I_1^2 - I_2^2$ 3. $4\theta^2 - 16\phi^2$ 4. $4I^2 - 9r^2$

5. $\frac{1}{4} - \theta^2$ 6. $\frac{\alpha^2}{\beta^2} - \frac{4\gamma^2}{9}$ 7. $1 - 225\omega^2$ 8. $\frac{1}{E_1^2} - \frac{1}{e^2}$

9. $81\theta^2\mu^2 - 1$ 10. $\frac{1}{X_C^2} - \frac{V^2}{Q^2}$ 11. $9c^2 - a^2 + 2ab - b^2$

SOLUTION:
$$\begin{aligned} 9c^2 - a^2 + 2ab - b^2 &= 9c^2 - (a^2 - 2ab + b^2) \\ &= [3c + (a - b)][3c - (a - b)] \\ &= (3c + a - b)(3c - a + b) \end{aligned}$$

12. $(\theta^2 + 4\theta\phi + 4\phi^2) - \omega^2$ 13. $36m^2 - 81p^2q^2 + 9a^2b^2 - 36abm$

14. $16I^2 - E^2 + \frac{14E}{X} - \frac{49}{X^2}$ 15. $100acl - 144l^2 + 25a^2 + 100c^2l^2$

10-13 PRODUCT OF TWO BINOMIALS HAVING A COMMON TERM

Type: $(x + a)(x + b) = x^2 + (a + b)x + ab$

The multiplication

$$\begin{array}{l} x + a \\ \underline{x + b} \\ x^2 + ax \\ \underline{\qquad\quad + bx + ab} \\ x^2 + ax + bx + ab \end{array}$$

when factored, results in $x^2 + (a + b)x + ab$.

This type of formula can be expressed as follows:

Rule

To obtain the product of two binomials having a common term, square the common term, multiply the common term by the algebraic sum of the second terms of the binomials, find the product of the second terms, and add the results.

example 32 Find the product of $x - 7$ and $x + 5$.

solution

$$(x - 7)(x + 5) = x^2 + (-7 + 5)x + (-7)(+5)$$
$$= x^2 - 2x - 35$$

example 33

$$(ir + 3)(ir - 6) = i^2r^2 + (+3 - 6)ir + (+3)(-6)$$
$$= i^2r^2 - 3ir - 18$$

Although the preceding examples have been written out in order to illustrate the method, the actual multiplication should be performed mentally. In Example 33, write the i^2r^2 term first. Then glance at the $+3$ and -6, note that their sum is -3 and their product is -18, and write down the complete product.

PROBLEMS 10-9

Mentally, multiply the following:

1. $(\theta + 4)(\theta + 3)$
2. $(Q + 1)(Q + 2)$
3. $(R + 1)(R - 2)$
4. $(\phi - 3)(\phi - 2)$
5. $(\theta + 6)(\theta + 3)$
6. $(2r + 3)(2r + 2)$
7. $(3\theta - 2)(3\theta + 1)$
8. $(4x + 2)(4x - 4)$
9. $(I - 3)(I - 4)$
10. $(\frac{1}{2}P + 2)(\frac{1}{2}P + 6)$
11. $(\alpha - 1)(\alpha - \frac{1}{4})$
12. $(\lambda + 6)(\lambda - \frac{1}{3})$
13. $(IR + \frac{1}{2})(IR - \frac{1}{3})$
14. $(2f + 12)(2f - \frac{1}{4})$
15. $(\alpha + \frac{2}{3})(\alpha + \frac{1}{3})$
16. $\left(\frac{1}{X} + 5\right)\left(\frac{1}{X} - 2\right)$
17. $\left(\frac{1}{\sqrt{LC}} - f\right)\left(\frac{1}{\sqrt{LC}} - 3f\right)$
18. $(vt - \frac{1}{12})(vt + \frac{1}{2})$
19. $(\alpha\beta^2 + \frac{1}{10})(\alpha\beta^2 + \frac{1}{5})$
20. $\left(I + \frac{6E}{R}\right)\left(I - \frac{4E}{R}\right)$

10-14 FACTORING TRINOMIALS OF THE FORM $a^2 + ba + c$

A trinomial of the form $a^2 + ba + c$ can be factored if it is the product of two binomials having a common term.

Rule

To factor a trinomial of the form $a^2 + ba + c$, find two numbers whose sum is b and whose product is c. Add each of them to the square root of the first term for the factors.

example 34 Factor $a^2 + 7a + 12$.

solution If this expression will factor, it will take the form

$$a^2 + 7a + 12 = (a + \quad)(a + \quad)$$

where the two blanks represent numbers whose product is 12 and whose sum is 7. The factors of 12 are

$$1 \times 12$$
$$2 \times 6$$
$$3 \times 4$$

The first two pairs will not do because the sum of neither pair is 7. The third pair gives the correct sum.

$$\therefore a^2 + 7a + 12 = (a + 3)(a + 4)$$

example 35 Factor $x^2 - 15x + 36$.

solution Since the 36 is positive, its factors must bear the same sign; also, since -15 is negative, it follows that both factors must be negative. The factors of 36 are

$$\begin{array}{c} 1 \times 36 \\ 2 \times 18 \\ 3 \times 12 \\ 4 \times 9 \\ 6 \times 6 \end{array}$$

Inspection of these factors shows that 3 and 12 are the required numbers.

$$\therefore x^2 - 15x + 36 = (x - 3)(x - 12)$$

example 36 Factor $e^2 - e - 56$.

solution Since we have -56, the two factors must have unlike signs. The sum of the factors must equal -1; therefore, the negative factor of -56 must have the greater absolute value. The factors of 56 are

$$\begin{array}{c} 1 \times 56 \\ 2 \times 28 \\ 4 \times 14 \\ 7 \times 8 \end{array}$$

Since the factors 7 and 8 differ in value by 1, we have

$$e^2 - e - 56 = (e + 7)(e - 8)$$

PROBLEMS 10-10
Factor:

1. $a^2 + 3a + 2$
2. $\theta^2 + 8\theta + 15$
3. $R^2 + 8R + 12$
4. $\alpha^2 - 9\alpha + 14$
5. $\beta^2 + 2\beta - 24$
6. $R^2 - 6R - 72$
7. $\theta^2 + 10\theta + 24$
8. $\omega^2 - 14\omega + 24$
9. $t^2 + 9t - 22$
10. $\alpha^2 t^2 - 13\alpha t + 36$
11. $Z^4 + 8Z^2 - 20$
12. $f^2 - 17f - 60$
13. $\pi^2 + \pi - 56$
14. $I^2R^2 + 4EIR + 3E^2$
15. $\omega^2 - \omega f - 6f^2$
16. $\theta^4 - 4\theta^2\phi - 12\phi^2$
17. $\theta^2 - \frac{5}{6}\theta + \frac{1}{6}$
18. $q^4 - \frac{1}{4}q^2 - \frac{1}{8}$
19. $\phi^4 + \dfrac{\phi^2}{10} - \dfrac{1}{50}$
20. $e^4 i^4 + 2e^2 i^2 P - 3P^2$

10-15 PRODUCT OF ANY TWO BINOMIALS

Type: $(ax + b)(cx + d)$

Up to the present, if it was desired to multiply $5x - 2$ by $3x + 6$, we multiplied in the following manner:

$$\begin{array}{r} 5x - 2 \\ 3x + 6 \\ \hline 15x^2 - 6x \qquad \\ + 30x - 12 \\ \hline 15x^2 + 24x - 12 \end{array}$$

Note that $15x^2$ is the product of the first terms of the binomials and the last term is the product of the last terms of the binomials. Also, the middle term is the sum of the products obtained by multiplying the first term of each binomial by the second term of the other binomial.

The preceding example can be written in the following manner:

$$\begin{array}{c} 5x - 2 \\ 3x + 6 \\ \hline 15x^2 + 24x - 12 \end{array}$$

The middle term $(+24x)$ is the sum of *cross products* $(5x)(+6)$ and $(3x)(-2)$, which is obtained by multiplying the first term of each binomial by the second term of the other.

The usual method of obtaining this product is indicated by the following solution:

$$(5x - 2)(3x + 6) = 15x^2 + 24x - 12$$

Rule

For finding the product of any two binomials,

1. The first term of the product is the product of the first terms of the binomials.
2. The second term is the algebraic sum of the product of the two outer terms and the product of the two inner terms.
3. The third term is the product of the last terms of the binomials.

example 37 Find the product of $(4e + 7j)(2e - 3j)$.

solution The only difficulty encountered in obtaining such products mentally is that of finding the second term.

$$(4e)(-3j) = -12ej$$
$$(7j)(2e) = 14ej$$
$$(-12ej) + (14ej) = +2ej$$
$$\therefore (4e + 7j)(2e - 3j) = 8e^2 + 2ej - 21j^2$$

example 38

Find the product $(7r^2 + 8Z)(8r^2 - 9Z)$.

solution

1. The first term of the product is $(7r^2)(8r^2) = 56r^4$.
2. Since $(7r^2)(-9Z) = -63r^2Z$ and $(8Z)(8r^2) = 64r^2Z$, the second term is $(-63r^2Z) + (64r^2Z) = +r^2Z$.
3. The third term is $(8Z)(-9Z) = -72Z^2$.

$$\therefore (7r^2 + 8Z)(8r^2 - 9z) = 56r^4 + r^2Z - 72Z^2$$

By repeated drills you should acquire skill enough that you can readily obtain such products mentally. This type of product is frequently encountered in algebra, and the ability to multiply rapidly will save you much time.

PROBLEMS 10-11

Multiply:

1. $(x + 2)(x - 5)$
2. $(IR - 4)(IR + 3)$
3. $(3\phi + 1)(2\phi + 3)$
4. $(2R + 6)(3R + 5)$
5. $(3j - 2)(4j + 2)$
6. $(7\lambda + 5)(2\lambda - 3)$
7. $(2\omega + 5)(3\omega - 1)$
8. $(7\theta + 3)(3\theta + 7)$
9. $(\frac{1}{2}\omega + 8)(\frac{1}{2}\omega - 4)$
10. $\left(\frac{3}{\theta} - 6\right)\left(\frac{2}{\theta} - 12\right)$
11. $(2Z + IR)(3Z + 5IR)$
12. $(I - 18)(I + 6)$
13. $(3X - 20)(5X + 2)$
14. $(12M - 3)(3M - 12)$
15. $(15\theta - 2)(\theta - 5)$
16. $(5 + 4p)(4 - 5p)$
17. $(5 - 3\pi)(7 - 2\pi)$
18. $(3\phi + 4)(5\phi + 3)$
19. $(3\alpha + 5\beta)(2\alpha + 7\beta)$
20. $(3x + 7)(4x - 5)$
21. $(2a - 7t)(2a - 5t)$
22. $(a + 0.5)(a - 0.3)$
23. $(\omega + 0.7f)(\omega - 0.2f)$
24. $(IR - 0.9)(IR + 1)$

25. $\left(\frac{x}{8} + \frac{\lambda}{4}\right)(2x - 16\lambda)$ **26.** $\left(8\delta - \frac{2}{3\eta}\right)\left(9\delta - \frac{1}{2\eta}\right)$

27. $\left(6Z + \frac{1}{2IR}\right)\left(4Z + \frac{1}{3IR}\right)$ **28.** $\left(12\pi L - \frac{2}{3\pi C}\right)\left(10\pi L - \frac{2}{3\pi C}\right)$

29. $(0.2p - 0.7q)(0.8p - 0.3q)$ **30.** $\left(4m + \frac{3r}{p}\right)\left(6m - \frac{2r}{3p}\right)$

10-16 FACTORING TRINOMIALS OF THE TYPE $ax^2 + bx + c$

The method of factoring trinomials of the type $ax^2 + bx + c$ is best illustrated by examples.

example 39 Factor $3a^2 + 5a + 2$.

solution It is apparent that the two factors are binomials and the product of the end terms must be $3a^2$ and 2. Therefore, the binomials to choose from are

$$(3a + 1)(a + 2)$$

and

$$(3a + 2)(a + 1)$$

However, the first factors when multiplied result in a product of $7a$ for the middle term. The second pair of factors when multiplied give a middle term of $5a$. Therefore,

$$3a^2 + 5a + 2 = (3a + 2)(a + 1)$$

example 40 Factor $6e^2 + 7e + 2$.

solution Again, the end terms of the binomial factors must be so chosen that their products result in $6e^2$ and 2. Both the last terms of the factors are of like signs, for the last term of the trinomial is positive. Also, both last terms of the factors must be positive, for the second term of the trinomial is positive. One of the several methods of arranging the work is as shown below. The tentative factors are arranged as if for multiplication:

Trial Factors	*Products*	
$(6e + 1)(e + 2)$	$= 6e^2 + 13e + 2$	Wrong
$(6e + 2)(e + 1)$	$= 6e^2 + 8e + 2$	Wrong
$(3e + 1)(2e + 2)$	$= 6e^2 + 8e + 2$	Wrong
$(3e + 2)(2e + 1)$	$= 6e^2 + 7e + 2$	Right

It is seen that any combination of the trial factors when multiplied results in the correct first and last term.

$$\therefore 6e^2 + 7e + 2 = (3e + 2)(2e + 1)$$

note This may seem to be a long process, but with practice, most of the factor trials can be tested mentally.

example 41 Factor $12i^2 - 17i + 6$.

solution The third term of this trinomial is $+6$; therefore, its factors must have like signs. Since the second term is negative, the cross products must be negative. Then it follows that both factors of 6 must be negative. Some of the combinations are as follows:

Trial Factors	*Products*	
$(2i - 3)(6i - 2)$	$= 12i^2 - 22i + 6$	Wrong
$(2i - 2)(6i - 3)$	$= 12i^2 - 18i + 6$	Wrong
$(3i - 3)(4i - 2)$	$= 12i^2 - 18i + 6$	Wrong
$(3i - 2)(4i - 3)$	$= 12i^2 - 17i + 6$	Right

$$\therefore 12i^2 - 17i + 6 = (3i - 2)(4i - 3)$$

example 42 Factor $8r^2 - 14r - 15$.

solution The factors of -15 must have unlike signs. The signs of these factors must be arranged so that the cross product of greater absolute value is minus, because the middle term of the trinomial is negative.

Trial Factors	*Products*	
$(8r + 3)(r - 5)$	$= 8r^2 - 37r - 15$	Wrong
$(4r + 5)(2r - 3)$	$= 8r^2 - 2r - 15$	Wrong
$(4r + 3)(2r - 5)$	$= 8r^2 - 14r - 15$	Right

example 43 Factor $6R^2 - 7R - 20$.

note Many students prefer the following method to the trial-and-error method of the foregoing examples.

solution Multiply and divide the entire expression by the coefficient of R^2. The result is

$$\frac{36R^2 - 42R - 120}{6}$$

Take the square root of the first term, which is $6R$, and let that be some other letter such as x. Then, if

$$6R = x$$

by substituting the value of $6R$ in the above expression, we obtain

$$\frac{x^2 - 7x - 120}{6}$$

This results in an expression with a numerator easy to factor. Thus,

$$\frac{x^2 - 7x - 120}{6} = \frac{(x + 8)(x - 15)}{6}$$

Substituting $6R$ for x in the last expression, we obtain

$$\frac{(6R + 8)(6R - 15)}{6}$$

Factoring the numerator,

$$\frac{2(3R + 4)3(2R - 5)}{6}$$

Canceling,

$$6R^2 - 7R - 20 = (3R + 4)(2R - 5)$$

note The denominator will *always* cancel out.

example 44 Factor $4E^2 - 8EI - 21I^2$.

solution Multiplying and dividing by the coefficient of E^2,

$$\frac{16E^2 - 32EI - 84I^2}{4}$$

Let the square root of the first term $4E = x$.

Then

$$\frac{x^2 - 8Ix - 84I^2}{4} = \frac{(x + 6I)(x - 14I)}{4}$$

Substituting for x,

$$\frac{(4E + 6I)(4E - 14I)}{4}$$

Factoring,

$$\frac{2(2E + 3I)2(2E - 7I)}{4}$$

Canceling,

$$4E^2 - 8EI - 21I^2 = (2E + 3I)(2E - 7I)$$

PROBLEMS 10-12

Factor:

1. $\omega^2 - 3\omega - 10$
2. $6\theta^2 + 7\theta + 2$
3. $8m^2 - 2m - 15$
4. $3I^2 - 14I + 8$

5. $6x^2 + 11x - 10$
6. $3\alpha^2 - 10\alpha + 7$
7. $9\phi^2 + 18\phi + 8$
8. $18L_1{}^2 + 31L_1 + 6$
9. $2\alpha^2 - \alpha\beta - 21\beta^2$
10. $10P^2 - 17PW + 3W^2$
11. $40m^2 + 2m - 21$
12. $20\lambda^2 - 22\lambda\phi + 6\phi^2$
13. $80l^2 + 14lw - 6w^2$
14. $18q^2 + 57qr + 35r^2$
15. $24\beta^4 - 30\beta^2\gamma + 9\gamma^2$
16. $42y^2z^2 + 11wyz - 20w^2$
17. $27l^2m^2 + 15lmw - 2w^2$
18. $2\mu^2 - 18\pi^2$
19. $6\psi^2 - 24\Omega^2$
20. $24m^4 - 43m^2p + 18p^2$
21. $15x^2 - 7\Delta x - 2\Delta^2$
22. $\alpha^2 - \frac{5\alpha}{6} + \frac{1}{6}$
23. $48\theta^2 + 5\theta + \frac{1}{8}$
24. $10Z^2 - \frac{3Z}{2} + \frac{1}{20}$
25. $0.18\theta^2 - 2$

10-17 SUMMARY

In this chapter, various cases of products and factoring have been treated separately in the different sections. Frequently, however, it becomes necessary to apply the principles underlying two or more cases to a single problem. It is very important, therefore, that you recognize the standard form for various types of algebraic expressions in order that you can apply the method of solution as needed. These forms are summarized in Table 10-1.

Problems 10-13 are included as a review of the entire chapter. If you can work all of them, you thoroughly understand the contents of this chapter. If not, a review of the doubtful parts is suggested, for a good working knowledge of special products and factoring makes it possible to do the following:

1. Multiply, divide, and factor very quickly in your head (mentally).
2. Find the solutions to problems which can be solved by (quick mental) factoring.

Table 10-1

General Type	*Factors*	*Section*
$ab + ac + ad$	$a(b + c + d)$	10-7
$a^2 + 2ab + b^2$	$(a + b)^2$	10-8
$a^2 - 2ab + b^2$	$(a - b)^2$	10-8
$a^2 - b^2$	$(a + b)(a - b)$	10-12
$a^2 + (b + c)a + bc$	$(a + b)(a + c)$	10-13
$acx^2 + (bc + ad)x + bd$	$(ax + b)(cx + d)$	10-15

PROBLEMS 10-13

Find the value of the following:

1. $(-4\omega L)^2$
2. $(-3\lambda^2\phi^3\omega)^3$
3. $\left(\frac{a^3b^3cd^2}{p^2q^3r}\right)^4$

4. $-\sqrt{64a^2x^4y^2z^6}$ 5. $\sqrt{\dfrac{144I^2R^2}{169F^2X_C{}^4}}$ 6. $\sqrt[3]{-64\alpha^6\beta^3\gamma^9}$

7. $-\sqrt[3]{\dfrac{125l^3m^6}{27x^{12}y^{15}z^3}}$ 8. $\sqrt{\dfrac{625I^4R^2P^2}{64E^2W^6}}$ 9. $-\sqrt[3]{-216\theta^3\phi^6\omega^6}$

10. $125a\sqrt[3]{\dfrac{a^5x^6z^8}{a^2x^3z^2}}$

Factor:

11. $IR^2 - Ir^2$ 12. $I^2R_1 + I^2R_2 + I^2R_3$

13. $\dfrac{3e^2}{8r_1} + \dfrac{5e^2}{8r_2} - \dfrac{7e^2}{8r_3}$ 14. $4.8\omega L_1 - 0.24\omega L_2 + 1.2\omega L_3$

15. $\frac{7}{16}xk - \frac{3}{16}xl - \frac{9}{16}xm$ 16. $\dfrac{2g_\mathrm{m}r_\mathrm{p}}{3} - 6g_\mathrm{m}R_\mathrm{p} + \dfrac{16g_\mathrm{m}R'_\mathrm{p}}{3}$

Mentally, find the products:

17. $(R + 12)^2$ 18. $(2\theta - 3\phi^2)^2$ 19. $(12I^2 + \frac{2}{3})^2$
20. $(0.5\omega M - 0.3\omega L)^2$ 21. $(\frac{5}{9}\beta - 3\lambda)^2$ 22. $(0.4X_C + 0.8X_L)^2$

Supply the missing term so that the three terms form a trinomial square:

23. $r^2 + ? + 9$ 24. $9\alpha^2 - ? + 4\beta^2$ 25. $? + 28Qr + 4r^2$

26. $64Z^6 - 32Z^3 + ?$ 27. $\mu^2 + \dfrac{\mu\lambda}{2} + ?$ 28. $? + \frac{3}{2}L^2M + \frac{9}{64}M^2$

Extract the square roots of the following:

29. $m^2 + 10m + 25$ 30. $\theta^2 + 14\,\theta\phi + 49\phi^2$

31. $16\alpha^2 + 80\alpha\beta + 100\beta^2$ 32. $F^2 - \dfrac{2Ff}{3} + \dfrac{f^2}{9}$

33. $\dfrac{\phi^2}{36} - \dfrac{\phi\lambda}{6} + \dfrac{\lambda^2}{4}$ 34. $\dfrac{4E^2}{25} - \dfrac{12EX}{5} + 9X^2$

Factor:

35. $24iR + 42IR$ 36. $6pq - 27pr$
37. $3ir^2 + 18ir + 27i$ 38. $16\pi^3Dr^2 + 48\pi^2CDr + 36\pi C^2D$

39. $768\theta^2\omega - 192\theta\phi\omega + 12\phi^2\omega$ 40. $\dfrac{12P^2}{EI} - \dfrac{144PW}{EI} + \dfrac{432W^2}{EI}$

Find the products:

41. $(\alpha + 2\beta)(\alpha - 2\beta)$

42. $(2IR - 3E)(2IR + 3E)$

43. $(Z - 12)(Z + 12)$

44. $(8\theta + 7\phi)(8\theta - 7\phi)$

45. $\left(\frac{24E}{IR} + 2P\right)\left(\frac{24E}{IR} - 2P\right)$

46. $(0.8\lambda + 0.3\Omega)(0.8\lambda - 0.3\Omega)$

Factor:

47. $Q^2 - 1$

48. $25 - f_o^2$

49. $4\omega^2L^2 - \frac{1}{16\omega^2C^2}$

50. $\frac{4}{25}\alpha^2\beta^4 - \frac{9}{16}\lambda^2$

51. $0.0025\psi^2 - 0.36\mu^2$

52. $0.01X_L^2 - 0.81X_C^2$

Find the quotients:

53. $(\lambda^2 - 4) \div (\lambda + 2)$

54. $(4L^2 - 9C^2) \div (2L + 3C)$

55. $(\frac{1}{9}\alpha^2 - \frac{4}{49}\beta^2) \div (\frac{1}{3}\alpha - \frac{2}{7}\beta)$

56. $(0.25\theta^2 - 0.04\delta^2) \div (0.5\theta - 0.2\delta)$

57. $(\frac{9}{25}e^2 - \frac{16}{81}i^2r^2) \div (\frac{3}{5}e + \frac{4}{9}ir)$

58. $\frac{L^2 + 2LM + M^2 - 25}{L + M + 5}$

Find the products:

59. $(\kappa + 2)(\kappa - 4)$

60. $(3 - Q)(4 - 2Q)$

61. $(0.2X_C - 3)(X_C + 0.5)$

62. $(0.1Z + 0.6R)(0.3Z - R)$

63. $(A - \frac{1}{3})(A + \frac{1}{5})$

64. $(5\lambda + 12)(\frac{1}{4}\lambda - \frac{1}{5})$

65. $(8\mu + 6g_m)(3\mu - 2g_m)$

66. $(2\pi fL + X_C)(2\pi fL - 3X_C)$

67. $(2R - r)(0.3R + 0.2r)$

68. $(0.5\alpha + 8\beta)(0.2\alpha + 0.4\beta)$

69. $\left(4\phi + \frac{2\theta}{3}\right)\left(6\phi - \frac{\theta}{2}\right)$

70. $\left(\frac{2v}{3} - \frac{4s}{t}\right)\left(\frac{v}{2} - \frac{6s}{t}\right)$

Factor (remove any common factors first):

71. $6z^2 + 11z + 3$

72. $12I_1^2 - 2I_1 - 4$

73. $\lambda^2 - 8\lambda + 15$

74. $e^2 - 0.2e - 0.03$

75. $x^2 - 2.6x + 1.2$

76. $A^2 - \frac{3A}{40} - \frac{1}{40}$

77. $12R^2 + 8RX - 15X^2$

78. $3\alpha^2\beta^2\gamma^2 + \alpha\beta\gamma\Omega - 10\Omega^2$

79. $2E^2 + 0.1EIR - 0.15I^2R^2$

80. $I^2 - 0.3Iq - 0.4q^2$

81. $\frac{X_C^2}{9} + \frac{2X_CZ}{3} + Z^2$

82. $a^2 + \frac{2a}{b} + \frac{1}{b^2}$

83. $16\pi^2 - 2\pi f - 5f^2$

84. $5\theta^2\omega - 5\phi^2\omega$

85. $3x^2 - 12$

86. $288f^2\lambda - \frac{2\lambda}{9}$

87. $\frac{3E^2}{2i} - \frac{18Ee}{i} + \frac{54e^2}{i}$

88. $\frac{x^3}{9Z} - \frac{x^2y}{6Z} + \frac{xy^2}{16Z}$

89. $\frac{a^2c}{2d} - \frac{145abc}{144d} + \frac{b^2c}{2d}$

90. $\frac{2ER_1{}^2}{3I} - \frac{1898ER_1R_2}{1350I} + \frac{2ER_2{}^2}{3I}$

11

algebraic fractions

Algebraic fractions play an important role in mathematics, especially in equations for electric and electronic circuits.

At this time, if you feel you have not thoroughly mastered arithmetical fractions, you are urged to review them. Despite the current emphasis on metrication and the increasing use of decimal fractions in measurement, a good foundation in arithmetical fractions is essential, for every rule and operation pertaining to them is applicable to algebraic fractions. It is a fact that anyone who really knows arithmetical fractions rarely has trouble with algebraic fractions.

11-1 THE DEGREE OF A MONOMIAL

The degree of a monomial is determined by the number of literal factors it has.

Thus, $6ab^2$ is a monomial of the third degree because $ab^2 = a \cdot b \cdot b$; $3mn$ is a monomial of the second degree. From these examples, it is seen that the degree of a monomial is the sum of the exponents of the literal factors (letters).

In such an expression as $5X^2Y^2Z$, we speak of the whole term as being of the fifth degree, X and Y as being of the second degree, and Z as being of the first degree.

The above definition for the degree of a monomial does not apply to letters in a denominator.

11-2 THE DEGREE OF A POLYNOMIAL

The degree of a polynomial is taken as the degree of its term of highest degree. Thus, $3ab^2 - 4cd - d$ is a polynomial of the *third degree* and $6x^2y + 5xy^2 + y^4$ is a polynomial of the *fourth degree*.

11-3 HIGHEST COMMON FACTOR

A factor of each of two or more expressions is a *common factor* of those expressions. For example, 2 is a common factor of 4 and 6; a^2 is a common factor of a^3, $(a^2 - a^2b)$, and $(a^2x^2 - a^2y)$.

The product of all the factors common to two or more numbers, or expressions, is called their *highest common factor*. That is, the highest common factor is the expression of highest degree that will divide each of them without a remainder. It is commonly abbreviated HCF.

example 1 Find the HCF of

$$6a^2b^3(c + 1)(c + 3)^2$$

and

$$30a^3b^2(c - 2)(c + 3)$$

solution 6 is the greatest integer that will divide both expressions. The highest power of a that will divide both is a^2. The highest power of b that will divide both is b^2. The highest power of $(c + 3)$ that will divide both is $(c + 3)$. $(c + 1)$ and $(c - 2)$ will not divide both expressions.

$$\therefore 6a^2b^2(c + 3) = \text{HCF}$$

Rule

To determine the HCF:

1. Determine all the prime factors of each expression.
2. Take the common factors of all the expressions and give to each the lowest exponent it has in any of the expressions.
3. The HCF is the product of all the common factors as obtained in step 2.

example 2 Find the HCF of

$$50a^2b^3c(x + y)^3(x - y)^4$$

and

$$75a^2bc^2(x + y)^2(x - y)$$

solution

$$50a^2b^3c(x + y)^3(x - y)^4 = 2 \cdot 5 \cdot 5a^2b^3c(x + y)^3(x - y)^4$$
$$75a^2bc^2(x + y)^2(x - y) = 3 \cdot 5 \cdot 5a^2bc^2(x + y)^2(x - y)$$
$$\therefore \text{HCF} = 5^2a^2bc(x + y)^2(x - y) = 25a^2bc(x + y)^2(x - y)$$

example 3 Find the HCF of

$$e^2 + er \qquad e^2 + 2er + r^2$$

and

$$e^2 - r^2$$

solution

$$e^2 + er = e(e + r)$$
$$e^2 + 2er + r^2 = (e + r)^2$$
$$e^2 - r^2 = (e + r)(e - r)$$
$$\therefore \text{HCF} = e + r$$

PROBLEMS 11-1

Find the HCF of:

1. 24, 40

2. 50, 125, 625

3. $4\theta^2\phi$, $12\theta\phi\omega$, $36\theta\phi^2\omega$

4. $16\lambda^2\omega$, $48\lambda^2\theta$, $36\lambda^2\phi$

5. $0.5a^3b^2c$, $0.25a^2b^2c^2$, $0.1a^2bc^3$

6. $39x^4y^2z^3$, $78x^3y^3z^3$, $156x^2y^4z^3$

7. $39I^2R$, $195I^2R^2$, $36IR$

8. $18\alpha\beta^2\gamma^3$, $162\alpha^2\beta^3\gamma$, $220\alpha\beta^3\gamma^2$

9. $X_L^2 - X_C^2$, $X_L^2 + X_LX_C$

10. $m^2 + 2mn + n^2$, $m^2 - n^2$

11. $E^2 - 2E + 1$, $E^2 - 1$, $3E^2 - 3E$

12. $12\pi + 4\phi$, $9\pi^2 + 6\pi\phi + \phi^2$, $9\pi^2 - \phi^2$

13. $L_1L_2 + 2\sqrt{L_1L_2}M + M^2$, $5\sqrt{L_1L_2} + 5M$

14. $9I^2R^2 - 24EIR + 16E^2$, $3I^2R^2 - 10EIR + 8E^2$, $15I^2R^2 - 17EIR - 4E^2$

15. $10I^2 + 25\frac{EI}{R} + 15\frac{E^2}{R^2}$, $30I + 45\frac{E}{R}$, $40I^2 + 40\frac{EI}{R} - 30\frac{E^2}{R^2}$

11-4 MULTIPLE

A number is a *multiple* of any one of its factors. For example, some of the multiples of 4 are 8, 16, 20, and 24. Similarly, some of the multiples of $a + b$ are $3(a + b)$, $a^2 + 2ab + b^2$, and $a^2 - b^2$. A *common multiple* of two or more numbers is a multiple of each of them. Thus, 45 is a common multiple of 1, 3, 5, 9, and 15.

11-5 LOWEST COMMON MULTIPLE

The smallest number that will contain each one of a set of factors is called their *lowest common multiple*. Thus, 48, 60, and 72 are all common multiples of 4 and 6, but the lowest common multiple of 4 and 6 is 12.

The lowest common multiple is abbreviated LCM.

example 4 Find the LCM of $6x^2y$, $9xy^2z$, and $30x^3y^3$.

solution

$$6x^2y = 2 \cdot 3 \cdot x^2y$$
$$9xy^2z = 3^2 \cdot xy^2z$$
$$30x^3y^3 = 2 \cdot 3 \cdot 5 \cdot x^3y^3$$

Because the LCM must contain *each* of the expressions, it must have 2, 3^2, and 5 as factors. Also, it must contain the literal factors of highest degree, or x^3y^3z.

$$\therefore \text{LCM} = 2 \cdot 3^2 \cdot 5 \cdot x^3y^3z = 90x^3y^3z$$

Rule

To determine the LCM of two or more expressions, determine all the prime factors of each expression. Find the product of all the different prime factors, taking each factor the greatest number of times it occurs in any one expression.

example 5 Find the LCM of

$$3a^3 + 6a^2b + 3ab^2$$
$$6a^4 - 12a^3b + 6a^2b^2$$
$$9a^3b - 9ab^3$$

solution

$$3a^3 + 6a^2b + 3ab^2 = 3a(a + b)^2$$
$$6a^4 - 12a^3b + 6a^2b^2 = 2 \cdot 3 \cdot a^2(a - b)^2$$
$$9a^3b - 9ab^3 = 3^2 \cdot ab(a + b)(a - b)$$
$$\therefore \text{LCM} = 2 \cdot 3^2 \cdot a^2b(a + b)^2(a - b)^2$$
$$= 18a^2b(a + b)^2(a - b)^2$$

PROBLEMS 11-2

Find the LCM of the following:

1. 12, 70, 210

2. 22, 154, 231

3. 40, 72, 180

4. a^3b^2c, a^2bc^3

5. $\theta^4\phi^3\lambda^2\omega$, $\theta^2\phi^2\lambda^3\mu\omega$

6. $2\alpha^4\beta^2$, $10\alpha^2\beta^2\gamma^3$, $15\alpha^3\beta^3\gamma$

7. $5m^3n^2p^2$, $20m^2np$, $45mnp^4$

8. I^2, $3IR$, $17I^2R^2$

9. $t - 3$, $t^2 - 5t + 6$

10. $X^2 - 11X + 30$, $X^2 - 9X + 20$

11. $\mu^2 + 3\mu$, $\mu^2 + 5\mu$, $\mu^2 + 8\mu + 15$

12. $6 + 4\psi$, $3 - 2\psi$, $9 - 12\psi + 4\psi^2$

13. $6\theta^2 + 7\theta - 3$, $44\theta^2 + 88\theta + 33$, $66\theta^2 + 11\theta - 11$

14. $4X_L^2 + 12X_LX_C + 8X_C^2$, $2X_L^2 + 10X_LX_C + 12X_C^2$, $X_L^2 + 4X_LX_C + 3X_C^2$

15. $8Q^2 - 38\dfrac{\omega LQ}{R} + 35\dfrac{\omega^2L^2}{R^2}$, $Q^2 - \dfrac{\omega^2L^2}{R^2}$, $2Q^2 - 9\dfrac{\omega LQ}{R} + 7\dfrac{\omega^2L^2}{R^2}$

11-6 DEFINITIONS

A fraction is an indicated division. Thus, we indicate 4 divided by 5 as $\frac{4}{5}$ (read four-fifths). Similarly, X *divided by* Y is written $\frac{X}{Y}$ (read X divided by Y or X over Y).

The quantity above the horizontal line is called the *numerator* and that below the line is called the *denominator* of the fraction. The numerator and denominator are often called the *terms* of the fraction.

11-7 OPERATIONS ON NUMERATOR AND DENOMINATOR

As in arithmetic, when fractions are to be simplified or affected by one of the four fundamental operations, we find it necessary to make frequent use of the following important principles:

1. The numerator and the denominator of a fraction can be multiplied by the same number or expression, except zero, without changing the value of the fraction.
2. The numerator and the denominator can be divided by the same number or expression, except zero, without changing the value of the fraction.

example 6

$$\frac{2}{3} = \frac{2 \times 3}{3 \times 3} = \frac{6}{9} = \frac{2}{3}$$

Also, $$\frac{6}{9} = \frac{6 \div 3}{9 \div 3} = \frac{2}{3} = \frac{6}{9}$$

example 7

$$\frac{x}{y} = \frac{x \cdot a}{y \cdot a} = \frac{ax}{ay} = \frac{x}{y}$$

Also, $$\frac{ax \div a}{ay \div a} = \frac{x}{y} \qquad \text{(where } a \neq 0\text{)}$$

No new principles are involved in performing these operations, for multiplying or dividing both numerator and denominator by the same number, except zero, is equivalent to multiplying or dividing the fraction by 1 in any form convenient for our use, such as

$$\frac{2}{2}, \frac{4}{4}, \frac{10}{10}$$

or $$\frac{-1}{-1}$$

It will be noted that, in the foregoing principles, multiplication and division of numerator and denominator by zero are excluded. When any expression is multiplied by zero, the product is zero. For example, $6 \times 0 = 0$. Therefore, if we multiplied both numerator and denominator of some fraction by zero, the result would be meaningless. Thus,

$$\frac{5}{6} \neq \frac{5 \times 0}{6 \times 0}$$

because $$\frac{5 \times 0}{6 \times 0} = \frac{0}{0}$$

Division by zero is meaningless. Some people say that any number divided by zero results in a quotient of infinity, denoted by ∞. If we accept this, we immediately impose a severe restriction on operations with even simple equations. For example, let us assume for the moment that any number divided by zero *does* result in infinity. Then if

$$\frac{4}{0} = \infty$$

by following Axiom 3, we should be able to multiply both sides of this equation by 0. If so, we obtain

$$4 = \infty \cdot 0$$

which we know is not sensible. Obviously, there is a fallacy here; therefore, we shall simply say at this time that *division by zero is not a permissible operation.*

11-8 EQUIVALENT FRACTIONS

Examples 6 and 7 show that when a numerator and a denominator are multiplied or divided by the same number, except zero, we change the *form* of the given fraction but *not* its value. Therefore, two fractions having the same value but not the same form are called *equivalent fractions.*

PROBLEMS 11-3

Supply the missing terms:

1. $\frac{3}{7} = \frac{?}{42}$

2. $\frac{7}{16} = \frac{?}{144}$

3. $\frac{1}{x} = \frac{?}{x^2y}$

4. $\frac{2\theta}{7\phi} = \frac{?}{35\phi\omega}$

5. $\frac{3ab}{25c} = \frac{?}{75cd}$

6. $\frac{\alpha}{\alpha + 3} = \frac{?}{(\alpha + 3)(\alpha - 2)}$

7. $\frac{t - 1}{t - 3} = \frac{?}{t^2 - 4t + 3}$

8. $\frac{7 + \theta}{\theta - 1} = \frac{?}{\theta^2 - 1}$

9. $\frac{i + \alpha}{\alpha - 3\beta} = \frac{?}{6\alpha - 18\beta}$

10. $\frac{x - 2y}{2x + y} = \frac{?}{2x^2 + 5xy + 2y^2}$

11. Change the fraction $\frac{3}{16}$ into an equivalent fraction whose denominator is 64.

12. Change the fraction $\frac{7}{25}$ into an equivalent fraction whose denominator is 150.

13. Change the fraction $\frac{\omega L}{R}$ into an equivalent fraction whose denominator is $R^3 - RX^2$

14. Change the fraction $\frac{L + 2}{L - 2}$ into an equivalent fraction whose denominator is $L^2 - 4$.

15. Change the fraction $\frac{Q}{EC + 1}$ into an equivalent fraction whose denominator is $2E^2C^2 - EC - 3$.

11-9 REDUCTION OF FRACTIONS TO THEIR LOWEST TERMS

If the numerator and denominator of a fraction have no common factor other than 1, the fraction is said to be in its lowest terms.

Thus, the fractions $\frac{2}{3}$, $\frac{3}{5}$, $\frac{x}{y}$, and $\frac{x + y}{x - y}$ are in their lowest terms, for the numerator and denominator of each fraction have no common factor except 1.

The fractions $\frac{4}{6}$ and $\frac{3x}{9x^2}$ are not in their lowest terms, for $\frac{4}{6}$ can be reduced to $\frac{2}{3}$ if both numerator and denominator are divided by 2. Similarly, $\frac{3x}{9x^2}$ can be reduced to $\frac{1}{3x}$ by dividing both numerator and denominator by $3x$.

Rule

To reduce a fraction to its lowest terms, factor the numerator and denominator into prime factors and cancel the factors common to both.

Cancellation as used in the rule really means that we actually *divide* both terms of the fraction by the *common factors*. Then, to reduce a fraction to its lowest terms, it is only necessary to divide both numerator and denominator by the highest common factor, which leaves an equivalent fraction.

example 8 Reduce $\frac{27}{108}$ to lowest terms.

solution $$\frac{27}{108} = \frac{\not{3} \cdot \not{3} \cdot \not{3}}{2 \cdot 2 \cdot \not{3} \cdot \not{3} \cdot \not{3}} = \frac{1}{4}$$

example 9 Reduce $\frac{24x^2yz^3}{42x^2yz^2}$ to lowest terms.

solution

$$\frac{24x^2yz^3}{42x^2yz^2} = \frac{\not{2} \cdot 2 \cdot 2 \cdot \not{3} \cdot x^2yz^3}{\not{2} \cdot \not{3} \cdot 7 \cdot x^2yz^2} = \frac{4z}{7}$$

Actually, the solution to Example 9 need not have been written out, for it can be seen by inspection that the HCF of both terms of the fraction is $6x^2yz^2$, which we divide into both terms to obtain the equivalent fraction $\frac{4z}{7}$.

Also, in reducing fractions, we may resort to direct cancellation as in arithmetic.

example 10 Reduce $\frac{x^2 - y^2}{x^3 - y^3}$ to lowest terms.

solution

$$\frac{x^2 - y^2}{x^3 - y^3} = \frac{(x+y)\cancel{(x-y)}}{\cancel{(x-y)}(x^2 + xy + y^2)} = \frac{x+y}{x^2 + xy + y^2}$$

example 11 Reduce to lowest terms

$$\frac{r^2 - R^2}{r^2 + 3rR + 2R^2}$$

solution

$$\frac{r^2 - R^2}{r^2 + 3rR + 2R^2} = \frac{\cancel{(r+R)}(r-R)}{(r+2R)\cancel{(r+R)}} = \frac{r-R}{r+2R}$$

PROBLEMS 11-4

Reduce to lowest terms:

1. $\frac{36}{48}$ **2.** $\frac{72}{729}$ **3.** $\frac{12}{156}$

4. $\frac{15}{240}$ **5.** $\frac{a^3b^2}{a^4b^5}$ **6.** $\frac{3\theta^2\phi}{12\theta\phi^3}$

7. $\frac{125I^2R}{25IR^2}$ **8.** $\frac{32\theta\lambda^3\mu\phi^2}{80\theta^2\lambda\mu\phi^3}$ **9.** $\frac{x^2}{x^3 + xy^2}$

10. $\dfrac{7.5p + 0.5q}{2.5pq}$ 11. $\dfrac{a^2 + 2ab + b^2}{a^2 - b^2}$ 12. $\dfrac{4m - 4n}{m^2 - n^2}$

13. $\dfrac{2x^2 + 5xy + 3y^2}{6x + 9y}$ 14. $\dfrac{\alpha^2 + 3\alpha\beta - 10\beta^2}{2\alpha^2 + 11\alpha\beta + 5\beta^2}$ 15. $\dfrac{\pi^2\omega^2 - 9\lambda^2\omega^2}{3\pi^2\omega - 8\pi\lambda\omega - 3\lambda^2\omega}$

11-10 SIGNS OF FRACTIONS

As stated in Sec. 11-6, a fraction is an indicated division or an indicated quotient. Heretofore, all our fractions have been positive, but now we must take into account three signs in working with an algebraic fraction: the sign of the numerator, the sign of the denominator, and the sign preceding the fraction. By the law of signs in division, we have

$$+\frac{+12}{+6} = +\frac{-12}{-6} = -\frac{+12}{-6} = -\frac{-12}{+6} = +2$$

or, in general,

$$+\frac{+a}{+b} = +\frac{-a}{-b} = -\frac{+a}{-b} = -\frac{-a}{+b}$$

Careful study of the above examples will show the truths of the following important principles:

1. The sign before either term of a fraction can be changed if the sign before the fraction is changed.
2. If the signs of both terms are changed, the sign before the fraction must not be changed.

That is, we can change *any two* of the three signs of a fraction without changing the value of the fraction.

It must be remembered that, when a term of a fraction is a polynomial, changing the sign of the term involves changing the sign of *each term* of the polynomial.

Changing the signs of both numerator and denominator, as mentioned in the second principle above, can be explained by considering both terms as multiplied or divided by -1, which, as previously explained, does not change the value of the fraction.

Multiplying (or dividing) a quantity by -1 twice does not change the value of the quantity. Hence, multiplying each of the two factors of a product by -1 does not change the value of the product. Thus,

$$(a - 4)(a - 8) = (-a + 4)(-a + 8)$$
$$= (4 - a)(8 - a)$$

Also,

$$(a - b)(c - d)(e - f) = (b - a)(d - c)(e - f)$$

The validity of these illustrations should be checked by multiplication.

example 12 Change $-\dfrac{a}{b}$ to three equivalent fractions having different signs.

solution $-\dfrac{a}{b} = \dfrac{-a}{b} = \dfrac{a}{-b} = -\dfrac{-a}{-b}$

example 13 Change $\dfrac{a - b}{c - d}$ to three equivalent fractions having different signs.

solution

$$\frac{a - b}{c - d} = \frac{-a + b}{-c + d}$$
$$= -\frac{-a + b}{c - d} = -\frac{a - b}{-c + d}$$

example 14 Change $\frac{a-b}{c-d}$ to a fraction whose denominator is $d - c$.

solution $\frac{a-b}{c-d} = \frac{-a+b}{-c+d} = \frac{b-a}{d-c}$

PROBLEMS 11-5

Express as fractions with positive numerators:

1. $-\frac{-a}{x}$
2. $\frac{-IR}{E_1 - e}$
3. $\frac{-2\pi fL}{X_L - X_C}$
4. $-\frac{\sqrt{L_1L_2}}{\omega L}$
5. $\frac{-\omega L}{R_1 - R_2}$
6. $\frac{-\pi - \omega}{\alpha - \beta}$

Express as fractions with positive denominators:

7. $\frac{IR}{-E - e}$
8. $\frac{\mu E_g}{-(R_p + R_L)}$
9. $\frac{\pi R^2}{-(A_1 - A_2)}$
10. $-\frac{\theta + \phi}{-2\lambda^2}$

Reduce to lowest terms:

11. $\frac{a-b}{b-a}$
12. $\frac{I - i}{-(i^2 - I^2)}$
13. $\frac{\theta - \phi}{\phi^2 - \theta^2}$
14. $\frac{x^2 - 2xy + y^2}{y^2 - 2yx + x^2}$
15. $\frac{\pi^2 - 8\pi + 16}{20 - \pi - \pi^2}$
16. $-\frac{4s^2t^2 + 3stv - v^2}{2s^2t^2 + stv - v^2}$

11-11 COMMON ERRORS IN WORKING WITH FRACTIONS

It has been demonstrated that a fraction may be reduced to lower terms by dividing both numerator and denominator by the same number (Sec. 11-9). Mistakes are often made by canceling parts of numerator and denominator that are not factors. For example,

$$\frac{5+2}{7+2} = \frac{7}{9}$$

Here is a case in which both terms of the fraction are polynomials and the terms, even if alike, can never be canceled. Thus,

$$\frac{5+\not{2}}{7+\not{2}} \neq \frac{5}{7}$$

because canceling terms has changed the value of the fraction. Similarly, it would be incorrect to cancel the x's in the fraction $\frac{6a - x}{6b - x}$, for the x's are not factors. At the same time, it is incorrect to cancel the 6's because, although they are factors of terms in the fraction, they are not factors of the complete numerator and denominator. Therefore, it is apparent that $\frac{6a - x}{6b - x}$ cannot be reduced to lower terms, for neither term (numerator or denominator) can be factored.

It is permissible to cancel x's in the fraction $\frac{6x}{ax + 5x}$, because each term of the denominator contains the common factor x.

The denominator may be factored to give $\frac{6x}{x(a+5)}$, the result being that x is a factor in both terms of the fraction. Note, however, that the single x in the numerator cancels both x's in the denominator.

Thus, we cannot remove, or cancel, like *terms* from the numerator and denominator of a fraction. Only like *factors* can be removed, or canceled.

Another important fact to be remembered is that adding the same number to or subtracting the same number from both numerator and denominator changes the value of the fraction. That is,

$$\frac{3}{4} \neq \frac{3+2}{4+2} \qquad \text{because the latter equals } \frac{5}{6}$$

Likewise,

$$\frac{3}{4} \neq \frac{3-2}{4-2} \qquad \text{because the latter equals } \frac{1}{2}$$

Similarly, squaring or extracting the same root of numerator and denominator results in a different value. For example,

$$\frac{3}{4} \neq \frac{3^2}{4^2} \qquad \text{because the latter equals } \frac{9}{16}$$

Likewise,

$$\frac{16}{25} \neq \frac{\sqrt{16}}{\sqrt{25}} \qquad \text{because the latter equals } \frac{4}{5}$$

Students sometimes thoughtlessly make the error of writing 0 (zero) as the result of the cancellation of all factors. For example,

$$\frac{4x^2y(a+b)}{4x^2y(a+b)} = 1, \textit{ not } 0$$

Another serious, although common, mistake is forgetting that the fraction bar, or vinculum, is a sign of grouping, so that $-\frac{x-y}{x}$ really means $-\left(\frac{x-y}{x}\right)$, or $-\left(\frac{x}{x}-\frac{y}{x}\right)$, or $-\left(1-\frac{y}{x}\right)$, and it does not reduce to $-(1-y)$.

Note that the *vinculum* is a sign of grouping and, when a minus sign precedes a fraction having a polynomial numerator, all the signs of the numerator must be changed in order to complete the process of subtraction.

Thus, $-\frac{x-y}{x}$ simplifies to $\frac{y}{x} - 1$.

11-12 CHANGING MIXED EXPRESSIONS TO FRACTIONS

In arithmetic, an expression such as $3\frac{1}{3}$ is called a *mixed number;* $3\frac{1}{3}$ means $3 + \frac{1}{3}$. Similarly, in algebra, an expression such as $x + \frac{y}{z}$ is called a *mixed expression.* Because

$$4\frac{2}{3} = 4 + \frac{2}{3} = \frac{4}{1} + \frac{2}{3} = \frac{12}{3} + \frac{2}{3} = \frac{14}{3}$$

then, $$x + \frac{y}{z} = \frac{x}{1} + \frac{y}{z} = \frac{xz}{z} + \frac{y}{z} = \frac{xz+y}{z}$$

Also, $$3x^2 - 4x + \frac{3}{x^2-1}$$
$$= \frac{3x^2}{1} - \frac{4x}{1} + \frac{3}{x^2-1}$$
$$= \frac{3x^2(x^2-1)}{x^2-1} - \frac{4x(x^2-1)}{x^2-1} + \frac{3}{x^2-1}$$
$$= \frac{3x^4 - 3x^2 - 4x^3 + 4x + 3}{x^2-1}$$

11-13 REDUCTION OF A FRACTION TO A MIXED EXPRESSION

As would be expected, reducing a fraction to a mixed expression is the reverse of changing a mixed expression to a fraction. That is, a fraction can be changed to a mixed expression by dividing the numerator by the denominator and adding to the quotient thus obtained the remainder, which is written as a fraction.

example 15

Change $\dfrac{12x^3 + 16x^2 - 8x - 3}{4x}$ to a mixed expression.

solution Divide each term of the numerator by the denominator.

Thus,

$$\frac{12x^3 + 16x^2 - 8x - 3}{4x} = 3x^2 + 4x - 2 - \frac{3}{4x}$$

example 16 Change $\dfrac{a^2 + 1}{a - 2}$ to a mixed expression.

solution By division,

$$\begin{array}{r|l} a^2 \qquad + 1 & a - 2 \\ \cline{2-2} & a + 2 \\ a^2 - 2a & \\ \cline{1-1} 2a + 1 & \\ 2a - 4 & \\ \cline{1-1} 5 & \end{array}$$

$$\therefore \frac{a^2 + 1}{a - 2} = a + 2 + \frac{5}{a - 2}$$

PROBLEMS 11-6

Change the following mixed expressions to fractions:

1. $2\frac{1}{8}$ **2.** $5\frac{3}{16}$ **3.** $a + \frac{b}{c}$

4. $R - \frac{E}{I}$ **5.** $4 - \frac{5}{F}$ **6.** $\frac{3}{Q} - \frac{5}{Q^2}$

7. $4 + \frac{2}{\pi + 1}$ **8.** $\theta + \frac{\theta}{2\pi}$ **9.** $R - 1 - \frac{E}{I}$

10. $5 + \frac{5x - 30}{x^2 - 2x}$ **11.** $\frac{9}{x^2} - \frac{14}{2x} - 2$ **12.** $4 - \frac{4}{c} - \frac{8}{c^2}$

13. $1 + \frac{6}{R} - \frac{7}{R^2}$ **14.** $\frac{a + b}{4} - \frac{a - b}{8}$ **15.** $1 - \frac{4\lambda + 1}{9\lambda^2 - 1}$

16. $\frac{x - 1}{2x} - \frac{x^2 - 1}{3x^2}$ **17.** $\frac{45}{\theta^2} + \frac{14}{\theta} - \frac{\theta + 1}{\theta - 1}$ **18.** $2 - \frac{12Q - 2}{Q^2 - 1}$

19. $2a^2 - 1 - \frac{4}{a^2 - 3}$ **20.** $1 - \frac{50\omega\pi - 30\pi^2}{(5\omega - 3\pi)(3\omega + 5\pi)}$

Reduce the following fractions to mixed expressions:

21. $\frac{83}{16}$ **22.** $\frac{231}{32}$

23. $\frac{x^2 + y^2}{x^2}$ **24.** $\frac{32a^2 - 16a + 4}{4a}$

25. $\dfrac{R^3 + 6R^2 + 7R - 8}{R - 1}$

26. $\dfrac{x^2 + 5x + 6}{x - 1}$

27. $\dfrac{E^4 - e^4 - 1}{E + e}$

28. $\dfrac{6\phi^5 - \phi^4 + 4\phi^3 - 5\phi^2 - \phi + 20}{2\phi^2 - \phi + 3}$

29. $\dfrac{2x^3 + 2x^2 + x + 2}{x^2 + 1}$

30. $\dfrac{2\alpha^3 + \alpha\beta^2}{\alpha + \beta}$

11-14 REDUCTION TO THE LOWEST COMMON DENOMINATOR

The *lowest common denominator* (LCD) of two or more fractions is the lowest common multiple of their denominators.

example 17 Reduce $\frac{1}{3}$ and $\frac{3}{5}$ to their LCD.

solution The LCM of 3 and 5 is 15. To change the denominator of $\frac{1}{3}$ to 15, we must multiply the 3 by 5 (15 ÷ 3). So that the value of the fraction will not be changed, we must also multiply the numerator by 5. Hence,

$$\frac{1}{3} = \frac{1}{3} \times \frac{5}{5} = \frac{5}{15}$$

For the second fraction, we must multiply the denominator by 3 in order to obtain a new denominator of 15 (15 ÷ 5). Again we must also multiply the numerator by 3 to maintain the original value of the fraction. Hence,

$$\frac{3}{5} = \frac{3}{5} \times \frac{3}{3} = \frac{9}{15}$$

example 18 Reduce $\dfrac{4a^2b}{3x^2y}$ and $\dfrac{6cd^2}{4xy^2}$ to their LCD.

solution The LCM of the two denominators is $12x^2y^2$. This is the LCD.

For the first fraction the LCD is divided by the denominator. That is,

$$12x^2y^2 \div 3x^2y = 4y$$

Multiplying both numerator and denominator by $4y$, we have

$$\frac{4a^2b}{3x^2y} = \frac{4a^2b}{3x^2y} \cdot \frac{4y}{4y} = \frac{16a^2by}{12x^2y^2}$$

For the second fraction we follow the same procedure.

$$12x^2y^2 \div 4xy^2 = 3x$$

Multiplying both numerator and denominator by $3x$, we have

$$\frac{6cd^2}{4xy^2} = \frac{6cd^2}{4xy^2} \cdot \frac{3x}{3x} = \frac{18cd^2x}{12x^2y^2}$$

Rule

To reduce fractions to their LCD:

1. Factor each denominator into its prime factors and find the LCM of the denominators. This is the LCD.
2. For each fraction, divide the LCD by the denominator and multiply both numerator and denominator by the quotient thus obtained.

example 19 Reduce $\dfrac{3x}{x^2 - y^2}$ and $\dfrac{4y}{x^2 - xy - 2y^2}$ to their LCD.

solution

$$\frac{3x}{x^2 - y^2} = \frac{3x}{(x + y)(x - y)}$$

$$\frac{4y}{x^2 - xy - 2y^2} = \frac{4y}{(x + y)(x - 2y)}$$

The LCM of the two denominators, and therefore the LCD, is $(x + y)(x - y)(x - 2y)$.

For the first fraction, the LCD divided by the denominator is:

$$(x + y)(x - y)(x - 2y) \div (x + y)(x - y) = x - 2y.$$

$$\therefore \frac{3x}{(x + y)(x - y)} = \frac{3x(x - 2y)}{(x + y)(x - y)(x - 2y)}$$

For the second fraction, the LCD divided by the denominator is:

$$(x + y)(x - y)(x - 2y) \div (x + y)(x - 2y) = x - y.$$

$$\therefore \frac{4y}{(x + y)(x - 2y)} = \frac{4y(x - y)}{(x + y)(x - 2y)(x - y)}$$

To check the solution, the fractions having the LCD can be changed into the original fractions by cancellation.

PROBLEMS 11-7

Convert the following sets of fractions to equivalent sets having their LCD:

1. $\frac{1}{2}, \frac{3}{7}, \frac{2}{5}$
2. $\frac{3}{16}, \frac{5}{8}, \frac{7}{32}$
3. $\frac{3}{4}, \frac{7}{16}, \frac{5}{12}$
4. $\frac{1}{x}, \frac{1}{y}$
5. $\frac{\theta}{\phi}, \frac{\lambda}{\omega}$
6. $\frac{1}{ir}, \frac{1}{\omega}, \frac{i}{\omega}$
7. $\frac{e}{r}, \frac{1}{ir}, ei$
8. $\frac{Q}{L_1}, \frac{1}{L_2}, \frac{\sqrt{L_1L_2}}{M}$
9. $\frac{1}{a - b}, \frac{1}{a + b}$
10. $\frac{x}{y}, \frac{2x + y}{x - y}$
11. $\frac{3}{\phi - \pi}, \frac{4}{\phi + \pi}$
12. $\frac{3\phi}{1 - \phi^2}, \frac{2}{\phi + 1}, \frac{2}{1 - \phi}$
13. $\frac{a}{c + d}, \frac{b}{c - d}, \frac{a - b}{d - c}$
14. $\frac{1}{2M + 2}, \frac{5}{3M - 3}, \frac{3M - 1}{1 - M^2}$
15. $\frac{\pi^2 - \phi^2}{\pi\phi}, \frac{\pi\phi - \phi^2}{\pi\phi - \pi^2}$
16. $\frac{R + 3Z}{4R^2 + 12RZ + 8Z^2}, \frac{R + Z}{4R^2 + 20RZ + 24Z^2}, \frac{R + 2Z}{R^2 + 4RZ + 3Z^2}$

11-15 ADDITION AND SUBTRACTION OF FRACTIONS

The sum of two or more fractions having the same denominator is obtained by adding the numerators and writing the result over the common denominator.

example 20 $\frac{2}{7} + \frac{1}{7} + \frac{5}{7} = \frac{2 + 1 + 5}{7}$

$$= \frac{8}{7}$$

example 21

$$\frac{3e}{R+r}+\frac{e}{R+r}+\frac{5e}{R+r}=\frac{3e+e+5e}{R+r}$$
$$=\frac{9e}{R+r}$$

To subtract two fractions having the same denominator, subtract the numerator of the subtrahend from the numerator of the minuend and write the result over their common denominator.

example 22 $\frac{4}{5}-\frac{3}{5}=\frac{4-3}{5}=\frac{1}{5}$

example 23 $\frac{a}{x}-\frac{b}{x}=\frac{a-b}{x}$

example 24 $\frac{a}{x}-\frac{b-c}{x}=\frac{a-b+c}{x}$

Note that *the vinculum is a sign of grouping* and that, when a minus sign precedes a fraction having a polynomial numerator, all the signs in the numerator must be changed in order to complete the process of subtraction.

We thus have the following rules:

Rule

To add or subtract fractions having unlike denominators:

1. Reduce them to equivalent fractions having their LCD.
2. Combine the numerators of these equivalent fractions, in parentheses, give each the sign of the fraction. This is the numerator of the result.
3. The denominator of the result is the LCD.
4. Simplify the numerator by removing parentheses and combining terms.
5. Reduce the fraction to the lowest terms.

example 25 Simplify $\frac{a-5}{6x}-\frac{2a-5}{16x}$

solution

$$\frac{a-5}{6x}-\frac{2a-5}{16x}=\frac{8(a-5)}{48x}-\frac{3(2a-5)}{48x}$$
$$=\frac{8(a-5)-3(2a-5)}{48x}$$
$$=\frac{8a-40-6a+15}{48x}$$
$$=\frac{2a-25}{48x}$$

check Let $a=6$, $x=1$.

$$\frac{a-5}{6x}=\frac{1}{6} \qquad \frac{2a-5}{16}=\frac{7}{16}$$
$$\frac{1}{6}-\frac{7}{16}=\frac{8-21}{48}=-\frac{13}{48}$$

Also, $\frac{2a-25}{48}=\frac{12-25}{48}=-\frac{13}{48}$

Solution is correct.

example 26

Simplify $x^2-xy+y^2-\frac{2y^3}{x+y}$

solution

$$x^2-xy+y^2-\frac{2y^3}{x+y}$$
$$=\frac{(x+y)x^2}{x+y}-\frac{(x+y)xy}{x+y}+\frac{(x+y)y^2}{x+y}-\frac{2y^3}{x+y}$$
$$=\frac{x^3+x^2y-x^2y-xy^2+xy^2+y^3-2y^3}{x+y}$$
$$=\frac{x^3-y^3}{x+y}$$

PROBLEMS 11-8

Perform the following indicated additions and subtractions:

1. $\frac{1}{2} + \frac{2}{5} - \frac{3}{7}$
2. $\frac{7}{32} - \frac{5}{8} + \frac{3}{16}$
3. $\frac{3}{4} - \frac{7}{16} - \frac{5}{12}$
4. $\frac{5a}{4} - \frac{a}{5} + \frac{7a}{3}$
5. $\frac{7IR}{8} + \frac{2IR}{3} - \frac{3IR}{16}$
6. $\frac{1}{I} + \frac{1}{i}$
7. $\frac{\alpha}{\beta} - \frac{\gamma}{\delta}$
8. $\frac{3p}{4q} - \frac{p}{6q} - \frac{5p}{30q}$
9. $\frac{a}{x} - \frac{b}{y} - \frac{c}{z}$
10. $\frac{3}{r_1} + \frac{2}{r_2} + \frac{5}{r_1 r_2}$
11. $\frac{10}{I^2} - \frac{3}{R} + \frac{4}{I^2 R}$
12. $\frac{3\alpha}{\phi\lambda} + \frac{2\phi}{\alpha\lambda} + \frac{6\lambda}{\alpha\phi}$
13. $\frac{3I - i}{2} + \frac{5I + 2i}{3}$
14. $\frac{a + 4}{7} - \frac{a - 1}{3}$
15. $\frac{2}{\alpha - \beta} + \frac{1}{\alpha + \beta}$
16. $\frac{3}{2e + 4} + \frac{5}{e + 2}$
17. $\frac{5}{L_1 - 2} - \frac{2}{L_1 + 6}$
18. $\frac{a}{c + d} + \frac{b}{c - d} - \frac{a - b}{d - c}$
19. $\frac{1}{2\theta + 2} - \frac{5}{3\theta - 3} + \frac{3\theta - 1}{1 - \theta^2}$
20. $\frac{8}{\alpha^2 - 9} - \frac{2}{\alpha^2 - 5\alpha + 6}$
21. $\frac{2}{I^2 + 7I} - \frac{3}{I} + \frac{3}{I - 7}$
22. $\frac{11R_1 - 2}{3R_1^2 - 3} - \frac{5R_1 + 1}{2R_1^2 - 2}$
23. $\frac{21}{14 - \pi} - \frac{35 - 2\pi^2}{\pi^2 - 11\pi - 42}$
24. $\frac{2L - 4M}{2L - 2M} - \frac{3M^2 - 3LM}{L^2 - 2LM + M^2}$
25. $\frac{\theta + \phi}{\theta - \phi} - \frac{\theta - \phi}{\theta + \phi} + \frac{4\theta\phi}{\theta^2 - \phi^2}$
26. $\frac{2X_C}{2X_C + 3X_L} - \frac{3X_L}{2X_C - 3X_L} + \frac{8X_L^2}{4X_C^2 - 9X_L^2}$
27. $\frac{E - 1}{E^2 - 9E + 20} - \frac{E + 1}{E^2 - 11E + 30}$
28. $a + b - \frac{a^2 - b^2}{a - b} + 1$
29. $\frac{\omega^2 + 3\omega + 9}{\theta^2 - 3\omega + 9} - \frac{54}{\omega^3 + 27} - \frac{\omega - 3}{\omega + 3}$
30. $\frac{\theta + 3\pi}{4\theta^2 + 12\theta\pi + 8\pi^2} + \frac{\theta + 2\pi}{\theta^2 + 4\theta\pi + 3\pi^2} - \frac{\theta + \pi}{4\theta^2 + 20\theta\pi + 24\pi^2}$

11-16 MULTIPLICATION OF FRACTIONS

The methods of multiplication of fractions in algebra are identical with those in arithmetic.

The product of two or more fractions is the product of their numerators divided by the product of their denominators.

example 27 $\frac{2}{3} \times \frac{3}{5} = \frac{6}{15}$

example 28 $\frac{a}{b} \cdot \frac{x}{y} = \frac{ax}{by}$

When a factor occurs one or more times in *any* numerator and in *any* denominator of the product of two or more fractions, it can be canceled the same number of times from both. This process results in the product of the given fractions in lower terms.

example 29 Multiply $\frac{6x^2y}{7b}$ by $\frac{21b^2c}{24xy^2}$.

solution $\frac{6x^2y}{7b} \cdot \frac{21b^2c}{24xy^2} = \frac{3bcx}{4y}$

example 30 Simplify

$$\frac{2a^2 - ab - b^2}{a^2 + 2ab + b^2} \cdot \frac{a^2 - b^2}{4a^2 + 4ab + b^2}.$$

solution

$$\frac{2a^2 - ab - b^2}{a^2 + 2ab + b^2} \cdot \frac{a^2 - b^2}{4a^2 + 4ab + b^2}$$
$$= \frac{(2a + b)(a - b)}{(a + b)(a + b)} \cdot \frac{(a + b)(a - b)}{(2a + b)(2a + b)}$$
$$= \frac{(a - b)(a - b)}{(a + b)(2a + b)} = \frac{a^2 - 2ab + b^2}{2a^2 + 3ab + b^2}$$

It is very important that you understand clearly what we are allowed to cancel in the numerators and the denominators. The *whole* of an expression is always canceled, *never one term*. For example, in the expression $\frac{8a}{a - 5}$, it is not permissible to cancel the a's and obtain $\frac{8}{-5}$. It must be remembered that the denominator $a - 5$ denotes *one quantity*. Because of the parentheses, we would not cancel the a's if the expression were written $\frac{8a}{(a - 5)}$. However, the parentheses are not needed; for the *vinculum, which is also a sign of grouping, serves the same purpose*. We will consider this again in the next chapter.

11-17 DIVISION OF FRACTIONS

As with multiplication, the methods of division of fractions in algebra are identical with those of arithmetic. Therefore, to divide by a fraction, invert the divisor fraction and proceed as in the multiplication of fractions.

example 31 $\frac{5}{2} \div \frac{2}{3} = \frac{5}{2} \cdot \frac{3}{2} = \frac{15}{4}$

example 32
$$\frac{ab^2}{xy} \div \frac{a^2b}{xy^2} = \frac{ab^2}{xy} \cdot \frac{xy^2}{a^2b}$$
$$= \frac{by}{a}$$

example 33

$$\frac{x}{y} \div \left(a + \frac{b}{c}\right) = \frac{x}{y} \div \frac{ac + b}{c}$$
$$= \frac{x}{y} \cdot \frac{c}{ac + b} = \frac{cx}{y(ac + b)} = \frac{cx}{acy + by}$$

Students often ask why we must invert the divisor and multiply by the dividend in dividing fractions. As an example, suppose we have $\frac{a}{b} \div \frac{x}{y}$. The dividend is $\frac{a}{b}$, and the divisor is $\frac{x}{y}$. Now

$$\text{Quotient} \times \text{divisor} = \text{dividend}$$

Therefore, the quotient must be a number such that, when multiplied by $\frac{x}{y}$, it will give $\frac{a}{b}$ as a product. Then,

$$\left(\frac{a}{b}\cdot\frac{y}{x}\right)\cdot\frac{x}{y}=\frac{a}{b}$$

Hence, the quotient is $\frac{a}{b}\cdot\frac{y}{x}$, which is the dividend multiplied by the inverted divisor.

PROBLEMS 11-9

Simplify:

1. $\frac{2}{3}\times\frac{5}{7}\times\frac{21}{40}$

2. $\frac{12}{35}\times\frac{5}{24}\times\frac{42}{21}$

3. $\frac{5}{16}\times\frac{6}{25}\times\left(\frac{-4}{15}\right)$

4. $\frac{5}{9}\div\frac{15}{18}$

5. $\frac{7}{8}\div\frac{7}{32}$

6. $-\frac{2}{3}\left(-\frac{5}{16}\div\frac{15}{64}\right)$

7. $\frac{4x^3}{5y}\times\frac{15y^4}{3x^2}$

8. $3p\left(\frac{5r}{6p^2}\times\frac{7pr}{15}\right)$

9. $\frac{40\theta\phi^2\omega}{21\theta^2\phi^3\omega^2}\div\frac{10\theta^3\phi\omega^2}{21\theta\phi^2\omega}$

10. $\frac{\pi r^2h}{3}\div 2\pi r$

11. $\frac{\omega L}{R}\div 2\pi fL$

12. $\left(\frac{m^2+4m}{m}\right)\left(\frac{m^2}{m^3+4m^2}\right)$

13. $\frac{4}{x-y}\div\frac{x^2-y^2}{x^2+2xy+y^2}$

14. $\frac{4\theta^2-1}{\theta^3-16\theta}\div\frac{2\theta-1}{\theta-4}$

15. $\frac{25x^2-y^2}{9x^2z-4z}\div\frac{5x-y}{3xz-2z}$

16. $\frac{I^2-4i^2}{Ii+2i^2}\cdot\frac{2i}{I-2i}$

17. $\frac{\lambda^2-2\lambda\mu+\mu^2}{4\lambda-4\mu}\cdot\frac{4\lambda+4\mu}{\phi^3-3\phi^2+2\phi}\cdot\frac{\phi^3-\phi^2}{\phi\lambda^2-\phi\mu^2}$

18. $\frac{F^2+2F+1}{P^3-PZ^2}\cdot\frac{P^2-Z^2}{5F^3+10F^2+5F}\cdot\frac{F^2P-10FP+25P}{F^2-110F+525}$

19. $\frac{\theta^2-2\theta-3}{-6\phi^2}\cdot\frac{5\theta\phi^6+25\phi^6}{\theta^2\phi-8\theta+\theta\phi-8}\cdot\frac{48\phi^2-6\theta\phi^3}{5\theta^2\phi^3+10\theta\phi^3-75\phi^3}$

20. $\frac{R^2-r^2}{r^2+Rr}\cdot\frac{R(R-r)}{(R-r)^2}\div\frac{R^2-3Rr+2r^2}{Rr-2r^2}$

21. $\frac{a^2-6a+8}{a^2}\cdot\frac{7a^4+7a^3}{a^2-11a+28}\div\frac{a^2-a-2}{2a^2-14a}$

22. $\frac{16I^4R^2-9}{4(I^2R+\frac{3}{4})}\cdot\frac{I^4R^2-3I^2R-28}{2I^4R^2-32}\div\frac{8I^4R^2-62I^2R+42}{8I^2R-32}$

23. $\left(4-\frac{4}{c}-\frac{8}{c^2}\right)\left(\frac{3c^4-6c^3}{2c^2-2c-4}\right)\left(\frac{2c^2+8c}{3c^3+6c^2-24c}\right)$

24. $\left(m - \dfrac{m^2}{m}\right)\left(\dfrac{m^2 - n^2}{m^2 + mn}\right)\left(\dfrac{m + n}{m^2 + mn}\right)$

25. $\left(\dfrac{5\phi^5 - 5\phi^4}{\phi^2 - \phi - 20}\right)\left(\dfrac{\phi^2 + 11\phi + 28}{5\phi - 5}\right) \div \left(\dfrac{\phi^4 + 9\phi^3 + 14\phi^2}{\phi^2 - 3\phi - 10}\right)$

26. $\left(\dfrac{I^2 + I - 6}{I^4 - 9I^2}\right)\left(I^2 + 4I + \dfrac{12I}{I - 3}\right) \div \left(\dfrac{I^2 - I - 2}{I^2 - 6I + 9}\right)$

27. $\left(\dfrac{\omega L + R}{2} + \dfrac{\omega L - R}{4}\right)\left(\dfrac{4}{9\omega^2 L^2 + 6\omega LR + R^2}\right)$

28. $\left(\dfrac{45}{\theta^2} + \dfrac{14}{\theta} + 1\right)\left(\dfrac{3\theta^3 + 6\theta^2}{\theta^2 + 18\theta + 81}\right)\left(\dfrac{\theta^2 + 13\theta + 36}{\theta^2 + 9\theta + 20}\right)\left(\dfrac{1}{3\theta + 3}\right)$

29. $\left(\dfrac{6m^2 - 2m}{-9m^2 + 4m + 2}\right)\left(\dfrac{2}{m^2} + \dfrac{10}{m} + 12\right)\left(\dfrac{4m + 1}{9m^2 - 1} - 1\right)\left(\dfrac{m^3}{4m^2 + 2m}\right)$

30. $\left(\dfrac{1}{f^2} + \dfrac{2}{f} + 1\right)\left(\dfrac{f^3 - f^2}{f^2 - 5f - 6}\right)\left(2 - \dfrac{12f - 2}{f^2 - 1}\right)$

11-18 COMPLEX FRACTIONS

A *complex fraction* is one with one or more fractions in its numerator, denominator, or both. The name is an unfortunate one. There is nothing complex or intricate about such compounded fractions, as we shall see.

Rule

To simplify a complex fraction, reduce both numerator and denominator to simple fractions; then perform the indicated division.

example 34 Simplify $\dfrac{\dfrac{1}{3} + \dfrac{1}{5}}{4 - \dfrac{1}{5}}$.

solution

$$\frac{\dfrac{1}{3} + \dfrac{1}{5}}{4 - \dfrac{1}{5}} = \frac{\dfrac{5 + 3}{15}}{\dfrac{20 - 1}{5}}$$

$$= \frac{\dfrac{8}{15}}{\dfrac{19}{5}} = \frac{8}{15} \times \frac{5}{19} = \frac{8}{57}$$

example 35 Simplify $\dfrac{5 - \dfrac{1}{a + 1}}{3 + \dfrac{2}{a + 1}}$.

solution

$$\frac{5 - \dfrac{1}{a + 1}}{3 + \dfrac{2}{a + 1}} = \frac{\dfrac{5(a + 1) - 1}{a + 1}}{\dfrac{3(a + 1) + 2}{a + 1}}$$

$$= \frac{\dfrac{5a + 4}{a + 1}}{\dfrac{3a + 5}{a + 1}}$$

$$= \frac{5a + 4}{a + 1} \cdot \frac{a + 1}{3a + 5}$$

$$= \frac{5a + 4}{3a + 5}$$

note It is evident that if the same factor occurs in both numerators of a complex fraction, the factors can be canceled. Also, if a factor occurs in both denominators, it can be canceled. Thus, $(a + 1)$ could have been canceled in Example 35 after the numerators and denominators were reduced from mixed expressions to simple fractions.

example 36 Simplify $\dfrac{\dfrac{a}{b}+\dfrac{a+b}{a-b}}{\dfrac{a}{b}-\dfrac{a-b}{a+b}}$.

solution

$$\frac{\dfrac{a}{b}+\dfrac{a+b}{a-b}}{\dfrac{a}{b}-\dfrac{a-b}{a+b}}=\frac{\dfrac{a(a-b)+b(a+b)}{b(a-b)}}{\dfrac{a(a+b)-b(a-b)}{b(a+b)}}$$

$$=\frac{\dfrac{a^2-ab+ab+b^2}{b(a-b)}}{\dfrac{a^2+ab-ab+b^2}{b(a+b)}}$$

$$=\frac{\dfrac{a^2+b^2}{b(a-b)}}{\dfrac{a^2+b^2}{b(a+b)}}=\frac{a+b}{a-b}$$

PROBLEMS 11-10

Simplify:

1. $\dfrac{2+\frac{1}{3}}{\frac{1}{3}-3}$

2. $\dfrac{2}{\frac{1}{7}+\frac{1}{2}}$

3. $\dfrac{(\frac{5}{8})^2-\frac{16}{25}}{\frac{5}{8}+\frac{4}{5}}$

4. $\dfrac{\theta+\dfrac{1}{\phi}}{\theta-\dfrac{1}{\phi}}$

5. $\dfrac{Q}{\dfrac{1}{\omega L_1}+\dfrac{1}{\omega L_2}}$

6. $\dfrac{\dfrac{i^2}{8}-8}{1+\dfrac{i}{8}}$

7. $\dfrac{I}{I-\dfrac{E}{r}}$

8. $\dfrac{5\theta+\dfrac{2\lambda}{5\phi}}{\dfrac{2\lambda}{5\theta}+5\phi}$

9. $\dfrac{\dfrac{E^2}{e^2}-1}{\dfrac{E^2+e^2}{2Ee}+1}$

10. $\dfrac{\dfrac{\lambda+\pi}{\lambda^2+\pi^2}-\dfrac{1}{\lambda+\pi}}{\dfrac{1}{\lambda+\pi}-\dfrac{\lambda}{\lambda^2+\pi^2}}$

11. $\dfrac{\dfrac{l}{l+w}}{1+\dfrac{w}{l-w}}$

12. $\dfrac{\omega+2-\dfrac{15}{\omega}}{1-\dfrac{8}{\omega}+\dfrac{15}{\omega^2}}$

13. $\dfrac{1-\dfrac{a-b}{a+b}}{1+\dfrac{a-b}{a+b}}$

14. $\dfrac{\dfrac{\theta}{\theta+\phi}-\dfrac{\theta}{\theta-\phi}}{\dfrac{\theta}{\theta+\phi}+\dfrac{\theta}{\theta-\phi}}$

15. $\dfrac{\dfrac{I-i}{I+i}+\dfrac{I+i}{I-i}}{\dfrac{I-i}{I+i}-\dfrac{I+i}{I-i}}$

16. $\dfrac{L_1}{Q-\dfrac{1}{Q+\dfrac{1}{Q}}}-\dfrac{L_1}{Q+\dfrac{1}{Q-\dfrac{1}{Q}}}$

12

fractional equations

An equation containing a fraction in which the unknown occurs in a denominator is called a *fractional equation*. Equations of this type are encountered in many problems involving electric and radio circuits. Simple fractional equations, wherein the unknown appeared only as a factor, were studied in earlier chapters.

12-1 FRACTIONAL COEFFICIENTS

A number of problems lead to equations containing *fractional coefficients*. This type of equation is included in this chapter because the methods of solution apply also to fractional equations.

example 1 $\frac{3x}{4} + \frac{3}{2} = \frac{5x}{8}$ and $\frac{x}{2} + \frac{x}{3} = 5$ are equations having fractional coefficients.

example 2 $\frac{60}{x} - 3 = \frac{60}{4x}$ and $\frac{x-2}{x} = \frac{4}{5}$ are fractional equations.

You are familiar with the methods of solving simple equations that do not contain fractions. An equation involving fractions can be changed to an equation containing no fractions by canceling the denominators and then solved as heretofore. To accomplish this we have the following rule:

Rule

To solve an equation containing fractions:

1. First clear the equation of fractions by multiplying every term by the LCD of the whole equation. (This will permit canceling all denominators.)
2. Solve the resulting equation.

example 3 Given $\frac{5x}{12} - 13 = \frac{x}{18}$. Solve for x.

solution

Given $\qquad \frac{5x}{12} - 13 = \frac{x}{18}$

M: 36, the LCD,

$$\frac{36 \cdot 5x}{12} - 36 \cdot 13 = \frac{36x}{18}$$

Canceling, $\qquad \frac{\overset{3}{\cancel{36}} \cdot 5x}{\cancel{12}} - 36 \cdot 13 = \frac{\overset{2}{\cancel{36}}x}{\cancel{18}}$

Simplifying, $\qquad 15x - 468 = 2x$

Collecting terms, $\qquad 13x = 468$

D: 13, $\qquad x = 36$

check Substitute 36 for x in the original equation:

$$\frac{5 \cdot 36}{12} - 13 = \frac{36}{18}$$

Clearing fractions, $15 - 13 = 2$

$$2 = 2$$

example 4

Given $\frac{e - 4}{9} = \frac{e}{10}$. Solve for e.

solution

Given	$\frac{e - 4}{9} = \frac{e}{10}$
M: 90, the LCD,	$\frac{90(e - 4)}{9} = \frac{90e}{10}$
Canceling,	$\frac{\overset{10}{\cancel{90}}(e - 4)}{\cancel{9}} = \frac{\overset{9}{\cancel{90}}e}{\cancel{10}}$
Simplifying,	$10(e - 4) = 9e$
or	$10e - 40 = 9e$
Collecting terms,	$10e - 9e = 40$
or	$e = 40$

check Substitute 40 for e in the original equation:

$$\frac{40 - 4}{9} = \frac{40}{10}$$

Clearing fractions, $4 = 4$

Note that when the fractions were cleared and the equation written in simplified form in the above solution, the resulting equation was

$$10(e - 4) = 9e$$

which is equivalent to multiplying each member by the denominator of the other member and expressing the resulting equation with no denominators. This is called *cross multiplication*. You will see the justification of this if each member is expressed as a fraction having the LCD. Although the method is convenient, it must be remembered that *cross multiplication is permissible only when each term of a member of an equation has the same denominator.*

PROBLEMS 12-1

Solve the following equations:

1. $\frac{\phi}{2} - \frac{\phi}{4} = 2$

2. $\frac{x}{3} = \frac{x}{6} + 4$

3. $\frac{3\alpha}{2} + \frac{\alpha}{4} = 10 + \frac{\alpha}{2}$

4. $I - \frac{1}{4} = \frac{2I}{5} - \frac{1}{16}$

5. $\omega - \frac{4\omega}{7} = 2\omega - \frac{11}{16}$

6. $\frac{1}{3} + \frac{Z}{5} = \frac{Z}{3}$

7. $\frac{6 + 3\phi}{4} + \frac{12 - 2\phi}{15} = \frac{6\phi}{5} - \frac{37}{60}$

8. $\frac{F}{6} + \frac{F - 3}{18} = \frac{3 + 3F}{12}$

note If a fraction is negative, the sign of each term of the numerator must be changed after removing the denominator. (See Sec. 11-10.) Remember that *the vinculum is a sign of grouping.*

9. $3 - \frac{1 + \lambda}{2} = \frac{2\lambda - 3}{3}$

10. $\frac{4I + 3}{5} - \frac{I - 5}{10} = \frac{I}{3}$

11. $\frac{\omega + 2}{2} - \frac{\omega - 3}{3} = 0$

12. $x - \frac{3 + 4x}{5} + \frac{2x - 3}{6} - \frac{5x - 4}{15} = 0$

13. $\frac{1}{16}(3\theta - 10) - \frac{1}{8}(5\theta - 6) = \frac{1}{2}(7\theta + 16)$

note $\frac{1}{16}(3\theta - 10) = \frac{3\theta - 10}{16}$

14. $\frac{2}{3}(z + 1) - \frac{3}{4}(z + 2) = \frac{1}{6}(z + 1)$

15. $\frac{1}{2}\left(\frac{5}{16} + \frac{1}{4}m\right) + 3 = \frac{1}{8}\left(3m - \frac{1}{3}\right)$

12-2 EQUATIONS CONTAINING DECIMALS

An equation containing decimals is readily solved by first clearing the equation of the decimals. This is accomplished by multiplying both members by a power of 10 that corresponds to the largest number of decimal places appearing in any term.

example 5 Solve $0.75 - 0.7a = 0.26$.

solution

Given	$0.75 - 0.7a = 0.26$
M: 100,	$75 - 70a = 26$
Collecting terms,	$70a = 49$
D: 70,	$a = 0.7$

check Substitute 0.7 for a in the original equation:

$$0.75 - 0.7 \cdot 0.7 = 0.26$$
$$0.75 - 0.49 = 0.26$$
$$0.26 = 0.26$$

If decimals occur in any denominator, multiply both numerator and denominator of the fraction by a power of 10 that will reduce the decimals to integers.

example 6

Solve $\frac{5m - 1.33}{0.02} - \frac{m}{0.05} = 1083.5$.

solution Given $\frac{5m - 1.33}{0.02} - \frac{m}{0.05} = 1083.5$.

Multiplying numerator and denominator of each fraction by 100,

$$\frac{500m - 133}{2} - \frac{100m}{5} = 1083.5$$

The equation is then solved and checked by the usual methods.

PROBLEMS 12-2

Solve the following equations:

1. $0.4Q = 16$
2. $0.05e = 0.20$
3. $0.8\theta = 1.6 + 0.4\theta$
4. $0.125x - 0.02 = 0.035x + 0.025$
5. $0.3r + 4 = 0.7r - 8$
6. $\phi + 2.6 - 0.2\phi = 1.4 + 0.3\phi$
7. $16.5 - 1.5(2R - 0.5) - 15.6 + 2.1(R + 0.3) = 0.03$
8. $0.2 - 0.5(E - 2) - E = 18.7 + 0.8(E + 4)$
9. $\dfrac{0.5b}{6} - \dfrac{0.2b - 0.5}{30} = \dfrac{0.3b + 0.3}{15}$
10. $\dfrac{0.5(\theta - 5)}{3.75} = \dfrac{0.3(\theta + 5)}{7.5} - \dfrac{0.2(3\theta - 2)}{5}$
11. $\dfrac{1.3a - 1.5}{30} = \dfrac{0.4a + 0.3}{5}$
12. $\dfrac{0.8r_i - 0.1}{3} - \dfrac{0.2r_i - 0.5}{5} + \dfrac{0.6r_i + 1.5}{15} - 0.25r_i = 2.75$
13. $\dfrac{\lambda - 2}{0.05} - 70 = \dfrac{\lambda - 4}{0.08}$
14. $\dfrac{0.2(\omega - 1)}{0.5(\omega + 5)} - \dfrac{0.3(1 - \omega)}{0.7(\omega + 5)} - \dfrac{29}{140} = 0$
15. $(0.7\alpha - 0.7)(0.2 + \alpha) = (1 - 1.4\alpha)(0.1 - 0.5\alpha)$

12-3 FRACTIONAL EQUATIONS

Fractional equations are solved in the same manner as equations containing fractional coefficients (Sec. 12-1). That is, every term of the equation must be multiplied by the LCD.

example 7 Solve $\dfrac{x + 2}{3x} - \dfrac{2x^2 + 3}{6x^2} = \dfrac{1}{2x}$.

solution

Given $$\frac{x + 2}{3x} - \frac{2x^2 + 3}{6x^2} = \frac{1}{2x}$$

M: $6x^2$, the LCD,

$$\frac{6x^2(x + 2)}{3x} - \frac{6x^2(2x^2 + 3)}{6x^2} = \frac{6x^2}{2x}$$

Canceling, $$\frac{\overset{2x}{\cancel{6x^2}}(x + 2)}{\cancel{3x}} - \frac{\cancel{6x^2}(2x^2 + 3)}{\cancel{6x^2}} = \frac{\overset{3x}{\cancel{6x^2}}}{\cancel{2x}}$$

Rewriting, $2x(x + 2) - (2x^2 + 3) = 3x$

Simplifying, $2x^2 + 4x - 2x^2 - 3 = 3x$

Collecting terms, $4x - 3x = 3$

or $x = 3$

check Substituting 3 for x in the original equation,

$$\frac{3 + 2}{9} - \frac{18 + 3}{54} = \frac{1}{6}$$

That is, $$\frac{30}{54} - \frac{21}{54} = \frac{9}{54}$$

example 8 Solve

$$\frac{8a+2}{a-2} - \frac{2a-1}{3a-6} + \frac{3a+2}{5a-10} + 5 = 15$$

solution Given $\dfrac{8a+2}{a-2} - \dfrac{2a-1}{3a-6} + \dfrac{3a+2}{5a-10} + 5 = 15.$

Factoring denominators,

$$\frac{8a+2}{a-2} - \frac{2a-1}{3(a-2)} + \frac{3a+2}{5(a-2)} + 5 = 15$$

M: $15(a-2)$, the LCD,

$$\frac{15(a-2)(8a+2)}{a-2} - \frac{15(a-2)(2a-1)}{3(a-2)} + \frac{15(a-2)(3a+2)}{5(a-2)} + 15(a-2)(5) = 15(a-2)(15)$$

Canceling,

$$\frac{15\cancel{(a-2)}(8a+2)}{\cancel{a-2}} - \frac{\overset{5}{\cancel{15}}\cancel{(a-2)}(2a-1)}{\cancel{3}\cancel{(a-2)}} + \frac{\overset{3}{\cancel{15}}\cancel{(a-2)}(3a+2)}{\cancel{5}\cancel{(a-2)}} + 15(a-2)(5) = 15(a-2)(15)$$

Rewriting, $15(8a+2) - 5(2a-1) + 3(3a+2) + 15(a-2)(5) = 15(a-2)(15)$

Simplifying, $120a + 30 - 10a + 5 + 9a + 6 + 75a - 150 = 225a - 450$

Collecting terms,

$$\begin{aligned} 120a - 10a + 9a + 75a - 225a &= -30 - 5 - 6 + 150 - 450 \\ -31a &= -341 \\ a &= 11 \end{aligned}$$

Check the solution by the usual method.

PROBLEMS 12-3

Solve the following equations:

1. $\dfrac{3}{I} + \dfrac{5}{I} = 4$

2. $2 - \dfrac{2}{E} = \dfrac{10}{E}$

3. $\dfrac{16}{q} - 5 = \dfrac{3}{q} - \dfrac{2}{q}$

4. $\dfrac{3}{5R} - \dfrac{1}{15} + \dfrac{7}{5R} + \dfrac{2}{5} = 1$

5. $\dfrac{1}{\phi} - 1 - \dfrac{3}{2\phi} = 1 - \dfrac{1}{\phi}$

6. $\dfrac{5}{3x} + \dfrac{13}{12} + \dfrac{2}{x} = 2$

7. $\dfrac{12-\omega}{\omega}-\dfrac{4}{\omega}=\dfrac{6}{\omega}$

8. $\dfrac{4}{8+2L}=\dfrac{3}{20-2L}$

9. $\dfrac{40-\pi}{24\pi}+\dfrac{5}{6}-\dfrac{40+\pi}{8\pi}=0$

10. $\dfrac{10}{W}-3=\dfrac{2-W}{W}$

11. $\dfrac{40+e_o}{8e_o}-\dfrac{5}{6}-\dfrac{40-e_o}{24e_o}=0$

12. $\dfrac{6m-17}{3m+3}-\dfrac{2m-5}{9+m}=0$

13. $\dfrac{6}{x-1}-\dfrac{5}{1-x}-\dfrac{8}{x-1}+\dfrac{x}{1-x}=0$

14. $\dfrac{3}{5+R}+\dfrac{R}{R+2}=\dfrac{R+4}{R+5}$

15. $\dfrac{27-\alpha}{\alpha+1}+\alpha=1+\alpha$

16. $\dfrac{5+R}{5-R}-\dfrac{16R}{25-R^2}+\dfrac{5-R}{5+R}+2=0$

17. $\dfrac{\omega+3}{\omega-8}-\dfrac{5-\omega}{\omega+1}=\dfrac{2\omega^2-2}{\omega^2-7\omega-8}$

18. $\dfrac{2\phi+7}{6\phi-4}-\dfrac{17\phi+7}{9\phi^2-4}-\dfrac{3\phi-5}{9\phi+6}=0$

19. $\dfrac{9\alpha+17}{\alpha^2-2\alpha-48}-\dfrac{2\alpha+1}{2\alpha-16}+\dfrac{2\alpha-1}{2\alpha+12}=0$

20. $\dfrac{a-7}{a+2}-\dfrac{6}{a+3}=\dfrac{a^2-a-42}{a^2+5a+6}$

21. A can do a piece of work in 8 h, and B can do it in 6 h; how long will it take them to do it together?

SOLUTION: Let n = number of hours it will take them to do it together. Now A does $\frac{1}{8}$ of the job in 1 h; therefore, A will do $\frac{n}{8}$ in n h. Also, B does $\frac{1}{6}$ of the job in 1 h; therefore, B will do $\frac{n}{6}$ in n h. Then they will do $\frac{n}{8}+\frac{n}{6}$ in n h.

The entire job will be completed in n h, which we may represent by $\frac{8}{8}$ or $\frac{6}{6}$ of itself, which is 1.

$$\therefore \frac{n}{8}+\frac{n}{6}=1$$

M: 24, the LCD,

$$3n+4n=24$$
$$7n=24$$
$$n=3\tfrac{3}{7}\text{ h}$$

22. A technician can install a television transmission line in 5 h, and the technician's helper can do it in 8 h. In how many hours should they be able to do it if they work together?

23. A water tank can be filled in 1 h and 10 min if one pipe is used. If a different pipe is used, it takes 1 h and 45 min to fill the tank. How long will it take to fill the tank if both pipes are used?

24. *A* can do a piece of work in *a* days, and *B* can do it in *b* days. Derive a general formula for the number of days it would take both together to do the work.

SOLUTION: Let $x =$ number of days it will take both together.

Now *A* will do $\frac{x}{a}$ of the job in x days. Also, *B* will do $\frac{x}{b}$ of the job in x days.

Then $$\frac{x}{a} + \frac{x}{b} = 1$$

M: ab, $$bx + ax = ab$$

Factoring, $$x(a + b) = ab$$

D: $(a + b)$, $$x = \frac{ab}{a + b}$$

ALTERNATE SOLUTION: Let $x =$ number of days it will take both together. Then $\frac{1}{x} =$ part that both together can do in 1 day; $\frac{1}{a} =$ part that *A* alone can do in 1 day; and $\frac{1}{b} =$ part that *B* can do in 1 day.

Now, $$\frac{1}{a} + \frac{1}{b} = \frac{1}{x}$$

M: abx, $$bx + ax = ab$$

Factoring, $$x(b + a) = ab$$

D: $(a + b)$, $$x = \frac{ab}{a + b}$$

25. *A* can do a piece of work in *a* days, *B* in *b* days, and C in c days. Derive a general formula for the number of days it would take them to do it together.

26. A tank can be filled by one of two pipes in 3 h and by the other of the two in 5 h. It can be emptied by the drain pipe in 6 h. If all three pipes are open, how long will it take to fill the tank?

27. Three circuits are connected to a storage battery. Circuit 1 completely discharges the battery in 20 h, circuit 2 in 15 h, and circuit 3 in 12 h. All circuits are connected to the battery in parallel. In how many hours will the battery be discharged?

28. A tank can be filled by one of two pipes in x h and by the other of the two in y h; it can be emptied by a drain pipe in z h. Derive a general formula for the number of hours required to fill the tank with all pipes open.

29. A bottle contains 1 liter (l) of a mixture of equal parts of acid and water. How much water must be added to make a mixture that will be one-tenth acid?

SOLUTION: Let n = number of liters of water to be added.

$$1\text{ l} = \text{amount of original mixture}$$

and $$0.5\text{ l} = \text{amount of acid}$$

Hence $$n + 1 = \text{amount of new one-tenth acid mixture}$$

Now, $$\frac{1}{10} = \frac{\text{amount of acid}}{\text{total mixture}}$$

Then $$\frac{1}{10} = \frac{0.5}{n + 1}$$

$$n + 1 = 5$$
$$n = 4\text{ l of water to be added}$$

30. How much metal containing 25% copper must be added to 20 kg of pure copper to obtain an alloy having 50% copper?

SOLUTION: Let x = desired amount of metal containing 25% copper.

Then $$0.25x = \text{amount of copper in this metal}$$
$$20 + 0.25x = \text{amount of copper in mixture}$$
$$x + 20 = \text{total weight of mixture}$$
$$0.5(x + 20) = \text{amount of copper in mixture}$$
$$0.5x + 10 = 20 + 0.25x$$
$$x = 40\text{ kg}$$

31. How much 10% nickel alloy must be added to 10 kg of 30% nickel alloy to form a 12% nickel alloy?

32. A full radiator contains 50 liters of a 30% mixture of antifreeze. How much antifreeze is required to obtain a 45% mixture?

SOLUTION: The radiator now contains 50 l of 30% antifreeze = 15 l. We want it to contain 50 l of 45% antifreeze = 22.5 l. But to get the mixture we want, we must drain off some quantity of 30% mixture and replace it with 100% antifreeze. Let the volume replaced be x liters:

$$15 - 0.3x + x = 22.5$$
$$x = 10.7\text{ l}$$

33. A diesel engine driving a 100-kW generator for an isolated communications center has a 200-l cooling system which, during the summer, contains a 20% antifreeze solution. At 50¢ per liter, what is the total cost of increasing the cold-weather protection by making the coolant 55% antifreeze?

34. A fighter plane traveling at 900 km/h leaves its base at 9:00 A.M. to overtake a bomber which departed from the same base at 7:00 A.M. and is traveling at 475 km/h. How much time is required for the fighter to overtake the bomber?

35. The sum of two numbers is 625. When the larger is divided by the smaller, the quotient is 24. Find the numbers.

36. The numerator of a fraction is 54 greater than the denominator. When 9 is subtracted from each term, the quotient is 4. What is the value of the fraction?

37. The sum of three consecutive numbers is $4\frac{1}{2}$. Find the numbers.

38. A certain number, plus 23, is divided by the same number plus 12. The quotient is $\frac{4}{3}$. What is the number?

39. The perimeter of a stock room is 20 m. The room is four times as long as it is wide. What are its dimensions?

40. A screened room is two-thirds as wide as it is long. If it had been 3 m wider and 3 m shorter, its area would have been 3 m² larger. What are its dimensions?

12-4 LITERAL EQUATIONS

Equations in which some or all of the numbers are replaced by letters are called *literal equations;* they were studied in Chap. 5. Having attained more knowledge of algebra, such as factoring and fractions, we are now ready to proceed with the solution of more difficult literal equations, or formulas. No new methods are involved in the actual solutions—we are prepared to solve a more complicated equation simply because we have available more tools with which to work. Again, we point out that the ability to solve formulas is of utmost importance.

example 9 Given $I = \dfrac{E}{R + r}$, solve for r.

solution

Given $\quad I = \dfrac{E}{R + r}$

M: $(R + r)$,

$$I(R + r) = E$$

Removing parentheses,

$$IR + Ir = E$$

S: IR, $\quad Ir = E - IR$

D: I, $\quad r = \dfrac{E - IR}{I}$

example 10 Given $S = \dfrac{RL - a}{R - 1}$, solve for L.

solution

Given: $\quad \dfrac{RL - a}{R - 1} = S$

M: $(R - 1)$,

$$RL - a = S(R - 1)$$

A: a, $\quad RL = S(R - 1) + a$

D: R, $\quad L = \dfrac{S(R - 1) + a}{R}$

example 11 Given $\frac{a}{x-b} = \frac{2a}{x+b}$, solve for x.

solution

Given $$\frac{a}{x-b} = \frac{2a}{x+b}$$

M: $(x^2 - b^2)$, the LCD,

$$\frac{(x^2-b^2)a}{x-b} = \frac{(x^2-b^2)2a}{x+b}$$

Canceling, $$\frac{\overset{x+b}{\cancel{(x^2-b^2)}}a}{\cancel{x-b}} = \frac{\overset{x-b}{\cancel{(x^2-b^2)}}2a}{\cancel{x+b}}$$

Rewriting,

$$(x+b)a = (x-b)2a$$

Removing parentheses,

$$ax + ab = 2ax - 2ab$$

Collecting terms,

$$ax - 2ax = -2ab - ab$$

or $$-ax = -3ab$$

M: -1, $$ax = 3ab$$

D: a, $$x = 3b$$

note The last two steps can be combined into one step by dividing $-ax = -3ab$ by $-a$ to obtain $x = 3b$.

check Substitute $3b$ for x in the given equation:

$$\frac{a}{3b-b} = \frac{2a}{3b+b}$$

Simplifying, $$\frac{a}{2b} = \frac{2a}{4b}$$

or $$\frac{a}{2b} = \frac{a}{2b}$$

PROBLEMS 12-4

Given:	Solve for:
1. $Y_d = \frac{LbV_d}{2aV_0}$	V_0, V_d
2. $\boldsymbol{E}_{\max} = \frac{V + V_{pt}}{\omega}$	V_{pt}, $\omega\boldsymbol{E}_{\max}$
3. $I = \frac{E_b - e}{R}$	E_b, e
4. $C = \frac{\omega_{01}}{R_1 + R_2}$	R_1, ω_{01}
5. $C_2 = \frac{V_3 - V_2}{\omega^2 L V_3}$	V_2, L
6. $I_1 = \frac{V_1 - I_2(R+s)}{R}$	V_1, s, R
7. $\alpha = \frac{R_t - R_0}{R_0 t}$	R_t, t
8. $I_{\lambda_2} = \frac{V_{e_2} + V_\lambda - V_2}{R_b}$	V_λ, V_2

Given:	Solve for:
9. $e = \dfrac{Er}{R + r}$	r, R
10. $\mu = \dfrac{g_m}{g'_m - g_m}$	g'_m, g_m
11. $\omega^2 C_1 C_2 R_3 = \dfrac{1}{R_1 + R_2}$	R_1
12. $\mu = \dfrac{2G_L + g_p - 2G_2}{G_2 - G_L}$	G_2, G_L, g_p
13. $\dfrac{V_0}{I_0} = \dfrac{R_0}{1 - \mu\beta}$	β
14. $\beta_m = \dfrac{m\pi a}{a + b}$	a, b
15. $C_0 = \dfrac{a - b}{a + b}$	a, b
16. $\gamma = \dfrac{I_n}{I_n + I_p}$	I_p, I_n
17. $\dfrac{E}{I} = \dfrac{Z_1Z_2 + Z_2Z_3 + Z_3Z_1}{Z_3}$	Z_1, Z_2, Z_3
18. $Z_0 = \dfrac{R_aR}{(\mu + 1)R + R_a}$	R, R_a, μ
19. $\dfrac{V}{V_1} = \dfrac{AR_y}{(A + 1)R_x + R_y}$	R_x, R_y, A
20. $B_c = \dfrac{\pi\sqrt{2}DFf_b}{\sqrt{2}D + F}$	D, F
21. $Z_{ab}{}^2 = \dfrac{X_s{}^2R}{X_p - X_s - R}$	X_p, R
22. $Z_1 = \dfrac{(\mu + 1)R_1R + R_a(R_1 + R)}{R_a + R}$	R_a, R
23. $I_2R - V_n = \left(\dfrac{R_1}{R_1 + R_2}\right) V_n$	V_n, R_1
24. $2C_2R_3 = \sqrt{2} - C_1R_1\left(\dfrac{R_2}{R_3} - 1\right)$	R_1, R_2, C_1
25. $F = \dfrac{9}{5}C + 32$	C

Given:	Solve for:
26. $r = \dfrac{\mu E_g - PR_p}{P}$	P, μ
27. $V_1 = \dfrac{BI_0R_0}{R + R_0}$	R, R_0
28. $f_{out} = \dfrac{C_1 f_{in}}{C_1 + C_2}$	f_{in}, C_1, C_2
29. $K = \dfrac{\mu m N}{g(R_H + r)}$	R_H, r
30. $r_p = \dfrac{GR_{pg}}{g_m R_{pg} - G}$	G, g_m
31. $\alpha = \dfrac{Z_1 + Z_2 - R}{Z_1(1 - k) + Z_2}$	R, Z_1, k
32. $X = \dfrac{K}{(f_1 - f_2) - (f_0 - f_2)}$	f_1, K
33. $F_{12} = \left(\dfrac{2f}{\alpha}\right)\left(1 + \dfrac{F_s}{F_2}\right)$	F_2, f
34. $\mu\beta = \dfrac{2N}{2L + N}$	L, N
35. $H_2S = \left(\dfrac{1}{R_1}\right)\left(\dfrac{S}{S + \alpha}\right)$	α, S
36. $\dfrac{n_2 - n_1}{n_1} = \dfrac{-h\nu}{kT}$	n_1
37. $e_1 = \mu e_g\left(\dfrac{R_p}{R_p + Z_1}\right)$	Z_1, R_p
38. $F = 1 + 2\left(\dfrac{T_s}{T_a}\right)\left(\dfrac{1}{X}\right)$	T_a, X, T_s
39. $X = \left(1 - \dfrac{C_v}{C_0}\right)\left(\dfrac{f_c}{f}\right)$	C_v, C_0
40. $\dfrac{C_3}{C_1 + C_2} = \dfrac{R_3}{\dfrac{1}{R_1} + \dfrac{1}{R_2}}$	C_2, R_1
41. $\mu = \dfrac{\omega s}{2}\left(\dfrac{1}{v_0'} - \dfrac{1}{v_m'}\right)$	v_0', v_m'
42. $\alpha = 1 + \dfrac{1}{\mu_0}\left(1 + 1.5\dfrac{d_2}{d_1}\right)$	d_1

Given:	Solve for:
43. $\dfrac{V - v_0}{v_0} = \dfrac{R_2}{R_1}\left(\dfrac{i_1 + i_2}{i_1}\right)$	v_0, i_1
44. $Z_{am}^{\ 2} = R\dfrac{(X_p - X_s)Z_{ab}^{\ 2}}{Z_{ab}^{\ 2} + X_s^{\ 2}}$	$Z_{ab}^{\ 2}$, X_p
45. $\mu_1 = \dfrac{G(\mu_2\beta_2 - 1)}{G\beta_1(\mu_2\beta_2 - 1) - \mu_2}$	G, β_1
46. $C_g = C_{gf} + C_{gp}\left(1 + \dfrac{\mu R_b}{r_p + r_b}\right)$	R_b, C_{gp}
47. $\sigma_0 = 2\pi\lambda^2\left(\dfrac{\gamma_1}{\gamma_1 + \gamma_f}\right)\left(\dfrac{2I_f + 1}{2I_1 + 1}\right)$	I_1, I_f
48. $K_\varepsilon^{\ 2}\left(1 + \dfrac{\tan^2 K_a}{\varepsilon_p^{\ 2}}\right) = -\alpha^2$	$\varepsilon_p^{\ 2}$
49. $n' = \dfrac{\lambda}{\pi d_0}\left(\dfrac{1 - d_1}{d_1 - d_0}\right)$	λ, d_1
50. $I_2 = \dfrac{ER_0}{R_1R_0 + R_1R_2 + R_2R_0}$	ER_0, R_1
51. $R = -\left(\dfrac{1}{k}\right)\left(\dfrac{Z_1R_2}{Z_2\alpha} + \dfrac{R_2}{\alpha} + \dfrac{Z(1 - \alpha)}{\alpha}\right)$	R_2, Z, α
52. $\dfrac{E_b - E_c}{\mu} = E_c + E_s\left(\dfrac{R_p}{R_1 + R_p}\right)$	R_1, E_s
53. $\dfrac{r_1}{r_1 + r_2} = \dfrac{r_3}{r_3 + r_4}$	r_1, r_3, r_4
54. $\dfrac{S^2}{N^2} = \dfrac{\alpha F}{2f\left(1 + \dfrac{F_s}{F_2}\right)}$	F_s, F_2
55. $V_{out} = \dfrac{Q}{C_f}\left(\dfrac{1}{1 + \dfrac{1}{G} + \dfrac{C_d}{C_{fg}}}\right)$	G
56. $T_m = \dfrac{T}{\dfrac{\omega_{32}k\nu_{12}T_m}{\omega_{21}h\nu_{12}} - 1}$	T, h
57. $\dfrac{P_L}{2p} = \dfrac{\omega\varepsilon_2 p_2(\tan\delta)}{2CN(p_1 + p_2)}$	p_2
58. $\alpha_2 = \dfrac{FC}{(\Omega_1 - B)(\Omega_2 - B) + c^2}$	Ω_1

Given:	Solve for:
59. $\frac{1}{R_p} = \frac{1}{R_1} + \frac{1}{R_2}$	R_p, R_1, R_2
60. $i_s = \frac{v}{L\left(S_s + \frac{R}{L}\right)}$	L, R
61. $\frac{E_0}{E} = \frac{\mu}{\mu + 1 + \frac{R_a}{R_3}}$	R_3, R_a, μ
62. $\frac{\omega_{01} L}{\frac{R_1 R_2}{R_1 + R_2}} = 1$	L, R_1
63. $M = \frac{k}{1 + \frac{N}{4\pi} k} H_0$	π, k
64. $HS = \frac{\frac{1}{C}}{S + \frac{1}{R_c C}}$	C, R_c
65. $d = b + \frac{2b}{\frac{X}{X'} + \frac{X'}{X}}$	b
66. $(G_2)(p) = \frac{A(p + \omega_1)}{p + \omega - \frac{AC_2}{C_1 + C_2} p}$	C_2
67. $\frac{E_0}{E} = \frac{\mu R_1 + R_a}{\mu R_1 + R_a + (R_s + R_1)\left(1 + \frac{R_a}{R_3}\right)}$	R_a, R_s, μ
68. $\frac{E_0}{E} = \frac{h_{fe} + 1 + \frac{h_{ie}}{R_B}}{h_{fe} + 1 + h_{ie}\left(\frac{1}{R_B} + \frac{1}{R_E}\right)}$	E, R_B
69. $R_0 = \left(\frac{1}{1 + \mu\frac{R_1}{R_a}}\right)\left(\frac{1}{\frac{1}{R_s + R_1} + \frac{1}{R_2}}\right)$	R_a, R_2, μ

Given: Solve for:

70. $R_{\text{in}} = R_E \left[\dfrac{h_{fe} + 1 + h_{ie}\left(\dfrac{1}{R_E} + \dfrac{1}{R_B}\right)}{1 + \dfrac{h_{ie}}{R_B}} \right]$ R_B, R_E

71. $R_i = R_1 \dfrac{\mu + R_a\left(\dfrac{1}{R_1} + \dfrac{1}{R_2} + \dfrac{1}{R_3}\right)}{1 + R_a\left(\dfrac{1}{R_1} + \dfrac{1}{R_2} + \dfrac{1}{R_3}\right)}$ R_a

72. $MH = \dfrac{4\pi r^2}{T^2\left(1 + \dfrac{\alpha}{\frac{1}{2}\pi - \alpha}\right)}$ α, π

73. $\dfrac{\alpha - \dfrac{\pi}{\alpha - \beta}}{\alpha + \dfrac{\pi}{\alpha - \beta}} - 1 = \dfrac{\alpha}{\beta}$ π, β

74. The force between two magnetic poles of strength S_1 and S_2 at a separation of d cm is

$$F = \frac{10 S_1 S_2}{d^2} \qquad \text{micronewtons } (\mu\text{N})$$

When the poles are separated by a distance of 60 cm, a force of 15 μN exists between them. $S_2 = 90$ units. What is the value of S_1?

75. The force acting to close the air gap of a simple electromagnetic relay is

$$F = \frac{B^2 A}{2\mu} \qquad \text{N}$$

What will be the value of A, the cross-sectional area of the gap, in square meters, which will permit a flux density B of 64×10^3 webers per square meter (Wb/m^2) to exert a force F of 96 N? (μ, the permeability of air, is $4\pi \times 10^{-7}$ SI units.)

76. When two impedances Z_1 and Z_2 are connected in parallel, the resultant joint impedance Z_p is

$$Z_p = \frac{Z_1 Z_2}{Z_1 + Z_2}$$

Solve for Z_2

77. Using the formula given in Prob. 76, what is the value of Z_2 when $Z_p = 3\ \Omega$ and $Z_1 = 6\ \Omega$?

78. $\frac{N_p}{N_s} = \frac{E_p}{E_s}$; $E_p = 100$, $E_s = 20$, $N_p = 400$. What is the value of N_s?

79. $\frac{V_1}{V_2} = \frac{R_1}{R_2}$; $V_1 = 16.2$ V, $V_2 = 34$ V, $R_1 = 47.7\ \Omega$. What is R_2?

80. Corresponding temperature readings in Fahrenheit degrees (°F) can be obtained from a Celsius thermometer by the use of the formula $F = \frac{9}{5}C + 32$, where C is the temperature in Celsius degrees. When the temperature is 77°F, what is the Celsius temperature?

81. Use the formula given in Prob. 80 to find the temperature at which the Fahrenheit and Celsius temperatures are equal, that is, at which $F = C$.

82. $L_t = L_0 + L_0\alpha t$. If $L_t = 15$, $\alpha = 8.33 \times 10^{-2}$, and $t = 6$, what is the value of L_0?

83. $R_t = R_0(1 + 0.0042t)\ \Omega$. What is the resistance R_0 at 0°C if, at a temperature $t = 59.5$°C, the resistance $R_t = 40\ \Omega$?

84. $P = \frac{LI^2}{2}$. The energy P stored in a circuit is 1250 joules (J). If the current $I = 2.5$ A, find the value of the coefficient of self-induction L.

85. When two capacitors C_1 and C_2 are connected in series, the resultant total capacitance can be computed by means of the equation

$$\frac{1}{C_s} = \frac{1}{C_1} + \frac{1}{C_2}$$

If $C_s = 2$ pF and $C_2 = 6$ pF, what is the value of C_1?

86. The joint conductance $\frac{1}{R_p}$ siemens of three resistances connected in parallel is expressed by

$$\frac{1}{R_p} = \frac{1}{R_1} + \frac{1}{R_2} + \frac{1}{R_3}$$

Solve for R_p.

87. A lens formula is $\frac{1}{f} = \frac{1}{p} + \frac{1}{q}$. What is the value of p when $q = 80$ and $f = 50$?

88. Use the lens equation given in Prob. 87 to find the image distance q when the focal length $f = 10$ cm and the object distance $p = 40$ cm.

89. $P = \frac{E^2}{R}$. (a) How is the value of P changed when E is doubled? (b) How is the value of P changed when R is doubled?

90. A source of emf consists of n cells in parallel, and each cell has an emf of E V and an internal resistance of $r\,\Omega$. The current that flows through a load of $R\,\Omega$ is given by the relation

$$I = \frac{E}{R + \frac{r}{n}} \quad \text{A}$$

Solve for r and R.

91. Use the formula stated in Prob. 90 to find the value of R when $E = 2.1$ V, $r = 0.6\,\Omega$, $I = 2$ A, and $n = 4$ cells.

92. Use the formula stated in Prob. 90 to find n in terms of I and E when $R = 32\,\Omega$ and $r = 0.1\,\Omega$.

93. A source of emf consists of n cells in series, and each cell has an emf of E V and an internal resistance of $r\,\Omega$. The current flowing through a load of $R\,\Omega$ is given by the relation

$$I = \frac{nE}{R + nr} \quad \text{A}$$

Solve for R and n.

94. Use the formula stated in Prob. 93 to find the number of identical cells of internal resistance $r = 0.6\,\Omega$ each, if they provide an emf of $E = 2.1$ V each, when they drive a current $I = 2$ A through a load $R = 4.5\,\Omega$.

95. When a signal voltage e_g is impressed on the grid of a vacuum tube which has an amplification factor of μ, the resulting plate current i_p flowing in the output circuit, which consists of the plate resistance r_p in series with the load circuit r_b, is

$$i_p = \frac{\mu e_g}{r_p + r_b} \quad \text{A}$$

Solve for μ and r_p.

96. Use the formula stated in Prob. 95 to find the value of r_b if $i_p = 500$ mA, $\mu = 5 \times 10^5$, $e_g = 20$ V, and $r_p = 10$ kΩ.

97. Does $\frac{IR + E}{R} = I + E$? Explain your answer.

98. If $I = \frac{E}{R_1 + R_2 + R_3}$, does $R_3 = \frac{E}{R_1 + R_2 + I}$? Explain your answer.

99. $S = V_0 t + \frac{1}{2}gt^2$. What is the value of the initial velocity V_0 in terms of S, g, and t?

100. Using the formula stated in Prob. 99, what is the acceleration due to gravity g if the initial velocity $V_0 = 3$ m/s, $S = 520.5$ m, and $t = 10$ s?

101. A radiosonde is dropped from an airplane and falls freely until its parachute opens. Ten seconds after the parachute opens, it has fallen an additional 1 km. What was its velocity when the parachute opened?

102. If

$$\frac{a}{b} = \frac{a - \dfrac{x}{a-b}}{a + \dfrac{x}{a-b}} - 1$$

what is the value of x when $b = 4.62$ and $a = 3$?

103. The incremental plate resistance R_b of a vacuum tube is equal to the quotient obtained by dividing the plate voltage swing by the plate current swing. That is,

$$R_b = \frac{E_{max} - E_{min}}{I_{max} - I_{min}}$$

Solve for E_{max} and I_{min}.

104. Use the formula stated in Prob. 103 to find the value of E_{min} when $R_b = 250\ \Omega$, $E_{max} = 475$ V, $I_{max} = 500$ mA, and $I_{min} = 300$ mA.

105. $I_p = \dfrac{E_p + \mu e_g + m}{R_p}$. What is the value of E_p when $I_p = 50$ mA, $\mu = 50$, $e_p = 50$ V, $m = -250$, and $R_g = 50$ kΩ?

106. $E = L\dfrac{I_1 - I_2}{t}$. What is the change in current when a voltage $E = 1$ kV is induced in an inductance $L = 5$ H in time $t = 0.5$ s?

107. $I = C\dfrac{E_1 - E_2}{t}$. What is the change of voltage which will produce a current flow of $I = 0.05$ A during the discharge of a 15-μF capacitor in 0.0294 s?

108. $R_a = \dfrac{R_1 R_3}{R_1 + R_2 + R_3}$. Three resistances $R_1 = ?$, $R_2 = 3\ \Omega$, and $R_3 = 2.14\ \Omega$ are connected in delta to produce an equivalent Y-circuit branch $R_a = 0.6\ \Omega$. Find R_1.

109. In transistor parameters, $\beta = \dfrac{\alpha}{1 - \alpha}$. Solve for α in terms of β.

110. Using the formula stated in Prob. 109, what is α when $\beta = 284.7$?

13

ohm's law—parallel circuits

Most of the systems employed for the distribution of electric energy consist of parallel circuits; that is, a source of emf is connected to a pair of conductors, known as *feeders*, and various types of load are connected across the feeders. A simple distribution circuit consisting of a motor and a bank of five lamps is represented schematically in Fig. 13-1 and pictorially in Fig. 13-2. The motor and the lamps are said to be in *parallel*, and it is evident that the current supplied by the generator divides between the motor and the lamps.

In this chapter you will analyze parallel circuits and solve parallel circuit problems. The solution of a parallel circuit generally consists in reducing the entire circuit to a single equivalent resistance that could replace the original circuit without any change in the supply voltage or current.

Fig. 13-1 Schematic Diagram of a Generator G connected to a Motor M in Parallel with a Bank of Five Lamps L

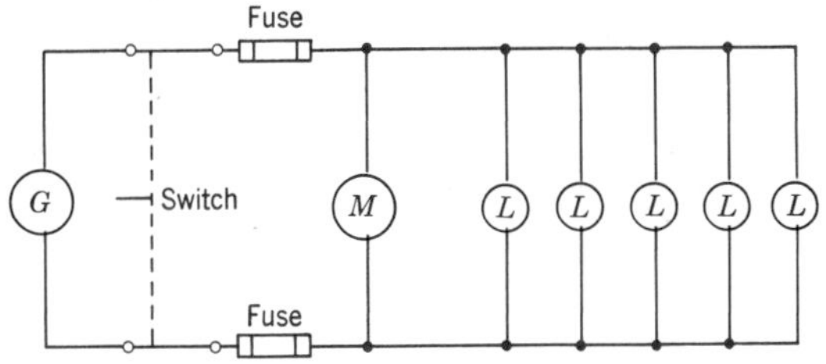

13-1 TWO RESISTANCES IN PARALLEL

The schematic diagram of Fig. 13-3 and the accompanying circuit shown in Fig. 13-4 represent two resistors R_1 and R_2 connected

Fig. 13-2 Illustration of Circuit Shown Schematically in Fig. 13-1

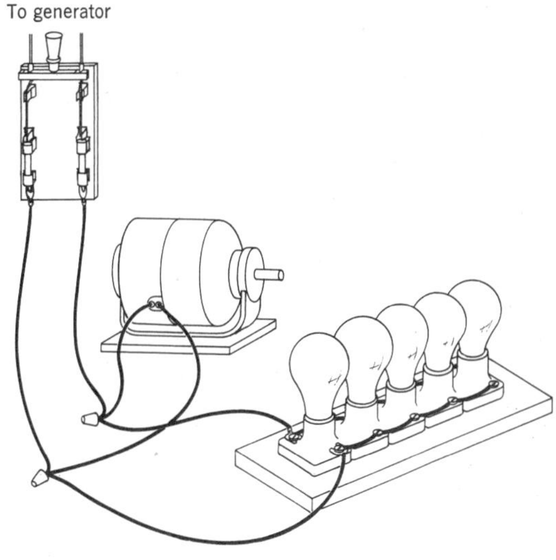

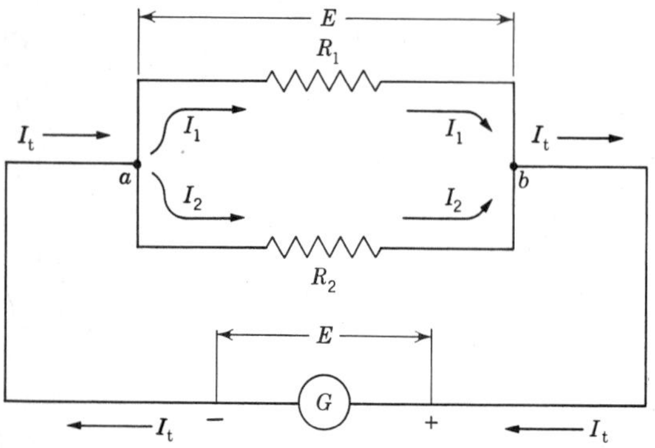

Fig. 13-3 Resistors R_1 and R_2 Connected in Parallel across Generator G, Which Maintains a Potential of E V

in parallel across a source of voltage E. An examination of the circuit arrangement brings out two important facts:

1. The same voltage exists across the two resistors.
2. The total current I_t delivered by the generator enters the paralleled resistors at junction a, divides between the resistors, and leaves the parallel circuit at junction b. Thus, the sum of the currents I_1 and I_2, which flow through R_1 and R_2, respectively, is equal to the total current I_t.

By making use of these facts and applying Ohm's law, it is easy to derive equations that show how paralleled resistances combine. From 1 above,

$$I_1 = \frac{E}{R_1} \qquad I_2 = \frac{E}{R_2} \qquad \text{and} \qquad I_t = \frac{E}{R_p}$$

where R_p is the joint resistance of R_1 and R_2, or the equivalent resistance of the parallel combination. From 2 above,

$$I_t = I_1 + I_2 \tag{1}$$

Substituting in Eq. (1) the value of the currents,

$$\frac{E}{R_t} = \frac{E}{R_1} + \frac{E}{R_2}$$

D: E,
$$\frac{1}{R_p} = \frac{1}{R_1} + \frac{1}{R_2} \tag{2}$$

Equation (2) states that the total conductance (Sec. 7-5) of the circuit is equal to the sum of the parallel conductances of R_1 and R_2; that is,

$$G_t = G_1 + G_2 \tag{3}$$

It is evident, therefore, that, when resistances are connected in parallel, each additional resistance represents another path (conductance) through which current will flow. Hence, increasing the number of resistances in parallel increases the total conductance

Fig. 13-4 Illustration of Schematic Circuit Shown in Fig. 13-3

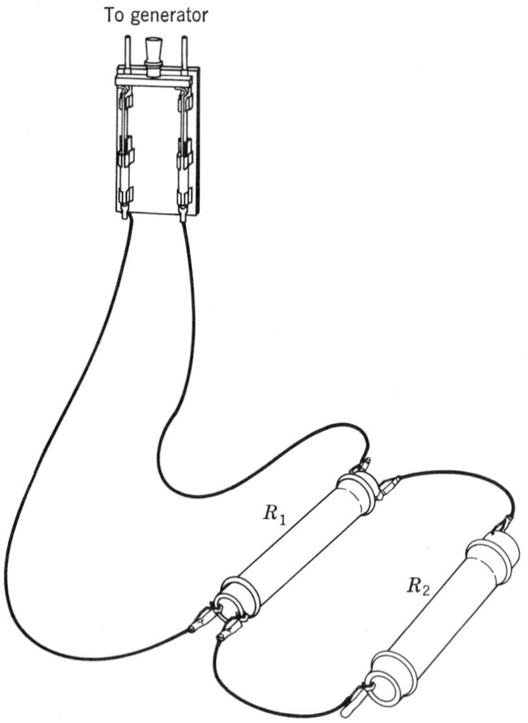

of the circuit and thus decreases the equivalent resistance of the circuit.

example 1 What is the joint resistance of the circuit of Fig. 13-3 if $R_1 = 6\ \Omega$ and $R_2 = 12\ \Omega$?

solution 1 Given $R_1 = 6\ \Omega$ and $R_2 = 12\ \Omega$. $R_p = ?$

Substituting the known values in Eq. (2),

$$\frac{1}{R_p} = \frac{1}{6} + \frac{1}{12} = 0.1667 + 0.0833$$

or $$\frac{1}{R_p} = 0.250$$

Solving for R_p,

$$R_p = \frac{1}{0.250} = 4.0\ \Omega$$

solution 2 A more convenient formula for the joint resistance of two parallel resistances is obtained by solving Eq. (2) for R_p. Thus,

$$R_p = \frac{R_1 R_2}{R_1 + R_2} \tag{4}$$

Hence, the joint resistance of two resistances in parallel is equal to their product divided by their sum.

Substituting the values of R_1 and R_2 in Eq. (4),

$$R_p = \frac{6 \times 12}{6 + 12} = \frac{72}{18} = 4.0\ \Omega$$

Thus, the paralleled resistors R_1 and R_2 are equivalent to a single resistance of $4.0\ \Omega$. Note that the joint resistance is *less* than either of the resistances in parallel.

example 2 (*a*) What is the joint resistance of the circuit of Fig. 13-3 if $R_1 = 21\ \Omega$ and $R_2 = 15\ \Omega$? (*b*) If the generator supplies 12 V across points *a* and *b*, what is the generator (line) current?

solution 1

(*a*) $$R_p = \frac{R_1 R_2}{R_1 + R_2} = \frac{21 \times 15}{21 + 15} = 8.75\ \Omega$$

(*b*) $$I_t = \frac{E}{R_t} = \frac{12}{8.75} = 1.371\ \text{A}$$

solution 2 Since 12 V exists across both resistors, the current through each can be found and added to obtain the total current. Thus,

Current through R_1,

$$I_1 = \frac{E}{R_1} = \frac{12}{21} = 0.571\ \text{A}$$

Current through R_2,

$$I_2 = \frac{E}{R_2} = \frac{12}{15} = 0.8\ \text{A}$$

Total current,

$$I_t = I_1 + I_2 = 0.571 + 0.8 = 1.371\ \text{A}$$

Hence,

$$R_p = \frac{E}{I_t} = \frac{12}{1.371} = 8.75\ \Omega$$

From the foregoing, it is evident that R_1 and R_2 could be replaced by a single resistor of $8.75\ \Omega$, connected between *a* and *b*, and the generator would be working under the same load conditions. Also, it is apparent that when a current enters a junction of resistors connected in parallel, the current divides between the branches in inverse proportion to their resistances; that is, the greatest current flows through the least resistance.

example 3 In the circuit of Fig. 13-3, $R_1 = 25\ \Omega$, $E = 220$ V, and $I_t = 14.3$ A. What is the resistance of R_2?

solution 1

Current through R_1,

$$I_1 = \frac{E}{R_1} = \frac{220}{25} = 8.8 \text{ A}$$

Since

$$I_t = I_1 + I_2$$

the current through R_2 is

$$I_2 = I_t - I_1 = 14.3 - 8.8 = 5.5 \text{ A}$$

Then

$$R_2 = \frac{E}{I_2} = \frac{220}{5.5} = 40 \ \Omega$$

solution 2 $R_p = \frac{E}{I_t} = \frac{220}{14.3} = 15.4 \ \Omega$

Solving Eq. (2) or (4) for R_2,

$$R_2 = \frac{R_1 R_p}{R_1 - R_p} = \frac{25 \times 15.4}{25 - 15.4} = 40 \ \Omega$$

PROBLEMS 13-1

1. Two 330-Ω resistors are connected in parallel. What is the equivalent resistance?

2. Two resistors, one of 1500 Ω and the other of 4700 Ω, are connected in parallel. What is the equivalent resistance of the combination?

3. What is the joint resistance of 68 kΩ in parallel with 82 kΩ?

4. What is the equivalent resistance of 27 kΩ in parallel with 1.5 kΩ?

5. What is the equivalent resistance of:

(*a*) Two 100-Ω resistors in parallel?

(*b*) Two 680-kΩ resistors in parallel?

(*c*) Two 3.9-kΩ resistors in parallel?

6. State a general formula for the total resistance R_p of two equal resistances of R Ω connected in parallel.

7. In the circuit of Fig. 13-3, how much generator voltage would be required to deliver a total current of 3.63 A through a parallel combination of $R_1 = 220 \ \Omega$ and $R_2 = 270 \ \Omega$?

8. How much power would be absorbed by the 270-Ω resistor of Prob. 7?

9. In the circuit of Fig. 13-3, $I_t = 20.3$ mA, $E = 220$ V, and $R_1 = 12$ kΩ. What is the resistance of R_2?

10. How much power is dissipated by R_1 of Prob. 9?

11. How much total power is drawn from the generator of Prob. 9?

12. In the circuit of Fig. 13-3, $R_1 = 18$ kΩ and the current through R_2 is 14.71 mA. A total current $I_t = 70.27$ mA flows through the parallel combination. What is the resistance of R_2?

13. How much power is expended in R_2 of Prob. 12?

14. How much power is drawn from the generator of Prob. 12?

15. What is the generated voltage of Prob. 12?

13-2 THREE OR MORE RESISTANCES IN PARALLEL

The procedure for deriving a general equation for the joint resistance of three or more resistances in parallel is the same as that of the preceding section. For example, Fig. 13-5 represents three resistors R_1, R_2, and R_3 connected in parallel across a source of voltage

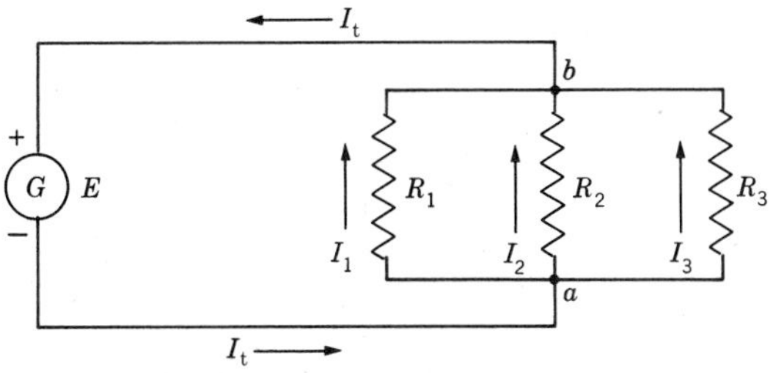

Fig. 13-5 Resistors R_1, R_2, and R_3 Connected in Parallel

E. The total line current I_t splits at junction a into currents I_1, I_2, and I_3, which flow through R_1, R_2, and R_3, respectively. Then

$$I_1 = \frac{E}{R_1} \qquad I_2 = \frac{E}{R_2}$$

$$I_3 = \frac{E}{R_3} \qquad I_t = \frac{E}{R_p}$$

where R_p is the joint resistance of the parallel combination.

Since $\qquad I_t = I_1 + I_2 + I_3$

by substituting,

$$\frac{E}{R_p} = \frac{E}{R_1} + \frac{E}{R_2} + \frac{E}{R_3}$$

D: E,

$$\frac{1}{R_p} = \frac{1}{R_1} + \frac{1}{R_2} + \frac{1}{R_3} \tag{5}$$

From Eq. (5), it is evident that the total conductance of the circuit is equal to the sum of the paralleled conductances of R_1, R_2, and R_3; that is,

$$G_p = G_1 + G_2 + G_3$$

In like manner, it can be demonstrated that the joint resistance R_p of any number of resistances connected in parallel is

$$\frac{1}{R_p} = \frac{1}{R_1} + \frac{1}{R_2} + \frac{1}{R_3} + \frac{1}{R_4} + \frac{1}{R_5} + \cdots$$

Or, in terms of conductances,

$$G_p = G_1 + G_2 + G_3 + G_4 + G_5 + \cdots$$

example 4 What is the joint resistance of the circuit of Fig. 13-5 if $R_1 = 5\ \Omega$, $R_2 = 10\ \Omega$, and $R_3 = 12.5\ \Omega$?

solution Substituting the known values in Eq. (5),

$$\frac{1}{R_p} = \frac{1}{5} + \frac{1}{10} + \frac{1}{12.5}$$

$$= 0.2 + 0.1 + 0.08$$

or $$\frac{1}{R_p} = 0.38$$

Solving for R_p,

$$R_p = \frac{1}{0.38} = 2.63\ \Omega$$

If Eq. (5) is solved for R_p, the result is

$$R_p = \frac{R_1R_2R_3}{R_1R_2 + R_1R_3 + R_2R_3} \tag{6}$$

It is seen that Eq. (6) is somewhat cumbersome for computing the joint resistance of three resistances connected in parallel. However, you should recognize such expressions for three or more resistances in parallel, for you will encounter them in the analysis of networks.

Finding the joint resistance of any number of resistors in parallel is facilitated by arbitrarily assuming a voltage to exist across the parallel combination. The currents through the individual branches that *would* flow if the assumed voltage were actually impressed are added to obtain the total line current. The assumed voltage divided by this total current results in the joint resistance of the combination.

The assumed voltage should always be a power of 10 in order that the slide rule operator can make full use of the reciprocal scales. In order to avoid decimal quantities, that is, currents of less than 1 A, the assumed voltage should be numerically greater than the highest resistance of any parallel branch.

example 5 Three resistances $R_1 = 10\ \Omega$, $R_2 = 15\ \Omega$, and $R_3 = 45\ \Omega$ are connected in parallel. Find their joint resistance.

solution Assume $E_a = 100$ V to exist across the combination.

Current through R_1,

$$I_1 = \frac{E_a}{R_1} = \frac{100}{10} = 10\ \text{A}$$

Current through R_2,

$$I_2 = \frac{E_a}{R_2} = \frac{100}{15} = 6.67\ \text{A}$$

Current through R_3,

$$I_3 = \frac{E_a}{R_3} = \frac{100}{45} = 2.22\ \text{A}$$

Total current,

$$I_t = 18.89\ \text{A}$$

Joint resistance,

$$R_p = \frac{E_a}{I_t} = \frac{100}{18.89} = 5.3\ \Omega$$

PROBLEMS 13-2

1. What is the equivalent resistance of 10 Ω, 15 Ω, and 30 Ω connected in parallel?
2. What is the joint resistance of 150 Ω, 470 Ω, and 470 Ω connected in parallel?
3. Three resistors of 12 Ω, 330 Ω, and 8.2 Ω are connected in parallel. What is the joint resistance?
4. Three resistors of 10 Ω, 100 Ω, and 1000 Ω are connected in parallel. Find the joint resistance of the combination.
5. What is the equivalent resistance of 22 Ω, 15 Ω, 33 Ω, and 47 Ω connected in parallel?
6. Four resistors of 8.2 Ω, 1.5 Ω, 2.7 Ω, and 3.3 Ω are connected in parallel. What is the equivalent resistance of the combination?
7. What is the joint resistance of:
   (*a*) Three 6.3-kΩ resistors in parallel?
   (*b*) Four 68-kΩ resistors in parallel?
8. What is the joint resistance of:
   (*a*) Three 100-kΩ resistors connected in parallel?
   (*b*) Four 100-kΩ resistors connected in parallel?
   (*c*) Five 100-kΩ resistors connected in parallel?
9. State a general formula for the resistance R_p of n equal resistance of R Ω connected in parallel.
10. In the circuit of Fig. 13-5, the total current $I_t = 18.03$ A, $R_1 = 100\ \Omega$, $R_2 = 150\ \Omega$, and $E = 475$ V. What is the resistance of R_3?
11. If the values of Prob. 10 are used, what is the power delivered to the 150-Ω resistor?

12. What would be the resistance in Prob. 10 if the 150-Ω resistor were shorted out?
13. In the circuit of Fig. 13-5, $R_1 = 12\ \Omega$, $R_2 = 18\ \Omega$, $I_3 = 4.545$ A, and $E = 100$ V. Find (*a*) the value of R_3 to two significant figures and (*b*) the total power delivered to the circuit.
14. In the circuit of Fig. 13-5, $R_2 = 510\ \Omega$, $R_3 = 270\ \Omega$, $I_t = 4.38$ A, and $I_1 = 1.52$ A. Find the value of R_1 to two significant figures.
15. In the circuit of Fig. 13-5, $R_1 = R_2 = 5$ kΩ, and R_3 is disconnected. $I_t = 0.40$ A. What must be the value of R_3 connected into the circuit to result in a total current of 0.50 A?
16. A 10-kΩ 100-W resistor, a 15-kΩ 50-W resistor, and a 100-kΩ 10-W resistor are connected in parallel.
    (*a*) What is the maximum voltage which may be applied without exceeding the rating of any resistor?
    (*b*) What is the total current drawn by the combination when the voltage of part (*a*) is applied?

13-3 COMPOUND CIRCUITS

The solution of circuits containing combinations of series and parallel branches generally consists in reducing the parallel branches to equivalent series circuits and combining these with the series branches. No set rules can be formulated for the solution of all types of such circuits, but from the examples that follow you will be able to build up your own methods of attack.

example 6 Find the total resistance of the circuit represented in Fig. 13-6.

solution Note that the parallel branch of Fig. 13-6 is the circuit of Example 1. Since the equivalent series resistance of the parallel branch is

$$\frac{R_2R_3}{R_2 + R_3}$$

the circuit reduces to two resistances in series, the total resistance of which is

$$R_t = R_1 + \frac{R_2R_3}{R_2 + R_3} = 5 + \frac{6 \times 12}{6 + 12} = 9.0\ \Omega$$

Fig. 13-6 Series-Parallel Circuit of Example 6

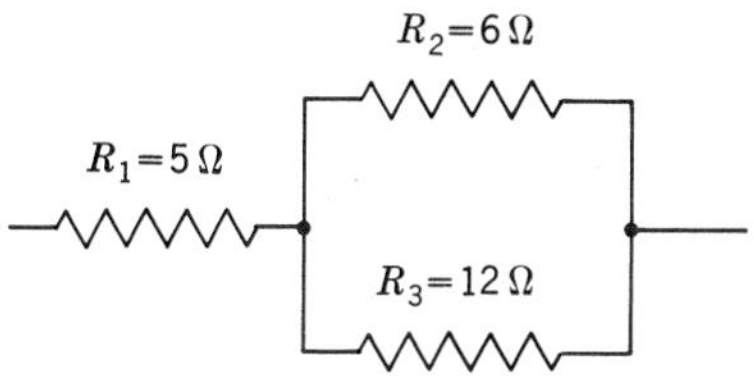

example 7 Find the total resistance of the circuit represented in Fig. 13-7.

solution This circuit is similar to that shown in Fig. 13-6, but with an additional parallel branch. By utilizing the expression for the joint resistance of two resistances in parallel, the entire circuit reduces to three resist-

Fig. 13-7 Circuit of Example 7 Consisting of One Resistance in Series with Two Parallel Branches

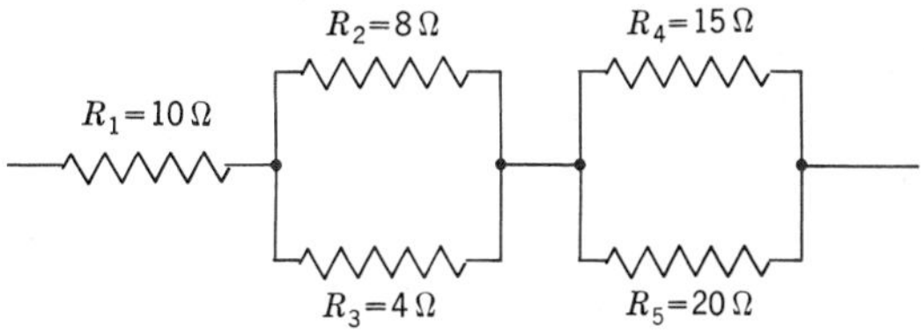

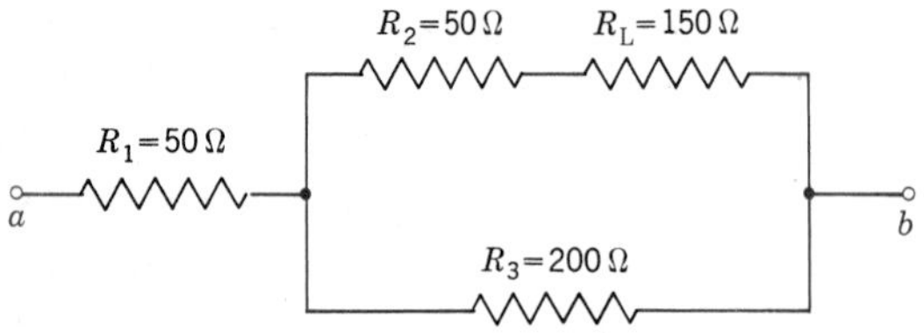

Fig. 13-8 Circuit of Example 8

ances in series, the total resistance of which is

$$R_t = R_1 + \frac{R_2R_3}{R_2 + R_3} + \frac{R_4R_5}{R_4 + R_5}$$
$$= 10 + \frac{8 \times 4}{8 + 4} + \frac{15 \times 20}{15 + 20}$$
$$= 21.2\ \Omega$$

example 8 Find the total resistance between points *a* and *b* in Fig. 13-8.

solution Since R_2 and R_L are in series, they must be added before being combined with R_3. Again, by utilizing the expression for the joint resistance of two resistances in parallel, the entire circuit reduces to two resistances in series. Thus, the total resistance is

$$R_t = R_1 + \frac{R_3(R_2 + R_L)}{R_3 + (R_2 + R_L)}$$
$$= 50 + \frac{200(50 + 150)}{200 + 50 + 150}$$
$$= 150\ \Omega$$

Note that the circuit of Fig. 13-8 is identical with that of Fig. 13-9. The latter is the customary method for representing T networks, often encountered in communication circuits, where R_L is the load or receiving resistance.

Fig. 13-9 Circuit of Example 8 Illustrated in T-Network Form

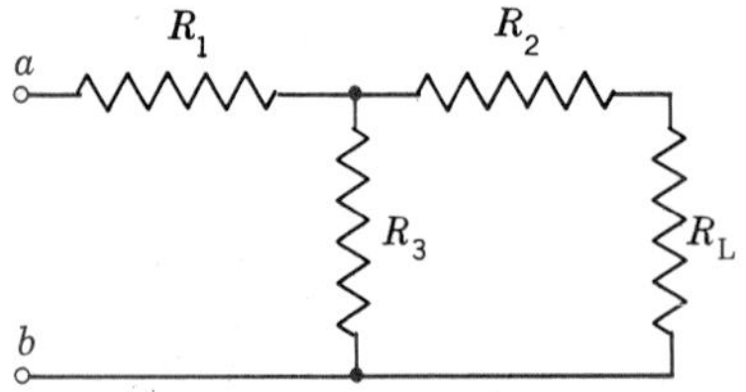

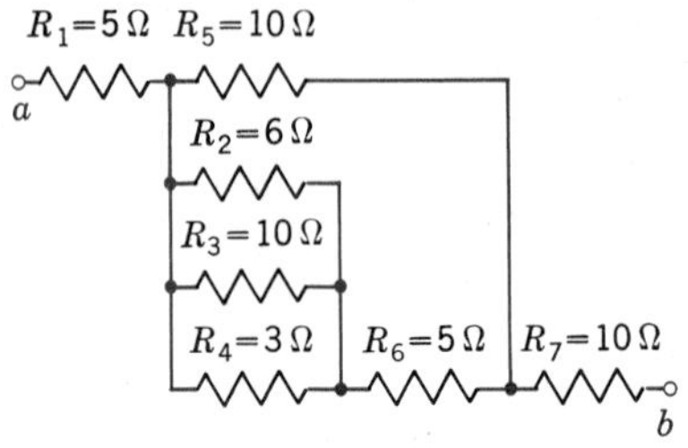

Fig. 13-10 Circuit of Example 9

example 9 Find the resistance between points *a* and *b* in Fig. 13-10.

solution In many instances a circuit diagram that *appears* to be complicated can be better understood and analyzed by redrawing it in a more simplified form. For example, Fig. 13-11 represents the circuit of Fig. 13-10.

First find the equivalent series resistance of the parallel group formed by R_2, R_3, and R_4 and add this resistance to R_6, which will result in the resistance R_{cd} between points c and d. Now combine R_{cd} with R_5, which is in parallel, to give an equivalent series resistance R_{ef} between points e and f. The circuit is now reduced to an equivalence of R_1, R_{ef}, and R_7 in series, which are added to obtain the total resistance R_{ab} between

Fig. 13-11 Simplified Circuit of Example 9

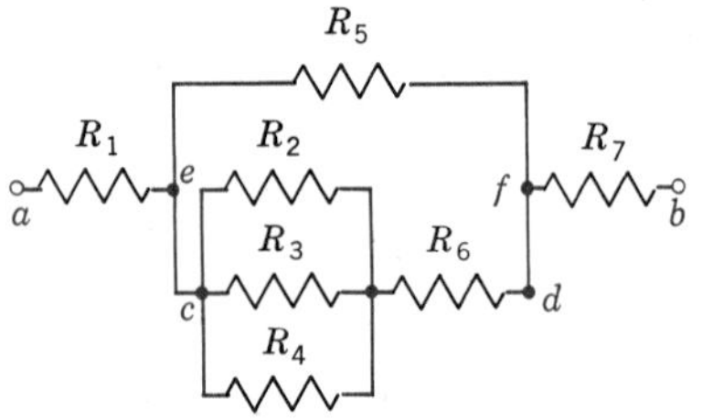

points a and b. The joint resistance of R_2, R_3, and R_4 is 1.67 Ω, which, when added to R_6, results in a resistance R_{cd} = 6.67 Ω between c and d. The equivalent series resistance R_{ef} between points e and f, formed by R_{cd} and R_5 in parallel, is 4.0 Ω. Therefore, the resistance R_{ab} between points a and b is

$$R_{ab} = R_1 + R_{ef} + R_7 = 19\ \Omega$$

PROBLEMS 13-3

1. In the circuit of Fig. 13-12, R_1 = 510 Ω, R_2 = 300 Ω, R_3 = 470 Ω, and E_G = 230 V. What is the total current I_t of the circuit?
2. In Prob. 1, how much power is expended in R_3?
3. In Prob. 1, if R_1 is short-circuited, how much power is expended in R_2?
4. In Prob. 1, what will be the total current I_t if R_2 is open-circuited?
5. In the circuit of Fig. 13-12, R_1 = 62 kΩ, R_2 = 15 kΩ, and I_t = 3.26 mA and the voltage across R_3 is 27.9 V. Find (a) E_G, (b) R_3, (c) R_t, (d) I_2, (e) I_3.
6. In Prob. 5, how much current will the generator supply if R_3 is short-circuited?
7. In the circuit of Fig. 13-12, R_t = 5.562 kΩ, R_1 = 3.9 kΩ, E_G = 1000 V, and I_2 = 135.4 mA. Find (a) voltage across R_1, (b) voltage across R_2, (c) resistance of R_2 to two significant figures, (d) resistance of R_3 to two significant figures, (e) total current I_t, (f) current through R_3, (g) total power expended in the circuit.
8. In Prob. 7, if R_1 is short-circuited, (a) how much power will be expended in R_2 and (b) how much current will flow through R_3?
9. In the circuit of Fig. 13-9, R_1, R_2, and R_3 are all 200-Ω resistors and R_L = 470 Ω. What is the effective resistance between points a and b?
10. In the circuit of Fig. 13-9, $R_1 = R_2 = R_3$ = 300 Ω, and R_L = 600 Ω. What is the resistance between points a and b?
11. In the circuit of Fig. 13-9, $R_1 = R_2 = R_L$ = 300 Ω and R_3 = 600 Ω. What is the resistance between points a and b?
12. In the circuit of Fig. 13-13, $R_1 = R_2 = R_4 = R_5$ = 10 Ω and $R_3 = R_L$ = 600 Ω. If a voltage of 30 V exists across R_L, what is the total current I_t?

Fig. 13-12 R_1 Connected in Series with R_2 and R_3 in Parallel

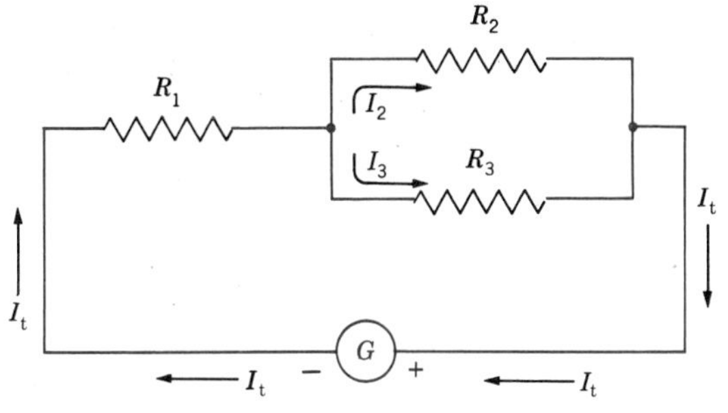

Fig. 13-13 Circuit of Prob. 12

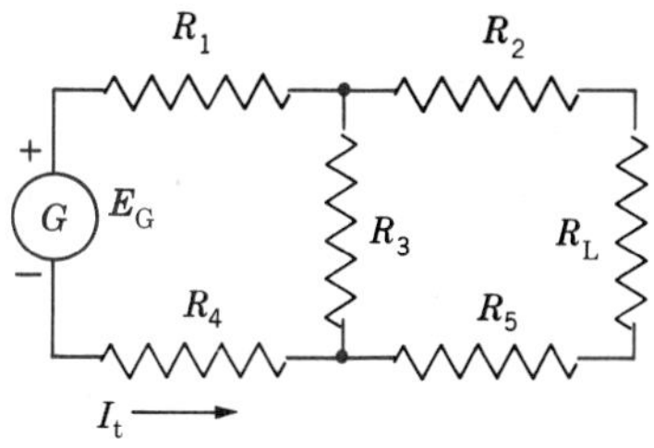

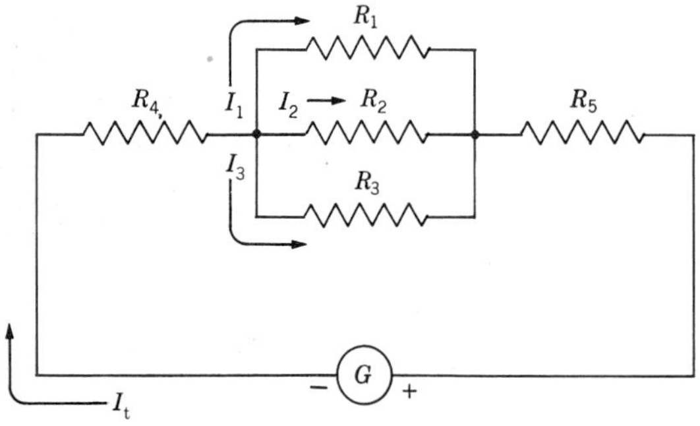

Fig. 13-14 Series-Parallel Circuit of Prob. 13

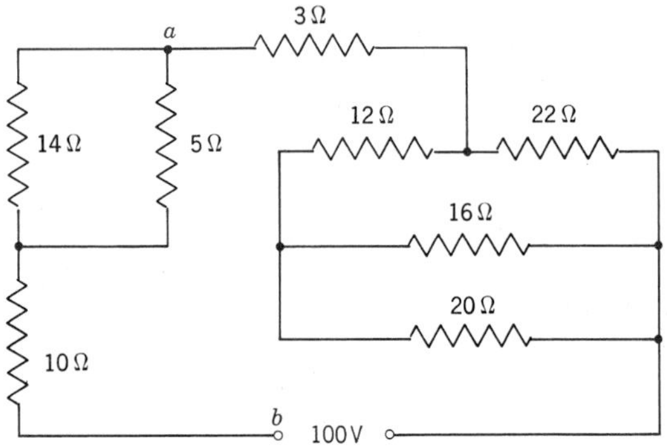

Fig. 13-15 Circuit of Prob. 14

13. In the circuit of Fig. 13-14, the generator voltage $E_G = 3500$ V, $R_4 = 1.5$ kΩ, $R_2 = 6.8$ kΩ, $I_2 = 52.9$ mA, $R_3 = 2.7$ kΩ, and $I_t = 273$ mA. Find to two significant figures (*a*) resistance of R_1, (*b*) resistance of R_5, and (*c*) power expended in R_3.

14. In the circuit represented in Fig. 13-15, find the total current I_t.

15. If, in Fig. 13-15, points *a* and *b* are short-circuited, find the total power expended.

16. What is the total current I_t in the circuit shown in Fig. 13-16?

17. In the circuit of Prob. 16, what is the current flow through the 5-Ω resistor?

18. What would be the power expended in the circuit of Fig. 13-16 if points *a* and *b* were short-circuited?

Fig. 13-16 Circuit of Prob. 16

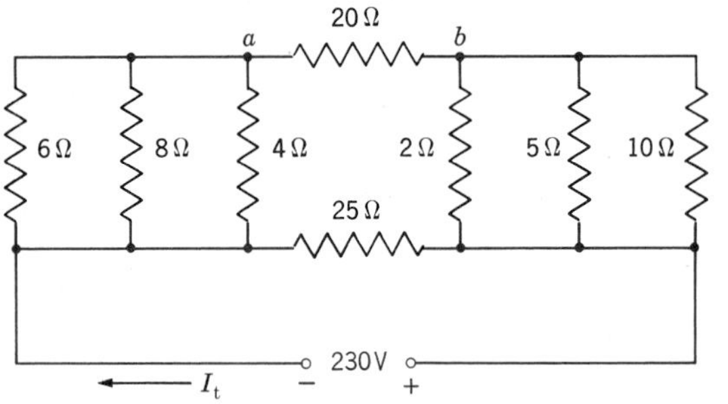

14 meter circuits

Chapters 8 and 13 dealt with the study of Ohm's law as applied to series and parallel circuits, and in Chap. 9 consideration was given to the effects of resistance in current-carrying conductors. The principles and methods learned therein are applied in the present chapter to circuits relating to *dc instruments* used for servicing electrical, radio, and other electronic equipment.

14-1 DIRECT-CURRENT INSTRUMENTS—BASIC METER MOVEMENT

The most common measuring instruments used with electric and electronic circuits are the *voltmeter* and the *ammeter.* As the names imply, a voltmeter is an instrument used to measure voltage and an ammeter is a current-measuring instrument.

The great majority of meters used with direct currents employ the D'Arsonval movement illustrated in Fig. 14-1. This movement utilizes a coil of wire mounted on jeweled bearings between the pole pieces of a permanent magnet. When direct current flows through the coil, a magnetic field is set up around the coil, thereby producing a force which, in conjunction with the magnetic field of the permanent magnet, causes the coil to rotate from the no-current position. Since the arc of rotation is proportional to the amount of current passing through the coil, a pointer can be attached to the coil and the deflection of the pointer over a calibrated scale can be used to indicate values of current.

The *sensitivity* of a current-indicating meter is the amount of current necessary to cause full-scale deflection of the pointer. For example, an instrument of wide usage is the

Fig. 14-1 D'Arsonval Meter Movement (*Courtesy of Weston Electrical Instrument Corporation*)

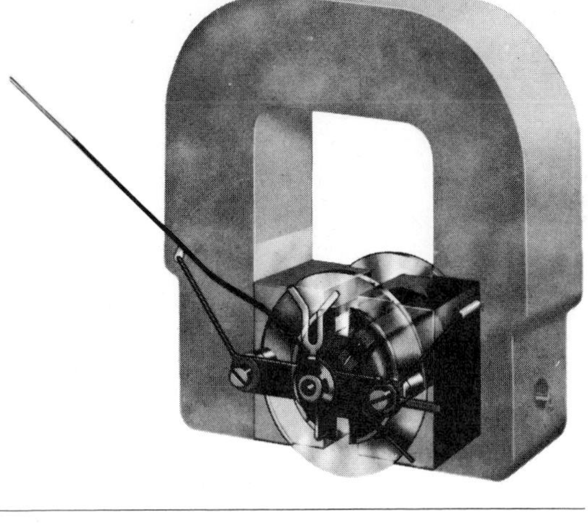

Fig. 14-2 0–1 Milliammeter (*Courtesy of Triplett Electrical Instrument Company*)

0–1 milliammeter illustrated in Fig. 14-2. This meter has a sensitivity of 1 mA because, when a current of 1 mA flows through the meter, the pointer indicates full-scale deflection. This particular meter has an internal resistance of 55 Ω. Other meter movements have different sensitivities with various values of internal resistance.

14-2 MULTIRANGE CURRENT METERS

Instead of utilizing a number of meters to make various current measurements, it is common practice to select a meter movement with sufficient sensitivity and, with the aid of one or more shunts, extend the range of the meter and therefore its usefulness. A shunt, in this application, is a resistor that is shunted (connected in parallel) across the meter coil as shown in Fig. 14-3.

Fig. 14-3 Total Current I_t Consists of Current I_s, Which Flows through Shunt Resistor R_s, and the Meter Current I_m, Which Flows through the Coil of the Meter. That Is, $I_t = I_s + I_m$

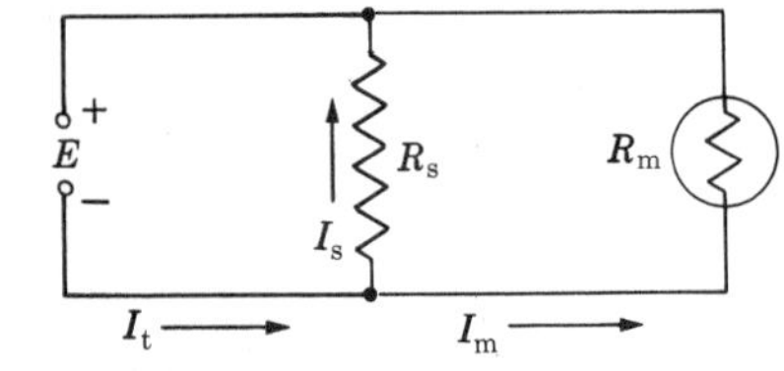

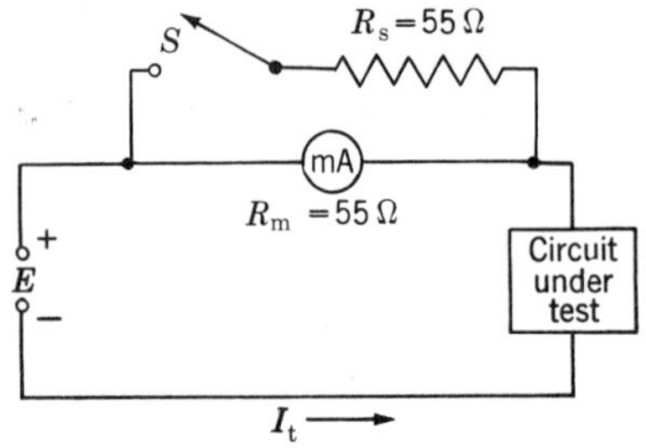

Fig. 14-4 Total Current I_t Flows through the Milliammeter, Which Indicates a Full-Scale Deflection of 1 mA

A meter such as illustrated in Fig. 14-2, with a resistance of 55 Ω, is connected to measure the circuit current of Fig. 14-4. In this condition the switch *S* is open and the meter indicates a full-scale deflection of 1 mA. In Fig. 14-5 the switch *S* is closed, thereby shunting the 55-Ω resistor R_s across the meter. Since the meter resistance and shunt resistance are equal, the circuit current I_t divides equally between them and the meter reads 0.5 mA.

In Fig. 14-4, with the switch open, the meter would indicate actual values of current. In Fig. 14-5, with the switch closed, circuit current would be obtained by multiplying the meter readings by a factor of 2 or by re-marking the scale as shown in Fig. 14-6.

example 1 A 0–1 milliammeter has an internal resistance of 70 Ω. Design a circuit that will allow this meter to be used as a

Fig. 14-5 Total Current I_t Divides Equally between Meter Resistance R_m and Shunt Resistance R_s. $I_t = I_m + I_s = 1$ mA and $I_s = I_m = 0.5$ mA

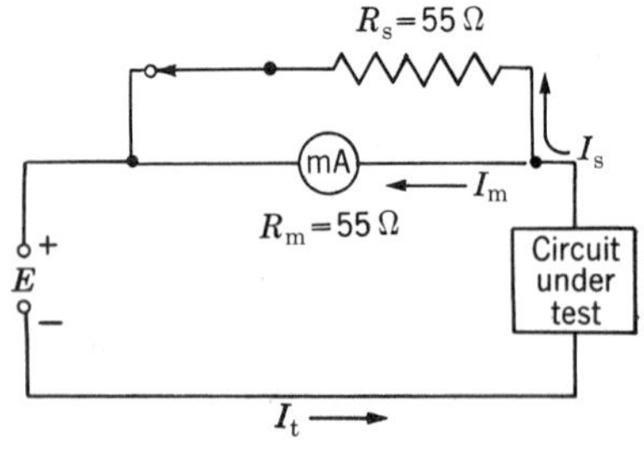

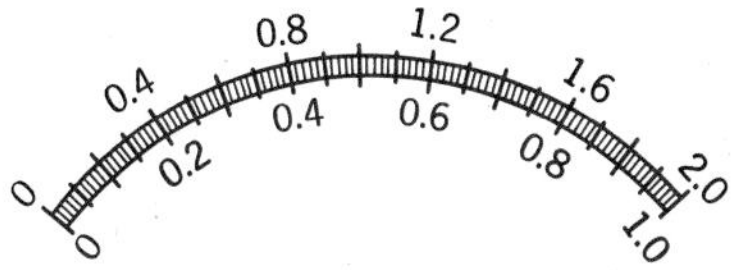

Fig. 14-6 Multirange Meter Scale

multirange meter having the ranges 0–1, 0–10, and 0–100 mA and 0–1 A.

solution The circuit is shown in Fig. 14-7. The switch S is used for range selection by switching in the proper shunt resistor. In its present position no shunt resistor is used and therefore the meter is connected to measure within its basic range of 0–1 mA. At full-scale deflection the voltage across the meter will be

$$E_m = I_t R_m = 0.001 \times 70 = 7 \times 10^{-2}\ \text{V}$$

Since whatever shunt resistor is in use will be in parallel with the resistance of the meter R_m, the same voltage will appear across the shunt resistance. That is,

$$E_m = E_s = 7 \times 10^{-2}\ \text{V}$$

When the 0- to 10-mA range is used, the switch S will connect R_{s10} in parallel with the meter and therefore its internal resistance R_m. For full-scale deflection 1 mA must flow through the meter coil, which leaves 9 mA to flow through R_{s10}. For this condition the value of R_{s10} must be

$$R_{s10} = \frac{E_s}{I_s} = \frac{7 \times 10^{-2}}{9 \times 10^{-3}} = 7.78\ \Omega$$

Similarly, when the 0- to 100-mA range is placed in operation by switching to shunt resistor R_{s100}, full-scale deflection 1 mA still must flow through the meter coil, leaving 99 mA to flow through R_{s100}. Then,

$$R_{s100} = \frac{E_s}{I_s} = \frac{7 \times 10^{-2}}{99 \times 10^{-3}} = 0.707\ \Omega$$

Likewise, when the 0- to 1-A (0- to 1000-mA) range is used, 999 mA must flow through the shunt resistor for full-scale deflection.

$$\therefore R_{s1000} = \frac{E_s}{I_s} = \frac{7 \times 10^{-2}}{999 \times 10^{-3}} = 0.0701\ \Omega$$

It will be noted that only basic Ohm's law was used in Example 1. This was done to emphasize the usefulness of the law. Also, special seldom-used formulas are difficult to remember and handbooks for ready reference are not always available on the job. Actually, you can find the resistance of a meter shunt by using your knowledge of current distribution in parallel circuits. For the 0- to 10-mA range of Example 1, the 70-Ω meter movement must carry 1 mA and the shunt resistor must carry 9 mA. Since the shunt carries nine times the meter current, the shunt resistance must be one-ninth the resistance of the meter, or $\frac{1}{9} \times 70 = 7.78\ \Omega$.

Similarly, for the range of 0 to 100 mA, the meter movement still must carry 1 mA, leaving 99 mA to flow through the shunt. Therefore, the resistance of the shunt will be one ninety-ninth of the resistance of the meter movement, or $\frac{1}{99} \times 70 = 0.707\ \Omega$.

Fig. 14-7 Circuit for Extending Range of 0–1 Milliammeter. Test Leads from Jacks Are Connected in Series with Circuit in Which Current Is to Be Measured

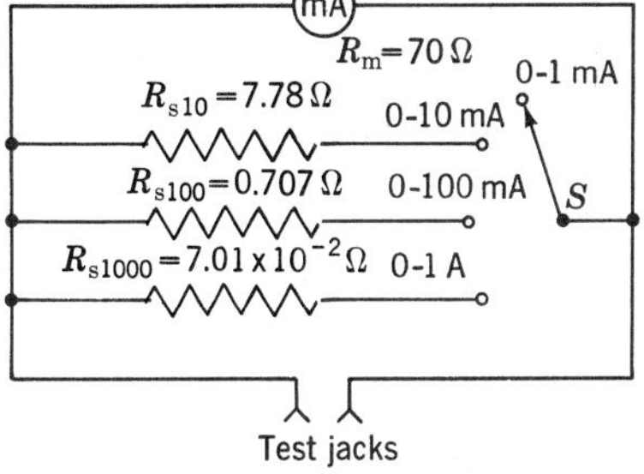

Now that the principles of meter shunts are understood, it is left as an exercise for you to show that

$$R_s = \frac{R_m}{N-1} \quad \Omega \qquad (1)$$

where R_s = shunt resistance, Ω
R_m = meter resistance, Ω
N = ratio obtained by dividing new full-scale reading by basic full-scale reading, both readings in same units

The ratio N is known as the *multiplying power* of the shunt resistor, that is, the factor by which the basic meter scale is multiplied when the shunt resistor R_s is connected in parallel with the meter resistance R_m. From Eq. (1),

$$N = \frac{R_m}{R_s} + 1$$

example 2 By what factor must the scale readings be multiplied when a resistance of 100 Ω is connected across a meter movement of 400 Ω?

solution $N = \frac{R_m}{R_s} + 1 = \frac{400}{100} + 1 = 5$

14-3 SHUNTING METHODS

Although mechanical details are not shown in Fig. 14-7, it is necessary to use a shorting switch in this type of circuit to avoid damage to the meter movement. When switching from one shunt to another, the new shunt must be connected before contact with the shunt in use is broken. If this is not done, the entire circuit current will flow through the meter movement while the switch is moving from one contact to another.

By another method of switching, illustrated in Fig. 14-8, shunts are connected into the

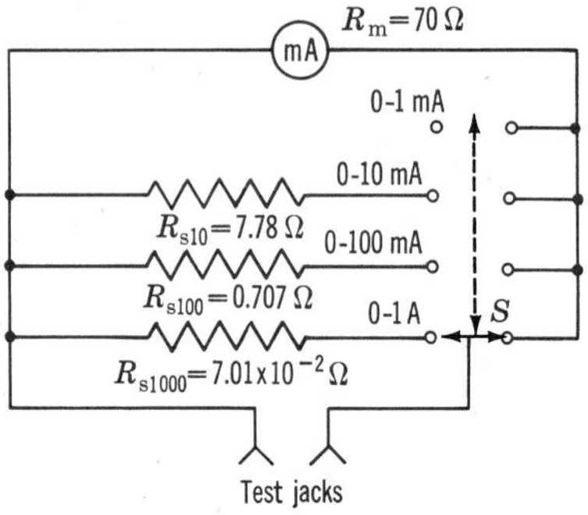

Fig. 14-8 Method of Switching Shunts

circuit by the two-pole rotary switch which makes connections between two sets of contacts. With this arrangement, the meter movement is protected by an open circuit when switching from one shunt to another.

Still another method of employing shunts is shown in Fig. 14-9. This is known as the *Ayrton*, or *universal*, shunt. In addition to other advantages, it provides a safe and convenient method of switching from one range to another. The total shunt resistance, which is permanently connected across the meter, generally has the same resistance as the meter movement. The value of the resistance for each range shunt can be computed by dividing the total circuit resistance

Fig. 14-9 Multicurrent Test Meter Using Universal Shunt

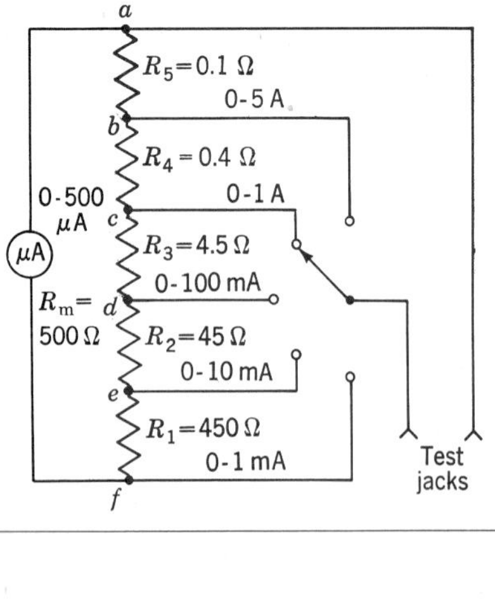

$R_{a-f} + R_m$ by the multiplying power N. This is demonstrated by the development which follows:

$$\begin{aligned} R_{a-e}(I_t - I_m) &= (R_{e-f} + R_m)I_m \\ &= (R_{a-f} - R_{a-e} + R_m)I_m \\ R_{a-e}I_t - R_{a-e}I_m &= R_{a-f}I_m - R_{a-e}I_m + R_mI_m \\ R_{a-e}I_t &= R_{a-f}I_m + R_mI_m \\ R_{a-e} &= \frac{I_m}{I_t}(R_{a-f} + R_m) \\ R_{a-e} &= \frac{1}{N}(R_{a-f} + R_m) \qquad (2) \end{aligned}$$

where R_{a-e} = portion of Ayrton shunt which is connected in shunt for the meter connection at point e ($R_{a-e} = R_2 + R_3 + R_4 + R_5$ in Fig. 14-9.)

R_{a-f} = total Ayrton shunt (In Fig. 14-9, $R_1 + R_2 + R_3 + R_4 + R_5$.)

R_m = meter movement resistance

N = multiplier for switch setting (In Fig. 14-9, $N = 2$ for setting at f, $N = 20$ for setting at e, and so on.)

For example, the 0–500 microammeter movement has a resistance R_m of 500 Ω and the total shunt resistance R_{a-f} connected across the meter is 500 Ω. When the switch is on the 0- to 1-mA position, the multiplying power N is 2.

For the 0- to 10-mA range, N would be 20 because 10 mA is 20 times the original full scale of 0.5 mA. Therefore, the required shunt for this range is

$$R_{a-e} = \frac{R_{a-f} + R_m}{N} = \frac{500 + 500}{20} = 50\ \Omega$$

Since the entire shunt resistance is 500 Ω

$$R_1 = R_{a-f} - R_{a-e} = 500 - 50 = 450\ \Omega$$

When the switch is connected to the 0- to 100-mA range, N becomes 200 and R_1 and R_2 in series (R_{d-f}) form the shunt. That is,

$$R_{a-d} = \frac{R_{a-f} + R_m}{N} = \frac{500 + 500}{200} = 5\ \Omega$$

note $R_{a-d} = \dfrac{2R_m}{N}$ when $R_{a-f} = R_m$

Since

$$R_1 = 450\ \Omega \quad \text{and} \quad R_{a-d} = 5\ \Omega$$

then

$$\begin{aligned} R_2 &= R_{a-f} - (R_1 + R_{a-d}) \\ &= 500 - (450 + 5) \\ &= 45\ \Omega \end{aligned}$$

The values of the remaining shunts are computed in the same manner.

PROBLEMS 14-1

1. A 0–1 milliammeter has an internal resistance of 53 Ω. What shunt resistance is required to extend the meter range to 0–50 mA?
2. A meter movement with a sensitivity of 100 μA has an internal resistance of 1250 Ω. How much shunt resistance is required to result in a 0- to 10-mA range?
3. The meter in Prob. 1 is being used as a multicurrent instrument. The shunt for the 0- to 50-mA range is burned out, but a spool of No. 30 enamel-covered copper wire is on hand. How much of this wire is needed to wind a substitute shunt?

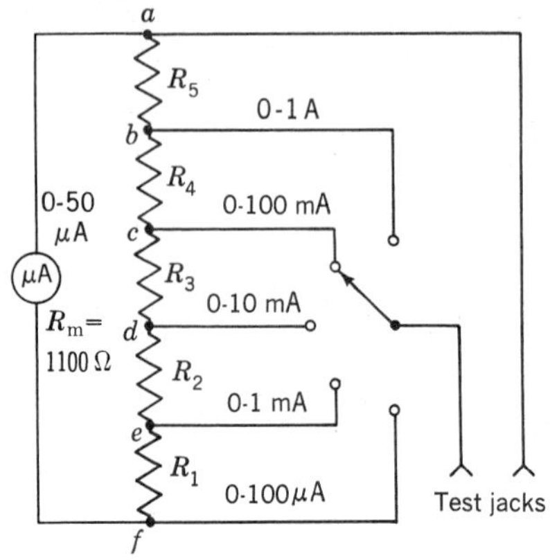

Fig. 14-10 Multicurrent Meter Circuit of Prob. 6

4. A 0–1 milliammeter has an internal resistance of 46 Ω. If this meter is shunted with a 0.939-Ω resistor, by what must the meter readings be multiplied to obtain the correct values of current?

5. It is desired to use the milliammeter illustrated in Fig. 14-2 as a multicurrent meter. What values of shunts are required for the following ranges: (*a*) 0–10 mA, (*b*) 0–100 mA, (*c*) 0–1 A, (*d*) 0–10 A?

6. In the circuit of Fig. 14-10, the total shunt resistance is equal to the resistance of the meter movement. Find the values of R_1, R_2, R_3, R_4, and R_5.

7. A 0–1 milliammeter is available. Design an Ayrton shunt to permit it to be used for the following ranges: (*a*) 0–10 mA, (*b*) 0–100 mA, (*c*) 0–1 A, (*d*) 0–10 A. The meter resistance is 1500 Ω.

14-4 VOLTMETERS

In Fig. 14-11, a voltage of 1 V is impressed across a circuit consisting of a 0–1 milliammeter in series with a variable resistor. The resistor is so adjusted that the circuit is limited to 1 mA; therefore, the meter indicates a full-scale deflection, or a reading of 1 mA.

Fig. 14-11 Basic Circuit of Milliammeter Used to Indicate Voltage

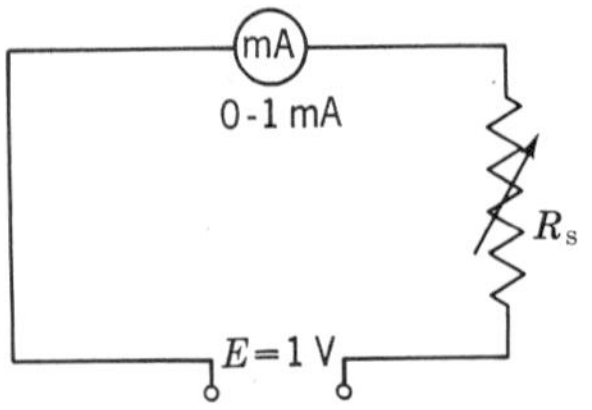

If the resistor is unchanged and the voltage is reduced to 0.5 V, then the circuit current will be reduced to one-half its original value and the meter will read 0.5 mA. Even though the meter deflection is the result of current flow, actually the meter can be used as a 0–1 voltmeter, indicating 1 V in the first instance and 0.5 V when the voltage is reduced.

Similarly, if the resistor is adjusted to a higher safe value so that the application of 150 V causes full-scale deflection, the instrument can be used as a 0–150 voltmeter. In that case voltage values will be obtained by multiplying the basic scale readings by a factor of 150 or by substituting a new scale as shown in Fig. 14-12.

Fig. 14-12 Panel Voltmeter (*Courtesy of Weston Electrical Instrument Corporation*)

example 3 It is desired to use the milliammeter of Fig. 14-2 as a 0–10 voltmeter. What resistance R_{mp} must be connected in series with the instrument to accomplish this?

solution The additional series resistance is called a *multiplier* resistance, and its value must be such that, when it is added to the resistance of the meter movement, the total resistance will limit the current through the instrument to 1 mA when 10 V is applied. The circuit is shown in Fig. 14-13. R_{mp} is the multiplier resistance, and $R_m = 55\ \Omega$ is the resistance of the meter movement.

If 10 V is to be applied across the two series resistances as shown in Fig. 14-13, in order to limit the current to 1 mA, 0.055 V must appear across the meter because

Fig. 14-13 Voltmeter Circuit of Example 3

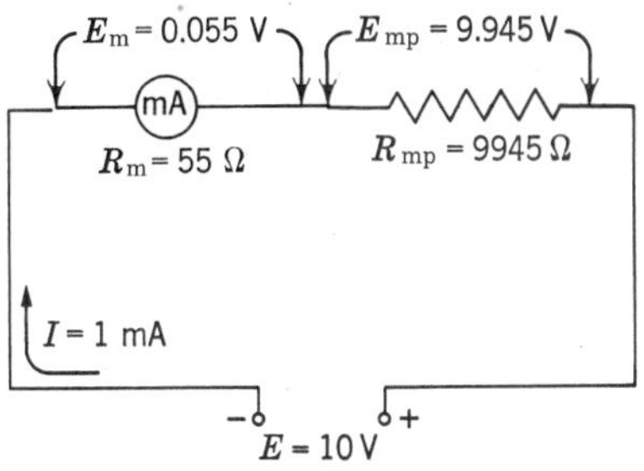

$$E_m = IR_m = 10^{-3} \times 55 = 0.055\ \text{V}$$

The remaining voltage, which is $10 - 0.055 = 9.945$ V, must appear across R_{mp}. Accordingly,

$$R_{mp} = \frac{E_{mp}}{I} = \frac{9.945}{10^{-3}} = 9945\ \Omega$$

If a 10 000-Ω resistor is used as a multiplier, with 10 V applied to the jacks, and if an observer could discern the difference, the voltage reading would be in error by only 0.05 V. (What percent error does this represent?)

example 4 A 0–50 microammeter, with a resistance of 1140 Ω, is to be used as a 0–100 voltmeter. What value of multiplier resistance is needed?

solution For full-scale deflection the voltage across the meter must be limited to

$$E_m = IR_m = 50 \times 10^{-6} \times 1140 = 0.057\ \text{V}$$

The remaining voltage across the multiplier is $100 - 0.057 = 99.943$ V, which results in

$$R_{mp} = \frac{E_{mp}}{I} = \frac{99.943}{50 \times 10^{-6}} = 1\ 998\ 860\ \Omega$$

Naturally, a 2-MΩ resistor would be used.

14-5 VOLTMETER SENSITIVITY

The *sensitivity* of a voltmeter is expressed in the number of ohms in the multiplier for each volt of range. For example, the voltmeter of Example 3 has a range of 10 V and a multiplier of 10 000 Ω, resulting in a sensitivity of 1000 Ω/V. The voltmeter of Example 4 has a sensitivity of 20 000 Ω/V.

14-6 VOLTMETER LOADING EFFECTS

The sensitivity of a voltmeter is a good indication of its accuracy. This is particularly true when the voltages in the low-current circuits often encountered in electronic equipment are measured. For example, a 0–150 voltmeter with a sensitivity of 200 Ω/V would give excellent service, say as a power switchboard meter, at an economical cost. However, it would not be satisfactory for some other applications. In Fig. 14-14, two 60-kΩ resistors are connected in series across 120 V. In this condition, 60 V will appear across each resistor. If the voltmeter is connected across R_2 as shown in Fig. 14-15, the joint resistance R_p of R_2 and R_{mp} becomes

$$R_p = \frac{R_2 R_{mp}}{R_2 + R_{mp}} = 20\,000\ \Omega$$

The total resistance of the circuit is now

$$R_t = R_1 + R_p = 60\,000 + 20\,000$$
$$= 80\,000\ \Omega = 80\ \text{k}\Omega$$

This results in a circuit current of

$$I_t = \frac{E}{R_t} = \frac{120}{80\,000} = 1.5 \times 10^{-3}\ \text{A}$$

Therefore, the voltage existing across R_2 due to the shunting effect of the voltmeter is

$$E_p = I_t R_p = 1.5 \times 10^{-3} \times 20\,000$$
$$= 30\ \text{V}$$

Fig. 14-14 The Current through the Resistors Is 1 mA, and the Voltage across Each Resistor Is 60 V

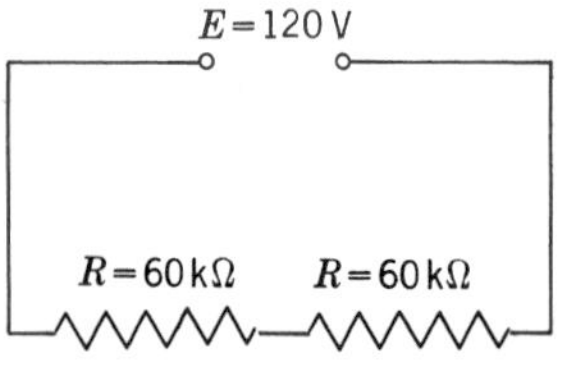

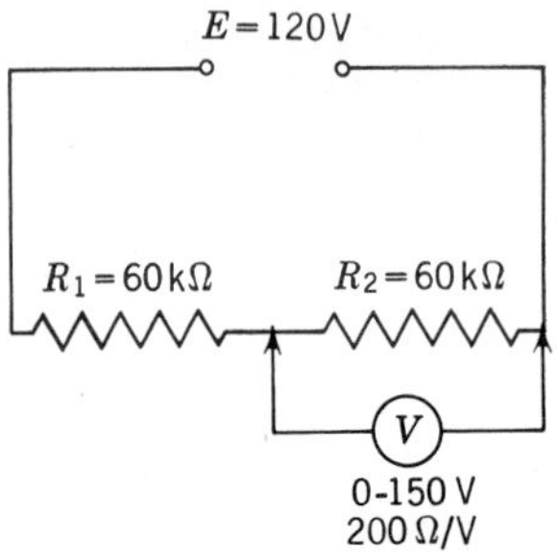

Fig. 14-15 A 30 000-Ω Voltmeter Connected across R_2. Total Circuit Current Is Now 1.5 mA, and the Voltage across R_2 Is 30 V

It is left as an exercise for you to show that if the voltmeter of Example 4 is used to measure the voltage across R_2, the reading will be 59.1 V.

14-7 MULTIRANGE VOLTMETERS

Using a single multiplier provides only one voltmeter range. Similar to the usage of current-measuring instruments, it has become practice to increase the usefulness of an instrument by selecting a meter movement of sufficient sensitivity and, with the use of several multipliers, use the instrument as a multirange voltmeter. Such an arrangement is shown in Fig. 14-16.

14-8 OHMMETERS

Owing to the fact that a change in the resistance of a circuit will cause a change in the

Fig. 14-16 A 0–50 Microammeter Used with Multipliers for Multirange Voltmeter

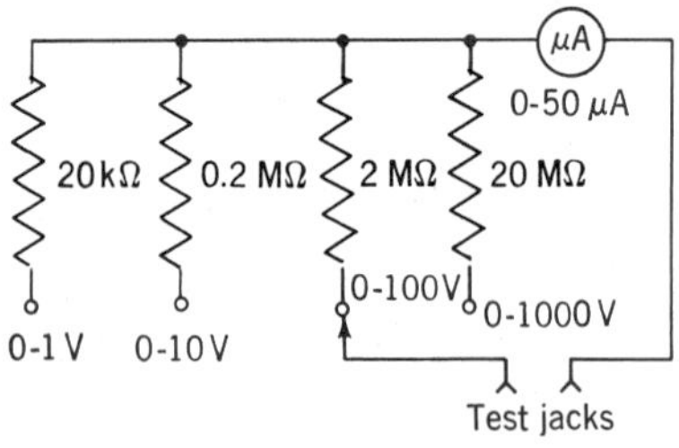

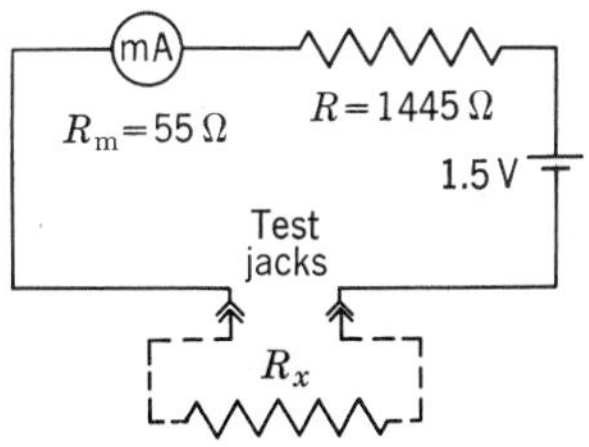

Fig. 14-17 A 0–1 Milliammeter Used in Ohmmeter Circuit

current in that circuit, a current-measuring instrument can be calibrated to indicate values of resistance required for a given change in current. Such a calibrated instrument is called an *ohmmeter.*

In the schematic diagram of Fig. 14-17, the 0–1 milliammeter of Fig. 14-2 is connected in series with a 1.5-V battery and a resistance of 1445 Ω. Since the total resistance of the circuit is 1500 Ω, if the test jacks are short-circuited, the meter will read full scale. If the short circuit is removed and a resistance R_x of 1500 Ω is connected across the jacks, the meter will indicate half-scale deflection because now the total circuit resistance is 3000 Ω. Therefore, at full-scale deflection the meter scale could be marked 0 Ω of external circuit resistance, and at half scale it could be marked 1500 Ω. Similarly, other values of known resistance could be used to calibrate the scale throughout its range. Also, unknown resistances can be used to calibrate the scale by making use of the relation

$$R_x = R_c \frac{I_1 - I_2}{I_2} \quad \Omega \tag{3}$$

where R_x = unknown resistance, Ω
R_c = circuit resistance when test jacks are short-circuited, Ω
I_1 = current when test jacks are short-circuited, A
I_2 = current when R_x is connected in circuit, A

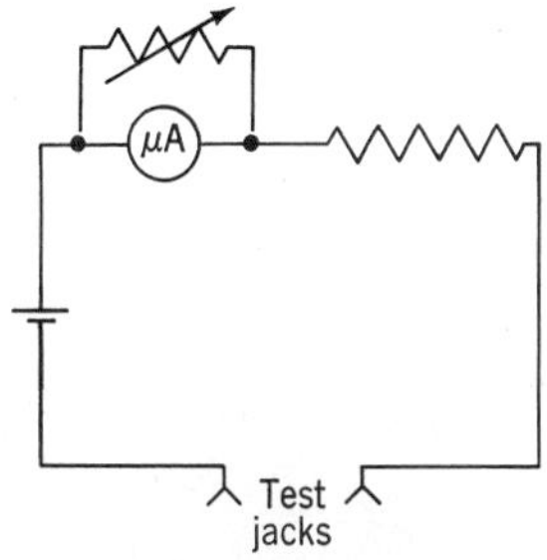

Fig. 14-18 Ohmmeter Circuit with Variable Shunt Resistance

Use your knowledge of Ohm's law and Axiom 5 (Sec. 5-2) to derive Eq. (3).

As a provision for compensating for battery aging and maintaining calibration, variable resistors controlled from the instrument panel are connected in ohmmeter circuits by either of two methods as illustrated in Figs. 14-18 and 14-19. In either case the test leads are short-circuited and the resistor control is adjusted until the meter reads full scale, or 0 Ω. An example of such a control is the "Ω ADJ" on the instrument shown in Fig. 14-20.

Since zero resistance between the test jacks results in maximum current and larger values of resistance result in less current, certain types of ohmmeter scales are marked with numbers increasing from right to left as illustrated on the ohms scale in Fig. 14-20.

In practice, the use of the ordinary ohmmeter should be limited from about one-tenth of to ten times the center-scale resistances reading because of the small deflection

Fig. 14-19 Ohmmeter Circuit with Variable Series Resistance

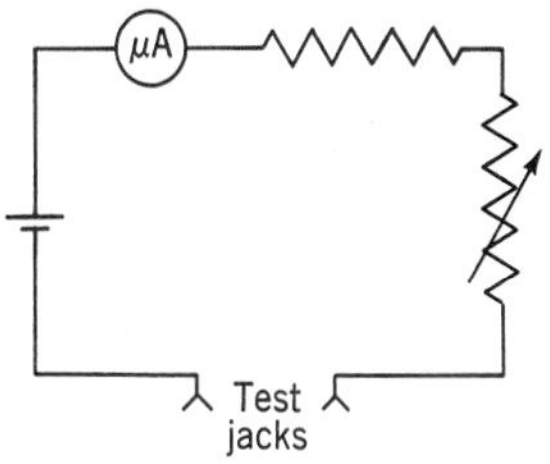

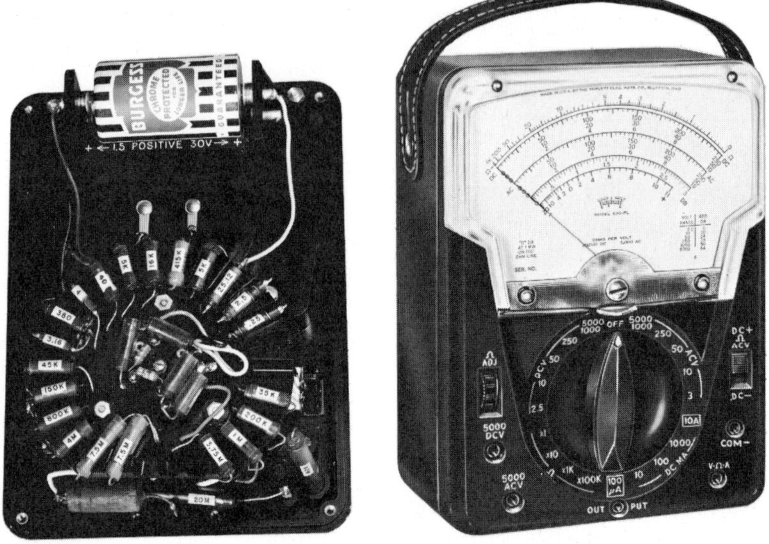

Fig. 14-20 Multimeter. See the Arrangement of Shunts and Multipliers on the Selector Switch (*Courtesy of Triplett Electrical Instrument Company*)

changes at the ends of the scale. For this reason multirange ohmmeters are employed for changing midscale values, and the ranges generally are designed to multiply the basic scale by some power of 10.

14-9 MULTIMETERS

For the purposes of convenience and economy, meters combining the functions and desired ranges of ammeters, voltmeters, and ohmmeters are incorporated into one instrument called a multimeter, one type of which is illustrated in Fig. 14-20. If the test leads are plugged into the proper pin jacks and the rotary switch is switched to the proper function and range, the instrument can be utilized for several functions.

PROBLEMS 14-2

1. In the circuit of Fig. 14-21: (*a*) What voltages are across R_1 and R_2? (*b*) A 0–100 voltmeter with a sensitivity of 1000 Ω/V is connected across R_1. What is the reading of the voltmeter?

2. In the circuit of Fig. 14-21:
   (*a*) A 0–100 voltmeter with a sensitivity of 20 000 Ω/V is connected across R_1. What is the voltmeter reading?
   (*b*) What will the voltmeter read if connected across points A and B?
   (*c*) When the voltmeter is connected across points A and B, what current flows through R_2?

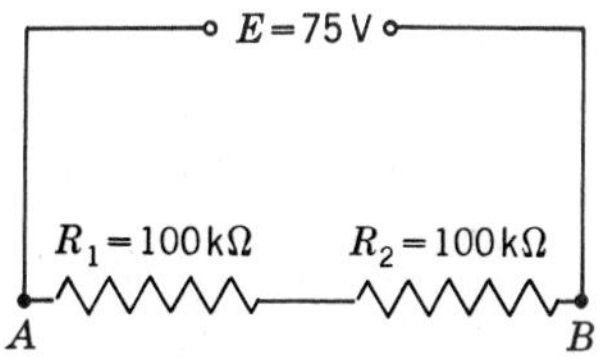

Fig. 14-21 Circuit of Probs. 1 and 2

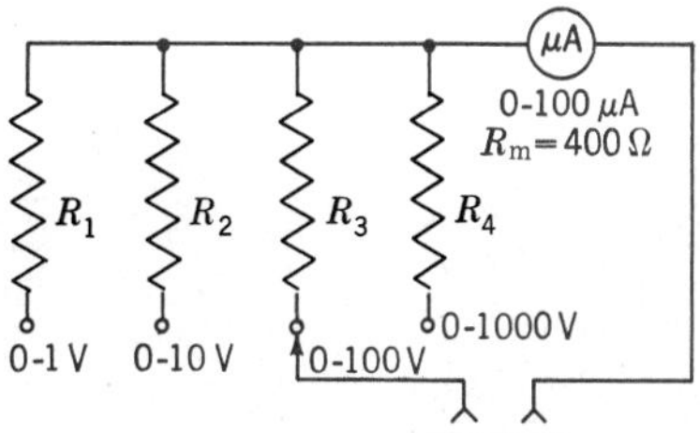

Fig. 14-22 Multirange Voltmeter Circuit of Prob. 3

3. What are the values of the multiplier resistors R_1, R_2, R_3, and R_4 in Fig. 14-22?

4. Refer to Eq. (1). Did you show that $R_s = \dfrac{R_m}{N-1}\ \Omega$?

5. Refer to the end of Sec. 14-6. Did you show that the voltmeter reading will be 59.1 V?

6. Refer to Eq. (3). Did you show that $R_x = R_c \dfrac{I_1 - I_2}{I_2}$?

15

divider circuits and wheatstone bridges

In this chapter consideration is given to voltage and current divider circuits. Computations involving voltages and currents in these circuits are simply applications of Ohm's law to series and parallel circuits.

The source of power for radio and television receivers, amplifiers, and similar electronic equipment generally consists of a filtered direct voltage which has been obtained from a rectified alternating voltage. For reasons of economy and design considerations, rectifier power supplies are usually so designed that only the highest voltage desired is available at the output. In most applications, however, other voltages are needed. For example, power tubes sometimes require higher voltages than voltage amplifier tubes require. Screen grids may require yet other voltages. Also, bias voltages are often required. These voltages can be made available from single sources of voltage by the use of *voltage dividers.*

15-1 VOLTAGE DIVIDERS

That several values of voltage can exist around a circuit was first demonstrated in Sec. 8-8 and Figs. 8-12 and 8-13. A similar situation exists when tapped resistors, or resistors in series, are connected across the output of a power supply as illustrated in Fig. 15-1. This represents a simple *voltage divider.*

Since the resistors are of equal value, one-third of the 300-V output voltage will appear across each resistor. Therefore, since terminal D is at zero or ground potential, terminal C will be $+100$ V with respect to D, terminal B will be $+200$ V, and terminal A will be $+300$ V.

Fig. 15-1 Voltage Divider Consisting of Three 25-kΩ Resistors Connected across 300-V Power Supply

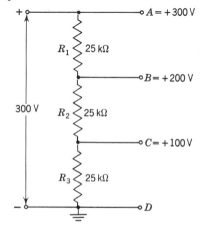

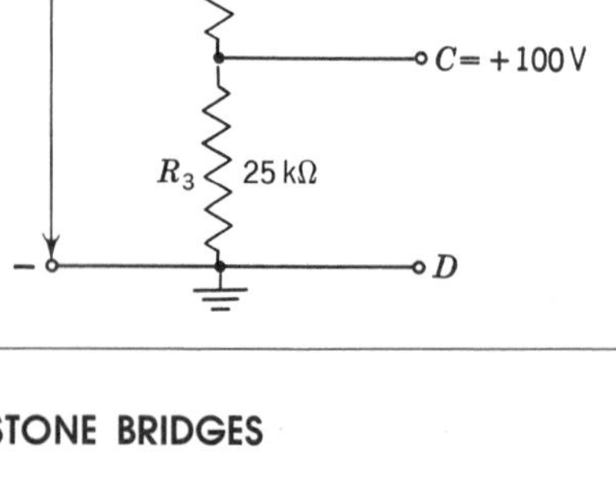

In addition to serving as a voltage divider, the total resistance connected across the output of a power supply generally serves as a *load resistor* and as a *bleeder*. The latter serves to "bleed off" the charge of the filter capacitors after the rectifier is turned off. As a compromise between output voltage regulation and efficiency of operation, the total value of the voltage divider resistance is so designed that the bleeder current will be about 10% of the full-load current. The bleeder current in Fig. 15-1 with no loads connected to the various voltage divider terminals is

$$I = \frac{E}{R_1 + R_2 + R_3} = \frac{300}{75\,000}$$
$$= 4.00 \text{ mA}$$

The grounded point of a voltage divider is generally used as the reference point for circuit voltages supplied by the voltage divider. In Fig. 15-1, this is at grounded terminal *D*.

If the power supply output voltage is grounded at no other point, the voltage divider can be grounded at an intermediate point so as to obtain both positive and negative voltages. For example, if the voltage divider resistors of Fig. 15-1 are grounded as shown in Fig. 15-2, the voltage relations

Fig. 15-2 Voltage Divider Grounded at C

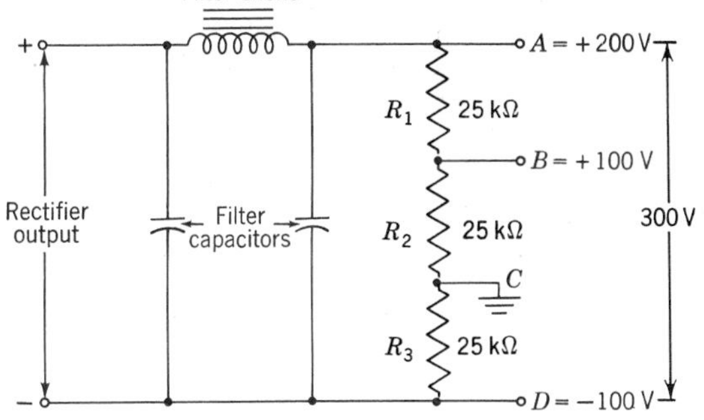

change. Terminal *D* is now −100 V with respect to ground, *B* is +100 V, and *A* is +200 V.

15-2 VOLTAGE DIVIDERS WITH LOADS

The voltage dividers of Figs. 15-1 and 15-2 have no loads connected to them; only the bleeder current of 4 mA flows through the voltage divider resistors. When loads are connected to the various terminals, the resulting additional currents must be taken into consideration because they affect the operating voltages. For example, assume a load of $R_4 = 50\,000\ \Omega$ connected between terminals *C* and *D* of Fig. 15-1. Under these conditions, the resistance between terminals *C* and *D* is

$$R_{CD} = \frac{R_3 R_4}{R_3 + R_4} = \frac{25\,000 \times 50\,000}{25\,000 + 50\,000}$$
$$= 16\,700\ \Omega = 16.7\text{ k}\Omega$$

The total resistance between terminals *A* and *D* is

$$R_{AD} = R_1 + R_2 + R_{CD} = 66\,700\ \Omega$$

resulting in a total current of

$$I_t = \frac{E}{R_{AD}} = 4.50 \text{ mA}$$

The voltage across terminals *B* and *D* is

$$E_{BD} = I_t R_{BD}$$
$$= 188 \text{ V} \qquad \text{(instead of 200 V)}$$

and across terminals *C* and *D* it is

$$E_{CD} = I_t R_{CD}$$
$$= 75 \text{ V} \qquad \text{(instead of 100 V)}$$

The circuit is shown in Fig. 15-3.

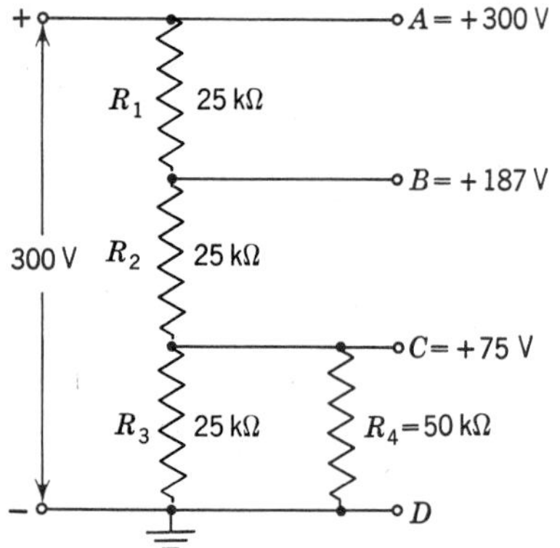

Fig. 15-3 Load of 50 kΩ Connected across Terminals C and D

Show that, if an additional load of $R_5 = 50$ kΩ is connected across terminals *B* and *D*, the terminal voltages would be as illustrated in Fig. 15-4.

example 1 Design a voltage divider circuit for a 250-V power supply. The connected loads are 60 mA at 250 V and 40 mA at 150 V. Allow a 10% bleeder current.

solution The circuit is shown in Fig. 15-5. The total load current is 100 mA; therefore, bleeder current is 10 mA, which flows

Fig. 15-4 Loads $R_4 = R_5 = 50$ kΩ Connected to Voltage Divider

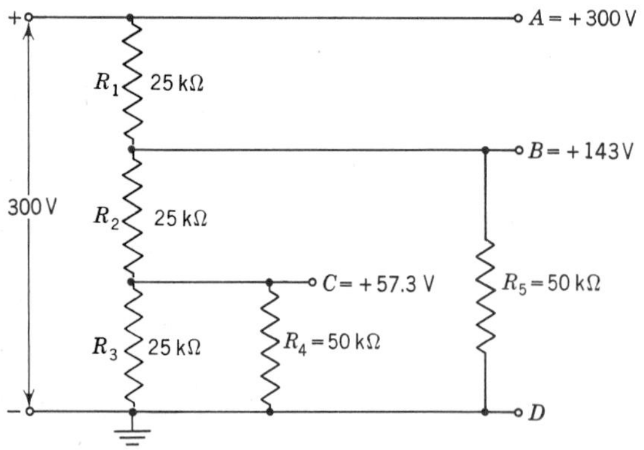

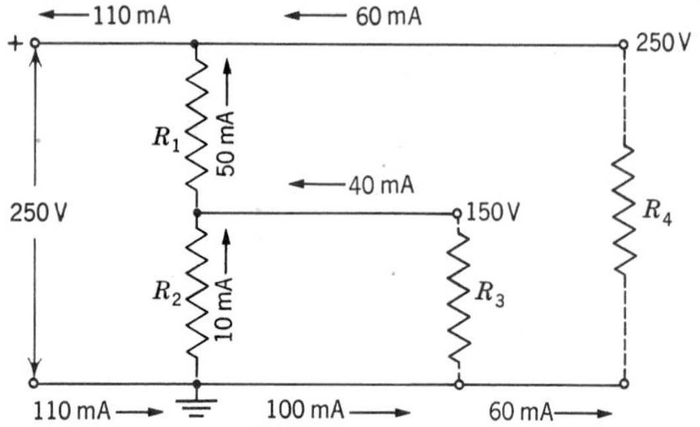

Fig. 15-5 Circuit of Example 1

through R_2. Since the voltage across R_2 is 150 V,

$$R_2 = \frac{150}{10 \times 10^{-3}} = 15\,000\ \Omega = 15\ \text{k}\Omega$$

The current flowing through R_1 is $40 + 10 = 50$ mA, and the voltage across R_1 is $250 - 150 = 100$ V. Then

$$R_1 = \frac{100}{50 \times 10^{-3}} = 2000\ \Omega$$

example 2 What are the values of the voltage divider resistors in Fig. 15-6 if the bleeder current is 10% of the total load current?

Fig. 15-6 Voltage Divider of Example 2

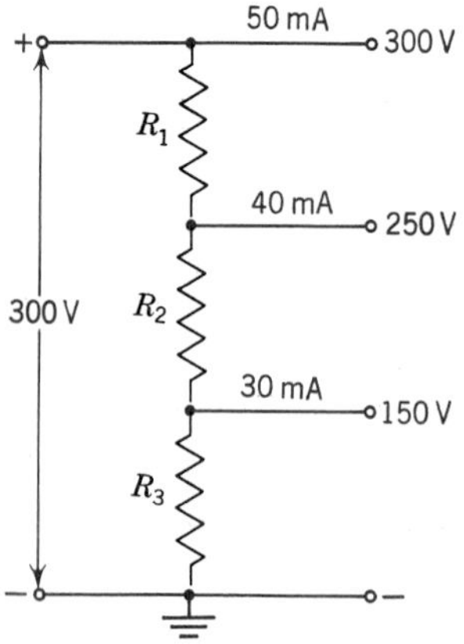

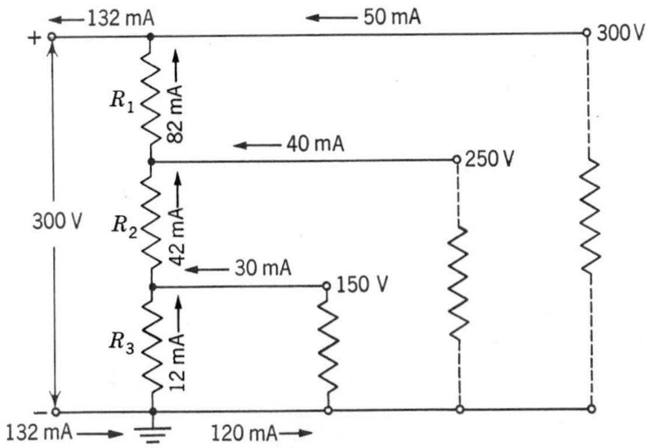

Fig. 15-7 Complete Circuit of Example 2

solution The total load current I_L is

$$I_L = 50 + 40 + 30 = 120 \text{ mA}$$

The bleeder current is

$$I_B = 0.1 \times 120 = 12 \text{ mA}$$

The complete circuit is shown in Fig. 15-7. The voltage across R_3 is 150 V, and only the bleeder current of 12 mA flows through this resistor. Therefore,

$$R_3 = \frac{150}{12 \times 10^{-3}} = 12.5 \text{ k}\Omega$$

The 30-mA load current of the 150-V load terminal combines with the bleeder current of 12 mA for a total of 42 mA through R_2, across which is 100 V. Therefore,

$$R_2 = \frac{100}{42 \times 10^{-3}} = 2.38 \text{ k}\Omega$$

Similarly, 82 mA flows through R_1, across which is 50 V. Then

$$R_1 = \frac{50}{82 \times 10^{-3}} = 610 \; \Omega$$

note Resistors $R_1 = 610 \; \Omega$, $R_2 = 2380 \; \Omega$, and $R_3 = 12\,500 \; \Omega$ are not readily available commercially. Try substituting standard preferred values of $R_1 = 560 \; \Omega$, $R_2 = 2.4 \text{ k}\Omega$, and $R_3 = 12 \text{ k}\Omega$ for the computed values, and determine how this would affect the loads.

example 3 Find the values of the voltage divider resistors of Fig. 15-8. The −50-V bias terminal draws no current, and the bleeder current is 10% of the total load current.

Fig. 15-8 Voltage Divider of Example 3

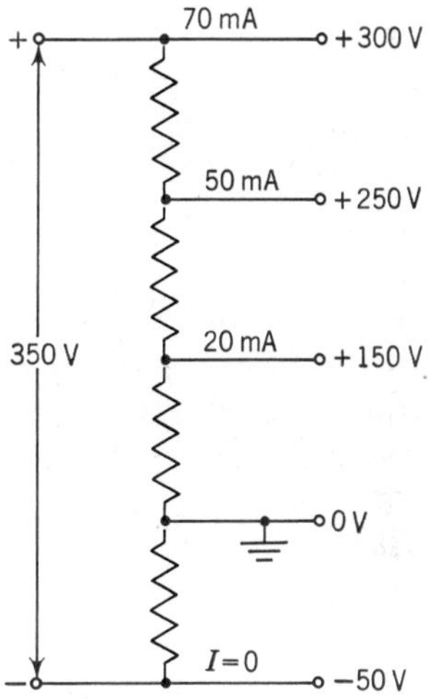

solution The total load current I_L is

$$I_L = 70 + 50 + 20 = 140 \text{ mA}$$

The bleeder current is

$$I_B = 0.1 I_L = 0.1 \times 140 = 14 \text{ mA}$$

The complete circuit is illustrated in Fig. 15-9. There is a voltage of 50 V across R_4, and the total current of 154 mA flows through this resistor. Therefore

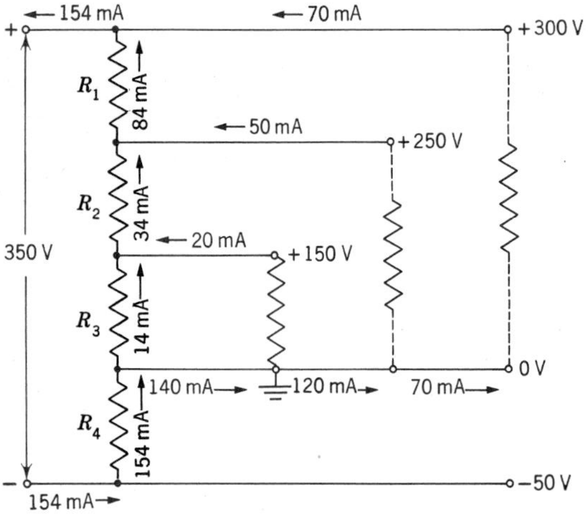

Fig. 15-9 Complete Circuit of Example 3

$$R_4 = \frac{50}{154 \times 10^{-3}} = 325\ \Omega$$

Since R_3 carries only the bleeder current and the voltage across this resistor is 150 V,

$$R_3 = \frac{150}{14 \times 10^{-3}} = 10.7\ \text{k}\Omega$$

In like manner,

$$R_2 = \frac{100}{34 \times 10^{-3}} = 2.94\ \text{k}\Omega$$

and

$$R_1 = \frac{50}{84 \times 10^{-3}} = 595\ \Omega$$

note As a problem, substitute the commercially available preferred values of $R_1 = 620\ \Omega$, $R_2 = 3\ \text{k}\Omega$, $R_3 = 11\ \text{k}\Omega$, and $R_4 = 300\ \Omega$ for the computed values, and determine how the loads would be affected.

PROBLEMS 15-1

1. The vertical attenuator of an oscilloscope is illustrated in Fig. 15-10. With an input voltage of 60 V, what voltages appear between the switch positions and the input to the vertical amplifier?
note No current flows from the circuit.

2. The horizontal hold control of a television receiver is shown in Fig. 15-11. What range of control voltage is available from the potentiometer to the horizontal hold control?
note The horizontal hold draws no current from the circuit.

3. What is the power dissipated by each of the resistors and the potentiometer of Prob. 2?

Fig. 15-10 Circuit of Prob. 1

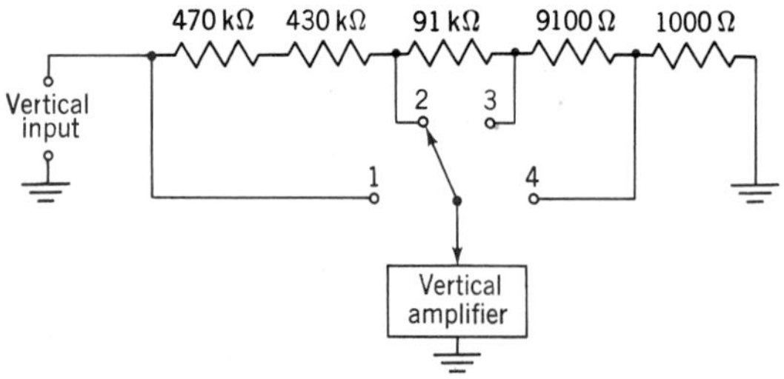

Fig. 15-11 Circuit of Prob. 2

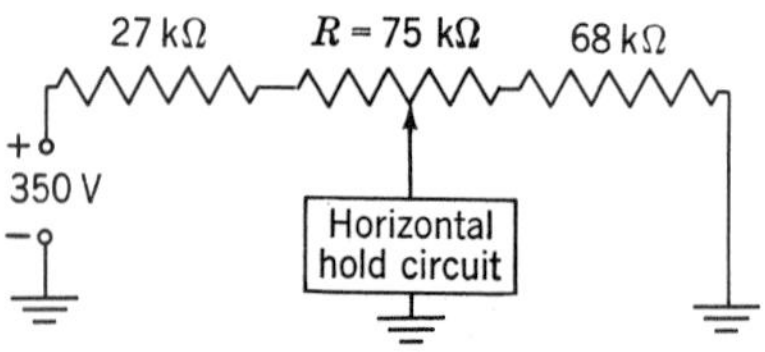

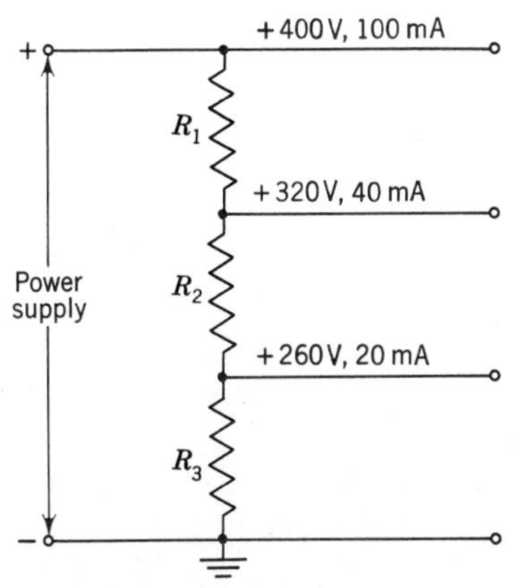

Fig. 15-12 Circuit of Probs. 4 and 5

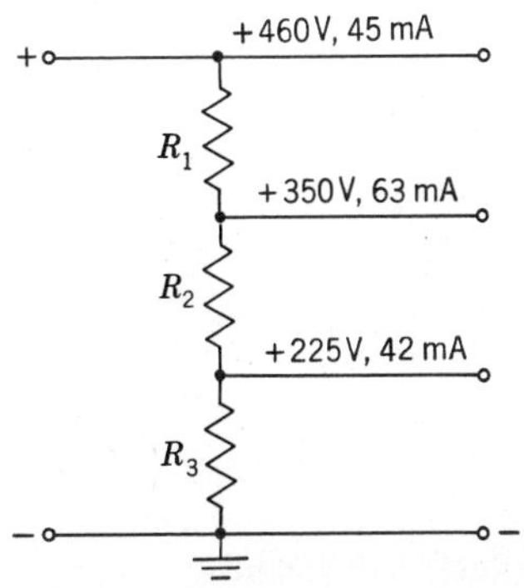

Fig. 15-13 Circuit of Probs. 6, 7, and 8

4. Determine the values of the voltage divider resistors of Fig. 15-12 if a total of 180 mA is drawn from the power supply.

5. What is the total power expended in the voltage divider of Prob. 4, and what power is dissipated by each of the resistors?

6. What are the values of the voltage divider resistors of Fig. 15-13 if the bleeder current is 10% of the total load current?

7. What is the power dissipated by each of the resistors in Prob. 6?

8. What is the total power delivered by the voltage source in Prob. 6?

9. What are the values of the voltage divider resistors of Fig. 15-14 if the bleeder current is 10 mA?

10. What wattage ratings should be used for the resistors in Prob. 9?

11. What is the total power delivered by the voltage source of Prob. 9?

Fig. 15-14 Circuit of Probs. 9, 10, and 11

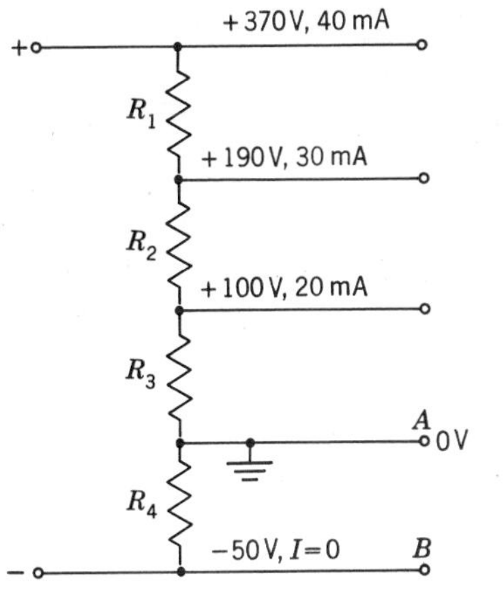

12. If the biasing resistor of R_4 of Fig. 15-14 became open-circuited, what would be the voltage between terminals A and B?
13. Referring to Sec. 15-2, did you show that Fig. 15-4 is the result when Fig. 15-3 is changed by the addition of a 50-kΩ load?
14. Referring to Example 2, did you try substituting standard 5% preferred values into the voltage divider of Fig. 15-7?
15. Referring to Example 3, did you try substituting standard 5% preferred values into the voltage divider of Fig. 15-9?

15-3 CURRENT DIVIDERS

We have seen that voltage dividers are employed to develop voltage drops across series resistors. Each voltage drop is proportional to the resistance value related to the total series resistance. When resistors are connected in parallel, the voltage is the same across each, but the current is divided in *inverse* proportion.

example 4 In Fig. 15-15, since

$$E = \text{voltage drop across } R_1 = I_1R_1$$

and $\quad E = \text{voltage drop across } R_2 = I_2R_2$

and $\quad I_t = I_1 + I_2$

Therefore

$$I_1R_1 = I_2R_2$$

and $\quad I_1R_1 = (I_t - I_1)R_2 = I_tR_2 - I_1R_2$

Collecting like terms,

$$I_1R_1 + I_1R_2 = I_tR_2$$
$$I_1(R_1 + R_2) = I_tR_2$$

so that $\quad I_1 = I_t\left(\frac{R_2}{R_1 + R_2}\right) \quad (1)$

Note carefully that the numerator in Eq. (1) is R_2 and *not* R_1. You should now prove to your own satisfaction that

$$I_2 = I_t\left(\frac{R_1}{R_1 + R_2}\right) \quad (2)$$

Fig. 15-15 Current Divider Circuit of Example 4

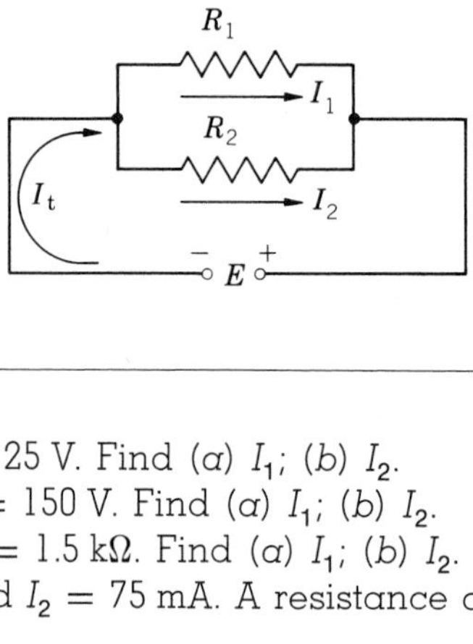

PROBLEMS 15-2

1. In Fig. 15-15, $R_1 = 100\ \Omega$ and $R_2 = 300\ \Omega$. $E = 25$ V. Find (*a*) I_1; (*b*) I_2.
2. In Fig. 15-15, $R_1 = 2.2$ kΩ and $R_2 = 4.7$ kΩ. $E = 150$ V. Find (*a*) I_1; (*b*) I_2.
3. In Fig. 15-15, $I_t = 150$ mA. $R_1 = 680\ \Omega$ and $R_2 = 1.5$ kΩ. Find (*a*) I_1; (*b*) I_2.
4. In Fig. 15-15, it is required that $I_1 = 25$ mA and $I_2 = 75$ mA. A resistance of 200 Ω has been established for R_1.
   (*a*) What value must be used for R_2?
   (*b*) What emf must be applied to achieve the required output?
5. In Fig. 15-15, the source of emf is replaced with a constant current source capable of delivering 5 A under all conditions. $R_1 = 100$ kΩ is a shunt resistor permanently connected across the constant current source. $R_2 = 1.2$ kΩ is a load driven by the shunted constant current source. Find I_2.

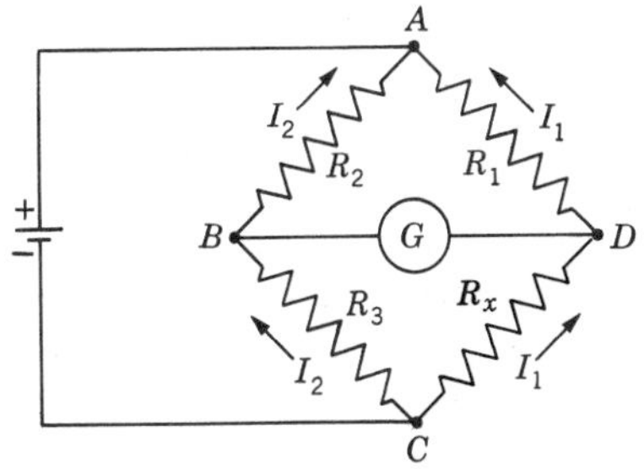

Fig. 15-16 Schematic Diagram of Wheatstone Bridge

15-4 WHEATSTONE BRIDGE CIRCUITS

The accuracy of resistance measurements by the voltmeter ammeter method is limited, mainly because of errors in the meters and the difficulty of reading the meters precisely. Probably the most widely used device for precise resistance measurement is the Wheatstone bridge, the circuit diagram of which is shown in Fig. 15-16.

Resistors R_1, R_2, and R_3 are known values, and R_x is the resistance to be measured. In most bridges, R_1 and R_2 are adjustable in ratios of 1:1, 10:1, 100:1, etc., and R_3 is adjustable in small steps. In measuring a resistance, R_3 is adjusted until the galvanometer reads zero, and in this condition the bridge is said to be "balanced." Since the galvanometer reads zero, it is evident that the points B and D are exactly at the same potential; that is, the voltage drop from A to B is the same as from A to D. Expressed as an equation,

$$E_{AD} = E_{AB}$$

or

$$I_1R_1 = I_2R_2 \qquad (3)$$

Similarly, the voltage drop across R_x must be equal to that across R_3; hence,

$$I_1R_x = I_2R_3 \qquad (4)$$

Dividing Eq. (4) by Eq. (3),

$$\frac{I_1R_x}{I_1R_1} = \frac{I_2R_3}{I_2R_2}$$

$$\therefore \frac{R_x}{R_1} = \frac{R_3}{R_2} \qquad (5)$$

Equation (5) is the fundamental equation of the Wheatstone bridge. By solving it for the only unknown R_x, the value of the resistance under measurement can be computed.

example 5 In the circuit of Fig. 15-16, $R_1 = 10\ \Omega$, $R_2 = 100\ \Omega$, and $R_3 = 13.9\ \Omega$. If the bridge is balanced, what is the value of the unknown resistance?

solution Solving Eq. (5) for R_x,

$$R_x = \frac{R_1R_3}{R_2}$$

Substituting the known values,

$$R_x = \frac{10 \times 13.9}{100} = 1.39\ \Omega$$

Locating the point at which a telephone cable or a long control line is grounded is simplified by the use of two circuits that are modifications of the Wheatstone bridge. These are the Murray loop and the Varley loop.

Figure 15-17 represents the method of locating the grounded point in a cable by

Fig. 15-17 Murray Loop

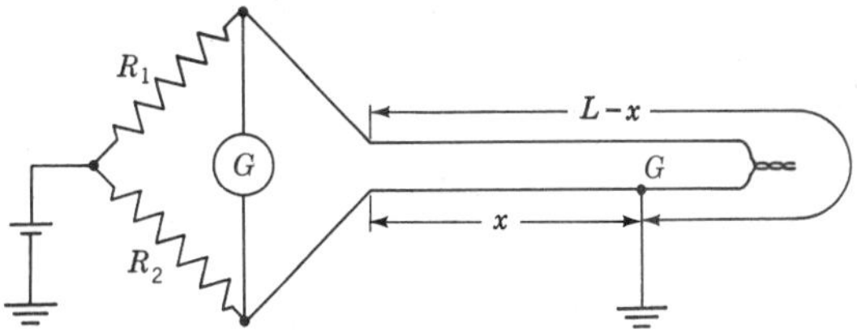

using a Murray loop. A spare ungrounded cable is connected to the grounded cable at a convenient location beyond the grounded point G. This forms a loop of length L, one part of which is the distance x from the point of measurement to the grounded point G. The other part of the loop is then $L - x$. These two parts of the loop form a bridge with R_1 and R_2, which are adjusted until the galvanometer shows no deflection. Because this results in a balanced bridge circuit,

$$\frac{R_2}{R_1} = \frac{x}{L - x} \tag{6}$$

Solving for x,

$$x = \frac{R_2}{R_1 + R_2} L \tag{7}$$

example 6 A Murray loop is connected as in Fig. 15-17 to locate a ground in a cable between two cities 60 km apart. The lines forming the loop are identical. With the bridge balanced, $R_1 = 645\ \Omega$ and $R_2 = 476\ \Omega$. How far is the grounded point from the test end?

solution Substituting the known values in Eq. (7),

$$x = \frac{476}{645 + 476} \times 2 \times 60 = 50.95 \text{ km}$$

If the two cables forming the loop are not the same size, the relations of Eq. (7) can be used to compute the resistance R_x of the grounded cable from the point of measurement to the grounded point. Then if R_L is the resistance of the entire loop.

$$R_x = \frac{R_2}{R_1 + R_2} R_L \tag{8}$$

example 7 A Murray loop is connected as in Fig. 15-17. The grounded cable is No. 19 wire, and wire of a different size is used to complete the loop. The resistance of the entire loop is 126 Ω, and when the bridge is balanced, $R_1 = 342\ \Omega$ and $R_2 = 217\ \Omega$. How far is the ground from the test end?

solution Substituting the known values in Eq. (8),

$$R_x = \frac{217}{342 + 217} \times 126 = 48.9\ \Omega$$

Since No. 19 wire has a resistance of 24.6 Ω/km, 48.9 Ω represents 1.852 km of wire between the test end and the grounded point.

PROBLEMS 15-3

1. In the Wheatstone bridge of Fig. 15-16, $R_1 = 0.001\ \Omega$, $R_2 = 1\ \Omega$, $R_3 = 52.4\ \Omega$. What is the value of the unknown resistance?
2. In the Wheatstone bridge of Fig. 15-16, the ratio of $R_2\!:\!R_3$ is 100:1. R_1 is 6.28 Ω. What is the unknown resistor?
3. In the Wheatstone bridge, the ratio of $R_1\!:\!R_2$ is 1000 and R_x is believed to be 22.6 Ω. At what setting of R_3 may a balance be expected?
4. A ground exists on one conductor of a lead-covered No. 19 pair. A Murray loop is used to locate the fault by connecting the pair together at the far end (Fig. 15-17). When the bridge circuit is balanced, $R_1 = 33.3\ \Omega$ and $R_2 = 21.7\ \Omega$. If the cable is 2.2 km long, how far from the test end is the cable grounded?

5. Several No. 8 wires run between two cities located 65 km apart. One wire becomes grounded, and a Murray loop is used in one city to locate the fault by connecting two of the wires in the other city. When the bridge is balanced, $R_1 = 716\ \Omega$ and $R_2 = 273\ \Omega$. How far from the test end is the wire grounded?
6. A No. 6 wire, which is known to be grounded, is made into a loop by connecting a wire of different size at its far end. The resistance of the loop thus formed is 5.62 Ω. When a Murray loop is connected and balanced, the value of R_1 is 16.8 Ω and that of R_2 is 36.2 Ω. How far from the test end does the ground exist?
7. As a research project, discover the details of the Varley loop and develop its equation, which is similar to that for the Murray loop.

16

graphs

A graph is a pictorial representation of the relationship between two or more quantities. Everyone is familiar with various types of graphs or graphic charts. They are used extensively in magazines, newspapers, annual reports, and trade journals published for engineers, manufacturers, and others concerned with relative values. It is difficult to conceive how engineers could dispense with them.

We have already used simple graphic representations in Chap. 3, and here we will develop a few of the uses of straight-line graphs. In later chapters we will use graphs in working out the solutions of problems and in quickly presenting information in varied forms.

The notions presented here are fundamental to the use of all graph forms, and we are paving the way for some important and interesting topics which will follow in later chapters.

16-1 LOCATING POINTS ON A GRAPH

The accurate location of points is vital, and the manner of marking points can help or seriously hinder in arriving at a correct solution to a problem. One of the most common methods of locating a point is by using a large dot (Fig. 16-1). But this is the poorest form of location, and Fig. 16-1 illustrates why: with a large dot, do you draw the line

Fig. 16-1 Illustration of Errors Introduced by the Use of Large Dots to Locate Points.
(a) Instead of a Single Fine Line, a Broad Range of Possibilities Is Presented.
(b) Shall We Join the Outside Edges of the Dots?
(c) Should We Join from Top to Top (or Bottom to Bottom)?
(d) Should We Just Pick a Line That Somehow Touches Both Dots Somewhere?

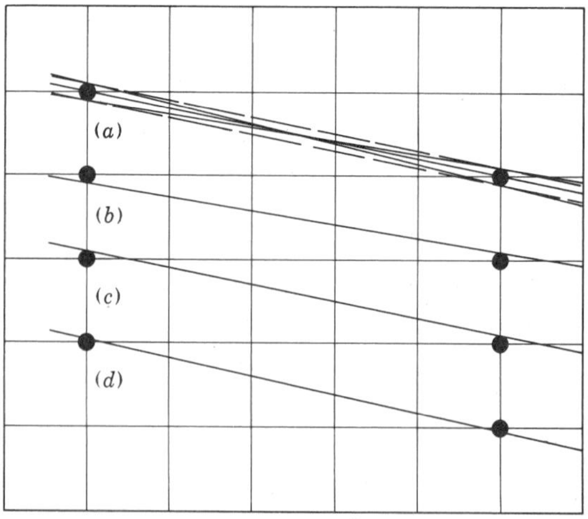

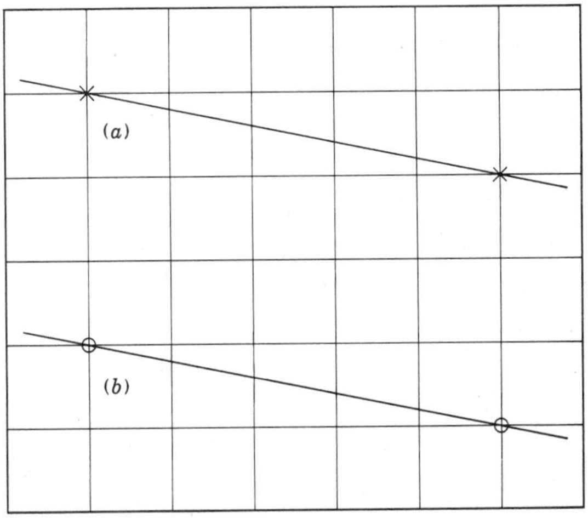

Fig. 16-2 Illustration of the Correct Method of Locating Points: After the Line Is Drawn, the Small Point Locations Are Still Indicated, But Only a Single Line Can Be Drawn between the Points

through the center, through the top, or through the bottom? Can you be sure where the center is? The possibility of introducing errors is great, and you should study the variations of error illustrated in the various parts of Fig. 16-1.

A more acceptable way to mark a point is to use an ×, with the intersection marking the spot, or else a circled dot, ⊙, with the tiny point marking the spot and the circle attracting your attention to it. These correct methods are illustrated in Fig. 16-2, and they should be used in all your graph-drawing practice.

A second important item to watch always is the placing of the points. If there is a choice, the points should be far apart, so that the line joining them spans the most important area of the graph. Thus, any error in locating the points themselves is minimized. If the points are located close together and an error is made in locating either one point or both points, then other useful locations "outside" the points plotted will be subject

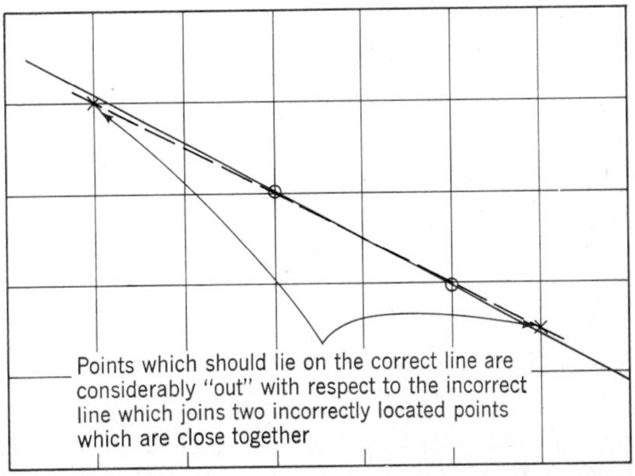

Points which should lie on the correct line are considerably "out" with respect to the incorrect line which joins two incorrectly located points which are close together

Fig. 16-3 Illustration of the Error Introduced When Points Are Plotted Close Together. If the Points Are Slightly Incorrect, Then Useful Points "Outside" the Plotted Area Are Even Further Off, and the Error Is Enlarged

to greater error. This fault is illustrated in Fig. 16-3, in which the two circled dots have been plotted slightly off their desired locations. The line joining them comes some distance away from the × points, which should lie on the line. In Fig. 16-4, the two circled dots are again plotted slightly off their desired loca-

Fig. 16-4 Illustration of the Reduction in Error When Incorrectly Plotted Points Are Far Apart, So That the Line Joining Them Spans the Working Area of the Graph. The error in Locating Each Circled Dot Is the Same As the Error in Fig. 16-3, But the × Locations are Closer to the Incorrect Line Which Joins the Plotted Points

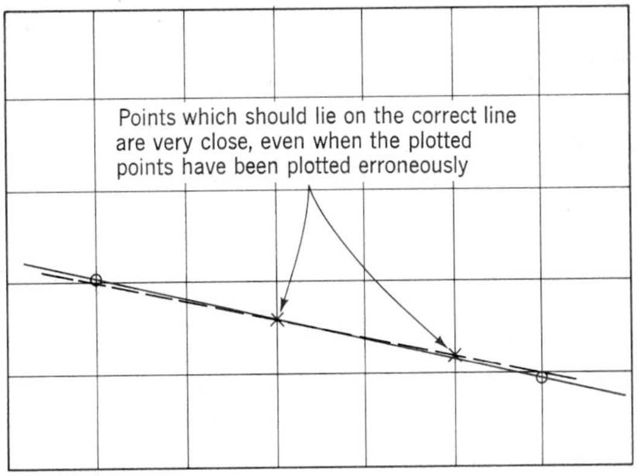

Points which should lie on the correct line are very close, even when the plotted points have been plotted erroneously

tions, but, since they are widely separated, the amount of error of intermediate points is less.

16-2 SOLVING PROBLEMS BY MEANS OF GRAPHS

In many instances, there arise problems involving relationships that, though readily solved by usual arithmetical or algebraic methods, are more clearly understood when solved graphically. It is also true that there are many problems which can be solved graphically with less labor than is required for the purely mathematical solutions. The following illustrative examples will show how some problems can be worked graphically.

example 1 Steamship *A* sails from New York at 6 A.M., steaming at an average speed of 10 knots (kn). (A knot is a measure of speed and is one nautical mile per hour.) The same day, at 9 A.M., steamship *B* sails from New York, steering the same course as *A* but steaming at 15 kn. (*a*) How long will it take *B* to overtake *A*? (*b*) What will be the distance from New York at that time?

solution Choose convenient scales on graph paper, and plot the distance in nautical miles covered by each vessel against the time in hours, as shown in Fig. 16-5. This is conveniently accomplished by making a table like Table 16-1.

Fig. 16-5 Graph of Example 1

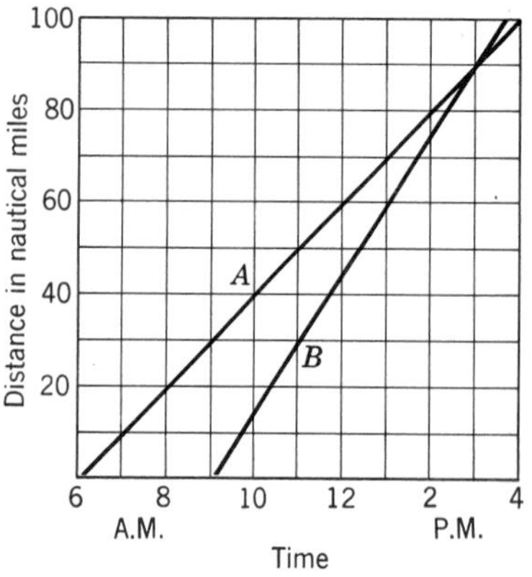

Table 16-1

Time, o'clock	Distance Covered by *A*, nmi	Distance Covered by *B*, nmi
6 A.M.	0	0
8	20	0
10	40	15
12	60	45
2 P.M.	80	75
4	100	105

It will be noted that the graphs of the two distances intersect at 90 nmi, or at 3 P.M. This means the two ships will be 90 nmi from New York at 3 P.M. Because they are both steering the same course, *B* will overtake *A* at this time and distance.

The graphic solution furnishes us with other information. For example, by measuring the vertical distance between the graphs, we can determine how far apart the ships will be at any time. Thus, at 11 A.M. the ships will be 20 nmi apart, at 1 P.M. they will be 10 nmi apart, etc.

example 2 Ship *A* is 200 nmi at sea, and ship *B* is in port. At 8 A.M., *A* starts toward the port, making a speed of 20 kn. At the same time, *B* leaves port at a speed of 30 kn to intercept *A*. After traveling 2 h, *B* is delayed for 1 h and 40 min at the lightship. *B* then continues on its course to intercept *A*. (*a*) At what time will the two ships meet? (*b*) How far will they be from port at that time?

solution Figure 16-6 is a graph showing the conditions of the problem. The graph is constructed as in Example 1. A table of

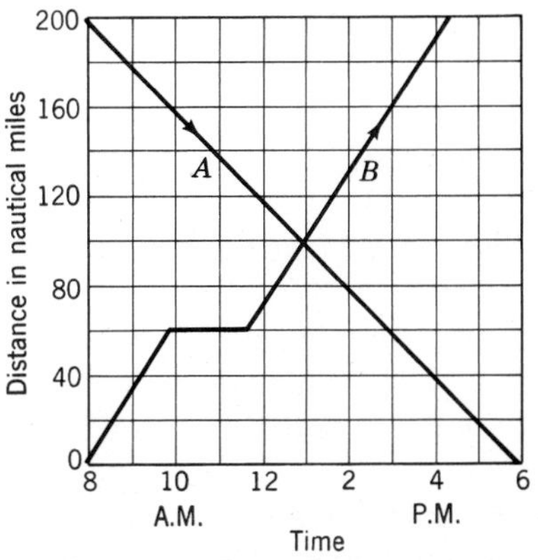

Fig. 16-6 Graph of Example 2

distances against time is made up, a convenient scale is chosen, and the points are plotted and joined with a straight line.

The intersection of the graphs shows that the ships will meet 100 nmi from port at 1 P.M. Why is there a horizontal portion in the graph of *B*'s distance from port? If *A* and *B* continue their speeds and courses, at what time will *A* reach port? At what time will *B* arrive at *A*'s 10 A.M. position? What will be the distance between the ships at that time?

PROBLEMS 16-1

1. A circuit consists of a 10-Ω resistor R_c connected across a variable emf E_v. Plot the current I through the resistor against the voltage E across the resistor as E_v is varied in 10-V steps from 0 to 100 V. What conclusion do you draw from this graph?
2. A circuit consists of a 50-Ω resistor R_L connected across the variable emf E_v of Prob. 1. On the same graph sheet as your solution to Prob. 1, plot the current I through R_L against the voltage E as E_v is varied between 0 and 100 V. What conclusion do you draw from the pair of graphs?
3. The distance s covered by a moving object is equal to the product of its velocity v and the time t during which the object is moving; that is, $s = vt$. Plot the distance in kilometers traveled by an automobile averaging 55 km/h against time for every hour from 9 A.M. to 6 P.M. What conclusions do you draw from the graph?
4. A variable resistor R_v is connected across a generator which maintains a constant voltage E_c of 120 V. Plot the current I through the resistor as its resistance is varied in 5-Ω steps between 5 and 50 Ω. What conclusions do you draw from this graph?

Solve these problems graphically:

5. Train *A* leaves a city at 8 A.M. traveling at the rate of 80 km/h. Two hours later train *B* leaves the same city, on the same track, traveling at the rate of 120 km/h.
   (*a*) At what time does train *B* overtake *A*?
   (*b*) How far from the starting point will the trains be at the time of part (*a*)?
   (*c*) How far apart will the trains be 2 h after *B* starts?

6. Two people start toward each other from points 144 km apart, the first traveling at 96 km/h and the second at 64 km/h.
   (a) How long will it be before they meet?
   (b) How far will each have traveled when they meet?
   (c) How far apart will they be after 30 min of travel?
7. *A* owns a motor that consumes 10 kWh per day, and *B* owns a motor that consumes 30 kWh per day. Beginning on the first day of a 30-day month, *A*'s motor runs continuously. *B*'s motor runs for 1 day, is idle for 4 days, then runs for 2 days, is idle for 6 days, and then runs continuously for the rest of the month. On what days of the month will *A*'s and *B*'s power bills be the same?
8. The owner of a radio store decides to pay the salespeople according to either of two plans. The first plan provides for a fixed salary of $25 per week plus a commission of $3 for each radio sold. According to the second plan, a salesperson may take a straight commission of $4 for each radio set sold. Determine at which point the second plan becomes more attractive for an energetic salesperson.

16-3 COORDINATE NOTATION

Let us suppose you are standing on a street corner and a stranger asks you for directions to some prominent building. You tell the stranger to go four blocks east and five blocks north. By these directions, you have automatically made the street intersection a *point of reference*, or *origin*, from which distances are measured. From this point you could count distances to any point in the city, using the blocks as a unit of distance and pairs of directions (east, north, west, or south) for locating the various points.

To draw a graph, we had to use two lines of reference, or *axes*. These correspond to the streets meeting at right angles. Also, in fixing a point on a graph, it was necessary to locate that point by pairs of numbers. For example, when we plot distance against time, we need one number to represent the time and another number to represent the distance covered in that time.

So far, only positive numbers have been used for graphs. To restrict graphs to positive values would impose just as severe a handicap as if we were to restrict algebra to positive numbers. Accordingly, a system must be established for plotting pairs of numbers, either or both of which may be positive or negative. In such a system, a sheet of squared paper is divided into four sections, or quadrants, by drawing two intersecting axes at right angles to each other. The point O, at the intersection of the axes, is called the *origin*. The horizontal axis is generally known as the *x axis* and the vertical axis is called the *y axis*.

There is nothing new about measuring distances along the *x* axis; it is the basic system described in Sec. 3-5 and shown in Fig. 3-3. That is, we agree to regard distances along the *x* axis to the *right* of the origin as *positive* and those to the *left* as *negative*. Also, we consider distances along the *y* axis as *positive* if *above* the origin and *negative* if *below* the origin. In effect, we have simply added to our method of graphical representation as originally outlined in Fig. 3-3.

With this system of representation, which is called a system of *rectangular coordinates*, we are able to locate any pair of numbers regardless of the signs. Because this system was developed by the French mathematician René Descartes, you will often hear it referred to as the system of *Cartesian coordinates*.

example 3 Referring to Fig. 16-7,

- Point A is in the first quadrant. Its x value is $+3$, and its y value is $+4$.
- Point B is in the second quadrant. Its x value is -4, and its y value is $+5$.
- Point C is in the third quadrant. Its x value is -5, and its y value is -2.
- Point D is in the fourth quadrant. Its x value is $+5$, and its y value is -3.

Thus, every point on the surface of the paper corresponds to a pair of coordinate numbers that completely describe the point. The two signed numbers that locate a point are called the *coordinates* of that point. The x value is called the *abscissa* of the point, and the y value is called the *ordinate* of the point. In describing a point in terms of its coordinates, the abscissa is always stated first. Thus, to locate point A in Fig. 16-7, we write $A = (3,4)$, meaning that, to locate point A, we count three divisions to the right of the origin along the x axis and up four divisions along the y axis. In like manner, we com-

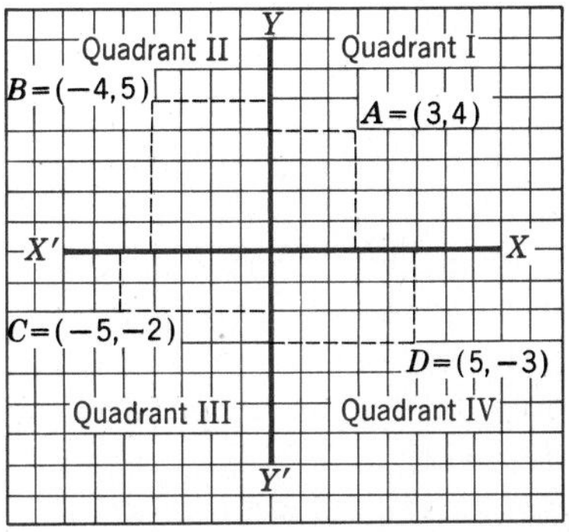

Fig. 16-7 System of Rectangular Coordinates

pletely describe point B by writing $B = (-4,5)$. Also,

$$C = (-5,-2) \quad \text{and} \quad D = (5,-3)$$

Fig. 16-8 Graph of Prob. 4

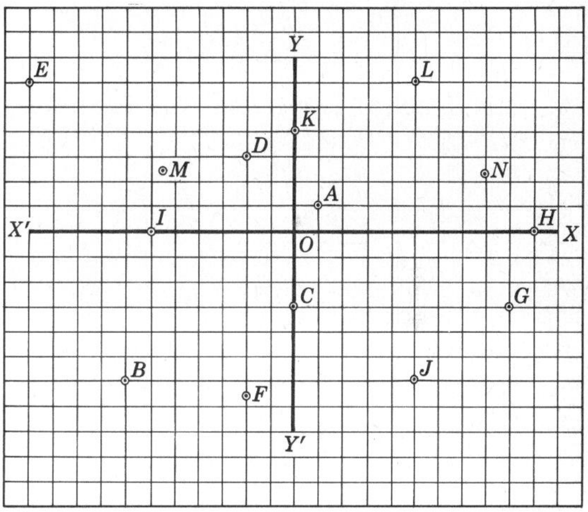

PROBLEMS 16-2

1. On a map, which lines correspond to the x axis, latitude or longitude?
2. Plot the following points: $(2,3)$, $(-6,-1)$, $(3,-7)$, $(0,-6)$, $(0,0)$, $(-8,0)$.
3. Plot the following points: $(-1.5,10)$, $(-6.5,-7.5)$, $(3.6,-4)$, $(0,2.5)$, $(6.5,8.5)$, $(3.5,0)$.
4. Using Fig. 16-8, give the coordinates of the points $A, B, C, D, E, F, G, H, I, J, K, L, M$, and N.
5. Plot the following points: $A = (-1,-2)$, $B = (5,-2)$, $C = (5,4)$, $D = (-1,4)$. Connect these points in succession. What kind of figure is $ABCD$? Draw the

diagonals DB and CA. What are the coordinates of the point of intersection of the diagonals?

16-4 GRAPHS OF LINEAR EQUATIONS

A relation between a pair of numbers, not necessarily connected with physical quantities such as those in foregoing exercises, can be expressed by a graph.

Consider the following problem: The sum of two numbers is equal to 5. What are the numbers? Immediately it is evident there is more than one pair of numbers that will fulfill the requirements of the problem. For example, if only positive numbers are considered, we have, by addition,

0	1	2	3	4	5
5	4	3	2	1	0
5	5	5	5	5	5

Similarly, if negative numbers are included, we can write

−1	− 2	− 3	−4	− 5	− 6
+6	+ 7	+ 8	+9	+10	+11
5	5	5	5	5	5

and so on, indefinitely.

Also, if fractions or decimals are considered, we have

1.5	−3.75	−1.63	− 8.36
3.5	+8.75	+6.63	+13.36
5	5	5	5

and so on, indefinitely.

It follows that there are an infinite number of pairs of numbers whose sum is 5.

Let x represent any possible value of one of these numbers, and let y represent the corresponding value of the second number. Then

$$x + y = 5$$

For any value assigned to x, we can solve for the corresponding value of y. Thus, if $x = 1, y = 4$. Also, if $x = 2, y = 3$. Likewise, if $x = -4$, $y = 9$, because, by substituting -4 for x in the equation, we obtain

$$-4 + y = 5$$

or

$$y = 9$$

In this manner, there may be obtained an unlimited number of values for x and y that satisfy the equations, some of which are listed at the bottom of the page.

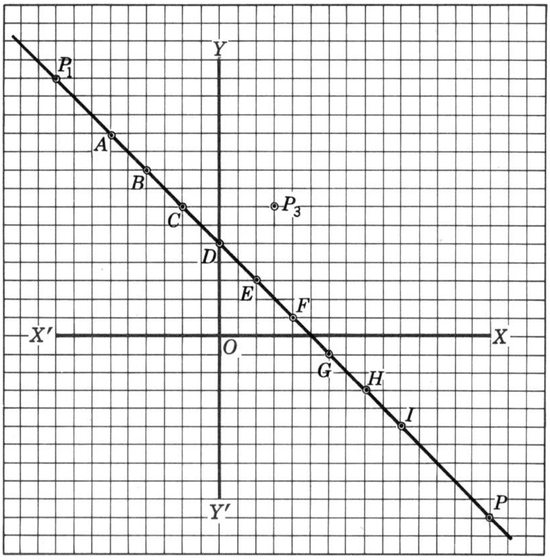

Fig. 16-9 Graph of the Equation $x + y = 5$

If $x =$	−6	−4	−2	0	2	4	6	8	10
Then $y =$	11	9	7	5	3	1	−1	−3	−5
Coordinates of	A	B	C	D	E	F	G	H	I

With the tabulated pairs of numbers as coordinates, the points are plotted and connected in succession as shown in Fig. 16-9. The line drawn through these points is called the *graph of the equation* $x + y = 5$.

Regardless of what pairs of numbers (coordinates) are chosen from the graph, it will be found that each pair satisfies the equation. For example, the point P has coordinates $(15,-10)$; that is, $x = 15$ and $y = -10$. These numbers satisfy the equation because $15 - 10 = 5$. Likewise, the point P_1 has coordinates $(-9,14)$ that also satisfy the equation because $-9 + 14 = 5$. The point P_3 has coordinates (3,7). This point is not on the line, nor do its coordinates satisfy the equation, for $3 + 7 \neq 5$. The straight line, or graph, can be extended in either direction, always passing through points whose coordinates satisfy the conditions of the equation. This is as would be expected, for there are an infinite number of pairs of numbers called *solutions* that, when added, are equal to 5.

PROBLEMS 16-3

1. Graph the equation $x - y = 8$ by tabulating and plotting five pairs of values for x and y that satisfy the equation. Can a straight line be drawn through these points? Plot the point (4,4). Is it on the graph of the equation? Do the coordinates of this point satisfy the equation? From the graph, when $x = 0$, what is the value of y? When $y = 0$, what is the value of x? Do these pairs of values satisfy the equation?
2. Graph the equation $2x + 3y = 6$ by tabulating and plotting at least five pairs of values for x and y that satisfy the equation. Can a straight line be drawn through these points? Plot the point $(-15,12)$. Is this point on the graph of the equation? Do the coordinates satisfy the equation? Plot the point $(10,-5)$. Is this point on the graph of the equation? Do the coordinates satisfy the equation? From the graph, when $x = 0$, what is the value of y? When $y = 0$, what is the value of x? Do these pairs of values satisfy the equation?

16-5 VARIABLES

When two variables, such as x and y, are so related that a change in x causes a change in y, then y is said to be a *function* of x. By assigning values to x and then solving for the value of y, we make x the *independent variable* and y the *dependent variable.*

The above definitions are applicable to all types of equations and physical relations. For example, in Fig. 16-5, distance is plotted against time. The distance covered by a body moving at a constant velocity is given by

$$s = vt$$

where s = distance
v = velocity
t = time

In this equation and therefore in the resulting graph, the distance is the dependent variable because it depends upon the amount of time. The time is the independent variable, and the velocity is a constant.

Similarly, in Prob. 1 of Problems 16-1, the formula $I = \dfrac{E}{R}$ is used to obtain values for plotting the graph. Here the resistance R is the constant, the voltage E is the independent variable, and the current I is the dependent variable.

In Prob. 4 of Problems 16-1, the same formula $I = \frac{E}{R}$ is used to obtain coordinates for the graph. Here the voltage E is a constant, the resistance R is the independent variable, and the current I is the dependent variable.

From these and other examples, it is evident, as will be shown in Sec. 16-6, that the graph of an equation having variables of the first degree is a straight line. This fact does not apply to variables in the denominator of a fraction as in the case above where R is a variable. However, $I = \frac{E}{R}$ is not an equation of the first degree as far as R is concerned because, by the law of exponents, $I = ER^{-1}$.

It is general practice to plot the independent variable along the horizontal, or x axis, and the dependent variable along the vertical, or y axis.

In plotting the graph of an equation, it is convenient to solve the equation for the dependent variable first. Values are then assigned to the independent variable in order to find the corresponding values of the dependent variable.

If an equation or formula contains more than two variables, after choosing the dependent variable, we must decide which one is to be the independent variable for each separate investigation, or graphing. For example, consider the formula

$$X_L = 2\pi fL$$

where X_L = inductive reactance of an inductor, Ω
f = frequency, Hz
L = inductance, H
$2\pi = 6.28\ldots$

In this case, we can vary either the frequency f or the inductance L in order to determine the effect upon the inductive reactance X_L, but we must not vary both at the same time. Either f must be fixed at some constant value and L varied, or L must be fixed. A little thought will show the difficulty of plotting, on a plane, the variations X_L if f and L are varied simultaneously.

16-6 THE GRAPH-EQUATION RELATIONSHIPS

Each of the equations that have been plotted is of the *first degree* (Sec. 11-1) and contains *two unknowns*. From their graphs the following important facts are obtained:

1. The graph of an equation of the first degree is a straight line.
2. The coordinates of every point on the graph satisfy the conditions of the equation.
3. The coordinates of every point not on the graph do not satisfy the conditions of the equation.

Because the graph of every equation of the first degree results in a straight line, as stated under 1 above, first-degree equations are called *linear equations*. Also, because such equations have an infinite number of solutions, they are called *indeterminate* equations.

As x changes in value in such an equation, the value of y also changes. Hence, x and y are called *variables*.

Now consider Fig. 16-9, the graph of $x + y = 5$. This equation may be written in the form $y = -x + 5$, where y is called the dependent variable, because its value depends upon the value of x, and x is called the independent variable, because we may assign to it any value we choose.

Notice in the graph first of all that the y intercept, the point where the curve cuts the y axis, is at the point $x = 0$, $y = 5$, and this value is revealed in the equation $y = -x + 5$ because, at the y axis, $x = 0$ and y then equals 5.

Second, note the slope of the line. For every step in the x direction (positive to the right), there is a downward (negative) step in the y direction. By definition, the slope of a line is the ratio of the change in the y values between two points to the corresponding change in x values between the same two points:

$$\text{Slope} = \frac{\Delta y}{\Delta x}$$

where the symbol Δ (Greek letter delta) means "the change in."

Figure 16-9 has been redrawn in Fig. 16-10 to show the changes in x and y between two arbitrarily selected points B and H. The slope of the graph equals

$$\frac{\Delta y}{\Delta x} = \frac{-12}{+12} = -1$$

Fig. 16-10 Figure 16-9 Redrawn to Show $m = \dfrac{\Delta y}{\Delta x}$

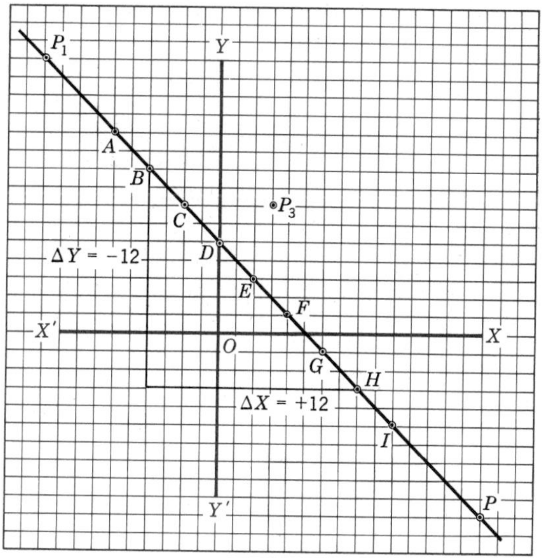

Now see in the equation $y = -x + 5$ that the slope, -1, is indicated in the coefficient of x.

Therefore, when we write the original equation $x + y = 5$ in standard form

$$y = -x + 5$$

the slope of the line is the coefficient of the x term and the y intercept is the constant term.

The general form of equation for a straight line is

$$y = mx + b$$

where y = dependent variable
x = independent variable
m = slope of the curve (straight line)
b = value of the y intercept

16-7 METHODS OF PLOTTING

To graph a linear equation of two variables,

1. Convert the equation to the standard form $y = mx + b$ to indicate quickly the values of the slope m and the y intercept b.
2. Choose a suitable value for x, substitute it into the standard form equation, and solve for the corresponding value of y. This results in one solution, or one set of coordinates.
3. Choose another value for x, and again solve for y. This second x value should be reasonably well spaced from the first (see Figs. 16-3 and 16-4).
4. Plot the two points whose coordinates were calculated in steps 2 and 3. Connect them with a fine straight line.
5. Check the resulting graph by solving for and plotting a third point. This third point must lie on the same straight line or its extension.

example 4

Graph the equation $2x - 5y = 10$.

solution

1. Rewrite the equation in the standard form: $y = \frac{2}{5}x - 2$.
2. Always plot first the value of y when $x = 0$. This value is immediately obtained from the "-2" of the equation, which shows the y intercept. This inspection results in a point, which we shall call A, whose coordinates are $(0,-2)$.
3. Now choose some value of x. Any value will serve, but one which cancels the denominator of the fractional coefficient will be the best choice. Let $x = 5$ and, by solving the equation, obtain $y = 0$. This gives the second point, B, at $(5,0)$. (Sometimes it may be more convenient to choose, as the second point, the value of $y = 0$ and solve for x.)
4. Choose another value of x in order to solve for the third (check) point. Let $x = -10$. Then $y = -6$, and this gives point C at $(-10,-6)$.
5. Draw the line of the equation by joining the three points. The points and the finished graph are shown in Fig. 16-11.

Fig. 16-11 Graph of the Equation $2x - 5y = 10$

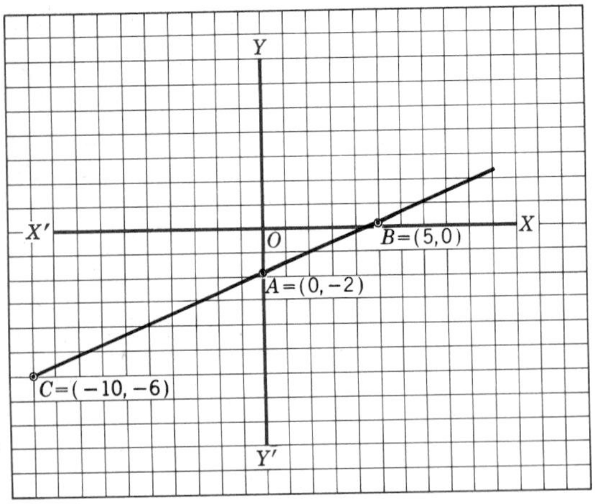

When x was set equal to zero, the resulting point A had coordinates that located the point where the graph crossed the y axis. This point is called the y *intercept*. Likewise, when y was set equal to zero, the resulting point B had coordinates that located the point where the graph crossed the x axis. This point is called the x *intercept*. Not only are these easy methods of locating two points with which to graph the equation but also these two points give us the exact location of the intercepts. The intercepts are important, as will be shown later.

The x intercept is often referred to as the *root* or *zero* of the equation.

An alternative method of plotting straight-line graphs is to use the information obtained from the standard form $y = mx + b$. If we locate the y intercept b immediately and then step over and up (or down) in accordance with the slope m, we can locate additional points. If, for example, $y = 2x + 9$, then the y intercept is at $+9$ and the slope is $+2:1$.

Follow the development of the graph in Fig. 16-12. First plot the y intercept, $+9$. Then step one unit in the positive x direction and two units in the positive y direction and plot the first point. Next, since $+\frac{2}{1} = \frac{-2}{-1}$, again starting at the y intercept, step one unit in the negative x direction and two units in the negative y direction and plot the second point. If these two points are too close together to be reliable, space them better by moving greater distances in the x and y directions while keeping the ratio $\frac{\Delta y}{\Delta x}$ equal to $2:1$ $(=m)$. Finally, join the two points so located with a straight line which passes through the third, or test, point, the y intercept.

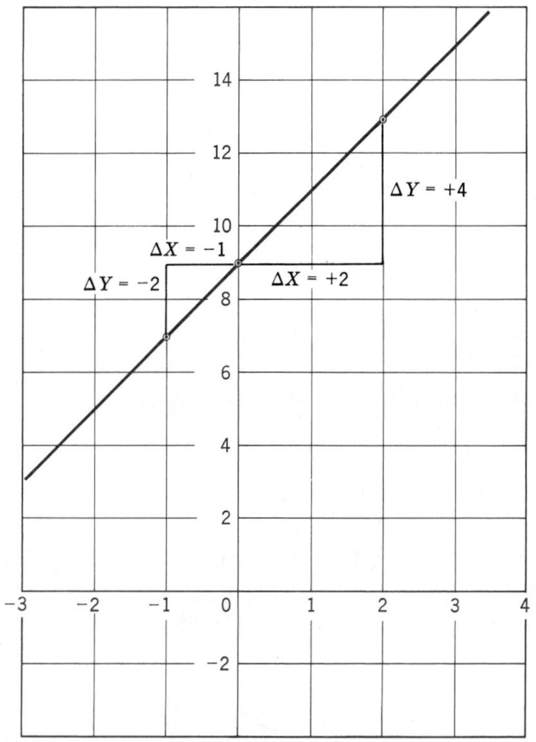

Fig. 16-12 Alternative Method of Plotting a Straight Line: First, Locate y Intercept, Given by the Constant in the Standard Form Equation. Then Step Off Δx and Δy So That $\frac{\Delta y}{\Delta x} = m$, or Slope, Also Given in the Standard Form of the Equation, First in the $+x$ Direction and Then in the $-x$ Direction

---

PROBLEMS 16-4

Graph the following equations and determine the x and y intercepts:

1. $5x + 4y = 12$ **2.** $2x - y = 8$ **3.** $x - 3y = 3$ **4.** $2x + y = 9$

5. Plot the following equations on the same sheet of graph paper (same axes), and carefully study the results: (*a*) $x - y = -8$; (*b*) $x - y = -5$; (*c*) $x - y = 0$; (*d*) $x - y = 4$; (*e*) $x - y = 8$.
Are the graphs parallel? Note that all left members of the given equations are identical. Solve each of these equations for y and write them in a column, thus:

$$
\begin{aligned}
&(a)\ y = x + 8\\
&(b)\ y = x + 5\\
&(c)\ y = x + 0\\
&(d)\ y = x - 4\\
&(e)\ y = x - 8
\end{aligned}
$$

In each equation, does the last term of the right member represent the y intercept?

When the equations are solved for y, as above, each coefficient of x is $+1$. All the graphs slant to the right because the coefficient of each x is positive. Each time an x increases one unit, note that the corresponding y increases one unit. That is because the coefficient of x in each equation is 1.

6. Plot the following equations on the same sheet of graph paper (same axes), and carefully study the results.

(a) $4x - 2y = -30$; (b) $4x - 2y = -16$; (c) $4x - 2y = 0$; (d) $4x - 2y = 12$; (e) $4x - 2y = 30$; (f) $8x - 4y = 60$.

Are all the graphs parallel? Again note that all left members are identical. Does the graph of Eq. (f) fall on that of Eq. (e)? Note that (e) and (f) are *identical equations*. Why?

Solve each of these equations, except (f), for y and write them in a column, thus:

$$
\begin{aligned}
&(a)\ y = 2x + 15\\
&(b)\ y = 2x + 8\\
&(c)\ y = 2x + 0\\
&(d)\ y = 2x - 6\\
&(e)\ y = 2x - 15
\end{aligned}
$$

In each equation, does the last term of the right member represent the y intercept? When linear equations are written in this form, this last term is known as the *constant term*.

Are all the coefficients of the x's positive? That is why all the graphs slant upward to the right. Lines slanting in this manner are said to have *positive slopes*.

Each time an x increases or decreases one unit, note that y respectively increases or decreases two units. That is because the coefficient of each x is 2. If a graph has a *positive slope*, an increase or decrease in x always results in a corresponding increase or decrease in y. In these equations, each line has a slope of $+2$, the coefficient of each x.

7. Plot the following equations on the same set of axes: (a) $x + 2y = 18$; (b) $x + 2y = 10$; (c) $x + 2y = 0$; (d) $x + 2y = -14$; (e) $x + 2y = -22$; (f) $3x + 6y = -66$.

Are all the graphs parallel? How should you have known they would be parallel without plotting them?

Does the graph of (*f*) fall on that of (e)? How should you have known (e) and (*f*) would plot the same graph without actually plotting them? Solve each equation for y as in Probs. 5 and 6. Does the constant term denote the y intercept in each case? Is the coefficient of each x equal to $-\frac{1}{2}$? The minus sign means that each graph has a *negative slope;* that is, the lines slant downward to the right. Thus, when x increases, y decreases, and vice versa. The $\frac{1}{2}$ slope means that, when x varies one unit, y is changed $\frac{1}{2}$ unit. Therefore, the variations of x and y are completely described by saying the slope is $-\frac{1}{2}$.

8. Plot the following equations on the same set of axes: (*a*) $x - 4y = 0$; (*b*) $x - 2y = 0$; (*c*) $x - y = 0$; (*d*) $2x - y = 0$; (e) $4x - y = 0$; (*f*) $4x + y = 0$; (*g*) $2x + y = 0$; (*h*) $x + y = 0$; (*i*) $x + 2y = 0$; (*j*) $x + 4y = 0$. Solve the equations for y, as before, and carefully analyze your results.

16-8 EQUATIONS DERIVED FROM GRAPHS

Often we obtain a set of readings relating two variables and want to know whether there is any definite relationship between the variables. This investigation makes use of both the graph showing the relationship and our understanding of the standard form of a straight-line equation

$$y = mx + b$$

1. Plot the observed values carefully on a graph. If a straight-line relationship is indicated, draw it.
2. Sometimes one or more points appear to be off the trend. There may or may not be errors in these readings. For the present, we will *assume* that they are errors.
3. If the trend is a straight line, but some points are off, try to draw the line so that there is an equal number of floating points above and below the line. (Use a transparent straightedge.)
4. The straight-line result must now obey the law $y = mx + b$.

example 5 Given the following set of readings, draw the graph and determine the law relating the variables:

x	-2	2	4	6
y	-7	5	11	17

solution First, plot the points as they have been given, and try them with a straightedge for a straight-line relationship. Since, in this case, Fig. 16-13, a straight line is indicated, draw the line joining the points. The y intercept is seen to be -1. This gives the value of b in the standard form. Then, to determine the slope m, choose any two convenient points, reasonably spaced, say (2,5) and (6,17). The difference between the points in the y direction is $17 - 5 = 12$.

Fig. 16-13 Graph of Example 5

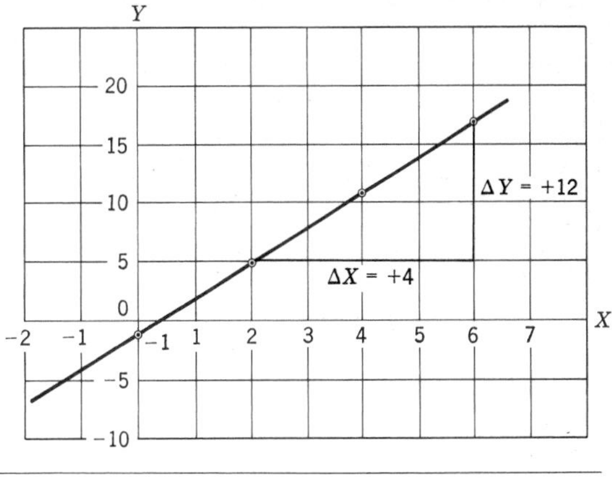

The difference between the points in the x direction is $6 - 2 = 4$.

Then the slope $m = \frac{\Delta y}{\Delta x} = \frac{17 - 5}{6 - 2} = \frac{+12}{+4}$
$= +3$

and the relationship is $y = 3x - 1$.

example 6 Given the readings relating P and V, determine the law relating them:

P	-4	-2	2	6	10
V	17	11	-5	-22	-38

solution Plot the points and test for a straight-line relationship. Because some of the points are not quite on the line, draw the straight line which will balance the floating points, (Fig. 16-14). Now the V intercept is seen to be $+2$, and the equation relating P and V will be of the form $V = mP + 2$. To evaluate the slope m, choose any two convenient points on the line, and arrive at m:

$$m = \frac{\Delta V}{\Delta P} = \frac{-38 - (-22)}{10 - 6}$$
$$= \frac{-16}{+4} = -4$$

and the relationship is seen to be

$$V = -4P + 2$$

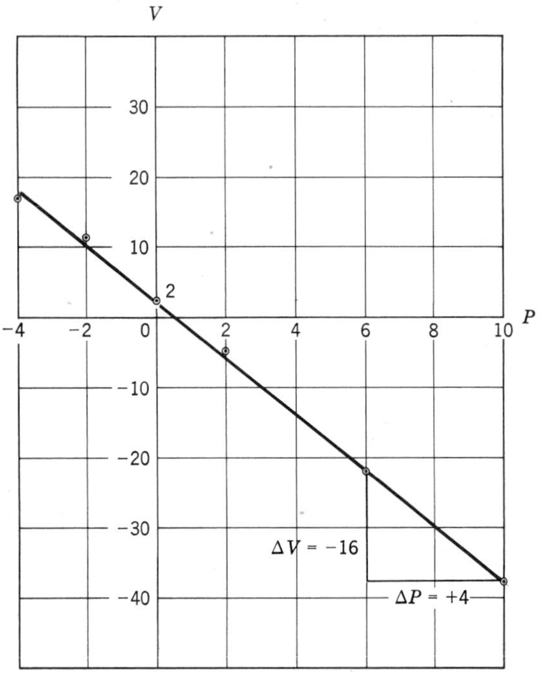

Fig. 16-14 Graph of Example 6

Referring to Examples 5 and 6, see how m may be found algebraically by realizing that $\Delta y = y_2 - y_1$, the difference of the values of y when going from point 1 to point 2, and $\Delta x = x_2 - x_1$, the difference of the values of x going from point 1 to point 2. Then

$$m = \frac{\Delta y}{\Delta x} = \frac{y_2 - y_1}{x_2 - x_1}$$

Always call your starting point 1 and your finishing point 2. That will yield the correct *sign* as well as the correct *value* of the slope.

PROBLEMS 16-5

1. What is the $y = mx + b$ form equation for the graph of Fig. 16-11?

2. A series of readings shows values of y for predetermined values of x:

x	5	10	15	20	25	30
y	100	200	300	400	500	600

Plot values of y against values of x and determine the values of the constants m and b which connect x and y in the form $y = mx + b$.

3. A laboratory experiment relates x and y as follows:

x	10	20	30	40	50	60
y	2.35	3.5	4.6	5.75	6.9	8.0

What is the equation, in the form $y = \alpha x + \theta$, which relates x and y?

4. The following is a series of readings relating s and t:

t	50	125	210	250	360	435
s	0.36	0.30	0.23	0.20	0.11	0.05

Plot s against t and, assuming s and t are connected by a law of the form $s = u + qt$, find u and q.

5. The following is a set of laboratory readings relating R and T:

T	30	75	150	210	270	300	360	390	425	450
R	0.38	0.35	0.31	0.26	0.22	0.195	0.16	0.13	0.12	0.10

Plot the graph of R versus T and determine the formula which relates them.

6. A comparison of Celsius (C) and Fahrenheit (F) temperatures is given in the following table:

°C	0	10	38	60	100
°F	32	50	100	140	212

Plot °F against °C.

(*a*) Determine from the graph the relationship between the two temperature scales in the form, $F = \theta C + \phi$.

(*b*) From the graph, what is the Fahrenheit equivalent of 25°C?

(*c*) From the graph, what is the Celsius equivalent of 165°F?

7. The readings of current flow I through a certain resistor as the emf E is changed are given in the following table:

E	10	20	30	40	50	60	70	80	90	100	V
I	0.2125	0.4255	0.638	0.851	1.062	1.278	1.49	1.702	1.915	2.125	A

(*a*) From the graph, what is the ratio $\dfrac{\text{change in voltage}}{\text{change in current}} \left(\dfrac{\Delta E}{\Delta I}\right)$?

(b) What is the ratio $\frac{\text{change in current}}{\text{change in voltage}}\left(\frac{\Delta I}{\Delta E}\right)$?

(c) From Ohm's law, what is the resistance of the resistor?

(d) What conclusions do you draw from your answers to questions (a), (b), and (c)?

8. The following is a series of readings of the avalanche breakdown of a Zener diode:

E	−14.2	−14.4	−14.6	−14.7	−14.8	−14.9	−15	−15.1	−15.2	−15.3	−15.4	−15.5	V
I	0	0	0	−10	−18.9	−28.2	−37.4	−46.8	−56	−65.2	−74.6	−83.9	mA

Plot the graph of I versus E and determine:

(a) What $\frac{\Delta I}{\Delta E}$ is after the voltage goes more negative than 14.6 V.

(b) What the ratio is for voltages less negative than 14.6 V.

9. When the control grid of a 6SN7GTB tube is biased at −6 V, the readings of plate current in milliamperes for selected plate voltages are

E_p	125	143	165	185	200	215	232	244	253	263	275	V
I_p	1	2	4	6	8	10	12	14	16	18	20	mA

Plot I_p against E_p.

(a) Over what range of voltages may the plate resistance of the tube be considered to be constant?

(b) What is your interpretation of other parts of the tube characteristic curve?

(c) What is $\frac{\Delta E_p}{\Delta I_p}$, that is, what is the change in plate voltage with respect to the change in plate current when the grid voltage E_g is constant over the straight-line portion of the graph?

(d) What does your electronic tube manual show as the value of R_p for the 6SN7GTB tube?

10. The readings of current versus applied voltage for a tunnel diode are as follows:

V_1	0.002	0.008	0.011	0.016	0.02	0.023	0.027	0.03	0.04	0.07	0.095	0.105	0.115
I_1	0.1	0.2	0.3	0.4	0.5	0.6	0.7	0.8	0.9	1.0	0.9	0.8	0.7

0.125	0.135	0.145	0.160	0.20	0.32	0.39	0.42	0.43	0.45	0.46	0.47	0.48	V
0.6	0.5	0.4	0.3	0.2	0.1	0.2	0.3	0.4	0.5	0.7	0.8	0.9	mA

only one point, there can be only one common set of values or one common solution that satisfies both equations.

Two equations, each with two variables, are called *inconsistent equations* when their plotted lines are parallel to each other. Because parallel lines do not intersect, there is no common solution for two or more inconsistent equations.

Considerable care must be used in graphing equations, for a deviation in the graph of either equation will cause the intersection to be in the wrong place and hence will lead to an incorrect solution.

PROBLEMS 17-1

Solve the following pairs of equations graphically, and check your solutions by substituting them into each of the original equations:

1. $x + 4y = 14$
$x - 4y = -2$

2. $6x - y = 15$
$2x + 5y = 21$

3. $x + y = 8$
$x - y = 2$

4. $x + 2y = 26$
$4x - y = 32$

5. $9E + 2I = 34$
$6E + 5I = -14$

6. $l - 8m = 0$
$l + m = 45$

7. $7\alpha + 3\beta = -23$
$5\beta + 4\alpha = -23$

8. $8F - f = 0$
$3f + 4F = 14$

9. $3I_1 + 7i = 50$
$5I_1 - 2i = 15$

10. $2Z_2 + 6Z_1 = 7$
$4Z_2 - 3Z_1 = 9$

17-2 SOLUTION OF SIMULTANEOUS LINEAR EQUATIONS BY ADDITION AND SUBTRACTION

It has been shown in preceding sections that an unlimited number of pairs of values of variables satisfy one linear equation. Also, it can be determined graphically whether there is one pair of values, or solution, that will satisfy two given linear equations. The solution of two simultaneous linear equations can also be found by algebraic methods, as illustrated in the following examples:

example 1 Solve the equations $x + y = 6$ and $x - y = 2$.

solution Given

$$x + y = 6 \qquad (a)$$
$$x - y = 2 \qquad (b)$$

Add (*a*) and (*b*),

$$2x = 8 \qquad (c)$$

D: 2 in (*c*), $x = 4$

Substitute this value of x in (*a*),

$$4 + y = 6$$

Collect terms, $y = 2$

The common solution for (*a*) and (*b*) is

$$x = 4 \qquad y = 2$$

check Substitute in (*a*), $4 + 2 = 6$
Substitute in (*b*), $4 - 2 = 2$

In Example 1 the coefficients of y in Eqs. (*a*) and (*b*) are the same except for sign. That being so, y can be *eliminated* by adding these equations, and the resulting sum is an equation in one unknown. This method of solution is called *elimination by addition.*

Because the coefficients of x are the same in Eqs. (a) and (b) of Example 1, x could have been eliminated by subtracting either equation from the other, and an equation containing only y as a variable would have been the result. This method of solution is called *elimination by subtraction.* The remaining variable x would have been solved for in the usual manner by substituting the value of y in either equation.

example 2

Solve the equations $3x - 4y = 13$ and $5x + 6y = 9$.

solution

Given	$3x - 4y = 13$	(a)
	$5x + 6y = 9$	(b)
M: 3 in (a),	$9x - 12y = 39$	(c)
M: 2 in (b),	$10x + 12y = 18$	(d)
Add (c) and (d)		
	$19x = 57$	(e)
D: 19 in (e),	$x = 3$	(f)
Substitute this value of x in (a),		
	$9 - 4y = 13$	(g)
Collect terms,	$-4y = 4$	(h)
D: -4 in (h),	$y = -1$	

The common solution for (a) and (b) is

$$x = 3 \qquad y = -1$$

check Substitute in (a), $9 + 4 = 13$
Substitute in (b), $15 - 6 = 9$

In Example 2 the coefficients of x and y in Eqs. (a) and (b) are not the same. The coefficients of y were made the same absolute value in Eqs. (c) and (d) in order to eliminate y by the method of addition.

example 3

Solve the equations $4a - 3b = 27$ and $7a - 2b = 31$.

solution

Given	$4a - 3b = 27$	(a)
	$7a - 2b = 31$	(b)
M: 7 in (a)	$28a - 21b = 189$	(c)
M: 4 in (b),	$28a - 8b = 124$	(d)
Subtract (d) from (c),		
	$-13b = 65$	(e)
D: -13 in (e),	$b = -5$	(f)
Substitute this value of b in (a),		
	$4a + 15 = 27$	(g)
Collect terms,	$4a = 12$	(h)
D: 4 in (h),	$a = 3$	(i)

The common solution for (a) and (b) is

$$a = 3 \qquad b = -5$$

check Substitute the values of the variables (a) and (b) as usual.

In Example 3 the coefficients of a and b in Eqs. (a) and (b) are not the same. The coefficients of a were made the same absolute value in Eqs. (c) and (d) in order to eliminate a by the method of subtraction.

Rule

To solve two simultaneous linear equations having two variables by the method of elimination by addition or subtraction:

1. If necessary, multiply each equation by a number that will make the coefficients of one of the variables of equal absolute value.
2. If these coefficients of equal absolute value have like signs, subtract one equation from the other; if they have unlike signs, add the equations.
3. Solve the resulting equation.

4. Substitute the value of the variable found in step 3 in one of the original equations, and then solve this resulting equation for the remaining variable.

5. Check the solution by substituting in both the original equations.

PROBLEMS 17-2

Solve for the unknowns by the method of addition and subtraction:

1. $2a + b = 9$
$4a - b = 6$

2. $E - 4I = 9$
$2E - 2I = 6$

3. $5Z + 2R = 16$
$3Z - R = 3$

4. $4E + 3I = -1$
$5E + I = 7$

5. $R_1 - 3R_2 = -8$
$3R_1 + R_2 = 6$

6. $5\theta + 4\phi = 12$
$\theta - 2\phi = 8$

7. $s + t = 0$
$3s + 7t = 8$

8. $2\alpha - \beta = 3$
$4\beta + 3\alpha = 10$

9. $5M + L = 11$
$3M + 2L = 8$

10. $4p - 3q = 5$
$9p - 8q = 0$

11. $3I_1 - 4I_2 = 17$
$I_1 + 3I_2 = -3$

12. $3Z_1 + Z_2 = 14$
$Z_1 + 2Z_2 = 13$

13. $E + 3e = 11$
$4E + 7e = 29$

14. $I + 3i = 25$
$4i + I = 31$

15. $3\lambda - 7\pi - 19 = 0$
$2\lambda - \pi = 9$

16. $5\alpha + 1 = -3\beta$
$7\beta + 3\alpha - 15 = 0$

17. $0.3E + 0.2e = -0.9$
$0.5E = -1.9 - 0.3e$

18. $0.9X_L + 0.04X_C = 9.4$
$0.05X_L + 2.5 = 0.3X_C$

19. $0.03I - 0.54 = -0.02i$
$21 - i = I$

20. $0.4L + 1.6 = 0.9X$
$0.7X + 0.2 = 0.6L$

21. Solve the problems of Problems 17-1 by the method of addition and subtraction, and confirm the answers obtained by the graphical method.

17-3 SOLUTION BY SUBSTITUTION

Another common method of solution is called *elimination by substitution.*

example 4

Solve the equations $16x - 3y = 10$ and $8x + 5y = 18$.

solution

Given
$$16x - 3y = 10 \qquad (a)$$
$$8x + 5y = 18 \qquad (b)$$

Solve (a) for x in terms of y,

$$x = \frac{10 + 3y}{16} \qquad (c)$$

Substitute this value of x in (b),

$$8\left(\frac{10 + 3y}{16}\right) + 5y = 18 \qquad (d)$$

M: 16 in (d),

$$8(10 + 3y) + 80y = 288 \qquad (e)$$

Expand (e),

$$80 + 24y + 80y = 288 \qquad (f)$$

Collect terms in (f),

$$104y = 208 \qquad (g)$$

D: 104 in (g), $\quad y = 2 \qquad (h)$

Substitute value of y in (a),

$$16x - 6 = 10 \qquad (i)$$

Collect terms in (i), $16x = 16$ $\qquad (j)$

D: 16 in (j), $x = 1$ $\qquad (k)$

check Usual method.

Not only is the method of substitution a very useful one; it also serves to emphasize that the values of the variables are the same in both equations.

The method of solving by substitution can be stated as follows:

Rule

To solve by substitution:

1. Solve one of the equations for one of the variables in terms of the other variable.
2. Substitute the resultant value of the variable, found in step 1, in the remaining equation.
3. Solve the equation obtained in step 2 for the second variable.
4. In the simplest of the original equations, substitute the value of the variable found in step 3 and solve the resulting equation for the remaining unknown variable.

PROBLEMS 17-3

Solve by the method of substitution:

1. $2E - I = 4$
$2E + 3I = 12$

2. $a + 2b = 6$
$3a - 10 = 2b$

3. $4I = -2 - 2i$
$3I + 12 = 2i$

4. $\pi - 8\omega = 0$
$\pi + \omega = 45$

5. $5\alpha - 8\beta = 0$
$8\alpha - 13\beta = -1$

6. $5I_1 + 7I_2 = 74$
$5I_2 - 7I_1 = 0$

7. $3 + 4E = 15e$
$2 - 9e = -2E$

8. $3X_L + 20 = 8X_C$
$3X_C - 44 = -8X_L$

9. $4\theta - 164 = 10\phi$
$3\theta - 2\phi = 68$

10. $3\lambda_1 + 11 = 4\lambda_2$
$3\lambda_2 = 9 + 2\lambda_1$

11. $3f + 5F = -9$
$17 - 4f = -3F$

12. $18 - 6I_1 = 8I_2$
$5I_1 + 4I_2 - 22 = 0$

13. $16 - 2\gamma = 3\delta$
$\delta - 52 = -4\gamma$

14. $2\pi - 8 = \omega$
$2\omega + 3\pi = 5$

15. $5 + \varepsilon = 2\psi$
$3\varepsilon + 4\psi = 20$

16. $4X_L - 9X_C = -16$
$7X_C + 2 = 6X_L$

17. $0.6\theta + 1.7\phi = 3.5$
$1.4\theta - 3.9 = 0.3\phi$

18. $0.6I + 0.8i = 2.6$
$7.0 - 0.5I = -0.3i$

19. $1.2a - 2b = 1$
$1.4a - 1.5b = 1.5$

20. $0.6L + 0.2M = 2040$
$0.5L + 0.3M = 1860$

21. Solve Probs. 1 to 20 graphically, and confirm the answers obtained algebraically.

17-4 SOLUTION BY COMPARISON

In this method, we solve for the value of the same variable in each equation in terms of the other variable and place these values equal to each other. The result is an equation having only one unknown.

example 5 Solve the equations $x - 4y = 14$ and $4x + y = 5$.

solution

Given $\qquad x - 4y = 14 \qquad (a)$

$\qquad 4x + y = 5 \qquad (b)$

Solve (a) for x in terms of y,

$$x = 14 + 4y \qquad (c)$$

Solve (b) for x in terms of y,

$$x = \frac{5 - y}{4} \qquad (d)$$

Equate values of x in (c) and (d),

$$14 + 4y = \frac{5 - y}{4} \qquad (e)$$

M: 4 in (e), $56 + 16y = 5 - y$ (f)

Collect terms in (f)

$$17y = -51 \qquad (g)$$

D: 17 in (g), $y = -3$

Substitute the value of y in (a),

$$x + 12 = 14$$

Collect terms, $x = 2$

check Usual method.

PROBLEMS 17-4

Solve by the method of comparison:

1. $3I + 2i = 5$
$I + i = 2$

2. $3Z - 2R = 7$
$Z + 2R = 5$

3. $\lambda + 2\pi = -2$
$15\lambda - 106 = 4\pi$

4. $4E + 3e = 15$
$2E + 11e = 36$

5. $4x + 2y = 20$
$1 + 2y = 3x$

6. $5L_1 + 24 = 6L_2$
$9L_2 - 22 = 4L_1$

7. $2\alpha - 5\beta - 7 = 0$
$7\alpha - 2\beta - 40 = 0$

8. $2M - 24Q = 0$
$3M - 20Q = 16$

9. $0.7p - 0.6q = 6.3$
$0.9p - 1.3 = -0.2q$

10. $2.8I - 2.7i = 19.9$
$6 + 5i = 2.1I$

11. Solve Probs. 1 to 10 graphically and by the other algebraic methods.

17-5 FRACTIONAL FORM

Simultaneous linear equations having fractions with numerical denominators are readily solved by first clearing the fractions from the equations and then solving by any method considered most convenient.

example 6

Solve the equations $\frac{x}{4} + \frac{y}{3} = \frac{7}{12}$ and $\frac{x}{2} - \frac{y}{4} = \frac{1}{4}$.

solution

Given

$$\frac{x}{4} + \frac{y}{3} = \frac{7}{12} \qquad (a)$$

$$\frac{x}{2} - \frac{y}{4} = \frac{1}{4} \qquad (b)$$

M: 12, the LCD, in (a),

$$3x + 4y = 7$$

M: 4, the LCD, in (b),

$$2x - y = 1$$

The resulting equations contain no fractions. Inspection of them shows that solution by addition is most convenient. The solution is

$$x = 1 \qquad y = 1$$

PROBLEMS 17-5

Solve the following sets of equations:

1. $\frac{a}{7} + \frac{4b}{7} = 2$

$\frac{a}{4} - b = -\frac{1}{2}$

2. $\frac{A}{3} + \frac{B}{5} = -\frac{3}{5}$

$\frac{3A}{34} - \frac{2B}{17} = -\frac{1}{2}$

3. $\frac{6\theta}{13} + \frac{8\phi}{13} = 2$

$\frac{\theta}{7} - \frac{3\phi}{35} = 2$

4. $2E - \frac{15e}{26} = 3\frac{1}{13}$

$\frac{13E}{33} - \frac{8e}{99} = 1$

5. $\varepsilon + \eta = 45$

$\frac{\varepsilon}{8} - \eta = 0$

6. $\frac{I}{3} + \frac{i}{5} = -\frac{1}{15}$

$\frac{7i}{30} + \frac{I}{10} = \frac{1}{2}$

7. $\frac{X_L}{4} - \frac{X_C}{8} = \frac{1}{2} = \frac{X_L}{12} + \frac{X_C}{8}$

8. $\frac{Z_1 + 2Z_2}{24} - \frac{Z_2 - 5}{4} = \frac{Z_1 + Z_2 + 1}{36}$

$\frac{Z_1 - 2}{12} = \frac{5 + Z_2}{3} - \frac{2Z_2 + 6}{6}$

9. $\frac{\lambda - \theta}{3} + \frac{5\lambda}{6} = \frac{5 - \theta}{6} - \frac{1 + \lambda}{4}$

$\frac{\lambda + 2}{5} = \theta - \frac{1}{4}$

10. $\frac{4I - i}{15} + \frac{1}{8} = 2i - \frac{12I}{5}$

$i - I = \frac{3}{16}$

17-6 FRACTIONAL EQUATIONS

When variables occur in denominators, it is generally easier to solve without clearing the equations of fractions.

example 7

Solve the equations $\frac{5}{x} - \frac{6}{y} = -\frac{1}{2}$ and $\frac{2}{x} - \frac{3}{y} = -1$.

solution

Given $\frac{5}{x} - \frac{6}{y} = -\frac{1}{2}$ (a)

$\frac{2}{x} - \frac{3}{y} = -1$ (b)

M: 2 in (a), $\frac{10}{x} - \frac{12}{y} = -1$ (c)

M: 5 in (b), $\frac{10}{x} - \frac{15}{y} = -5$ (d)

Subtract (d) from (c),

$$\frac{3}{y} = 4$$

$$y = \tfrac{3}{4}$$

Substitute $\frac{3}{4}$ for y in (b),

$$\frac{2}{x} - 4 = -1$$

Collect terms, $\frac{2}{x} = 3$

$$\therefore x = \tfrac{2}{3}$$

check Usual method.

PROBLEMS 17-6

Solve the following sets of equations:

1. $\frac{1}{R} + \frac{1}{Z} = \frac{7}{12}$

   $\frac{1}{R} - \frac{1}{Z} = \frac{1}{12}$

2. $\frac{2}{E_1} + \frac{3}{E_2} = \frac{13}{6}$

   $\frac{1}{E_1} + \frac{1}{E_2} = \frac{1}{6}$

3. $\frac{2}{X_L} - \frac{3}{X_C} = \frac{7}{55}$

   $\frac{1}{X_L} + \frac{1}{X_C} = \frac{27}{55} - \frac{1}{X_L}$

4. $\frac{6}{p} = \frac{1}{3}$

   $\frac{5}{q} - \frac{3}{p} = \frac{1}{4}$

5. $\frac{7}{\theta} + \frac{1}{\phi} = \frac{51}{80}$

   $\frac{4}{\phi} - \frac{4}{\theta} = \frac{11}{20}$

6. $\frac{4}{a-1} = \frac{3}{1-b}$

   $\frac{7}{2a-39} = \frac{5}{2b-5}$

7. $\frac{G-5}{5} - \frac{Y+3}{3} = -1$

   $\frac{18}{Y-1} = \frac{27}{G-12}$

8. $\frac{\lambda + 3\pi}{7} + 1 = \pi$

   $\frac{2}{\lambda} - \frac{4}{\pi} = 0$

9. $\frac{1}{M} + \frac{1}{L_1} = \frac{4}{15}$

   $\frac{19}{15} - \frac{3}{M} = \frac{2}{L_1}$

10. $\frac{1}{\pi} + \frac{1}{\lambda} = 3\frac{31}{35}$

    $\frac{1}{2\pi} + \frac{1}{4\lambda} = 1\frac{19}{35}$

17-7 LITERAL EQUATIONS IN TWO UNKNOWNS

The solution of literal simultaneous equations involves no new methods of solution. In general, it will be found that the addition or subtraction method will suffice for most cases.

example 8 Solve the equations $ax + by = c$ and $mx + ny = d$.

solution

Given $\quad ax + by = c \qquad (a)$

$\qquad mx + ny = d \qquad (b)$

First eliminate x.

M: m in (a),

$$amx + bmy = cm \qquad (c)$$

M: a in (b),

$$amx + any = ad \qquad (d)$$

Subtract (d) from (c),

$$bmy - any = cm - ad \qquad (e)$$

Factor (e),

$$y(bm - an) = cm - ad \qquad (f)$$

D: $(bm - an)$ in (f).

$$y = \frac{cm - ad}{bm - an}$$

Now go back to (a) and (b), and eliminate y.

M: n in (a),

$$anx + bny = cn \qquad (g)$$

M: b in (b),

$$bmx + bny = bd \qquad (h)$$

Subtract (h) from (g),

$$anx - bmx = cn - bd \qquad (i)$$

Factor (i),

$$x(an - bm) = cn - bd \qquad (j)$$

D: $(an - bm)$ in (j),

$$x = \frac{cn - bd}{an - bm}$$
$$= \frac{bd - cn}{bm - an}$$

example 9 Solve the equations

$$\frac{a}{x} + \frac{b}{y} = \frac{1}{xy} \quad \text{and} \quad \frac{c}{x} + \frac{d}{y} = \frac{1}{xy}$$

solution Given

$$\frac{a}{x} + \frac{b}{y} = \frac{1}{xy} \qquad (a)$$
$$\frac{c}{x} + \frac{d}{y} = \frac{1}{xy} \qquad (b)$$

First eliminate y, although it makes no difference which variable is eliminated first.

M: xy, the LCD, in (a),

$$ay + bx = 1 \qquad (c)$$

M: xy, the LCD, in (b),

$$cy + dx = 1 \qquad (d)$$

M: c in (c),

$$acy + bcx = c \qquad (e)$$

M: a in (d),

$$acy + adx = a \qquad (f)$$

Subtract (f) from (e),

$$bcx - adx = c - a \qquad (g)$$

Factor (g),

$$x(bc - ad) = c - a \qquad (h)$$

D: $(bc - ad)$ in (h),

$$x = \frac{c - a}{bc - ad}$$

Now go back to (a) and (b) to eliminate x, and find

$$y = \frac{b - d}{bc - ad}$$

PROBLEMS 17-7

Given:	Solve for
1. $4\alpha - \beta = P$ $\beta + 2\alpha = Q$	α and β
2. $3\pi + 2\lambda = x$ $2\pi - \lambda = y$	π and λ
3. $E + IR = a$ $3E + 7IR = b$	E and IR
4. $4L_1 + 3L_2 = C$ $3L_1 - 2L_2 = C$	L_1 and L_2
5. $6\theta + 5\phi = \alpha$ $3\phi - 4\theta = \beta$	θ and ϕ
6. $5r + 3R = Z_1$ $3r + 7R = Z_2$	R and r
7. $0.04X_C + 0.3X_L = Z_1$ $0.02X_C + 0.3X_L = Z_2$	X_C and X_L

Given:	Solve for
8. $\frac{R_L}{4} + \frac{R_p}{3} = R_T$ $\frac{R_L}{2} - \frac{R_p}{4} = R_1$	R_L and R_p
9. $\frac{1}{R_1} + \frac{1}{R_2} = \frac{1}{R_p}$ $\frac{3}{R_1} + \frac{2}{R_2} = \frac{1}{R_t}$	R_1 and R_2
10. $\frac{1}{3}(Z_1 - Z_2) = Z_1 - Z_2 - X_C$ $\frac{2}{5}Z_1 - Z_2 = 0$	Z_1 and Z_2

17-8 EQUATIONS CONTAINING THREE UNKNOWNS

In the preceding examples and problems, two equations were necessary to solve for two unknown variables. For problems involving three variables, three equations are necessary. The same methods of solution apply.

example 10 Solve the equations

$$2x + 3y + 5z = 0 \qquad (a)$$
$$6x - 2y - 3z = 3 \qquad (b)$$
$$8x - 5y - 6z = 1 \qquad (c)$$

solution Choose a variable to be eliminated. Let it be x.

M: 3 in (a),

$$6x + 9y + 15z = 0 \qquad (d)$$
$$6x - 2y - 3z = 3 \qquad (b)$$

Subtract (b) from (d),

$$11y + 18z = -3 \qquad (e)$$

M: 4 in (a),

$$8x + 12y + 20z = 0 \qquad (f)$$
$$8x - 5y - 6z = 1 \qquad (c)$$

Subtract (c) from (f),

$$17y + 26z = -1 \qquad (g)$$

This gives Eqs. (e) and (g) in two variables y and z. Solving them, we obtain $y = 3$, $z = -2$.

Substitute these values into (a),

$$2x + 9 - 10 = 0 \qquad (h)$$

Collect terms, $2x = 1$ (i)

D: 2 in (i), $x = \frac{1}{2}$

check Substitute the values of the variables in the equations.

PROBLEMS 17-8

Solve:

1. $\theta + 3\phi + 4\pi = 14$
$\theta + 2\phi + \pi = 7$
$2\theta + \phi + 2\pi = 2$

2. $X_L - X_C + R = 2$
$X_C + R + X_L = 6$
$X_L - R + X_C = 0$

3. $R_1 + 2R_2 + R_3 = 9$
$R_2 + R_3 + 2R_1 = 16$
$2R_3 + R_1 + R_2 = 3$

4. $a - 2b + c = 3$
$a + b + 2c = 1$
$2a - b + c = 2$

5. $\frac{1}{R_L} - \frac{1}{R_p} - \frac{1}{R_1} = \frac{1}{120}$
$\frac{1}{R_L} + \frac{1}{R_p} - \frac{1}{R_1} = \frac{49}{120}$
$\frac{1}{R_p} - \frac{1}{R_1} - \frac{1}{R_L} = \frac{-31}{120}$

6. $\frac{1}{a} - \frac{1}{b} - \frac{1}{c} = 1$
$\frac{1}{b} - \frac{1}{a} - \frac{1}{c} = 1$
$\frac{1}{c} - \frac{1}{a} - \frac{1}{b} = 1$

7. $0.1r - 0.1R + 0.6R_L = 4.1$
$2r + 3R + 6R_L = 70$
$\frac{3}{40}r + \frac{1}{20}R - \frac{1}{40}R_L = \frac{1}{2}$

8. $E_1 - E_2 - E_3 = \alpha$
$E_3 - E_1 - E_2 = \beta$
$E_2 - E_3 - E_1 = \gamma$

9. $s - t = 8$
$2v - 6 = s - 2$
$3v - 12 = 3t$

10. $a + 5 = c$
$7b = 3c - 1$
$2b - a = c - 9$

17-9 METHODS OF SOLUTION OF PROBLEMS

In working a problem involving more than one unknown, it is convenient to solve it by setting up a system of simultaneous equations according to the statements of the problem.

example 11 When a certain number is increased by one-third of another number the result is 23. When the second number is increased by one-half of the first number, the result is 29. What are the numbers?

solution Let x = first number, y = second number.

Then $$x + \frac{1}{3}y = 23 \qquad (a)$$

Also, $$y + \frac{1}{2}x = 29 \qquad (b)$$

Solving the equations, we obtain $x = 16$, $y = 21$.

check When 16, the first number, is increased by one-third of 21, we have

$$16 + 7 = 23$$

When 21, the second number, is increased by one-half of 16, we have

$$21 + 8 = 29$$

example 12 Two airplanes start from Omaha at the same time. The plane traveling west has a speed of 130 km/h faster than that of the plane traveling east. At the end of 4 h they are 2600 km apart. What is the speed of each plane?

solution Let x = rate of plane flying west and y = rate of plane flying east.

Then $$x - y = 130 \qquad (a)$$

Since

Rate × time = distance

then $4x$ = distance traveled by plane flying west

and $4y$ = distance traveled by plane flying east

Hence, $$4x + 4y = 2600 \qquad (b)$$

Solving Eqs. (a) and (b), we obtain

$$x = 390 \text{ km/h}$$
$$y = 260 \text{ km/h}$$

check Substitute these values into the statements of the example.

Often it is possible to derive a formula from known data and thereby eliminate terms which are not desired or cannot be used conveniently in some investigation.

example 13 The effective voltage E of an alternating voltage is equal to 0.707 times its maximum value E_{max}. That is,

$$E = 0.707E_{max} \tag{1}$$

Also, the average value E_{av} is equal to 0.637 times the maximum value. That is,

$$E_{av} = 0.637E_{max} \tag{2}$$

It is desired to express the effective value E in terms of the average value E_{av}.

solution E_{max} must be eliminated.

Solving Eq. (1) for E_{max},

$$E_{max} = \frac{E}{0.707}$$

Solving Eq. (2) for E_{max},

$$E_{max} = \frac{E_{av}}{0.637}$$

By Axiom 5, $\dfrac{E}{0.707} = \dfrac{E_{av}}{0.637}$

Solving for E, $E = 1.11E_{av}$ (3)

Equation (3) shows that the effective value of an alternating voltage is 1.11 times the average value of the voltage.

example 14 You know that in a dc circuit $P = EI$ and also that $P = I^2R$. Derive a formula for E in terms of I and R.

solution It is evident that P must be eliminated. Because both equations are equal to P, we can equate them (Axiom 5) and obtain

$$EI = I^2R$$

D: I,

$$E = IR \tag{4}$$

example 15 The quantity of electricity Q, in coulombs, in a capacitor is equal to the product of the capacitance C and the applied voltage E. That is,

$$Q = CE \tag{5}$$

The total voltage across capacitors C_a and C_b connected in series is $E = E_a + E_b$. Find C in terms of C_a and C_b.

solution Solve for E, E_a, and E_b. Thus

$$E = \frac{Q}{C}$$

$$E_a = \frac{Q}{C_a}$$

and $E_b = \dfrac{Q}{C_b}$

Then, since $E = E_a + E_b$

By substitution

$$\frac{Q}{C} = \frac{Q}{C_a} + \frac{Q}{C_b}$$

D: Q, $\dfrac{1}{C} = \dfrac{1}{C_a} + \dfrac{1}{C_b}$

M: CC_aC_b, the LCD

$$C_aC_b = CC_b + CC_a$$

Transposing,

$$CC_a + CC_b = C_aC_b$$

D: $(C_a + C_b)$ $C = \dfrac{C_aC_b}{C_a + C_b}$ (6)

This is the formula for the resultant capacitance C of two capacitors C_a and C_b connected in series.

PROBLEMS 17-9

1. The sum of two currents is I_t A, and their difference is I_d A. What are the currents?
2. Find two numbers whose sum is 19 and whose difference is 5.
3. If 1 is added to each term of a fraction, the value of the fraction becomes 0.75, and if 1 is subtracted from each term, the value of the fraction becomes 0.5. What is the fraction?
4. In a right triangle, the acute angles are complementary (that is, they add up to 90°). What are the angles if their difference is 40°?
5. The difference between the two acute angles of a right angle is $\alpha°$. Find the angles.
6. The sum of the three angles of any triangle is 180°. Find the three angles of a particular triangle if the smallest angle is one-third the middle angle and the largest is 5° larger than the middle one.
7. A TV repair technician goes to the parts dealer for an assortment of common resistors and capacitors. The sales clerk replies: "We have two such assortments: 20 resistors and 8 capacitors for $3.60, or 60 resistors and 40 capacitors for $14.00. Both assortments come under the same discount schedule." "I'll take the larger selection," says the technician, "if you'll figure out the price of one resistor and one capacitor."
   Help the sales clerk.
8. A takes 1 h longer than B to walk 30 km, but if A's pace were doubled, then A would take 2 h less than B. Find their rates of walking.
9. In 3 h, L drives 40 km farther than Q does in 2 h. In 6 h, Q drives 180 km more than L does in 4 h. Find their average rates of driving.
10. $v = gt$ and $s = \frac{1}{2}gt^2$. Solve for v in terms of s and t.
11. $C = \dfrac{Q}{V}$ and $W = \dfrac{QV}{2}$. Solve for W in terms of C and Q.
12. $I = \dfrac{E}{R}$ and $P = I^2R$. If $P = 2.7$ kW and $E = 180$ V, find the current I and the resistance R.
13. $v = u + at$ and $s = \frac{1}{2}(u + v)t$. Find the distance s in terms of initial velocity u and acceleration a and time t.
14. Use the information of Prob. 13 to show that $v^2 = u^2 + 2as$.
15. $\mu = \dfrac{\Delta e_b}{\Delta e_c}$, $r_p = \dfrac{\Delta e_b}{\Delta i_b}$, and $g_m = \dfrac{\Delta i_b}{\Delta e_c}$. Solve for μ in terms of g_m and r_p.
16. $R = 2D_L fL$ and $Q = \dfrac{2\pi fL}{R}$. Solve for D_L in terms of π and Q.
17. $R = \omega LQ$, $Q = \dfrac{\omega L}{r}$, and $\omega^2 = \dfrac{1}{LC}$. Solve for R in terms of L, C, and r.
18. $I = \dfrac{E}{R}$ and $I_1 = \dfrac{E}{R + R_1}$. Solve for R in terms of R_1, I, and I_1.

19. $Q = It$ coulombs (C), and $I = \frac{CE}{t}$ A. Solve for Q in terms of C and E.

20. Given $P = EI$ W, $I = \frac{E}{R}$ A, and $H = 0.24I^2Rt$. Solve for H in terms of P and t.

21. Use the data of Prob. 20 to find H when $E = 30$ V over a time $t = 10$ s if the heater resistance $R = 300\ \Omega$.

22. Given $I_aR_a = I_bR_b$, $\frac{Q_a}{Q_b} = \frac{C_a}{C_b}$, $I_a = \frac{Q_a}{t}$, and $I_b = \frac{Q_b}{t}$, show that $R_aC_a = R_bC_b$.

23. $I_p = \frac{\mu E_g}{R + R_p}$ and $E_p = I_pR$. Solve for R in terms of R_p, μ, E_p, and E_g.

24. Use the data of Prob. 23 to find E_p when $\mu = 50$, $E_g = 5$ V, $I_p = 12.5$ mA, and $R_p = 10$ kΩ.

25. The three-Varley method of cable fault location yields the following relationships:

$$R_AV_1 + R_AR_Y = R_AV_2 - R_YR_B$$
$$R_AV_2 + R_AR_X = R_AV_3 - R_XR_B$$
$$R_AV_1 + R_AR_T = R_AV_3 - R_TR_B$$

Solve these equations for R_Y, R_X, and R_T, and show that $R_T = R_Y + R_X$.

26. Given $E = I_x(R + R_x)$, $E = I_a(R + R_a)$, and $E = IR$. Show that

$$R_x = R_a \times \frac{\dfrac{I - I_x}{I_x}}{\dfrac{I - I_a}{I_a}}$$

27. Given three star-delta transformation equations:

$$R_a = \frac{R_1R_3}{R_1 + R_2 + R_3}$$

$$R_b = \frac{R_1R_2}{R_1 + R_2 + R_3}$$

$$R_c = \frac{R_2R_3}{R_1 + R_2 + R_3}$$

Solve for R_1, R_2, and R_3 in terms of R_a, R_b, and R_c.

28. If three resistors R_1, R_2, and R_3 are connected in parallel so that the total circuit current I_t is divided into I_1, I_2, and I_3, respectively, determine relationships similar to Eq. (1) in Chap. 15 for I_1, I_2, and I_3 in terms of R_1, R_2, R_3, and I_t.

18

determinants

In Chap. 17 we learned four methods of solving simultaneous equations of the second order, and we used some of those methods to solve equation sets of the third order. Indeed, some of the methods we learned are limited to solving simultaneous equations of the second order, while others may be used to solve third-, fourth-, fifth-, or even higher-order systems.

However, after about the third order, the method of repeated addition and subtraction, with its attendant multiplication, becomes tedious. In this chapter we shall investigate a "mechanical" method of solving simultaneous equations. This method, known as the method of determinants, is usually not introduced until students are well along in advanced mathematics, so we are not going to study all the fascinating developments which the whole subject of determinants may involve. (That would take a separate book of its own.) Instead, we are going to see how determinants may be put to work for us in order to simplify our solutions to simultaneous equations.

18-1 SECOND-ORDER DETERMINANTS

In Sec. 17-2, we learned how to solve pairs of simultaneous equations by the method of addition and subtraction. Let us apply this method to a pair of *general equations:*

$$\begin{aligned} a_1x + b_1y &= c_1 \\ a_2x + b_2y &= c_2 \end{aligned} \qquad (1)$$

where a_1, a_2, b_1, b_2, c_1, and c_2 represent any numbers, positive or negative, integers or fractions, or zero. Let us solve these general equations for x:

$$a_1x + b_1y = c_1 \qquad (a)$$

$$a_2x + b_2y = c_2 \qquad (b)$$

M: b_2 in (a), $\quad a_1b_2x + b_1b_2y = b_2c_1 \qquad (c)$

M: b_1 in (b), $\quad a_2b_1x + b_1b_2y = b_1c_2 \qquad (d)$

Subtract (d) from (c),

$$(a_1b_2 - a_2b_1)x = b_2c_1 - b_1c_2 \qquad (e)$$

Solve for x, $\quad x = \dfrac{b_2c_1 - b_1c_2}{a_1b_2 - a_2b_1} \qquad (2)$

It is left as an exercise for you to prove similarly that

$$y = \frac{a_1c_2 - a_2c_1}{a_1b_2 - a_2b_1} \tag{3}$$

Observe that we have kept the literal factors in alphabetical order for convenience in checking.

Note several interesting facts about these two solutions:

1. Their denominators are identical, and they contain only the coefficients of x and y.
2. The numerator for the solution of y contains no y coefficients.
3. The numerator for the solution of x contains no x coefficients.

For a few minutes, let us consider just the denominator: $a_1b_2 - a_2b_1$. We are going to define a new, alternative method of writing this expression.

$$a_1b_2 - a_2b_1 = \begin{vmatrix} a_1 & b_1 \\ a_2 & b_2 \end{vmatrix}$$

This arrangement is called the *determinant* of the denominator. It is a mechanical statement made up of two horizontal *rows* and two vertical *columns* of two elements each, and it is a *second-order determinant*. Whenever this form appears, it is understood to mean $a_1b_2 - a_2b_1$. To obtain this evaluation of the determinant, we perform diagonal multiplication, first of all downward to the right to obtain

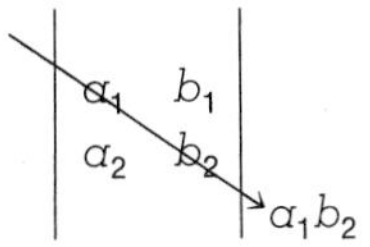

(this is, by definition, positive multiplication)

and, second, we multiply upward to the right to obtain

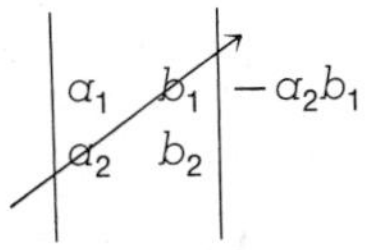

(this is, by definition, negative multiplication)

Rule

1. The diagonal multiplication in determinants derives its sign from the direction of the multiplication, and not primarily from any algebraic signs of the elements being multiplied.
2. After the individual steps of multiplication, with the appropriate sign of the multiplication affixed, the products are added algebraically to form the evaluation of the determinants.

example 1 Evaluate the determinant

$$\begin{vmatrix} -3 & 2 \\ 5 & 1 \end{vmatrix}$$

solution Perform the signed diagonal multiplication:

$$+(-3)(1) - (5)(2) = -3 - 10 = -13$$

PROBLEMS 18-1

Evaluate the following determinants:

1. $\begin{vmatrix} 4 & 1 \\ 2 & 1 \end{vmatrix}$ **2.** $\begin{vmatrix} 1 & 8 \\ 3 & -12 \end{vmatrix}$ **3.** $\begin{vmatrix} 2 & -8 \\ 3 & 5 \end{vmatrix}$ **4.** $\begin{vmatrix} -1 & 4 \\ 3 & 12 \end{vmatrix}$

5. $\begin{vmatrix} 3 & -4 \\ 3 & -4 \end{vmatrix}$ **6.** $\begin{vmatrix} 3 & -2 \\ 1 & 2 \end{vmatrix}$ **7.** $\begin{vmatrix} 9 & 4 \\ 15 & -6 \end{vmatrix}$ **8.** $\begin{vmatrix} -3 & -7 \\ -4 & -2 \end{vmatrix}$

9. $\begin{vmatrix} 0.8 & 0.2 \\ 0.5 & 0.1 \end{vmatrix}$ **10.** $\begin{vmatrix} -0.06 & 0.02 \\ 0.05 & -1.6 \end{vmatrix}$ **11.** $\begin{vmatrix} a & b \\ a & b \end{vmatrix}$ **12.** $\begin{vmatrix} a & b \\ x & y \end{vmatrix}$

13. $\begin{vmatrix} b & a \\ y & x \end{vmatrix}$ **14.** $\begin{vmatrix} a & x \\ b & y \end{vmatrix}$ **15.** $\begin{vmatrix} b & y \\ a & x \end{vmatrix}$ **16.** $\begin{vmatrix} y & x \\ b & a \end{vmatrix}$

18-2 SOLUTION OF EQUATIONS

Consider Eqs. (2) and (3), the solutions for x and y in the general equations (1):

$$x = \frac{b_2c_1 - b_1c_2}{a_1b_2 - a_2b_1} \qquad y = \frac{a_1c_2 - a_2c_1}{a_1b_2 - a_2b_1}$$

or, in determinant form:

$$x = \frac{\begin{vmatrix} b_2 & c_2 \\ b_1 & c_1 \end{vmatrix}}{\begin{vmatrix} a_1 & b_1 \\ a_2 & b_2 \end{vmatrix}} \tag{4}$$

$$y = \frac{\begin{vmatrix} a_1 & c_1 \\ a_2 & c_2 \end{vmatrix}}{\begin{vmatrix} a_1 & b_1 \\ a_2 & b_2 \end{vmatrix}} \tag{5}$$

Let us see how the determinant form may be developed directly from the original equations without performing the intervening additions and subtractions. Given the original equations:

$$\begin{aligned} a_1x + b_1y &= c_1 \\ a_2x + b_2y &= c_2 \end{aligned} \tag{1}$$

First, produce the determinant of the denominator by setting, in order, the coefficients of the unknowns:

$$\begin{vmatrix} a_1 & b_1 \\ a_2 & b_2 \end{vmatrix}$$

Second, using the denominator determinant as a base, develop the determinant of the numerator of the solution for x by replacing the column of x coefficients by the corresponding column of constants (the right-hand sides of the equations). Then complete the new determinant by putting in the column of the y coefficients in its original position:

$$\begin{vmatrix} c_1 & b_1 \\ c_2 & b_2 \end{vmatrix}$$

Confirm that this determinant is identical in value with

$$\begin{vmatrix} b_2 & c_2 \\ b_1 & c_1 \end{vmatrix}$$

given as Eq. (4), but easier to develop automatically.

Third, still using the denominator determinant as a starting place, develop the determinant of the numerator of y by replacing the column of y coefficients by the column of constants and leaving the column of x coefficients in its original position:

$$\begin{vmatrix} a_1 & c_1 \\ a_2 & c_2 \end{vmatrix}$$

Last, put these three determinants together to form the full solution statements:

$$x = \frac{\begin{vmatrix} c_1 & b_1 \\ c_2 & b_2 \end{vmatrix}}{\begin{vmatrix} a_1 & b_1 \\ a_2 & b_2 \end{vmatrix}} \tag{4}$$

$$y = \frac{\begin{vmatrix} a_1 & c_1 \\ a_2 & c_2 \end{vmatrix}}{\begin{vmatrix} a_1 & b_1 \\ a_2 & b_2 \end{vmatrix}} \qquad (5)$$

Rule

To solve two simultaneous equations having two variables by the method of determinants:

1. Form the denominator determinant by using the coefficients of the unknowns in their correct rows and columns.
2. Form the x numerator determinant by replacing the column of x coefficients in the denominator determinant by the column of constants.
3. Form the y numerator determinant by replacing the column of y coefficients in the denominator by the column of constants.
4. Combine the three determinants so formed to produce the pair of solution equations.

example 2 Solve the simultaneous equations

$$3p + 2q = 8$$
$$5p + q = 11$$

solution The denominator determinant is

$$\begin{vmatrix} 3 & 2 \\ 5 & 1 \end{vmatrix}$$

Using this determinant as a base, the determinant for the numerator of p must be $\begin{vmatrix} 8 & 2 \\ 11 & 1 \end{vmatrix}$ and the determinant for the numerator of q must be $\begin{vmatrix} 3 & 8 \\ 5 & 11 \end{vmatrix}$. Thus,

$$p = \frac{\begin{vmatrix} 8 & 2 \\ 11 & 1 \end{vmatrix}}{\begin{vmatrix} 3 & 2 \\ 5 & 1 \end{vmatrix}} \quad \text{and} \quad q = \frac{\begin{vmatrix} 3 & 8 \\ 5 & 11 \end{vmatrix}}{\begin{vmatrix} 3 & 2 \\ 5 & 1 \end{vmatrix}}$$

When evaluating these determinants, *always* evaluate the denominator first. (The reason will be explained soon.) The value of the denominator is

$$+(3)(1) - (5)(2) = -7$$

The numerator of p has the value

$$+(8)(1) - (11)(2) = -14$$

$$p = \frac{-14}{-7} = 2$$

The numerator of q has the value $+(3)(11) - (5)(8) = -7$, and

$$q = \frac{-7}{-7} = 1$$

18-3 CONSISTENCY OF EQUATIONS

In solving systems of second-order simultaneous equations, there are three main possibilities:

1. The equations may represent straight lines which intersect. These are said to be *independent equations.* They are in no way related to each other except that the unknowns have similar symbols, A, b, x, θ, etc., and one pair of values constitutes the whole solution.
2. The equations may represent superimposed lines. These are said to be *dependent equations.* They are related to each other, and every solution of the one is also a solution of the other. There is an endless number of solutions.
3. The equations may represent parallel lines. These are said to be *inconsistent equations.* They differ only in the constant terms (the y intercepts), and there is no solution for one equation which satisfies the other.

The values of the denominator and the numerators quickly show us into which classification any system of simultaneous equations falls:

1. To be independent, the denominators may not equal zero.
2. To be dependent, the denominator is zero and the numerators equal zero.
3. To be inconsistent, the denominator is zero and at least one of the numerators does not equal zero.

This is why we evaluate the denominator first. If it is zero, there is no single set of values which will constitute the entire solution, and, in electronics problems, there is no use investigating further.

PROBLEMS 18-2

Solve these systems of simultaneous equations by using determinants:

1. $4a - 3b = 10$
$3a + b = 14$

2. $4x + y = 15$
$2x + 3y = 15$

3. $2\theta + \pi = 22$
$3\theta - 5\pi = 20$

4. $R_1 + 3R_2 = 23$
$R_1 - 3R_2 = 5$

5. $I + 4i = -5$
$2I + i = 4$

6. $3E + 2E_g = 1$
$E_g + E = -2$

7. $4r_p + 3r_L = 3$
$6r_p - 9r_L = 0$

8. $4X_C + 3X_L = 2.9$
$30X_L = 17 - 8X_C$

9. $0.5R_1 + 0.2R_2 = 315$
$0.6R_1 - 54 = 0.03R_2$

10. $Z_1 = 9300 - Z_2$
$192 + 0.06Z_2 = 0.04Z_1$

18-4 THIRD-ORDER DETERMINANTS

When solving sets of three simultaneous equations, naturally, we arrive at third-order determinants consisting of three columns and three rows of three elements each, such as

$$\begin{vmatrix} 3 & 1 & 2 \\ 2 & 6 & 5 \\ 4 & 8 & 1 \end{vmatrix} \qquad \begin{vmatrix} a_1 & b_1 & c_1 \\ a_2 & b_2 & c_2 \\ a_3 & b_3 & c_3 \end{vmatrix}$$

Now, when we multiply on the diagonal, we find a slight complication. Multiplying the main diagonal is simple:

$$\begin{vmatrix} 3 & 1 & 2 \\ 2 & 6 & 5 \\ 4 & 8 & 1 \end{vmatrix} = +(3)(6)(1) = +18$$

but the next diagonal gets complicated:

$$\begin{vmatrix} 3 & 1 & 2 \\ 2 & 6 & 5 \\ 4 & 8 & 1 \end{vmatrix} = +(1)(5)(4) = +20$$

and also the next:

$$\begin{vmatrix} 3 & 1 & 2 \\ 2 & 6 & 5 \\ 4 & 8 & 1 \end{vmatrix} = +(2)(8)(2) = +32$$

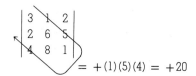

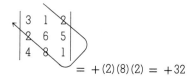

And you can see that the negative diagonals will be just as complicated. So we devise a method of notation which gets around this complication and enables us to perform straight-line multiplication. First, we set down the determinant in its usual form, with three columns and three rows. Then, to the right of this determinant, we repeat the first two columns. This process straightens out the diagonals

$$= -(1)(2)(1) = -2$$

$$
\begin{vmatrix}
3 & 1 & 2 & 3 & 1 \\
2 & 6 & 5 & 2 & 6 \\
4 & 8 & & 4 & 8
\end{vmatrix}
$$

$$= +(3)(6)(1) = +18$$

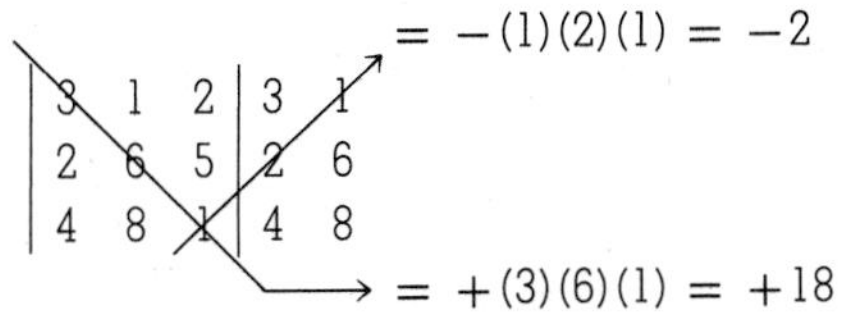

and we obtain, with a complete program of diagonal multiplication, the value of the determinant $= -100$.

example 3 Evaluate the determinant

$$
\begin{vmatrix}
2 & -1 & 4 \\
1 & 6 & 5 \\
7 & -3 & -2
\end{vmatrix}
$$

solution Rewrite the determinant and repeat the first two columns outside to the right:

$$
\begin{vmatrix}
2 & -1 & 4 \\
1 & 6 & 5 \\
7 & -3 & -2
\end{vmatrix}
\begin{matrix}
2 & -1 \\
1 & 6 \\
7 & -3
\end{matrix}
$$

Then perform the diagonal multiplication, signed, as for second-order determinants and obtain

$$-24 - 35 - 12 - 168 + 30 - 2 = -211$$

example 4 Solve the third-order set of simultaneous equations:

$$
\begin{aligned}
a + 2b + \ c &= 7 \\
2a + \ b + 2c &= 2 \\
a + 3b + 4c &= 14
\end{aligned}
$$

solution First, write and evaluate the denominator determinant:

$$
\begin{vmatrix}
1 & 2 & 1 \\
2 & 1 & 2 \\
1 & 3 & 4
\end{vmatrix}
\begin{matrix}
1 & 2 \\
2 & 1 \\
1 & 3
\end{matrix} = -9
$$

Second, develop the determinant for the numerator of a, replacing the column of a coefficients by the column of constants, and evaluate it:

$$
\begin{vmatrix}
7 & 2 & 1 \\
2 & 1 & 2 \\
14 & 3 & 4
\end{vmatrix}
\begin{matrix}
7 & 2 \\
2 & 1 \\
14 & 3
\end{matrix} = 18
$$

Third, combine the denominator and numerator to evaluate a:

$$a = \frac{18}{-9} = -2$$

You should immediately prove that $b = 4$ and $c = 1$.

PROBLEMS 18-3
Evaluate these third-order determinants:

1.
$$
\begin{vmatrix}
1 & 1 & 1 \\
2 & -1 & -1 \\
3 & 2 & -5
\end{vmatrix}
$$

2.
$$
\begin{vmatrix}
1 & 3 & 1 \\
5 & 40 & 6 \\
-2 & -25 & -3
\end{vmatrix}
$$

3.
$$
\begin{vmatrix}
2 & 3 & 32 \\
5 & -2 & 0 \\
4 & -8 & -41
\end{vmatrix}
$$

4.
$$
\begin{vmatrix}
-3 & -2 & 3 \\
0 & -7 & 2 \\
0 & 7 & -4
\end{vmatrix}
$$

5.
$$
\begin{vmatrix}
4 & 6 & 8 \\
-10 & -3 & 4 \\
2 & 12 & -20
\end{vmatrix}
$$

6.
$$
\begin{vmatrix}
3 & 8 & 3.2 \\
12 & 20 & 16.5 \\
-16 & -12 & -7.8
\end{vmatrix}
$$

Solve these simultaneous equations by using determinants:

7. $x + y + z = 15$
$2x - y - z = 0$
$3x + 2y - 5z = 14$

8. $R_1 + R_2 + R_3 = 3$
$5R_1 - 2R_2 + 6R_3 = 40$
$-2R_1 + 3R_2 - 3R_3 = -25$

9. $2\alpha + 3\beta + 2\gamma = 32$
$5\alpha - \gamma = 2\beta$
$4\alpha - 8\beta = 3\gamma - 41$

10. $3r + 5p - 2q = -3$
$p + q = 4r$
$3p - 7q + 2r = -42$

11. $4E + 6e + 8(IR) = 6$
$4(IR) - 10E - 3e = -5$
$12e - 20(IR) + 12E = 5$

12. $12I_1 + 20I_2 + 10I_3 = 16.5$
$8I_2 - 6I_3 + 3I_1 = 3.2$
$20I_3 - 16I_1 - 12I_2 = -7.8$

18-5 MINORS

The method of diagonal multiplication works perfectly for both second- and third-order determinants. Unfortunately, it will not work for higher-order systems. Thus, if we are required to evaluate by determinants a fourth- or fifth-order set of equations such as might arise from the solution of a complicated circuit (see Chap. 22), we must work out another useful system. Since we can do this without difficulty, we will not try to prove the statement above. (Even many "higher mathematics" texts say simply: Do not use diagonal multiplication for fourth-order determinants or higher.)

This is how minors come about: Let us evaluate the general third-order determinant:

$$\left|\begin{array}{ccc|cc} a_1 & b_1 & c_1 & a_1 & b_1 \\ a_2 & b_2 & c_2 & a_2 & b_2 \\ a_3 & b_3 & c_3 & a_3 & b_3 \end{array}\right|$$
$$= a_1b_2c_3 + a_3b_1c_2 + a_2b_3c_1 - a_3b_2c_1 - a_1b_3c_2 - a_2b_1c_3 \qquad (6)$$

Consider the terms which involve the value a_1. These may be collected to yield $a_1(b_2c_3 - b_3c_2)$, which in turn could be written

$$a_1\begin{vmatrix} b_2 & c_2 \\ b_3 & c_3 \end{vmatrix}$$

where the new second-order determinant is called the *minor of the element* a_1.

We can develop this minor from the original third-order determinant by selecting the element a_1, crossing out the other elements of the row and column which contain a_1, and writing the minor with the elements remaining.

$$\begin{vmatrix} \bigcirc\!\!\!\!\!\!a_1 & \cancel{b_1} & \cancel{c_1} \\ \cancel{a_2} & b_2 & c_2 \\ \cancel{a_3} & b_3 & c_3 \end{vmatrix}$$

yields $$\begin{vmatrix} b_2 & c_2 \\ b_3 & c_3 \end{vmatrix}$$

Rule

To find the *minor* of any element in a determinant, select the element, cross out the row and column containing that element, and write the lower-order determinant which contains all the other elements that remain.

Thus, in the third-order determinant of Eq. (6), the minor of the element b_3 is

$$\begin{vmatrix} a_1 & c_1 \\ a_2 & c_2 \end{vmatrix}$$

example 5 Evaluate the minor of 2 in the determinant

$$\begin{vmatrix} 1 & 4 & 0 \\ 3 & 1 & 5 \\ 5 & 6 & 2 \end{vmatrix}$$

solution Striking out the elements in the row and column containing the 2 yields

$$\begin{vmatrix} 1 & 4 \\ 3 & 1 \end{vmatrix} = +1 - 12 = -11$$

PROBLEMS 18-4

Write and evaluate the *minors* of the indicated elements:

1. $\begin{vmatrix} 2 & 3 & 2 \\ -4 & -1 & 3 \\ 5 & 2 & \textcircled{6} \end{vmatrix}$

2. $\begin{vmatrix} 3 & 1 & -1 \\ \textcircled{8} & -2 & 2 \\ -13 & -3 & -1 \end{vmatrix}$

3. $\begin{vmatrix} 3 & 2 & 5 \\ \textcircled{-2} & -3 & 4 \\ 6 & 5 & 0 \end{vmatrix}$

4. $\begin{vmatrix} -3 & -7 & 16 \\ -8 & 2 & 84 \\ 2 & 3 & \textcircled{-26} \end{vmatrix}$

5. $\begin{vmatrix} 2 & 3 & -5 \\ 3 & 0 & 4 \\ 0 & \textcircled{-2} & 7 \end{vmatrix}$

6. $\begin{vmatrix} -8 & \textcircled{-13} & 10 \\ 0 & 2 & 5 \\ 2 & 10 & -20 \end{vmatrix}$

7. $\begin{vmatrix} 0 & 0 & 4 \\ 0 & 2 & 4 \\ -4 & \textcircled{10} & 0 \end{vmatrix}$

8.* $\begin{vmatrix} 3 & 6 & -3 & 2 \\ 2 & -2 & 2 & -1 \\ 5 & \textcircled{25} & 0 & 3 \\ 0 & 5 & -5 & 1 \end{vmatrix}$

* ***hint*** The minor of any element of a fourth-order determinant will be a third-order determinant which may itself be evaluated by the diagonal method or by second-step cofactors, which are discussed in the following section.

18-6 COFACTORS

A simple step converts the *minor* into a *cofactor*. When evaluating a complete determinant by the method of cofactors, we first find the minors of all the elements in any given row or column. Then we convert these minors into cofactors by assigning them algebraic signs according to this simple rule:

Rule

Each element of a determinant, regardless of its actual algebraic value, has a cofactor sign according to its place in the determinant. The signs are found by a checkerboard arrangement:

$$\begin{vmatrix} + & - & + \\ - & + & - \\ + & - & + \end{vmatrix}$$

The only thing to remember is to always start the upper left-hand corner (the element in row 1 and column 1) with a + sign. All the rest follows automatically, regardless of the number of elements in the determinant.

example 6 Evaluate the following determinant by means of cofactors:

$$\begin{vmatrix} 1 & 4 & 0 \\ -3 & 1 & 5 \\ 5 & 6 & -2 \end{vmatrix}$$

solution Choose any convenient row or column, and, one after the other, set down the individual elements of that row or column, together with their minors:

$$4\begin{vmatrix} -3 & 5 \\ 5 & -2 \end{vmatrix} \quad 1\begin{vmatrix} 1 & 0 \\ 5 & -2 \end{vmatrix} \quad 6\begin{vmatrix} 1 & 0 \\ -3 & 5 \end{vmatrix}$$

Then assign the cofactor signs according to the checkerboard plan:

$$-4\begin{vmatrix} -3 & 5 \\ 5 & -2 \end{vmatrix} \quad +1\begin{vmatrix} 1 & 0 \\ 5 & -2 \end{vmatrix} \quad -6\begin{vmatrix} 1 & 0 \\ -3 & 5 \end{vmatrix}$$

Evaluate each minor, multiply its value by the element of which it is the minor, and add algebraically according to the cofactor signs and the actual algebraic sign of the multiplications:

$$-4(6 - 25) + 1(-2 - 0) - 6(5 - 0)$$
$$= 76 - 2 - 30 = 44$$

You should immediately evaluate the same third-order determinant by the cofactors of the elements of each other row and column in turn. The answer must always be 44.

example 7 Solve this set of simultaneous equations by means of cofactors:

$$\begin{aligned} 2p + 10q + 5r &= 9 \\ -3p + 9q + 4r &= -3 \\ 7p - 6q - r &= 17 \end{aligned}$$

solution Using the information now at hand, we may immediately set up the determinant form of solution:

$$p = \frac{\begin{vmatrix} 9 & 10 & 5 \\ -3 & 9 & 4 \\ 17 & -6 & -1 \end{vmatrix}}{\begin{vmatrix} 2 & 10 & 5 \\ -3 & 9 & 4 \\ 7 & -6 & -1 \end{vmatrix}}$$

$$q = \frac{\begin{vmatrix} 2 & 9 & 5 \\ -3 & -3 & 4 \\ 7 & 17 & -1 \end{vmatrix}}{\begin{vmatrix} 2 & 10 & 5 \\ -3 & 9 & 4 \\ 7 & -6 & -1 \end{vmatrix}}$$

$$r = \frac{\begin{vmatrix} 2 & 10 & 9 \\ -3 & 9 & -3 \\ 7 & -6 & 17 \end{vmatrix}}{\begin{vmatrix} 2 & 10 & 5 \\ -3 & 9 & 4 \\ 7 & -6 & -1 \end{vmatrix}}$$

Always evaluate the denominator first. To solve by means of cofactors, we choose any row or column in the denominator determinant, evaluate their minors, and multiply by the elements, adding algebraically and using the checkerboard signs.

$$\begin{vmatrix} 2 & 10 & 5 \\ -3 & 9 & 4 \\ 7 & -6 & -1 \end{vmatrix}$$
$$= -(-3)\begin{vmatrix} 10 & 5 \\ -6 & -1 \end{vmatrix} + (9)\begin{vmatrix} 2 & 5 \\ 7 & -1 \end{vmatrix} - (4)\begin{vmatrix} 2 & 10 \\ 7 & -6 \end{vmatrix}$$
$$= 3(-10 + 30) + 9(-2 - 35) - 4(-12 - 70)$$
$$= 60 - 333 + 328$$
$$= 55$$

Since the denominator is not zero, we should evaluate the numerators, in turn, of p, q, and r.

Numerator of

$$p = \begin{vmatrix} 9 & 10 & 5 \\ -3 & 9 & 4 \\ 17 & -6 & -1 \end{vmatrix}$$
$$= +(17)\begin{vmatrix} 10 & 5 \\ 9 & 4 \end{vmatrix} - (-6)\begin{vmatrix} 9 & 5 \\ -3 & 4 \end{vmatrix} + (-1)\begin{vmatrix} 9 & 10 \\ -3 & 9 \end{vmatrix}$$
$$= +110$$

Therefore, $p = \frac{110}{55} = 2$. You should now prove that $q = -1$ and $r = 3$.

18-7 USEFUL PROPERTIES OF DETERMINANTS

The evaluation of determinants by the methods of diagonal multiplication or cofactors will yield the correct answers if you keep close watch on your arithmetic and the algebraic signs of positive and negative diagonals or of the checkerboard cofactor signs. There are, however, a few very useful properties of determinants which will simplify the process of evaluation. These properties are described briefly below, and it is left to you to perform the diagonal multiplication or cofactor evaluation methods to confirm them immediately when you meet them.

1. When all the elements of any row (or column) are zero, the value of the determinant is zero:

$$\begin{vmatrix} a_1 & b_1 & 0 \\ a_2 & b_2 & 0 \\ a_3 & b_3 & 0 \end{vmatrix} = 0$$

example 8 Evaluate the determinant:

$$\begin{vmatrix} 2 & 4 & -3 \\ 0 & 0 & 0 \\ -4 & 6 & 1 \end{vmatrix}$$

solution

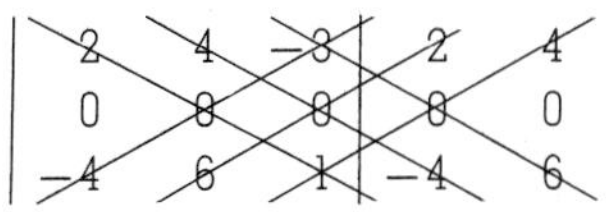

Each diagonal multiplication introduces a factor of zero. Therefore, each diagonal product is zero, and the value of the determinant is zero.

2. When all the elements to the right (or left) of the principal diagonal are zero, the value of the determinant is the product of the elements of the principal diagonal:

$$\begin{vmatrix} a_1 & 0 & 0 \\ a_2 & b_2 & 0 \\ a_3 & b_3 & c_3 \end{vmatrix} = a_1 b_2 c_3$$

(It is left to you to prove that this is also true for fourth-order determinants.)

example 9 Evaluate the determinant:

$$\begin{vmatrix} 3 & 8 & 5 \\ 0 & -2 & 7 \\ 0 & 0 & -5 \end{vmatrix}$$

solution

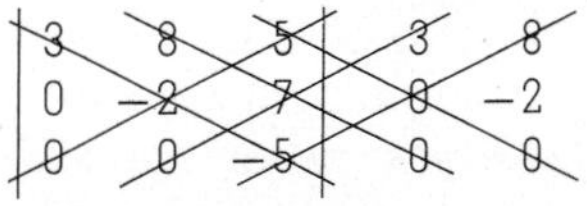

All the diagonal multiplications except the first, through the principal diagonal, are zero. Therefore the value of the determinant is $(3)(-2)(-5) = 30$.

3. Interchanging all the rows and columns gives the identical result, both absolute value and algebraic sign:

$$\begin{vmatrix} a_1 & a_2 & a_3 \\ b_1 & b_2 & b_3 \\ c_1 & c_2 & c_3 \end{vmatrix} = \begin{vmatrix} a_1 & b_1 & c_1 \\ a_2 & b_2 & c_2 \\ a_3 & b_3 & c_3 \end{vmatrix}$$

4. Interchanging two rows (or columns) gives the same absolute value but the opposite algebraic sign:

$$\begin{vmatrix} c_1 & b_1 & a_1 \\ c_2 & b_2 & a_2 \\ c_3 & b_3 & a_3 \end{vmatrix} = -\begin{vmatrix} a_1 & b_1 & c_1 \\ a_2 & b_2 & c_2 \\ a_3 & b_3 & c_3 \end{vmatrix}$$

5. When the corresponding elements of any two rows (or columns) are identical or proportional, the value of the determinant is zero:

$$\begin{vmatrix} a_1 & ka_1 & c_1 \\ a_2 & ka_2 & c_2 \\ a_3 & ka_3 & c_3 \end{vmatrix} = 0 \qquad (k \text{ may} = 1)$$

example 10 Evaluate the determinant:

$$\begin{vmatrix} 3 & 5 & 6 \\ 2 & -1 & 4 \\ 7 & 4 & 14 \end{vmatrix}$$

solution

$$\left|\begin{array}{ccc|cc} 3 & 5 & 6 & 3 & 5 \\ 2 & -1 & 4 & 2 & -1 \\ 7 & 4 & 14 & 7 & 4 \end{array}\right.$$

Diagonal multiplication yields are zero value. Observation of the first and third columns shows that col 3 = 2 × col 1.

6. A common factor of any row (or column) may be factored out as a common factor of the whole determinant:

$$\begin{vmatrix} a_1 & b_1 & kc_1 \\ a_2 & b_2 & kc_2 \\ a_3 & b_3 & kc_3 \end{vmatrix} = k \begin{vmatrix} a_1 & b_1 & c_1 \\ a_2 & b_2 & c_2 \\ a_3 & b_3 & c_3 \end{vmatrix}$$

example 11 Evaluate the determinant

$$\begin{vmatrix} 3 & 6 & 2 \\ -2 & 8 & 5 \\ 40 & 30 & -70 \end{vmatrix}$$

solution

$$\begin{vmatrix} 3 & 6 & 2 \\ -2 & 8 & 5 \\ 40 & 30 & -70 \end{vmatrix} = 10 \left|\begin{array}{ccc|cc} 3 & 6 & 2 & 3 & 6 \\ -2 & 8 & 5 & -2 & 8 \\ 4 & 3 & -7 & 4 & 3 \end{array}\right.$$
$$= 10(-253) = -2530$$

7. When the elements of any row (or column) are increased by a constant times the corresponding elements of any other row (or column), the value of the determinant is unchanged. (k may equal 1, −1, or any other positive or negative integer or fraction):

$$\begin{vmatrix} a_1 & b_1 & ka_1 + c_1 \\ a_2 & b_2 & ka_2 + c_2 \\ a_3 & b_3 & ka_3 + c_3 \end{vmatrix} = \begin{vmatrix} a_1 & b_1 & c_1 \\ a_2 & b_2 & c_2 \\ a_3 & b_3 & c_3 \end{vmatrix}$$

example 12 Evaluate the determinant

$$\begin{vmatrix} 2 & 8 & 3 \\ 3 & 7 & -6 \\ -1 & 2 & 1 \end{vmatrix}$$

solution If the spaces filled by the elements 8, 3, and −6 can be converted to zeros, the evaluation of the determinant will be the product of the elements of the principal axis. Or if any two spaces in any row or column can be adjusted to zero, the evaluation becomes a single element times its cofactor.

Using the principle introduced above, let us attempt to eliminate the element 3. We will multiply each element of the third row by −3 and add the result to the corresponding elements of the first row:

$$\begin{vmatrix} 2 & 8 & 3 \\ 3 & 7 & -6 \\ -1 & 2 & 1 \end{vmatrix}$$

$$= \begin{vmatrix} 2 + (-3)(-1) & 8 + (-3)(2) & 3 + (-3)(1) \\ 3 & 7 & -6 \\ -1 & 2 & 1 \end{vmatrix}$$

$$= \begin{vmatrix} 5 & 2 & 0 \\ 3 & 7 & -6 \\ -1 & 2 & 1 \end{vmatrix}$$

Then, to eliminate the −6, we will multiply

the third row by 6 and add the results to the second row:

$$\begin{vmatrix} 5 & 2 & 0 \\ 3 & 7 & -6 \\ -1 & 2 & 1 \end{vmatrix} = \begin{vmatrix} 5 & 2 & 0 \\ 3 + (6)(-1) & 7 + (6)(2) & -6 + (6)(1) \\ -1 & 2 & 1 \end{vmatrix} = \begin{vmatrix} 5 & 2 & 0 \\ -3 & 19 & 0 \\ -1 & 2 & 1 \end{vmatrix}$$

This determinant may be evaluated by the product of the element 1 and its cofactor:

$$\begin{vmatrix} 5 & 2 & 0 \\ -3 & 19 & 0 \\ -1 & 2 & 1 \end{vmatrix} = +1\begin{vmatrix} 5 & 2 \\ -3 & 19 \end{vmatrix} = 95 + 6 = 101$$

You should test this solution by the diagonal multiplication of the original determinant. Alternatively, the simplification may continue by removal of the element 2 in the first row. If we add to the first row the product of $-\frac{2}{19}$ (second row),

$$\begin{vmatrix} 5 + (-\frac{2}{19})(-3) & 2 + (-\frac{2}{19})(19) & 0 + (-\frac{2}{19})(0) \\ -3 & 19 & 0 \\ -1 & 2 & 1 \end{vmatrix} = \begin{vmatrix} 5\frac{6}{19} & 0 & 0 \\ -3 & 19 & 0 \\ -1 & 2 & 1 \end{vmatrix}$$

Evaluation by the principal diagonal yields

$$(5\tfrac{6}{19})(19)(1) = 101$$

With practice, the addition of a fraction in the form $-\frac{a_x}{a_y}a_y$ will reveal itself as a valuable tool.

8. When the elements of any row (or column) may be written as sums, the determinant may be written as the sum of two determinants with the rows (or columns) of the sum elements in their corresponding places:

$$\begin{vmatrix} a_1 & b_1 & p_1 + q_1 \\ a_2 & b_2 & p_2 + q_2 \\ a_3 & b_3 & p_3 + q_3 \end{vmatrix} = \begin{vmatrix} a_1 & b_1 & p_1 \\ a_2 & b_2 & p_2 \\ a_3 & b_3 & p_3 \end{vmatrix} + \begin{vmatrix} a_1 & b_1 & q_1 \\ a_2 & b_2 & q_2 \\ a_3 & b_3 & q_3 \end{vmatrix}$$

Now apply these fundamental properties of determinants in the solution of the following problems and in problems like them in later chapters.

PROBLEMS 18-5

Evaluate the following determinants by means of the *cofactors* of the indicated rows or columns:

1. $\begin{vmatrix} 5 & 41 & 6 \\ 2 & 1 & -2 \\ -4 & 8 & 3 \end{vmatrix}$ Row 1

2. $\begin{vmatrix} 2 & 1 & -1 \\ 4 & -3 & -1 \\ 3 & -2 & 2 \end{vmatrix}$ Col 2

3. $\begin{vmatrix} 5 & 2 & 3 \\ 4 & -3 & 12 \\ 0 & 5 & -8 \end{vmatrix}$ Row 3

4. $\begin{vmatrix} -26 & 3 & 2 \\ 84 & 2 & -10 \\ 16 & -7 & 4 \end{vmatrix}$ Col 1

5. $\begin{vmatrix} 3 & 0 & 21.7 \\ 2 & 3 & 15.3 \\ 0 & -2 & 1.9 \end{vmatrix}$ Col 3

6. $\begin{vmatrix} 2 & 4 & 10 \\ -8 & -16 & -13 \\ 0 & 16 & 2 \end{vmatrix}$ Row 2

7. $\begin{vmatrix} 3 & 2 & -3 & 2 \\ 2 & -3 & 2 & -1 \\ 5 & 4 & 0 & 3 \\ 0 & 8 & -5 & 1 \end{vmatrix}$ Col 2

hint The cofactors of elements in a fourth-order determinant will themselves be third-order determinants which may be evaluated by diagonals or by cofactors.

8. $\begin{vmatrix} 2 & 16 & 12 & -10 & -2 \\ 5 & 2 & 2 & 3 & -9 \\ 11 & 0 & 0 & 5 & 4 \\ 5 & 0 & 2 & 15 & 4 \\ 0 & -4 & 10 & -8 & 0 \end{vmatrix}$ Row 3

Solve by using determinants and cofactors:

9. $5I_1 + 2I_2 + 6I_3 = 41$
$2I_1 + 3I_2 - 2I_3 = 1$
$-4I_1 - I_2 + 3I_3 = 8$

10. $2\theta + \phi - \lambda = 3$
$3\theta - 2\phi + 2\lambda = 8$
$4\theta - 3\phi - \lambda = -13$

11. $3\alpha + 2\beta + 3\gamma = 5$
$-2\alpha - 3\beta + 12\gamma = 4$
$6\alpha + 5\beta - 8\gamma = 0$

12. $2I_1 + 3I_2 + 2I_3 = -26$
$-8I_1 + 2I_2 - 10I_3 = 84$
$-3I_1 - 7I_2 + 4I_3 = 16$

13. $3R_1 + 4R_3 = 21.7$
$2R_1 + 3R_2 - 5R_3 = 15.3$
$7R_3 - 2R_2 = 1.9$

14. $2x + 4y + 10z = 10$
$-8x - 16y + 5z = -13$
$16y - 20z = 2$

15. $3\varepsilon + 2\eta - 3\kappa + 2\lambda = 6$
$2\varepsilon - 3\eta + 2\kappa - \lambda = -2$
$5\varepsilon + 4\eta + 3\lambda = 25$
$8\eta - 5\kappa + \lambda = 5$

16. $2I_1 + 16I_2 + 12I_3 - 10I_4 - 2I_5 = 100$
$5I_1 + 2I_2 + 2I_3 + 3I_4 - 9I_5 = 0$
$11I_1 + 5I_4 + 4I_5 = 100$
$5I_1 + 2I_3 + 15I_4 + 4I_5 = 100$
$-4I_2 + 10I_3 - 8I_4 = 0$

17. Solve selected problems from Chap. 17 by means of determinants.

18. Use determinants for the solution of appropriate problems throughout the remainder of this book.

19

batteries

In order to avoid confusion in previous discussions of electric circuits, all sources of electromotive force have been considered to be sources of constant potential, and nothing has been said of their internal resistances. At the same time, no mention has been made of the actual sources of the emf. In this chapter we will consider both of these factors. First of all, electrical devices which produce electric energy, as well as those which consume energy, have a certain amount of internal resistance which materially affects their operation. The application of Ohm's law to the internal resistance of batteries is the feature topic of this chapter. And despite the prevalence of utility power supply, batteries are still useful, indeed necessary, sources of portable power. For this reason, the electronics technician should be aware of the problems which arise in the use of batteries.

19-1 ELECTROMOTIVE FORCE

A battery is a device which converts chemical energy into electric energy. Essentially, it consists of a cell, or several cells connected in series or parallel, conveniently packaged. The emf of the battery is the total voltage developed by the chemical action. However, this total voltage is not all available for doing useful work in an external circuit, because some of it is needed to overcome the internal resistance of the battery itself. The voltage which is supplied to the external circuit is known as the terminal voltage; that is,

Terminal voltage = emf − internal voltage drop

19-2 BATTERIES

The word *battery* is taken to mean two or more *cells* connected to each other, although a single cell is often referred to as a battery.

Figure 19-1 represents a circuit by which the voltage existing across the cell can be read with the resistance connected across the battery or with the resistance disconnected from the circuit.

The emf of a cell is the total amount of

Fig. 19-1 High-Resistance Voltmeter Used for Measuring Electromotive Force of Cell

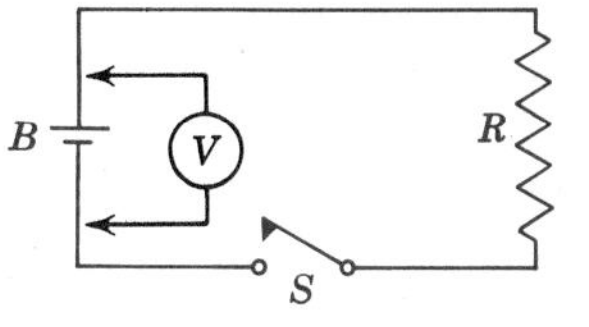

voltage developed by the cell. For all practical purposes the emf of a cell can be read with a high-resistance voltmeter connected across the cell when it is not supplying current to any other circuit, as is the case with the switch S, Fig. 19-1, open.

When a cell supplies current to an external circuit, as with the switch in Fig. 19-1 closed, it will be found that the voltmeter no longer reads the open-circuit voltage (emf) of the cell. The reason for this is that part of the emf is used in forcing current through the resistance of the cell and the remainder is used in forcing current through the external circuit. Expressed as an equation,

$$E = E_t + Ir \tag{1}$$

where E is the emf of the cell or group of cells and E_t is the voltage measured across the terminals while forcing a current I through the internal resistance r. Since I also flows through the external circuit of resistance R, Eq. (1) can be written

$$E = IR + Ir$$

or

$$E = I(R + r) \tag{2}$$

example 1 A cell whose internal resistance is 0.15 Ω delivers 0.50 A to a resistance of 2.85 Ω. What is the emf of the cell?

solution Given $r = 0.15\ \Omega$, $R = 2.85\ \Omega$, and $I = 0.50$ A.

From Eq. (2), $E = 0.50(2.85 + 0.15) = 1.5$ V

example 2 Figure 19-2 represents a cell with an emf of 1.2 V and an internal resistance r of 0.2 Ω connected to a resistance R of 5.8 Ω. How much current flows in the circuit?

solution Solving Eq. (2) for the current,

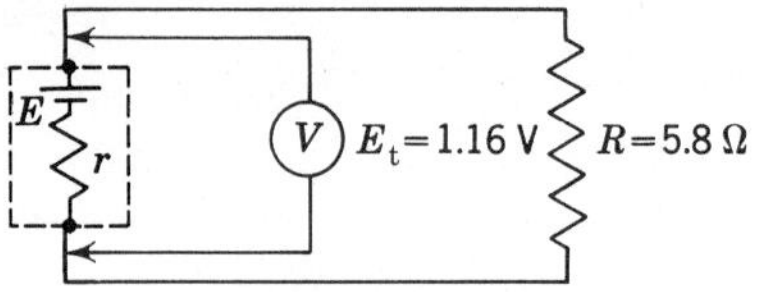

Fig. 19-2 Circuit of Example 2

$$I = \frac{E}{R + r} \tag{3}$$

$$= \frac{1.2}{5.8 + 0.2} = 0.2 \text{ A}$$

Note the significance of Eq. (3). It says that the current which flows in a circuit is proportional to the emf of the circuit and inversely proportional to the *total* resistance of the circuit. This is Ohm's law for the *complete circuit.*

example 3 A cell with an emf of 1.6 V delivers a current of 2 A to a circuit of 0.62 Ω. What is the internal resistance of the cell?

solution Solving Eq. (2) for the internal resistance,

$$r = \frac{E - IR}{I} \tag{4}$$

$$= \frac{1.6 - 2 \times 0.62}{2} = 0.18\ \Omega$$

Therefore, the significance of Eq. (4) is that a voltage equal to $E - IR$ is sending the current I through the internal resistance r.

Since Eq. (4) can be rearranged to

$$r = \frac{E}{I} - R$$

and

$$\frac{E}{I} = R_t$$

Eq. (4) can be written

$$r = R_t - R$$

or

$$R_t = R + r \tag{5}$$

Equation (5) states simply that the resistance of the entire circuit is equal to the resistance of the external circuit plus the internal resistance of the source of the emf.

PROBLEMS 19-1

1. A battery taken off the shelf gives a voltmeter reading of 9 V. When connected across a 24-Ω circuit, it drives a current of 360 mA. What is the internal resistance of the battery?
2. A 24-cell battery measures 38.4 V on open circuit. If the total internal resistance is 7.2 Ω, how much current will flow through a 430-Ω circuit?
3. A 6-V battery drives a current of 1 A through a 5.6-Ω load. What is the internal resistance of the battery?
4. With the circuit of Prob. 3, how much power is absorbed by the internal resistance of the battery?
5. With the circuit of Prob. 3, (*a*) how much power is delivered to the load, and (*b*) what is the efficiency of the circuit?

19-3 CELLS IN SERIES

If *n* identical cells are connected in series, the emf of the combination will be *n* times the emf of each cell. Similarly, the total internal resistance of the circuit will be *n* times the internal resistance of each cell. By modifying Eq. (2), the expression for the current through an external resistance of *R* Ω is

$$I = \frac{nE}{R + nr} \tag{6}$$

example 4 Six cells, each with an emf of 2.1 V and an internal resistance of 0.1 Ω, are connected in series, and a resistance of 3.6 Ω is connected across the combination. (*a*) How much current flows in the circuit? (*b*) What is the terminal voltage of the group?

solution Figure 19-3 is a diagram of the circuit. The resistance *nr* represents the total internal resistance of all cells in series.

(*a*) $I = \dfrac{nE}{R + nr} = \dfrac{6 \times 2.1}{3.6 + 6 \times 0.1} = 3.0 \text{ A}$

(*b*) The terminal voltage of the group is equal to the total emf minus the voltage drop across the internal resistance. From Eq. (1),

$$\begin{aligned} E_t &= nE - Inr \\ &= 6 \times 2.1 - 3 \times 6 \times 0.1 \\ &= 10.8 \text{ V} \end{aligned}$$

Since the terminal voltage exists across the external circuit, a more simple relation is

$$E_t = IR = 3 \times 3.6 = 10.8 \text{ V}$$

19-4 CELLS IN PARALLEL

If *n* identical cells are connected in parallel, the emf of the group will be the same as the emf of one cell and the internal resistance of the group will be equal to the internal resistance of one cell divided by the number of cells in parallel, that is, to $\frac{r}{n}$. By modifying

Fig. 19-3 Circuit of Example 4

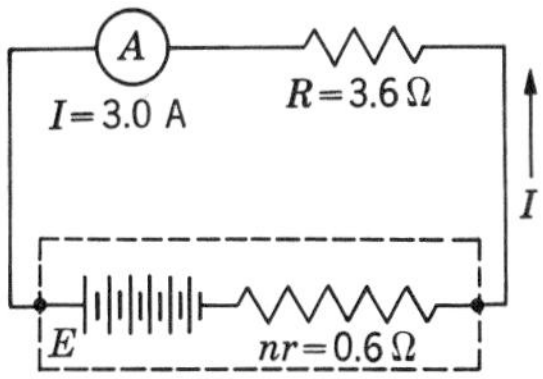

Eq. (2), the expression for the current through an external resistance of $R\ \Omega$ is

$$I = \frac{E}{R + \frac{r}{n}} \qquad (7)$$

example 5 Three cells, each with an emf of 1.4 V and an internal resistance of 0.15 Ω, are connected in parallel, and a resistance of 1.35 Ω is connected across the group. (*a*) How much current flows in the circuit? (*b*) What is the terminal voltage of the group?

solution Figure 19-4 is a diagram of the circuit. The resistance $\frac{r}{n}$ represents the internal resistance of the group.

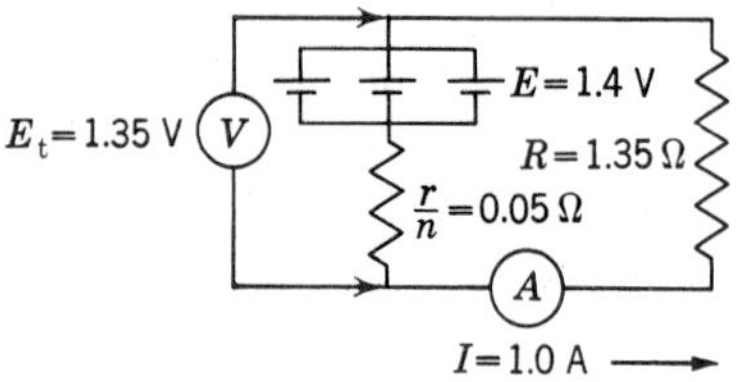

Fig. 19-4 Circuit of Example 5

(*a*) $I = \frac{E}{R + \frac{r}{n}} = \frac{1.4}{1.35 + \frac{0.15}{3}} = 1.0\ \text{A}$

(*b*) $E_t = IR = 1.0 \times 1.35 = 1.35\ \text{V}$

PROBLEMS 19-2

1. The emf of a cell is 1.5 V; the internal resistance of the cell is 0.15 Ω. When current is supplied to a load, the voltage drop across the internal resistance is 0.2 V.
(*a*) What is the terminal voltage?
(*b*) What is the current flow?
(*c*) What is the connected load?

2. A cell whose emf is 1.4 V is supplying 1.5 A to a 0.733-Ω circuit.
(*a*) What is the internal resistance of the cell?
(*b*) How much power is lost in the cell?

3. A cell of emf 1.6 V develops a terminal voltage of 1.48 V when delivering 250 mA to an external circuit.
(*a*) What is the internal resistance of the cell?
(*b*) How much power is expended in the cell?
(*c*) What is the resistance of the external circuit?
(*d*) How much power is absorbed by the load circuit?
(*e*) What is the efficiency of the power transfer?

4. A high-resistance voltmeter reads 2 V when connected across the terminals of an open-circuit cell. What will the meter read when a 5-A current is delivered to a 0.22-Ω load if the internal resistance of the cell is 0.18 Ω?

5. Using the data and results of Prob. 4, how much current would flow if the cell itself were short-circuited?

6. A cell with an emf of 2 V and an internal resistance of 0.1 Ω is connected to a load consisting of a variable resistor.
(*a*) Plot the power delivered to the load as the load resistance is varied in 0.01-Ω steps from 0.05 to 0.15 Ω. What conclusion do you draw from this graph?

(*b*) Plot the efficiency of power transfer over the same resistance range. What conclusion do you draw?

7. Six identical cells, each of emf 1.5 V and internal resistance 0.1 Ω, are connected in series across a load resistor, and they deliver a circuit current of 1.0 A.

(*a*) What is the resistance of the load?

(*b*) How much power is absorbed by the battery?

(*c*) How much current would flow if the battery were short-circuited?

8. If the cells in Prob. 7 are connected in parallel, how much power will be delivered to the load?

9. Ten cells of emf 1.5 V and internal resistance 0.6 Ω each are connected in series across a load of 33 Ω.

(*a*) How much current will flow in the circuit?

(*b*) What will be the terminal voltage of the battery?

(*c*) How much power will be delivered to the load?

10. If the cells of Prob. 9 are connected in parallel across the same load, how much current will flow?

11. Twelve identical cells are hooked up so that four groups of three cells each in series are connected in parallel as shown in Fig. 19-5. The emf of each cell is 1.6 V, and each cell has an internal resistance of 0.2 Ω. If the load *R* is 0.85 Ω and the measured current flow through *R* is 4.8 A:

(*a*) What is the terminal voltage of the battery?

(*b*) What is the emf of each cell?

(*c*) How much power is expended in each cell?

Fig. 19-5 Circuit of Prob. 11

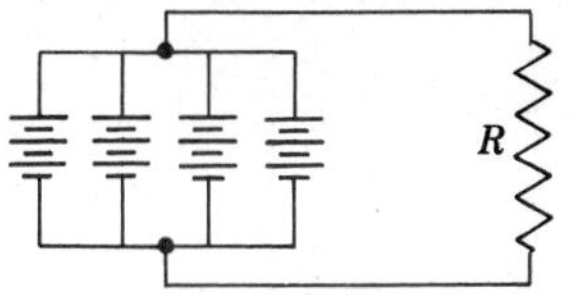

12. The cells of Prob. 11 are so arranged that there are two-cells-per-series groups (six groups in parallel).

(*a*) How much power is dissipated in *R*?

(*b*) How much current flows through each cell?

13. Each cell of a six-cell storage battery has an emf of 2.0 V and an internal resistance of 0.01 Ω. The battery is to be charged from a 14-V line.

(*a*) How much resistance must be connected in series with the battery to limit the charging current to 15 A?

(*b*) What current would flow if the battery were disconnected from the charging circuit and short-circuited?

14. Sixteen storage batteries of three cells each are to be charged in series from a 115-V line. Each cell has an emf of 2.1 V and an internal resistance of 0.02 Ω.

(*a*) How much resistance must be connected in series with the battery to limit the charging current to 10 A?
(*b*) How much power is dissipated in the entire circuit?
(*c*) How much power is dissipated in the series charging resistance?
(*d*) What current would flow if the batteries were disconnected from the charging circuit and short-circuited?

15. Six identical cells connected in series deliver 4 A to a circuit of 2.7 Ω. When two of the same cells are connected in parallel, they deliver 5 A to an external resistance of 0.375 Ω. What are the emf and internal resistance of each cell?

SOLUTION:

Let E = emf of each cell
r = internal resistance of each cell
I = current in external circuit
R = resistance of external circuit

For the series connection,

$6E$ = emf of six cells in series

and

$6r$ = internal resistance of six cells in series

Substituting in Eq. (2),

$$6E = 4(2.7 + 6r) = 10.8 + 24r \qquad (a)$$

For the parallel connection,

E = emf of cells in parallel

and

$\frac{r}{2}$ = internal resistance of two cells in parallel

Substituting in Eq. (2),

$$E = 5\left(0.375 + \frac{r}{2}\right)$$

or

$$2E = 3.75 + 5r \qquad (b)$$

Solve Eqs. (*a*) and (*b*) simultaneously to obtain

$$E = 2.0 \text{ V}$$

and

$$r = 0.05\ \Omega$$

16. Ten identical cells connected in series send a current of 3 A through a 1-Ω circuit. When three of these cells are connected in parallel, they send a current of 6 A through an external resistance of 0.1 Ω. What are the emf and internal resistance of each cell?

17. Five cells connected in series send a current of 5 A through a resistance of 0.4 Ω. When four of these cells are connected in parallel, they send 1 A through 1.35 Ω. What are the emf and internal resistance of each cell?

18. Twelve cells in series, each with an emf of 2.0 V, send a certain current through a 2.4-Ω circuit. The same current flows through a 0.24-Ω circuit when five of these cells are connected in parallel. What is the value of the current and what is the internal resistance of each cell?

19. A cell with an internal resistance of 0.035 Ω sends a current of 3 A through an external circuit. Another cell, with the same emf but with an internal resistance of 0.385 Ω, sends a current of 2 A through the external circuit when substituted for the first cell. What is the emf of the cells and what is the resistance of the external circuit?

20. A cell sends a current of 20 A through an external circuit of 0.04 Ω. When the resistance of the external circuit is increased to 3.96 Ω, the current is 0.4 A. What is the emf and what is the internal resistance of the cell?

20

exponents and radicals

In earlier chapters, examples and problems have been limited to those containing exponents and roots that consisted of integers. In this chapter the study of exponents and radicals is extended to include new operations that will enable you to solve electrical formulas and equations of a type hitherto omitted. In addition, new ideas that will be of fundamental importance in your study of alternating currents are introduced.

20-1 FUNDAMENTAL LAWS OF EXPONENTS

As previously explained, if n is a positive integer, a^n means that a is to be taken as a factor n times. Thus, a^4 is defined as being a shortened form of notation for the product $a \cdot a \cdot a \cdot a$. The number a is called the *base*, and the number n is called the *exponent*.

For the purpose of review, the fundamental laws for the use of *positive-integer exponents* are listed below:

$$a^m \cdot a^n = a^{m+n} \qquad \text{(Sec. 4-3)} \quad (1)$$

$$a^m \div a^n = a^{m-n} \qquad (\text{when } n < m) \qquad \text{(Sec. 4-9)} \quad (2)$$

$$= \frac{1}{a^{n-m}} \qquad (\text{when } n > m)$$

$$(a^m)^n = a^{mn} \qquad \text{(Sec. 6-11)} \quad (3)$$

$$(ab)^m = a^m b^m \qquad \text{(Sec. 6-12)} \quad (4)$$

$$\left(\frac{a}{b}\right)^m = \frac{a^m}{b^m} \qquad (b \neq 0) \qquad (5)$$

20-2 ZERO EXPONENT

If a^0 is to obey the law of exponents for multiplication as stated under Eq. (1) of the preceding section, then

$$a^m \cdot a^0 = a^{m+0} = a^m$$

Also, if a^0 is to obey the law of exponents for division, then

$$\frac{a^m}{a^0} = a^{m-0} = a^m$$

Therefore, the zero power of any number, except zero, is defined as being equal to 1, for 1 is the only number that, when used to multiply another number, does not change the value of the multiplicand.

20-3 NEGATIVE EXPONENTS

If a^{-n} is to obey the multiplication law, then

$$\frac{a^n}{a^n} = a^{n-n} = a^0 = 1$$

In Sec. 4-11, it was shown that a *factor* can be transferred from one term of a fraction to the other if the sign of its exponent is changed, that is, from numerator to denominator, or vice versa.

PROBLEMS 20-1

Making use of the five fundamental laws of exponents, write the results of the indicated operations:

1. $a^4 \cdot a^3$	**2.** $\pi^2 \cdot \pi^5$	**3.** $x^2 \cdot x$	**4.** $\theta^3 \cdot \theta^7$
5. $p^q p^r$	**6.** $\lambda^{2x} \cdot \lambda^{5x}$	**7.** $I^\alpha \cdot I^\beta$	**8.** $m^{x+y} \cdot m^{x-y}$
9. $x^8 \div x^5$	**10.** $\alpha^{5.3} \div \alpha^{2.7}$	**11.** $X^{5y} \div X^2$	**12.** $e^{\pi+2} \div e^3$
13. $\theta^{\alpha+\beta} \div \theta^{\alpha-\beta}$	**14.** $\psi^{\alpha+\beta} \div \psi^{\alpha-\gamma}$	**15.** $(I^3)^3$	**16.** $(f^2)^5$
17. $(x^2y^3)^3$	**18.** $(IR^2t)^3$	**19.** $(a^x)^4$	**20.** $(a^4)^x$
21. $(-x^l y^m z^p)^4$	**22.** $(-a^\pi b^\lambda)^3$	**23.** $\left(\frac{E}{R}\right)^2$	**24.** $\left(\frac{R_1R_2}{R_3}\right)^3$
25. $\left(\frac{\omega^3}{2\pi f^2}\right)^6$	**26.** $\left(\frac{Z_1{}^2}{Z_3Z_4}\right)^2$	**27.** $\left(\frac{-X_C{}^2}{X_L}\right)^3$	**28.** $\left(\frac{\pi D^2}{4}\right)^4$
29. $\left(\frac{\alpha^{3x}}{\alpha^{x+2}}\right)^2$	**30.** $\left(\frac{a^{3\pi}}{a^{5\lambda}}\right)^{4\gamma}$		

Express with all positive exponents:

31. I^2R^{-1}	**32.** $x^{-3}y^{-2}$	**33.** $y^{-\pi}z^{3\lambda}$	**34.** $16L_1^{-2}L_2^{-2}$
35. $\theta^4\phi^{-3}\lambda^{-2x}$	**36.** $(\pi R^2)^{-2i}$	**37.** $\frac{a^{-3}b}{c^{-1}}$	**38.** $\left(\frac{Z_1Z_2}{Z_4}\right)^{-3}$
39. $\frac{3I^3R^{-2}}{12I^2r^{-3}}$	**40.** $\frac{\alpha^3}{2(4\beta\gamma)^{-2}}$		

20-4 FRACTIONAL EXPONENTS

The meaning of a base affected by a fractional exponent is established by methods similar to those employed in determining meanings for zero or negative exponents. If we assume that Eq. (1) of Sec. 20-1 holds for fractional exponents, we should obtain, for example,

$$a^{\frac{1}{2}} \cdot a^{\frac{1}{2}} = a^{\frac{1}{2}+\frac{1}{2}} = a^1 = a$$

Also, $$a^{\frac{1}{3}} \cdot a^{\frac{1}{3}} \cdot a^{\frac{1}{3}} = a^{\frac{1}{3}+\frac{1}{3}+\frac{1}{3}} = a^1 = a$$

That is, $a^{\frac{1}{2}}$ is one of two equal factors of a, and $a^{\frac{1}{3}}$ is one of three equal factors of a. Therefore, $a^{\frac{1}{2}}$ is the square root of a, and $a^{\frac{1}{3}}$ is the cube root of a. Hence,

$$a^{\frac{1}{2}} = \sqrt{a}$$

and $$a^{\frac{1}{3}} = \sqrt[3]{a}$$

Likewise, $$a^{\frac{2}{3}} \cdot a^{\frac{2}{3}} \cdot a^{\frac{2}{3}} = a^{\frac{2}{3}+\frac{2}{3}+\frac{2}{3}} = a^{\frac{6}{3}} = a^2$$

Hence, $$(a^{\frac{2}{3}})^3 = a^2$$

or $$a^{\frac{2}{3}} = \sqrt[3]{a^2}$$

In a fractional exponent, the denominator

denotes the root and the numerator denotes the power of the base.

In general, $a^{\frac{m}{n}} = \sqrt[n]{a^m}$

example 1 $a^{\frac{3}{5}} = \sqrt[5]{a^3}$

example 2 $(-8)^{\frac{1}{3}} = \sqrt[3]{-8} = -2$

PROBLEMS 20-2

Find the value of:

1. $16^{\frac{1}{2}}$
2. $(-27)^{\frac{1}{3}}$
3. $16^{\frac{1}{4}}$
4. $-(-32)^{\frac{1}{5}}$
5. $(-64a^6b^3c^{12})^{\frac{1}{3}}$
6. $(L_1{}^4L_2{}^4)^{\frac{1}{2}}$
7. $(I^4R^2)^{\frac{3}{2}}$
8. $(\theta^3\pi^6)^{\frac{2}{3}}$
9. $\left(\frac{27\lambda^9}{\omega^{12}}\right)^{\frac{2}{3}}$
10. $\left(\frac{r^{12}R^8}{16E^4}\right)^{\frac{3}{4}}$

Express with radical signs:

11. $9^{\frac{1}{2}}$
12. $8\alpha^{\frac{1}{3}}$
13. $(8\alpha)^{\frac{1}{3}}$
14. $6^{\frac{2}{3}}$
15. $\theta^{\frac{3}{4}}\lambda^{\frac{3}{4}}$
16. $x^{\frac{2}{3}}y^{\frac{3}{2}}$

Express with fractional exponents:

17. $\sqrt{a^3}$
18. $\sqrt[3]{x^2}$
19. $\sqrt[3]{16E}$
20. $\sqrt[3]{a^2b^4c^6}$
21. $\sqrt[3]{\theta^2\omega^4}$
22. $\alpha\sqrt[5]{\beta^2}$
23. $\sqrt[5]{\alpha^2\beta^2}$
24. $4L\sqrt{\omega^3}$
25. $2\pi\sqrt[3]{16f^3}$
26. $5\alpha^2\sqrt[5]{-32\alpha^3\beta^7}$

20-5 RADICAND

The meaning of the radical sign was explained in Sec. 2-11. The number under the radical sign is called the *radicand*.

20-6 SIMPLIFICATION OF RADICALS

The form in which a radical expression is written can be changed without altering the numerical value of the expression. Such a change is desirable for many reasons. For example, addition of several fractions containing different radicals in the denominators would be more difficult than addition with the radicals removed from the denominators. Similarly, it will be shown later that

$$\frac{1}{\sqrt{3}} = \frac{\sqrt{3}}{3}$$

It is apparent that the value to several decimal places could be computed more easily from the second fraction than from the first.

Because we are chiefly concerned with radicals involving a square root, only that type will be considered.

20-7 REMOVING A FACTOR FROM THE RADICAND

Since, in general, $\sqrt{ab} = \sqrt{a} \cdot \sqrt{b}$, the following is evident:

Rule
A radicand can be separated into two factors one of which is the greatest perfect square it contains. The square root of this factor can then be written as the coefficient of a radical the other factor of which is the radicand.

example 3 $\sqrt{27} = \sqrt{9 \cdot 3}$
$= \sqrt{9} \cdot \sqrt{3}$
$= \pm 3\sqrt{3}$

example 4 $\sqrt{8} = \sqrt{4 \cdot 2}$
$= \sqrt{4} \cdot \sqrt{2}$
$= \pm 2\sqrt{2}$

example 5 $\sqrt{75} = \sqrt{25 \cdot 3}$
$= \sqrt{25} \cdot \sqrt{3}$
$= \pm 5\sqrt{3}$

example 6 $\sqrt{200a^5b^3c^2d}$
$= \sqrt{100a^4b^2c^2} \cdot \sqrt{2abd}$
$= \pm 10a^2bc\sqrt{2abd}$

PROBLEMS 20-3
Simplify by removing factors from the radicand:

1. $\sqrt{8}$
2. $\sqrt{32}$
3. $\sqrt{18}$
4. $\sqrt{24}$
5. $\sqrt{50}$
6. $\sqrt{20}$
7. $\sqrt{80}$
8. $\sqrt{28}$
9. $\sqrt{720}$
10. $\sqrt{27x^4}$
11. $\sqrt{12\theta^2\phi^4}$
12. $\sqrt{99A^3D}$
13. $5\sqrt{96I^2R}$
14. $3\pi\sqrt{72r^3z^5\pi^3}$
15. $6\omega\sqrt{63f^4F^3T^5}$
16. $7x\sqrt{147xy^2z^3D^3}$
17. $3\alpha^2\sqrt{242\alpha^5\beta^7\gamma^8}$
18. $8\sqrt{567X_L^2Z_1^4}$
19. $2r^3\sqrt{588\pi^4L^4X_L^2}$
20. $5\theta\sqrt{289\theta^5\lambda^7}$

20-8 SIMPLIFYING RADICALS CONTAINING FRACTIONS

Since $\sqrt{\dfrac{4}{9}} = \pm\dfrac{2}{3}$

and $\dfrac{\sqrt{4}}{\sqrt{9}} = \pm\dfrac{2}{3}$

then $\sqrt{\dfrac{4}{9}} = \pm\dfrac{\sqrt{4}}{\sqrt{9}}$

Also, $\sqrt{\dfrac{16}{25}} = \pm\dfrac{4}{5}$

and $\dfrac{\sqrt{16}}{\sqrt{25}} = \pm\dfrac{4}{5}$

then $\sqrt{\dfrac{16}{25}} = \dfrac{\sqrt{16}}{\sqrt{25}}$

Or, in general terms,

$$\sqrt{\frac{a}{b}} = \frac{\sqrt{a}}{\sqrt{b}}$$

The above relation permits simplification of radicals containing fractions by removing the radical from the denominator. This process, by which the denominator is made a rational number, is called *rationalizing the denominator*.

Rule
To rationalize the denominator:
1. Multiply both numerator and denominator by a number that will make the resulting denominator a perfect square.

2. Simplify the resulting radical by removing factors from the radicands.

example 7 $\sqrt{\frac{2}{5}} = \sqrt{\frac{2}{5}\cdot\frac{5}{5}}$

$= \sqrt{\frac{10}{25}}$

$= \frac{\sqrt{10}}{\sqrt{25}}$

$= \pm\frac{\sqrt{10}}{5}$

example 8 $\sqrt{\frac{1}{2}} = \sqrt{\frac{1}{2}\cdot\frac{2}{2}}$

$= \sqrt{\frac{2}{4}}$

$= \frac{\sqrt{2}}{\sqrt{4}}$

$= \pm\frac{\sqrt{2}}{2}$

example 9 $\frac{3}{\sqrt{6}} = \frac{3}{\sqrt{6}}\cdot\frac{\sqrt{6}}{\sqrt{6}}$

$= \pm\frac{1}{2}\sqrt{6}$

example 10 $\sqrt{\frac{3a}{5x}} = \sqrt{\frac{3a}{5x}\cdot\frac{5x}{5x}}$

$= \sqrt{\frac{15ax}{25x^2}}$

$= \frac{\sqrt{15ax}}{\sqrt{25x^2}}$

$= \pm\frac{1}{5x}\sqrt{15ax}$

PROBLEMS 20-4

Simplify the following:

1. $\sqrt{\frac{1}{3}}$
2. $\sqrt{\frac{1}{7}}$
3. $\sqrt{\frac{2}{5}}$
4. $\sqrt{\frac{4}{7}}$
5. $\sqrt{\frac{3}{4}}$
6. $\sqrt{\frac{7}{15}}$
7. $\frac{8}{\sqrt{2}}$
8. $\frac{9}{\sqrt{3}}$
9. $\frac{1}{\sqrt{\lambda}}$
10. $\frac{21\sqrt{35}}{\sqrt{7}}$
11. $\sqrt{\frac{9}{16\theta}}$
12. $\sqrt{\frac{Q}{R}}$
13. $\theta\sqrt{\frac{\lambda}{\theta}}$
14. $\pi\sqrt{\frac{X_L}{2\pi fL}}$
15. $\sqrt{\frac{\alpha^2}{\gamma}}$
16. $\frac{2F}{f_0}\sqrt{\frac{f_0}{F}}$
17. $\frac{\pi R^2}{A}\sqrt{\frac{A}{\pi}}$
18. $\sqrt{\frac{E-e}{E+e}}$
19. $\sqrt{X_L{}^2 - \left(\frac{X_L}{4}\right)^2}$
20. $\sqrt{R^2 + \left(\frac{R}{3}\right)^2}$
21. $\sqrt{Q^4 - \left(\frac{Q}{3}\right)^4}$

20-9 ADDITION AND SUBTRACTION OF RADICALS

Terms that are the same except in respect to their coefficients are called *similar terms.* Likewise, *similar radicals* are defined as radicals that have the same index and the same radicand and differ only in their coefficients. For example, $-2\sqrt{5}$, $3\sqrt{5}$, and $\sqrt{5}$ are similar radicals.

Similar radicals can be added or subtracted in the same way that similar terms are added or subtracted.

example 11

$$3\sqrt{6} - 4\sqrt{6} - \sqrt{6} + 8\sqrt{6} = 6\sqrt{6}$$

example 12

$$\sqrt{12} + \sqrt{27} = 2\sqrt{3} + 3\sqrt{3} = 5\sqrt{3}$$

Note that, in the simplification of radicals, the positive root is assumed.

example 13

$$\sqrt{48x} + \sqrt{\frac{x}{3}} + \sqrt{3x} = 4\sqrt{3x} + \frac{1}{3}\sqrt{3x} + \sqrt{3x} = \frac{16}{3}\sqrt{3x}$$

If the radicands are alike, then factors removed are assumed to be positive roots. If the radicands are not alike and cannot be reduced to a common radicand, then the radicals are dissimilar terms and addition and subtraction can only be indicated. Thus the following statement can be made:

Rule

To add or subtract radicals:

1. Reduce them to their simplest form.
2. Combine similar radicals, and assume positive square roots of factors removed from the radicands.
3. Indicate addition or subtraction of dissimilar radicals.

PROBLEMS 20-5

Simplify:

1. $5\sqrt{3} - 2\sqrt{3}$

2. $3\sqrt{5} + 2\sqrt{20}$

3. $5\sqrt{5} - \sqrt{80}$

4. $\sqrt{63} - \sqrt{28}$

5. $m\sqrt{3} - p\sqrt{3} + q\sqrt{3}$

6. $\alpha\sqrt{2} + \beta\sqrt{8} - \gamma\sqrt{50}$

7. $5\sqrt{48} + 2\sqrt{108} - \sqrt{12}$

8. $2\sqrt{\frac{1}{3}} + \sqrt{\frac{1}{3}}$

9. $7\sqrt{5} - \frac{15}{\sqrt{5}} - 16\sqrt{\frac{5}{16}}$

10. $6\sqrt{27} + 5\sqrt{32}$

11. $4\sqrt{\frac{1}{8}} + 6\sqrt{\frac{1}{2}} + 2\sqrt{2}$

12. $\frac{R_1}{3} + \sqrt{\frac{16R_1^2}{3}}$

13. $\sqrt{\frac{4}{5}} - \sqrt{\frac{9}{15}}$

14. $\sqrt{\frac{\varepsilon + \eta}{\varepsilon - \eta}} + \sqrt{\frac{\varepsilon - \eta}{\varepsilon + \eta}}$

15. $\sqrt{\frac{\pi}{8}} - \sqrt{\frac{\pi}{32}}$

16. $\sqrt{\frac{7R^2}{16E}} + \sqrt{\frac{M^2E}{28}} - 4\sqrt{\frac{63}{16E}}$

20-10 MULTIPLICATION OF RADICALS

Obtaining the product of radicals is the inverse of removing a factor, as will be shown in the following examples:

example 14 $3\sqrt{3}\cdot 5\sqrt{4} = 15\sqrt{3\cdot 4}$
$= 15\cdot 2\sqrt{3}$
$= 30\sqrt{3}$

example 15

$$\begin{aligned} 4\sqrt{3a}\cdot 2\sqrt{6a} &= 8\sqrt{3a\cdot 6a} \\ &= 8\sqrt{18a^2} \\ &= 8\sqrt{9\cdot 2a^2} \\ &= 24a\sqrt{2} \end{aligned}$$

example 16 Multiply $3\sqrt{2} + 2\sqrt{3}$ by $4\sqrt{2} - 3\sqrt{3}$.

solution

$$\begin{array}{lll} & 3\sqrt{2} + 2\sqrt{3} & \\ & 4\sqrt{2} - 3\sqrt{3} & \\ \hline 24 & + 8\sqrt{6} & \\ & - 9\sqrt{6} - 18 & \\ \hline 24 & - \sqrt{6} - 18 = 6 - \sqrt{6} & \end{array}$$

PROBLEMS 20-6

Perform the indicated operations:

1. $\sqrt{2}\cdot\sqrt{3}$
2. $\sqrt{12}\cdot\sqrt{3}$
3. $2\sqrt{10}\cdot\sqrt{2}$
4. $8\sqrt{5}\cdot 4\sqrt{15}$
5. $2\sqrt{8}\cdot 3\sqrt{5}$
6. $\sqrt{6}\cdot\sqrt{24}$
7. $\sqrt[3]{2}\cdot\sqrt[3]{4}$
8. $\sqrt{\frac{7}{16}}\cdot\sqrt{\frac{21}{3}}$
9. $(\sqrt{A} - D)^2$
10. $(\varepsilon + \sqrt{3})(\varepsilon - \sqrt{3})$
11. $(\sqrt{\alpha} - \sqrt{\alpha - 7})^2$
12. $(3 + \sqrt{5})^2$
13. $\sqrt{(\theta - \phi)^2}$
14. $(2\sqrt{5} + 3\sqrt{2})(\sqrt{5} + 5\sqrt{2})$
15. $\sqrt{6\pi}\cdot\sqrt{12\alpha^2\pi}$
16. $\sqrt{2(x^2 - 4x + 4)}\cdot\sqrt{\dfrac{8}{4x^2 + 16x + 16}}$
17. $(-1 - \sqrt{3})(3 - 3\sqrt{3})$
18. $(4 + 2\sqrt{3})(2 - \sqrt{3})$
19. $\left(\dfrac{36 - 9\sqrt{5}}{2}\right)\left(\dfrac{2\sqrt{5} + 8}{11}\right)$
20. $\dfrac{(\sqrt{\alpha} - \sqrt{\beta})(\alpha + 2\sqrt{\alpha\beta} + \beta)}{\alpha - \beta}$

20-11 DIVISION

An indicated root whose value is irrational but whose radicand is rational is called a *surd*. Thus, $\sqrt[3]{3}$, $\sqrt{2}$, $\sqrt[4]{5}$, $\sqrt{3}$, etc., are surds. If the indicated root is the square root, then the surd is called a *quadratic surd*. For example, $\sqrt{2}$, $\sqrt{5}$, $\sqrt{6}$, $\sqrt{15}$ are quadratic surds. Then, by extending the definition, such expressions as $3 + \sqrt{2}$ and $\sqrt{3} - 6$ are called *binomial quadratic surds*.

It is important that you become proficient in the multiplication and division of binomial quadratic surds. One method of solving ac circuits, which will be discussed later, makes wide use of these particular operations. Multiplication of such expressions was covered in the preceding section. However, a new method is necessary for division.

Consider the two expressions $a - \sqrt{b}$ and $a + \sqrt{b}$. They differ only in the sign between the terms. These expressions are *conjugates;*

that is, $a - \sqrt{b}$ is called the conjugate of $a + \sqrt{b}$, and $a + \sqrt{b}$ is called the conjugate of $a - \sqrt{b}$. Remember this meaning of "conjugate," for it is the same with reference to certain circuit characteristics.

To divide a number by a binomial quadratic surd, rationalize the divisor (denominator) by multiplying both dividend (numerator) and divisor by the conjugate of the divisor.

example 17

$$\frac{1}{3+\sqrt{2}} = \frac{3-\sqrt{2}}{(3+\sqrt{2})(3-\sqrt{2})} = \frac{3-\sqrt{2}}{7}$$

example 18

$$\frac{1}{3\sqrt{3}-1} = \frac{3\sqrt{3}+1}{(3\sqrt{3}-1)(3\sqrt{3}+1)} = \frac{3\sqrt{3}+1}{26}$$

example 19

$$\frac{3-\sqrt{2}}{4+\sqrt{2}} = \frac{(3-\sqrt{2})(4-\sqrt{2})}{(4+\sqrt{2})(4-\sqrt{2})} = \frac{14-7\sqrt{2}}{14} = \frac{2-\sqrt{2}}{2}$$

note In each of the foregoing examples the resulting denominator is a rational number. In general, the product of two conjugate surd expressions is a rational number. This important fact is widely used in the solution of ac problems.

PROBLEMS 20-7

Perform the indicated division:

1. $\dfrac{2\sqrt{10}}{\sqrt{8}}$
2. $\dfrac{3}{3-\sqrt{2}}$
3. $\dfrac{8}{3+\sqrt{7}}$
4. $\dfrac{7}{2\sqrt{3}-2}$
5. $\dfrac{9}{3-3\sqrt{3}}$
6. $\dfrac{x+\sqrt{y}}{x-\sqrt{y}}$
7. $\dfrac{x-\sqrt{y}}{x+\sqrt{y}}$
8. $\dfrac{3-\sqrt{5}}{2+\sqrt{5}}$
9. $\dfrac{3+2\sqrt{3}}{2+2\sqrt{3}}$
10. $\dfrac{\sqrt{R}+\sqrt{Z}}{\sqrt{R}-\sqrt{Z}}$
11. $\dfrac{\sqrt{2}+3}{\sqrt{3}+2}$
12. $\dfrac{50+j35}{8+j5}$

hint Maintain order j35, j5, etc. and treat terms containing algebraic symbol j as if they were radicals.

20-12 THE OPERATOR j

In our studies so far, we have met with several mathematical symbols which actually indicate *commands;* +, −, ×, ÷, and $\sqrt{\ }$ are all symbols which actually tell us to perform some specific operation. In Sec. 3-5, for instance, we saw that the minus sign is equivalent to a rotation of a quantity through 180°, and, by definition, this rotation is in the positive, or counterclockwise, direction.

Now we must meet the operator j, which also provides a rotation, not of 180°, but of

90°. You have noticed that all the algebraic symbols used so far in this book are printed in *italic* (slanting) type. The operator j, however, is printed in roman (regular) type to distinguish it as an operator and to constantly remind the student that it is not just another algebraic symbol. The use of j is an extremely useful notation in the solution of electronic circuits, and although it is a simple, straightforward idea—*just rotate through 90° in a counterclockwise* (ccw) *direction*—it is essential that we understand exactly how to operate with it. In Fig. 20-1, the line OA, which lies on the x axis and is a units long, can be operated on by the operator j to become ja, a line of the same length as before but now rotated CCW through 90° to lie on the y axis.

Note how the rotated quantity is described: first is given the symbol of the operator j, and then the quantity which has been operated upon, a. Thus, when a is "j'd", it becomes ja. This practice of placing the operator first draws attention to the fact that we are not dealing with some quantity j multiplied by some other quantity a, but that the j operator is operating on the quantity a. The algebraic symbol ja represents for us the geometric symbol of a line rotated through 90° in a counterclockwise direction.

Fig. 20-1 Representation of a Quantity Affected by the Operator j

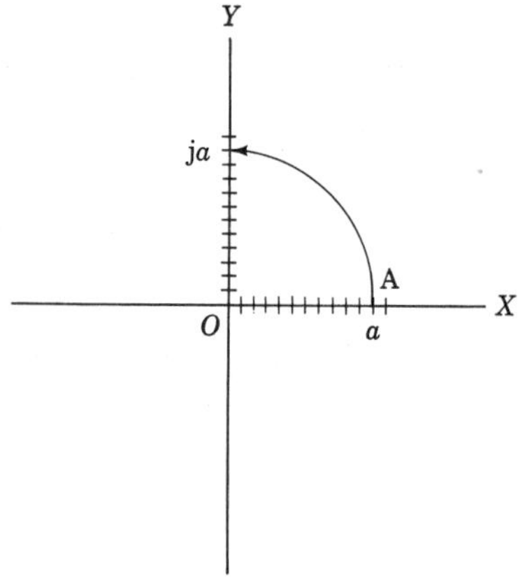

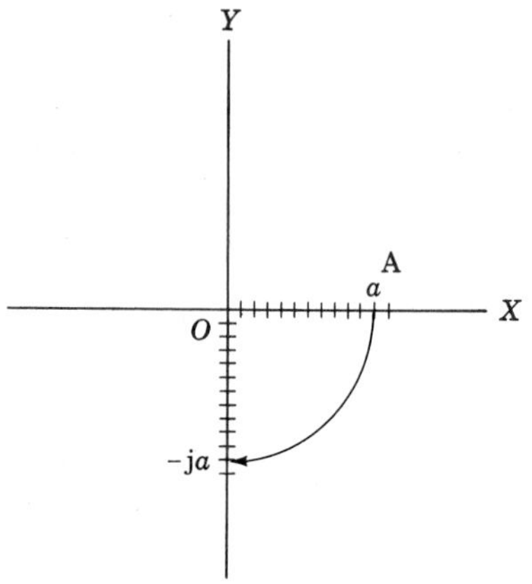

Fig. 20-2 Representation of Quantity Affected by the Operator −j

Any quantity operated upon by j will rotate through 90° in a counterclockwise direction, and, similarly, any quantity operated upon by −j will rotate through 90° in a clockwise direction. (See Fig. 20-2.)

Figure 20-3 relates four different quantities by way of review: $A = 5$, $B = \mathrm{j}5$, $C = -7$, and $D = -\mathrm{j}3$.

A quantity may be j-operated more than once. If we start with a quantity ja, as in Fig. 20-1, and j it again, we cause it to rotate through an additional 90° CCW, as shown in Fig. 20-4.

j(ja) may be written jj$a$, or, more simply, j^2a. Similarly, j^3 indicates that a quantity has been operated on three times in succession; that is, it has been rotated through 90° CCW

Fig. 20-3 Comparison of Quantities: $A = +5$, $B = +\mathrm{j}5$, $C = -7$, and $D = -\mathrm{j}3$

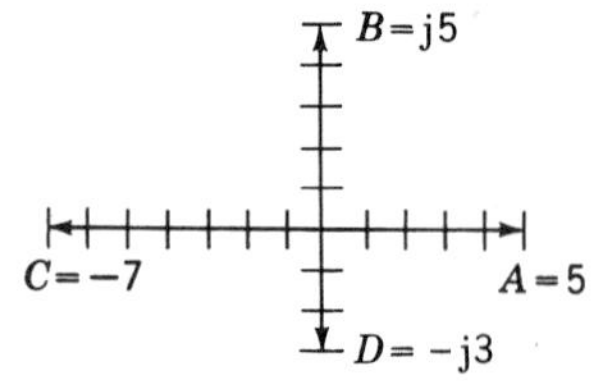

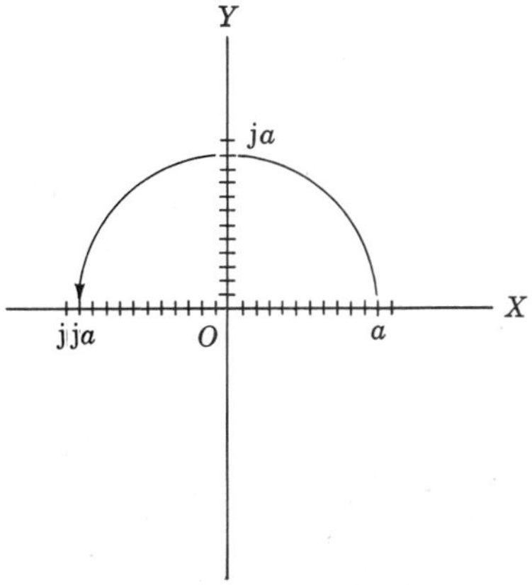

Fig. 20-4 Representation of Repeated Operation by j

three times in succession. Figures 20-5 and 20-6 indicate repeated rotations resulting from repeated operations by j and −j.

Note, in passing, a very interesting point about j^2a: j-ing a twice in succession brings it to be the same point as a single operation with a minus sign. From this graphic illustration, you can see that

$$j^2 = -1$$

and $$j = \sqrt{-1}$$

This added relationship, $j = \sqrt{-1}$, is an extremely interesting one, because so far, in the removal of factors from radicands, all the radicands have been positive numbers. In Sec. 20-13, we will use the important relationship $j = \sqrt{-1}$ to factor negative radicands and to determine (or, at least represent) the square roots of negative numbers.

Fig. 20-5 Repeated Rotation of Numbers in Counterclockwise Direction

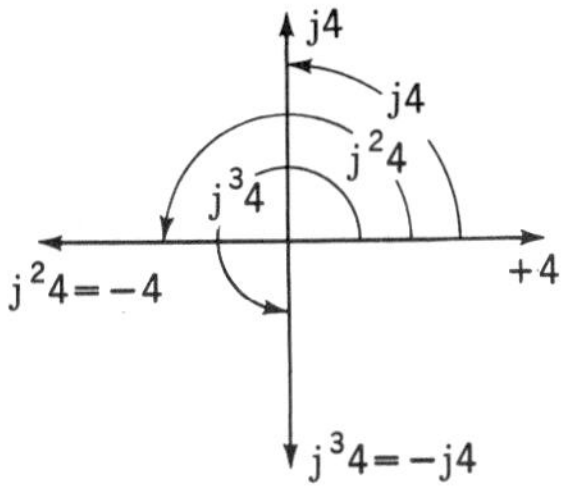

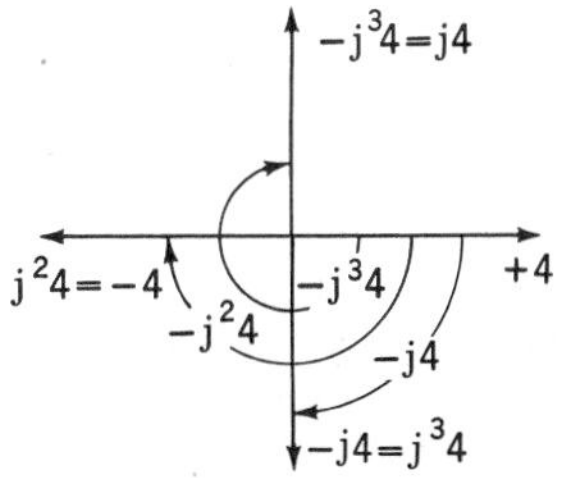

Fig. 20-6 Repeated Rotation of Numbers in Clockwise Direction

First, however, let us continue with the fascinating relationships exhibited by repeated operations with j. Since $j^2 = -1$, then j^3 must equal $j(-1)$, or $-j$, and j^4 must equal j^2j^2, that is, $(-1)(-1) = +1$. The truth of these statements can be justified by the following considerations:

$$\sqrt{-1} \cdot \sqrt{-1} = -1$$

That is, $$j \cdot j = -1$$
$$\therefore j^2 = -1$$

Also, $$\sqrt{-1} \cdot \sqrt{-1} \cdot \sqrt{-1} = -1 \cdot \sqrt{-1} = -j$$

That is, $$j \cdot j \cdot j = j^3$$
$$\therefore j^3 = -j$$

Also, $$\sqrt{-1} \cdot \sqrt{-1} \cdot \sqrt{-1} \cdot \sqrt{-1} = (\sqrt{-1} \cdot \sqrt{-1})(\sqrt{-1} \cdot \sqrt{-1}) = (-1)(-1) = 1$$

That is, $$j \cdot j \cdot j \cdot j = j^4$$
$$\therefore j^4 = 1$$

Similarly, it can be shown that successive multiplication by each $+j$ rotates the number 90° in a counterclockwise direction.

If we consider successive multiplication by $-j$, we have

$$(-\sqrt{-1})(-\sqrt{-1}) = -1$$

That is, $$(-j)(-j) = j^2$$
$$\therefore (-j)^2 = -1$$

Also, $(-\sqrt{-1})(-\sqrt{-1})(-\sqrt{-1})$
$= (-1)(-\sqrt{-1} = \sqrt{-1}$

That is, $(-j)(-j)(-j) = (j^2)(-j)$
$= (-1)(-j) = j$
$\therefore (-j)^3 = j$

To demonstrate that $(-j)^4 = 1$ and $\frac{1}{j} = -j$ is left as an exercise for you.

Note the convenience of the graphic method of representation of the j operations, Figs. 20-5 and 20-6. This method is an advantageous one because, if we can *visualize* a graph or diagram when we come up against certain types of numbers and equations, we often have a better understanding of the manner in which the quantities vary or are related.

One special note must be drawn to your attention: Long before the operator j was found to have practical application in electrical and electronics calculations, mathematicians used the symbol i to represent $\sqrt{-1}$. When electrical theory adopted the symbol i for instantaneous current flow in a circuit, we switched the mathematicians' i to j for our symbol of rotation through 90° ccw. Sometimes in your reading you will meet i instead of j, but you will know what it really means: "Rotate the quantity operated upon by 90° in a counterclockwise direction."

As a mathematical definition, j is sometimes referred to as the "complex operator," but, as we have seen, there is nothing particularly complex about j.

20-13 INDICATED SQUARE ROOTS OF NEGATIVE NUMBERS

So far, in the removal of factors from radicands, all the radicands have been positive numbers. Also, we have extracted the square roots of positive numbers only. How shall we proceed to factor negative radicands, and what is the meaning of the square root of a negative number?

According to our laws for multiplication, no number multiplied by itself or raised to any even power will produce a negative result. For example, what does $\sqrt{-25}$ mean when we know of no number that, when multiplied by itself, will produce -25?

The indicated square root of a negative number is known as an *imaginary number*. It is probable that this name was assigned before mathematicians could visualize such a number and that the word "imaginary" was originally used to distinguish such numbers from the so-called "real numbers" previously studied. In any event, calling such a number imaginary might be considered unfortunate, because in working with circuits such numbers become very real in the physical sense. If you accidently touch a large capacitor that is highly charged, you are likely to be killed by some of those "imaginary" volts. This will be discussed later.

To avoid the difficulty of operations with the indicated square roots of negative numbers, or imaginary numbers, it becomes necessary to introduce a new type of number. That is, we agree that every imaginary number can be expressed as the product of a positive number and $\sqrt{-1}$.

example 20 $\sqrt{-25} = \sqrt{(-1)25}$
$= \sqrt{-1}\sqrt{25}$
$= \sqrt{-1} \cdot 5$

As we saw in Sec. 20-12, $\sqrt{-1}$ may be represented by the operator j, and we may now rewrite $\sqrt{-1} \cdot 5$ as j5.

example 21 $\sqrt{-16} = \sqrt{(-1)16}$
$= \sqrt{-1}\sqrt{16}$
$= \sqrt{-1} \cdot 4 = j4$

example 22 $\sqrt{-X^2} = \sqrt{(-1)X^2}$
$= \sqrt{-1}\sqrt{X^2}$
$= \sqrt{-1} \cdot X = \mathrm{j}X$

example 23

$$-\sqrt{-4X^2} = -\sqrt{(-1)4X^2} = -\sqrt{-1}\sqrt{4X^2}$$
$$= -\sqrt{-1} \cdot 2X = -\mathrm{j}2X$$

PROBLEMS 20-8

Express the following by using the operator j:

1. $\sqrt{-36}$ **2.** $\sqrt{-64}$ **3.** $\sqrt{-144}$ **4.** $\sqrt{-\theta^2}$

5. $-\sqrt{-z^2}$ **6.** $-\sqrt{-49\omega^2}$ **7.** $\sqrt{-I^4X^2}$ **8.** $\sqrt{\dfrac{-Q^4}{\omega^2L^2}}$

9. $-5\sqrt{-49}$ **10.** $2\sqrt{-48}$ **11.** $\sqrt{\dfrac{-16}{121}}$ **12.** $-\sqrt{\dfrac{169}{-\alpha^2}}$

13. $\sqrt{\dfrac{-32}{75}}$ **14.** $-\sqrt{-\lambda^2\pi}$ **15.** $-\sqrt{\dfrac{-E^2}{P}}$

16. Did you demonstrate that $(-\mathrm{j})^4 = 1$?

17. Did you demonstrate that $\dfrac{1}{\mathrm{j}} = -\mathrm{j}$?

20-14 COMPLEX NUMBERS

If a "real" number is united to an "imaginary" number by a plus or a minus sign, the expression thus obtained is called a *complex number*. Thus, $3 - \mathrm{j}4$, $a + \mathrm{j}b$, $R + \mathrm{j}X$, etc., are complex numbers. At this time, we shall consider, not their graphical representation, but simply how to perform the four fundamental operations algebraically. Figure 20-7 shows the representation of the complex number $a + \mathrm{j}b$.

20-15 ADDITION AND SUBTRACTION OF COMPLEX NUMBERS

Combining a real number with an imaginary number cannot be accomplished by the usual methods of addition and subtraction; these processes can only be expressed. For example, if we have the complex number $5 + \mathrm{j}6$, this is as far as we can simplify it at this time. We should not attempt to add 5 and j6 arithmetically, for these two numbers are at right angles to each other, and such an operation would be meaningless. However, we *can* add and subtract complex numbers by treating them as ordinary binomials.

Fig. 20-7 Representation of a Complex Number $a + \mathrm{j}b$. a Lies in OX, b Is Rotated through 90° Counterclockwise. The Point p Represents the "Sum" of a and jb

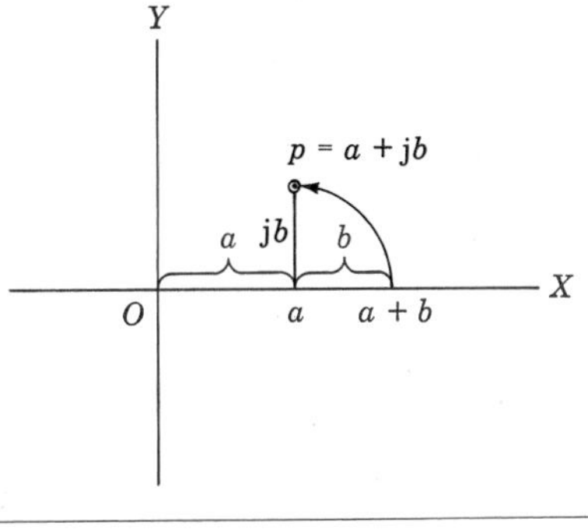

example 24 Add 3 + j7 and 4 − j5.

solution

$$\begin{array}{r} 3 + j7 \\ 4 - j5 \\ \hline 7 + j2 \end{array}$$

example 25 Subtract −15 − j6 from −5 + j8.

solution

$$\begin{array}{r} -5 + j8 \\ -15 - j6 \\ \hline 10 + j14 \end{array}$$

PROBLEMS 20-9

Find the indicated sums:

1. $\begin{array}{r} 3 + j12 \\ 2 + j8 \\ \hline \end{array}$ **2.** $\begin{array}{r} 14 + j3 \\ 12 + j3 \\ \hline \end{array}$ **3.** $\begin{array}{r} 25 + j8 \\ 16 - j10 \\ \hline \end{array}$ **4.** $\begin{array}{r} 96 - j22 \\ 32 - j5 \\ \hline \end{array}$

5. $\begin{array}{r} 47 - j3 \\ 125 + j8 \\ \hline \end{array}$ **6.** $\begin{array}{r} 32 \\ 5 + j6 \\ \hline \end{array}$ **7.** $\begin{array}{r} 20 + j3 \\ - j5 \\ \hline \end{array}$ **8.** $\begin{array}{r} 26 - j6 \\ 31 \\ \hline \end{array}$

9 to **16.** Subtract the lower complex number from the upper in each of the above problems.

20-16 MULTIPLICATION OF COMPLEX NUMBERS

As in addition and subtraction, complex numbers are treated as ordinary binomials when multiplied. However, when writing the result, we must not forget that $j^2 = -1$.

example 26 Multiply 4 − j7 by 8 + j2.

solution

$$\begin{array}{r} 4 - \ j7 \\ 8 + \ j2 \\ \hline 32 - j56 \qquad \\ + \ j8 - j^2 14 \\ \hline 32 - j48 - j^2 14 \end{array}$$

Since $j^2 = -1$, the product is

$$\begin{aligned} 32 - j48 - (-1)(14) &= 32 - j48 + 14 \\ &= 46 - j48 \end{aligned}$$

example 27 Multiply 7 + j3 by 6 + j2.

solution

$$\begin{array}{r} 7 + \ j3 \\ 6 + \ j2 \\ \hline 42 + j18 \qquad \\ + j14 + j^2 6 \\ \hline 42 + j32 + j^2 6 = 36 + j32 \end{array}$$

20-17 DIVISION OF COMPLEX NUMBERS

As in the division of binomial quadratic surds, we simplify an indicated division by rationalizing the denominator in order to obtain a "real" number as divisor (Sec. 20-11). We do this by multiplying by the conjugate in the usual manner.

example 28

$$\begin{aligned} \frac{10}{1 + j2} = \frac{10(1 - j2)}{(1 + j2)(1 - j2)} &= \frac{10(1 - j2)}{1 - j^2 4} \\ &= \frac{10(1 - j2)}{5} \\ &= 2(1 - j2) \end{aligned}$$

example 29

$$\frac{5+j6}{3-j4} = \frac{(5+j6)(3+j4)}{(3-j4)(3+j4)} = \frac{15+j38+j^2 24}{9-j^2 16} = \frac{-9+j38}{25}$$

example 30

$$\frac{a+jb}{a-jb} = \frac{(a+jb)(a+jb)}{(a-jb)(a+jb)} = \frac{a^2+j2ab+j^2b^2}{a^2-j^2b^2} = \frac{a^2+j2ab-b^2}{a^2+b^2}$$

PROBLEMS 20-10

Find the indicated products:

1. $(3)(1-j3)$ **2.** $(6+j2)(2+j3)$ **3.** $(8-j9)(6+j3)$

4. $(3-j5)(6-j7)$ **5.** $(\theta+j\phi)(\theta+j\phi)$ **6.** $(R-jX_C)(R+jX_L)$

Find the quotients:

7. $\frac{1}{1+j1}$ **8.** $\frac{10}{1-j3}$ **9.** $\frac{1+j1}{1-j1}$ **10.** $\frac{1-j1}{1+j1}$

11. $\frac{8}{8+j8}$ **12.** $\frac{3+j2}{6-j5}$ **13.** $\frac{6}{6-jx}$ **14.** $\frac{\theta+j\phi}{\theta-j\phi}$

15. $\frac{R+j\omega X}{R-j\omega X}$ **16.** $\frac{j3}{2+j3}$ **17.** $\frac{j\phi}{\theta-j\phi}$ **18.** $\frac{1+j\frac{\omega}{\omega_o}}{1-j\frac{\omega}{\omega_o}}$

19. $\frac{R}{\frac{1}{j\omega C}+R+j\omega L}$

20. Write in the form $a+jb$: $\frac{(1+j\omega\tau_1)(1+j\omega\tau_2)}{\mu_o-\beta}$

20-18 RADICAL EQUATIONS

An equation in which the unknown occurs in a radicand is called an *irrational* or *radical equation*. To solve such an equation, arrange it in such a manner that the radical is the only term in one member of the equation. Then eliminate the radical by squaring both members of the equation.

example 31 Given $\sqrt{3x} = 6$; solve for x.

solution

$$\sqrt{3x} = 6$$

Squaring, $3x = 36$

D: 3, $x = 12$

check Substituting 12 for x in the given equation,

$$\sqrt{3 \cdot 12} = 6$$
$$\sqrt{36} = 6$$
$$6 = 6$$

example 32 Given $\sqrt{2x+3} = 7$; solve for x.

solution

$$\sqrt{2x+3} = 7$$

Squaring, $\quad 2x + 3 = 49$

S: 3, $\quad 2x = 46$

D: 2, $\quad x = 23$

check

$$\sqrt{2 \cdot 23 + 3} = 7$$
$$\sqrt{49} = 7$$
$$7 = 7$$

example 33 The time for one complete swing of a simple pendulum is given by

$$t = 2\pi\sqrt{\frac{L}{g}}$$

where t = time, s
L = length of pendulum
g = force due to gravity
Solve the equation for g and for L.

solution

Given $\quad t = 2\pi\sqrt{\frac{L}{g}} \quad (a)$

Squaring (a), $\quad t^2 = 4\pi^2 \frac{L}{g} \quad (b)$

M: g in (b), $\quad gt^2 = 4\pi^2 L \quad (c)$

D: t^2 in (c), $\quad g = \frac{4\pi^2 L}{t^2} \quad (d)$

Rewrite (c), $\quad 4\pi^2 L = gt^2 \quad (e)$

D: $4\pi^2$ in (e), $\quad L = \frac{gt^2}{4\pi^2}$

example 34 Given $E = I_pZ_p + j\omega MI_s$ and $I_sZ_s = -j\omega MI_p$. Show that

$$E = I_p\left[Z_p + \frac{(\omega M)^2}{Z_s}\right]$$

solution Since I_s does not appear in the final equation, it must be eliminated. Solving the given equations for I_s,

$$I_s = \frac{E - I_pZ_p}{j\omega M} \quad (a)$$

$$I_s = \frac{-j\omega MI_p}{Z_s} \quad (b)$$

Equating the right members of (a) and (b),

$$\frac{E - I_pZ_p}{j\omega M} = \frac{-j\omega MI_p}{Z_s}$$

M: $j\omega M$ $\quad E - I_pZ_p = \frac{-j^2\omega^2M^2I_p}{Z_s}$

Substituting -1 for j^2 in the right member,

$$E - I_pZ_p = \frac{\omega^2M^2I_p}{Z_s}$$

A: I_pZ_p, $\quad E = I_pZ_p + \frac{(\omega M)^2I_p}{Z_s}$

Factoring the right member,

$$E = I_p\left[Z_p + \frac{(\omega M)^2}{Z_s}\right]$$

PROBLEMS 20-11

Solve the following equations:

1. $\sqrt{x} = 2$
2. $\sqrt{R} = 6$
3. $\sqrt{\gamma} = 3$
4. $\sqrt{i} + 1 = 4$
5. $\sqrt{Z} - 5 = 20$
6. $\sqrt{\theta + 3} = 7$
7. $\sqrt{M - 3} = 8$
8. $2\sqrt{\theta - 2} = 6$
9. $4\sqrt{\lambda + 3} - 2 = 6$

10. $\sqrt{\dfrac{7K+4}{2}} = 4$

11. $3\sqrt{\phi + 3} = 2\sqrt{3\phi - 12}$

Given:	Solve for:
12. $E = \sqrt{\dfrac{\eta\phi}{\omega^2\theta}}$	ϕ
13. $i_s = \rho\sqrt{2P_rP_s}$	P_r
14. $\dfrac{i_s}{i_n} = \sqrt{\dfrac{\rho P_s}{e(\Delta f)}}$	P_s
15. $\dfrac{S}{N} = \alpha\sqrt{\eta\tau}$	η
16. $\lambda = \dfrac{4\pi}{\gamma Q}\sqrt{\dfrac{KFTS(\Delta f)}{NP_0}}$	$\dfrac{S}{N}$, P_0
17. $\dfrac{V}{C} = \sqrt{\dfrac{1}{\dfrac{w-a}{w} + \dfrac{\varepsilon_1 a}{w}}}$	w
18. $\gamma = \sqrt{\dfrac{1 - \mu_x\eta E}{\omega X}}$	μ_x
19. $Y_n = G\sqrt{\left(\dfrac{n^2-1}{n}\right)^2 Q_2 + 1}$	Q_2
20. $Z_t = R\sqrt{1 + \left(\dfrac{f}{f_0}\right)^4}$	f_0
21. $G_a = \sqrt{G_1 + \dfrac{G_1}{R_{eq} + \dfrac{G_L}{g_m^2}}}$	g_m^2

22. At a resonant frequency of f Hz, the inductive reactance X_L of a circuit of L H is $X_L = \omega L\ \Omega$ and the capacitive reactance X_C of a circuit with a capacitance of C F is $X_C = \dfrac{1}{\omega C}\ \Omega$. $\omega = 2\pi f$. At the resonant frequency, with both inductance and capacitance in the circuit, $X_L = X_C$. Solve for the resonant frequency f in terms of π, L, and C.

23. Use the formula for the resonant frequency derived in Prob. 22 to find the value of C in picofarads when $f = 1.4$ MHz and $L = 51.7\ \mu$H.

24. Use the formula derived in Prob. 22 to find the value of f when $C = 47$ nF and $L = 15$ nH.

25. $f = \dfrac{1}{\sqrt{2\pi \dfrac{LC_aC_b}{C_a + C_b}}}$. Solve for C_a.

26. In a conductor through which current I flows, the power P_m existing in the magnetic field about the line is $\dfrac{LI^2}{2}$ W, where L is the inductance of the line per unit length. An equal power P_e exists in the electrostatic field of the line, equal to $\dfrac{CE^2}{2}$ W, where C is the capacitance of the line per unit length. If the surge impedance Z_o of the line is $\dfrac{E}{I}$ Ω, show that $Z_o = \sqrt{\dfrac{L}{C}}$

27. Given $\Delta = \dfrac{4}{\pi}\sqrt{1 + \left(\dfrac{\pi\tau\omega}{4}\right)^2}$, show that

$$\omega = \pm \frac{1}{\pi\tau}\sqrt{(\pi\Delta + 4)(\pi\Delta - 4)}$$

28. Given $\sqrt{\dfrac{1}{\tau_1\tau_2} - \dfrac{1}{4\tau_2{}^2}} = 786$ and $\dfrac{1}{2\tau_2} = 78.6$, solve for τ_1.

29. Show that $KE_p^{\frac{3}{2}} = KE_p\sqrt{E_p}$. This is a convenient transformation for the slide rule operator.

30. A West Coast semiconductor products manufacturer, in a design for a 100-W 10-MHz power amplifier, equates the actual output circuit to its equivalent:

$$\frac{R_L\left(\dfrac{1}{j\omega C_7}\right)}{R_L + \dfrac{1}{j\omega C_7}} = R'_L + \frac{1}{j\omega C'_7}$$

(*a*) Show that

$$C_7 = \frac{1}{\omega R_L}\sqrt{\frac{R_L}{R'_L} - 1}$$

and

$$C'_7 = C_7\left[1 + \left(\frac{1}{\omega C_7 R_L}\right)^2\right]$$

(*b*) If $R_L = 50$ Ω, $R'_L = 12.5$ Ω, and $\omega = 2\pi \times 10^7$, show that $C_7 = 552$ pF and $C'_7 = 738$ pF.

21

quadratic equations

In preceding chapters the study of equations has been limited mainly to equations which contain the unknown quantity in the first degree. This chapter is concerned with equations of the second degree, which are called quadratic equations.

21-1 DEFINITIONS

In common with polynomials (Sec. 11-2), the degree of an equation is defined as the degree of the term of highest degree in it. Thus, if an equation contains the square of the unknown quantity and no higher degree, it is an equation of the second degree, or a *quadratic equation.*

A quadratic equation that contains terms of the second degree only of the unknown is called a *pure quadratic equation.* For example,

$$x^2 = 25 \qquad R^2 - 49 = 0$$
$$3x^2 = 12 \qquad ax^2 + c = 0$$

are pure quadratic equations.

A quadratic equation that contains terms of *both* the first and the second degree of the unknown is called an *affected* or a *complete quadratic equation.* Thus, $x^2 + 3x + 2 = 0$, $3x^2 + 11x = -2$, $ax^2 + bx + c = 0$, etc., are affected, or complete, quadratic equations.

When a quadratic equation is solved, values of the unknown that will satisfy the conditions of the equation are found.

A value of the unknown that will satisfy the equation is called a *solution* or a *root* of the equation.

21-2 SOLUTION OF PURE QUADRATIC EQUATIONS

As stated in Sec. 10-5, every number has two square roots that are equal in magnitude but opposite in sign. Hence, all quadratic equations have two roots. In pure quadratic equations, the absolute values of the roots are equal but of opposite sign.

example 1 Solve the equation $x^2 - 16 = 0$.

solution

Given $\qquad x^2 - 16 = 0$
A: 16, $\qquad x^2 = 16$
$\sqrt{\ }$ (see note on page 272), $x = \pm 4$

check Substituting in the equation either $+4$ or -4 for the value of x, because either squared results in $+16$, we have

$$(\pm 4)^2 - 16 = 0$$

or

$$16 - 16 = 0$$

note Hereafter, the radical sign will mean "take the square root of both members of the preceding or designated equation."

example 2

Solve the equation $5R^2 - 89 = 91$.

solution

Given	$5R^2 - 89 = 91$
A: 89,	$5R^2 = 180$
D: 5,	$R^2 = 36$
$\sqrt{}$,	$R = \pm 6$

check

$$5(\pm 6)^2 - 89 = 91$$
$$5 \times 36 - 89 = 91$$
$$180 - 89 = 91$$
$$91 = 91$$

example 3 Solve the equation

$$\frac{I+4}{I-4} + \frac{I-4}{I+4} = \frac{10}{3}$$

solution

Given $\quad \dfrac{I+4}{I-4} + \dfrac{I-4}{I+4} = \dfrac{10}{3}$

Clearing fractions,

$$3(I+4)(I+4) + 3(I-4)(I-4) = 10(I-4)(I+4)$$

Expanding,

$$3I^2 + 24I + 48 + 3I^2 - 24I + 48 = 10I^2 - 160$$

Collecting terms,	$-4I^2 = -256$
D: -4,	$I^2 = 64$
$\sqrt{}$,	$I = \pm 8$

check By the usual method.

PROBLEMS 21-1

Solve the following:

1. $E^2 - 25 = 0$
2. $s^2 - 49 = 0$
3. $i^2 + 36 = 225$
4. $\theta^2 - 0.25 = 0$
5. $5\omega^2 - 180 = 0$
6. $\phi^2 - 0.0004 = 0.0012$
7. $\lambda^2 - \frac{9}{121} = 0$
8. $49I^2 - 144 = 0$
9. $5\mu^2 = 3\frac{1}{5}$
10. $5x^2 - 0.0308 = 0.0817$
11. $2(m+1) - m(m-3) - 5m = 0$
12. $\dfrac{28}{R^2 - 9} = \dfrac{R+3}{R-3} - 1 + \dfrac{R-3}{R+3}$
13. $\dfrac{3\lambda - 18}{6} + \dfrac{90 + 9\lambda - 4\lambda^2}{3\lambda} = 0$
14. $6\alpha(4\alpha - 3) + 3(6\alpha - 16) = 0$
15. $X_C = \dfrac{24 - X_C + (X_C - 1)^3}{2 + X_C^2} - 2$

21-3 COMPLETE QUADRATIC EQUATIONS—SOLUTION BY FACTORING

As an example, let it be assumed that all that is known about two expressions x and y is that $xy = 0$. We know that it is impossible to find the value of either unless the value of the other is known. However, we do know that, if $xy = 0$, *either* $x = 0$ *or* $y = 0$, for the product of two numbers can be zero if, and only if, one of the numbers is zero.

example 4

Solve the equation $x(5x - 2) = 0$.

solution Here we have the product of two numbers x and $(5x - 2)$, equal to zero, and in order for the equation to be satisfied one of the numbers must be equal to zero. Therefore, $x = 0$ or $5x - 2 = 0$. Solving the latter equation, we have $x = \frac{2}{5}$. Hence,

$$x = 0 \quad \text{or} \quad x = \tfrac{2}{5}$$

check If $x = 0$,

$$x(5x - 2) = 0(5 \cdot 0 - 2) = 0(-2) = 0$$

If $x = \frac{2}{5}$,

$$x(5x - 2) = \tfrac{2}{5}(5 \cdot \tfrac{2}{5} - 2) = \tfrac{2}{5}(2 - 2) = 0$$

It is evident that the roots of a complete quadratic may be of unequal absolute value and may or may not have the same signs. It is incorrect to say $x = 0$ *and* $x = \frac{2}{5}$, for x cannot be equal to both 0 and $\frac{2}{5}$ at the same time. This will be more apparent in the following examples.

example 5

Solve the equation $(x - 5)(x + 3) = 0$.

solution Again, we have the product of two numbers, $(x - 5)$ and $(x + 3)$, equal to zero. Hence, either

$$x - 5 = 0 \quad \text{or} \quad x + 3 = 0$$
$$\therefore x = 5 \quad \text{or} \quad x = -3$$

check If $x = 5$,

$$(x - 5)(x + 3) = (5 - 5)(5 + 3) = 0(8) = 0$$

If $x = -3$,

$$(x - 5)(x + 3) = (-3 - 5)(-3 + 3) = (-8)0 = 0$$

example 6

Solve the equation $x^2 - x - 6 = 0$.

solution

Given	$x^2 - x - 6 = 0$
Factoring	$(x - 3)(x + 2) = 0$
Then, if $x - 3 = 0$,	$x = 3$
Also, if $x + 2 = 0$,	$x = -2$

$\therefore x = 3$ or -2

check If $x = 3$,

$$x^2 - x - 6 = 3^2 - 3 - 6 = 9 - 3 - 6 = 0$$

If $x = -2$,

$$x^2 - x - 6 = (-2)^2 - (-2) - 6 = 4 + 2 - 6 = 0$$

example 7

Solve the equation $(E - 3)(E + 2) = 14$.

solution Given $(E - 3)(E + 2) = 14$.

Expanding,	$E^2 - E - 6 = 14$
S: 14,	$E^2 - E - 20 = 0$
Factoring,	$(E - 5)(E + 4) = 0$
Then, if $E - 5 = 0$,	$E = 5$
Also, if $E + 4 = 0$,	$E = -4$

$\therefore E = 5$ or -4

check If $E = 5$,

$$(E - 3)(E + 2) = (5 - 3)(5 + 2) = (2)(7) = 14$$

If $E = -4$,

$$(E - 3)(E + 2) = (-4 - 3)(-4 + 2) = (-7)(-2) = 14$$

PROBLEMS 21-2

Solve by factoring:

1. $\alpha^2 + 5\alpha + 4 = 0$
2. $e^2 + 2e - 15 = 0$
3. $R^2 + 14 = 9R$
4. $x^2 = 5x - 6$
5. $\lambda^2 = 2 - \lambda$
6. $\psi^2 = 17\psi - 60$
7. $E^2 + 40 = 22E$
8. $26 + 11L - L^2 = 0$
9. $\dfrac{2Q - 13}{Q - 5} = \dfrac{7Q - 5}{5Q - 7}$
10. $\dfrac{8}{\kappa} + \kappa + 2 = \dfrac{2}{\kappa} - 3$
11. $\alpha + 32 + \dfrac{20}{\alpha} = 5 - \dfrac{30}{\alpha}$
12. $\dfrac{160}{I^2} = \dfrac{26}{I} - 1$
13. $\dfrac{1}{Z - 4} - 1 = \dfrac{-2}{Z - 2}$
14. $\dfrac{2F - 6}{17 - F} = 1 - \dfrac{2}{F - 2}$
15. $\dfrac{4}{2i + 2} + \dfrac{i}{3i + 7} - \dfrac{11}{4i + 4} = 0$

21-4 SOLUTION BY COMPLETING THE SQUARE

Some quadratic equations are not readily solved by factoring, but frequently such quadratic equations are readily solved by another method known as *completing the square.*

In Problems 10-5, missing terms were supplied in order to form a perfect trinomial square. This is the basis for the method of completing the square. For example, in order to make a perfect square of the expression $x^2 + 10x$, 25 must be added as a term to obtain $x^2 + 10x + 25$, which is the square of the quantity $x + 5$.

example 8

Solve the equation $x^2 - 10x - 20 = 0$.

solution Inspection of the given equation shows that it cannot be factored with integral numbers. Therefore, the solution will be accomplished by the method of completing the square.

Given $\qquad x^2 - 10x - 20 = 0$

A: 20, $\qquad x^2 - 10x = 20$

Squaring one-half the coefficient of x and adding to both members,

$$x^2 - 10x + 25 = 20 + 25$$

Collecting terms,

$$x^2 - 10x + 25 = 45$$

Factoring, $\qquad (x - 5)^2 = 45$

$\sqrt{\ }$, $\qquad x - 5 = \pm 6.71$

A: 5, $\qquad x = 5 \pm 6.71$

or $\qquad x = 11.71$ or -1.71

The above answers are correct to three significant figures. The values of x are more precisely stated by maintaining the radical sign in the final roots. That is, if

$$(x - 5)^2 = 45$$

$\sqrt{\ }$, $\qquad x - 5 = \pm\sqrt{45}$

or $\qquad x - 5 = \pm 3\sqrt{5}$

A: 5, $\qquad x = 5 \pm 3\sqrt{5}$

That is, $\qquad x = 5 + 3\sqrt{5}$ or $5 - 3\sqrt{5}$

example 9

Solve the equation $3x^2 - x - 1 = 0$.

solution

Given $\qquad 3x^2 - x - 1 = 0$

D: 3 (because the coefficient of x^2 must be 1),

$$x^2 - \tfrac{1}{3}x - \tfrac{1}{3} = 0$$

Transposing the constant term,

$$x^2 - \tfrac{1}{3}x = \tfrac{1}{3}$$

Squaring one-half the coefficient of x and adding to both members,

$$x^2 - \tfrac{1}{3}x + \tfrac{1}{36} = \tfrac{1}{3} + \tfrac{1}{36}$$

Collecting terms,

$$x^2 - \tfrac{1}{3}x + \tfrac{1}{36} = \tfrac{13}{36}$$

Factoring,

$$(x - \tfrac{1}{6})^2 = \tfrac{13}{36}$$

$\sqrt{}$,

$$x - \frac{1}{6} = \pm\frac{\sqrt{13}}{6}$$

$$\therefore x = \frac{1 + \sqrt{13}}{6} \text{ or } \frac{1 - \sqrt{13}}{6}$$

To summarize the method, we have the following:

Rule

To solve by completing the square:

1. If the coefficient of the square of the unknown is not 1, divide both members of the equation by the coefficient.
2. Transpose the constant terms (those not containing the unknown) to the right member.
3. Find one-half the coefficient of the unknown of the first degree, square the result, and add this square to both members of the equation. This makes the left member a perfect trinomial square.
4. Take the square root of both members of the equation and write the ± sign before the square root of the right member.
5. Solve the resulting simple equation.

PROBLEMS 21-3

Solve by completing the square:

1. $x^2 - 8x + 12 = 0$

2. $\alpha^2 - 4\alpha - 45 = 0$

3. $E^2 - 15E + 54 = 0$

4. $\Omega^2 + 5\Omega + 6 = 0$

5. $i^2 - 27i = -50$

6. $63 - a^2 = 2a$

7. $\theta^2 + 2 = 3\theta$

8. $e^2 - 6 = e$

9. $M^2 = 22M + 48$

10. $24E^2 = 2E + 1$

11. $3 + \theta = \theta^2 - 3$

12. $17I - 42 = I^2 + 2I - 16$

13. $\phi = \dfrac{60}{\phi} + 4$

14. $1 + \dfrac{12}{f} + \dfrac{35}{f^2} = 0$

15. $\dfrac{7(R - 4)}{R - 3} - (R - 2) = \dfrac{R - 4}{2}$

16. $\dfrac{Z - 1}{Z + 1} = \dfrac{Z - 2}{Z + 2} - 6$

21-5 STANDARD FORM

Any quadratic equation can be written in the general form

$$ax^2 + bx + c = 0$$

This is called the *standard form* of the quadratic equation. When it is written in this way, a represents the coefficient of the term containing x^2, b represents the coefficient of the term containing x, and c represents the constant term. Note that all terms of the equation, when written in standard form, are in the left member of the equation.

example 10 Given $2x^2 + 5x - 3 = 0$. In this equation, $a = 2$, $b = 5$, and $c = -3$.

example 11 Given $R^2 - 5R - 6 = 0$. In this equation, $a = 1$, $b = -5$, and $c = -6$.

example 12 Given $9E^2 - 25 = 0$. In this equation, $a = 9$, $b = 0$, and $c = -25$.

21-6 THE QUADRATIC FORMULA

Because the standard form

$$ax^2 + bx + c = 0$$

represents *any* quadratic equation, it follows that the roots of $ax^2 + bx + c = 0$ represent the roots of *any* quadratic equation. Therefore, if the standard quadratic equation can be solved for the unknown, the values, or roots, thereby obtained will serve as a formula for finding the roots of *any* quadratic equation.

This formula is derived by solving the standard form by the method of completing the square as follows:

Given $\qquad ax^2 + bx + c = 0$

Divide by a (Rule 1):

$$x^2 + \frac{bx}{a} + \frac{c}{a} = 0$$

Transpose the constant term (Rule 2):

$$x^2 + \frac{bx}{a} = -\frac{c}{a}$$

Add the square of one-half the coefficient of x to both members (Rule 3):

$$x^2 + \frac{bx}{a} + \frac{b^2}{4a^2} = \frac{b^2}{4a^2} - \frac{c}{a}$$

Factor the left member, and add terms in the right member:

$$\left(x + \frac{b}{2a}\right)^2 = \frac{b^2 - 4ac}{4a^2}$$

Take the square root of both members:

$$x + \frac{b}{2a} = \pm\frac{\sqrt{b^2 - 4ac}}{2a}$$

Subtract $\frac{b}{2a}$:

$$x = -\frac{b}{2a} \pm \frac{\sqrt{b^2 - 4ac}}{2a}$$

Collect terms of the right member:

$$x = \frac{-b \pm \sqrt{b^2 - 4ac}}{2a}$$

This equation is known as the quadratic formula.

Instead of attempting to solve a quadratic equation by factoring or by completing the square, we now make use of the quadratic formula. Upon becoming proficient in the use of the formula, you will find this method a convenience.

example 13
Solve the equation $5x^2 + 2x - 3 = 0$.

solution Comparing this equation with the standard form

$$ax^2 + bx + c = 0$$

we have $a = 5$, $b = 2$, and $c = -3$. Substituting in the quadratic formula

$$x = \frac{-b \pm \sqrt{b^2 - 4ac}}{2a}$$

$$= \frac{-2 \pm \sqrt{2^2 - 4 \cdot 5 \cdot (-3)}}{2 \cdot 5}$$

$$\text{Hence, } x = \frac{-2 \pm \sqrt{64}}{10}$$

$$= \frac{-2 \pm 8}{10}$$

$$= \frac{-2 + 8}{10} \quad \text{or} \quad \frac{-2 - 8}{10}$$

$$\therefore x = \tfrac{3}{5} \text{ or } -1$$

check Substitute the values of x in the given equation.

note It must be remembered that the expression $\sqrt{b^2 - 4ac}$ is the square root of the *quantity* $(b^2 - 4ac)$ *taken as a whole.*

example 14

Solve the equation $\dfrac{3}{5 - R} = 2R$.

solution Clearing the fractions results in $2R^2 - 10R + 3 = 0$. Comparing this equation with the standard form

$$ax^2 + bx + c = 0,$$

we have $a = 2$, $b = -10$, and $c = 3$. Substituting in the quadratic formula.

$$x = \frac{-b \pm \sqrt{b^2 - 4ac}}{2a}$$

$$R = \frac{-(-10) \pm \sqrt{(-10)^2 - 4 \cdot 2 \cdot 3}}{2 \cdot 2}$$

Hence,

$$R = \frac{10 \pm \sqrt{76}}{4}$$

Factoring the radicand,

$$R = \frac{10 \pm 2\sqrt{19}}{4}$$

Dividing both terms of the fraction by 2,

$$R = \frac{5 \pm \sqrt{19}}{2}$$

$$= \frac{5 + \sqrt{19}}{2} \quad \text{or} \quad \frac{5 - \sqrt{19}}{2}$$

$$\therefore R = 4.68 \text{ or } 0.320$$

These final answers are correct to three significant figures. Check the solution by the usual method.

21-7 TESTING SOLUTIONS

Now that we can obtain solutions to quadratic equations by means of the quadratic formula, there will be two possible answers so long as $b^2 - 4ac \neq 0$. One of these answers we may call α:

$$\alpha = \frac{-b + \sqrt{b^2 - 4ac}}{2a}$$

and the other we may call β:

$$\beta = \frac{-b - \sqrt{b^2 - 4ac}}{2a}$$

By suitable combinations of α and β, we can achieve two useful relationships the proof of which we leave to you as an exercise:

$$\alpha + \beta = \frac{-b}{a} \qquad (1)$$

$$\alpha \cdot \beta = \frac{c}{a} \qquad (2)$$

Whenever you obtain answers to quadratic equations by means of the formula (or any other means), you may quickly test them for accuracy. The sum of the two answers must equal $-\dfrac{b}{a}$, and the product of the two must equal $\dfrac{c}{a}$.

example 15

Solve the equation $6x^2 - 2x - 4 = 0$, and test the answers.

solution Using the quadratic formula, $x = 1$ or $x = -\frac{2}{3}$. Applying the tests:

$$\alpha + \beta = 1 - \frac{2}{3} = +\frac{1}{3}$$

$$-\frac{b}{a} = -\frac{-2}{6} = +\frac{1}{3}$$

and

$$\alpha \cdot \beta = (1)\left(-\frac{2}{3}\right) = -\frac{2}{3}$$

$$\frac{c}{a} = \frac{-4}{6} = -\frac{2}{3}$$

The tests show that the solutions obtained are correct. You should make a habit of applying the tests to every solution to quadratic equations that you obtain.

PROBLEMS 21-4

Solve the following equations by using the quadratic formula, and apply the tests of Eqs. (1) and (2):

1. $\theta^2 = 4 - 3\theta$

2. $\lambda^2 + 7\lambda = 18$

3. $2I + 35 = I^2$

4. $\alpha^2 - 4\alpha + 3 = 0$

5. $15 - 14q = 8q^2$

6. $3I^2 - 7I + 2 = 0$

7. $5 = 6Z^2 - 3Z$

8. $5(R + 2) = 2R(R - 1)$

9. $24 - \frac{2}{m} - \frac{1}{m^2} = 0$

10. $\frac{2}{I_1} + \frac{3}{I_1} = \frac{1}{I_1^2} - 14$

11. $\frac{4 - R_1}{1 - R_1} - \frac{12}{3 - R_1} = 0$

12. $\frac{2}{\lambda + 3} = \frac{3}{\lambda - 2} - 1$

13. $\frac{7}{\beta - 3} - \frac{1}{2} = \frac{\beta - 2}{\beta - 4}$

14. $\frac{36}{(I + 3)^2} - \frac{I + 2}{I + 3} = 1$

15. $7i + 5 = \frac{21i^3 - 16}{3i^2 - 4}$

16. $4 - E - \frac{1}{2E} = -\frac{E^2 + 25}{7E}$

21-8 THE GRAPH OF A QUADRATIC EQUATION —THE PARABOLA

In Chap. 16 we spent some time on the drawing of graphs, especially graphs of unity-power (first-degree) equations, or linear graphs. Graphs of quadratic equations may also be drawn, and in this section we will investigate the common methods of producing such graphs and also a method of predicting the shape of graphs just from the equation itself in the same way that we learned to use the standard form $y = mx + b$ to predict the slope and y intercept of linear graphs.

All the quadratic equations we have studied so far in this chapter have contained only one unknown, but that is because we were looking at special cases. In the algebraic solution of quadratics, the standard form $ax^2 + bx + c = 0$ is sufficient, because we want to know the values of x which will satisfy this standard form equation. However, to draw a graph requires two variables, an independent one x and a dependent one y, so we rewrite the standard equation:

$$y = ax^2 + bx + c$$

Then, by plotting values of y for given values of x, we can draw the complete graph. Note that the algebraic solutions so far in this chapter have simply let $y = 0$, that is, the algebraic solutions have given us the x intercepts for the equation of the general form

$$y = ax^2 + bx + c$$

example 16
Graph the equation $x^2 - 10x + 16 = 0$.

solution Set the equation equal to y:

$$y = x^2 - 10x + 16$$

Make a table of the values of y corresponding to assigned values of x (see Eq. Fig. 21-1).
Plotting the corresponding values of x and y as pairs of coordinates and drawing a smooth curve through the points results in the graph shown in Fig. 21-1.

From Fig. 21-1 it is apparent that the graph has two x intercepts at $x = 2$ and $x = 8$. That is, when $y = 0$, the graph crosses the x axis at $x = 2$ and $x = 8$. This is to be expected, for when $y = 0$, the given equation

$$x^2 - 10x + 16 = 0$$

can be solved algebraically to obtain $x = 2$ or 8. Hence, it is evident that the points at which the graph crosses the x axis denote the values of x when $y = 0$, which are the roots of the equation.

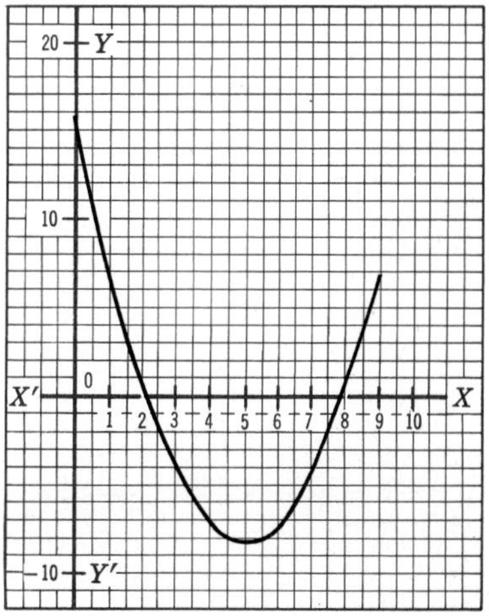

Fig. 21-1 Graph of the Equation $y = x^2 - 10x + 16$

Another interesting fact regarding this graph is that the curve goes through a *minimum value*. Suppose it is desired to solve for the coordinates of the point of minimum value. First, if the equation is changed to standard form, we obtain $a = 1$, $b = -10$, and $c = 16$. If the value of $\frac{-b}{2a}$ is computed, the result is the x value, or abscissa, of the minimum point on the curve. That is,

$$x = -\frac{b}{2a} = -\frac{-10}{2 \times 1} = \frac{10}{2} = 5$$

Substituting this value of x in the original equation,

$$y = x^2 - 10x + 16$$
$$y = 5^2 - 10 \times 5 + 16 = -9$$

Thus, the point (5,−9) is where the curve

Eq. Fig. 21-1

If $x =$	0	1	2	3	4	5	6	7	8	9	10
Then $x^2 =$	0	1	4	9	16	25	36	49	64	81	100
$10x =$	0	10	20	30	40	50	60	70	80	90	100
$x^2 - 10x =$	0	−9	−16	−21	−24	−25	−24	−21	−16	−9	0
$\therefore y = x^2 - 10x + 16 =$	16	7	0	−5	−8	−9	−8	−5	0	7	16

passes through a minimum value. That is, the dependent variable y is a minimum and equal to -9 when x, the independent variable, is equal to 5.

A third point of interest is that the parabola, as the graph of the quadratic is called, is symmetrical about its turning point, which lies midway between the two intercepts. Indeed, this can be seen from a revision of the quadratic formula:

$$x = \frac{-b \pm \sqrt{b^2 - 4ac}}{2a}$$

may appropriately be written

$$x = \frac{-b}{2a} \pm \frac{\sqrt{b^2 - 4ac}}{2a}$$

from which we can see that, with the turning point at $\frac{-b}{2a}$, the values of the x intercepts, or roots, of the graph will be offset from the x value of the turning point by amounts equal to $\pm\frac{\sqrt{b^2 - 4ac}}{2a}$.

Look now at some of the main possibilities concerning the appearance of parabolas:

1. They may open upward or downward (Fig. 21-2).
2. They may be symmetrical about the y axis or about some line parallel to the y axis (Fig. 21-3).
3. They may (a) cut the x axis in two places, (b) touch the x axis (cut it in one place), or (c) not touch the x axis at all (Fig. 21-4).

It is possible to decide many of these possibilities from the values of a particular quadratic equation. This general equation $y = ax^2 + bx + c$ offers many possibilities and a restriction:

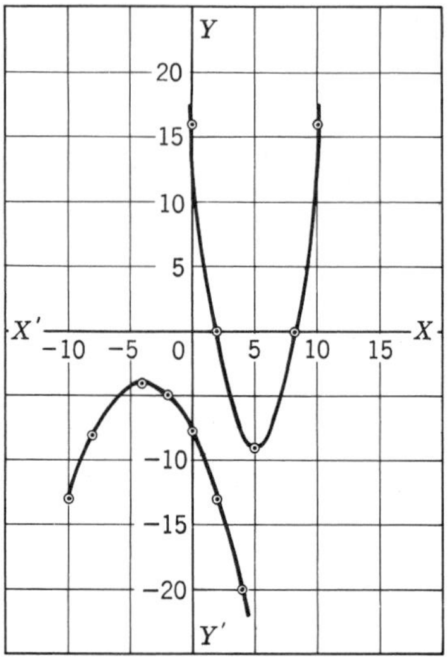

Fig. 21-2 Quadratic Graph May Open Upward or Downward

1. a, the coefficient of the square term, may be any number, positive or negative, but *not* zero. (Why?)
2. b, the coefficient of the unity-power term,

Fig. 21-3 Quadratic Graphs May be Symmetrical about the y Axis or about a Line Parallel to the y Axis

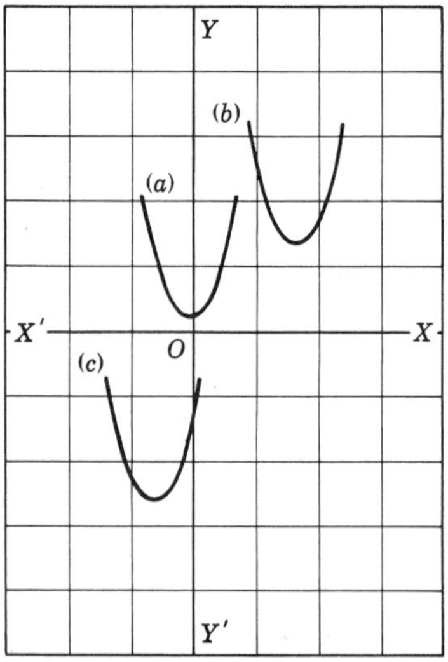

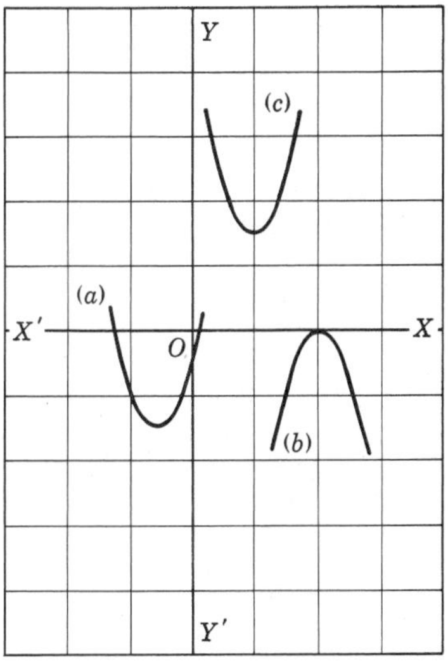

Fig. 21-4 Quadratic Graphs May Cut the x Axis in Two Places, in One Place, or Not at All

may be any number, positive or negative, or zero.

3. c, the constant term, may be any number, positive or negative, or zero.

Now, what is the effect of these algebraic possibilities on the graph? You should, at this point, arm yourself with graph paper, and confirm the following statements:

Rule

The effect of a on the graph of the quadratic equation:

The value of a in the quadratic equation governs the steepness of the parabola. When a is large, the parabola is very steep, approaching a needlelike shape. When a is small, the parabola is shallow, approaching a dished shape.

Let $b = c = 0$ and plot the comparison graph $y = x^2$, in which the value of a is 1. Then plot various graphs of $y = ax^2$, letting a equal, in succession, 2, 5, 10, $\frac{1}{2}$, and $\frac{1}{4}$. If these are all plotted on the same graph sheet, with different colors or dashed lines, etc., then the effect of the value of a will be impressed on your mind forever.

Rule

The effect of a on the appearance of the parabola:

The algebraic sign of a determines the opening of the parabola. $+a$ causes the curve to open upward, and the turning point is the *minimum* value. $-a$ causes the parabola to open downward, and the turning point is the *maximum* value.

You have already plotted a number of graphs with $+a$. Now plot a few graphs of $y = -ax^2$, letting $a = 2, 5, 10, \frac{1}{2}$, and $\frac{1}{4}$.

Rule

The effect of c on the appearance of the parabola:

The constant c in the quadratic equation determines the y intercept, and therefore the amount of vertical shift of the parabola. When c is positive, the curve is raised to cut the y axis above the x axis. When c is negative, the curve cuts the y axis below the x axis.

Let $a = 1$ and $b = 0$ and vary the value of c in the equation $y = x^2 + c$. Draw the curves when $c = +5$ and -5, and compare with the standard parabola $y = x^2$.

Rule

The effect of b on the appearance of the parabola:

The factor b in the quadratic equation determines the rotational shift of the turning point of the graph. When b is positive, the turning point shifts in a positive (ccw) direction about its "original" position, and when b is negative, the turning point shifts in a negative (cw) direction about its original position.

Let $a = 1$, $c = 0$, and vary the value of b in the equation $y = x^2 + bx$. Draw the curves when $b = +2, +5, +10, -2, -5$, and

−10. Repeat these curves with $a = -1$, and draw curves for $y = -x^2 - bx$.

example 17

Plot the curve $y = 27 - 3x - 4x^2$.

solution Predict, first of all, what effect the various coefficients will have on the graph:

1. The value of a is 4, so that the curve will be reasonably steep.
2. The algebraic sign of a is minus, so that the curve will open downward.
3. The constant term is +27, so that the curve cuts the y axis at +27, well above the x axis. Since the curve opens downward and the y intercept is above the x axis, the curve will cut the x axis in two places. That is, the special equation $27 - 3x - 4x^2 = 0$ will have two definite solutions.
4. The value of b is −3, so that the turning point will be shifted from the "ideal" value of $x = 0$, $y = 27$ in the clockwise direction. The turning point will then be at a value of y greater than 27 and at some value of x to the left, or minus, side of the y axis.

With these predictions, together with a sketch of the probable appearance of the curve (Fig. 21-5), you may assign values to x and calculate the corresponding values of y (see table below).

Plotting the corresponding values of x and y as pairs of coordinates and drawing a smooth curve through them results in the graph shown in Fig. 21-5.

From the graph of the equation $y = 27 - 3x - 4x^2$, Fig. 21-5, it is observed:

1. The roots (solution) of the equation are denoted by the x intercepts. These are $x = -3$ and $x = 2.25$. They can be checked algebraically to obtain

$$27 - 3x - 4x^2 = 0$$

Factoring, $$(3 + x)(9 - 4x) = 0$$

$$x = -3 \text{ or } 2.25$$

Fig. 21-5 Graph of the Equation $y = 27 - 3x - 4x^2$

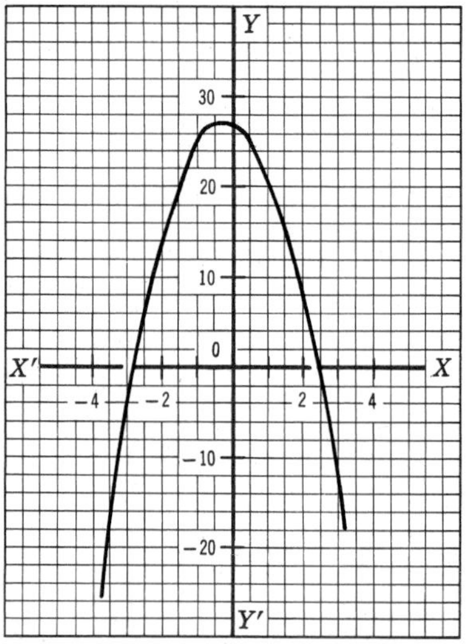

If $x =$	−4	−3	−2	−1	0	1	2	3
Then $3x =$	−12	−9	−6	−3	0	3	6	9
$27 - 3x =$	39	36	33	30	27	24	21	18
$x^2 =$	16	9	4	1	0	1	4	9
$4x^2 =$	64	36	16	4	0	4	16	36
$\therefore y = 27 - 3x - 4x^2 =$	−25	0	17	26	27	20	5	−18

2. The parabola opens *downward* because the coefficient of x^2 is negative ($a = -4$.)
3. Because the parabola opens downward, the graph goes through a *maximum* value. The point of maximum value is found in the same manner as the minimum point of Example 16. That is,

$$x = \frac{-b}{2a} = \frac{-(-3)}{2(-4)} = -\frac{3}{8}$$

Substituting $-\frac{3}{8}$ for x in the original equation,

$$y = 27 - 3(-\tfrac{3}{8}) - 4(-\tfrac{3}{8})^2 = 27.6$$

Thus, the dependent variable y is a maximum and equal to 27.6 when x, the independent variable, is equal to $-\frac{3}{8}$.

example 18 Graph the equations

$$y = x^2 - 8x + 12 \qquad (a)$$
$$y = x^2 - 8x + 16 \qquad (b)$$
$$y = x^2 - 8x + 20 \qquad (c)$$

solution

1. Based on an analysis of the values of a, b, and c, predict the probable appearance of the curve.
2. As before, and for each equation, make up a table of values of y corresponding to chosen values of x. Using these x and y values as pairs of coordinates, plot the graphs of the equations. These graphs are shown in Fig. 21-6.
   The coefficients of the equations are the same except for the values of the constant term c.

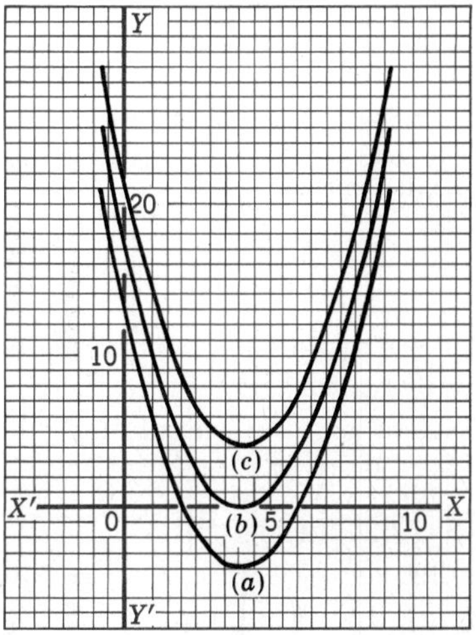

Fig. 21-6 Graphs of the Equations of Example 18

From the graphs of the equations of Example 18, it is observed that:

1. The curve of (a) intercepts the x axis at $x = 2$ and $x = 6$, and the roots of the equation are thus denoted as $x = 2$ or 6. This checks with the algebraic solution.
2. The curve of (b) just *touches* the x axis at $x = 4$. Solving (b) algebraically shows that the roots are *equal*, both roots being 4.
3. The curve of (c) does not intersect or touch the x axis. Solving (c) algebraically results in the imaginary roots $x = 4 \pm j2$.
4. All curves pass through minimum values at points having equal x values. This is as expected, for the x value of a maximum or a minimum is given by $x = \dfrac{-b}{2a}$, and these values are equal in each of the given equations.
5. Checking the y values of the minima, it is seen that they must be affected by the constant terms, for, as previously mentioned, the other coefficients of the equations are the same.

21-9 GRAPHICAL SOLUTIONS

From the foregoing comments, it must now be obvious that quadratic equations can be solved graphically by letting the equation $ax^2 + bx + c = 0$ take the more general form $ax^2 + bx + c = y$ or, more commonly, $y = ax^2 + bx + c$. Then the two x intercepts of the graph will give the roots of the original equation. It is for this reason that you will often hear the solutions to a quadratic equation referred to as the *zeros* of the equation—they occur when $y = 0$.

PROBLEMS 21-5

Select problems from Problems 21-1, 21-2, 21-3, and 21-4 and solve them graphically to confirm the algebraic solutions. Predict what the graphs will look like before plotting calculated values of x and y.

21-10 THE DISCRIMINANT

The quantity $b^2 - 4ac$ under the radical in the quadratic formula is called the *discriminant* of the quadratic equation. The two roots of the equation are

$$x = \frac{-b + \sqrt{b^2 - 4ac}}{2a}$$

and

$$x = \frac{-b - \sqrt{b^2 - 4ac}}{2a}$$

Now, if $b^2 - 4ac = 0$, it is apparent that the two roots are equal. Also, if $b^2 - 4ac$ is *positive*, each of the roots is a *real* number. But if $b^2 - 4ac$ is *negative*, the roots are *imaginary*. Therefore, there is a direct relationship between the value of the discriminant and the roots, and hence the graph, of a quadratic equation.

For example, the discriminants of the equations of Example 18 in the preceding section are

$$b^2 - 4ac = (-8)^2 - 4 \cdot 1 \cdot 12 = 16$$
$$b^2 - 4ac = (-8)^2 - 4 \cdot 1 \cdot 16 = 0$$
$$b^2 - 4ac = (-8)^2 - 4 \cdot 1 \cdot 20 = -16$$

Upon checking these values with the curves of Fig. 21-6 and also checking the values of the discriminants found in the preceding exercises with their respective curves, it is evident that the roots of a quadratic equation are:

1. Real and unequal if and only if $b^2 - 4ac$ is positive.
2. Real and equal if and only if $b^2 - 4ac = 0$.
3. Imaginary and unequal if and only if $b^2 - 4ac$ is negative.
4. Rational if and only if $b^2 - 4ac$ is a perfect square.

21-11 MAXIMA AND MINIMA

As previously stated, in the general quadratic equation $ax^2 + bx + c = 0$, the relation $x = \frac{-b}{2a}$ gives the value of the independent variable x at which the dependent variable y will be maximum or minimum. Then by substituting this value of x, the independent variable, in the equation, the corresponding value of y can be obtained. Also, it has been shown that the function will be maximum if a, the coefficient of x^2, is negative because the curve opens downward. Similarly, if the coefficient of x^2 is positive, the curve will pass through a minimum because the curve opens upward.

This knowledge facilitates the solution of many problems that heretofore would have involved considerable labor.

example 19 A source of emf E, with an internal resistance r, is connected to a load of variable resistance R. What will be the value of R with respect to r when maximum power is being delivered to the load?

solution The circuit can be represented as shown in Fig. 21-7. By Ohm's law, the current flowing through the circuit is

$$I = \frac{E}{r + R} \qquad (a)$$

The power delivered to the external circuit is

$$P = VI = I^2R \qquad (b)$$

where V is the terminal voltage of the source and is

$$V = E - Ir \qquad (c)$$

Now the terminal voltage V will decrease as the current I increases. Therefore, the power P supplied to the load is a function of the two variables V and I. Substituting Eq. (c) in Eq. (b),

$$P = (E - Ir)I = EI - I^2r$$

that is,

$$P = -rI^2 + EI \qquad (d)$$

Equation (d) is a quadratic in I, where

$$a = -r \qquad \text{and} \qquad b = E$$

Fig. 21-7 Circuit of Example 19

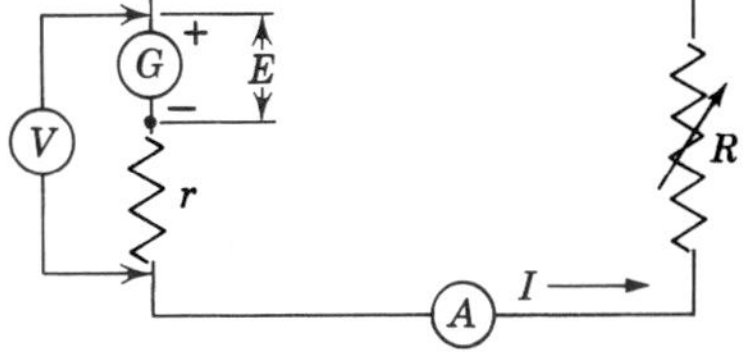

Then, since, for maximum conditions, $I = \frac{-b}{2a}$,

$$I = \frac{-b}{2a} = \frac{-E}{2(-r)} = \frac{E}{2r} \qquad (e)$$

which is the value of the current through the circuit when maximum power is being delivered to the load. Substituting Eq. (e) in Eq. (a),

$$\frac{E}{2r} = \frac{E}{r + R} \qquad (f)$$

Solving Eq. (f) for R,

$$R = r \qquad (g)$$

Equation (g) shows that maximum power will be delivered to any load when the resistance of that load is equal to the internal resistance of the source of emf. This is one of the important concepts in electronics engineering. For example, we are concerned with obtaining maximum power output from several types of power amplifier. We obtain it when the amplifier load resistance matches the plate resistance of the associated vacuum tube. Also, maximum power is delivered to an antenna circuit when the impedance of the antenna is made to match that of the transmission line that feeds it.

In Fig. 21-8, power delivered to the load is

Fig. 21-8 Power Delivered to Load Plotted against Load Resistance

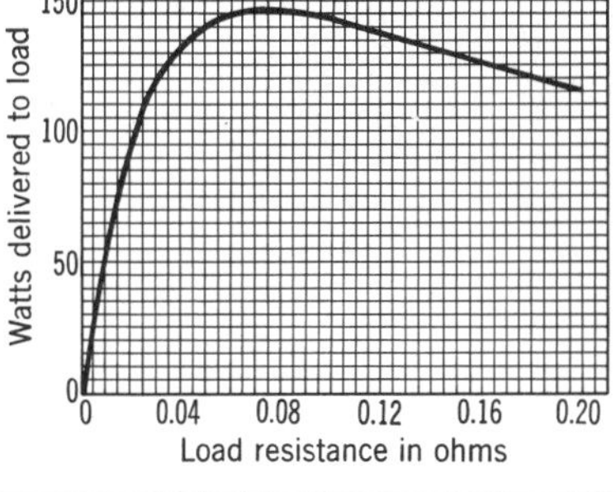

plotted against values of the load resistance R_L when a storage battery with an emf E of 6.6 V and an internal resistance $r = 0.075\ \Omega$ is used. The circuit is as shown in Fig. 21-9.

It is apparent that, when the battery or any other source of emf is delivering maximum power, half the power is lost within the battery. Under these conditions, therefore, the efficiency is 50%.

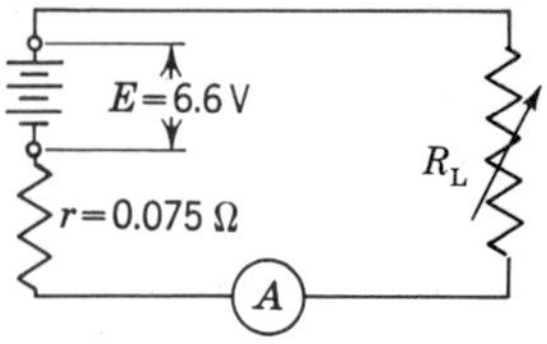

Fig. 21-9 Load Resistance R_L Is Varied to Obtain Power Values Plotted in Fig. 21-8

PROBLEMS 21-6

1. Graph the following equations all on the same sheet with the same axes:
(a) $x^2 - 6x - 16 = 0$ (b) $x^2 - 6x - 7 = 0$
(c) $x^2 - 6x = 0$ (d) $x^2 - 6x + 5 = 0$
(e) $x^2 - 6x + 9 = 0$ (f) $x^2 - 6x + 12 = 0$
(g) $x^2 - 6x + 15 = 0$
Does changing the constant term change only the vertical positions of the graphs and the solutions of the equations? Do all graphs pass through minimum values at the same value of x?

2. Solve the equations of Prob. 1 algebraically. Do these solutions check with the graphs of the equations? Test your solutions by means of the quadratic tests.

3. Compute the discriminant for each equation of Prob. 1. Do you see any connection between the value and the graph?

4. Compute the minimum value of the dependent variable y for each equation of Prob. 1. Does the value check with the graph?

5. What do you see from the graphs of Prob. 1 when x is equal to zero?

21-12 SUMMARY

Several methods are available for solving quadratic equations. All quadratic equations can be solved by factoring, by completing the square, by use of the quadratic formula, or by graphical methods. However, some of these methods involve unnecessary work for certain forms or types of quadratic equations; therefore, one tries to choose the most convenient method for a particular equation. For example, a pure quadratic equation is readily solved merely by reducing the equation to its simplest form and extracting the square root of both members of the equation in order to obtain the two roots, which are equal in absolute value but of opposite sign (Sec. 21-2).

In practical problems involving complete quadratic equations the numerical coefficients are such that you will seldom be able to solve the equation readily by factoring. Also, solution by completing the square sometimes can become a chore. Probably the most widely used method is solution by use of the quadratic formula, which, if you forget it, can be found in most handbooks and put to use whenever needed.

Solution by graphical methods allows you to visualize the variation of quantities and serves to check computations. In any event, through solving many problems, you will develop your own methods of attack.

In solving problems involving quadratic equations, care must be used because two answers (roots) are obtained. In all cases both roots will satisfy the mathematics of the equation, but in some cases only one root will satisfy the conditions of the problem. Therefore, we reject the obviously impossible or the impractical answers and retain the ones that are consistent with the physical conditions of the problem.

example 20 The square of a certain number plus four times the number is 12. Find the number.

solution

Let $x =$ the number

Then $x^2 =$ the square of the number

and $4x =$ four times the number

From the problem

$$x^2 + 4x = 12$$

S: 12,

$$x^2 + 4x - 12 = 0$$

Factoring,

$$(x + 6)(x - 2) = 0$$

Then $x = -6$ or 2

Both roots satisfy the equation and the condition of the problem; therefore, both answers are correct.

example 21 Find the dimensions of a right triangle if its hypotenuse is 40 ft and the base exceeds the altitude by 8 ft.

solution In any right triangle, Fig. 21-10, $c^2 = a^2 + b^2$.

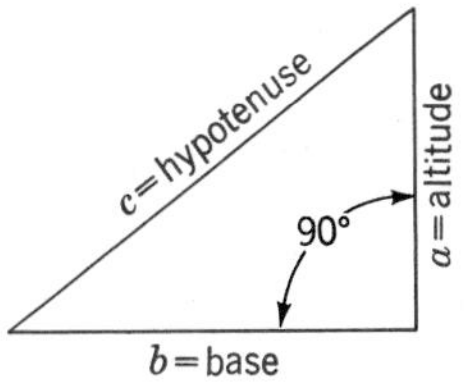

Fig. 21-10 In Any Right Triangle, $c^2 = a^2 + b^2$

Since $c = 40$

and $a = b - 8$

then $1600 = (b - 8)^2 + b^2$

Are both the roots of this equation consistent with the physical conditions of the problem?

example 22 A storage battery has an emf of 6.3 V and an internal resistance of 0.015 Ω. The battery is used to drive a dynamotor that requires 300 W. What current will the battery deliver to the dynamotor, and what will be the voltage reading across the battery terminals while this current is supplied?

solution The circuit is represented in Fig. 21-11.

Let $P =$ power consumed by dynamotor $= 300$ W

$E_B =$ voltage across battery terminals when dynamotor is delivering 300 W

Fig. 21-11 Circuit of Example 22

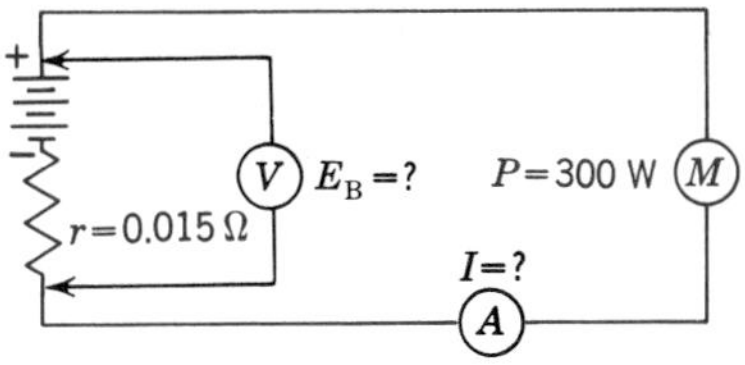

Since $I = \dfrac{P}{E_B}$

then $I = \dfrac{300}{E_B}$

Now $E_B = 6.3 - rI$

Substituting for r,

$$E_B = 6.3 - 0.015I$$

Substituting for I,

$$E_B = 6.3 - 0.015 \times \frac{300}{E_B}$$

Multiplying, $E_B = 6.3 - \dfrac{4.5}{E_B}$

Clearing fractions,

$$E_B^2 = 6.3E_B - 4.5$$

Transposing,

$$E_B^2 - 6.3E_B + 4.5 = 0$$

This equation is a quadratic in E_B; hence, $a = 1$, $b = -6.3$, and $c = 4.5$. Substituting these values in the quadratic formula,

$$E_B = \frac{-(-6.3) \pm \sqrt{(-6.3)^2 - 4 \cdot 1 \cdot 4.5}}{2 \cdot 1}$$

or $E_B = \dfrac{6.3 \pm \sqrt{21.7}}{2}$

$$\therefore E_B = 5.48 \text{ V or } 0.82 \text{ V}$$

$$I = \frac{300}{E_B} = \frac{300}{5.48} = 54.7 \text{ A}$$

Why was 5.48 V chosen instead of 0.82 V in the above solution?

PROBLEMS 21-7

1. Compute the discriminant, and tell what it shows, in each of these equations:
   (*a*) $x^2 - 8x + 12 = 0$
   (*b*) $9x^2 - 42x + 49 = 0$
   (*c*) $4x^2 - 20x + 30 = 0$

2. Find two positive consecutive even numbers whose product is 288.

3. Find two positive consecutive odd numbers whose product is 483.

4. Can the sides of a right triangle ever be consecutive integers? If so, find the integers.

5. Find the dimensions of a rectangular parking lot whose area is 22 400 m² and whose perimeter is 600 m.

6. Separate 156 into two parts such that one part is the square of the other.

7. One number is 20 less than another, and the difference of their squares is 9200. What are the numbers?

8. $F = \dfrac{Wv^2}{32r}$
   (*a*) Solve for v.
   (*b*) If W is doubled and r is halved, what happens to F?
   (*c*) What is W if $F = 12$, $r = 1\frac{1}{3}$, and $v = 16$?

9. Given $P = \dfrac{kE^2}{nR}$. Solve for E. If k and n are doubled and P and R are held constant, what happens to E?

10. $R_t = \dfrac{r}{\left(\dfrac{d_o}{d_i}\right)^2} - 1$. Solve for $\dfrac{d_o}{d_i}$.

11. $P = \dfrac{R(r^2 + x^2)}{r(Rr + Xx)}$. Solve for r and x.

12. The following relations exist in the Wien bridge:

$$\omega^2 = \frac{1}{R_1 R_2 c_1 c_2} \qquad \text{and} \qquad \frac{c_1}{c_2} = \frac{R_b - R_2}{R_a R_1}$$

Solve for c_1 and c_2 in terms of resistance components and ω.

13. Kinetic energy (KE) is equal to one-half the product of mass m in kilograms and the square of velocity v in meters per second; that is, $\text{KE} = \frac{1}{2}mv^2$ joules. Find the value of v when $\text{KE} = 1.1 \times 10^6$ J and $m = 2.2$ kg.

14. A ball rolls down a slope and travels a distance $d = 6t + \frac{1}{2}t^2$ meters in t seconds. Solve for t.

15. The distance through which an object will fall in t seconds is $s = \frac{1}{2}gt^2$ meters, where $g = 9.81$ m/s^2. The velocity v attained after t seconds is $v = gt$ m/s. Solve for the velocity in terms of g and s.

16. If an object is thrown straight up with a velocity of v m/s, its height t seconds later is given by $h = vt - 4.9t^2$ meters. If a rocket were fired upward with a velocity of 1176 m/s, neglecting air resistance:

(*a*) At what time would its height be 15 km on the way up?
(*b*) At what time would its height be 15 km on the way down?
(*c*) At what time would it attain its maximum height?
(*d*) What maximum height would it attain?

Attempt these solutions both graphically and algebraically.

17. Use the formula for height in Prob. 16 to derive a formula for maximum height attained for any initial velocity v.

18. In an ac series circuit containing resistance R in ohms and inductance L in henrys, the current I may be computed from the formula

$$I = \frac{E}{\sqrt{R^2 + \omega^2 L^2}} \qquad \text{A}$$

where E is the emf in volts applied across the circuit. Find the value of R to three significant figures if $E = 282$ V, $I = 2$ A, $\omega = 2\pi f$, $f = 60$ Hz, and $L = 0.264$ H.

19. In an ac circuit containing $R\,\Omega$ resistance and $X_C\,\Omega$ reactance, the impedance is

$$Z = \sqrt{R^2 + X_C^2} \qquad \Omega$$

Find the value of R if $Z = 130\ \Omega$ and $X_C = 120\ \Omega$.

20. The susceptance of an ac circuit containing $R\,\Omega$ resistance and $X\,\Omega$ reactance is

$$B = \frac{X}{R^2 + X^2} \qquad \text{siemens (S)}$$

Find the value of R to three significant figures when $B = 0.008$ S and $X = 100\ \Omega$.

21. The equivalent noise resistance R_n of a pentode tube in terms of cathode current I_k, anode current I_a, screen-grid current I_{sg}, and mutual transconductance g_m, is

$$R_n = \frac{2.5}{g_m}\left(\frac{I_a}{I_k}\right)^2 + \frac{20 I_{sg} I_a}{g_m{}^2 I_k}$$

Show that the ratio of anode current to cathode current is

$$\frac{-20 I_{sg}}{5 g_m} + \frac{1}{5 g_m}\sqrt{400 I_{sg}{}^2 + 10 R_n g_m{}^3}$$

22. Did you prove that $\alpha + \beta = \dfrac{-b}{a}$?

23. Did you prove that $\alpha \cdot \beta = \dfrac{c}{a}$?

24. Find the two combinations of resistance of R_2 and R_3 that will satisfy the circuit conditions of Fig. 21-12.

25. The circuit conditions as shown in Fig. 21-13 existed when the generator G was supplying current to the circuit. When the generator was disconnected, an ohmmeter connected between points A and B read 60 Ω.
(a) What was the circuit current?
(b) What was the generator voltage?
(c) What is the value of each resistor?

Fig. 21-12 Circuit of Prob. 24

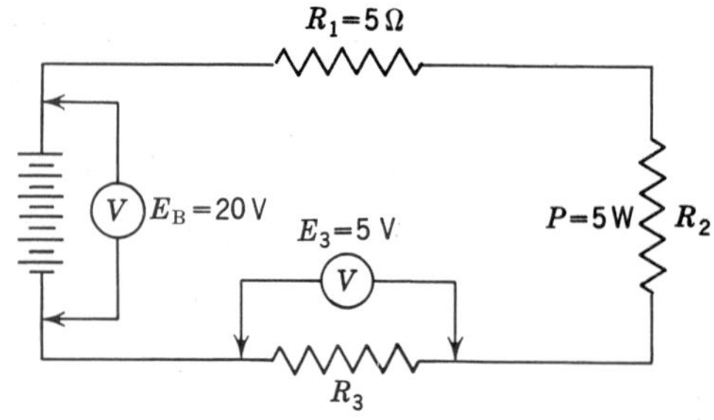

Fig. 21-13 Circuit of Prob. 25

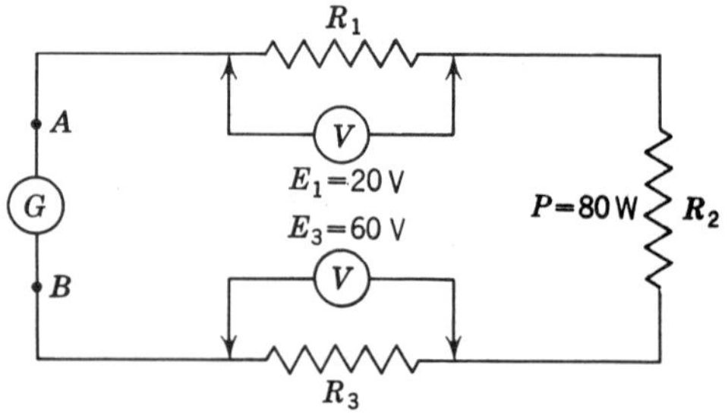

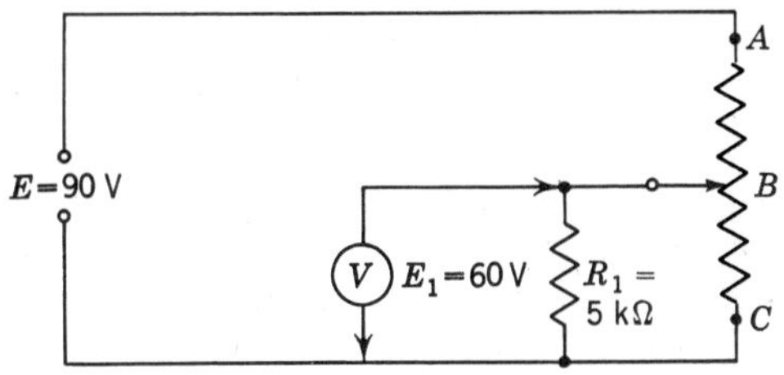

Fig. 21-14 Circuit of Prob. 26

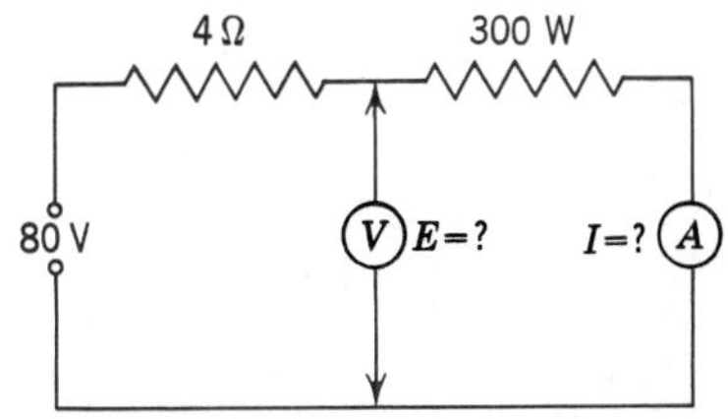

Fig. 21-15 Circuit of Prob. 27

26. In the circuit of Fig. 21-14, the resistor *ABC* represents a potentiometer with a total resistance (*A* to *C*) of 25 000 Ω. R_1 = 5000 Ω, across which is 60 V.
(*a*) What is the resistance from *A* to *B*?
(*b*) How much current flows from *B* to *C*?

27. What are the meter readings in the circuit of Fig. 21-15?

28. When two capacitors C_1 and C_2 are connected in series, the total capacitance C_t of the combination is always less than either of the two capacitors. That is,

$$C_t = \frac{C_1 C_2}{C_1 + C_2}$$

Suppose we have a tuning capacitor that varies from 200 to 300 pF; that is, it has a *change* in capacitance of 100 pF. What value of fixed capacitor should be connected in series with the tuning capacitor to limit the total *change* of circuit capacitance to 50 pF?

22

network simplification

An understanding of Kirchhoff's laws, plus the ability to apply them in analyzing circuit conditions, will give you a better insight into the behavior of circuits. Furthermore, you will be able to solve circuit problems that, with only a knowledge of Ohm's law, would be very difficult in some cases and impossible in others.

22-1 DIRECTION OF CURRENT FLOW

As stated in Sec. 8-1, the most generally accepted concept of an electric current is that it consists of a motion of electrons from a negative toward a more positive point in a circuit. That is, a positively charged body is taken to be one that is deficient in electrons, whereas a negatively charged body carries an excess of electrons. When the two are joined by a conductor, electrons flow from the negative charge to the positive charge. Hence, if two such points in a circuit are *maintained* at a difference of potential, a *continuous* flow of electrons, or current, will take place from negative to positive. Therefore, in the consideration of Kirchhoff's laws, current will be thought of as flowing from the negative terminal of a source of emf, through the external circuit, and back to the positive terminal of the source. Thus, in Fig. 22-1, the current flows away from the negative terminal of the battery, through R_1 and R_2, and back to the positive terminal of the battery. Note that point *b* is positive with respect to point *a* and that point *d* is positive with respect to point *c*.

22-2 STATEMENT OF KIRCHHOFF'S LAWS

In 1847, G. R. Kirchhoff extended Ohm's law by two important statements which have become known as Kirchhoff's laws. These laws can be stated as follows:

1. The algebraic sum of the currents at any junction of conductors is zero.

Fig. 22-1 Current *I* Flowing from − to + through the Connected Circuit

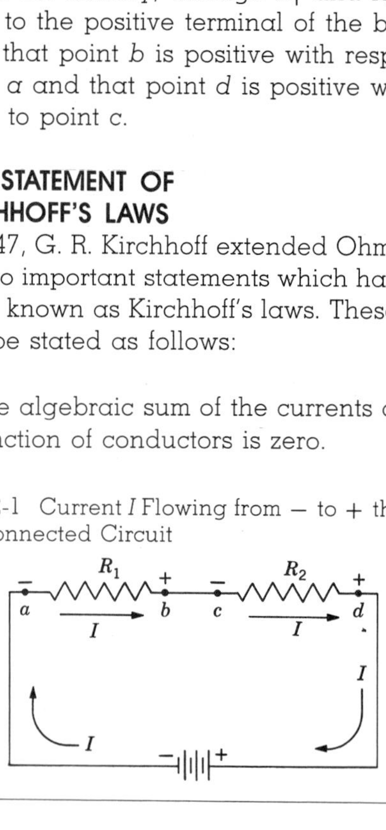

That is, at any point in a circuit, there is as much current flowing away from the point as there is flowing toward it.

2. The algebraic sum of the emf's and voltage drops around any closed circuit is zero.

That is, in any closed circuit, the applied emf is equal to the voltage drops around the circuit.

These laws are straightforward and need no proof here, for the first is self-evident from the study of parallel circuits, and the second was stated in different words in Sec. 8-8. When properly applied, they enable us to set up equations for any circuit and solve for the unknown circuit components, voltages, or currents as required.

22-3 APPLICATION OF SECOND LAW TO SERIES CIRCUITS

The second law is considered first because of its applications to problems with which you are already familiar.

Figure 22-2 represents a 20-V generator connected to three series resistors. The validity of Kirchhoff's second law was demonstrated in Sec. 8-8: that is, in any closed circuit the applied emf is equal to the sum of

Fig. 22-2 The Sum of the Voltage Drops across the Resistors Is Equal to the Applied EMF.

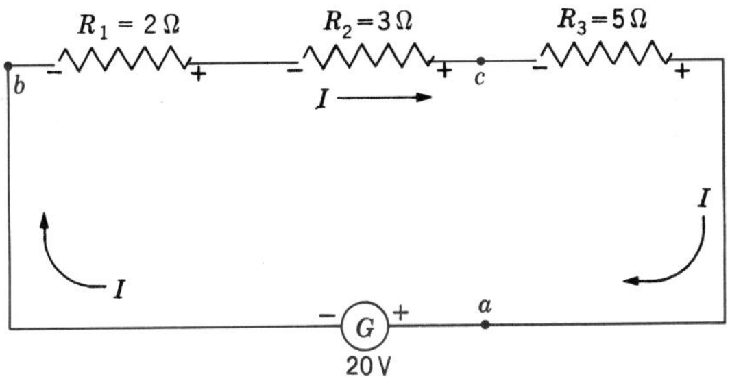

the voltage drops around the circuit. Thus, neglecting the internal resistance of the generator and the resistance of the connecting wires in Fig. 22-2,

$$E = IR_1 + IR_2 + IR_3 \tag{1}$$

or $\quad 20 = 2I + 3I + 5I$

Hence, $\quad I = 2\text{ A}$

Equation (1) is satisfactory for a circuit containing one source of emf. By considering the circuit from a different viewpoint, however, the voltage relations around the circuit become more understandable. For example, by starting at any point in the circuit, such as point *a*, we proceed completely around the circuit in the direction of current flow, remembering that, when current passes through a resistance, there is a voltage drop that represents a loss and therefore is subtractive. Also, in going around the circuit, sources of emf represent a gain in voltage if they tend to aid current flow and therefore are additive. By this method, according to the second law, the algebraic sum of all emf's and voltage drops around the circuit is zero. For example, in starting at point *a* in Fig. 22-2 and proceeding around the circuit in the direction of current flow, the first thing encountered is the positive terminal of a source of emf of 20 V. Because this causes current to flow in the direction we are going, it is written +20. This is easily remembered, for the positive terminal was the first one encountered; therefore, write it plus. Next comes R_1, which is responsible for a *drop* in voltage due to the current I passing through it. Hence, this voltage drop is written $-IR_1$ or $-2I$, for R_1 is known to be $2\,\Omega$. R_2 and R_3 are treated in a similar manner because both represent voltage *drops*. This completes the trip around the circuit, and by equating the algebraic sum of the emf and voltage drops to zero,

$$20 - 2I - 3I - 5I = 0 \qquad (2)$$

or

$$I = 2 \text{ A}$$

Note that Eq. (2) is simply a different form of Eq. (1). If the polarities of the sources of emf are marked, they will serve as an aid in remembering whether to add or subtract. In going around the circuit, if the first terminal of a source of emf is positive, the emf is added; if negative, the emf is subtracted.

The point at which to start around the circuit is purely a matter of choice, for the algebraic sum of all voltages around the circuit is equal to zero. For example, starting at point b,

$$-2I - 3I - 5I + 20 = 0$$
$$I = 2 \text{ A}$$

Starting at point c,

$$-5I + 20 - 2I - 3I = 0$$
$$I = 2 \text{ A}$$

example 1 Find the amount of current flowing in the circuit represented in Fig. 22-3 if the internal resistance of battery E_1 is 0.3 Ω, that of E_2 is 0.2 Ω, and that of E_3 is 0.5 Ω

solution Figure 22-4 is a diagram of the circuit in which the internal resistances are

Fig. 22-3 Circuit of Example 1

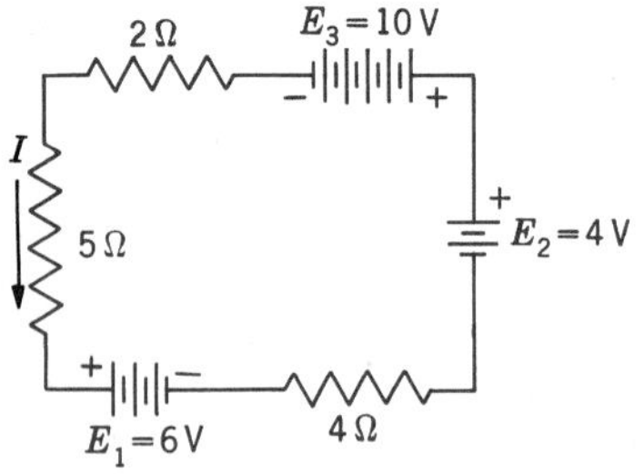

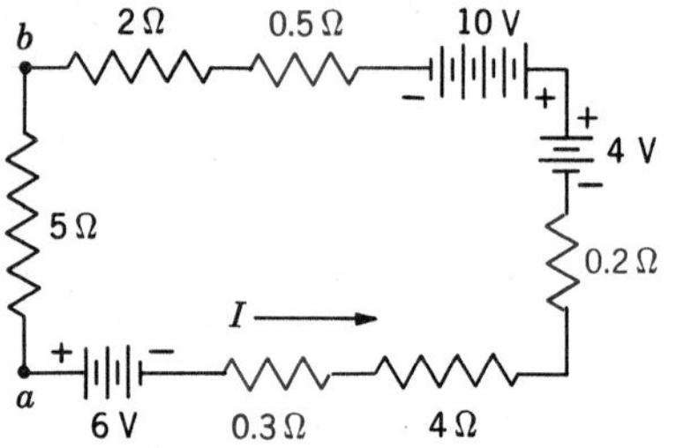

Fig. 22-4 Circuit of Example 1 Illustrating Internal Resistances of the Batteries

represented in color as an aid in setting up the circuit equation. Beginning at point a and going around the circuit in the direction of current flow,

$$6 - 0.3I - 4I - 0.2I - 4 + 10 - 0.5I - 2I - 5I = 0$$

Hence,

$$I = 1 \text{ A}$$

In more complicated circuits the direction of the current is often in doubt. However, this need cause no confusion, for the direction of current flow can be *assumed* and the circuit equation written in the usual manner. If the current results in a negative value when the equation is solved, the negative sign denotes that the assumed direction was wrong. As an example, let it be assumed that the current in the circuit of Fig. 22-4 flows in the direction from a to b. Then, starting at point a and going around the circuit in the assumed direction,

$$-5I - 2I - 0.5I - 10 + 4 - 0.2I - 4I - 0.3I - 6 = 0$$
$$\therefore I = -1 \text{ A}$$

As stated above, the minus sign shows that the assumed direction of the current was wrong; therefore, the current flows in the direction from b to a.

PROBLEMS 22-1

1. Three resistors, R_1 = 22 kΩ, R_2 = 39 kΩ, and R_3 = 33 kΩ, are connected in parallel across a 12-V power supply whose internal resistance is 1.8 kΩ. How much current is drawn from the source?

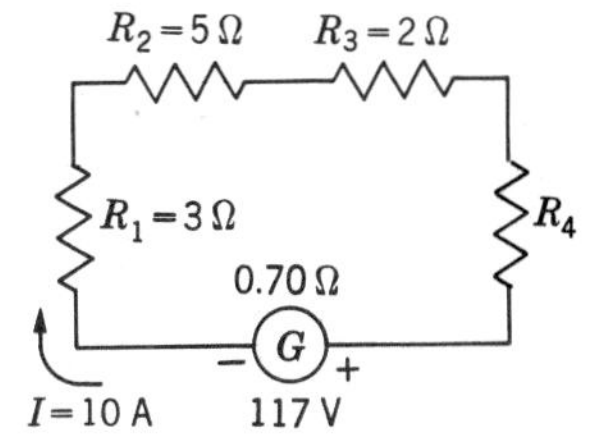

Fig. 22-5 Circuit of Prob. 4

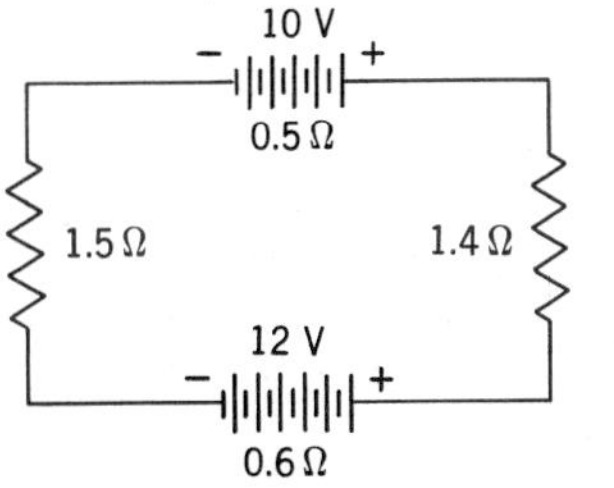

Fig. 22-6 Circuit of Prob. 7

2. The resistors in Prob. 1 are replaced by new values $R_1 = 2.2\ \text{k}\Omega$, $R_2 = 3.9\ \text{k}\Omega$, and $R_3 = 3.3\ \text{k}\Omega$. How much current will be drawn from the source?
3. Three resistors, $R_1 = 68\ \text{k}\Omega$, $R_2 = 22\ \text{k}\Omega$, and $R_3 = 18\ \text{k}\Omega$, are connected in series across a signal generator whose internal resistance is 600 Ω. If 0.500 mA flows through the circuit, what is the terminal voltage of the generator?
4. What is the value of R_4 in Fig. 22-5?
5. A motor that draws 16 A at 234 V is connected to a generator through two No. 8 copper feeders each of which is 500 ft long. What is the generator terminal voltage?
6. A generator with a terminal voltage of 117 V is supplying 63 A to a load through two feeders each 1500 ft long. If the feeders are No. 0 copper, what is the voltage across the load?
7. (*a*) How much current flows in the circuit of Fig. 22-6?
   (*b*) What is the terminal voltage of the 12-V battery?
8. (*a*) How much current flows in the circuit of Fig. 22-7?
   (*b*) What is the terminal voltage of the generator?
9. A current of 5 A flows through the circuit of Fig. 22-8. What is the value of R?

Fig. 22-7 Circuit of Prob. 8

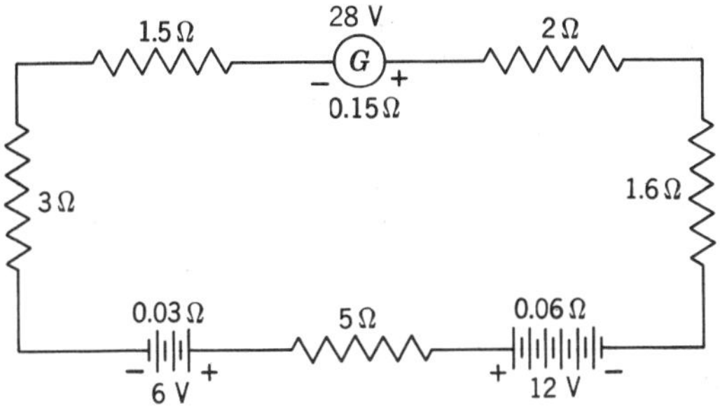

Fig. 22-8 Circuit of Prob. 9

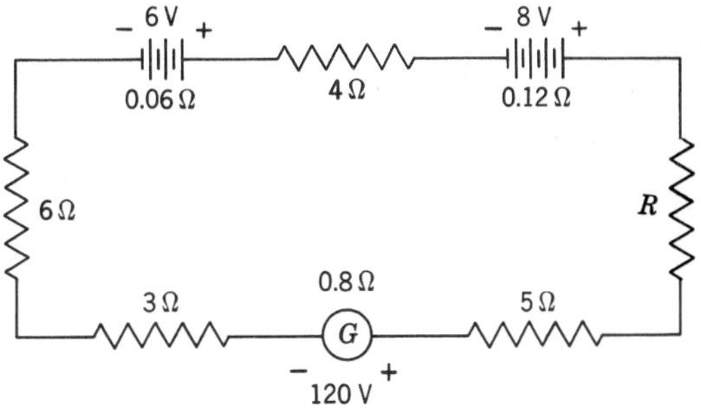

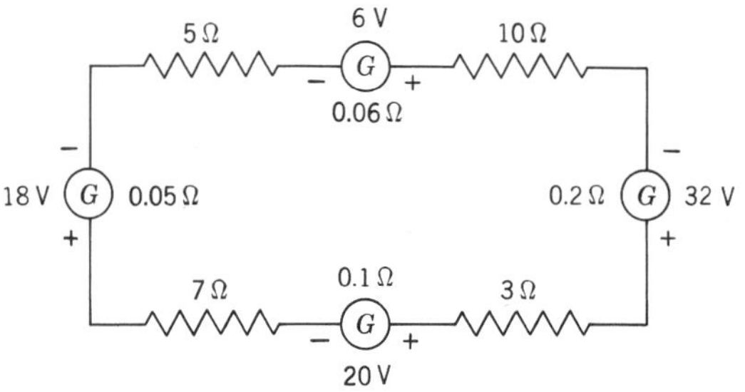

Fig. 22-9 Circuit of Prob. 10

10. How much current flows in the circuit of Fig. 22-9?

22-4 SIMPLE APPLICATIONS OF BOTH LAWS

Although the circuits of the following examples can be solved by Ohm's law, they are included here because you are familiar with such circuits. You will have no trouble in solving circuits that appear to be complicated if you understand the applications of Kirchhoff's laws to simple circuits, for all circuits are combinations of the fundamental series and parallel circuits.

example 2 A generator supplies 7 A to two resistances of 40 and 30 Ω connected in parallel. Neglecting the internal resistance of the generator and the resistance of the connecting wires, find the generator voltage and the current through each resistance.

solution Figure 22-10 is a diagram of the circuit. From our knowledge of parallel circuits, it is evident that the line current I divides at junction c into the branch currents I_1 and I_2. Similarly, I_1 and I_2 combine at junction f to form the line current I. Therefore,

$$I = I_1 + I_2$$

which is the same as

$$I - I_1 - I_2 = 0 \qquad (3)$$

These are algebraic expressions for Kirchhoff's first law and, when used in conjunction with the second law, facilitate solution of circuits.

If we start at point a and go around the circuit in the direction of current flow, the equation for the voltages around path $abcdefa$ is

$$E - 40I_1 = 0$$

$$I_1 = \frac{E}{40} \qquad (4)$$

The equation for the voltages around path $abcghfa$ is

$$E - 30I_2 = 0$$

$$I_2 = \frac{E}{30} \qquad (5)$$

Fig. 22-10 Circuit of Example 2

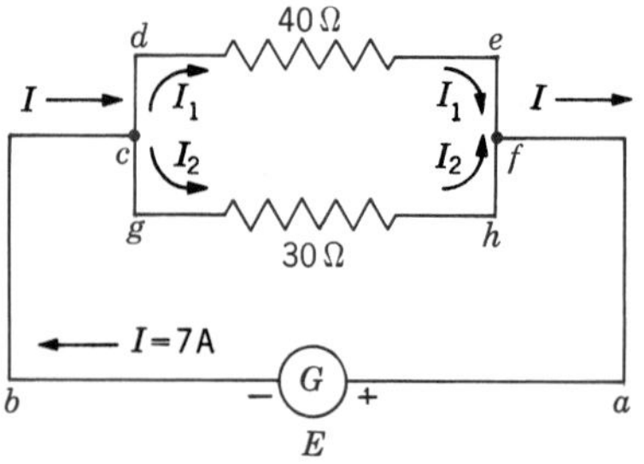

Substituting the known values in Eq. (3),

$$7 - \frac{E}{40} - \frac{E}{30} = 0$$
$$E = 120 \text{ V}$$

$I_1 = 3$ A and $I_2 = 4$ A are found from Eqs. (4) and (5), respectively.

example 3 Two 6-V batteries, each with an internal resistance of 0.05 Ω, are connected in parallel to a load resistance of 9.0 Ω. How much current flows through the load resistance?

solution Figure 22-11 is a diagram of the circuit. In the circuit, two identical sources of emf are connected in parallel to supply the line current I to the load resistance. Again,

$$I = I_1 + I_2$$

or $$I - I_1 - I_2 = 0$$

Starting at junction *a*, the equation for the voltages around path *abcdefa* is

Fig. 22-11 Circuit of Example 3

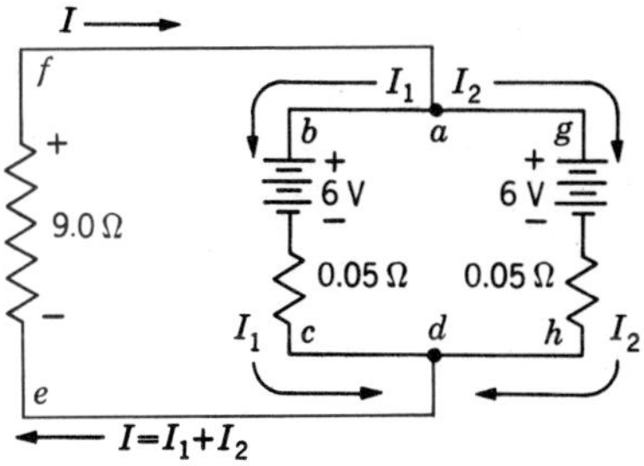

$$6 - 0.05I_1 - 9I = 0$$

Solving for I_1, $\quad I_1 = 120 - 180I \qquad (6)$

Starting at junction *a*, the equation for the voltages around path *aghdefa* is

$$6 - 0.05I_2 - 9I = 0$$

Solving for I_2, $\quad I_2 = 120 - 180I \qquad (7)$

As would be expected, I_1 and I_2 are equal. Substituting the values of I_1 and I_2 in Eq. (3),

$$I - (120 - 180I) - (120 - 180I) = 0$$

Hence, $\quad I = 0.6648$ A

The foregoing solution assumes three unknowns I, I_1, and I_2. However, in writing the equations for the voltages around any path, only two unknowns can be used, for $I = I_1 + I_2$. Thus, around path *abcdefa*,

$$6 - 0.05I_1 - 9(I_1 + I_2) = 0$$

Collecting terms, $\quad 9.05I_1 + 9I_2 = 6 \qquad (8)$

Voltages around path *aghdefa*,

$$6 - 0.05I_2 - 9(I_1 + I_2) = 0$$

Collecting terms, $\quad 9I_1 + 9.05I_2 = 6 \qquad (9)$

Since Eqs. (8) and (9) are simultaneous equations, they can be solved for I_1 and I_2. Hence,

$$I_1 = 0.3324 \text{ A}$$
$$I_2 = 0.3324 \text{ A}$$

and $\quad I = I_1 + I_2 = 0.6648$ A

PROBLEMS 22-2

1. A power supply supplies a total of 1.46 A to two resistors of 75 and 43 Ω connected in parallel. What is the terminal voltage of the power supply?
2. A battery supplies 5.53 A to three resistors of 2 Ω, 2.7 Ω, and 3 Ω connected in parallel. What is the terminal voltage of the battery?
3. A generator with an internal resistance of 0.05 Ω supplies 15.2 A to three

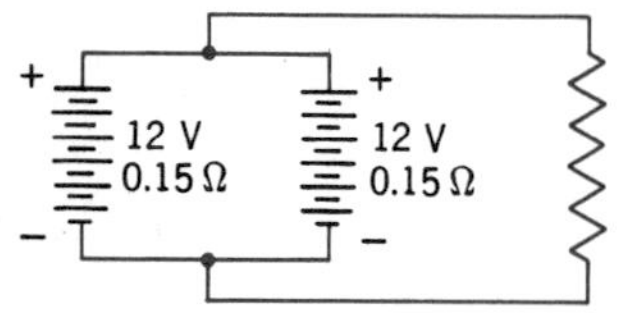

Fig. 22-12 Circuit of Probs. 5 and 6

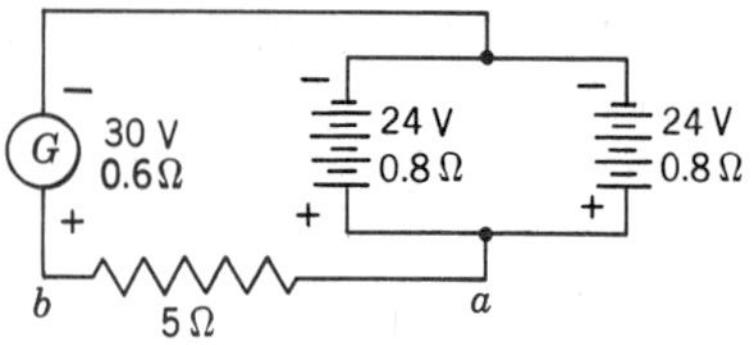

Fig. 22-13 Circuit of Probs. 7 and 8

resistors of 8, 4, and 10 Ω connected in parallel. What is the generator terminal voltage?

4. A battery supplies 9.7 A to four resistors of 110, 50, 100, and 200 Ω connected in parallel. What is the voltage across the resistors?
5. (*a*) What is the value of the current in the circuit of Fig. 22-12?
   (*b*) How much power is expended in each of the batteries?
6. How much power would be expended in each battery in the circuit of Fig. 22-12 if the load resistance were changed from 10 to 0.5 Ω?
7. (*a*) What is the generator current in the circuit of Fig. 22-13?
   (*b*) In what direction does the current flow?
8. (*a*) What is the value of the generator current in the circuit of Fig. 22-13 if the generator emf voltage is decreased to 12 V?
   (*b*) In what direction does the current flow?

22-5 FURTHER APPLICATIONS OF KIRCHHOFF'S LAWS

In preceding examples and problems if two sources of emf have been connected to the same circuit, the values of emf and internal resistance have been equal. However, there are many types of circuits that contain more than one source of power, each with a different emf and different internal resistance.

example 4 Figure 22-14 represents two batteries connected in parallel and supplying current to a resistance of 2 Ω. One battery has an emf of 6 V and an internal resistance of 0.15 Ω, and the other battery has an emf of 5 V and an internal resistance of 0.05 Ω. Determine the current through the batteries and the current in the external circuit. Neglect the resistance of the connecting wires.

Fig. 22-14 Circuit of Example 4

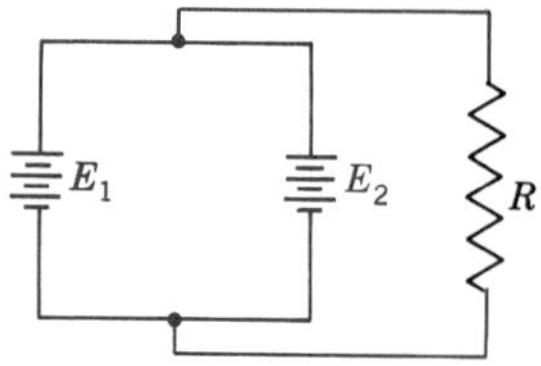

solution Draw a diagram of the circuit representing the internal resistance of the batteries, and label the circuit with all the known values as shown in Fig. 22-15. Label

Fig. 22-15 Circuit of Example 4 Labeled with Known Values

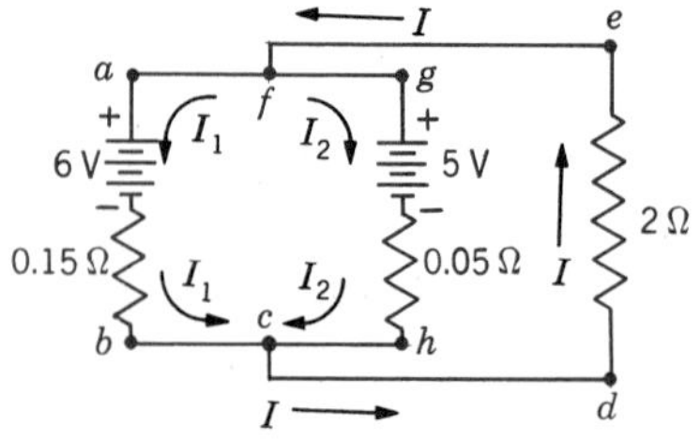

the unknown currents, and mark the direction in which each current is assumed to flow. There are three currents of unknown value in the circuit, I_1, I_2, and the current I which flows through the external circuit. However, because $I = I_1 + I_2$, the unknown currents can be reduced to two unknowns by considering a current of $I_1 + I_2$ A flowing through the external circuit.

For the path *abcdefa*,

$$6 - 0.15I_1 - 2(I_1 + I_2) = 0$$

Collecting terms,

$$2.15I_1 + 2I_2 = 6 \tag{10}$$

For the path *ghcdefg*,

$$5 - 0.05I_2 - 2(I_1 + I_2) = 0$$

Collecting terms,

$$2I_1 + 2.05I_2 = 5 \tag{11}$$

Equations (10) and (11) are simultaneous equations that, when solved, result in

$$I_1 = 5.64 \text{ A}$$

and

$$I_2 = -3.07 \text{ A}$$

The negative sign of the current I_2 denotes that this current is flowing in a direction opposite to that assumed. The value of the line current is

$$I = I_1 + I_2 = 5.64 + (-3.07)$$
$$= 2.57 \text{ A}$$

Try checking this solution by changing the direction of I_2 in Fig. 22-15 and rewriting the voltage equations accordingly, remembering that now, at junction *f*, for example, $I + I_2 - I_1 = 0$. This will demonstrate that it is immaterial which way the arrows point, for the signs preceding the current values, when found, determine whether or not the assumed directions are correct. As previously mentioned, however, it must be remembered that going through a resistance in a direction opposite to the current arrow represents a voltage (rise) which must be added, whereas going through a resistance in the direction of the current arrow represents a voltage (drop) which must be subtracted.

example 5 Figure 22-16 represents a network containing three unequal sources of emf. Find the current flowing in each branch.

solution Assume a direction I_1, I_2, and I_3, and label them as shown in the circuit diagram.

Although three unknown currents are involved, they can be reduced to two unknowns by expressing one current in terms of the other two. This is accomplished by applying Kirchhoff's first law to some junction such as c. By considering current flow toward a junction as positive and that flowing away from a junction as negative,

$$I_1 + I_3 - I_2 = 0$$
$$I_3 = I_2 - I_1 \tag{12}$$

Since there are now only two unknown currents I_1 and I_2, Kirchhoff's second law may be applied to any two different closed loops in the network.

For path *abcda*,

$$4 - 0.1I_1 + 6 - 0.2I_2 - 2I_1 = 0$$

Fig. 22-16 Circuit of Example 5

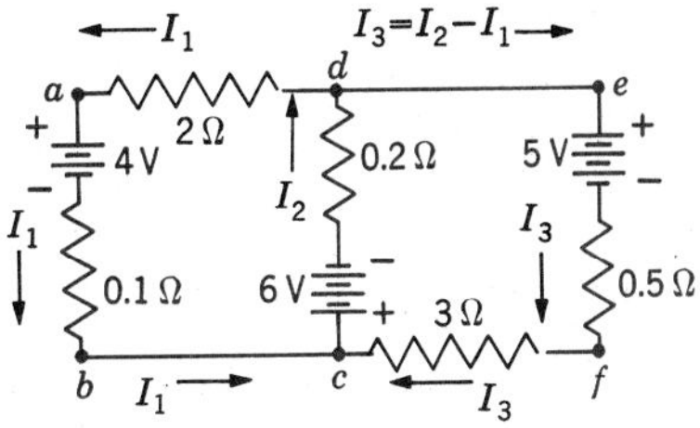

Collecting terms,

$$2.1I_1 + 0.2I_2 = 10 \qquad (13)$$

For path *efcde,*

$$5 - 0.5(I_2 - I_1) - 3(I_2 - I_1) + 6 - 0.2I_2 = 0$$

Collecting terms,

$$3.5I_1 - 3.7I_2 = -11 \qquad (14)$$

Equations (13) and (14) are simultaneous equations that, when solved, result in

$$I_1 = 4.109 \text{ A}$$

and

$$I_2 = 6.860 \text{ A}$$

Substituting in Eq. (12),

$$I_3 = 6.860 - 4.109 = 2.751 \text{ A}$$

The assumed directions of current flow are correct because all values are positive. The solution can be checked by applying Kirchhoff's second law to a path not previously used. When the current values are substituted in the equation for this path an identity should result. Thus, for path *adefcba,*

$$2I_1 + 5 - 0.5(I_2 - I_1) - 3(I_2 - I_1) + 0.1I_1 - 4 = 0$$

Collecting terms,

$$5.6I_1 - 3.5I_2 = -1 \qquad (15)$$

The substitution of the numerical values of I_1 and I_2 in Eq. (15) verifies the solution within reasonable limits of accuracy.

22-6 OUTLINE FOR SOLVING NETWORKS

In common with all other problems, the solution of a circuit or a network should not be started until the conditions are analyzed and it is clearly understood what is to be found. Then a definite procedure should be adopted and followed until the solution is completed.

In order to facilitate solutions of networks by means of Kirchhoff's laws, the following procedure is suggested:

1. Draw a large, neat diagram of the network, and arrange the circuits so that they appear in their simplest form.
2. Letter the diagram with all the known values such as sources of emf, currents, and resistances. Carefully mark the polarities of the known emf's.
3. Assign a symbol to each unknown quantity.
4. Indicate with arrows the assumed direction of current flow in each branch of the network. The number of unknown currents can be reduced by assigning a direction to all but one of the unknown currents at a junction. Then, by Kirchhoff's first law, the remaining current can be expressed in terms of the others.
5. Using Kirchhoff's second law, set up as many equations as there are unknowns to be determined. So that each equation will contain some relation that has not been expressed in another equation, each circuit path followed should cover some part of the circuit not used for other paths.
6. Solve the resulting simultaneous equations for the values of the unknown quantities.
7. Check the values obtained by substituting them in a voltage equation that has been obtained by following a circuit path not previously used.

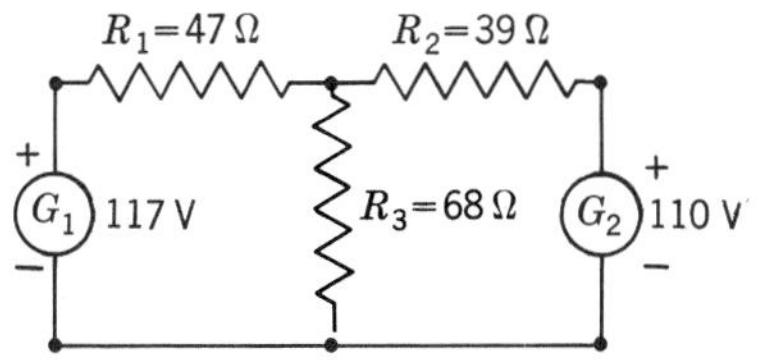

Fig. 22-17 Circuit of Probs. 1 and 2

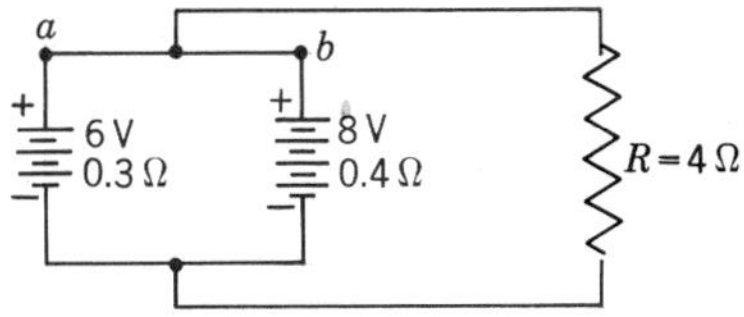

Fig. 22-18 Circuit of Probs. 3 and 4

PROBLEMS 22-3

1. In the circuit of Fig. 22-17, (*a*) how much current flows through R_3 and (*b*) how much power is expended in R_2?
2. In the circuit of Fig. 22-17, R_3 becomes short-circuited.
   (*a*) How much current flows through the short circuit?
   (*b*) How much power is supplied by generator G_1?
3. In the circuit of Fig. 22-18, (*a*) how much current flows through R and (*b*) how much current flows through the batteries when R is open-circuited and in what direction?
4. In the original circuit of Fig. 22-18, R is shunted by a resistor of 1 Ω.
   (*a*) How much power is expended in the shunting resistor?
   (*b*) What is the terminal voltage of the 6-V battery?
5. In the circuit of Fig. 22-19, if the internal resistance of the generator is neglected, (*a*) how much power is being supplied by the generator and (*b*) what is the voltage across R?
6. In the circuit of Fig. 22-19, the generator has an internal resistance of 0.15 Ω. If the connections of the generator are reversed, (*a*) how much power will be dissipated in R and (*b*) what will be the terminal voltage of the 10-V battery?
7. In the circuit of Fig. 22-20, (*a*) how much power is dissipated in R_4 and (*b*) what is the voltage across R_1?
8. If R_1 is short-circuited in the circuit of Fig. 22-20, (*a*) what is the voltage across R_4 and (*b*) how much power is dissipated in the battery?

Fig. 22-19 Circuit of Probs. 5 and 6

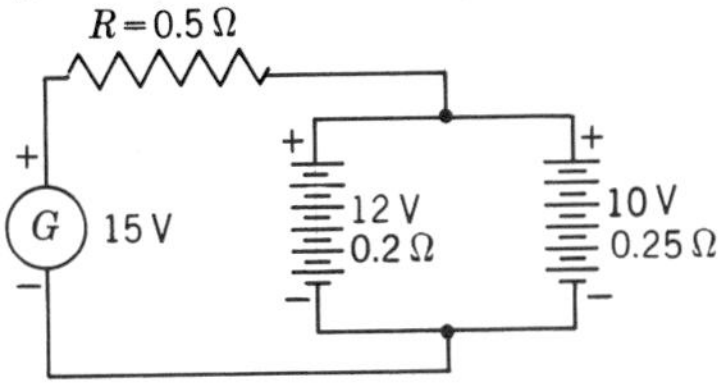

Fig. 22-20 Circuit of Probs. 7 and 8

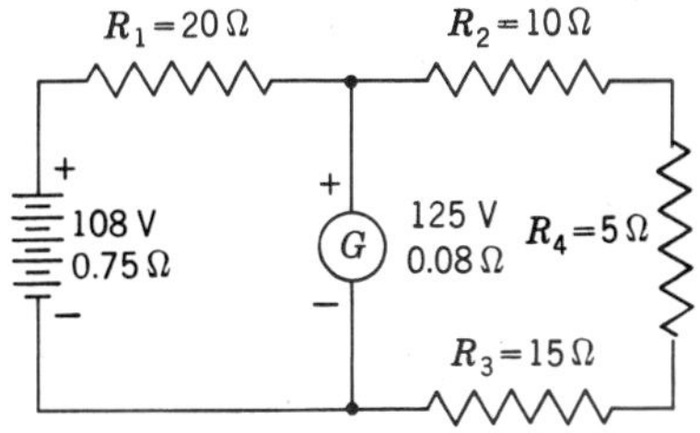

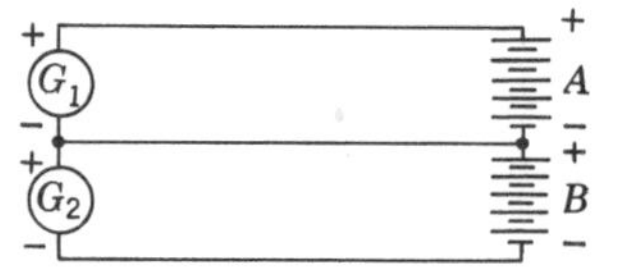

Fig. 22-21 Circuit of Prob. 9

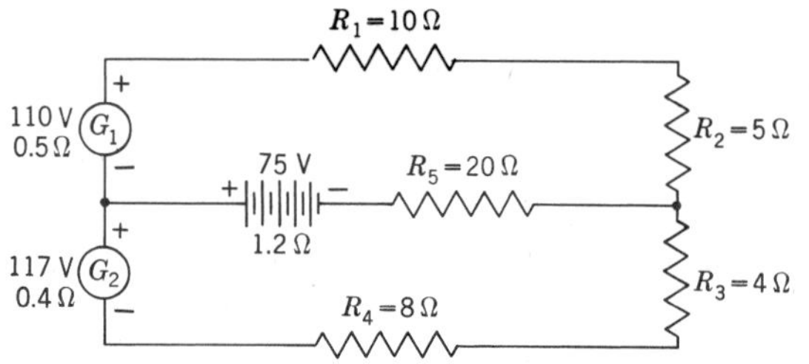

Fig. 22-22 Circuit of Probs. 10 and 11

9. In the circuit of Fig. 22-21, battery A has an emf of 114 V and an internal resistance of 1.5 Ω. Battery B has an emf of 108 V and an internal resistance of 1 Ω. Each generator has an emf of 122 V and an internal resistance of 0.05 Ω. The resistance of each feeder is 0.02 Ω.
(a) How much current flows through battery A?
(b) How much power is expended in battery B?

10. In the circuit of Fig. 22-22, (a) how much power is expended in R_5 and (b) how much power is expended in generator G_2?

11. If the connections of the battery are reversed in Fig. 22-22, (a) what is the voltage across R_5 and (b) how much power is expended in the entire circuit?

12. Figure 22-23 represents a bank of batteries supplying power to loads R_a and R_b, with R_1, R_2, and R_3 representing the lumped line resistance. R_b is disconnected, and R_a draws 50 A. Neglecting the internal resistance of the generator and batteries, (a) what is the voltage across R_2 and (b) how much current is flowing in the batteries and in what direction?

13. R_b is connected in the circuit of Fig. 22-23 and draws 75 A. If R_a draws 50 A, (a) what is the voltage across R_b and (b) how much power is expended in R_2?

14. In the circuit of Fig. 22-23 the loads are adjusted until R_a draws 150 A and R_b draws 25 A. How much power is lost in R_2?

Fig. 22-23 Circuit of Probs. 12, 13, and 14

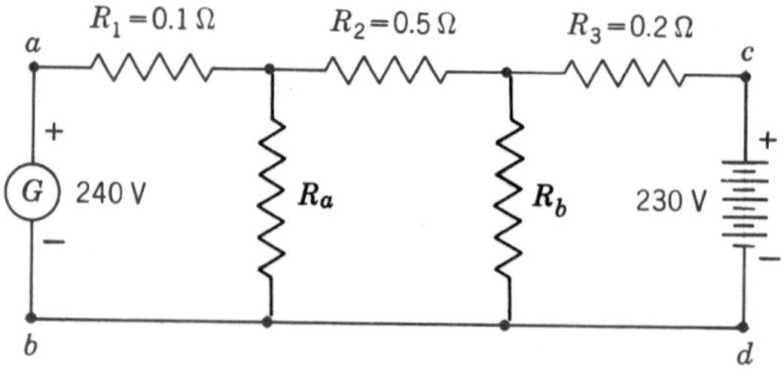

22-7 EQUIVALENT STAR AND DELTA CIRCUITS

example 6 Determine the currents through the branches of the network of Fig. 22-24 and find the equivalent resistance between points *a* and *c*.

solution Assume directions for all the currents, and label them on the figure. By Kirchhoff's second law.

Path *efabce*,

$$3I - 3I_1 + 4I - 4I_2 = 10 \qquad (16)$$

Path *efadce*,

$$2I_1 + 5I_2 = 10 \qquad (17)$$

Path *adba*,

$$2I_1 + 6I_1 - 6I_2 - 3I + 3I_1 = 0 \qquad (18)$$

Collecting like terms,

Equation (16) becomes

$$7I - 3I_1 - 4I_2 = 10 \qquad (19)$$

Equation (17) becomes

$$2I_1 + 5I_2 = 10 \qquad (20)$$

Equation (18) becomes

$$-3I + 11I_1 - 6I_2 = 0 \qquad (21)$$

Equations (19), (20), and (21) permit us to write

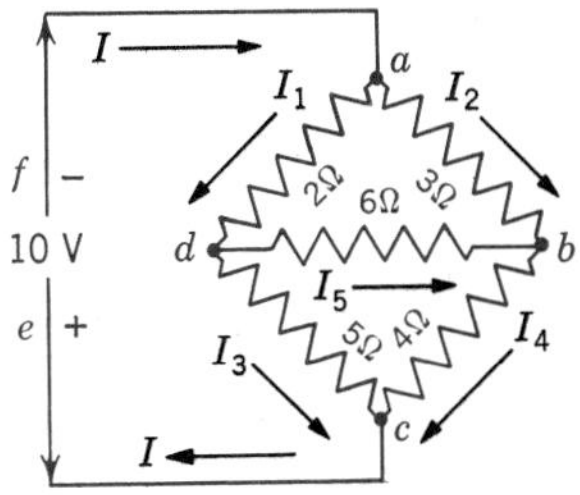

Fig. 22-24 Circuit of Example 6

$$I = \frac{\begin{vmatrix} 10 & -3 & -4 \\ 10 & 2 & 5 \\ 0 & 11 & -6 \end{vmatrix}}{\begin{vmatrix} 7 & -3 & -4 \\ 0 & 2 & 5 \\ -3 & 11 & -6 \end{vmatrix}} = 2.879 \text{ A}$$

The equivalent resistance between points *a* and *c* is

$$\frac{E}{I} = \frac{10}{2.879} = 3.47\ \Omega$$

By expressing the branch currents in terms of other currents and labeling the circuit accordingly, this problem can be solved with a smaller number of equations. This is left as an exercise for you.

You will note, from the solution of Example 5, that the solution by Kirchhoff's laws of networks containing such configurations can become complicated. There are many cases, however, in which such networks can be replaced with more convenient equivalent circuits.

The three resistors R_1, R_2, and R_3 in Fig. 22-25*a* are said to be connected in *delta* (Greek letter Δ). R_a, R_b, and R_c in Fig. 22-25*b* are connected in *star*, or Y.

If these two circuits are to be made equiva-

Fig. 22-25 (a) Resistors Connected in Delta (b) Resistors Connected in Star or Y

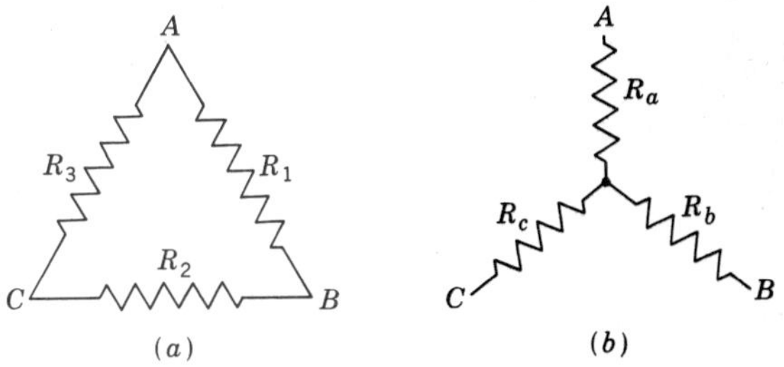

lent, then the resistance between terminals A and B, B and C, and A and C must be the same in each circuit. Hence, in Fig. 22-25a the resistance from A to B is

$$R_{AB} = \frac{R_1(R_2 + R_3)}{R_1 + R_2 + R_3} \tag{22}$$

In Fig. 22-25b the resistance from A to B is

$$R_{AB} = R_a + R_b \tag{23}$$

Equating Eqs. (22) and (23),

$$R_a + R_b = \frac{R_1R_2 + R_1R_3}{R_1 + R_2 + R_3} \tag{24}$$

Similarly,

$$R_b + R_c = \frac{R_1R_2 + R_2R_3}{R_1 + R_2 + R_3} \tag{25}$$

and

$$R_a + R_c = \frac{R_1R_3 + R_2R_3}{R_1 + R_2 + R_3} \tag{26}$$

Equations (24), (25), and (26) are simultaneous and, when solved, result in

$$R_a = \frac{R_1R_3}{R_1 + R_2 + R_3} = \frac{R_1R_3}{\Sigma R_\Delta} \tag{27}$$

$$R_b = \frac{R_1R_2}{R_1 + R_2 + R_3} = \frac{R_1R_2}{\Sigma R_\Delta} \tag{28}$$

and

$$R_c = \frac{R_2R_3}{R_1 + R_2 + R_3} = \frac{R_2R_3}{\Sigma R_\Delta} \tag{29}$$

Since Σ (Greek letter sigma) is used to denote "the summation of,"

$$\Sigma R_\Delta = R_1 + R_2 + R_3$$

example 7 In Fig. 22-25a, $R_1 = 2\ \Omega$, $R_2 = 3\ \Omega$, and $R_3 = 5\ \Omega$. What are the values of the resistances in the equivalent Y circuit of Fig. 22-25b?

solution

$$\Sigma R_\Delta = 2 + 3 + 5 = 10\ \Omega$$

Substituting in Eq. (27),

$$R_a = \frac{2 \times 5}{10} = 1\ \Omega$$

Substituting in Eq. (28),

$$R_b = \frac{2 \times 3}{10} = 0.6\ \Omega$$

Substituting in Eq. (29),

$$R_c = \frac{3 \times 5}{10} = 1.5\ \Omega$$

Fig. 22-26 Circuits of Example 8

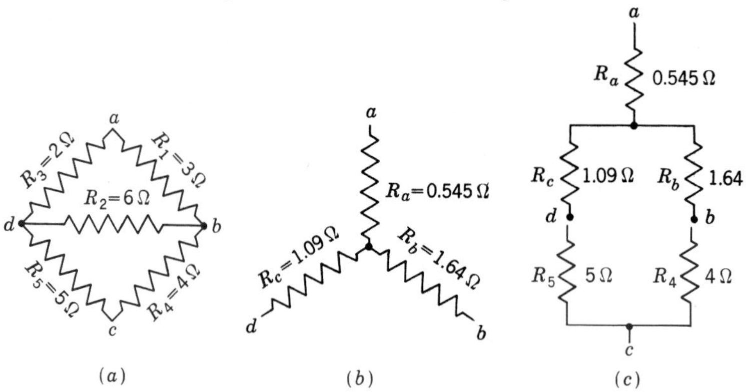

example 8 Determine the equivalent resistance between points *a* and *c* in the circuit of Fig. 22-26*a*.

solution Convert one of the delta circuits of Fig. 22-26*a* to its equivalent Y circuit. Thus, for the delta *abd*,

$$\Sigma R\Delta = 3 + 6 + 2 = 11\ \Omega$$

The equivalent Y resistances, which are shown in Fig. 22-26*b*, are

$$R_a = \frac{3 \times 2}{11} = 0.545\ \Omega$$

$$R_b = \frac{3 \times 6}{11} = 1.64\ \Omega$$

and $$R_c = \frac{2 \times 6}{11} = 1.09\ \Omega$$

The equivalent Y circuit is connected to the remainder of the network as shown in Fig. 22-26*c* and is solved as an ordinary series-parallel combination. Thus,

$$R_{ac} = R_a + \frac{(R_c + R_5)(R_b + R_4)}{R_c + R_5 + R_b + R_4}$$

$$= 0.545 + \frac{(1.09 + 5)(1.64 + 4)}{1.09 + 5 + 1.64 + 4}$$

$$= 3.47\ \Omega$$

Note that the values of Fig. 22-26 are the same as those of Fig. 22-24.

The equations for converting a Y circuit to its equivalent delta circuit are obtained by solving Eqs. (27), (28), and (29) simultaneously. This results in

$$R_1 = \frac{\Sigma R_Y}{R_c} \qquad (30)$$

$$R_2 = \frac{\Sigma R_Y}{R_a} \qquad (31)$$

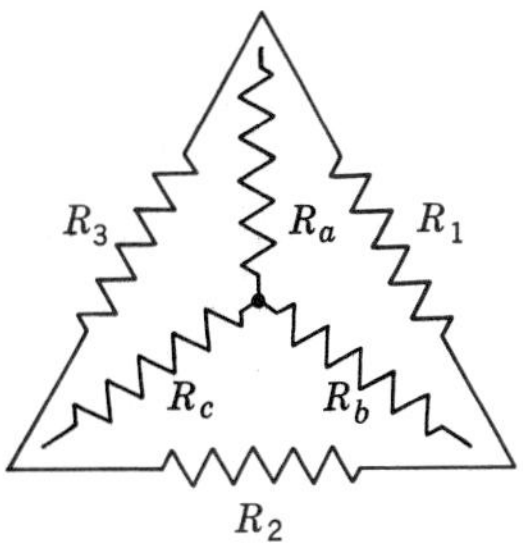

Fig. 22-27 Resistance Equivalents

$$R_3 = \frac{\Sigma R_Y}{R_b} \qquad (32)$$

where $\Sigma R_Y = R_aR_b + R_bR_c + R_aR_c$

A convenient method for remembering the Δ to Y and Y to Δ conversions is illustrated in Fig. 22-27.

In converting from Δ to Y, each equivalent Y resistance is equal to the product of the two *adjacent* Δ resistances divided by the sum of the Δ resistances. For example, R_1 and R_3 are adjacent to R_a; therefore,

$$R_a = \frac{R_1R_3}{\Sigma R_\Delta}$$

In converting from Y to Δ, each equivalent Δ resistance is found by dividing ΣR_Y by the *opposite* Y resistance. For example, R_1 is opposite R_c; therefore,

$$R_1 = \frac{\Sigma R_Y}{R_c}$$

note A comparison of the Y network of Fig. 22-25*b* with the network formed by R_1, R_2, and R_3 of Fig. 13-9 will show the common interchangeability of the names T and Y and π (pi) and Δ (delta) in electronics circuitry.

PROBLEMS 22-4

1. In the Δ circuit of Fig. 22-25*a*, $R_1 = 12\ \Omega$, $R_2 = 15\ \Omega$, and $R_3 = 18\ \Omega$. Determine the resistances of the equivalent Y circuit.

2. In the Δ circuit of Fig. 22-25*a*, $R_1 = 120\ \Omega$, $R_2 = 240\ \Omega$, and $R_3 = 300\ \Omega$. Determine the resistances of the equivalent Y circuit.

3. In the π circuit of Fig. 22-25*a*, $R_1 = R_2 = R_3 = 500\ \Omega$. Determine the resistances of the equivalent T circuit.

4. In the Y circuit of Fig. 22-25*b*, $R_a = 8\ \Omega$, $R_b = 16\ \Omega$, and $R_c = 40\ \Omega$. Determine the resistances of the equivalent Δ circuit.

5. In the T circuit of Fig. 22-25*b*, $R_a = 4.7\ \text{k}\Omega$, $R_b = 3.3\ \text{k}\Omega$, and $R_c = 1.8\ \text{k}\Omega$. Determine the resistances of the equivalent π circuit.

6. In the T circuit of Fig. 22-25*b*, $R_a = R_b = R_c = 1.5\ \text{k}\Omega$. Determine the resistances of the equivalent π circuit.

In Probs. 7 to 17, solve the circuits by both the Δ to Y conversion and Kirchhoff's laws:

7. In the circuit of Fig. 22-28, $R_1 = 20\ \Omega$, $R_2 = 10\ \Omega$, $R_3 = 45\ \Omega$, $R_4 = 12\ \Omega$, $R_5 = 15\ \Omega$, and $E = 1.5$ V. What is the value of I?

8. How much current flows through R_5 of Prob. 7?

9. How much current is flowing through R_2 of Prob. 7?

10. In the circuit of Fig. 22-28, $R_1 = 25\ \Omega$, $R_2 = 10\ \Omega$, $R_3 = 15\ \Omega$, $R_4 = 50\ \Omega$, $R_5 = 30\ \Omega$, and $E = 50$ V. What is the value of I?

11. How much current is flowing through R_2 of Prob. 10?

12. In the circuit of Fig. 22-28, $R_1 = ?$, $R_2 = 10\ \Omega$, $R_3 = 15\ \Omega$, $R_4 = 12\ \Omega$, $R_5 = 8\ \Omega$, $E = 32$ V, and $I = 2.39$ A. What is the resistance of R_1?

13. Determine the value of the current I in Fig. 22-29 if $E = 100$ V.

14. How much current flows through R_4 of Prob. 13?

15. How much current flows through R_5 of Prob. 13?

16. How much current flows through the load resistance R_L in Fig. 22-30?

17. How much current does the signal generator G supply to the circuit of Fig. 22-31?

Fig. 22-28 Circuit of Probs. 7 to 12

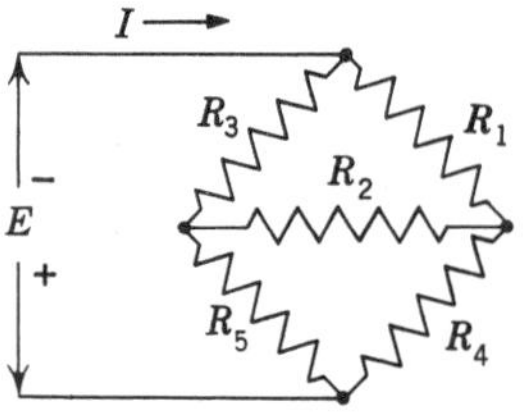

Fig. 22-29 Circuit of Probs. 13, 14, and 15

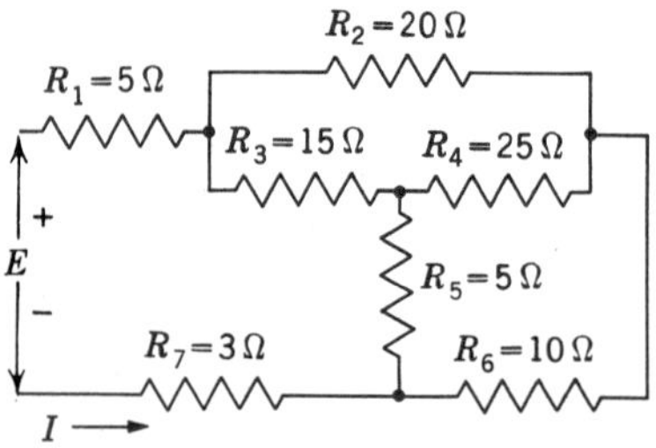

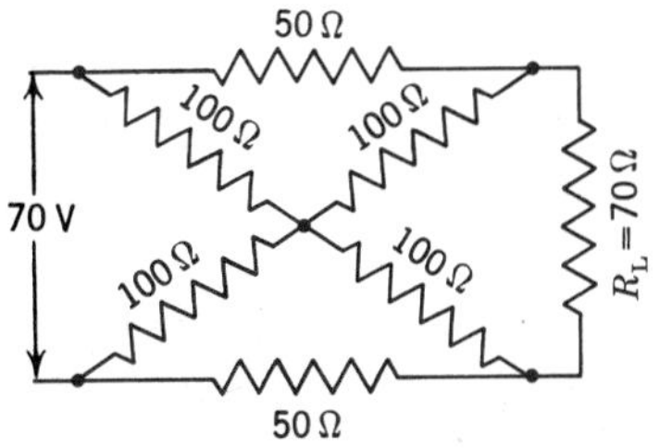

Fig. 22-30 Circuit of Prob. 16

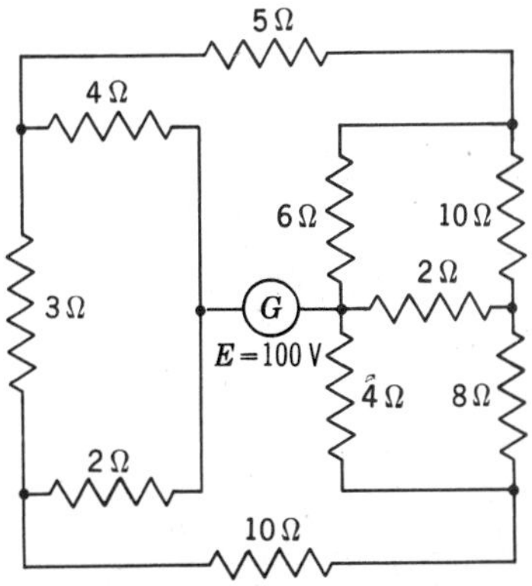

Fig. 22-31 Circuit of Prob. 17

22-8 THEVENIN AND NORTON EQUIVALENTS

Often a knowledge of the actual components inside a power supply circuit is immaterial so long as we can measure the open-circuit output voltage and the short-circuit output current. From these easy measurements, we can picture a model of the circuit which will behave in exactly the same way as the original so far as any external connected circuit is concerned. Figure 22-32 illustrates this idea.

A voltmeter with extremely high resistance can make a reasonable measurement of the open-circuit emf which the power supply generates. And an ammeter with very low resistance can make a reasonable measurement of the short-circuit current which the power supply can deliver. Two such models of power supplies are available to us:

Thevenin's theorem suggests that the power supply of Fig. 22-32 can be pictured as consisting of a simple equivalent source of constant emf E_{Th} in series with an equivalent resistance R_{Th}. Figure 22-33 shows the Thevenin equivalent of the circuit of Fig. 22-32. Obviously,

$$E_{\text{Th}} = E_0 \tag{33}$$

$$R_{\text{Th}} = \frac{E_0}{I_{\text{sc}}} \tag{34}$$

$$I_L = \frac{E_{\text{Th}}}{R_{\text{Th}} + R_L} \tag{35}$$

Fig. 22-32(*a*) Any Power Supply Will Deliver a Particular Open-Circuit (No Load) EMF, E_0, Which May Be Measured by an Infinite Resistance Voltmeter
(*b*) Any Power Supply Will Deliver a Short-Circuit (Maximum Load) Current, I_{sc}, Which May Be Measured by a Zero Resistance Ammeter

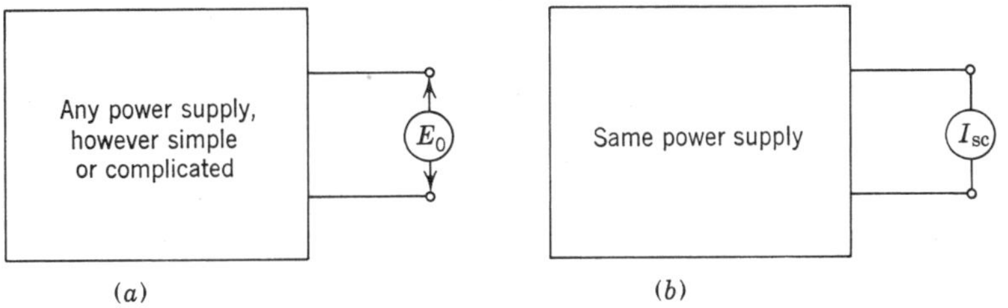

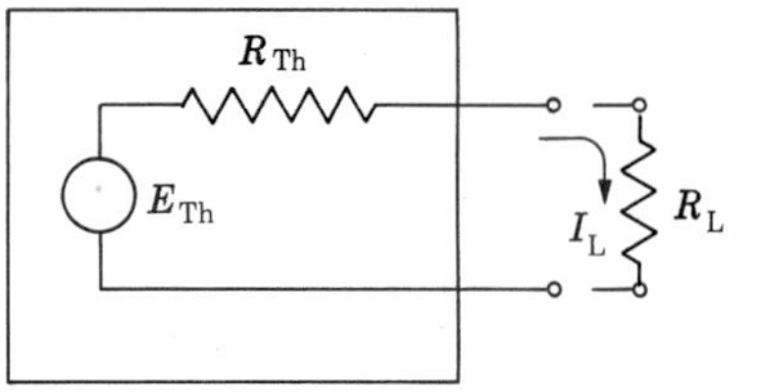

Fig. 22-33 Thevenin's Equivalent of Power Supply of Fig. 22-32; That Is, a Source of Constant EMF E_{Th} in Series with Internal Resistance R_{Th}

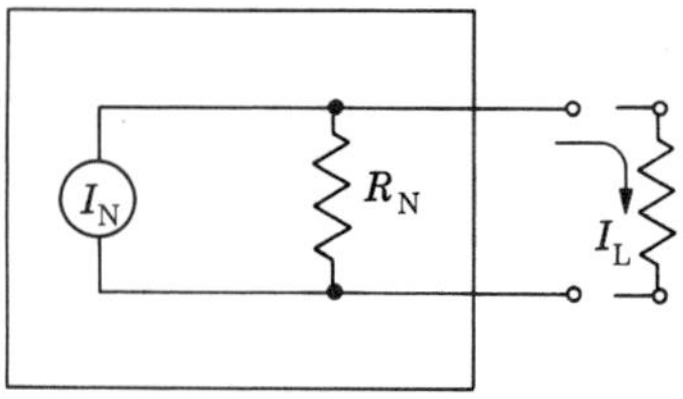

Fig. 22-34 Norton's Equivalent of Power Supply of Fig. 22-32; That Is, a Source of Constant Current I_N in Parallel with Internal Resistance R_N

Norton's theorem suggests that the power supply of Fig. 22-32 can be pictured as consisting of a simple equivalent source of constant current I_N in parallel with an equivalent resistance R_N. Figure 22-34 shows the Norton equivalent of the circuit of Fig. 22-32. You can see that

$$R_N = R_{Th} \tag{36}$$

$$I_N = \frac{E_{Th}}{R_{Th}} \tag{37}$$

$$I_L = \frac{R_N}{R_N + R_L} \cdot I_N \tag{38}$$

You should apply your knowledge of parallel resistances to prove Eq. (38).

The solution to network problems may sometimes be simplified by applying one or the other of these two theorems.

example 9 Use Thevenin's theorem to solve the current flow through the load resistor R in Fig. 22-14. $E_1 = 6$ V with an internal resistance of 0.15 Ω, $E_2 = 5$ V with an internal resistance of 0.05 Ω, and $R = 2$ Ω.

solution Redraw the figure to show R as the load to be connected and the rest of the circuit as a power supply (Fig. 22-35).

Fig. 22-35(*a*) Redrawn from Fig. 22-14 for Thevenin's Solution
(*b*) Redrawn from Fig. 22-14 for Thevenin's Equivalent Circuit
(*c*) Redrawn from Fig. 22-14 for Norton's Equivalent Circuit

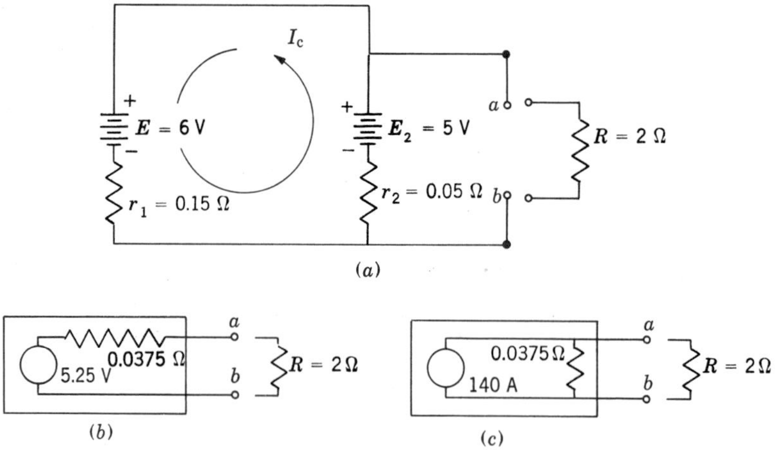

Determine the open-circuit voltage which would appear across the terminals *ab* of Fig. 22-35. A high-resistance voltmeter would measure

$$E_{Th} = E_0 = E_2 + I_c r_2$$

The circulating current I_c is found by applying Ohm's law to the internal circuit:

$$I_c = \frac{6 - 5}{0.15 + 0.05} = \frac{1}{0.20} = 5 \text{ A}$$

and $E_{Th} = 5 + 5(0.05) = 5 + 0.25$
$= 5.25 \text{ V}$

Determine the circuit resistance which would be seen by an ohmmeter connected to terminals *ab* with the sources of emf shorted and represented by their internal resistances. Under such circumstances, an ohmmeter would see r_1 and r_2 in parallel:

$$R_{Th} = \frac{0.15 \times 0.05}{0.15 + 0.05} = 0.0375\ \Omega$$

Thus, the Thevenin equivalent circuit (Fig. 22-35*b*) is a constant source of 5.25 V in series with 0.0375 Ω. Then the current through the 2-Ω "load" is

$$I_R = \frac{5.25}{2.0375} = 2.58 \text{ A}$$

(Compare with Example 4, Sec. 22-5.)

example 10 Solve the same problem by using Norton's theorem.

solution As before, determine the equivalent internal resistance of the "power supply" as seen by a connected load:

$$R_N = R_{Th} = 0.0375\ \Omega$$

Then determine the current which the "power supply" would drive through a short circuit across terminals *ab*. This may be done by using a Thevenin open-circuit approach and finding

$$I_N = \frac{E_{Th}}{R_{Th}} = \frac{5.25}{0.0375} = 140 \text{ A}$$

from which

$$I_R = \frac{0.0375}{2.0375} \times 140 = 2.58 \text{ A}$$

Alternatively, determine from first principles what the short-circuit current through *ab* would be. Figure 22-36 shows this approach. Using Kirchhoff's laws,

$$0.15(I_{sc} + I_5) + 0.05I_5 = 1$$
$$0.015(I_{sc} + I_5) = 6$$

from which $I_{sc} = 140 \text{ A}$
and $I_R = 2.58 \text{ A}$

Fig. 22-36 Determination of $I_N = I_{SC}$ for Fig. 22-35

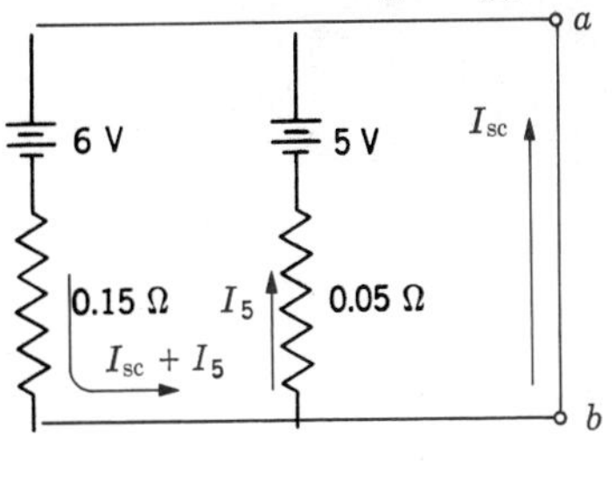

22-9 OUTLINE FOR THEVENIN AND NORTON SOLUTIONS

The following systematic procedure will simplify the utilization of these two circuit simplification theorems:

1. Determine the *leg* of a circuit through which the current flow is to be determined and redraw the circuit, omitting that part.
2. The balance of the circuit is considered

a power supply whose terminals are eventually to deliver current to the part omitted. Often it is helpful to letter all connecting points in the original circuit to make sure that the equivalent has been drawn correctly.

3. Determine the voltage which would be indicated by a voltmeter connected across the open-circuit terminals of the "power supply." This is E_{Th}.
4. Short-circuit all the internal sources of emf, leaving them represented by their internal resistances, and determine the resistance which would be indicated by an ohmmeter connected across the open-circuit terminals of the power supply. This is $R_{\text{Th}} = R_{\text{N}}$.
5. Determine $I_{\text{N}} = \dfrac{E_{\text{Th}}}{R_{\text{Th}}}$, or
6. Determine the value of I_{N} as the current which the power supply would drive through an ammeter connected across its terminals.
7. Use Eq. (35) or (38) to determine the current flow through the reconnected "load."

example 11 Determine the current I_5 through the 6-Ω resistor of Fig. 22-24.

solution Redraw the circuit. Omit the 6-Ω bridging resistor and let the balance of the circuit be a power supply which will later serve the 6-Ω load (Fig. 22-37).
When terminals *bd* are open-circuited, the 10-V source will drive currents I_a and I_b through the power supply internal circuitry, thereby producing voltage drops across the 4- and 5-Ω resistors with the polarities indicated:

$$I_a = \frac{10}{7} = 1.43\text{ A}$$

$$I_b = \frac{10}{7} = 1.43\text{ A}$$

$$V_4 = 1.43 \times 4 = 5.72\text{ V}$$
$$V_5 = 1.43 \times 5 = 7.15\text{ V}$$

A voltmeter across terminals *bd* will measure

$$E_{\text{Th}} = 7.15 - 5.72 = 1.43\text{ V}$$

When the 10-V internal source is shorted, its internal resistance being zero, an ohmmeter across terminals *bd* will measure

$$R_{\text{Th}} = \frac{3 \times 4}{3 + 4} + \frac{2 \times 5}{2 + 5}$$
$$= 1.715 + 1.43$$
$$= 3.145\ \Omega$$

Fig. 22-37(*a*) Redrawn from Fig. 22-24 for Thevenin's Solution of Current I_S Through Resistor across Points *bd*
(*b*) Thevenin's Equivalent Circuit for (*a*)

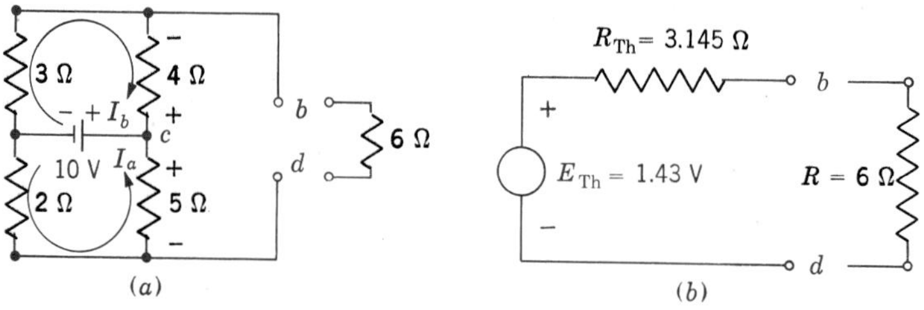

Then the Thevenin equivalent to the power supply is a constant 1.43 V in series with 3.145 Ω.

$$I_5 = \frac{1.43}{6 + 3.145} = 156 \text{ mA}$$

(Compare with Example 6, Sec. 22-7.)

PROBLEMS 22-5

1. A power supply delivers an open-circuit emf of 120 V. An ammeter connected across its terminals measures a short-circuit current of 150 A.
   (*a*) What is the Thevenin circuit equivalent to the power supply so far as any connected load is concerned?
   (*b*) What is the Norton equivalent to the power supply?

2. A power supply delivers an open-circuit emf of 6 V. An ammeter connected across its terminals measures a short-circuit current of 220 mA.
   (*a*) What is the Thevenin equivalent circuit?
   (*b*) What is the Norton equivalent circuit?

3. What is the Thevenin equivalent circuit of the "power supply" portion of the circuit of Fig. 22-28 for the solution of the current through R_2 if $R_1 = 20\ \Omega$, $R_2 = 10\ \Omega$, $R_3 = 45\ \Omega$, $R_4 = 12\ \Omega$, $R_5 = 15\ \Omega$, and $E = 1.5$ V? What is the current flow through R_2?

4. What is the Thevenin equivalent circuit of the power supply portion of the circuit of Fig. 22-28 for the solution of the current through R_2 if $R_1 = 25\ \Omega$, $R_2 = 10\ \Omega$, $R_3 = 15\ \Omega$, $R_4 = 50\ \Omega$, $R_5 = 30\ \Omega$, and $E = 50$ V? What is the current through R_3?

5. What is the Norton equivalent circuit of the power supply portion of the circuit of Prob. 3 for the solution of the current through R_5? What is the current through R_5?

6. What is the Thevenin equivalent circuit of the power supply portion of the circuit of Prob. 4 for the solution of the current through R_1? What is the current through R_1?

23

angles

This chapter deals with the study of angles as an introduction to the branch of mathematics called *trigonometry*. The word "trigonometry" is derived from two Greek words meaning "measurement" or "solution" of triangles.

Trigonometry is both algebraic and geometric in nature. It is not confined to the solution of triangles but forms a basis for more advanced subjects in mathematics. A knowledge of the subject paves the way for a clear understanding of ac and related circuits.

23-1 ANGLES

In trigonometry, we are concerned primarily with the many relations that exist among the sides and angles of triangles. In order to understand the meaning and measurement of angles, it is essential that you thoroughly understand these correlations.

Fig. 23-1 Formation of Angles

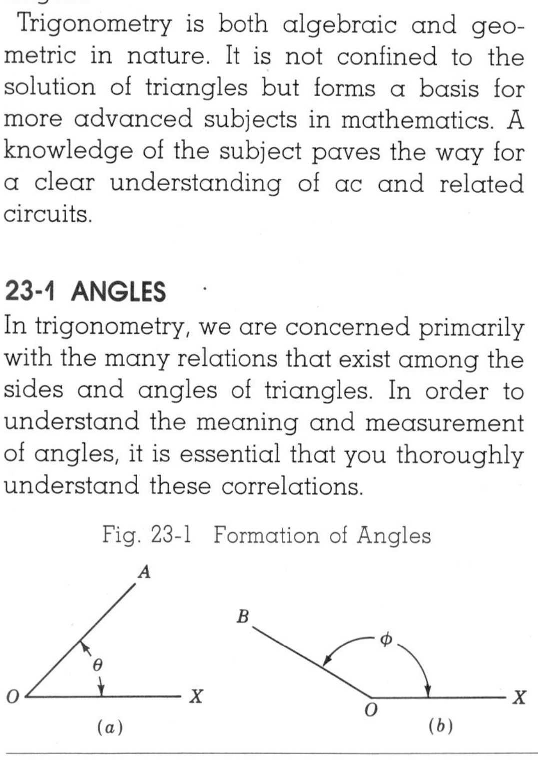

An angle is formed when two straight lines meet at a point. In Fig. 23-1*a*, lines *OA* and *OX* meet at the point O to form the angle *AOX*. Similarly, in Fig. 23-1*b*, the angle *BOX* is formed by lines *OB* and *OX* meeting at the point O. This point is called the *vertex* of the angle, and the two lines are called the *sides* of the angle. The size, or magnitude, of an angle is a measure of the difference in directions of the sides. Thus, in Fig. 23-1, angle *BOX* is a larger angle than *AOX*. The lengths of the sides of an angle have no bearing on the size of the angle.

In geometry it is customary to denote an angle by the symbol $\angle$. If this notation is used, "angle *AOX*" would be written $\angle AOX$.

An angle is also denoted by the letter at the vertex or by a supplementary letter placed inside the angle. Thus, angle *AOX* is correctly denoted by $\angle AOX$, $\angle O$, or $\angle \theta$. Also, *BOX* could be written $\angle BOX$, $\angle O$, or $\angle \phi$.

If equal angles are formed when one straight line intersects another, the angles are called *right angles*. In Fig. 23-2, angles *XOY*, ϕ, *X′OY′*, and α are all right angles.

An *acute angle* is an angle that is less than a right angle. In Fig. 23-3*a*, $\angle \alpha$ is an acute angle.

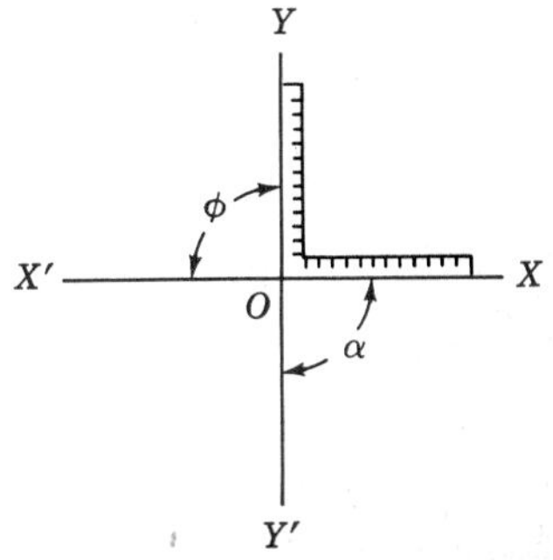

Fig. 23-2 Right Angles

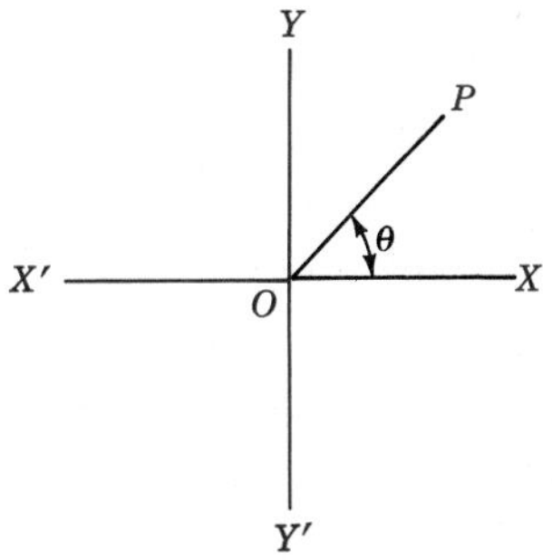

Fig. 23-4 Angle θ in Standard Position

An *obtuse angle* is an angle that is greater than a right angle. In Fig. 23-3*b*, $\angle \beta$ is an obtuse angle.

Two angles whose sum is one right angle are called *complementary* angles. Either one is said to be the "complement" of the other. Thus, in Fig. 23-3*c*, angles ϕ and θ are complementary angles; ϕ is the complement of θ, and θ is the complement of ϕ.

Two angles whose sum is two right angles (a straight line) are called *supplementary angles*. Either one is said to be the *supplement* of the other. Thus, in Fig. 23-3*d*, angles *b* and *a* are supplementary angles; *b* is the supplement of *a*, and *a* is the supplement of *b*.

23-2 GENERATION OF ANGLES

In the study of trigonometry, it becomes necessary to extend our concept of angles beyond the geometric definitions stated in Sec. 23-1. An angle should be thought of as being generated by a line (line segment or half ray) that starts in a certain initial position and rotates about a point called the *vertex* of the angle until it stops at its final position. The original position of the rotating line is called the *initial side* of the angle, and the final position is called the *terminal side* of the angle.

An angle is said to be in *standard position* when its vertex is at the origin of a system of rectangular coordinates and its initial side extends in the positive direction along the *x* axis. Thus, in Fig. 23-4, the angle θ is in standard position. The vertex is at the origin, and the initial side is on the positive *x* axis. The angle has been generated by the line *OP* revolving, or sweeping, from *OX* to its final position.

An angle is called a *positive angle* if it is generated by a line revolving counterclockwise. If the generating line revolves clockwise, the angle is called a *negative angle.*

Fig. 23-3 (a) Acute Angle, (b) Obtuse Angle, (c) Complementary Angles, (d) Supplementary Angles

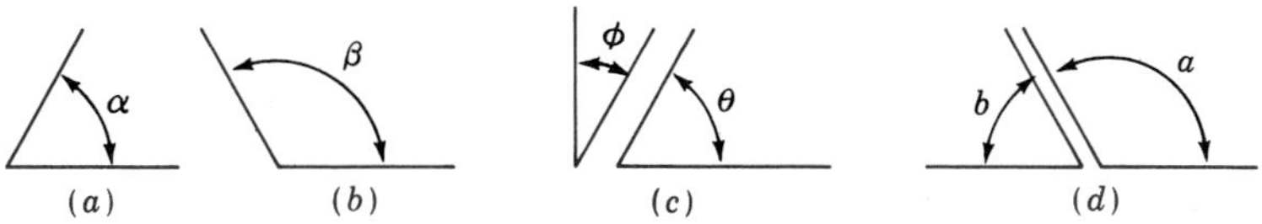

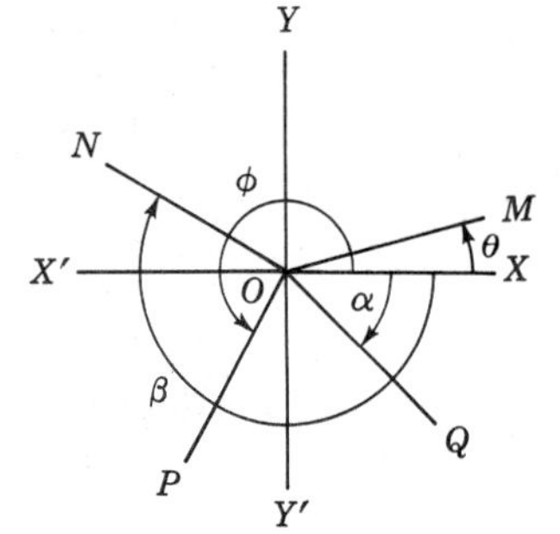

Fig. 23-5 Generation of Angles

In Fig. 23-5, all angles are in standard position. θ is a positive angle that was generated by the line *OM* revolving counterclockwise from *OX*. ϕ is also a positive angle whose terminal side is *OP*. α is a negative angle that was generated by the line *OQ* revolving in a clockwise direction from the initial side *OX*. β is also a negative angle whose terminal side is *ON*.

If the terminal side of an angle that is in standard position lies in the first quadrant, then that angle is said to be *an angle in the first quadrant*, etc. Thus, θ in Fig. 23-4 and θ in Fig. 23-5 are in the first quadrant. Similarly, in Fig. 23-5, β is in the second quadrant, ϕ is in the third quadrant, and α is in the fourth quadrant.

23-3 THE SEXAGESIMAL SYSTEM

There are several systems of angular measurement. The three most commonly used are the right angle, the circular (or natural) system, and the sexagesimal system. The right angle is almost always used as a unit of angular measure in plane geometry and is constantly used by builders, surveyors, etc. However, for the purposes of trigonometry, it is an inconvenient unit because of its large size.

The unit most commonly used in trigonometry is the *degree*, which is one-ninetieth

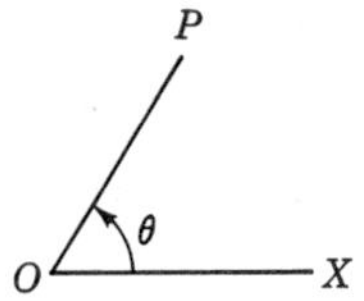

Fig. 23-6 Angle to Be Measured

of a right angle. The degree is defined as the angle formed by one three hundred sixtieth part of a revolution of the angle-generating line. The degree is divided into 60 equal parts called *minutes*, and the minute into 60 equal parts called *seconds*. The word "sexagesimal" is derived from a Latin word pertaining to the number 60.

Instead of dividing the degrees into minutes and seconds, we shall divide them decimally for convenience. For example, instead of expressing an angle of 43 degrees 36 minutes as 43°36′, we write 43.6°.

The actual measurement of an angle consists in finding how many degrees and a decimal part of a degree there are in the angle. This can be accomplished with a fair degree of accuracy by means of a *protractor*, which is an instrument for measuring or constructing angles.

To measure an angle *XOP*, as in Fig. 23-6, place the center of the protractor indicated by O at the vertex of the angle with, say, the line *OX* coinciding with one edge of the protractor as shown in Fig. 23-7. The magnitude

Fig. 23-7 Using Protractor to Measure Angle *XOP* of Fig. 23-6

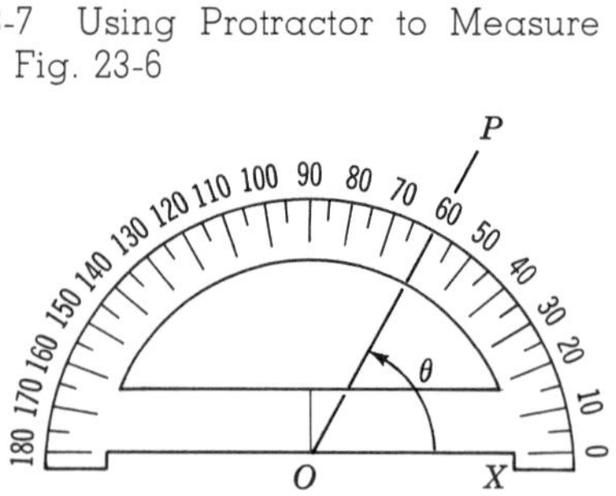

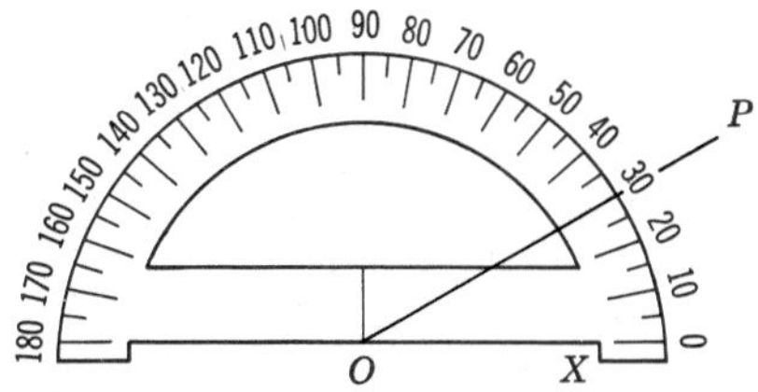

Fig. 23-8 Using Protractor to Construct Angle

of the angle, which is 60°, is indicated where the line *OP* crosses the graduated scale.

To construct an angle, say 30° from a given line *OX*, place the center of the protractor on the vertex O. Pivot the protractor about this point until *OX* is on a line with the 0° mark on the scale. In this position, 30° on the scale now marks the terminal side *OP* as shown in Figs. 23-8 and 23-9.

23-4 ANGLES OF ANY MAGNITUDE

In the study of trigonometry, it will be necessary to extend our concept of angles in order to include angles greater than 360°, either

Fig. 23-9 30° Angle Constructed by Protractor in Fig. 23-8

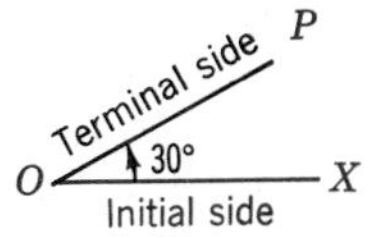

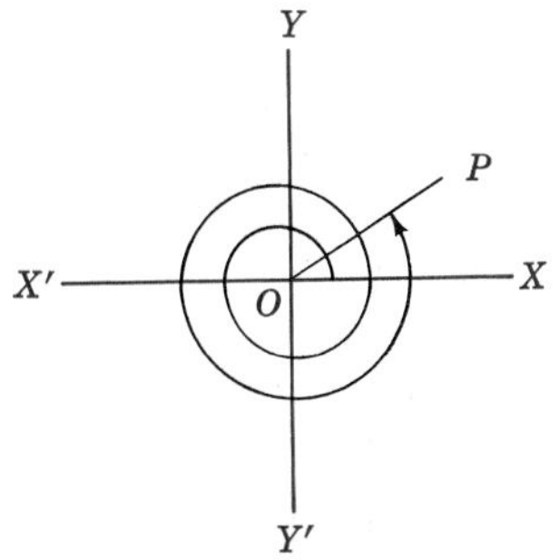

Fig. 23-10 Generation of 750° Angle

positive or negative. Thinking of an angle being generated, as explained in Sec. 23-2, permits consideration of angles of any size, for the generating line can rotate from its initial position in a positive or negative direction so as to produce an angle of any size, even greater than 360°. Figure 23-10 illustrates how an angle of +750° is generated. However, for the purpose of ordinary computation, we consider such an angle to be in the same quadrant as its terminal side with a magnitude equal to the remainder after the largest multiple of 360° it will contain has been subtracted from it. Thus, in Fig. 23-10, the angle is in the first quadrant and, geometrically, is equal to $750° - 720° = 30°$.

PROBLEMS 23-1

1. What is the complement of (*a*) 68°, (*b*) 23°, (c) 41°, (*d*) 170°, (e) 255°, (*f*) −10°?
2. What is the supplement of (*a*) 75°, (*b*) 153°, (c) 258°, (*d*) 270°, (e) 350°, (*f*) −150°?
3. Construct two complementary angles each in standard position on the same pair of axes.
4. Construct two supplementary angles each in standard position on the same pair of axes.
5. By using a protractor, construct the following angles and place them in standard position on rectangular coordinates. Indicate by arrows the di-

rection and amount of rotation necessary to generate these angles: (*a*) 45°, (*b*) 160°, (*c*) 220°, (*d*) 315°, (*e*) 405°, (*f*) −60°, (*g*) −315°, (*h*) −300°, (*i*) −390°, (*j*) −850°.

6. Through how many degrees does the minute hand of a clock turn in (*a*) 20 min, (*b*) 40 min?

7. Through how many right angles does the minute hand of a clock turn from 10:30 A.M. to 5:00 P.M. of the same day?

8. Through how many degrees per minute do (*a*) the second hand, (*b*) the minute hand, (*c*) the hour hand of a clock rotate?

9. A motor armature has a speed of 3600 rev/min. What is the angular velocity (speed) in degrees per second?

10. The shaft of the motor armature in Prob. 9 is directly connected to a pulley 300 mm in diameter. What is the pulley rim speed in meters per second?

23-5 THE CIRCULAR, OR NATURAL, SYSTEM

The circular, or natural, system of angular measurement is sometimes called *radian measure* or *π measure*. The unit of measure is the *radian*. [In this book the abbreviation for *radian* is "rad" when used with units (0.55 rad/s); but an angle of 0.55 radian is written symbolically with a Roman superscript "r" (0.55^{r}) to parallel the use of the degree symbol (288°).]

A radian is an angle that, when placed with its vertex at the center of a circle, intercepts an arc equal in length to the radius of the circle. Thus, in Fig. 23-11, if the length of the arc *AP* equals the radius of the circle, then angle *AOP* is equal to one radian. Figure 23-12 shows a circle divided into radians.

The circular system of measure is used extensively in electrical and electronics formulas and is almost universally used in the higher branches of mathematics.

From geometry, it is known that the circumference of a circle is given by the relation

$$C = 2\pi r \qquad (1)$$

where *r* is the radius of the circle. Dividing both sides of Eq. (1) by *r*, we have

$$\frac{C}{r} = 2\pi \qquad (2)$$

Fig. 23-11 Angle $AOP = 1^{\mathrm{r}}$

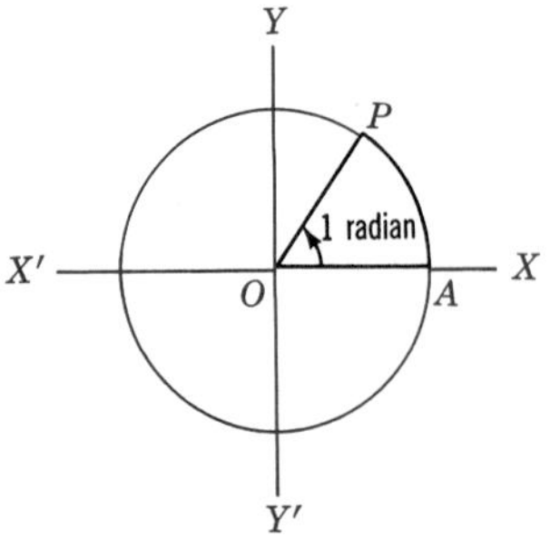

Fig. 23-12 Circle Divided into $2\pi^{\mathrm{r}}$

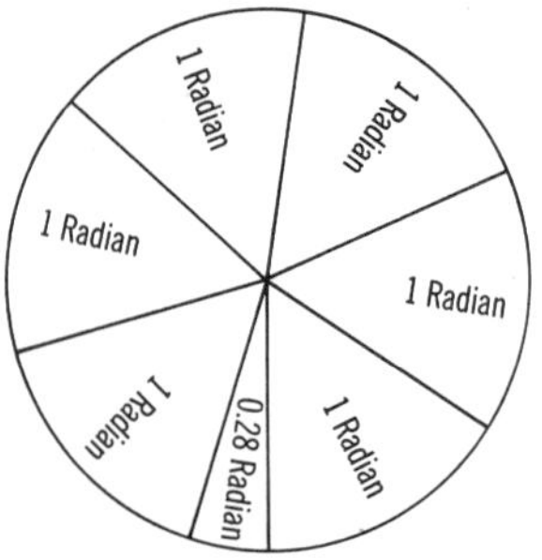

Now Eq. (2) says simply that the ratio of the circumference to the radius is 2π; that is, the length of the circumference is 2π times longer than the radius. Therefore, a circle must contain 2π radians ($2\pi^r$). Also, since the circumference subtends 360°, it follows that

$$2\pi^r = 360°$$
$$\pi^r = 180°$$

or $$1^r = \frac{180°}{\pi} = 57.2959° \cong 57.3° \qquad (3)$$

From Eq. (3), the following is evident:

- To reduce radians to degrees, multiply the number of radians by $\frac{180°}{\pi^r}$ ($\simeq 57.3$).
- To reduce degrees to radians, multiply the number of degrees by $\frac{\pi^r}{180°}$ ($\simeq 0.017\,45$).

Several types of slide rules have gage marks at 57.3 on scales C and D denoted by R (for radians). These marks are a convenience in converting from radians to degrees. Since 0.017 45 is the reciprocal of 57.3, the former number will be found on the reciprocal scales opposite the R gage marks. Similarly, if 180 on scale CF is set to π on DF, 0.017 45 will appear on scale D opposite the index of scale C. In this manner the rule is set up for multiplication by 0.017 45.

example 1 Reduce 1.7^r to degrees.

solution

$$1^r = 57.3°$$

Hence, $$1.7^r = 1.7 \times 57.3 = 97.4°$$

example 2 Convert 15.6° to radians.

solution

$$1° = 0.01745^r$$

Hence, $$15.6° = 15.6 \times 0.017\,45 = 0.272^r$$

PROBLEMS 23-2

1. Express the following angles in radians, first in terms of π and second as decimals: (*a*) 60°, (*b*) 120°, (*c*) 165°, (*d*) 225°, (*e*) 285°, (*f*) 5°.
2. Express the following angles in degrees: (*a*) 2^r, (*b*) 0.6^r, (*c*) $\frac{1^r}{\pi}$, (*d*) $\frac{2\pi^r}{3}$, (*e*) $\frac{5\pi^r}{6}$, (*f*) $0.610\,87^r$.
3. Through how many radians does the second hand of a clock turn between 6:35 A.M. and 9:20 A.M. of the same day?
4. Through how many radians does the hour hand of a clock turn in 40 min?
5. Through how many radians does the minute hand of a clock turn in 1 h 15 min?
6. What is the angular velocity in radians per second of (*a*) the second hand, (*b*) the minute hand, (*c*) the hour hand of a clock?
7. The speed of a rotating switch is 400 rev/min. What is the angular velocity of the switch in radians per second (rad/s)?
8. A radar antenna rotates at 6 rev/min. What is its angular velocity in radians per second?

9. A radar antenna has an angular velocity of π rad/s. What is its speed of rotation in revolutions per minute?
10. What is the approximate angular velocity of the earth in radians per minute (rad/min)?

23-6 GONS AND GRADS

In some parts of Europe, as a stage in furthering decimalization, the right angle is divided into one hundred equal parts known as *grads* (from the German) or, internationally, as *gons* (from the Greek). Each gon (grad) may be subdivided into 100 centigons (centigrads). Sometimes angles measured in this system will be written 20^g, sometimes 20 gon or 20 grad (no *s* for plural). The manner of notation will be a matter for future international agreement. Some calculators now available to electronics technicians offer alternative angle calculations in gons (grads).

$$1 \text{ right angle} = 90° = \frac{\pi^r}{2} = 100^g \qquad (4)$$

From Eq. (4), the following is evident:

- To reduce gons to degrees, multiply the number of gons by $\dfrac{90°}{100^g}$ $(= 0.9)$.
- To reduce degrees to gons, multiply the number of degrees by $\dfrac{100^g}{90°}$ $(=1.111)$.
- To reduce gons to radians, multiply the number of gons by $\dfrac{\pi/2^r}{100^g}$ $(\simeq 0.0157)$.
- To reduce radians to gons, multiply the number of radians by $\dfrac{100^g}{\pi/2^r}$ $(\simeq 63.7)$.

PROBLEMS 23-3

1. Express the following angles in gons: (*a*) 45°, (*b*) 30°, (*c*) 60°, (*d*) 120°, (e) 225°, (*f*) 315°.
2. Express the following angles in degrees: (*a*) 50^g, (*b*) 20^g, (*c*) 75^g, (*d*) 150^g, (e) 200^g, (*f*) 400^g.
3. Express the following angles in radians: (*a*) 50^g, (*b*) 20^g, (*c*) 75^g, (*d*) 150^g, (e) 200^g, (*f*) 400^g.
4. Express the following angles in gons: (*a*) $\dfrac{\pi^r}{4}$, (*b*) $\dfrac{\pi^r}{6}$, (*c*) $\dfrac{2\pi^r}{5}$, (*d*) $\dfrac{3\pi^r}{2}$, (e) $\dfrac{5\pi^r}{6}$, (*f*) 1.5708^r.

If you have an electronic calculator which gives angles in gons (grads), you may convert angles less than 90° as follows:

- Enter degrees, 45°; call for sin and read 0.707 11; convert to gons and call for arcsin to read 50^g.
- Enter gons, 50^g; call for sin and read 0.707 11; convert to rad and call for arcsin to read $0.785\ 40^r$.

23-7 SIMILAR TRIANGLES

Two triangles are said to be *similar* when their corresponding angles are equal. That is, similar triangles are identical in shape but may not be the same size. The important characteristic of similar triangles is that a direct proportionality exists between corresponding sides. The three triangles of Fig. 23-13 have been so constructed that their corresponding angles are equal. Therefore, the three triangles are similar, and their corresponding sides are proportional. This leads to the proportions

$$\frac{AB}{AC} = \frac{DE}{DF} = \frac{GH}{GI}$$

$$\frac{BC}{AB} = \frac{EF}{DE} = \frac{HI}{GH} \qquad \text{etc.}$$

As an example, if $AB = 0.5$ cm, $DE = 1$ cm, and $GH = 1.5$ cm, then DF is twice as long as AC and GI is three times as long as AC. Similarly, HI is three times as long as BC, and EF is twice as long as BC.

The properties of similar triangles are used extensively in measuring distance, such as the distances across bodies of water or other

Fig. 23-13 Similar Triangles

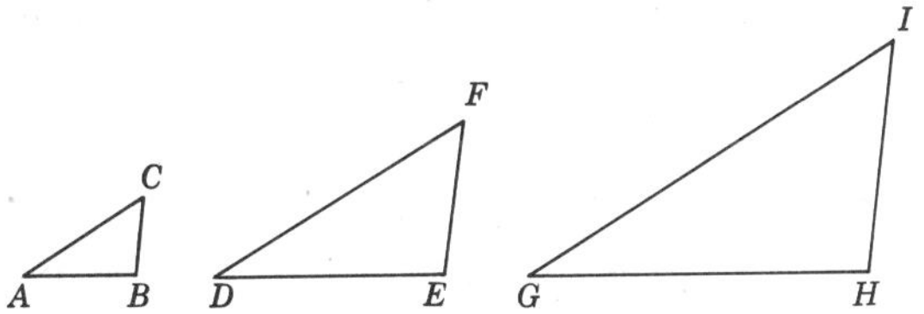

obstructions and the heights of various objects. In addition, the relationship between similar triangles forms the very basis of trigonometry.

Since the sum of the three angles of any triangle is 180°, it follows that if two angles of a triangle are equal to two angles of another triangle, the third angle of one must also be equal to the third angle of the other. Therefore, two triangles are similar if two angles of one are equal to two angles of the other.

If the numerical values of the necessary parts of a triangle are known, the triangle can be drawn to scale with the use of compasses, protractor, and ruler. The completed figure can then be measured with protractor and ruler to obtain the numerical values of the unknown parts. This is conveniently accomplished on squared paper.

note In the following problems the sides and angles of all triangles will be as represented in Fig. 23-14. That is, the angles will be represented by the capital letters A, B, and C and the sides opposite these angles will be the corresponding letters a, b, and c.

Fig. 23-14 Triangle for Probs. 3 to 10

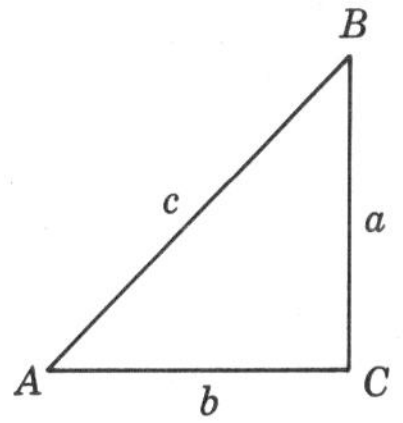

PROBLEMS 23-4

1. The sides of a triangular plot are 8, 12, and 16 m. The shortest side of a scale triangle is 3 m. How long are the other two sides of the smaller triangle?
2. Two triangles are similar. The sides of the first are 18, 30, and 36 in. The

longest side of the second is 20 mm. How long are the other two sides of the second triangle?

Solve the following triangles by graphical methods:

3. $b = 3$, $A = 53.1°$, $C = 90°$
4. $a = 15$, $b = 20$, $c = 25$
5. $b = 4$, $A = 80°$, $C = 80°$
6. $b = 5$, $c = 4.75$, $A = 110°$
7. $a = 10$, $B = 100°$, $C = 46.2°$
8. $a = 4.95$, $c = 7$, $B = 45°$
9. $a = 15.4$, $b = 20$, $C = 29.3°$
10. $a = 35$, $c = 35$, $A = 60°$

23-8 THE RIGHT TRIANGLE

If one of the angles of a triangle is a right angle, the triangle is called a *right triangle*. Then, since the sum of the angles of any triangle is 180°, a right triangle contains one right angle and two acute angles. Also, the sum of the acute angles must be 90°. This relation enables us to find one acute angle when the other is given. For example, in the right triangle shown in Fig. 23-15, if $\theta = 30°$, then $\phi = 60°$.

Since all right angles are equal, if an acute angle of one right triangle is equal to an acute angle of another right triangle, the two triangles are similar.

The side of a right triangle opposite the right angle is called the *hypotenuse*. Thus, in Fig. 23-15, the side c is the hypotenuse. When a right triangle is in standard position as in Fig. 23-15, the side a is called the *altitude* and the side b is called the *base*.

Another very important property of a right triangle is that the square of the hypotenuse is equal to the sum of the squares of the other two sides. That is,

$$c^2 = a^2 + b^2$$

Fig. 23-15 Right Triangle

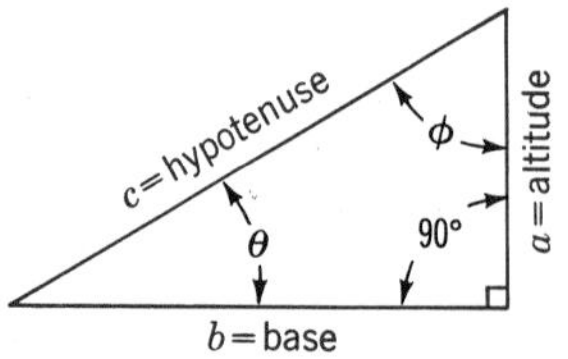

This relationship provides a means of computing any one of the three sides if two sides are given.

example 3 A chimney is 40 m high. What is the length of its shadow at a time when a vertical post 2 m high casts a shadow that is 2.1 m long?

solution BC in Fig. 23-16 represents the post, and EF represents the stack. Because the rays of the sun strike both chimney and post at the same angle, right triangles ABC and DEF are similar. Then, since

$$\frac{DF}{AC} = \frac{EF}{BC}$$

by substituting, $\dfrac{DF}{2.1} = \dfrac{40}{2}$

or $\qquad DF = 42 \text{ m}$

example 4 What is the length of a in the triangle of Fig. 23-17?

Fig. 23-16 Similar Right Triangles of Example 3

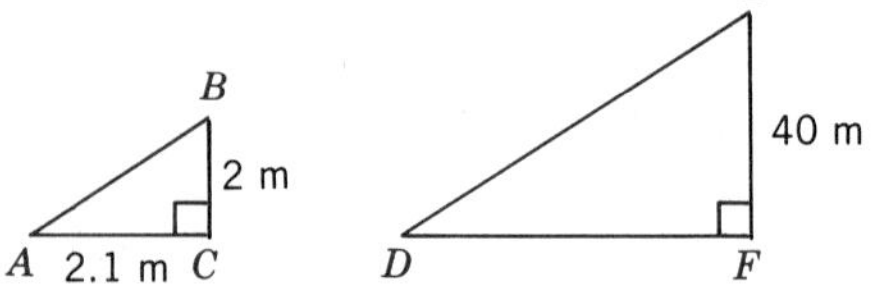

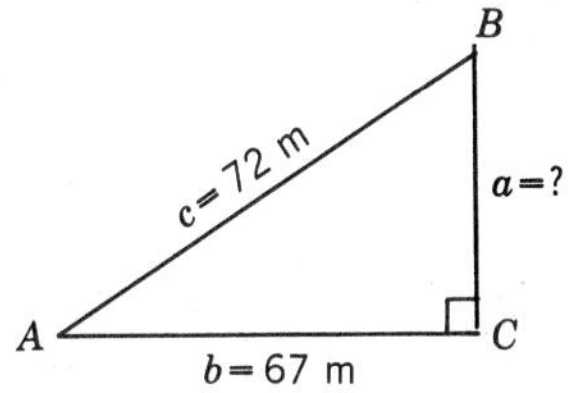

Fig. 23-17 Right Triangle of Example 4

solution

Given	$c^2 = a^2 + b^2$
Transposing,	$a^2 = c^2 - b^2$
$\sqrt{\ }$,	$a = \sqrt{c^2 - b^2}$
Substituting,	$a = \sqrt{72^2 - 67^2} = \sqrt{695}$
	$\therefore\ a = 26.4 \text{ m}$

PROBLEMS 23-5

In the following right triangles, solve for the indicated elements:

1. $a = 56$, $b = 15$, $A = 75°$. Find c and B.

2. $a = 24$, $c = 30$, $A = 53.1°$. Find b and B.

3. $b = 78$, $c = 80$, $B = 77°$. Find a and A.

4. An instrument plane flies north at the rate of 650 kn, and a hurricane hunter flies east at 1100 kn. If both planes start from the same place at the same time, how far apart will they be in 2 h?

5. In Fig. 23-18, if $AC = 18$ m, $BC = 24$ m, and $AE = 9$ m, find the length of DE.

6. In Fig. 23-18, if $AD = 30$ cm, $DB = 20$ cm, and $BC = 40$ cm, what is the length of DE?

7. In Fig. 23-18, $AE = 12$ m, $EC = 12$ m, and $AB = 46.5$ m. What is the length of DE?

8. The top of an antenna tower is 40 m above the ground. The tower is to be guyed at a point 6 m below its top to a point on the ground 18 m from the base of the tower. What is the length of the guy?

9. A transmitter antenna tower casts a shadow 270 m long at a time when a meterstick held upright with one end touching the ground casts a shadow 1.8 m long. What is the height of the tower?

10. The tower in Prob. 9 is to be guyed from its top with a 230-m guy wire. How far out from the base of the tower may the guy be anchored?

Fig. 23-18 Similar Right Triangles of Probs. 5, 6, and 7

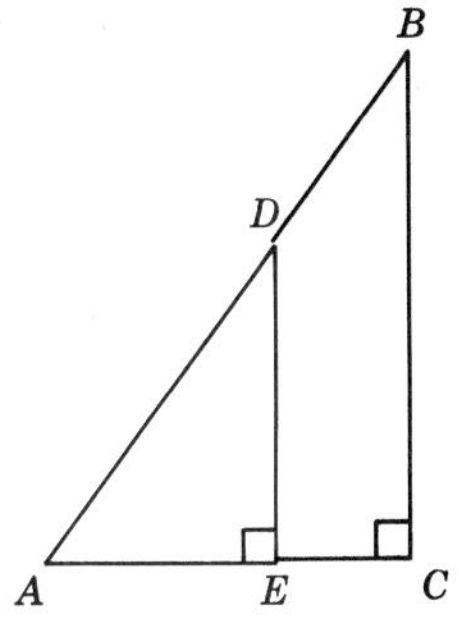

24

trigonometric functions

In the preceding chapter, it was shown that plane geometry furnishes two important properties of right triangles. These are

$$A + B = 90^\circ$$

and

$$a^2 + b^2 = c^2$$

The first relation makes it possible to find one acute angle when the other is known. By means of the second, any one side can be computed if the other two sides are known. These relations, however, furnish no methods for computing the magnitude of an acute angle when two sides are given. Also, using these relations, we cannot compute two sides of a right triangle if the other side and one acute angle are given. With only this amount of knowledge, we should be forced to resort to actual measurement by *graphical methods*.

The results obtained by such methods are unsuitable for use in many problems, for even with the greatest care and large-scale drawings the degree of accuracy is definitely limited. There is, then, an evident need for certain other relations in which the sides of a right triangle and the angles are united. Such relations form the foundation of trigonometry.

24-1 TRIGONOMETRIC FUNCTIONS ARE RATIOS

In Sec. 23-7, we saw that triangles may be similar regardless of their respective sizes. For example, in Fig. 24-1, the two triangles *ABC* and *DEF* are similar, and

$$\frac{AB}{AC} = \frac{DE}{DF} \qquad \frac{BC}{AC} = \frac{EF}{DF} \qquad \text{etc.}$$

Even if one of the pair of similar triangles is tilted (Fig. 24-2), the ratios still hold, since the triangles themselves have not changed in any of their dimensions. We may, however, have to look a little harder to see that this is so.

Consider the 30°-60°-90° triangle developed by bisecting an equilateral triangle (Fig. 24-3). First of all, you should confirm that, if the hypotenuse is 2 units long, then the base *AC* will be 1 unit long, and the

Fig. 24-1 Similar Triangles

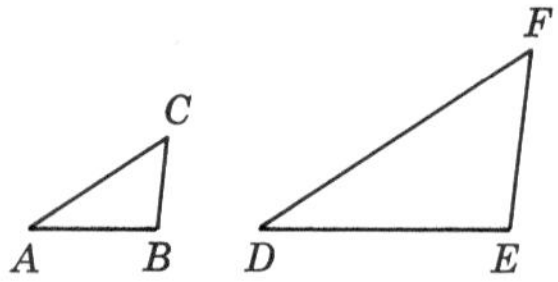

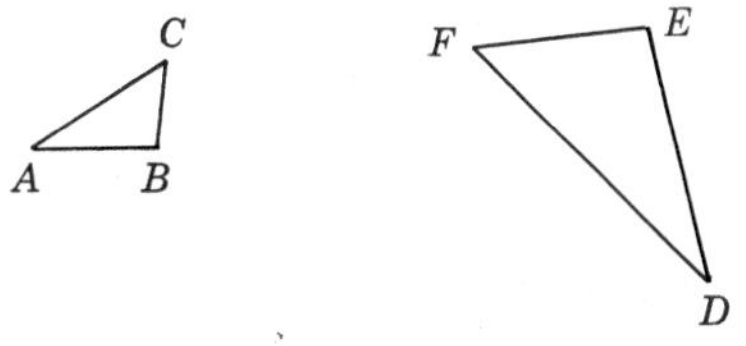

Fig. 24-2 Similar Triangles of Fig. 24-1, Except Triangle *DEF* Has Been Rotated

altitude *CB* will be $\sqrt{3}$ units long. Then consider the truth of the following statement:

In the 30°-60°-90° triangle, regardless of its size, the ratio of the base to the hypotenuse will always be 0.5000.

You should draw several 30°-60°-90° triangles of different sizes and prove to your complete satisfaction that this statement *must* always be true.

If the triangle were now rotated so that the side *CB* were the base and *AC* the altitude, the above statement would have to be adjusted. Therefore, we should rename the parts of the triangle so that there can be no possibility of misunderstanding a statement about it. The most convenient way to refer to a side of a triangle is to relate it to the angles in the triangle. For instance, the hypotenuse is always the longest side, it is always opposite the right angle, and it is always adjacent to (forms) each of the other two angles. We can always refer to it as simply the hypotenuse without introducing any possibility of being misunderstood.

Fig. 24-3 Equilateral Triangle Divided into Two Equal 30°-60°-90° Right Triangles

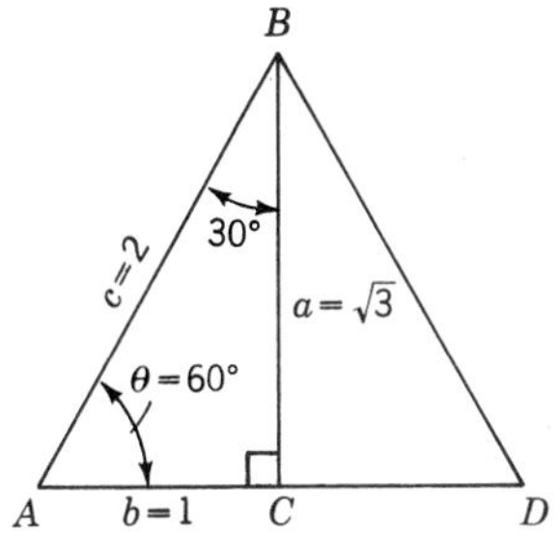

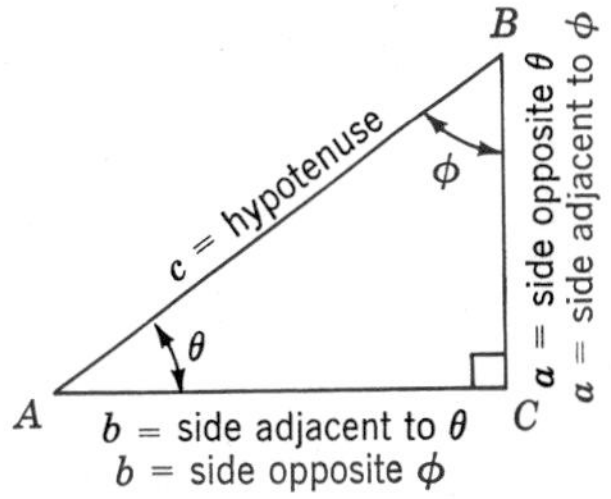

Fig. 24-4 Side-Angle Relationships in the Standard Right Triangle

In the 30°-60°-90° triangle with which we are dealing, the side *AC* is always the side *opposite* the 30° angle, and it is always the side *adjacent* to the 60° angle, regardless of the letter designation given it or the orientation of the triangle.

Similarly, the side *CB* is always opposite the 60° angle, and it is always adjacent to the 30° angle, regardless of the symbols used to identify the side or how the triangle is tilted. These side-angle relationships are illustrated in Fig. 24-4, and they must be memorized, because they will be used continuously henceforth.

For the rest of this chapter and the next, we shall be dealing only with right triangles. The hypotenuse is always the longest side and is opposite the right angle. The other two sides will be designated according to their relationships to the acute angles.

You should immediately confirm, using sketches as required, the truth of the following statements relating to the sides of the 30°-60°-90° triangle, first as they apply to the 30° angle and then as they apply to the 60° angle:

1. In the 30°-60°-90° triangle, regardless of its size or orientation, the ratio of the side opposite the 30° angle to the hypotenuse will always be 0.5000.
2. In the 30°-60°-90° triangle, regardless of

its size or orientation, the ratio of the side adjacent to the 30° angle to the hypotenuse will always be 0.866.

3. In the 30°-60°-90° triangle, regardless of its size or orientation, the ratio of the side opposite the 30° angle to the side adjacent to the 30° angle will always be 0.577.
4. In the 30°-60°-90° triangle, regardless of its size or orientation, the ratio of the side opposite the 60° angle to the hypotenuse will always be 0.866.
5. In the 30°-60°-90° triangle, regardless of its size or orientation, the ratio of the side adjacent to the 60° angle to the hypotenuse will always be 0.5000.
6. In the 30°-60°-90° triangle, regardless of its size or orientation, the ratio of the side opposite the 60° angle to the side adjacent to the 60° angle will always be 1.732.

It is left as an exercise for you to develop the three similar statements for the 45°-45°-90° triangle. (Why only three statements?)

Now, student, stop and look at these statements. See what they really mean. Make sure that their message is plain. When you fully understand the import of the relationships between sides of triangles, you will have trigonometry in the palm of your hand forever. We do not say that all of trigonometry is simple. But to grasp quickly the fact that the trigonometric functions are merely ratios of sides of triangles is to resolve most of the difficulties which stand in the way of students who have never properly understood how simple the functions of trigonometry really are.

The word "trigonometry" just means "measurement of triangles," and one of the most useful tools in the measurement of triangles is the ratios of sides.

"In the triangle, regardless of its size or orientation" means that, so long as the angles made by the sides are specified, the triangle itself may be formed by:

1. Three lines on a piece of paper
2. A ladder, the ground, and the wall of a house
3. An antenna mast, its shadow on the ground, and the line of sight from the end of the shadow to the top of the mast
4. The lines of sight between two surveyors and a distant landmark
5. A mast, a guy wire, and the ground between the foot of the mast and the guy anchor
6. Any other system which uses three straight lines to form three enclosed angles

The entire statement, "In the . . . triangle . . . will always be . . ." is quite a mouthful, far too lengthy for convenience, and it is often abbreviated. For instance, statement 1 above becomes

$$\frac{\text{opp } 30^\circ}{\text{hyp}} = 0.500$$

or

$$\frac{\text{opp}}{\text{hyp}}\, 30^\circ = 0.500$$

and all the other parts of the statement are understood to apply. Statement 2 becomes

$$\frac{\text{adj}}{\text{hyp}}\, 30^\circ = 0.866$$

and statement 3 becomes

$$\frac{\text{opp}}{\text{adj}}\, 30^\circ = 0.577$$

You should now write similar abbreviations for statements 4, 5, and 6 and check your work for the 45°-45°-90° triangle to show your own statements 7, 8, and 9 may be written

$$\frac{\text{opp}}{\text{hyp}} 45^\circ = 0.7071$$

$$\frac{\text{adj}}{\text{hyp}} 45^\circ = 0.7071$$

$$\frac{\text{opp}}{\text{adj}} 45^\circ = 1.000$$

example 1 A triangular piece of farm land is to be used as an "antenna farm." It is in the shape of a 30°-60°-90° triangle the shortest side of which is 600 m long (Fig. 24-5). What are the dimensions of the other two sides?

solution By using the ratios which have been discovered above and drawing a sketch of the triangle to show the relationships between the sides and angles, we find that the 600-m side must be adjacent to the 60° angle. Then we have

$$\frac{600}{\text{hyp}} 60^\circ = 0.500$$

from which

$$\text{hyp} = \frac{600}{0.5} = 1200 \text{ m}$$

Fig. 24-5 Triangle of Example 1

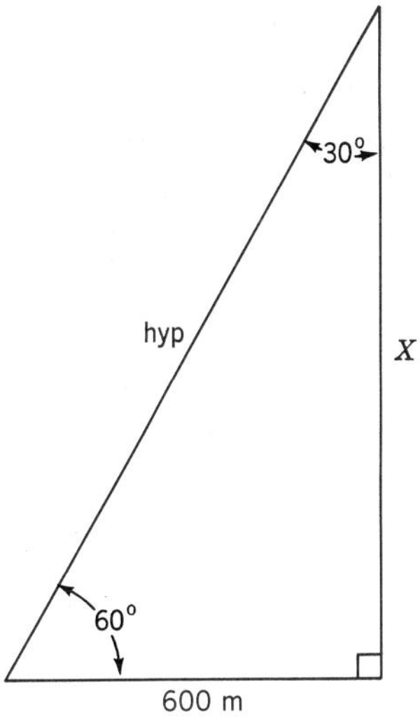

and $\frac{600}{\text{adj}} 30^\circ = 0.577$

from which

$$\text{adj } 30^\circ = \frac{600}{0.577} = 1040 \text{ m}$$

Even these abbreviations are more than we require for everyday use, and we now introduce the proper *trigonometric names* for the different ratios (*functions*). θ is the "general angle," just as x is the "general number."

1. The ratio $\frac{\text{opp}}{\text{hyp}} \theta$ is properly called sine θ, abbreviated to sin θ.
2. The ratio $\frac{\text{adj}}{\text{hyp}} \theta$ is properly called cosine θ, abbreviated to cos θ.
3. The ratio $\frac{\text{opp}}{\text{adj}} \theta$ is properly called tangent θ, abbreviated to tan θ.

It must be clearly understood that the names sine, cosine, and tangent are meaningless in themselves; you must relate them to angles of triangles. To say simply "cosine" means nothing. But "cos 60°" means, very specifically, the ratio of the side adjacent to the 60° angle of a 30°-60°-90° triangle to the hypotenuse of the same triangle.

In the general triangle, Fig. 24-6, it will be

Fig. 24-6 Standard Right Triangle, as Used in Electronics Problems

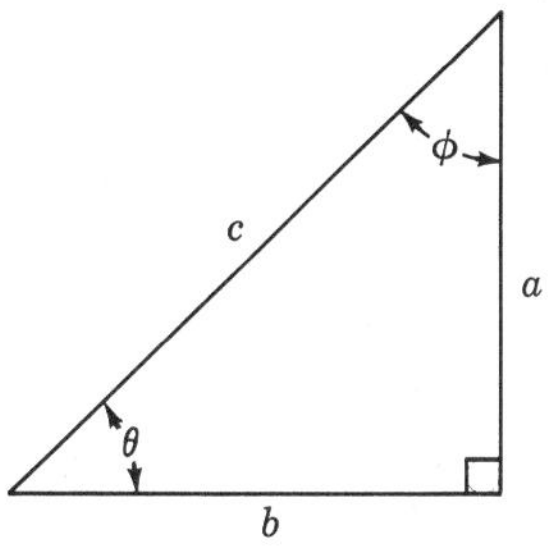

seen that there exist *six* possible trigonometric functions. Three of them we have already discovered, and the others are reciprocals of those three.

$$\frac{\text{opp}}{\text{hyp}}\,\theta = \sin\theta = \frac{a}{c}$$

$$\frac{\text{adj}}{\text{hyp}}\,\theta = \cos\theta = \frac{b}{c}$$

$$\frac{\text{opp}}{\text{adj}}\,\theta = \tan\theta = \frac{a}{b}$$

$$\frac{\text{hyp}}{\text{opp}}\,\theta = \text{cosecant}\,\theta \quad = \csc\theta = \frac{c}{a}$$

$$\frac{\text{hyp}}{\text{adj}}\,\theta = \text{secant}\,\theta \quad = \sec\theta = \frac{c}{b}$$

$$\frac{\text{adj}}{\text{opp}}\,\theta = \text{cotangent}\,\theta = \cot\theta = \frac{b}{a}$$

The cosecant, secant, and cotangent should always be thought of as the reciprocals of the sine, cosine, and tangent, respectively. This is shown easily by considering the reciprocal of sin θ:

$$\frac{1}{\sin\theta} = \frac{1}{\frac{a}{c}} = \csc\theta$$

You should confirm the other two reciprocal functions.

These definitions should be memorized so thoroughly that you can tell instantly any ratio of either acute angle of a right triangle, regardless of its position.

The sine, cosine, and tangent are the ratios most frequently used in practical work. If they are carefully learned, the others are easily remembered because they are reciprocals.

The fact that the numerical value of any one of the trigonometric functions (ratios) depends only upon the magnitude of the angle θ is of fundamental importance. This is established from a consideration of Fig. 24-7. The angle θ is generated by the line AD revolving about the point A. From the points B, B', and B'', perpendiculars are let fall to the initial line, or adjacent side, AX. These form similar triangles ABC, $AB'C'$, and $AB''C''$ because all are right triangles having a common acute angle θ (Sec. 23-8). Hence,

$$\frac{BC}{AB} = \frac{B'C'}{AB'} = \frac{B''C''}{AB''}$$

Each of these ratios defines the sine of θ. Similarly, it can be shown that this property is true for each of the other functions. Therefore, the size of the right triangle is immaterial, for only the *relative* lengths of the sides are of importance.

Each one of the six ratios will change in value whenever the angle changes in magnitude. Thus, it is evident that the ratios are really functions of the angle under consideration. If the angle is considered to be the independent variable, then the six functions (ratios) and the relative lengths of the sides of the triangles are dependent variables.

example 2 Calculate the functions of the angle θ in the right triangle of Fig. 24-6 if $a = 6$ mm and $c = 10$ mm.

solution Since $c^2 = a^2 + b^2$,

then $b = \sqrt{c^2 - a^2} = \sqrt{100 - 36}$
$= \sqrt{64} = 8$ mm

Fig. 24-7 The Values of the Functions Depend Only on the Size of the Angle

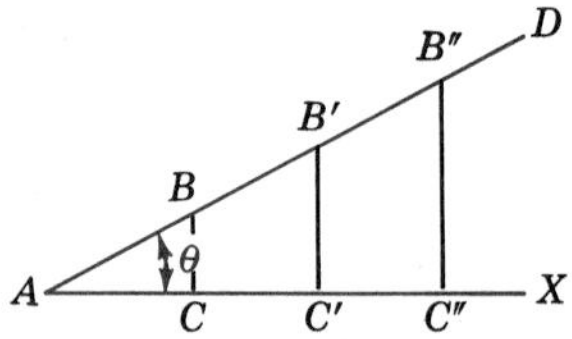

Applying the definitions of the six functions,

$$\sin\theta = \tfrac{6}{10} = \tfrac{3}{5} \qquad \cos\theta = \tfrac{8}{10} = \tfrac{4}{5}$$
$$\tan\theta = \tfrac{6}{8} = \tfrac{3}{4} \qquad \cot\theta = \tfrac{8}{6} = \tfrac{4}{3}$$
$$\sec\theta = \tfrac{10}{8} = \tfrac{5}{4} \qquad \csc\theta = \tfrac{10}{6} = \tfrac{5}{3}$$

What would be the values of the above functions if $a = 6$ m, $b = 8$ m, and $c = 10$ m?

24-2 FUNCTIONS OF COMPLEMENTARY ANGLES

By applying the definitions of the six functions to the angle ϕ in Fig. 24-8 and noting the positions of the adjacent and opposite sides for this angle, we obtain

$$\sin\phi = \frac{\text{opp}}{\text{hyp}} = \frac{b}{c} \qquad \csc\phi = \frac{\text{hyp}}{\text{opp}} = \frac{c}{b}$$
$$\cos\phi = \frac{\text{adj}}{\text{hyp}} = \frac{a}{c} \qquad \sec\phi = \frac{\text{hyp}}{\text{adj}} = \frac{c}{a}$$
$$\tan\phi = \frac{\text{opp}}{\text{adj}} = \frac{b}{a} \qquad \cot\phi = \frac{\text{adj}}{\text{opp}} = \frac{a}{b}$$

Upon comparing these with the original definitions given for the triangle of Fig. 24-2, we find the following relations:

$$\sin\phi = \cos\theta \qquad \cos\phi = \sin\theta$$
$$\tan\phi = \cot\theta \qquad \cot\phi = \tan\theta$$
$$\sec\phi = \csc\theta \qquad \csc\phi = \sec\theta$$

Since $\phi = 90° - \theta$, the above relations can be written

Fig. 24-8 Right Triangle for Determining Functions of Angle ϕ

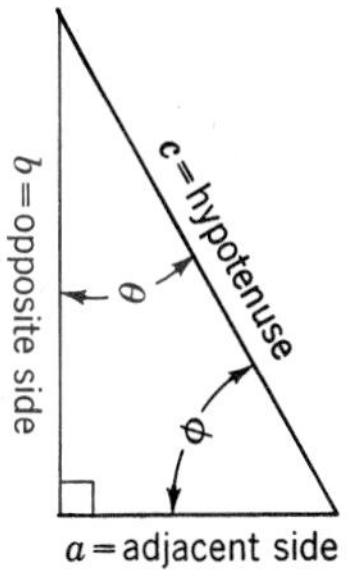

$$\sin(90° - \theta) = \cos\theta \qquad \cos(90° - \theta) = \sin\theta$$
$$\tan(90° - \theta) = \cot\theta \qquad \cot(90° - \theta) = \tan\theta$$
$$\sec(90° - \theta) = \csc\theta \qquad \csc(90° - \theta) = \sec\theta$$

The above can be stated in words as follows: *A function of an acute angle is equal to the cofunction of its complementary angle.* This enables us to find the function of every acute angle greater than 45° if we know the functions of all angles less than 45°. For example, sin 56° = cos 34°, tan 63° = cot 27°, cos 70° = sin 20°, etc.

24-3 CONSTRUCTION OF AN ANGLE WHEN ONE FUNCTION IS GIVEN

When the trigonometric function of an acute angle is given, the angle can be constructed geometrically by using the definition for the given function. Also, the magnitude of the resulting angle can be measured by the use of a protractor.

example 3 Construct the acute angle whose tangent is $\frac{9}{10}$.

solution Erect perpendicular lines AC and BC, preferably on cross-sectional paper. Measure off 10 units along AC and 9 units along BC. Join A and B and thus form the right triangle ABC. $\tan A = \frac{9}{10}$; therefore, A is the required angle. Measuring A with a protractor shows it to be an angle of approximately 42°. The construction is shown in Fig. 24-9.

Fig. 24-9 Construction of Acute Angle Whose Tangent Is $\frac{9}{10}$

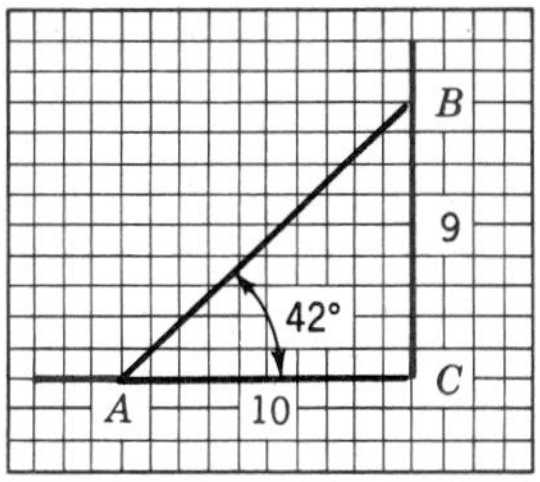

example 4 Find by construction the acute angle whose cosine is $\frac{3}{4}$.

solution Erect perpendicular lines AC and BC. Measure off three units along AC. (Let three divisions of the cross-sectional paper be equal to one unit for greater accuracy.) With A as a center and with a radius of 4 units, draw an arc to intersect the perpendicular at B. Connect A and B. $\cos A = \frac{3}{4}$; therefore A is the required angle. Measuring A with a protractor shows it to be an angle of approximately 41.4°. The construction is shown in Fig. 24-10.

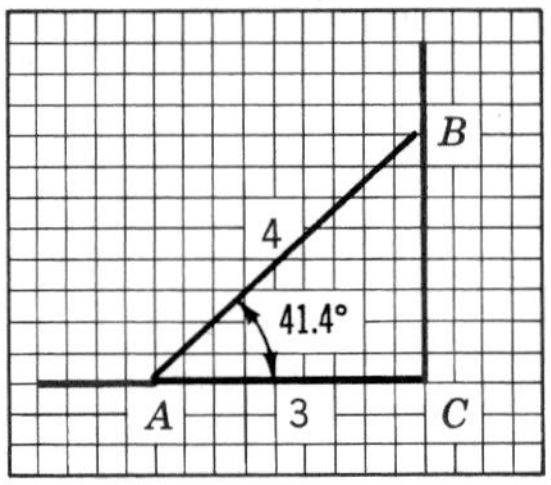

Fig. 24-10 Construction of Acute Angle Whose Cosine Is $\frac{3}{4}$

PROBLEMS 24-1

1. In Fig. 24-11, what are the values of the trigonometric functions for the angles θ and ϕ in terms of ratios of the sides, a, b, and c?

2. In Fig. 24-12, (a) $\sin\alpha = ?$ (b) $\sin\beta = ?$ (c) $\cot\beta = ?$ (d) $\sec\alpha = ?$ (e) $\tan\alpha = ?$

3. In Fig. 24-12, (a) $\frac{OP}{OR} = \tan?$ (b) $\frac{PR}{PO} = \sec?$ (c) $\frac{OR}{PR} = \cos?$ (d) $\frac{OP}{RP} = \sin?$ (e) $\frac{PR}{RO} = \csc?$

4. The three sides of a right triangle are 5, 12, and 13. Let α be the acute angle opposite the side 5 and let β be the other acute angle. Write the six functions of α and β.

5. In Fig. 24-13, if $X = R$, find the six functions of θ.

6. In Fig. 24-13, if $R = \frac{1}{2}Z$, find the sine, cosine, and tangent of θ.

7. In Fig. 24-13, if $X = 2R$, find the sine, cosine, and tangent of ϕ.

Fig. 24-12 Right Triangle of Probs. 2 and 3

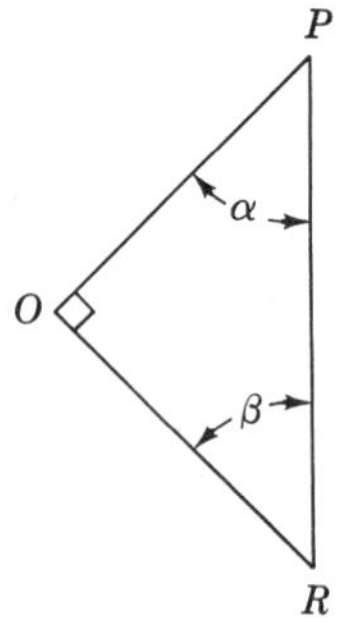

Fig. 24-11 Right Triangle of Prob. 1

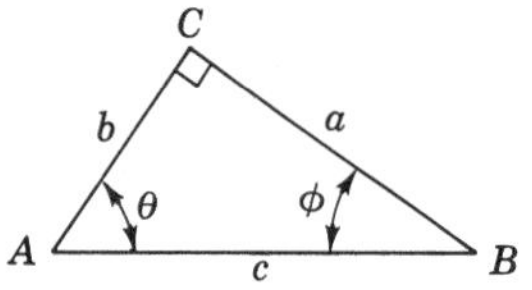

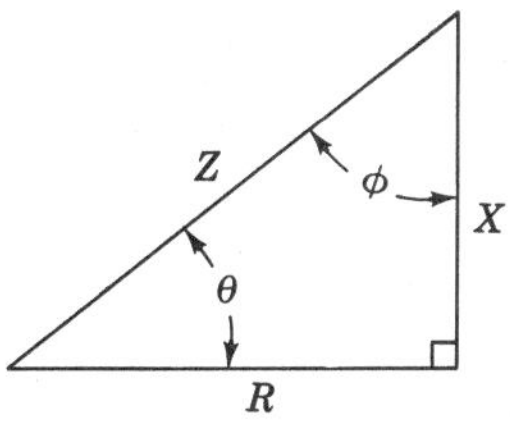

Fig. 24-13 Right Triangle of Probs. 5, 6, and 7

8. (*a*) $\sin\theta = \frac{2}{3}$, $\csc\theta = ?$ (*b*) $\sec\alpha = 2$, $\cos\alpha = ?$ (c) $\cot\beta = \frac{7}{8}$, $\tan\beta = ?$ (*d*) $\cos\phi = \frac{5}{16}$, $\sec\phi = ?$ (e) $\tan\phi = 12$, $\cot\phi = ?$ (*f*) $\csc\alpha = 4$, $\sin\alpha = ?$

9. The three sides of a right triangle are 6, 8, and 10. Write the six functions of the largest acute angle.

10. Write the other functions of an acute angle whose cosine is $\frac{4}{5}$.

11. In a right triangle, $c = 5$ cm and $\cos A = \frac{4}{5}$. Construct the triangle, and write the functions of the angle B.

12. State which of the following is greater if $\theta \neq 0°$ and is less than 90°: (*a*) $\sin\theta$ or $\tan\theta$, (*b*) $\cos\theta$ or $\cot\theta$, (c) $\sec\theta$ or $\tan\theta$, (*d*) $\csc\theta$ or $\cot\theta$.

24-4 FUNCTIONS OF ANY ANGLE

The notion of trigonometric functions has been introduced from the point of view of right triangles because this allows for an easy introduction which most students can follow with assurance. However, the total concept applies to far more than just right triangles and to far more than angles between 0° and 90°. In Chap. 27 we shall investigate a few interesting and useful relationships in nonright triangles. For the moment, we will concentrate on the trigonometric functions of any angle.

In Chap. 23 we found the concepts of angles were extended to include angles in any quadrant and both positive and negative angles. In Fig. 24-14 the line **r** is revolving about the origin of the rectangular coordinate system in a counterclockwise (positive) direction. This line, which generates the angle θ, is known as the *radius vector*. The initial side of θ is the positive x axis, and the terminal side is the radius vector. If a perpendicular is let fall from any point P along the radius vector, in any of the quadrants, a right triangle $xy\mathbf{r}$ will be formed with $\mathbf{r}$ as a hypotenuse of constant unit length and with x and y having lengths equal to the respective coordinates of P.

Fig. 24-14 Radius Vector **r** Generating Angles

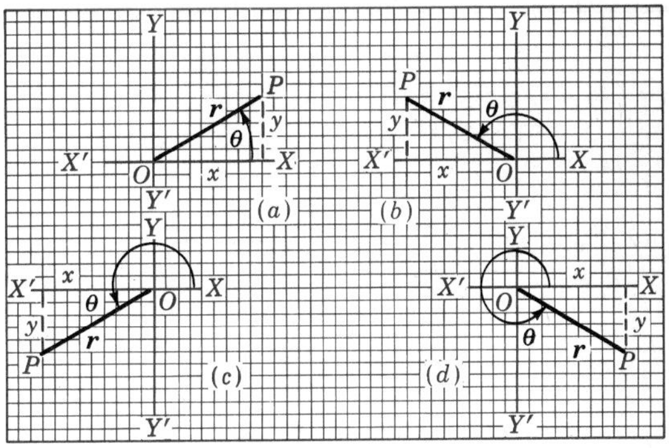

We then define the trigonometric functions of θ as follows:

$$\sin\theta = \frac{y}{r} = \frac{\text{ordinate}}{\text{radius}} \qquad \csc\theta = \frac{r}{y} = \frac{\text{radius}}{\text{ordinate}}$$

$$\cos\theta = \frac{x}{r} = \frac{\text{abscissa}}{\text{radius}} \qquad \sec\theta = \frac{r}{x} = \frac{\text{radius}}{\text{abscissa}}$$

$$\tan\theta = \frac{y}{x} = \frac{\text{ordinate}}{\text{abscissa}} \qquad \cot\theta = \frac{x}{y} = \frac{\text{abscissa}}{\text{ordinate}}$$

Since the values of the six trigonometric functions are entirely independent of the position of the point *P* along the radius vector, it follows that they depend only upon the position of the radius vector, or the size of the angle. Therefore, for every angle there is one, and only one, value of each function.

24-5 SIGNS OF THE FUNCTIONS

The signs of the functions of angles in various quadrants are very important. If you remember the signs of the abscissas (x values) and the ordinates (y values) in the four quadrants, you will encounter no trouble.

For angles in the first quadrant, as shown in Fig. 24-14*a*, the x and y values are positive. Since the length of the radius vector r is always considered positive, it is evident that all functions of angles in the first quadrant are positive. For angles in the second quadrant, as shown in Fig. 24-14*b*, the x values are negative and the y values are positive. Therefore, the sine and its reciprocal are positive and the other four functions are negative. Similarly, the signs of all the functions can be checked from their definitions

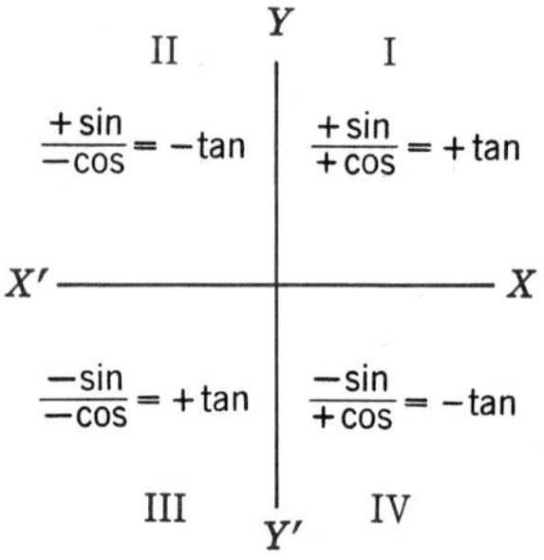

Fig. 24-15 Signs of Functions in Quadrants

as given in the preceding section. You should verify each part of Table 24-1.

If the proper signs for the sine and cosine are fixed in mind, the other signs will be remembered because of an important relation

$$\frac{\sin\theta}{\cos\theta} = \frac{\frac{y}{r}}{\frac{x}{r}} = \frac{y}{x}$$

Since $\qquad \tan\theta = \dfrac{y}{x}$

then $\qquad \dfrac{\sin\theta}{\cos\theta} = \tan\theta$

If the sine and cosine have like signs, the tangent is positive, and if they have unlike signs, the tangent is negative. Because the signs of the sine, cosine, and tangent always agree with signs of the respective reciprocals, the cosecant, secant, and cotangent, the signs for the latter are obtainable from the signs of the sine and cosine as outlined above. Figure 24-15 will serve as an aid in remembering the signs.

Table 24-1

Quadrant	$\sin\theta$	$\cos\theta$	$\tan\theta$	$\cot\theta$	$\sec\theta$	$\csc\theta$
I	+	+	+	+	+	+
II	+	−	−	−	−	+
III	−	−	+	+	−	−
IV	−	+	−	−	+	−

PROBLEMS 24-2

In what quadrant or quadrants is θ for each of the following conditions?

1. $\sin\theta$ is positive.
2. $\cos\theta$ is positive.
3. $\sin\theta$ is negative
4. $\tan\theta$ is negative.
5. $\cos\theta$ is negative.
6. $\sin\theta$ positive, $\cos\theta$ negative.
7. $\tan\theta$ and $\sin\theta$ both positive.
8. $\cot\theta$ negative, $\cos\theta$ negative.
9. $\tan\theta$ negative, $\cos\theta$ positive.
10. All functions of θ are positive.
11. $\tan\theta = 6$
12. $\cos\theta = -\frac{3}{4}$
13. Is there an angle whose cosine is negative and whose secant is positive?
14. When $\tan\theta = \frac{3}{4}$, find the value of

$$\frac{\sin\theta - \csc\theta}{\cot\theta - \sec\theta}$$

Give the signs of the sine, cosine, and tangent of each of the following angles:

15. 32°
16. 210°
17. 98°
18. 350°
19. −175°
20. $\frac{\pi^r}{3}$
21. $\frac{-3\pi^r}{4}$
22. −72°
23. 780°

Find the value of the radius vector r for each of the following positions of P, and then find the trigonometric functions of the angle θ ($\angle XOP$). Keep answers in fractional form.

24. $(-9,12)$

SOLUTION: Draw the radius vector r from O to $P = (-9,12)$ as shown in Fig. 24-16. Hence, θ is an angle in the second quadrant with a side adjacent that has an x value of -9 and a side opposite that has a y value of 12.

Fig. 24-16 Diagram of Prob. 24

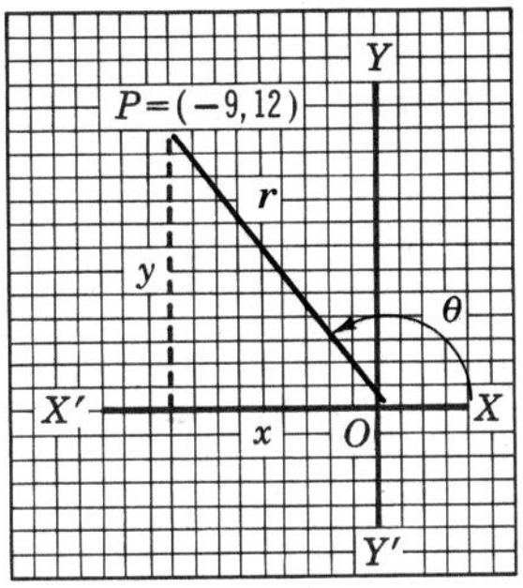

Then $$r = \sqrt{x^2 + y^2} = \sqrt{(-9)^2 + (12)^2} = 15$$

Hence, by definition,

$$\sin\theta = \frac{y}{r} = \frac{12}{15} = \frac{4}{5} \qquad \csc\theta = \frac{r}{y} = \frac{15}{12} = \frac{5}{4}$$

$$\cos\theta = \frac{x}{r} = \frac{-9}{15} = -\frac{3}{5} \qquad \sec\theta = \frac{r}{x} = \frac{15}{-9} = -\frac{5}{3}$$

$$\tan\theta = \frac{y}{x} = \frac{12}{-9} = -\frac{4}{3} \qquad \cot\theta = \frac{x}{y} = \frac{-9}{12} = -\frac{3}{4}$$

25. (12,−5)

SOLUTION: Draw the radius vector r from O to P as shown in Fig. 24-17. θ is an angle in the fourth quadrant with a side adjacent that has an x value of 12 and a side opposite that has a y value of −5. Then

$$r = \sqrt{x^2 + y^2} = \sqrt{12^2 + (-5)^2} = 13$$

Hence, by definition,

$$\sin\theta = \frac{y}{r} = -\frac{5}{13} \qquad \csc\theta = \frac{r}{y} = -\frac{13}{5}$$

$$\cos\theta = \frac{x}{r} = \frac{12}{13} \qquad \sec\theta = \frac{r}{x} = \frac{13}{12}$$

$$\tan\theta = \frac{y}{x} = -\frac{5}{12} \qquad \cot\theta = \frac{x}{y} = -\frac{12}{5}$$

26. (3,4) **27.** (12,5) **28.** (−3,4) **29.** (−4,−5)
30. (3,3) **31.** (4,−3) **32.** (−8,6) **33.** (−5,−3)
34. (8,8)

Fig. 24-17 Diagram of Prob. 25

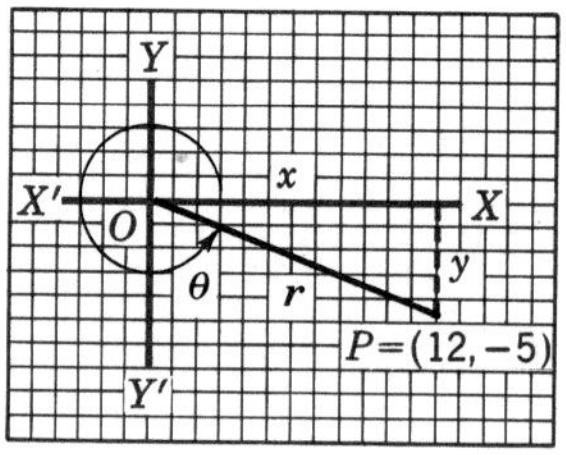

24-6 COMPUTATION OF THE FUNCTIONS

In Sec. 24-1, we developed the functions of 30°, 45°, and 60° by merely using simple notions about right triangles. These angles are very important and will be used often, so that they and their trigonometric functions are worthy of the time you spend in this development. At the same time, their use will make it easy for some students to quickly relearn trigonometry a few years hence if their work has been such that they do not require it immediately. In Chap. 25 we will extend our notions of trigonometric functions and develop and use the tables prepared by expert mathematicians for our use and convenience.

24-7 FUNCTIONS OF 0°

For an angle of 0°, the initial and terminal sides are both on OX. At any distance a from O, choose the point P as shown in Fig. 24-18. Then the coordinates of P are $(a,0)$. That is, the x value is equal to a units, and the y value is zero. Since the radius vector r is equal to a, by definition,

$$\sin 0° = \frac{y}{r} = \frac{0}{r} = 0 \qquad \csc 0° = \frac{r}{y} = \frac{a}{0} = \infty$$

$$\cos 0° = \frac{x}{r} = \frac{a}{a} = 1 \qquad \sec 0° = \frac{r}{x} = \frac{a}{a} = 1$$

$$\tan 0° = \frac{y}{x} = \frac{0}{a} = 0 \qquad \cot 0° = \frac{x}{y} = \frac{a}{0} = \infty$$

By $\frac{a}{0} = \infty$ is meant the value of $\frac{a}{y}$ as y approaches zero without limit. Thus, as y gets nearer and nearer to zero, $\frac{a}{y}$ gets larger and larger. Therefore, $\frac{a}{y}$ is said to *approach* infinity as y approaches zero. However, $\frac{a}{0}$ does not actually result in a quotient of infinity, for division by zero is meaningless.

Fig. 24-18 $\theta = 0°$, $x = a$, and $y = 0$

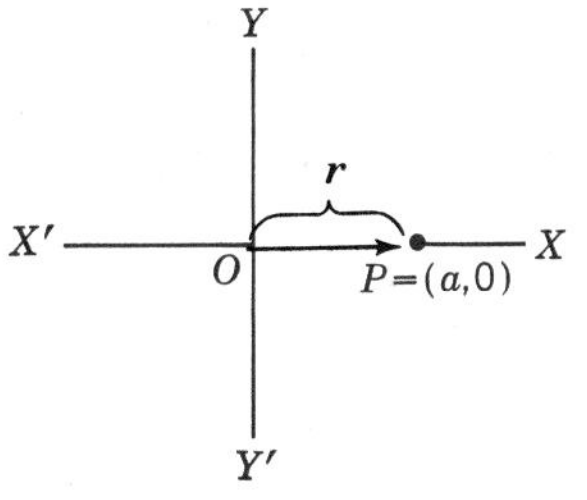

Determining the functions of 90°, 180°, and 270° is accomplished by the same method as that used for 0°. This is left as an exercise for you.

24-8 THE RANGES OF THE FUNCTIONS

As the radius vector r starts from OX and revolves about the origin in a positive (counterclockwise) direction, the angle θ is generated and varies in magnitude continuously from 0° to 360° through the four quadrants. Figure 24-19 illustrates the manner in which the sine, cosine, and tangent vary as the angle θ changes in value.

Quadrant I. As θ increases from 0° to 90°,

Fig. 24-19 Lengths of Lines Showing the Ranges of Sin θ, Cos θ, and Tan θ

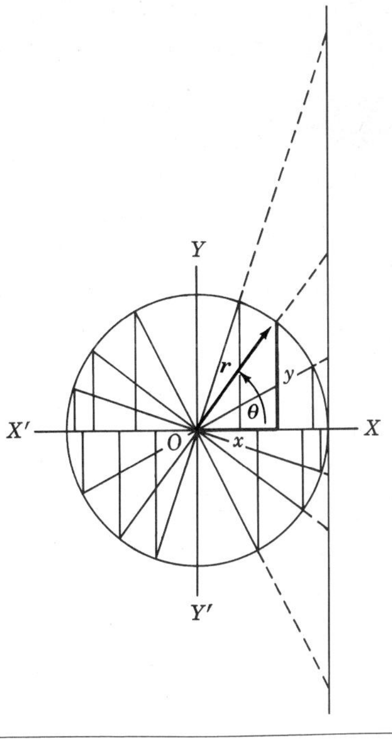

- x is positive and decreases from r to 0.
- y is positive and increases from 0 to r.

Therefore,

$\sin\theta = \dfrac{y}{r}$ is *positive* and increases from 0 to 1.

$\cos\theta = \dfrac{x}{r}$ is *positive* and decreases from 1 to 0.

$\tan\theta = \dfrac{y}{x}$ is *positive* and increases from 0 to ∞.

Quadrant II. As θ increases from 90° to 180°,

- x is negative and increases from 0 to $-r$.
- y is positive and decreases from r to 0.

Therefore,

$\sin\theta = \dfrac{y}{r}$ is *positive* and decreases from 1 to 0.

$\cos\theta = \dfrac{x}{r}$ is *negative* and increases from 0 to -1.

$\tan\theta = \dfrac{y}{x}$ is *negative* and decreases from $-\infty$ to 0.

Quadrant III. As θ increases from 180° to 270°,

- x is negative and decreases from $-r$ to 0.
- y is negative and increases from 0 to $-r$.

Therefore,

$\sin\theta = \dfrac{y}{r}$ is *negative* and increases from 0 to -1.

$\cos\theta = \dfrac{x}{r}$ is *negative* and decreases from -1 to 0.

$\tan\theta = \dfrac{y}{x}$ is *positive* and increases from 0 to ∞.

Quadrant IV. As θ increases from 270° to 360°,

- x is positive and increases from 0 to r.
- y is negative and decreases from $-r$ to 0.

Therefore,

$\sin\theta = \dfrac{y}{r}$ is *negative* and decreases from -1 to 0.

$\cos\theta = \dfrac{x}{r}$ is *positive* and increases from 0 to 1.

$\tan\theta = \dfrac{y}{x}$ is *negative* and decreases from $-\infty$ to 0.

Students often become confused in comparing the variations of the functions, when represented as lines, with their actual numerical value. For example, in quadrant II as the angle θ increases from 90 to 180°, we say that $\cos\theta$ *increases* from 0 to $-r$. Actually, the abscissa representing the cosine is getting *longer;* confusion results from not remembering that a negative number is always greater than zero in the defined negative direction. The *lengths* of the lines representing the functions, when compared with the radius vector, indicate only the *magnitude* of the function. The positions of the lines, with respect to the x or y axis, specify the signs of the functions.

24-9 LINE REPRESENTATION OF THE FUNCTIONS

By representing the functions as lengths of lines, we are able to obtain a mental picture of the manner in which the functions vary as the radius vector r revolves and generates angles. Since we are primarily concerned with the sine, cosine, and tangent, only these functions will be represented graphically.

In Fig. 24-20 the radius vector r, with a length of one unit, is revolving about the origin and generating the angle θ. Then, in each of the four quadrants,

$$\sin\theta = \frac{BC}{r} = \frac{BC}{1} = BC$$

and $$\cos\theta = \frac{OC}{r} = \frac{OC}{1} = OC$$

It is evident *that the sine of an angle can be represented by the ordinate (y value) of any point where the end of the radius vector coincides with the circumference of the circle.* Hence, the length BC represents $\sin\theta$ in all quadrants, as shown in Fig. 24-20. Note that the ordinate gives both the sign and the magnitude of the sine in any quadrant. Thus, in quadrants I and II, $\sin\theta = +0.6$; in quadrants III and IV, $\sin\theta = -0.6$. That is, when the radius vector is above the x axis, the ordinate and therefore the sine are positive. When the radius vector is below the x axis, the ordinate and therefore the sine are negative.

Similarly, *the cosine of an angle can be represented by the abscissa (x value) of any point where the end of the radius vector coincides with the circumference of the circle.* Hence, the length OC represents $\cos\theta$ in all quadrants, as shown in Fig. 24-20. The abscissa gives both the sign and the magnitude of the cosine in any quadrant. Thus, in quadrants I and IV, $\cos\theta = +0.8$; in quadrants II and III, $\cos\theta = -0.8$. That is, when the radius vector is to the right of the y axis, the abscissa and therefore the cosine are positive. When the radius vector is to the left of the y axis, the abscissa, and therefore the cosine, are negative.

In Fig. 24-20, the radius vector has been

Fig. 24-20 Line Representation of Functions

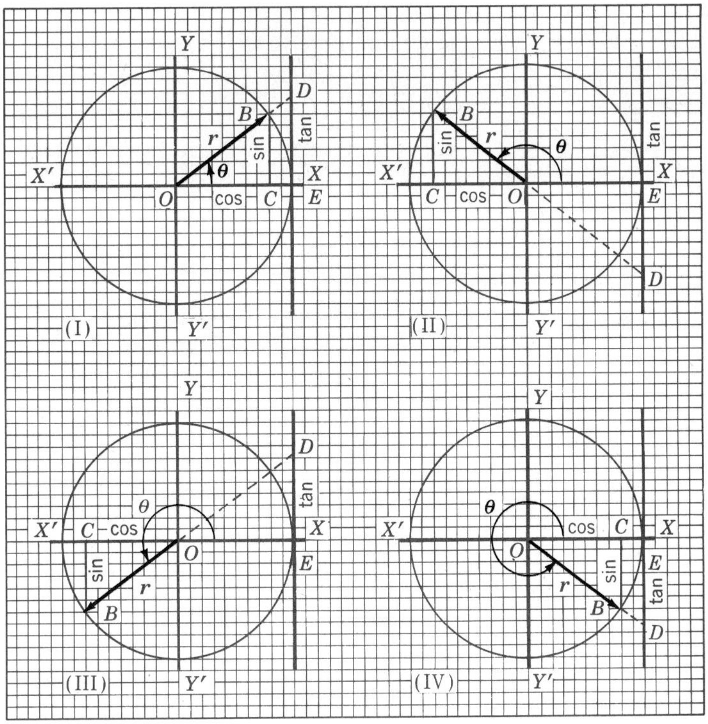

extended to intersect the tangent line DE which has been drawn tangent to the circle at the positive x axis. Since by construction, DE is perpendicular to OX, OBC and ODE are similar right triangles, for they have a common acute angle BOC. From the similar triangles,

$$\frac{BC}{OC} = \frac{DE}{OE}$$

Then, in each of the four quadrants,

$$\tan\theta = \frac{BC}{OC} = \frac{DE}{OE} = \frac{DE}{1} = DE$$

From the above, it is evident that *the tangent of an angle can be represented by the ordinate (y value) of any point where the extended radius vector intersects the tangent line.* The ordinate gives both the sign and the magnitude of the tangent in any quadrant. Thus, in quadrants I and III, $\tan\theta = +0.75$; in quadrants II and IV, $\tan\theta = -0.75$.

PROBLEMS 24-3

1. What is the least value $\sin\theta$ may have?
2. What is the least value $\cos\theta$ may have?
3. What is the greatest value $\csc\theta$ may have in the first quadrant?
4. What is the greatest value $\sec\theta$ may have in the fourth quadrant?
5. Can the secant and cosecant have values between -1 and $+1$?
6. What is the greatest value $\sin\theta$ may have in the (*a*) first quadrant, (*b*) second quadrant, (*c*) third quadrant, and (*d*) fourth quadrant?
7. What is the greatest value $\cos\theta$ may have in the (*a*) first quadrant, (*b*) second quadrant, (*c*) third quadrant, and (*d*) fourth quadrant?

25 trigonometric tables

For the purpose of making computations, it is evident that a table of trigonometric functions would be helpful. Such a table could be made by computing the functions of all angles by graphical methods. However, that would be laborious and the resulting functions would not be accurate.

Fortunately, mathematicians have calculated the values of the trigonometric functions by the use of advanced mathematics and have tabulated the results. These tables are known as *tables of natural functions* to distinguish them from *tables of the logarithms of the functions.* In Tables 9 to 12 of the Appendix are arranged the natural functions of angles for every one-tenth of a degree from 0° to 90°.

25-1 GIVEN AN ANGLE—TO FIND THE DESIRED FUNCTION

How to use the tables of natural functions is best illustrated by examples.

WHEN THE ANGLE IS GIVEN IN THE TABLES

example 1 Find the sine of 36.7°.

solution The angle 36° is found in the left column of Table 9. The sine of 36.7° is read in the 36° row and the column headed 0.7°. It is 0.597 63.

$$\sin 36.7° = 0.597\ 63$$

example 2 Find the cosine of 7.9°.

solution The angle 7° is in the left column of Table 10. The cosine of 7.9° is read in the 7° row and in the column headed 0.9°. It is 0.990 51.

$$\cos 7.9° = 0.990\ 51$$

example 3 Find the tangent of 79.1°.

solution Opposite 79° in Table 11, and in the column headed 0.1°, read 5.192 93.

$$\tan 79.1° = 5.192\ 93$$

WHEN THE ANSWER IS NOT GIVEN IN THE TABLES DIRECTLY

example 4 Find the sine of 26.42°.

solution 1 Since 26.42° is between 26.4° and

26.5°, its sine value must be between sin 26.4° and sin 26.5°.

$$\begin{aligned} \sin 26.5° &= 0.446\,20 \\ \underline{\sin 26.4°} &= \underline{0.444\,64} \\ \text{Difference} &= 0.001\,56 \end{aligned}$$

The *tabular difference* between these sine values is 0.001 56, and it is apparent that an increase of 0.1° from 26.4° to 26.5° causes the sine value to increase by 0.001 56. Therefore, an increase from 26.4° to 26.42° (which is an increase of 0.02°) must have a corresponding increase in sine value 0.2 times 0.001 56 = 0.000 312.

$$\begin{aligned} \sin 26.42° &= 0.444\,64 + 0.000\,312 \\ &= 0.444\,952 \end{aligned}$$

The sine of 26.42°, as written in Example 4, is another good example of how the retention of decimals might easily convey a false impression of accuracy. The tables from which the sine values were taken are correct to five significant figures. Therefore any sine value found by interpolation (the technical expression for the operation developed in Example 4) should not be expressed beyond five significant figures. Thus, it is correct to write

$$\sin 26.42° = 0.444\,95$$

solution 2 The trigonometric tables (Tables 9 to 11) contain, to the right of the main columns, additional columns headed "Proportional Parts." In these portions of the tables the arithmetic (the subtraction and subsequent multiplication in Example 4) has been done for you. The small numbers of the proportional parts combined with the five-figure numbers of the main tables provide the complete trigonometric value:

Find sin 26.42°.

From Table 9, sin 26.4°	= 0.444 64
From the same horizontal line, add the proportional part for 2	= 31
Adding, sin 26.42°	= 0.444 95

For further details relating to the use of the proportional parts, especially as they relate to repeated interpolation, please refer to the introductory note "Interpolation of Mathematical Tables" immediately preceding Table 9 in the Appendix.

example 5 Find the cosine of 53.77°.

solution cos 53.7°	= 0.592 01
From the same horizontal line, subtract (indicated by the shading of the proportional parts) the proportional part for 7	= 98
Subtracting, cos 53.77°	= 0.591 03

example 6 Find the tangent of 48.13°.

solution tan 48.1°	= 1.114 52
Add the proportional part for 3	= 1 19
Adding, tan 48.13°	= 1.115 71

PROBLEMS 25-1

1. Find the sine, cosine, and tangent of (*a*) 18°, (*b*) 68°, (*c*) 9.3°, (*d*) 52.5°, (*e*) 2.6°.

2. Find the sine, cosine, and tangent of (*a*) 12°, (*b*) 88.7°, (*c*) 70.2°, (*d*) 0.8°, (*e*) 20.1°.

3. Find the sine, cosine, and tangent of (*a*) 1.94°, (*b*) 57.36°, (*c*) 38.91°, (*d*) 40.28°, (*e*) 55.37°.
4. Find the sine, cosine, and tangent of (*a*) 7.39°, (*b*) 12.18°, (*c*) 32.65°, (*d*) 41.55°, (*e*) 3.17°.

25-2 INVERSE TRIGONOMETRIC FUNCTIONS

Frequently some form of notation is needed in order to express an angle in terms of one of its functions. For example, in Sec. 24-3 Example 3 dealt with an angle whose tangent was $\frac{9}{10}$. Similarly, in Example 4 of the same section, we considered an angle whose cosine was $\frac{3}{4}$.

If $\sin\theta = x$, then θ is an angle whose sine is x. It has been agreed to express such a relation by the notation

$$\theta = \sin^{-1} x \qquad \text{or} \qquad \theta = \arcsin x$$

Both are read "θ is equal to the angle whose sine is x" or "the inverse sine of x." For example, the tangent of 32.7° is 0.759 04. Stated as an inverse function, this would be written

$$37.2° = \arctan 0.759\,04$$

Similarly, in the case of a right triangle labeled as in Fig. 24-8, we should write $\theta = \arctan\dfrac{a}{b}$, $\theta = \arccos\dfrac{b}{c}$, etc. In this book, we shall not use the notation "$\theta = \sin^{-1} x$" (although it appears on some calculators), for we prefer not to use an exponent when no exponent is intended. Although this form of notation is used in a number of texts, you will find that nearly all recent mathematics and engineering texts are using the "$\theta = \arcsin x$" form of notation. Because more advanced mathematics employs trigonometric functions affected by exponents, it is evident that confusion would eventually result from utilizing the other notation for specifying the inverse functions.

25-3 GIVEN A FUNCTION—TO FIND THE CORRESPONDING ANGLE

As in Sec. 25-1 the use of the tables is best illustrated by examples.

WHEN THE FUNCTION IS GIVEN IN THE TABLES

example 7 Find the angle whose sine is 0.235 14.

solution Find 0.235 14 in Table 9. In the degrees column opposite the row in which 0.235 14 is located, read 13°. The column in which 0.235 14 is located is the 0.6° column. Thus, 0.235 14 is the sine of 13.6°. This could be written

$$13.6° = \arcsin 0.235\,14$$

example 8 Find θ if $\cos\theta = 0.033\,16$

solution Since the given cosine value is a very small number, we deduce that the corresponding angle must be large. Here again a knowledge of how the functions vary is an asset because it saves time in looking up angles whose functions are given. In the degrees column of Table 10 opposite the row in which 0.033 16 is located, read 88°; and 0.033 16 is in the 0.1°

column. Thus, 0.033 16 is the cosine of 88.1°. This can be written

$$\cos 88.1° = 0.033\ 16$$

or

$$88.1° = \arccos 0.033\ 16$$

example 9 Find θ if $\theta = \arctan 1.142\ 29$.

solution Since $\tan 45° = 1$ and the tangent value increases as the angle increases, it is evident that θ is somewhat larger than 45°. This knowledge enables us to begin searching for the given tangent value somewhere near its location.

In the degrees column of Table 11 the row in which 1.142 29 is located, read 48°; and 1.142 29 is in the 0.8° column. Thus, 1.142 29 is the tangent of 48.8°. That is,

$$48.8° = \arctan 1.142\ 29$$

WHEN THE FUNCTION IS NOT GIVEN IN THE TABLES

example 10 Find θ if $\theta = \arcsin 0.445\ 26$.

solution 1 Examination of the table shows that 0.445 26 is not given exactly in the sine values; therefore, we find the two consecutive sine values between which the given sine value lies. These are 0.444 64 and 0.446 20, which are the sines of 26.4° and 26.5°, respectively. Tabulating,

$$\begin{aligned} \sin 26.5° &= 0.446\ 20 \\ \sin 26.4° &= \underline{0.444\ 64} \\ \text{Difference} &= 0.001\ 56 \end{aligned}$$

The *tabular difference* between these sine values is 0.001 56, and it is apparent that an increase of 0.1° from 26.4° causes the sine value to increase 0.001 56. Now, the given sine value is 0.000 62 larger than the sine of the smaller angle taken from the table (0.445 26 − 0.444 64 = 0.000 62). Then, since

$$\frac{\text{Increase}}{\text{Difference}} = \frac{0.000\ 62}{0.001\ 56} = 0.4$$

the given sine value is 0.4 of the way from 0.444 64 to 0.446 20. Therefore, we assume that θ is 0.4 of the way from 26.4° to 26.5°. Hence,

$$\begin{aligned} \theta &= 26.4° + (0.4)(0.1°) \\ &= 26.4° + 0.04° = 26.44° \end{aligned}$$

(As in Example 4, it is meaningless to take the interpolated value to an excessive number of decimal places.)

solution 2 Find $\theta = \arcsin 0.445\ 26$.

The closest value to 0.445 26 in Table 9 is 0.444 64. Find the difference between the desired value and the tabular value:

$$\begin{aligned} \text{Desired} \quad & 0.445\ 26 \\ \underline{\text{Tabular}} \quad & \underline{0.444\ 64} = \sin 26.4° \\ \text{Difference} = {} & 0.000\ 62 \end{aligned}$$

On the same horizontal line in Table 9 as 0.444 64, find the difference 62 in the proportional parts column headed 4: add

$$\begin{aligned} & \underline{0.04} \\ \arcsin 0.445\ 26 = {} & 26.44° \end{aligned}$$

example 11 Find θ if $\cos\theta = 0.373\ 15$.

solution

Available in Table 10,		0.374 61	= cos 68.0°
Desired value ($\cos\theta$)	=	0.373 15	
Difference	=	0.001 46	

From the proportional parts columns on the same horizontal line as 68°, a difference of 146 = 9:

$$\begin{aligned} & \underline{.09°} \\ \theta = {} & 68.09° \end{aligned}$$

Since the *smaller* cosine value represents a *larger* angle, we subtract the desired cosine from the tabular value. The shading of the proportional parts in Table 10 reminds us of this reverse trend for cosines.

example 12 Find θ if $\theta = \arctan 0.591\,87$.

solution

Desired value ($\tan \theta$)	$= 0.591\,87$	
Available in Table 11	$0.591\,40$	$= \tan 30.6°$
Difference	$= 0.000\,47$	

From the proportional parts of Table 11, a difference of 47 represents 2:

$$\begin{array}{r} .02° \\ \hline \theta = 30.62° \end{array}$$

25-4 ACCURACY

The methods of interpolation illustrated here are for the use of those who require a greater degree of accuracy than that given by working with angles to the nearest tenth of a degree. In our considerations of ac circuits, we shall confine our accuracy to three significant figures and angles to the nearest tenth of a degree. This, except for isolated cases, will more than meet all practical requirements. Also, it reduces interpolation to an inspection of the values of the tabulated functions in order to determine which tenth of a degree to choose.

Inside the front cover of this book is a three-place table of sines, cosines, and tangents for each degree from 0° to 90°. With the confidence gained from working with the components that form all but the most precise circuits, you will find that this table will serve many of your needs.

You should study the tables at this point and satisfy yourself that, for angles up to about 6°, the values of $\sin \theta$ and $\tan \theta$ are within 0.55% of each other and, at 10°, the difference is only 1.52%. It is because of the closeness of the values of sin and tan that many slide rules incorporate an ST or SRT scale that gives as equal the sines and tangents of angles up to 5.73°. The percentage error in accepting the approximation is well within the tolerance of ordinary electronics components. 5.73° is the reasonable place to break the scales because it is at this point that the numerical values change from 10^{-2} to 10^{-1}, which offers a ready relationship between the ST and the D scales.

If you have a slide rule, you should take special pains to relate the S, T, and ST scales on your particular rule to the D, C, DI, or A and B scales, so that you will be able to perform with ease all the necessary operations of multiplying and dividing by the trigonometric functions. The use of the slide rule reduces the necessity of using the tables of functions except when an extremely high degree of accuracy is desired. Finding an angle corresponding to a given function or finding the function of a given angle may be accomplished by one setting of the cursor. It is in work involving trigonometric functions that the use of the slide rule and electronic calculator begins to be rewarding in saving time and labor.

PROBLEMS 25-2

1. Find the angles having the following values as sines:
(*a*) 0.453 99, (*b*) 0.116 67, (*c*) 0.878 82, (*d*) 0.644 12, (*e*) 0.037 34.

2. Find the angles having the following values as cosines:
(*a*) 0.965 93, (*b*) 0.190 81, (*c*) 0.998 72, (*d*) 0.866 90, (*e*) 0.343 17.

3. Find the angles whose tangents are (a) 12.429, (b) 0.008 73, (c) 0.842 08, (d) 1.651 20, (e) 0.482 34.

4. Find θ if:
(a) $\theta = \arctan 1.356\,37$
(b) $\theta = \arccos 0.486\,34$
(c) $\theta = \arcsin 0.273\,96$
(d) $\theta = \arccos 0.048\,85$
(e) $\theta = \arcsin 0.518\,03$

5. Find θ if:
(a) $\theta = \arccos 0.973\,74$
(b) $\theta = \arctan 0.009\,25$
(c) $\theta = \arcsin 0.963\,06$
(d) $\theta = \arctan 0.893\,15$
(e) $\theta = \arcsin 0.732\,66$

25-5 FUNCTIONS OF ANGLES GREATER THAN 90°

You have noted that the trigonometric functions have been tabulated only for angles of 0° to 90°. The signs and magnitudes for angles in all quadrants were considered in the preceding chapter, and it is evident that a table of functions for all angles will be needed. Because the existing tables are for angles in the first quadrant, there must be methods of expressing any angle in terms of an angle of the first quadrant in order to make use of the table of functions.

25-6 TO FIND THE FUNCTIONS OF AN ANGLE IN THE SECOND QUADRANT

In Fig. 25-1, let θ represent any angle in the second quadrant. From any point P on the radius vector $\mathbf{r}$, draw the perpendicular y to the horizontal axis. The acute angle that $\mathbf{r}$ makes with the horizontal axis is designated by ϕ. Then, since $\theta + \phi = 180°$, θ and ϕ are supplementary angles. Hence,

Fig. 25-1 θ and ϕ Are Supplementary Angles; $\theta + \phi = 180°$

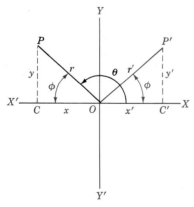

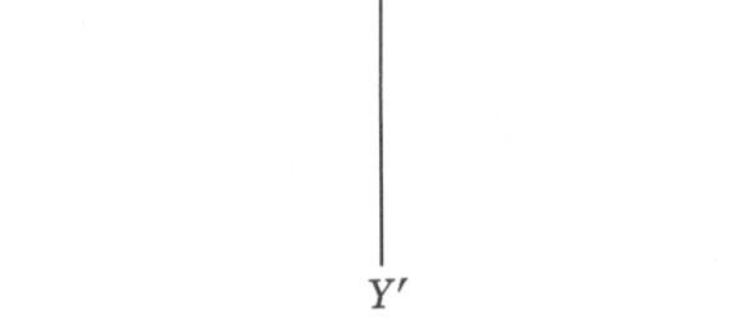

$$\phi = 180° - \theta$$

Now construct the angle XOP' in the first quadrant equal to ϕ, make $\mathbf{r}'$ equal to $\mathbf{r}$, and draw y' perpendicular to OX. Since the right triangles OPC and $OP'C'$ are equal, $x = -x'$ and $y = y'$. Then

$$\sin (180° - \theta) = \frac{y}{r} = \frac{y'}{r'} = \sin \phi$$

$$\cos (180° - \theta) = \frac{x}{r} = \frac{-x'}{r'} = -\cos \phi$$

$$\tan (180° - \theta) = \frac{y}{x} = \frac{y'}{-x'} = -\tan \phi$$

These relationships show that the function of an angle has the same absolute value as the same function of its supplement. That is, if two angles are supplementary, their sines are equal and their cosines and tangents are equal in magnitude but opposite in sign.

example 13

$$\begin{aligned} \sin 140° &= \sin (180° - 140°) \\ &= \sin 40° = 0.642\,79 \\ \cos 100° &= -\cos (180° - 100°) \\ &= -\cos 80° = -0.173\,65 \\ \tan 175° &= -\tan (180° - 175°) \\ &= -\tan 5° = -0.087\,49 \end{aligned}$$

25-7 TO FIND THE FUNCTION OF AN ANGLE IN THE THIRD QUADRANT

In Fig. 25-2, let θ represent any angle in the third quadrant and let ϕ be the acute angle that the radius vector $\mathbf{r}$ makes with the horizontal axis. Then

$$\phi = \theta - 180°$$

Now construct the angle XOP' in the first quadrant equal to ϕ, make $\mathbf{r}'$ equal to $\mathbf{r}$, and draw y and y' perpendicular to the horizontal axis. Since the right triangles OPC and $OP'C'$ are equal, $x = -x'$ and $y = -y'$. Then

$$\sin(\theta - 180°) = \frac{y}{\mathbf{r}} = \frac{-y'}{\mathbf{r}'} = -\sin\phi$$

$$\cos(\theta - 180°) = \frac{x}{\mathbf{r}} = \frac{-x'}{\mathbf{r}'} = -\cos\phi$$

$$\tan(\theta - 180°) = \frac{y}{x} = \frac{-y'}{-x'} = \tan\phi$$

These relationships show that the function of an angle in the third quadrant has the same absolute value as the same function of the acute angle between the radius vector and the horizontal axis. The signs of the functions are the same as for any angle in the third quadrant, as discussed in Sec. 24-5.

Fig. 25-2 θ Is in the Third Quadrant; $\phi = \theta - 180°$

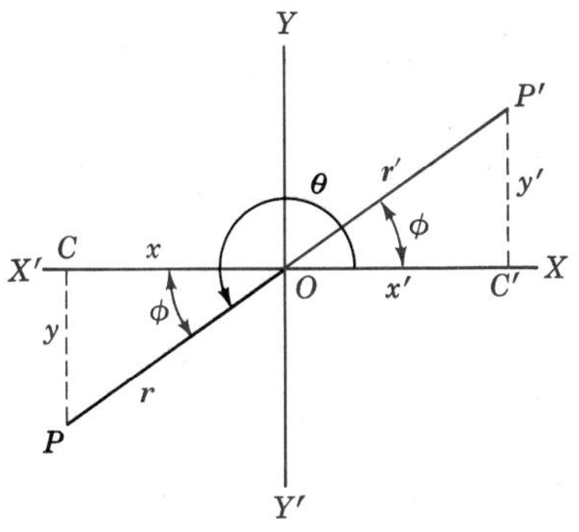

example 14

$$\begin{aligned}
\sin 200° &= -\sin(200° - 180°)\\
&= -\sin 20° = -0.342\,02\\
\cos 260° &= -\cos(260° - 180°)\\
&= -\cos 80° = -0.173\,65\\
\tan 234° &= \tan(234° - 180°)\\
&= \tan 54° = 1.376\,38
\end{aligned}$$

25-8 TO FIND THE FUNCTIONS OF AN ANGLE IN THE FOURTH QUADRANT

In Fig. 25-3, let θ represent any angle in the fourth quadrant and let ϕ be the acute angle that the radius vector $\mathbf{r}$ makes with the horizontal axis. Then

$$\phi = 360° - \theta$$

Now construct the angle XOP' in the first quadrant equal to ϕ, make $\mathbf{r}'$ equal to $\mathbf{r}$, and draw y and y' perpendicular to the horizontal axis. Since the right triangles OPC and $OP'C$ are equal, $y = -y'$. Then

Fig. 25-3 θ Is in the Fourth Quadrant; $\phi = 360° - \theta$

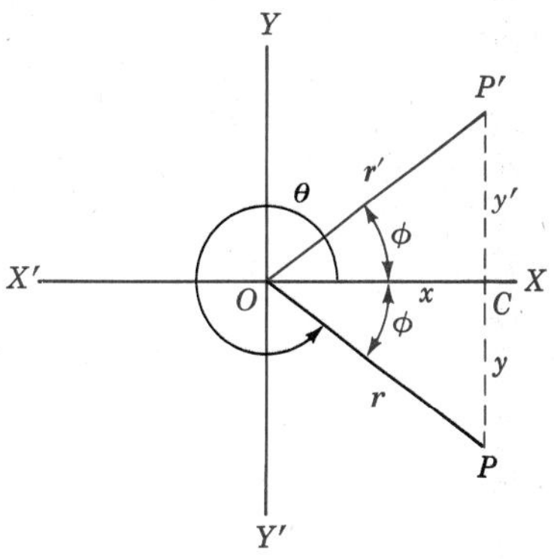

$$\sin(360° - \theta) = \frac{y}{r} = \frac{-y'}{r'} = -\sin\phi$$

$$\cos(360° - \theta) = \frac{x}{r} = \frac{x}{r'} = \cos\phi$$

$$\tan(360° - \theta) = \frac{y}{x} = \frac{-y'}{x} = -\tan\phi$$

These relationships show that the functions of an angle in the fourth quadrant have the same absolute value as the same functions of an acute angle in the first quadrant equal to $360° - \theta$. The signs of the functions, however, are those for an angle in the fourth quadrant, as discussed in Sec. 24-5.

example 15

$$\begin{aligned}
\sin 300° &= -\sin(360° - 300°) \\
&= -\sin 60° = -0.866\,03 \\
\cos 285° &= \cos(360° - 285°) \\
&= \cos 75° = 0.258\,82 \\
\tan 316° &= -\tan(360° - 316°) \\
&= -\tan 44° = -0.965\,69
\end{aligned}$$

25-9 TO FIND THE FUNCTIONS OF AN ANGLE GREATER THAN 360°

Any angle θ greater than 360° has the same trigonometric functions as θ minus an integral multiple of 360°. That is, a function of an angle larger than 360° is found by dividing the angle by 360° and finding the required function of the remainder. Thus θ in Fig. 25-4 is a positive angle of 955°. To find any function of 955°, divide 955° by 360°, which gives 2 with a remainder of 235°. Hence,

$$\begin{aligned}
\sin 955° &= \sin 235° = -\sin(235° - 180°) \\
&= -\sin 55° = -0.819\,15 \\
\cos 955° &= \cos 235° = -\cos(235° - 180°) \\
&= -\cos 55° = -0.573\,58 \\
\tan 955° &= \tan 235° = \tan(235° - 180°) \\
&= \tan 55° = 1.428\,15
\end{aligned}$$

Fig. 25-4 $\theta = 955°$

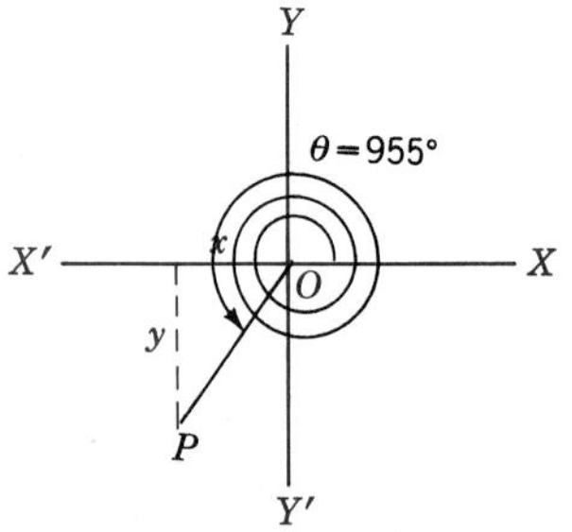

25-10 TO FIND THE FUNCTIONS OF A NEGATIVE ANGLE

In Fig. 25-5, let $-\theta$ represent a negative angle in the fourth quadrant made by the radius vector r and the horizontal axis. Construct the angle θ in the first quadrant equal to $-\theta$, make r' equal to r, and draw y and y' perpendicular to the horizontal axis. Since the right triangles OPC and $OP'C$ are equal, $y = -y'$. Then

$$\sin(-\theta) = \frac{y}{r} = \frac{-y'}{r'} = -\sin\theta$$

$$\cos(-\theta) = \frac{x}{r} = \frac{x}{r'} = \cos\theta$$

$$\tan(-\theta) = \frac{y}{x} = \frac{-y'}{x'} = -\tan\theta$$

These relationships are true for any values of $-\theta$ regardless of the quadrant or the magnitude of the angle.

Fig. 25-5 $-\theta$ Generated by Clockwise Rotation

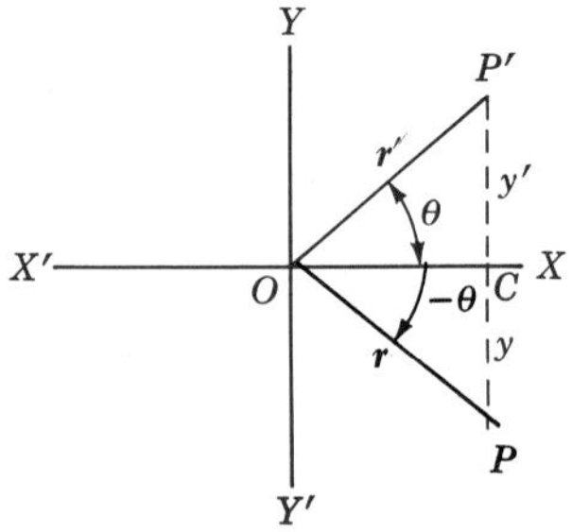

example 16

$$\sin(-65°) = -\sin 65° = -0.906\ 31$$
$$\cos(-150°) = -\cos 150° = -\cos(180° - 150°) = -\cos 30° = -0.866\ 03$$
$$\tan(-287°) = -\tan 287° = -\tan(360° - 287°) = -(-\tan 73°) = 3.270\ 85$$

25-11 TO REDUCE THE FUNCTIONS OF ANY ANGLE TO THE FUNCTIONS OF AN ACUTE ANGLE

It has been shown in the preceding sections that all angles can be reduced to terms of $(180° - \theta)$, $(\theta - 180°)$, $(360° - \theta)$, or θ. These results can be summarized as follows:

Rule

To find any function of any angle θ, take the same function of the acute angle formed by the terminal side (radius vector) and the *horizontal* axis and prefix the proper algebraic sign for that quadrant.

When finding the functions of angles, you should make a sketch showing the approximate location of the angle. This procedure will clarify the trigonometric relationships, and in addition, many errors will be avoided by using it.

example 17 Find the functions of 143°.

solution Construct the angle 143°, and mark the signs of the radius vector, abscissa, and ordinate, as shown in Fig. 25-6. (The radius vector is always positive.) Since 180° − 143° = 37° the acute angle for the functions is 37°. Hence,

Fig. 25-6 180° − 143° = 37°

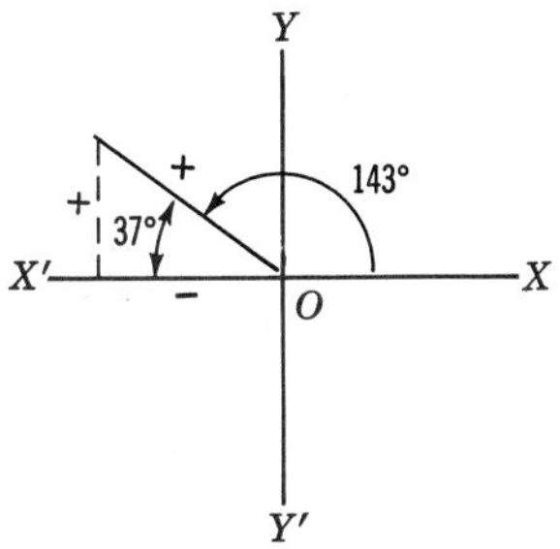

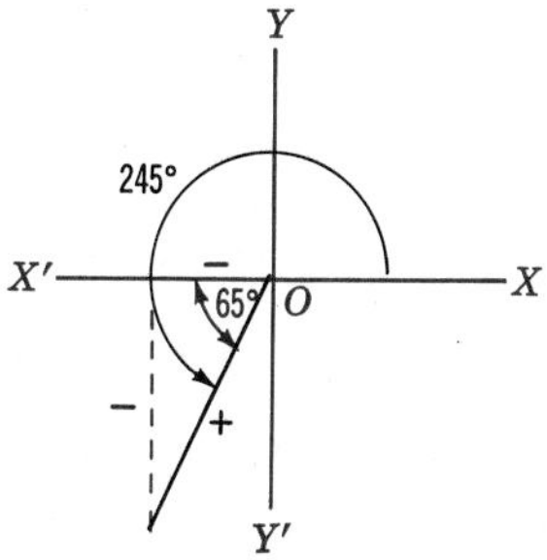

Fig. 25-7 245° − 180° = 65°

$$\sin 143° = \sin 37° = 0.601\ 82$$
$$\cos 143° = -\cos 37° = -0.798\ 64$$
$$\tan 143° = -\tan 37° = -0.753\ 55$$

example 18 Find the functions of 245°.

solution Construct the angle 245° as shown in Fig. 25-7. Since 245° − 180° = 65° the acute angle for the functions is 65°. Hence,

$$\sin 245° = -\sin 65° = -0.906\ 31$$
$$\cos 245° = -\cos 65° = -0.422\ 62$$
$$\tan 245° = \tan 65° = 2.144\ 51$$

example 19 Find the functions of 312°.

solution Construct the angle 312° as shown in Fig. 25-8. Since 360° − 312° = 48° the

Fig. 25-8 360° − 312° = 48°

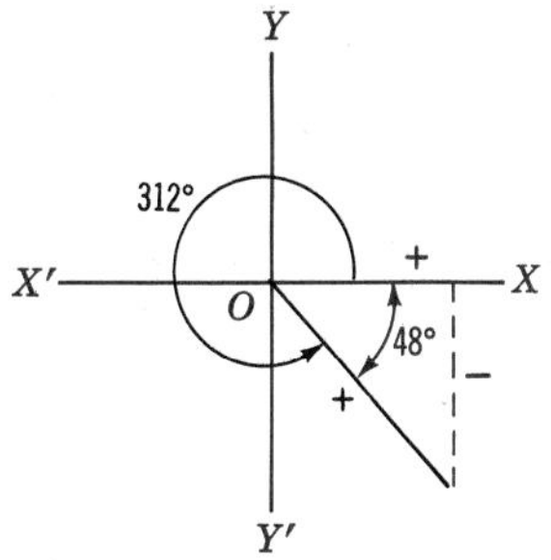

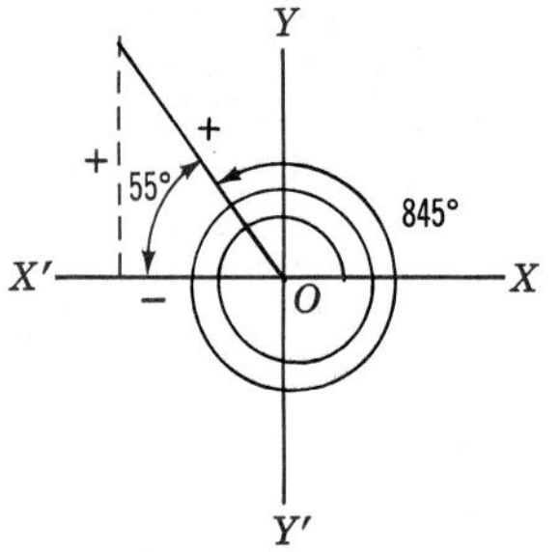

Fig. 25-9 Functions of 845° Are the Same as Those of 125°

acute angle for the functions is 48°. Hence,

$$\sin 312° = -\sin 48° = -0.743\ 14$$
$$\cos 312° = \cos 48° = 0.669\ 13$$
$$\tan 312° = -\tan 48° = -1.110\ 61$$

example 20 Find the functions of 845°.

solution $845° \div 360° = 2 + 125°$. Therefore, the functions of 125° will be identical with those of 845°. The construction is shown in Fig. 25-9. Since $180° - 125° = 55°$, the acute angle for the functions is 55°. Hence,

$$\sin 845° = \sin 55° = 0.819\ 15$$
$$\cos 845° = -\cos 55° = -0.573\ 58$$
$$\tan 845° = -\tan 55° = -1.428\ 15$$

example 21 Find the functions of −511°.

solution $-511° \div 360° = -(1 + 151°)$. Therefore, the functions of −151° will be identical with those of −511°. The construction is shown in Fig. 25-10. Since $180° - 151° = 29°$, the acute angle for the functions is 29°. Hence,

$$\sin(-151°) = -\sin 29° = -0.484\ 81$$
$$\cos(-151°) = -\cos 29° = -0.874\ 62$$
$$\tan(-151°) = \tan 29° = 0.554\ 31$$

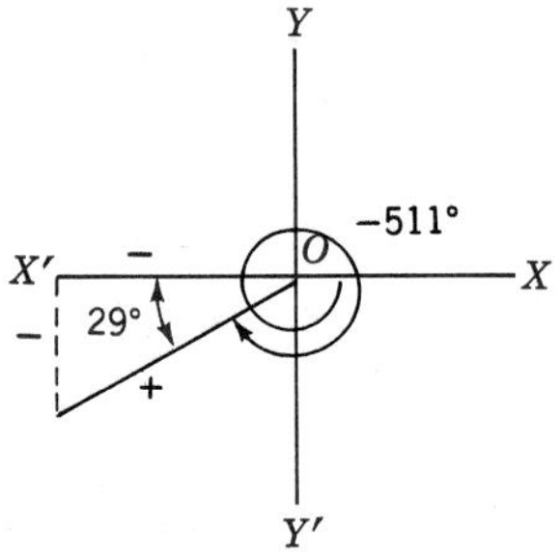

Fig. 25-10 Functions for −511° Are the Same as Those of −151°

25-12 ANGLES CORRESPONDING TO INVERSE FUNCTIONS

Now that we are able to express all angles as acute angles in order to use the table of functions from 0° to 90°, it has probably occurred to you that an important distinction exists between the direct trigonometric functions and the inverse trigonometric functions. The trigonometric functions of any given angle have only one value, whereas a given function corresponds to an infinite number of angles. For example, an angle of 30° has but one sine value, which is 0.500 00, but an angle whose sine is 0.500 00 (arcsin 0.500 00) may be taken as 30°, 150°, 390°, 480°, 510° etc.

To avoid confusion, it has been agreed that the values of arcsin θ and arctan θ which lie between +90° and −90°, in the first and fourth quadrants, are to be known as the *principal values* of arcsin θ and arctan θ. The principal value is often denoted by using a capital letter, as Arcsin θ. Thus,

	Arcsin 0.575 01	= 35.1°
and	Arcsin (−0.998 03)	= −86.4°
Also	Arctan 1.482 56	= 56°
and	Arctan (−0.069 93)	= −4°

The principal values of arccos θ are taken as the values between 0° and 180° and are denoted by Arccos θ. Thus,

Arccos 0.173 65 = 80°
and Arccos (−0.981 63) = 169°

Users of electronic calculators must take special note of the principal angles because functions of angles greater than 90° may lead to errors of interpretation.

example 22 (for electronic calculators)
Enter 120°, and call up sin 120° = 0.866 03.
Now call up arcsin 0.866 03, and read 60°.

example 23 (for electronic calculators)
Find cos 240° = −0.500 00.
Now call up arccos −0.500 00, and read 120°.

example 24 (for electronic calculators)
Find cos 300° = 0.500 00.
Now call up arccos 0.500 00, and read 60°.

example 25 (for electronic calculators)
Find tan 240° = 1.732 05.
Now call up arctan 1.732 05, and read 60°.

PROBLEMS 25-3

1. Find the sine, cosine, and tangent of (*a*) 107°, (*b*) 160°, (c) 130.1°, (*d*) 147.5°, (e) 176.2°.
2. Find the sine, cosine, and tangent of (*a*) 183°, (*b*) 235°, (c) 217.8°, (*d*) 180.9°, (e) 268.1°.
3. Find the sine, cosine, and tangent of (*a*) 280°, (b) 318°, (c) 349.9°, (*d*) 300.1°, (e) 359.5°.
4. Find the sine, cosine, and tangent of (*a*) 461°, (*b*) 510°, (c) 480.5°, (*d*) 523.2°, (e) 539.3°.
5. Find the sine, cosine, and tangent of (*a*) 905°, (*b*) −17.1°, (c) 940.7°, (*d*) −362.6°, (e) 1260.2°.
6. Find θ if:
   (*a*) θ = Arccos 0.969 02 (*b*) θ = Arcsin 0.582 12
   (c) θ = Arccos (−0.455 55) (*d*) θ = Arctan (−3.510 53)
   (e) θ = Arcsin (−0.377 84)
7. Find ϕ if:
   (*a*) ϕ = Arctan (−1.076 13) (*b*) ϕ = Arccos (−0.027 92)
   (c) ϕ = Arcsin 0.780 43 (*d*) ϕ = Arccos (−0.976 30)
   (e) ϕ = Arctan (−2.732 63)
8. The illumination on a surface that is not perpendicular to the rays of light from a light source is given by the formula

$$E = \frac{F \cos \theta}{d^2} \quad \text{lux}$$

where E = illumination at a point on the surface, lux
F = intensity of light output of source, lumens (lm)
d = distance of source of light to surface, m
θ = angle between incident light ray and a line perpendicular to the surface
Solve for F, d, and θ.

9. In the formula of Prob. 8, find the value of d if $F = 900$ lm, $\theta = 48°$, and $E = 300$ lux.

10. A 100-W lamp has a total light output of 1700 lm. Disregarding reflection, compute the illumination at a point on a surface 3 m from the lamp if the plane of the surface is at an angle of 30° to the incident rays.

11. In the formula of Prob. 8, at what angle of the plane of the surface to the incident ray will the illumination be the greatest?

12. The illumination on a horizontal surface from a source of light at a given vertical distance from the surface is given by the formula

$$E_{\mathrm{h}} = \frac{F}{h^2}\cos^3\theta \qquad \text{lux}$$

where E_{h} = illumination at a point on horizontal surface, lux
F = intensity of light output from source of light, lm
h = vertical distance from horizontal surface to source of light, m
θ = angle between incident ray and vertical line, as shown in Fig. 25-11.

note $\cos^3\theta$ means $(\cos\theta)$ raised to the third power.

Solve for F, h, and θ.

13. Use the formula of Prob. 12 to solve for E_{h} if $F = 3260$ lm, $h = 4$ m, and $\theta = 18°$.

14. Use the formula of Prob. 12 to solve for F if $E_{\mathrm{h}} = 330$ lux, $h = 4.5$ m, and $\theta = 50°$.

15. According to illumination experts, 1000 to 1500 lux of illumination on the printed page should be provided for study purposes. A 60-W, 850-lm lamp is suspended 2 m above a reading table. The reflector used projects 70% of the light downward. Does this produce a satisfactory amount of illumination on a book directly below the lamp?

16. To produce 1250 lux on the book in Prob. 15, what lumen rating lamp should be used?

17. Snell's law states that, when a wave of electromagnetic energy passes from one dielectric material to another, the ratio of the sines of the angles of

Fig. 25-11 Illumination at P from Source

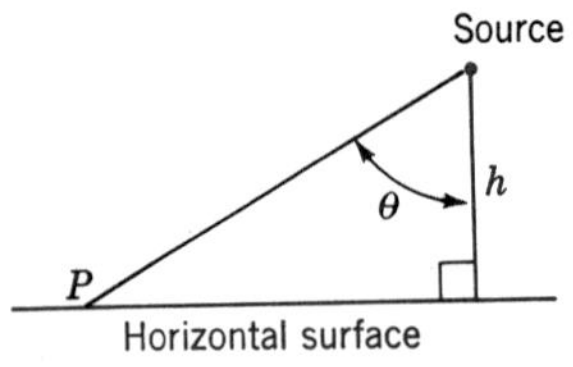

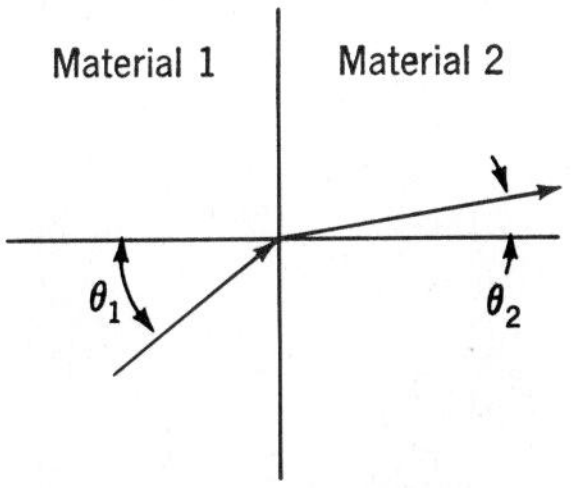

Fig. 25-12 Diagram for Prob. 17

incidence θ_1 and refraction θ_2 is inversely proportional to the square root of the ratio of the dielectric relative permittivities (Fig. 25-12). That is,

$$\frac{\sin \theta_1}{\sin \theta_2} = \sqrt{\frac{\varepsilon_2}{\varepsilon_1}}$$

If the angle of incidence $\theta_1 = 70°$, material 1 is lucite, $\varepsilon_1 = 2.6$, and material 2 is mica, $\varepsilon_2 = 5.4$, what is the angle of refraction θ_2?

26

solution of right triangles

One of the most important applications of trigonometry is the solution of triangles, both right and oblique. This chapter is concerned with the former. The right triangle is probably the most universally used geometric figure; with the aid of trigonometry, it is applied to numerous problems in measurement that otherwise might be impossible to solve.

A large percentage of the problems relating to the analysis of ac circuits and networks involves the solution of the right triangle in one form or another.

26-1 FACTS CONCERNING RIGHT TRIANGLES

Before we proceed with the actual solutions of right triangles, we will review the following useful facts regarding the properties of the right triangle:

1. The square of the hypotenuse is equal to the sum of the squares of the other two sides ($c^2 = a^2 + b^2$).
2. The acute angles are complements of each other; that is, the sum of the two acute angles is 90° ($A + B = 90°$).
3. The hypotenuse is greater than either of the other two sides and is less than their sum.
4. The greater angle is opposite the greater side, and the greater side is opposite the greater angle.

These facts will often be a material aid in checking computations made by trigonometric methods.

26-2 PROCEDURE FOR SOLUTION OF RIGHT TRIANGLES

Every triangle has three sides and three angles, and these are called the six *elements* of the triangle. To *solve* a triangle is to find the values of the unknown elements.

A triangle can be solved by two methods:

1. By constructing the triangle accurately from known elements with scale, protractor, and compasses. The unknown elements can then be measured with the scale and the protractor.
2. By computing the unknown elements from those that are known.

The first method has been used to some extent in preceding chapters. However, as previously discussed, the graphical method is cumbersome and has a limited degree of accuracy.

Trigonometry, combined with simple algebraic processes, furnishes a powerful tool for solving triangles by the second method listed above. Moreover, the degree of accuracy is limited only by the number of significant figures to which the elements have been measured and the number of significant figures available in the table of functions or the calculator used for the solution.

As pointed out in earlier chapters, every type of problem should be approached and solved in a planned and systematic manner. Only in this way are the habits of clear and ordered thinking developed, the principles of the problem understood, and the possibility of errors reduced to a minimum. With the foregoing in mind, we list the following suggestions for solving right triangles as a guide:

1. Make a reasonable sketch of the triangle, and mark the known (given) elements. This shows the relation of the elements, helps you choose the functions needed, and will serve as a check for the solution. List what is to be found.
2. To find an unknown element, select a formula that contains two known elements and the required unknown element. Substitute the known elements in the formula, and solve for the unknown.
3. As a rough check on the solution, compare the results with the drawing. To check the values accurately, note whether they satisfy relationships different from those already employed for the solution of the values being checked. A convenient check for the sides of a right triangle is the relation

$$a^2 = c^2 - b^2 = (c + b)(c - b)$$

4. In the computations, round off the numbers representing the lengths of sides to three significant figures and all angles to the nearest tenth of a degree. This means that the values of the functions employed in computations are to be used to only three significant figures. As previously stated, such accuracy is sufficient for ordinary *practical circuit* computations.

Heretofore, the right triangles used in figures for illustrative examples have been lettered in the conventional manner, as shown in Figs. 24-4, 24-11, etc. At this point the notation for the various elements will be changed to that of Fig. 26-1. In no way does this change of lettering have any effect on the fundamental relations existing among the elements of a right triangle, nor are any new ideas involved in connection with the trigonometric functions. Because certain ac problems will employ this form of notation, this is a convenient place to introduce it in order that you may become accustomed to solving right triangles lettered in this manner.

The following sections illustrate all the possible conditions encountered in the solution of right triangles.

26-3 GIVEN AN ACUTE ANGLE AND A SIDE NOT THE HYPOTENUSE

example 1 Given $R = 30.0$ and $\theta = 25.0°$. Solve for Z, X, and ϕ.

Fig. 26-1 Lettering of "Standard" Electrical Right Triangle

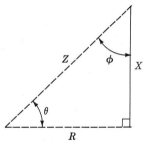

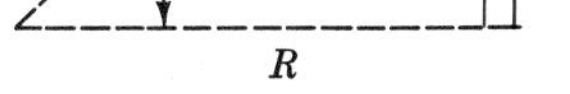

solution The construction is shown in Fig. 26-2.

$$\phi = 90° - \theta = 90° - 25° = 65°$$

An equation containing the two known elements and one unknown is

$$\tan\theta = \frac{X}{R}$$

Solving for X, $X = R\tan\theta$

Substituting the values of R and $\tan\theta$,

$$X = 30 \times 0.466 = 14.0$$

Also, since $\sin\theta = \frac{X}{Z}$

Solving for Z, $Z = \frac{X}{\sin\theta}$

Substituting the values of X and $\sin\theta$,

$$Z = \frac{14.0}{0.423} = 33.1$$

This solution can be checked by using some relation other than the relations already used. Thus, substituting values in

$$X^2 = (Z + R)(Z - R)$$

results in

$$14.0^2 = (33.1 + 30.0)(33.1 - 30.0)$$
$$196 = 63.1 \times 3.10 = 196$$

Since all results were rounded off to three significant figures, the check shows the solution to be correct for this degree of accuracy.

Fig. 26-2 Construction for Solution of Example 1

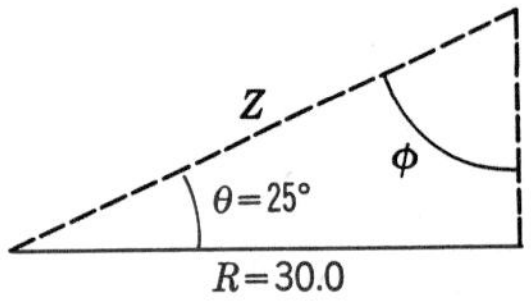

The value of Z can be checked by employing a function not used in the solution. Thus, since

$$R = Z\cos\theta$$

by substituting the values,

$$30 = 33.1 \times 0.906$$

Still another check could be made by use of an inverse function employing two of the elements found in the solution. For example,

$$\phi = \arccos\frac{X}{Z} = \arccos\frac{14.0}{33.1}$$
$$= \arccos 0.423 = 65°$$

example 2 Given $X = 106$ and $\theta = 36.4°$. Solve for Z, R, and ϕ.

solution The construction is shown in Fig. 26-3.

$$\phi = 90° - \theta = 90° - 36.4° = 53.6°$$

An equation containing two known elements and one unknown is

$$\sin\theta = \frac{X}{Z}$$

Solving for Z, $Z = \frac{X}{\sin\theta}$

Fig. 26-3 Triangle of Example 2

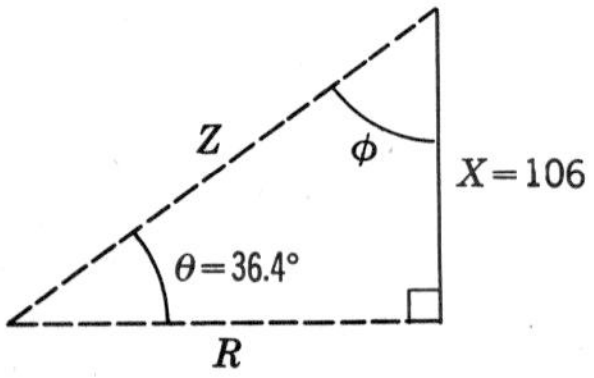

Substituting the values of X and $\sin\theta$,

$$Z = \frac{106}{0.593} = 179$$

Also, since $\cos\theta = \frac{R}{Z}$

solving for R, $R = Z\cos\theta$

Substituting the values of Z and $\cos\theta$,

$$R = 179 \times 0.805 = 144$$

Check the solution by one of the methods previously explained.

example 3 Given $R = 8.35$ and $\phi = 62.7°$. Find Z, X, and θ.

solution The construction is shown in Fig. 26-4.

$$\theta = 90° - \phi = 90° - 62.7° = 27.3°$$

When θ is found, the methods to be used in the solution of this example become identical with those of Example 1. Hence,

$$X = R\tan\theta = 8.35 \tan 27.3°$$
$$= 8.35 \times 0.516 = 4.31$$
$$Z = \frac{X}{\sin\theta} = \frac{4.31}{\sin 27.3°}$$
$$= \frac{4.31}{0.459} = 9.39$$

Fig. 26-4 Triangle of Example 3

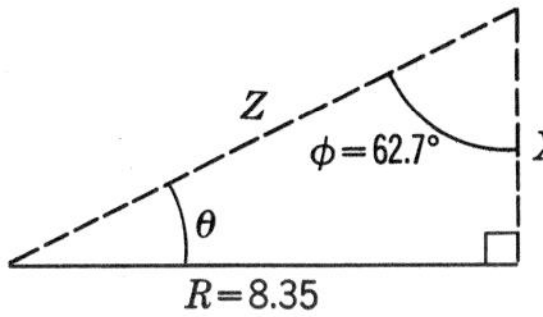

Check the solution by a method considered most convenient.

example 4 Given $X = 1290$ and $\phi = 41.9°$. Find Z, R, and θ.

solution The construction is shown in Fig. 26-5.

$$\theta = 90° - \phi = 90° - 41.9° = 48.1°$$

When θ is found, the methods to be used in the solution of this example become identical with those of Example 2. Hence,

$$Z = \frac{X}{\sin\theta} = \frac{1290}{\sin 48.1°} = \frac{1290}{0.744} = 1730$$
$$R = Z\cos\theta = 1730 \cos 48.1°$$
$$= 1730 \times 0.688 = 1190$$

Check the solution by a method considered most convenient.

With the exception of finding the unknown acute angle, which involves subtraction, any of the foregoing examples and the following problems can be solved with two movements on many slide rules.

Fig. 26-5 $X = 1290$, $\phi = 41.9°$

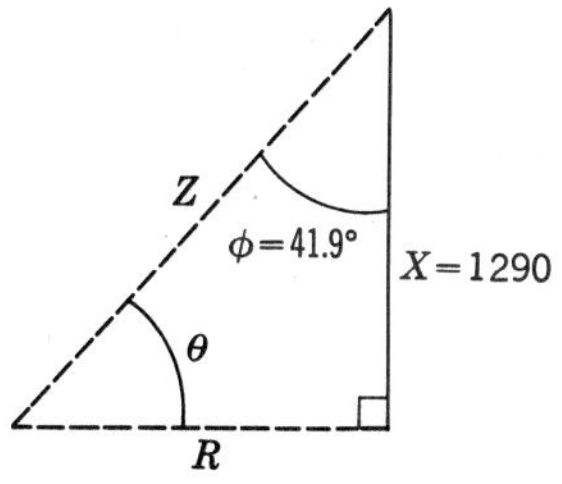

PROBLEMS 26-1

Solve the following right triangles for the unknown elements. Check each by making a construction and by substituting into a formula not used in the solution:

1. $R = 22.0,\ \theta = 34.7°$ **2.** $X = 4.39,\ \phi = 86.5°$ **3.** $X = 424,\ \phi = 45°$
4. $R = 8.10,\ \phi = 21°$ **5.** $R = 63.5,\ \theta = 24.9°$ **6.** $X = 1530,\ \theta = 73.5°$
7. $R = 8.85 \times 10^5,\ \theta = 27.7°$ **8.** $R = 222,\ \phi = 26.3°$ **9.** $X = 867,\ \theta = 57.3°$
10. $R = 0.230,\ \theta = 77°$ **11.** $X = 124,\ \theta = 51.1°$ **12.** $X = 0.0929,\ \theta = 6.4°$
13. $R = 0.105,\ \theta = 63.9°$ **14.** $R = \frac{2}{3},\ \theta = 51.9°$ **15.** $X = \frac{3}{8},\ \theta = 82.4°$
16. $R = \dfrac{1}{\sqrt{2}},\ \theta = 45°$

26-4 GIVEN AN ACUTE ANGLE AND THE HYPOTENUSE

example 5 Given $Z = 45.3$ and $\theta = 20.3°$. Find R, X, and ϕ.

solution The construction is shown in Fig. 26-6.

$$\phi = 90° - \theta = 90° - 20.3° = 69.7°$$

An equation containing two known elements and one unknown is

$$\cos\theta = \frac{R}{Z}$$

Solving for R, $\quad R = Z\cos\theta$

Substituting the values of Z and $\cos\theta$,

$$R = 45.3 \times 0.938 = 42.5$$

Another convenient equation is

Fig. 26-6 $Z = 45.3,\ \theta = 20.3°$

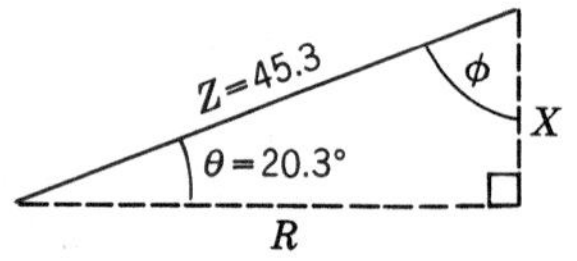

$$\sin\theta = \frac{X}{Z}$$

Solving for X, $\quad X = Z\sin\theta$

Substituting the values of Z and $\sin\theta$,

$$X = 45.3 \times 0.347 = 15.7$$

The solution can be checked by any of the usual methods.

example 6 Given $Z = 265$ and $\phi = 22.4°$. Find R, X, and θ.

solution The construction is shown in Fig. 26-7.

Fig. 26-7 $Z = 265,\ \phi = 22.4°$

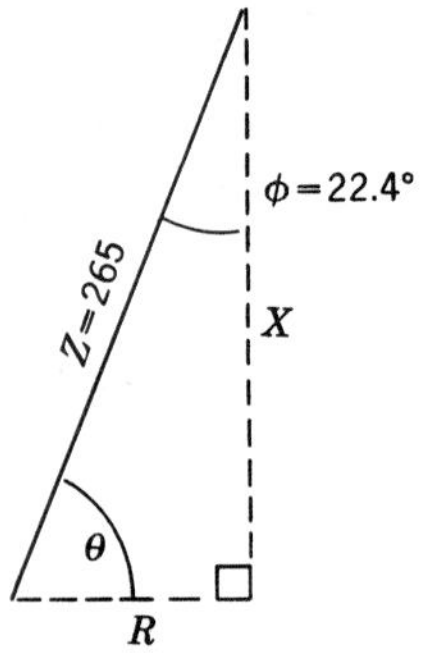

$\theta = 90° - \phi = 90° - 22.4° = 67.6°$

When θ is found, this triangle is solved by the methods used in Example 1. Hence,

$R = Z \cos\theta = 265 \cos 67.6° = 265 \times 0.381 = 101$
$X = Z \sin\theta = 265 \sin 67.6° = 265 \times 0.924 = 245$

Check the solution by one of the several methods.

PROBLEMS 26-2

Solve the following right triangles for the unknown elements. Check each by construction and by substituting into an equation not used in the solution.

1. $Z = 76.2$, $\phi = 75°$ **2.** $Z = 464$, $\theta = 23.6°$ **3.** $Z = 47.6$, $\theta = 69.1°$
4. $Z = 179$, $\phi = 77.7°$ **5.** $Z = 1 \times 10^4$, $\phi = 51.6°$ **6.** $Z = 60$, $\theta = 48.2°$
7. $Z = 0.948$, $\phi = 79.6°$ **8.** $Z = 610$, $\phi = 79.7°$ **9.** $Z = 5.10$, $\theta = 52.3°$
10. $Z = 0.342$, $\phi = 73.2°$

26-5 GIVEN THE HYPOTENUSE AND ONE OTHER SIDE

example 7 Given $Z = 38.3$ and $R = 23.1$. Find X, θ, and ϕ.

solution The construction is shown in Fig. 26-8.

An equation containing two known elements and one unknown is

$$\cos\theta = \frac{R}{Z}$$

Substituting the values of R and Z,

Fig. 26-8 Triangle of Example 7

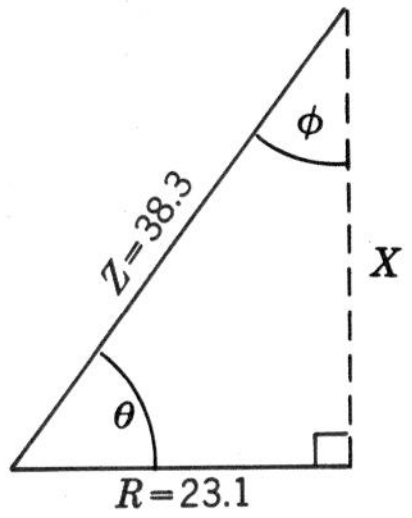

$$\cos\theta = \frac{23.1}{38.3} = 0.603$$

$$\therefore \theta = 52.9°$$
$$\phi = 90° - \theta = 90° - 52.9° = 37.1°$$

Then, since $\sin\theta = \frac{X}{Z}$

Solving for X, $X = Z \sin\theta$

Substituting the values of Z and $\sin\theta$,

$$X = 38.3 \times 0.798 = 30.6$$

example 8 Given $Z = 10.7$ and $X = 8.10$. Find R, θ, and ϕ.

solution The construction is shown in Fig. 26-9.

Fig. 26-9 Triangle of Example 8

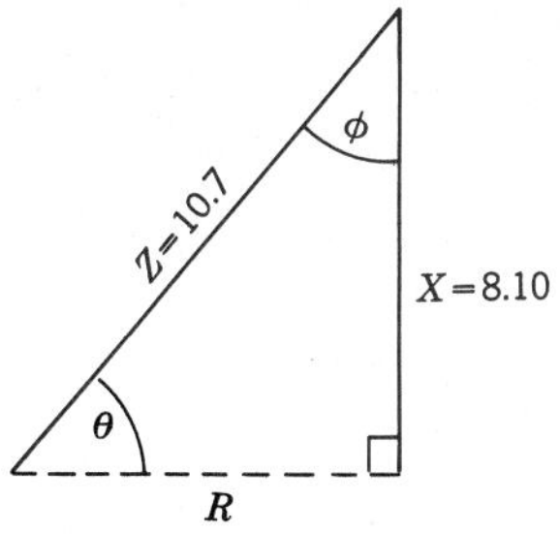

An equation containing two known elements and one unknown is

$$\sin\theta = \frac{X}{Z}$$

Substituting the values of X and Z,

$$\sin\theta = \frac{8.10}{10.7} = 0.757$$

$$\therefore \theta = 49.2°$$

$$\phi = 90° - \theta$$

$$= 90° - 49.2°$$

$$= 40.8°$$

Then, since $\cos\theta = \frac{R}{Z}$

Solving for R, $R = Z\cos\theta$

Substituting the values of Z and $\cos\theta$,

$$R = 10.7 \times 0.653 = 6.99$$

PROBLEMS 26-3

Solve the following right triangles and check each graphically and algebraically as in the preceding problems:

1. $Z = 229$, $X = 200$
2. $Z = 2160$, $R = 1200$
3. $Z = 47.6$, $R = 17$
4. $Z = 3100$, $R = 3060$
5. $Z = 0.742$, $R = 0.734$
6. $Z = 407$, $X = 57.0$
7. $Z = 1 \times 10^4$, $X = 6210$
8. $Z = 39.7$, $R = 11.4$
9. $Z = 1.08$, $R = 0.667$
10. $Z = 0.342$, $R = 0.327$

26-6 GIVEN TWO SIDES NOT THE HYPOTENUSE

example 9 Given $R = 76.0$ and $X = 37.4$. Find Z, θ, and ϕ.

solution The construction is shown in Fig. 26-10.

An equation containing two known elements and one unknown is

$$\tan\theta = \frac{X}{R}$$

Substituting the values of X and R,

$$\tan\theta = \frac{37.4}{76.0} = 0.492$$

$$\therefore \theta = 26.2°$$

$$\phi = 90° - \theta = 90° - 26.2° = 63.8°$$

$Z = 84.7$ can be found by one of the methods explained in the preceding sections.

Fig. 26-10 Triangle of Example 9

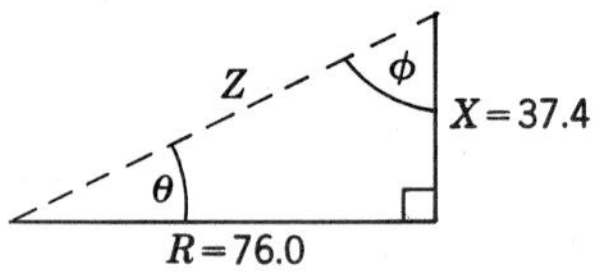

PROBLEMS 26-4

Solve the following right triangles and check as in the preceding problems:

1. $R = 35.5$, $X = 6.19$
2. $R = 11.5$, $X = 6.94$
3. $X = 5.30$, $R = 4.79$

4. $R = 76.3, X = 277$ **5.** $X = 20.3, R = 430$ **6.** $X = 50.6, R = 10.3$

7. $R = 5.43, X = 48.4$ **8.** $R = \frac{\sqrt{3}}{2}, X = \frac{1}{2}$ **9.** $X = 0.290, R = 0.280$

10. $X = 4.01, R = 5.25$

26-7 TERMS RELATING TO MISCELLANEOUS TRIGONOMETRIC PROBLEMS

If an object is higher than an observer's eye, the *angle of elevation* of the object is the angle between the horizontal and the line of sight to the object. This is illustrated in Fig. 26-11.

If an object is lower than an observer's eye, the *angle of depression* of the object is the angle between the horizontal and the line of sight to the object. This is illustrated in Fig. 26-12.

The *horizontal distance* between two points is the distance from one of the two points to a vertical line drawn through the other. Thus, in Fig. 26-13, the line AC is a vertical line through the point A and CB is a horizontal line through the point B. Then the horizontal distance from A to B is the distance between C and B.

The *vertical distance* between two points is the distance from one of the two points to the horizontal line drawn through the other. Thus, the vertical distance from A to B, in Fig. 26-13, is the distance between A and C.

Fig. 26-11 Angle of Elevation

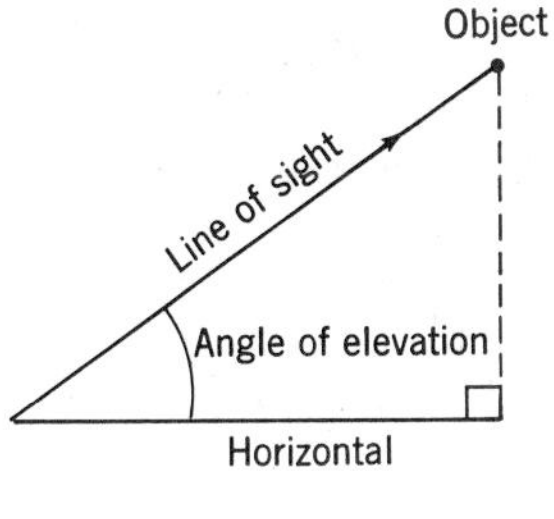

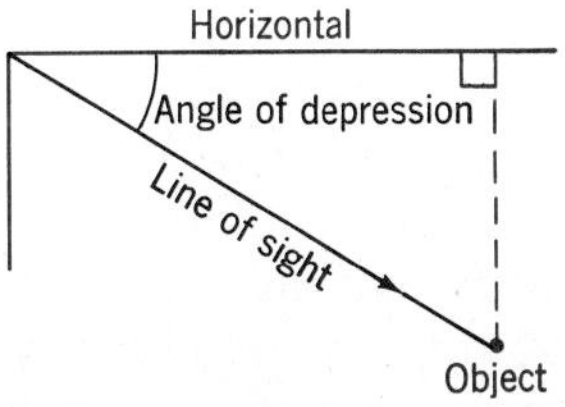

Fig. 26-12 Angle of Depression

Calculations of distance in the vertical plane are made by means of right triangles having horizontal and vertical sides. The horizontal side is usually called the *run*, and the vertical side is called the *rise* or *fall*, as the case may be.

The *slope* or *grade* of a line is the rise or fall divided by the run. Thus, if a road rises 5 m in a run of 100 m, the grade of the road is

$$5 \div 100 = 0.05 = 5\%$$

Fig. 26-13 Vertical and Horizontal Distances

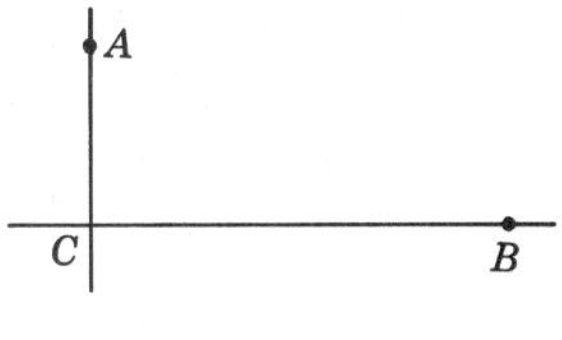

PROBLEMS 26-5

1. What is the angle of inclination of a stairway with the floor if the steps have a tread of 27 cm and a rise of 18 cm?
2. What angle does an A-frame rafter make with the horizontal if it has a rise of 3.75 m in a run of 1.5 m?

3. A transmission line rises 2.65 m in a run of 36 m. What is the angle of elevation of the line with the horizontal?
4. A radio tower casts a shadow 174 m long, and at the same time the angle of elevation of the sun is 41.7°. What is the height of the tower?
5. An antenna mast 120 m tall casts a shadow 62 m long. What was the angle of elevation of the sun?
6. At a horizontal distance of 85 m from the foot of a radio tower, the angle of elevation of the top is found to be 31°. How high is the tower?
7. A telephone pole 12.5 m high is to be guyed from its middle, and the guy is to make an angle of 45° with the ground. Allowing 1 m extra for splicing, how long must the guy wire be?
8. An extension ladder 15 m long rests against a vertical wall with its foot 4 m from the wall. (Do not use Pythagoras' theorem to solve.)
   (*a*) What angle does the ladder make with the ground?
   (*b*) How far up the wall does the ladder reach?
9. A ladder 15 m long can be placed so that it will reach a point on a wall 12 m above the ground. By tipping the ladder back without moving its foot, it will reach a point on another wall 10 m above the ground. What is the horizontal distance between the walls?
10. From the top of a cliff 58 m high, the angle of depression of a boat is 28.6°. How far out is the boat?
11. In order to find the width *BC* of a river, a distance *AB* was laid off along the bank, the point *B* being directly opposite a tree *C* on the opposite side, as shown in Fig. 26-14. If the angle *BAC* was observed to be 62.9° and *AB* was measured at 50 m, find the width of the river.
12. In order to measure the distance *AC* across a swamp, a surveyor lays off a line *AB* such that the angle *BAC* = 90°, as shown in Fig. 26-15. At point *B*, 240 m from *A*, the surveyor observes that angle *ABC* = 59.1°. Find the distance *AC*.

Fig. 26-14 Measuring across a River

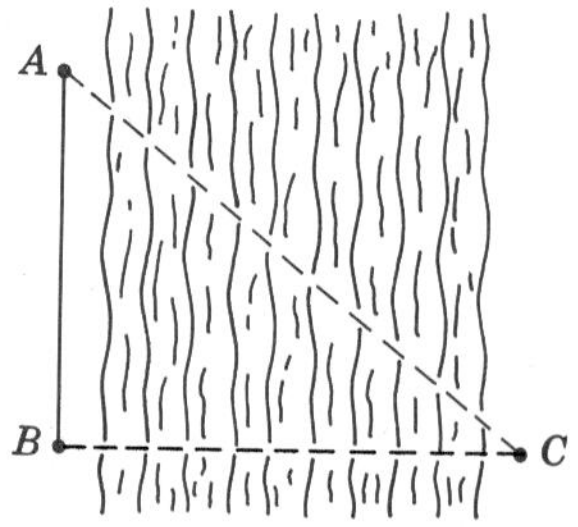

Fig. 26-15 Measuring across a Pond or Swamp

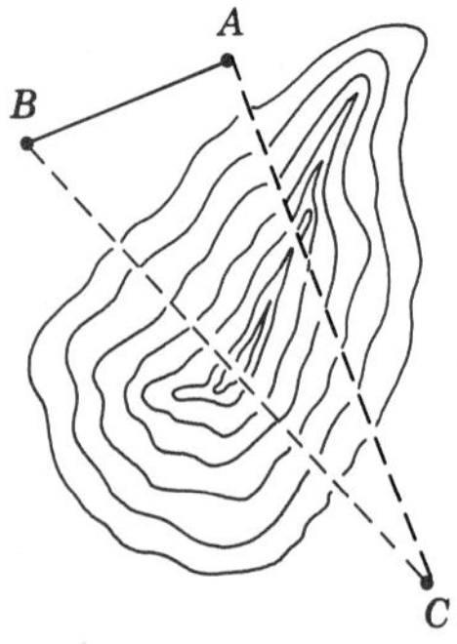

26-8 THE AREA OF TRIANGLES

A convenient use of trigonometry is the calculation of the area of a triangle. In Fig. 26-16, the area of the triangle *ABC*, from previous knowledge, is known to be

$$A = \tfrac{1}{2}ab$$

But $b = c \sin \phi$ and $a = c \sin \theta$, from which we can write

$$A = \tfrac{1}{2}ac \sin \phi$$

or

$$A = \tfrac{1}{2}bc \sin \theta$$

Fig. 26-16 Area of a Right Triangle

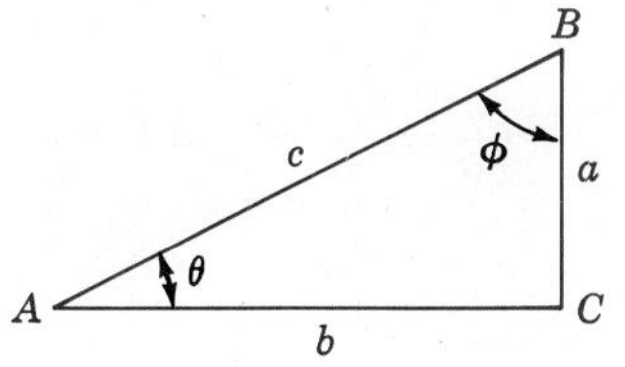

Either of these expressions may be stated:

The area of a triangle is one-half the product of any two sides times the sine of the angle between them.

You should prove that the formula holds for the more general case of the triangle of Fig. 26-17.

hint Draw an altitude perpendicular to the base.

Fig. 26-17 Area of an Oblique Triangle

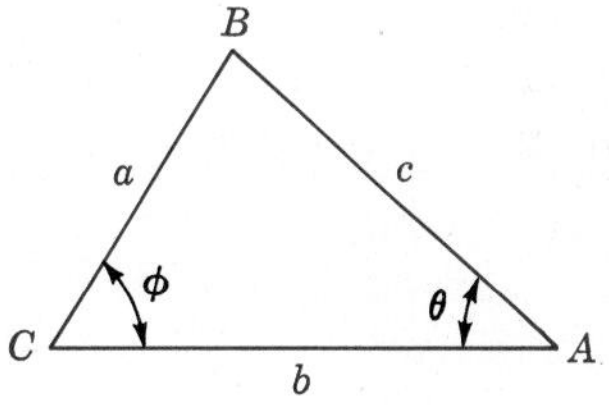

PROBLEMS 26-6

1. In the right triangle of Fig. 26-16, $a = 15$ m and $c = 40$ m.
  (*a*) What is the angle ϕ?
  (*b*) What is the area of the triangle by the sine formula?
  (*c*) What is the length of *b*?
  (*d*) What is the area by the formula $A = \tfrac{1}{2}$(base)(altitude)?

2. In the triangle of Fig. 26-17, $a = 77$ mm, $b = 96.4$ mm, $c = 72$ mm, $\phi = 47.5°$, and $\theta = 52°$.
  (*a*) What is the angle opposite side *b*?
  (*b*) What is the area of the triangle by the sine formula?
  (*c*) What is the length of altitude *h*?
  (*d*) What is the area by the formula $A = \tfrac{1}{2}$(base)(altitude)?

3. In the triangle of Fig. 26-17, $a = 100$ mm, $c = 86.5$ mm, and angle $CBA = 107°$. What is the area of the triangle?

27

trigonometric identities and equations

So far, our studies in trigonometry have been confined to the solution of *right triangles*, but there are times when other types of problems must be considered. In this chapter, we shall develop some useful relationships between the trigonometric functions, and also solve oblique triangles.

27-1 SIMPLE IDENTITIES

Consider the right triangle *ABC* (Fig. 27-1). From our studies in trigonometry we know that:

$$\sin\theta = \frac{X}{Z}$$

and

$$\cos\theta = \frac{R}{Z}$$

The ratio of these two functions is

Fig. 27-1 "Standard" Right Triangle

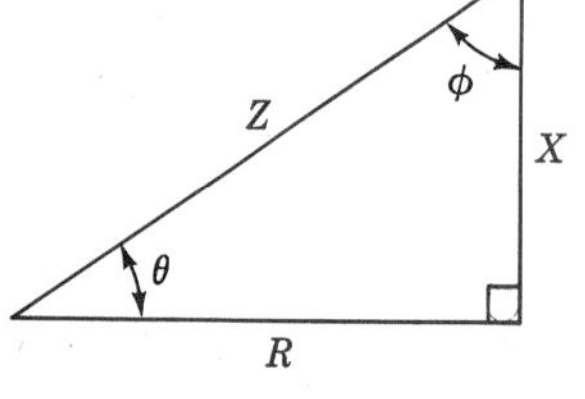

$$\frac{\sin\theta}{\cos\theta} = \frac{\frac{X}{Z}}{\frac{R}{Z}} = \frac{X}{R} = \tan\theta \qquad (1)$$

This interesting and useful relationship is the simplest of a group of trigonometric interrelationships called *identities*. We shall develop a few of the simpler identities and then tabulate them for convenience.

27-2 THE PYTHAGOREAN IDENTITIES

In the triangle of Fig. 27-1, we can readily see that

$$X^2 + R^2 = Z^2$$

the statement of Pythagoras' theorem. Dividing the entire equation by Z^2:

$$\frac{X^2}{Z^2} + \frac{R^2}{Z^2} = \frac{Z^2}{Z^2}$$

from which we can see that

$$(\sin\theta)^2 + (\cos\theta)^2 = 1$$

which is usually written (as in 12 of Problems 25-3)

$$\sin^2\theta + \cos^2\theta = 1 \tag{2}$$

This is the first of the interrelationships known as the *Pythagorean identities* because they are derived from Pythagoras' theorem. You should now repeat the process twice, dividing first by X^2 and then by R^2 to develop the other two Pythagorean identities:

$$1 + \cot^2\theta = \csc^2\theta \tag{3}$$
$$\tan^2\theta + 1 = \sec^2\theta \tag{4}$$

These relationships will prove quite useful in the advanced study of electronics because many of the mathematical descriptions of electrical and electronic phenomena are described by rather complicated combinations of trigonometric functions, and these may be often simplified by the use of identities. Here we shall confine ourselves to achieving some practice in manipulation of identities.

No set rule may be established about simplifying or proving identities. Usually, one side of the identity is manipulated until it is shown to be equal to the other side. Sometimes, each side is developed into the same equivalent in order to arrive at an obvious equality.

example 1 Show that $\dfrac{\tan^2\theta}{\sec^2\theta} + \dfrac{\cot^2\theta}{\csc^2\theta} = 1.$

solution

(*a*) One possible method of solution uses the fundamental relationships between the trigonometric functions:

$$\frac{\tan^2\theta}{\sec^2\theta} + \frac{\cot^2\theta}{\csc^2\theta} = \frac{\left(\dfrac{\sin\theta}{\cos\theta}\right)^2}{\left(\dfrac{1}{\cos\theta}\right)^2} + \frac{\left(\dfrac{1}{\tan\theta}\right)^2}{\left(\dfrac{1}{\sin\theta}\right)^2}$$
$$= \sin^2\theta + \sin^2\theta\left(\frac{\cos^2\theta}{\sin^2\theta}\right)$$
$$= 1$$

(*b*) An alternative solution is to start with the Pythagorean identities, which suggests itself from the square relationships in the problem:

$$\frac{\tan^2\theta}{\sec^2\theta} + \frac{\cot^2\theta}{\csc^2\theta} = \frac{\sec^2\theta - 1}{\sec^2\theta} + \frac{\csc^2\theta - 1}{\csc^2\theta}$$
$$= 1 - \frac{1}{\sec^2\theta} + 1 - \frac{1}{\csc^2\theta}$$
$$= 2 - (\cos^2\theta + \sin^2\theta)$$
$$= 2 - 1 = 1$$

PROBLEMS 27-1

Prove that the following equations are identities:

1. $\cos\theta\tan\theta = \sin\theta$

2. $(\sec\phi + \tan\phi)(\sec\phi - \tan\phi) = 1$

3. $\cos^2\lambda - \sin^2\lambda = 1 - 2\sin^2\lambda$

4. $\sin^4\alpha - \cos^4\alpha = \sin^2\alpha - \cos^2\alpha$

5. $\dfrac{2\tan\phi}{1 + \tan^2\phi} = 2\sin\phi\cos\phi$

6. $\dfrac{\cos^2\phi}{1 - \sin\phi} = 1 + \sin\phi$

7. $(1 + \tan^2\beta)\cos^2\beta = 1$

8. $\tan\theta + \cot\theta = \sec\theta\csc\theta$

9. $(\sin\theta + \cos\theta)^2 + (\sin\theta - \cos\theta)^2 = 2$

10. $1 - 2\sin^2\omega = 2\cos^2\omega - 1$

11. $\tan^2\psi - \sin^2\psi = \tan^2\psi\sin^2\psi$

12. $\dfrac{1 - 2\cos^2\alpha}{\sin\alpha\cos\alpha} = \dfrac{\sin^2\alpha - \cos^2\alpha}{\sin\alpha\cos\alpha}$

13. $\dfrac{1 - \tan^2\theta}{1 + \tan^2\theta} = \cos^2\theta - \sin^2\theta$

14. $\sec\phi - \cos\phi = \sqrt{(\tan\phi + \sin\phi)(\tan\phi - \sin\phi)}$

15. $\cot\theta\cos\theta = \csc\theta - \sin\theta$

16. $\dfrac{\sin\theta + \tan\theta}{\cot\theta + \csc\theta} = \sin\theta\tan\theta$

17. $\tan\lambda + \cot\lambda = \dfrac{\csc^2\lambda + \sec^2\lambda}{\csc\lambda\sec\lambda}$

18. $(\tan\alpha - \sin\alpha)^2 + (1 - \cos\alpha)^2 = (1 - \sec\alpha)^2$

19. $\dfrac{1 - \sin\omega}{1 + \sin\omega} = (\sec\omega - \tan\omega)^2$

20. $\dfrac{\tan\alpha + \tan\beta}{\cot\alpha + \cot\beta} = \tan\alpha\tan\beta$

27-3 LAW OF SINES

Consider the triangle *ABC* (Fig. 27-2). This is not a right triangle, and therefore we have no relationships which we can use to solve the triangle, that is, to relate the various sides and angles in order to find the unknown dimensions in an actual numerical problem. But if we were to develop within it our own right triangles, we might derive some useful relationships.

First of all, we redraw the triangle, Fig. 27-3, and from the vertex *B* we drop the altitude *h* perpendicular to the base *b*. This yields two right triangles, from which we develop the relationships:

Fig. 27-2 Nonright Triangle Cannot Be Solved by Simple Trigonometric Relationships

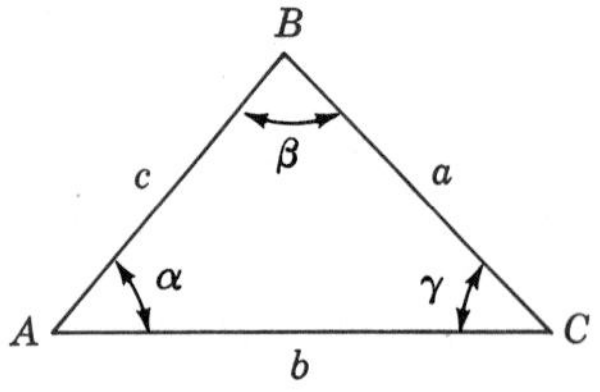

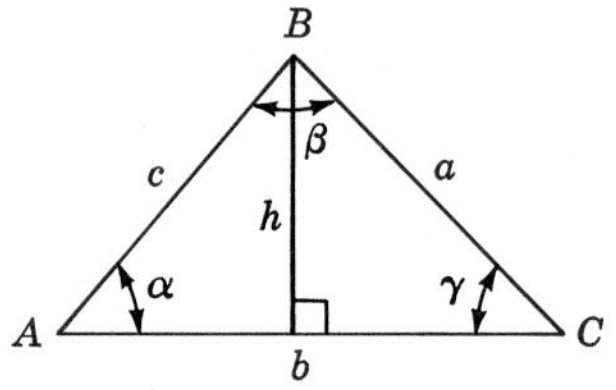

Fig. 27-3 Redrawn from Fig. 27-2 with Altitude *h* Perpendicular to Base *b*

$$h = c\sin\alpha \qquad \text{and} \qquad h = a\sin\gamma$$

Then, equating things equal to the same thing (Axiom 5, Sec. 5-2):

$$c\sin\alpha = a\sin\gamma$$

We rewrite this equation in the simple easy-to-remember form

$$\frac{a}{\sin\alpha} = \frac{c}{\sin\gamma} \tag{5}$$

which is also the most useful form for obtaining slide rule solutions when using this law of sines. You should immediately prove the more general statement:

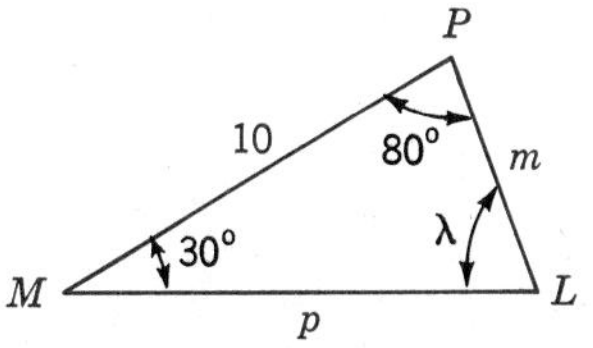

Fig. 27-4 Triangle of Example 2

$$\frac{a}{\sin \alpha} = \frac{b}{\sin \beta} = \frac{c}{\sin \gamma} \tag{6}$$

example 2 Given the triangle *MPL*, Fig. 27-4, find the values of m and p.

solution First of all, solve for

$$\lambda = 180° - (80° + 30°) = 70°$$

Then, using the law of sines,

$$\frac{10}{\sin 70°} = \frac{p}{\sin 80°} = \frac{m}{\sin 30°}$$

so that

$$p = \frac{10 \sin 80°}{\sin 70°} = 10.5$$

and

$$m = \frac{10 \sin 30°}{\sin 70°} = 5.32$$

note To be able to use the law of sines, it is necessary for us to know certain specific data: two sides and the angle opposite one of them or two angles and the side opposite one of them.

PROBLEMS 27-2

Referring to Fig. 27-2, solve the following triangles:

1. $a = 8.04$, $\alpha = 57°$, $\beta = 53°$
2. $a = 19$, $\beta = 80°$, $\gamma = 88°$
3. $b = 16.3$, $\alpha = 44°$, $\beta = 61°$
4. $c = 760$, $\alpha = 68°$, $\beta = 42°$
5. $b = 76$, $\alpha = 20°$, $\beta = 52°$
6. $b = 3.26$, $\alpha = 25°$, $\beta = 41°$
7. $c = 7.6$, $\beta = 60°$, $\gamma = 112°$
8. $a = 600$, $\beta = 17.6°$, $\gamma = 105.9°$
9. $b = 58$, $\alpha = 9.2°$, $\gamma = 115.3°$
10. $c = 635$, $\alpha = 15.5°$, $\beta = 26°$
11. Two observers who are 1500 m apart on a horizontal plane observe a radiosonde balloon in the same vertical plane as themselves and between themselves. The angles of elevation are 72° and 75°. Find the height of the balloon.
12. A 50-m antenna mast stands on the edge of the roof of the studio building. From a point on the ground at some distance from the base of the building, the angles of elevation of the top and bottom of the mast are respectively 76.5° and 54.5°. How high is the building?

27-4 LAW OF COSINES

Sometimes we are not given data suitable for solving a triangle by means of the law of sines. But another useful relationship can be readily developed. Using the triangle *ABC* of Fig. 27-2, copied as Fig. 27-5 and adjusted with an altitude h perpendicular to

Fig. 27-5 Redrawn from Fig. 27-2. Altitude h Divides Base b into Parts x and y

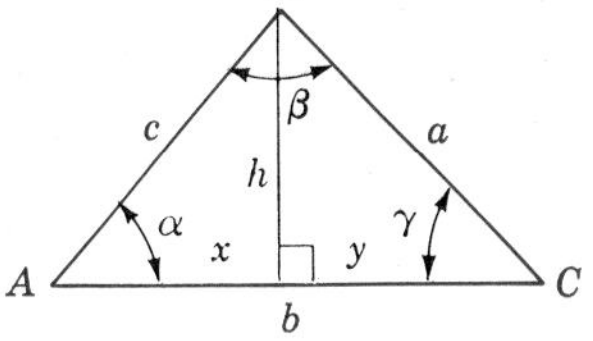

the base and rising to the vertex, so that the base is divided into parts x and y,

$$h^2 = c^2 - x^2 = a^2 - y^2$$

from which

$$\begin{aligned} a^2 &= c^2 - x^2 + y^2 \\ &= c^2 - x^2 + (b - x)^2 \\ &= c^2 - x^2 + b^2 - 2bx + x^2 \\ &= b^2 + c^2 - 2bx \end{aligned}$$

but $\quad x = c \cos \alpha$

and $\quad a^2 = b^2 + c^2 - 2bc \cos \alpha \qquad (7)$

See how straightforward this statement may be: "In any triangle, the square of any one side is equal to the sum of the squares of the other two sides minus twice their product times the cosine of the angle between them." You should prove that this statement holds true for right triangles, to become Pythagoras' theorem.

Like the law of sines, the law of cosines has a rhythm which makes it easy to memorize one part and simply rotate the other parts into duplicate statements. However, besides merely memorizing the result, you should prove that all parts of the full statement of the law of cosines are true:

$$\begin{aligned} a^2 &= b^2 + c^2 - 2bc \cos \alpha \\ b^2 &= a^2 + c^2 - 2ac \cos \beta \qquad (8) \\ c^2 &= a^2 + b^2 - 2ab \cos \gamma \qquad (9) \end{aligned}$$

The careful use of these three equations, together with what we have learned about the *signs* of the cosine, will enable us to prepare any triangle so that we may complete its solution by means of the law of sines.

example 3 Acute triangle. Solve the triangle of Fig. 27-6.

solution Using the law of cosines:

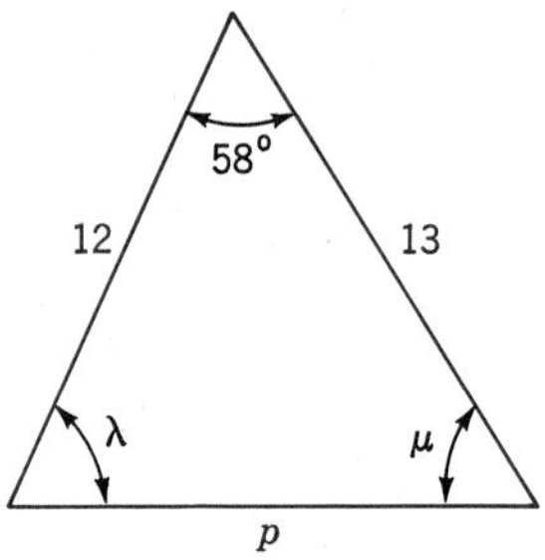

Fig. 27-6 Triangle of Example 3

$$\begin{aligned} p^2 &= 12^2 + 13^2 - 2 \times 12 \times 13 \times \cos 58° \\ &= 144 + 169 - 312 \cos 58° \\ &= 147.9 \\ p &= 12.2 \end{aligned}$$

Now, having at least two sides and the angle opposite one of them, we may, if we wish, complete the solution by means of the law of sines instead of repeating the cosine solution. Since this method is easier to set up on the slide rule:

$$\frac{12.2}{\sin 58°} = \frac{12}{\sin \mu}$$

from which $\quad \mu = 65°$

Similarly $\quad \lambda = \arcsin \dfrac{12 \sin 58°}{12.2} = 65°$

test $58° + 56.8° + 65° = 179.8°$

example 4 Oblique triangle. Solve the triangle of Fig. 27-7.

solution Since the information given is not sufficient to use the law of sines, check to see if the law of cosines may be applied.

Fig. 27-7 Triangle of Example 4

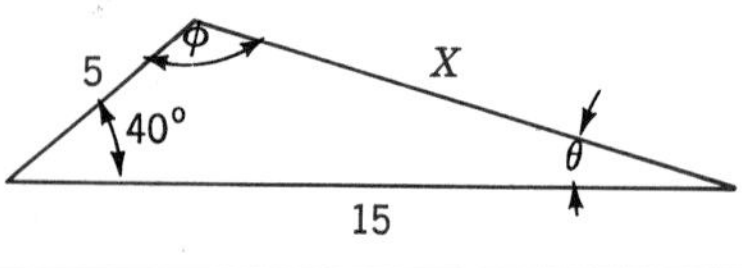

Knowing two sides and the angle between them is sufficient:

$$\begin{aligned} X^2 &= 5^2 + 15^2 - 2 \times 5 \times 15 \times \cos 40° \\ &= 135.1 \\ X &= 11.6 \end{aligned}$$

Then, using the law of sines,

$$\theta = \arcsin \frac{5 \sin 40°}{11.6} = 16.1°$$

and $$\phi = \arcsin \frac{15 \sin 40°}{11.6} = 56.1°$$

test $40° + 16.1° + 56.1° = 112.2°$ Oh. From Fig. 27-7, the side of length 15, being the longest side, *must* be opposite the largest angle, which we have calculated as 56.1°. Since this must be the largest angle, since it *could* be obtuse (greater than 90°, an angle in the second quadrant), and since all that our calculations guarantee is that $\phi = \arcsin\ 0.831$, perhaps ϕ is $180° - 56.1° = 123.9°$.
Testing this possibility,

$$40° + 16.1° + 123.9° = 180°$$

We have arrived at the correct solution.

Be sure to test your solutions.

example 5 The three sides of a triangle are given, and it is required to solve the angles. (Note that if just three angles are given, there are an infinite number of solutions.) Solve the triangle of Fig. 27-8.

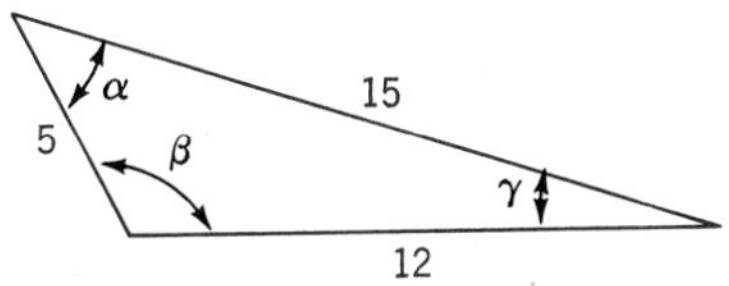

Fig. 27-8 Triangle of Example 5

solution If any angle in the triangle can be obtuse, it will be angle β. (Why?) We will defer solving for β for now. Consider the angle α. It is related, by the law of cosines, as follows:

$$12^2 = 5^2 + 15^2 - 2 \times 5 \times 15 \times \cos \alpha$$

from which

$$\alpha = \arccos \frac{5^2 + 15^2 - 12^2}{2 \times 5 \times 15} = 45.1°$$

You should confirm that

$$\gamma = \arccos \frac{12^2 + 15^2 - 5^2}{2 \times 12 \times 15} = 17.2°$$

Then $$\beta = 180° - (45.1° + 17.2°) = 117.7°$$

Alternatively, starting the solution for β,

$$15^2 = 5^2 + 12^2 - 2 \times 5 \times 12 \times \cos \beta$$

from which

$$\beta = \arccos(-0.466).$$

This negative cosine indicates immediately that β must be an angle between 90° and 180°, and we find it to be

$$180° - 62.3° = 117.7°$$

PROBLEMS 27-3
Referring to Fig. 27-2, solve the following triangles:

1. $b = 5.2,\ c = 8,\ \alpha = 63°$
2. $a = 544,\ b = 805,\ \gamma = 80°$
3. $a = 0.17,\ b = 0.785,\ \gamma = 132°$
4. $a = 2.6,\ c = 8.45,\ \beta = 48.8°$
5. $a = 1600,\ b = 3260,\ \gamma = 147.7°$
6. $b = 0.0945,\ c = 0.0980,\ \alpha = 5°$

7. $a = 3, b = 5, c = 7$
8. $a = 2000, b = 4000, c = 6000$
9. $a = 1280, b = 3260, c = 3935$
10. $a = 25, b = 30, c = 50.$
11. The diagonals of a parallelogram are 130 mm and 180 mm, and they intersect at an angle of 38°. What are the sides of the parallelogram?
12. Using the data of Prob. 11, but *not* your results, what is the area of the parallelogram? (After obtaining a solution, check it by means of a different computational method.)

27-5 THE SUM IDENTITIES

Often in the solution of antenna and modulation problems we come upon various combinations such as $\sin(\theta + \phi)$ and $\cos(\theta - \phi)$. It is often convenient to resolve these forms into the products of simple trigonometric functions.

Consider triangle PQR, Fig. 27-9, with the altitude h dividing the angle RPQ into two angles, α and β. Since the area of the whole triangle must be equal to the sum of the areas of the two component triangles,

$$\tfrac{1}{2}qr \sin(\alpha + \beta) = \tfrac{1}{2}qh \sin \alpha + \tfrac{1}{2}rh \sin \beta$$

from which

$$\sin(\alpha + \beta) = \frac{h}{r} \sin \alpha + \frac{h}{q} \sin \beta$$

which yields

$$\sin(\alpha + \beta) = \sin \alpha \cos \beta + \cos \alpha \sin \beta \quad (10)$$

Again, using the same triangle, Fig. 27-9, and the law of cosines,

Fig. 27-9 Triangle Adjusted for Development of the Sum Identities

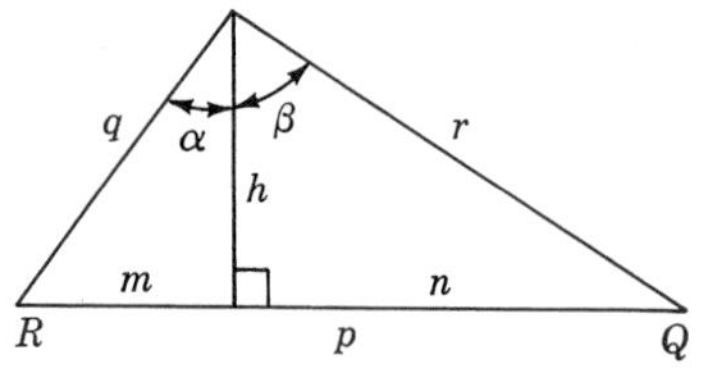

$$(m + n)^2 = q^2 + r^2 - 2qr \cos(\alpha + \beta)$$

from which

$$\cos(\alpha + \beta) = \frac{q^2 + r^2 - m^2 - n^2 - 2mn}{2qr}$$
$$= \frac{q^2 - m^2}{2qr} + \frac{r^2 - n^2}{2qr} - \frac{2mn}{2qr}$$
$$= \frac{2h^2}{2qr} - \frac{2mn}{2qr}$$
$$= \frac{h}{q} \cdot \frac{h}{r} - \frac{m}{q} \cdot \frac{n}{r}$$

which converts to

$$\cos(\alpha + \beta) = \cos \alpha \cos \beta - \sin \alpha \sin \beta \quad (11)$$

27-6 THE DIFFERENCE IDENTITIES

Sometimes, instead of functions of the sum of two angles, it is necessary to deal with the differences of two angles: In triangle PQR, Fig. 27-10, the line q divides the vertex into

Fig. 27-10 Triangle Adjusted for Development of the Difference Identities

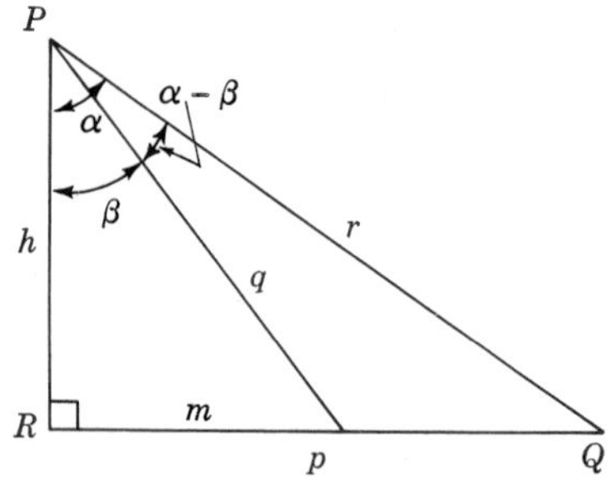

two angles, β and $\alpha - \beta$. As in the sum identity, the area of the whole triangle is equal to the sum of the parts:

$$\tfrac{1}{2}hr \sin \alpha = \tfrac{1}{2}hq \sin \beta + \tfrac{1}{2}qr \sin (\alpha - \beta)$$

from which

$$\sin (\alpha - \beta) = \frac{hr \sin \alpha - hq \sin \beta}{qr} = \frac{h}{q} \sin \alpha - \frac{h}{r} \sin \beta$$

which yields

$$\sin (\alpha - \beta) = \sin \alpha \cos \beta - \cos \alpha \sin \beta \quad (12)$$

And, as before, using the law of cosines:

$$(p - m)^2 = q^2 + r^2 - 2qr \cos (\alpha - \beta)$$

from which

$$\cos (\alpha - \beta) = \frac{q^2 + r^2 - p^2 - m^2 + 2mp}{2qr} = \frac{q^2 - m^2}{2qr} + \frac{r^2 - p^2}{2qr} + \frac{2mp}{2qr} = \frac{h}{q} \cdot \frac{h}{r} + \frac{m}{q} \cdot \frac{p}{r}$$

which yields

$$\cos (\alpha - \beta) = \cos \alpha \cos \beta + \sin \alpha \sin \beta \quad (13)$$

example 6 Simplify the expression

$$\sin (\theta + 45^\circ) + \cos (\theta + 45^\circ)$$

solution Using Eqs. (10) and (11) and substituting the equivalent product expressions:

$$\begin{aligned}\sin (\theta + 45^\circ) &+ \cos (\theta + 45^\circ)\\ &= \sin \theta \cos 45^\circ + \cos \theta \sin 45^\circ\\ &\quad + \cos \theta \sin 45^\circ - \sin \theta \sin 45^\circ\\ &= 0.7071 \sin \theta + 0.7071 \cos \theta\\ &\quad + 0.7071 \cos \theta - 0.7071 \sin \theta\\ &= 1.4142 \cos \theta\end{aligned}$$

Table 27.1 TRIGONOMETRIC IDENTITIES AND USEFUL RELATIONSHIPS

$$\tan \theta = \frac{\sin \theta}{\cos \theta} \qquad \cot \theta = \frac{\cos \theta}{\sin \theta}$$

$$\sin^2 \theta + \cos^2 \theta = 1$$

$$1 + \tan^2 \theta = \sec^2 \theta$$

$$1 + \cot^2 \theta = \csc^2 \theta$$

$$\frac{a}{\sin \alpha} = \frac{b}{\sin \beta} = \frac{c}{\sin \gamma}$$

$$a^2 = b^2 + c^2 - 2bc \cos \alpha$$

$$\sin (\theta + \phi) = \sin \theta \cos \phi + \cos \theta \sin \phi$$

$$\cos (\theta + \phi) = \cos \theta \cos \phi - \sin \theta \sin \phi$$

$$\sin (\theta - \phi) = \sin \theta \cos \phi - \cos \theta \sin \phi$$

$$\cos (\theta - \phi) = \cos \theta \cos \phi + \sin \theta \sin \phi$$

PROBLEMS 27-4

Using the sum and difference relationships, simplify:

1. $\sin (\theta + 30^\circ) + \cos (\theta + 30^\circ)$
2. $\sin (45^\circ - \theta) - \cos (45^\circ + \theta)$
3. $(\sin (\theta - 60^\circ) + \cos (\theta + 60^\circ)$
4. $\sin (\theta - 30^\circ) - \cos (\theta - 45^\circ)$

Given $\sin \theta = \frac{3}{5}$ and $\sin \phi = \frac{5}{12}$, evaluate:

5. $\cos(\theta + \phi)$
6. $\sin(\theta - \phi) - \cos(\theta - \phi)$
7. Use Eq. (10) to show that $\sin 2\theta = 2 \sin\theta \cos\theta$.
8. Use Eq. (11) to show that $\cos 2\theta = \cos^2\theta - \sin^2\theta$.
9. When a VHF direction-finding array is fed in modulation phase quadrature, the two fields about the antennas are

$$E_1 = K \cos\theta \cos pt \cos\omega t$$
$$E_2 = K \sin\theta \sin pt \cos\omega t$$

Show that the total field $E_t = E_1 + E_2 = K \cos\omega t \cos(pt - \theta)$.

10. Use Eqs. (11) and (13) to show that

$$\tfrac{1}{2}\cos(\omega t - pt) - \tfrac{1}{2}\cos(\omega t + pt) = \sin pt \sin\omega t$$

It is based on this relationship that an amplitude-modulated carrier wave is shown to consist of a fundamental and two sidebands. The equation of the modulated carrier wave is

$$e = E \sin\omega t + mE \sin\omega t \sin pt$$

where m is the depth of modulation, and your work in this problem shows the correctness of the substitution:

$$e = E \sin\omega t + \tfrac{1}{2}mE \cos(\omega t - pt) - \tfrac{1}{2}mE \cos(\omega t + pt)$$

where $E \sin\omega t$ represents the original carrier and the other two parts represent the difference and sum sideband frequencies whose amplitudes are each one-half that of the carrier.

28

elementary plane vectors

Many physical quantities can be expressed by specifying a certain number of units. For example, the volume of a tank may be expressed as so many cubic feet, the temperature of a room as a certain number of degrees, and the speed of a moving object as a number of linear units per unit of time such as miles per hour or feet per second. Such quantities are *scalar quantities,* and the numbers that represent them are called *scalars.* A scalar quantity is one that has only magnitude; that is, it is a quantity fully described by a number, but it does not involve any concept of direction.

28-1 VECTORS

Many other types of physical quantities need to be expressed more definitely than is possible by specifying magnitude alone. For example, the velocity of a moving object has a direction as well as a magnitude. Also, a force due to a push or a pull is not completely described unless the direction as well as the magnitude of the force is given. In addition, electric circuit analysis is built up around the idea of expressing the directions and magnitudes of voltages and currents. Those quantities which have both magnitude and direction are called *vector quantities.* A vector quantity is conveniently represented by a directed straight-line segment called a *vector,* whose length is proportional to the magnitude and whose head points in the direction of the vector quantity.

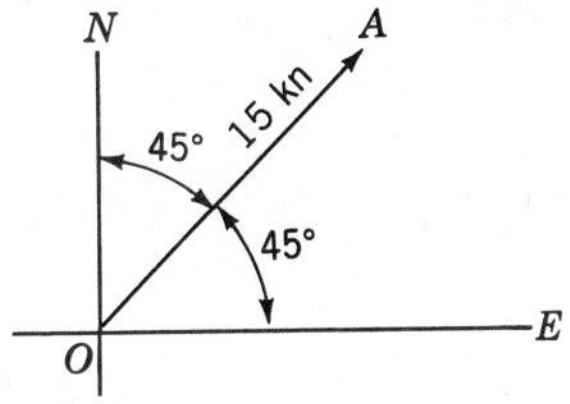

Fig. 28-1 Vector **OA** of Example 1

example 1 If a vessel steams northeast at a speed of 15 kn, its speed can be represented by a line whose length represents 15 kn, to some convenient scale, as shown in Fig. 28-1. The direction of the line represents the direction in which the vessel is traveling. Thus the line **OA** is a vector that completely describes the velocity of the vessel.

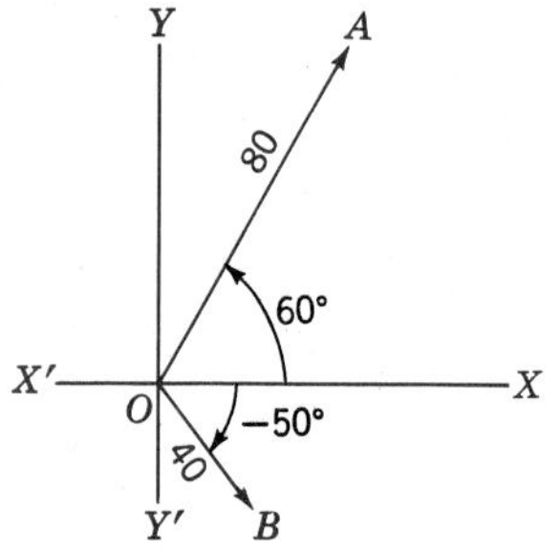

Fig. 28-2 Vector diagram of Example 2

example 2 In Fig. 28-2, the vector **OA** represents a force of 80 N pulling on a body at O in a direction of 60°. The vector **OB** represents a force of 40 N acting on the same body in a direction of 310° or −50°.

Two vectors are equal if they have the same magnitude and direction. Thus, in Fig. 28-3, vectors ***A*, *B*,** and ***C*** are equal.

28-2 NOTATION

As you progress in the study of vectors, you will find that vectors and scalars satisfy different algebraic laws. For example, a scalar when reduced to its simplest terms is simply a number and as such obeys all the laws of ordinary algebraic operations. Since a vector involves direction, in addition to magnitude, it does not obey the usual algebraic laws and therefore has an analysis peculiar to itself.

From the foregoing, it is apparent that it is desirable to have a notation that indicates clearly which quantities are scalars and which are vectors. Several methods of notation are used, but you will find little cause for confusion, for most authors specify and explain their particular system of notation.

A vector can be denoted by two letters, the first indicating the origin, or initial point, and the second indicating the head, or terminal point. This form of notation was used in Examples 1 and 2 of the preceding section. Sometimes a small arrow is placed over these letters to emphasize that the quantity considered is a vector. Thus, $\overrightarrow{OA}$ could be used to represent the vector from O to *A* as in Fig. 28-2. In most texts, vectors are indicated by boldface type; thus, ***A*** denotes the vector *A*. Other common forms of specifying a vector quantity, as, for example, the vector *A*, are $\bar{A}$, $\dot{A}$, $\underset{.}{A}$, and *A*.

28-3 ADDITION OF VECTORS

Scalar quantities are added algebraically. Thus

$$20 \text{ cents} + 8 \text{ cents} = 28 \text{ cents}$$

and

$$16 \text{ insulators} - 7 \text{ insulators} = 9 \text{ insulators}$$

Since vector quantities involve direction as well as magnitude, they cannot be added algebraically unless their directions are parallel. Figure 28-4 illustrates vectors **OA** and ***AB*.** Vector **OA** can be considered as a motion from O to *A*, and vector ***AB*** as a motion from *A* to *B*. Then the sum of the vectors represents the sum of the motions from O to *A*

Fig. 28-3 Vectors ***A*, *B*,** and ***C*** Are Equal.

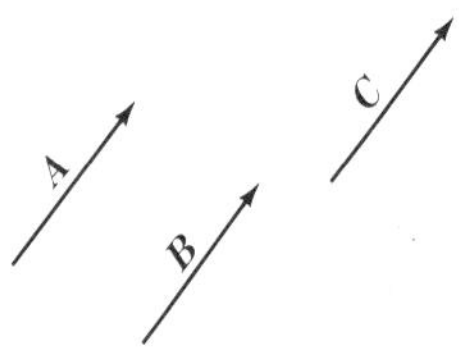

Fig. 28-4 Vector **OB** Is the Vector Sum of **OA** and ***AB***.

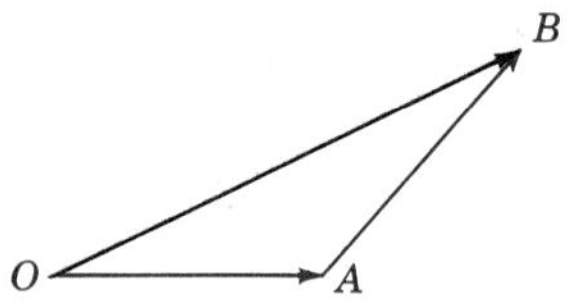

and from A to B, which is the motion from O to B. This sum is the vector ***OB;*** that is, the *vector sum* of **OA** and ***AB*** is **OB.** Therefore, the sum of two vectors is the vector joining the initial point of the first to the terminal point of the second if the initial point of the second vector is joined to the terminal point of the first vector as shown in Fig. 28-4.

In Fig. 28-5, vectors **OC** and **OD** are equal to vectors **OA** and ***AB,*** respectively, of Fig. 28-4. In Fig. 28-5, however, the vectors start from the same origin. That their sum can be represented by the diagonal of a parallelogram of which the vectors are adjacent sides is evident by comparing Figs. 28-4 and 28-5. This is known as the *parallelogram law* for the composition of forces, and it holds for the composition or addition of all vector quantities.

The addition of vectors that are not at right angles to each other will be considered in Sec. 28-7. At this time, it is sufficient to know that two forces acting simultaneously on a point, or an object, can be replaced by a single force called the *resultant.* That is, the resultant force will produce the same effect on the object as the joint action of the two forces. Thus, in Fig. 28-4 the vector **OB** is the resultant of vectors **OA** and ***AB.*** Similarly, in Fig. 28-5, the vector **OE** is the resultant of the vectors **OC** and **OD.** Note that **OB** = **OE.**

example 3 Three forces A, B, and C are acting on point O as shown in Fig. 28-6. Force A exerts 150 N at an angle of 60°, B exerts 100 N at an angle of 135°, and C exerts 150 N at an angle of 260°. What is the resultant force on point O?

Fig. 28-5 Resultant Vector **OE** Is the Vector Sum of **OC** and **OD**

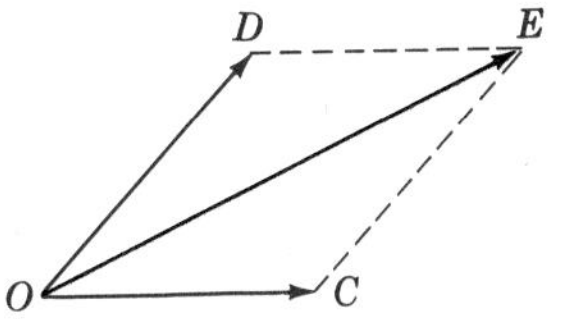

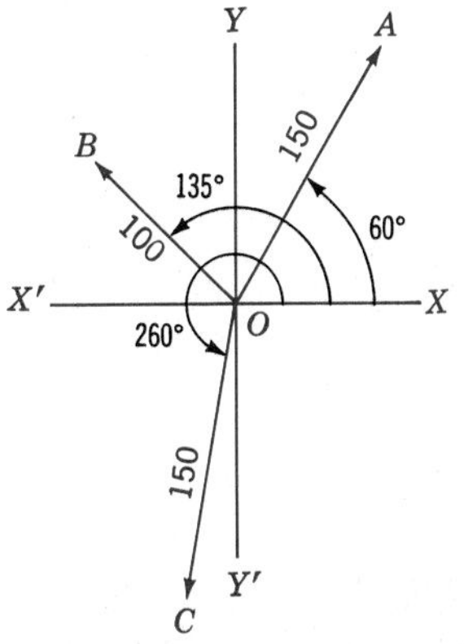

Fig. 28-6 Vector Diagram of Example 3

solution The resultant of vectors ***A, B,*** and **C** can be found graphically by two methods.

(a) First draw the vectors to scale. Find the resultant of any two vectors, such as **OA** and **OC,** by constructing a parallelogram with **OA** and **OC** as adjacent sides. Then the resultant of **OA** and **OC** will be the diagonal **OD** of the parallelogram $OADC$ as shown in Fig. 28-7. In effect, there are

Fig. 28-7 **OE** Is the Vector Sum of Vectors ***A, B,*** and **C**

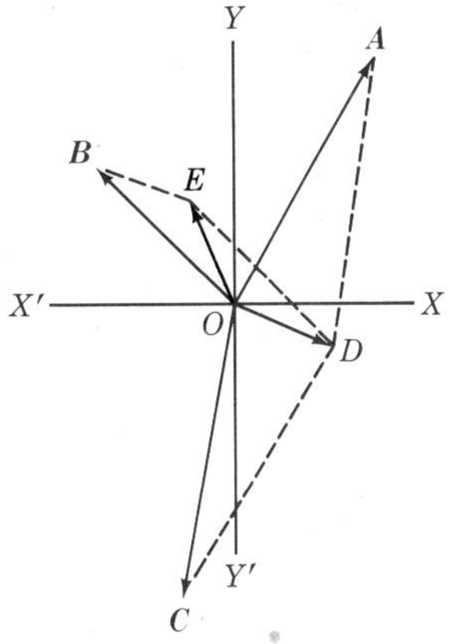

now but two forces, **OB** and **OD**, acting on point O. The resultant of these two forces is found as before by constructing a parallelogram with **OB** and **OD** as adjacent sides. The resultant force on point O is then the diagonal **OE** of the parallelogram *OBED*. By measurement with scale and protractor, *OE* is found to be 57 N acting at an angle of 112°.

(*b*) Draw the vectors to scale as shown in Fig. 28-8, joining the initial point of *B* to the terminal point of *A* and then joining the initial point of *C* to the terminal point of *B*. The vector drawn from the point O to the terminal point of *C* is the resultant force, and measurements show it to be the same as that found by the method illustrated in Fig. 28-7.

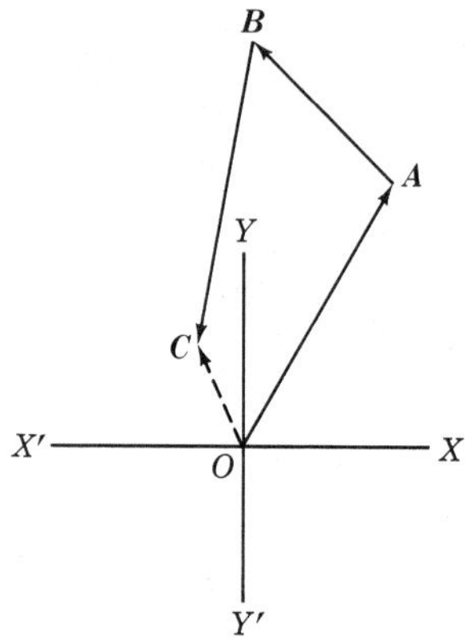

Fig. 28-8 **OC** Is the Vector Sum of Vectors ***A*, *B*,** and **C**

A figure such as *OABCO*, in Fig. 28-8, is called a *polygon of forces*. The vectors can be joined in any order as long as the initial point of one vector joins the terminal point of another vector and the vectors are drawn with the proper magnitude and direction. The length and direction of the line that is necessary to close the polygon, that is, the line from the original initial point to the terminal point of the last vector drawn, constitute a vector that represents the magnitude and the direction of the resultant.

PROBLEMS 28-1

1. to **4.** Find the magnitude and direction, with respect to the positive *x* axis, of the vectors shown in Figs. 28-9 to 28-12.

Fig. 28-9 Vector Diagram of Prob. 1

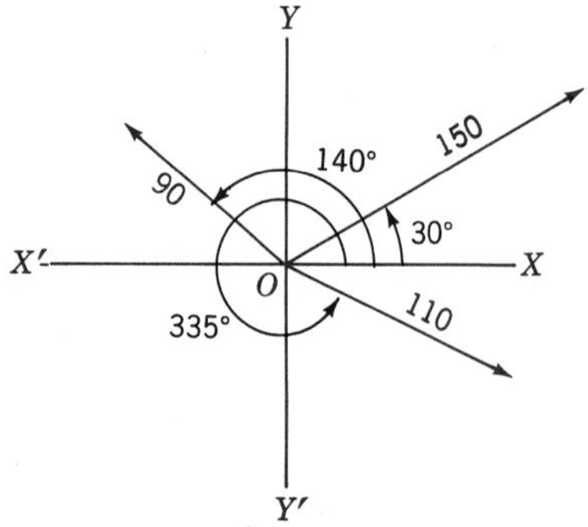

Fig. 28-10 Vector Diagram of Prob. 2

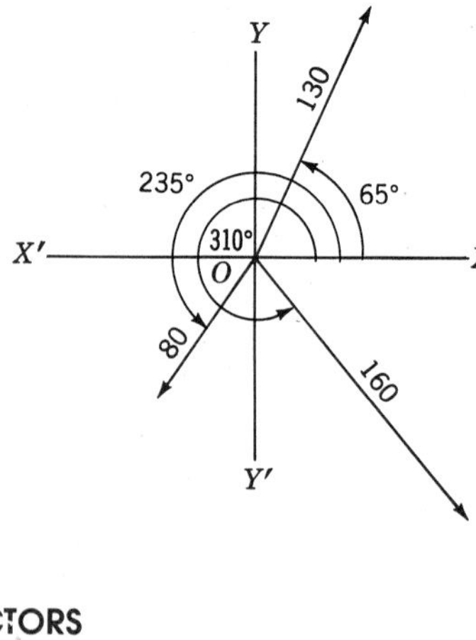

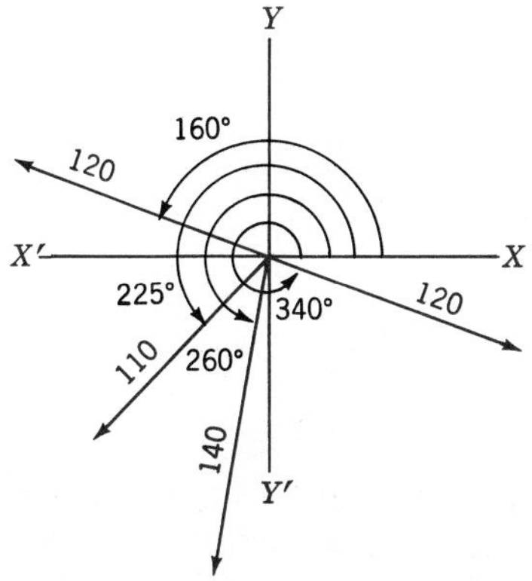

Fig. 28-11 Vector Diagram of Prob. 3

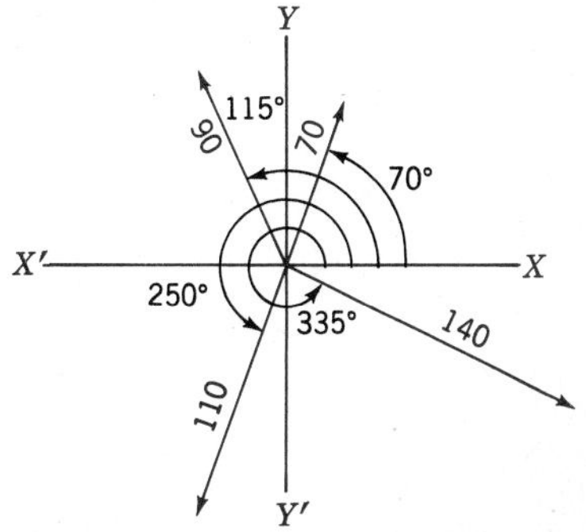

Fig. 28-12 Vector Diagram of Prob. 4

28-4 COMPONENTS OF A VECTOR

From what has been considered regarding combining or adding vectors, it follows that a vector can be resolved into components along any two specified directions. For example, in Fig. 28-4, the vectors ***OA*** and ***AB*** are components of the vector ***OB.*** If the directions of the components are so chosen that they are at right angles to each other, the components are called *rectangular components.*

By placing the initial point of a vector at the origin of the x and y axes, the rectangular components are readily obtained either graphically or mathematically.

example 4 A vector with a magnitude of 10 makes an angle of 53.1° with the horizontal. What are the vertical and horizontal components?

solution The vector is illustrated in Fig. 28-13 as the directed line segment ***OA.*** Its length drawn to scale represents the magnitude of 10, and it makes an angle of 53.1° with the x axis.

The *horizontal component* of ***OA*** is the horizontal distance (Sec. 26-7) from O to A and is found graphically by projecting the vector ***OA*** upon the x axis. Thus the vector ***OB*** is the horizontal component of ***OA.***

The *vertical component of* ***OA*** is the vertical distance from O to A and is found graphically by projecting the vector ***OA*** upon the y axis. Similarly, the vector ***OC*** is the vertical component of ***OA.*** Finding the horizontal and vertical components of ***OA*** by mathematical methods is simply a problem in solving a right triangle as outlined in Sec. 26-4. Hence,

$$\mathbf{OB} = 10 \cos 53.1° = 6$$

and $$\mathbf{OC} = \mathbf{BA} = 10 \sin 53.1° = 8$$

Fig. 28-13 Vertical and Horizontal Components of Vector

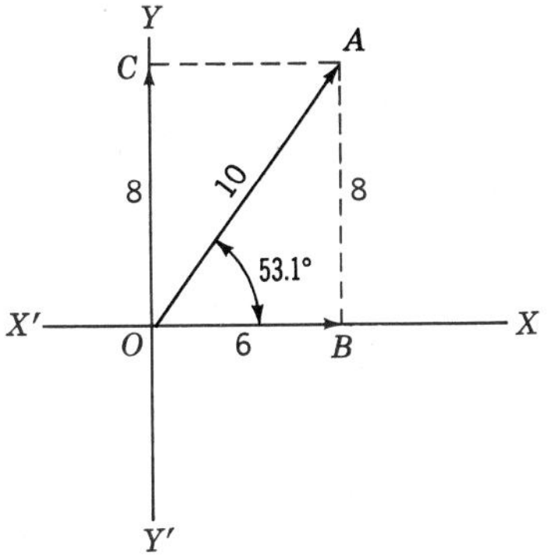

check $\theta = \arctan \frac{8}{6} = \arctan 1.33 = 53.1°$

$10^2 = 6^2 + 8^2 = 36 + 64 = 100$

The foregoing can be summarized as follows:

Rule

1. The horizontal component of a vector is the projection of the vector upon a horizontal line and equals the magnitude of the vector multiplied by the cosine of the angle made by the vector with the horizontal.
2. The vertical component of a vector is the projection of the vector upon a vertical line and equals the magnitude of the vector multiplied by the sine of the angle made by the vector with the horizontal.

example 5 An airplane is flying on a course of 40° at a speed of 400 km/h. How many kilometers per hour is the plane advancing in a due eastward direction? In a direction due north?

solution Draw the vector diagram as shown in Fig. 28-14. (Courses are measured from the north.) The vector **OB**, which is the horizontal component of **OA**, represents the velocity of the airplane in an eastward direction. The vector **OC**, which is the vertical component of **OA**, represents the velocity of the airplane in a northward direction. Again, the process of finding the magnitude of **OB** and **OC** resolves into a problem in solving the right triangle OBA. Hence,

Fig. 28-14 Vector Diagram of Example 5

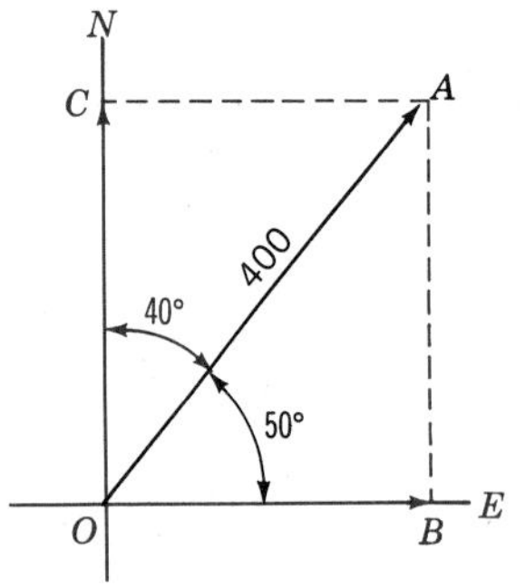

$$\mathbf{OB} = 400 \cos 50° = 257 \text{ km/h eastward}$$
$$\mathbf{OC} = \mathbf{BA} = 400 \sin 50° = 306.5 \text{ km/h northward}$$

If the vector diagram has been drawn to scale, an approximate check can be made by measuring the lengths of **OB** and **OC.** Such a check will disclose any large errors in the mathematical solution.

example 6 A radius vector of unit length is rotating about a point with a velocity of $2\pi^r$/s. What are its horizontal and vertical components (*a*) at the end of 0.15 s, (*b*) at the end of 0.35 s, (*c*) at the end of 0.75 s?

solution

(*a*) At the end of 0.15 s the rotating vector will have generated $2\pi \times 0.15 = 0.942^r$, or $0.942 \times 57.3° = 54°$ as shown in Fig. 28-15. The horizontal component, measured along the x axis, is

$$x = 1 \cos 54° = 0.588$$

The vertical component, measured along the y axis, is

Fig. 28-15 When $t = 0.15$ s, Angle $\theta = 54°$

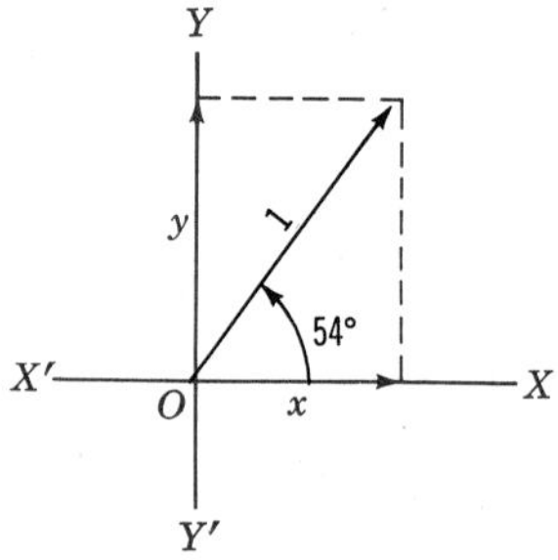

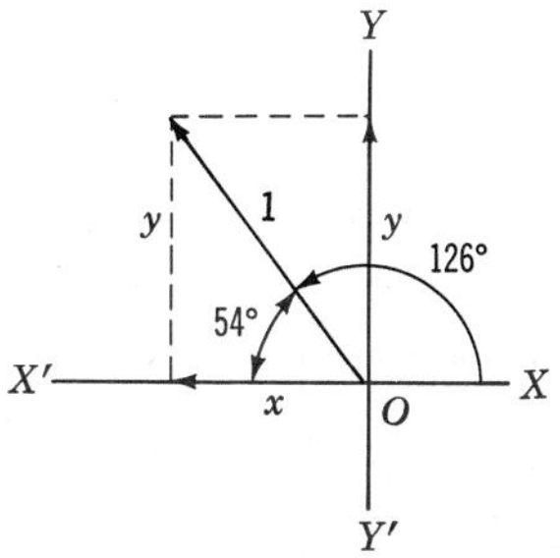

Fig. 28-16 When $t = 0.35$ s, Angle $\theta = 126°$

$$y = 1 \sin 54° = 0.809$$

Check the solution by measurement or any other method considered convenient.

(*b*) At the end of 0.35 s the rotating vector will have generated an angle of $2\pi \times 0.35 = 2.20^r$, or $2.20 \times 57.3° = 126°$ as shown in Fig. 28-16. The horizontal component, measured along the x axis, is

$$x = 1 \cos 126° = 1(-\cos 54°)$$
$$= -0.588 \quad \text{(Sec. 25-11)}$$

The vertical component, measured along the y axis, is

$$y = 1 \sin 126° = 1 \sin 54° = 0.809 \quad \text{(Sec. 25-11)}$$

Check by some convenient method.

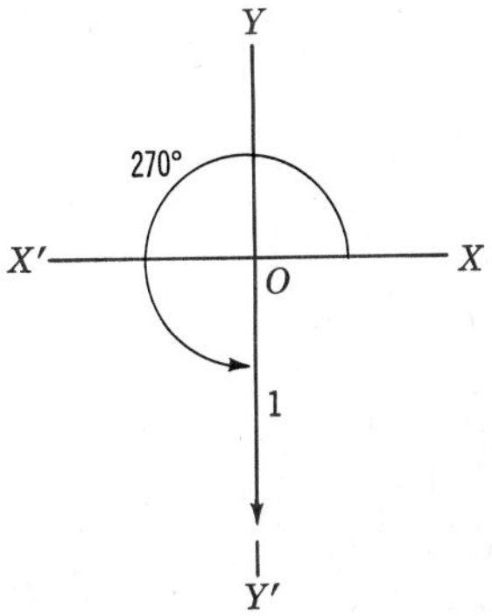

Fig. 28-17 When $t = 0.75$ s, Angle $\theta = 270°$

(*c*) At the end of 0.75 s the rotating vector will have generated $2\pi \times 0.75 = 4.71^r$, or $4.71 \times 57.3° = 270°$ as shown in Fig. 28-17. The horizontal component is

$$x = 1 \cos 270° = 0$$

The vertical component is

$$y = 1 \sin 270° = -1$$

PROBLEMS 28-2

Find the horizontal and vertical components, denoted by x and y, respectively, of the following vectors. Check the mathematical solution of each by drawing a vector diagram to scale.

1. 30 at 65.5° (This is commonly written $30\underline{/65.5°}$)

2. $99\underline{/22.8°}$ **3.** $0.865\underline{/87.2°}$ **4.** $1800\underline{/120°}$ **5.** $46.3\underline{/180°}$

6. $0.987\underline{/295.5°}$ **7.** $185.5\underline{/252.2°}$ **8.** $27.8275\underline{/90°}$ **9.** $30.8\underline{/157.3°}$

10. $1600\underline{/270°}$

11. The resultant of two forces acting at right angles is a force of 765 N which makes an angle of 17.8° with one of the forces. Find the component forces.

12. A test missile was fired at an angle of 82° from the horizontal. At a particular instant its velocity was 1950 km/h. Find its horizontal velocity at that instant in meters per second.
13. A jet fighter leaves its base and flies 1200 m southeast. How far east does it go?
14. Resolve a force of 250 N into two rectangular components one of which is 155 N.
15. The resultant of two forces acting at right angles is 199 N. One of the forces is 150 N. What is the other?

28-5 PHASORS

Early in this chapter we discovered the difference between scalar quantities, which involve magnitude only, and vectors, which involve both magnitude and direction. When electrical units are shown on paper, with the length of the line indicating the magnitude and the direction of the line indicating the phase relationship, they may be thought of as *vectors*. However, when an emf is impressed across a circuit, its *polarity* is not *direction* in the sense of vector definition. The paper representation as vectors serves a valuable purpose in our circuit calculations, but the electrical quantities are not true vectors. Since the angular separation of electrical units always represents *time* revealed as a *phase* relationship, scientists and engineers prefer to use the term *phasors* when discussing electrical "vectors."

On paper (in a "uniplanar" representation) there is no difference between phasors and vectors. The operations of conversion between rectangular and polar forms are the same. The summation of perpendicular components is the same. But since our purpose is to study the mathematics of electronics in an electronics environment and our communication is with electronics and scientific people, we will use the expressions *phasor* and *phasor summation* throughout the remaining chapters of this book.

28-6 PHASOR SUMMATION OF RECTANGULAR COMPONENTS

If two forces that are at right angles to each other are acting on a body, their resultant can be found by the usual methods of phasor summation as outlined in Sec. 28-3. However, the resultant can be obtained by geometric or trigonometric methods, for the problem is that of solving for the hypotenuse of a right triangle when the other two sides are given, as outlined in Sec. 26-6.

example 7 Two phasors are acting at a point. One with a magnitude of 6 is directed along the horizontal to the right of the point, and the other with a magnitude of 8 is directed vertically above the point. Find their resultant.

solution 1 In Fig. 28-18 the horizontal

Fig. 28-18 Addition of Rectangular Components

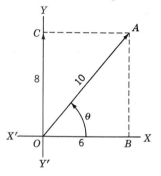

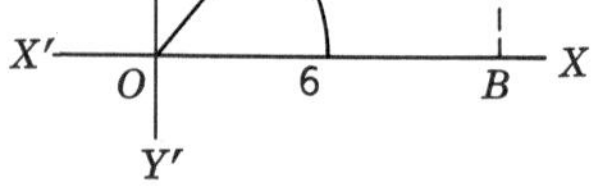

phasor, with a magnitude of 6, is shown as **OB.** The vertical phasor, with a magnitude of 8, is shown as **OC.** The resultant of these two phasors can be obtained graphically by completing the parallelogram of forces *OCAB*, as outlined in Sec. 28-3. Thus, the magnitude of the resultant will be represented by the length of **OA** in Fig. 28-18. The angle, or direction of the resultant, can be measured with the protractor.

Graphical methods have a limited degree of accuracy, as pointed out in earlier sections. They should be used as an approximate check for more precise mathematical methods.

solution 2 Since **BA** = **OC** in Fig. 28-18, then *OBA* is a right triangle the hypotenuse of which is the resultant **OA.** Therefore the magnitude of the resultant is

$$\mathbf{OA} = \sqrt{\mathbf{OB}^2 + \mathbf{BA}^2} = \sqrt{6^2 + 8^2} = 10$$

The angle, or direction of the resultant, is

$$\theta = \arctan \frac{\mathbf{BA}}{\mathbf{OB}} = \arctan \frac{8}{6} = \arctan 1.33 = 53.1°$$

Although the method of Solution 2 is accurate and mathematically correct, there are several operations involved. For example, in finding the magnitude, 6 and 8 must be squared, these squares must be added, and then the square root of this sum must be extracted. This involves four operations.

solution 3 Since *OBA* is a right triangle for which **OB** and **BA** are given, the hypotenuse (resultant) can be computed as explained in Sec. 26-6. Hence,

$$\tan \theta = \frac{BA}{OB} = 1.33$$

$$\therefore \theta = 53.1°$$

Then $$\mathbf{OA} = \frac{\mathbf{OB}}{\cos 53.1°} = \frac{6}{0.6} = 10$$

or $$\mathbf{OA} = \frac{\mathbf{BA}}{\sin 53.1°} = \frac{8}{0.8} = 10$$

The method of Solution 3 is to be preferred, owing to the minimum number of operations involved; in addition, this is the method used when the slide rule is used for solving the resultant. It is worthy of note that this solution can be completed with a total of three movements on many slide rules and without referring to a table of trigonometric functions.

It should be noted that Example 4 of Sec. 28-4 involves the same quantities as those used in the example of this section and that Figs. 28-13 and 28-18 are alike. In the earlier example a vector that is resolved into its rectangular components is given. In the example of this section, the same components are given as vectors which are added vectorially to obtain the vector of the first example. From this it is apparent that resolving a vector into its rectangular components and adding vectors that are separated by 90° are inverse operations. Basically, either problem resolves itself into the solution of a right triangle.

PROBLEMS 28-3

Find the resultants of the following sets of phasors.

1. $64.3\underline{/0°}$ and $415\underline{/90°}$

2. $10.6\underline{/0°}$ and $2.04\underline{/90°}$

3. $1.23\underline{/90°}$ and $1.47\underline{/0°}$

4. $45.4\underline{/0°}$ and $153\underline{/90°}$

5. $351\angle 0^\circ$ and $94.8\angle 90^\circ$
6. $459\angle 0^\circ$ and $405\angle 0^\circ$
7. $307\angle 0^\circ$ and $124\angle 180^\circ$
8. $5.27\angle 180^\circ$ and $6.0\angle 90^\circ$
9. $310\angle 270^\circ$ and $185\angle 90^\circ$
10. $323\angle 270^\circ$ and $323\angle 0^\circ$
11. $2.34\angle 180^\circ$ and $7.30\angle 270^\circ$
12. $84.2\angle 0^\circ$, $34.4\angle 90^\circ$, and $37\angle 90^\circ$
13. $23.5\angle 270^\circ$, $32\angle 90^\circ$, $26.5\angle 0^\circ$, and $51\angle 180^\circ$
14. $167\angle 270^\circ$, $252\angle 0^\circ$, $143.8\angle 180^\circ$, and $81.3\angle 90^\circ$
15. $12.1\angle 0^\circ$, $72.3\angle 270^\circ$, $51.9\angle 90^\circ$, $2.7\angle 270^\circ$, $8.6\angle 90^\circ$, and $31.6\angle 180^\circ$
16. Check your calculated answers graphically.

28-7 PHASOR SUMMATION OF NONRECTANGULAR COMPONENTS

Often we are called upon to resolve into a resultant a set of phasors which are not themselves perpendicular (Fig. 28-19). The best analytical method of arriving at a solution is to apply the methods already developed in this chapter.

The first step is to find the perpendicular components of each of the phasors to be added and determine their magnitudes and directions. These are shown in Fig. 28-19 as h_A and v_A, the components of phasor A, and h_B and v_B, the components of phasor B.

Second, these components are added algebraically. The horizontal components are added to obtain the resultant horizontal phasor, and then the vertical components are added to obtain the resultant vertical phasor:

$$h_R = h_A + h_B$$

and

$$v_R = v_A + v_B$$

taking into consideration the signs as well as the magnitudes of the components.

Finally, the resultant is the phasor summation of the new perpendicular components:

$$R = \sqrt{h_R{}^2 + v_R{}^2}$$

$$\theta_R = \arctan \frac{v_R}{h_R}$$

$$R = \frac{h_R}{\cos \theta} = \frac{v_R}{\sin \theta}$$

Fig. 28-19 Summation of Nonrectangular Phasors by Resolution into Rectangular Components

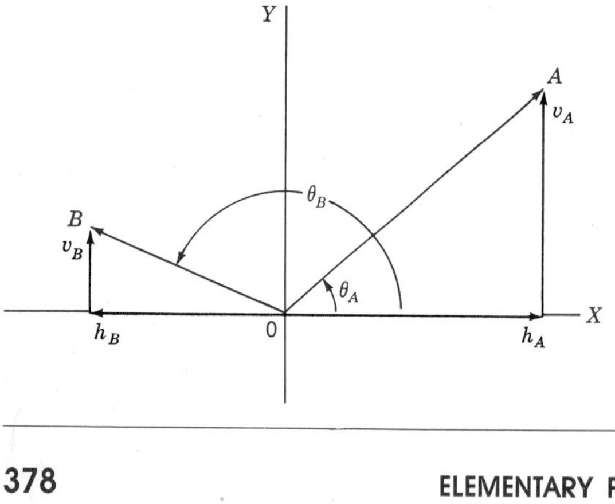

example 8 Find the resultant of two phasors $500\angle 36.9^\circ$ and $142\angle 135^\circ$.

solution Sketch the two phasors in the standard position (Fig. 28-20), and then resolve each phasor into its perpendicular components:

$$h_{500} = 500 \cos 36.9^\circ = 400$$
$$h_{142} = 142 \cos 135^\circ = -142 \cos 45^\circ = -100$$
$$v_{500} = 500 \sin 36.9^\circ = 300$$
$$v_{142} = 142 \sin 135^\circ = 142 \sin 45^\circ = 100$$

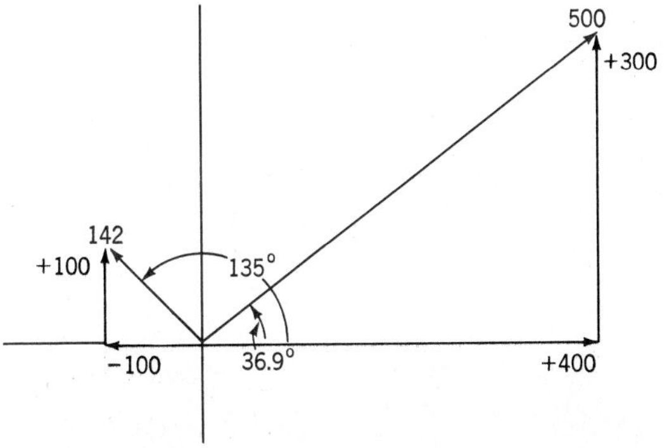

Fig. 28-20 Nonrectangular Phasor Summation of Example 8

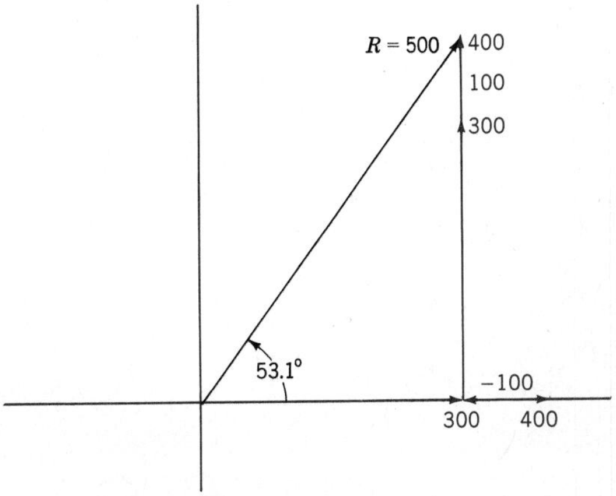

Fig. 28-21 Perpendicular Components of Fig. 28-20 Resolved into Resultant R

Add these components algebraically to obtain the new horizontal and vertical resultants:

$$h_R = +400 - 100 = 300$$
$$v_R = +300 + 100 = 400 \qquad \text{(Fig. 28-21)}$$

The angle θ_R, which R makes with the x axis, is

$$\theta = \arctan \frac{400}{300} = 53.1°$$

and the resultant R of the two resultant perpendicular components is

$$R = \frac{300}{\cos 53.1°}$$

or

$$R = \frac{400}{\sin 53.1°} = 500$$

This process of analysis of phasors into their components and synthesis of resultant components into a final phasor resultant may be applied to any number of phasors.

PROBLEMS 28-4

Find the resultants of the following sets of phasors. Check your solutions graphically:

1. $217\underline{/63.8°}$ and $110\underline{/40.3°}$

2. $799\underline{/48.7°}$ and $233\underline{/120.2°}$

3. $100\underline{/40.3°}$ and $39.6\underline{/315°}$

4. $7.65\underline{/17.8°}$ and $4.34\underline{/137.5°}$

5. $10.7\underline{/32.8°}$, $42.0\underline{/81.2°}$, and $61.2\underline{/221.4°}$

29 periodic functions

In Sec. 24-9, it was shown that the trigonometric functions could be represented by the ratios of lengths of certain lines to the unit radius vector. Also, in Sec. 24-8, the variation of the functions was represented by lines.

The complete variation of the functions is more clearly illustrated and better understood by plotting their continuous values on rectangular coordinates.

29-1 THE GRAPH OF THE SINE CURVE $y = \sin x$

The equation $y = \sin x$ can be plotted just as the graphs of algebraic equations are plotted, that is, by assigning values to the angle x (the independent variable), computing the corresponding value of y (the dependent variable), plotting the points whose coordinates are thus obtained, and drawing a smooth curve through the points. This is the same procedure as used for plotting linear equations in Chap. 16 and for plotting quadratic equations in Chap. 21.

The first questions that come to mind in preparing to graph this equation are, "What values shall be assigned to x? Shall they be in radians or degrees?" Either might be used, but it is more reasonable to use radians. In Sec. 23-5, it was shown that an angle measured in radians can be represented by the arc intercepted by this angle on the circumference of a circle of unit radius. Since, as previously mentioned, the functions of an angle can be represented by suitable lengths of lines, it follows that if an angle is expressed in radian measure, both the angle and its functions can be expressed in terms of a common unit of length. Therefore, we shall select a suitable unit of length and plot both x and y values in terms of this unit. Then to graph the equation $y = \sin x$, the procedure is as follows:

1. Assign values to x.
2. From the slide rule, the calculator, or the tables, determine the corresponding values of y (Table 29-1).
3. Take each pair of values of x and y as coordinates of a point, and plot the point.
4. Draw a smooth curve through the points.

It is not necessary to tabulate values of $\sin x$ between π and 2π radians (180 to 360°), for these values are negative but equal in magnitude to the sines of the angles between 0 and π radians (0 to 180°). The curve should

Table 29-1

x, degrees	*x, radians* (*π measure*)	*x, radians* (*unit measure*)	*y* (*sin x*)	*Point*
0	0	0	0	$P_0 = (0,0)$
30	$\frac{\pi}{6}$	0.52	0.50	$P_1 = (0.52,0.50)$
60	$\frac{\pi}{3}$	1.05	0.87	$P_2 = (1.05,0.87)$
90	$\frac{\pi}{2}$	1.57	1.00	$P_3 = (1.57,1.00)$
120	$\frac{2\pi}{3}$	2.09	0.87	$P_4 = (2.09,0.87)$
150	$\frac{5\pi}{6}$	2.62	0.50	$P_5 = (2.62,0.50)$
180	π	3.14	0	$P_6 = (3.14,0)$

be plotted with the angle and the function having the same unit or scale; that is, one unit on the y axis should be the same length as that representing 1 radian on the x axis. When the curve is so plotted, it is called a *proper sine curve*, as shown in Fig. 29-1. This wave-shaped curve is called the *sine curve* or *sinusoid*.

If additional values of x are chosen, both positive and negative, the curve continues indefinitely in both directions while repeating in value. Note that, as x increases from 0 to $\frac{\pi}{2}$ (or $\frac{1}{2}\pi$), $\sin x$ increases from 0 to 1; as x increases from $\frac{1}{2}\pi$ to π, $\sin x$ decreases from 1 to 0; as x increases from π to $\frac{3\pi}{2}$, $\sin x$ increases from 0 to -1; and as x increases from $\frac{3\pi}{2}$ to 2π, $\sin x$ decreases from -1 to 0. Thus the curve repeats itself for every multiple of 2π radians.

Fig. 29-1 Graph of the Equation $y = \sin x$

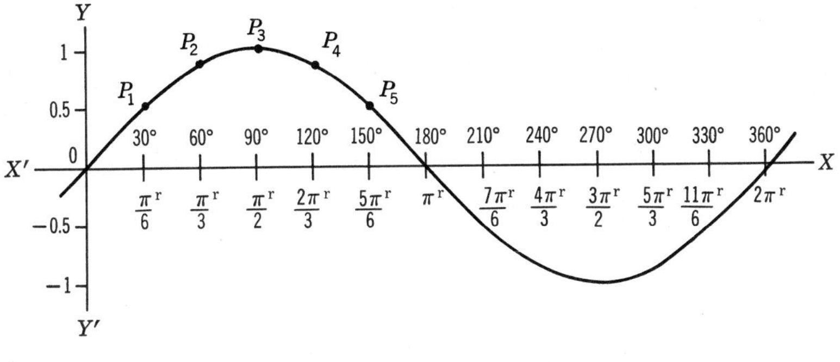

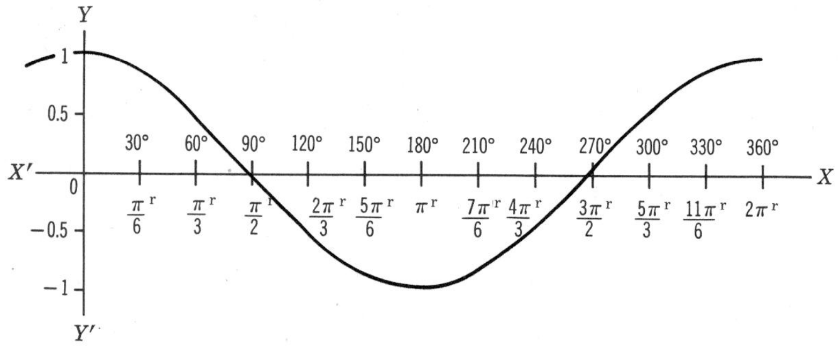

Fig. 29-2 Graph of the Equation $y = \cos x$

29-2 THE GRAPH OF THE COSINE CURVE $y = \cos x$

By following the procedure for plotting the sine curve, you can easily verify that the graph of $y = \cos x$ appears as shown in Fig. 29-2.

Note that, as x increases from 0 to $\frac{1}{2}\pi$, $\cos x$ decreases from 1 to 0; as x increases from $\frac{1}{2}\pi$ to π, $\cos x$ increases from 0 to -1; as x increases from π to $\frac{3\pi}{2}$ $\cos x$ decreases from -1 to 0; and as x increases from $\frac{3\pi}{2}$ to 2π, $\cos x$ increases from 0 to 1. If additional values of x are chosen, both positive and negative, the curve will repeat itself indefinitely in both directions. The cosine curve is identical in shape with the sine curve except that there is a difference of 90° between corresponding points on the two curves. Another similarity between these curves is that both curves repeat their values for every multiple of 2π radians ($2\pi^r$).

29-3 THE GRAPH OF THE TANGENT CURVE $y = \tan x$

The graph of the equation $y = \tan x$, shown in Fig. 29-3, has characteristics different from those of the sine or cosine curve. The curve slopes upward and to the right. At points where x is an odd multiple of $\frac{1}{2}\pi$, the curve is discontinuous. This is to be expected from the discussion of the tangent function in Sec. 24-8.

The tangent curve repeats itself at intervals of π radians (π^r), and is thus seen to be a series of separate curves, or branches, rather than a continuous curve.

PROBLEMS 29-1

1. Plot the equation $y = \sin x$ from -2π to $2\pi^r$.
2. Plot the equation $y = \cos x$ from -2π to $2\pi^r$.
3. Plot the equation $y = \cot x$ from 0 to $2\pi^r$.
4. Plot the equation $y = \sec x$ from 0 to $2\pi^r$.
5. Plot the equation $y = \csc x$ from 0 to $2\pi^r$.
6. Plot the equations $y = \sin^2 x$ and $y = \cos^2 x$ on the same coordinates and to the same scale. In computing points, remember that when a negative number is squared, the result is positive. Add the respective ordinates of

the curves for several different values of angle, and plot the results. What conclusion do you draw from these results?

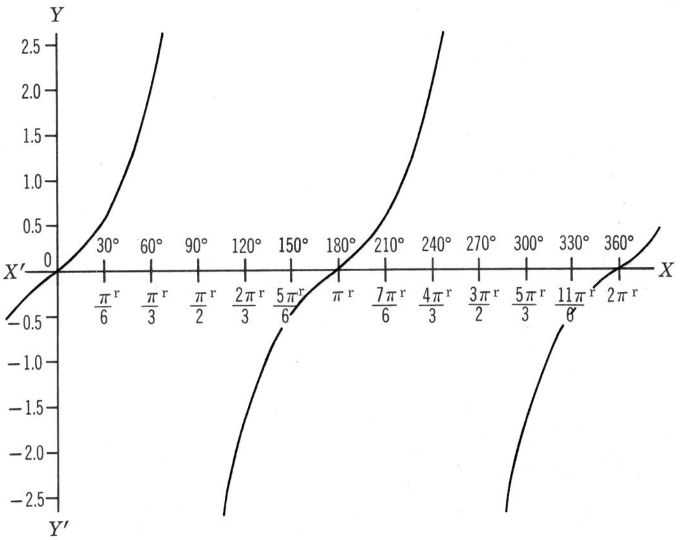

Fig. 29-3 Graph of the Equation $y = \tan x$

29-4 PERIODICITY

From the graphs plotted in the preceding figures and from earlier considerations of the trigonometric functions, it is evident that each trigonometric function repeats itself exactly in the same order and at regular intervals. A function that repeats itself periodically is called a *periodic function*. From this definition, it is apparent that the trigonometric functions are periodic functions.

Owing to the fact that many natural phenomena are periodic in character, the sine and cosine curves lend themselves ideally to graphical representation and mathematical analysis of these recurrent motions. For example, the rise and fall of tides, motions of certain machines, the vibrations of a pendulum, the rhythm of our bodily life, sound waves, and water waves are all familiar happenings that can be represented and analyzed by the use of these curves. An alternating current follows these variations, as will be shown in Chap. 30, and it is because of this fact that you must have a good grounding in trigonometry. It is essential that you understand the mathematical expressions for various periodic functions and especially their applications to ac circuits.

The tangent, cotangent, secant, and cosecant curves are not used to represent recurrent happenings, for although these curves are periodic, they are discontinuous for certain values of angles.

29-5 ANGULAR MOTION

The *linear velocity* of a point or object moving in a particular direction is the rate at which distance is traveled by the point or object. The unit of velocity is the distance

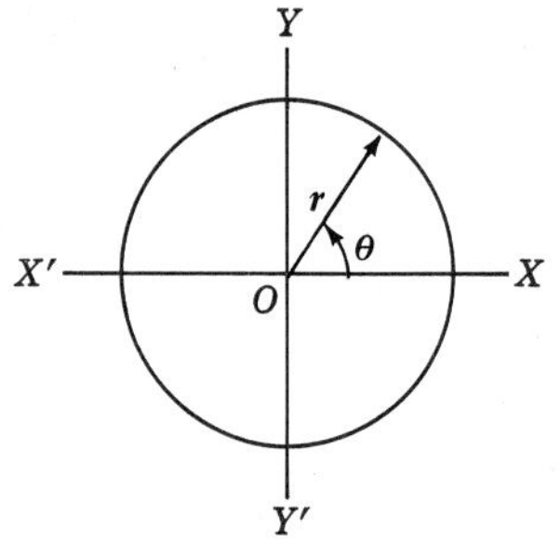

Fig. 29-4 Radius Vector r Generates Angle θ.

traveled in unit time when the motion of the point or object is uniform, such as miles per hour, feet per second, or centimeters per second.

The same concept is used to measure and define *angular velocity*. In Fig. 29-4 the radius vector r is turning about the origin in a counterclockwise direction to generate the angle θ. The *angular velocity* of such a rotating line is the rate at which an angle is generated by rotation. When the rotation is uniform, the unit of angular velocity is the angle generated per unit of time. Thus, angular velocity is measured in degrees per second or radians per second, the latter being the more widely used.

Angular velocity may be expressed in terms of revolutions per minute or revolutions per second. For example, if f is the number of revolutions per second of the vector of Fig. 29-4, then $2\pi f$ is the number of radians generated per second. The angular velocity in radians per second is denoted by ω (Greek letter omega). Thus, if the radius vector is rotating f revolutions per second,

$$\omega = 2\pi f \qquad \text{rad/s}$$

If the armature of a generator is rotating at 1800 rev/min, which is 30 rev/s, it has an angular velocity of

$$\omega = 2\pi f = 2\pi \times 30 = 188.4^{r}/\text{s}$$

where we have introduced ^{r} as the symbol for radians.

The total angle θ generated by a rotating line in t seconds at an angular velocity of ω^{r}/s is

$$\theta = \omega t \qquad \text{rad}$$

Thus the angle generated by the armature in 0.01 s is

$$\theta = \omega t = 188.4 \times 0.01 = 1.884^{r}$$

or
$$\theta = 1.884 \times \frac{180}{\pi} = 108°$$

example 1 A flywheel has a velocity of 300 rev/min. (*a*) What is its angular velocity? (*b*) What angle will be generated in 0.2 s? (*c*) How much time is required for the wheel to generate 628^{r}?

solution

(*a*)
$$f = \frac{300 \text{ rev/min}}{60} = 5 \text{ rev/s}$$

Then

$$\omega = 2\pi f = 2\pi \times 5 = 10\pi \text{ or } 31.4^{r}/\text{s}$$

(*b*)

$$\theta = \omega t = 10\pi \times 0.2 = 2\pi^{r}$$
$$\theta = 360°$$

(c) Since

$$\theta = \omega t$$

then

$$t = \frac{\theta}{\omega} = \frac{628}{10\pi} = 20 \text{ s}$$

PROBLEMS 29-2

1. What is the angular velocity, in terms of π^{r}/s, of (*a*) the hour hand of a clock, (*b*) the minute hand of a clock, and (c) the second hand of a clock?

2. Express the angular velocity of 1800 rev/min in (*a*) radians per second and (*b*) degrees per second.
3. If a satellite circles the earth in 80 min, what is its average angular velocity in (*a*) degrees per minute and (*b*) radians per second?
4. A revolution counter on an armature shaft recorded 900 revolutions in 30 s. What is the value of its angular velocity in (*a*) radians per minute and (*b*) degrees per minute?
5. The radius vector $\mathbf{r}$ of Fig. 29-4 is rotating at the rate of 3600 rev/s. What is the value of θ in radians at the end of (*a*) 0.01 s, (*b*) 0.001 s, and (*c*) 0.5 ms?
6. If the radius vector $\mathbf{r}$ of Fig. 29-4 is rotating at the rate of 1 rev/s, what is the value of sin ωt at the end of (*a*) 0.001 s, (*b*) 0.1 s, (*c*) 0.5 s, and (*d*) 0.95 s?

29-6 PROJECTION OF A POINT HAVING UNIFORM CIRCULAR MOTION

In Fig. 29-5 the radius vector $\mathbf{r}$ rotates about a point in a counterclockwise direction with a uniform angular velocity of 1 rev/s. Then every point on the radius vector, such as the end point P, rotates with uniform angular velocity. If the radius vector starts from 0°, at the end of $\frac{1}{12}$ s it will have rotated 30° or 0.5236^r, to P_1; at the end of $\frac{1}{6}$ s, it will have rotated to P_2 and generated an angle of 60°, or 1.047^r, etc.

The projection of the end point of the radius vector, that is, its ordinate value at any time, can be plotted as a curve. This is accomplished by extending the horizontal diameter of the circle to the right for use as an x axis along which time is to be plotted. Choose a convenient length along the x axis and divide it into as many intervals as there are angle values to be plotted. In Fig. 29-5, projections have been made every 30°, starting from 0°. Therefore, the x axis is divided into 12 divisions, and since one complete revolution takes place in 1 s, each division on the time axis will represent $\frac{1}{12}$ s, or 30° rotation.

Through the points of division on the time axis (x axis), construct vertical lines, and through the corresponding points (made by the end point of the radius vector at that particular time) draw lines parallel to the time axis. Draw a smooth curve through the points of intersection. Thus the resulting sine curve traces the ordinate of the end point of the radius vector for any time t, and from it

Fig. 29-5 Radius Vector Generating Sine Curve.

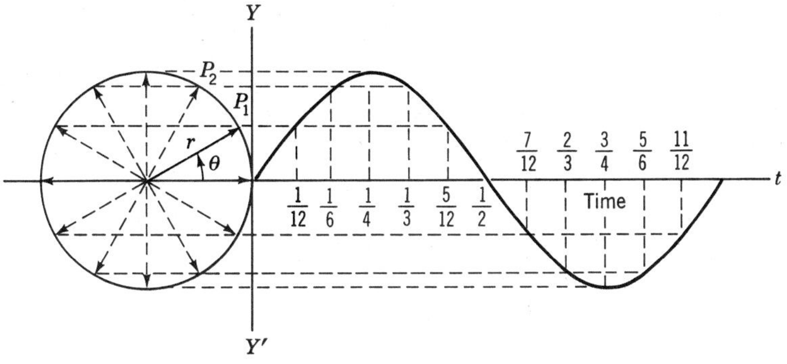

we could obtain the sine value for any angle generated by the radius vector.

As the vector continues to rotate, successive revolutions will generate repeating, or periodic, curves.

Since the y value of the curve is proportional to the sine of the generated angle and the length of the radius vector, we have

$$y = \boldsymbol{r} \sin \theta$$

Then, since the radius vector rotates through $2\pi^{r}$ in 1 s, the y value at any time t is

$$y = \boldsymbol{r} \sin 2\pi t$$

or $$y = \boldsymbol{r} \sin 6.28t$$

which is the equation of the sine curve of Fig. 29-5.

From the foregoing considerations, it is apparent that if a straight line of length $\boldsymbol{r}$ rotates about a point with a uniform angular velocity of ω^{r} per unit time, starting from a horizontal position when the time $t = 0$, the projection y of the end point upon a vertical straight line will have a motion that can be represented by the relation

$$y = \boldsymbol{r} \sin \omega t \qquad (1)$$

This equation is of fundamental importance in describing the motion of any object or quantity that varies periodically, or with simple harmonic motion. Thus the value of an alternating emf at any instant can be completely described in terms of such an equation, that is, if the motion or variation can be represented by a sine curve, it is said to be *sinusoidal* or to vary *sinusoidally*.

example 2 A crank 150 mm long, starting from 0°, turns in a counterclockwise direction at the rate of 1 revolution in 10 s. (*a*) What is the equation for the projection of the crank handle upon a vertical line at any instant? That is, what is the vertical distance from the crankshaft at any time? (*b*) What is the vertical distance from the handle to the shaft at the end of 3 s? (*c*) At the end of 8 s?

solution

(*a*) The general equation for the projection of the end point on a vertical line is

$$y = \boldsymbol{r} \sin \omega t \qquad (1)$$

where $\boldsymbol{r}$ = length of rotating object
ω = angular velocity, rad/s
t = time at any instant, s

Then, since the crank makes 1 revolution, or $2\pi^{r}$, in 10 s, the angular velocity is

$$\omega = \frac{2\pi}{10} = \frac{\pi}{5}, \text{ or } 0.628^{r}/\text{s}$$

Substituting the values of r and ω in Eq. (1),

$$y = 150 \sin 0.628t \text{ mm}$$

(*b*) At the end of 3 s, the crank will have turned through

$$0.628 \times 3 = 1.88^{r}$$

which is $1.88 \times \dfrac{180}{\pi} = 108°$. Substituting this value for $0.628t$ in Eq. (1) results in

$$y = 150 \sin 108° = 150 \times 0.951 = 142.6 \text{ mm}$$

which is the vertical distance of the handle from the shaft at the end of 3 s.

(*c*) At the end of 8 s the crank will have turned through

$$0.628 \times 8 = 5.02^{r}$$

which is $5.02 \times \frac{180}{\pi} = 288°$. Substituting this value for $0.628t$ in the above equation results in

$$y = 150 \sin 288° = 150 \times (-0.951) = -142.6 \text{ mm}$$

which is the vertical distance of the handle from the shaft at the end of 8 s. The negative sign denotes that the handle is *below* the shaft, that is, the distance is measured downward, whereas the distance in (*b*) above was taken as positive, or *above* the shaft.

If it is desired to express the projection of the end point of the radius vector upon the horizontal, the relation is

$$y = \mathbf{r} \cos \omega t \qquad (2)$$

which, when plotted, results in a cosine curve. Thus, in the foregoing example, the horizontal distance (Sec. 26-7) between the handle and shaft at the end of 8 s will be

$$y = 150 \cos 288° = 150 \times 0.309 = 46.4 \text{ mm}$$

29-7 AMPLITUDE

The graphs of Figs. 29-1, 29-2, and 29-5 have an equal amplitude of 1, that is, an equal vertical displacement from the horizontal axis. The value of the radius vector $\mathbf{r}$ determines the amplitude of a general curve, and for this reason the factor $\mathbf{r}$ in the general equation

$$y = \mathbf{r} \sin \omega t$$

is called the *amplitude factor*. Thus the amplitude of a periodic curve is taken as the maximum displacement, or value, of the curve. It is apparent that, if the length of the radius vector which generates a sine wave is varied, the amplitude of the sine wave will be varied accordingly. This is illustrated in Fig. 29-6.

29-8 FREQUENCY

When the radius vector makes one complete revolution, regardless of its starting point, it has generated one complete sine wave; hence, we say the sine wave has gone through one complete *cycle*. Thus the number of cycles occurring in a periodic curve

Fig. 29-6 A: $y = \sin \theta$, B: $y = 2 \sin \theta$, C: $y = 3 \sin \theta$

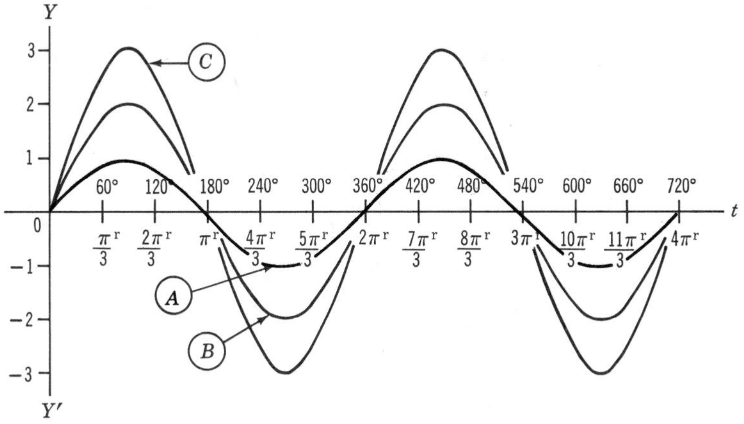

in a unit of time is called the *frequency* of the curve. For example, if the radius vector rotated 5 rev/s, the curve describing its motion would go through 5 cycles in 1 s of time. The frequency f in hertz is obtained by dividing the angular velocity ω by 360° when the latter is measured in degrees or by 2π when measured in radians. That is,

$$f = \frac{\omega}{2\pi} \qquad \text{Hz} \tag{3}$$

Curves for different frequencies are shown in Fig. 29-7.

In the equation $y = \boldsymbol{r} \sin \frac{1}{2}t$, since $\omega t = \frac{1}{2}t$, the angular velocity ω is 0.5 rad/s. That is, at the end of 2π, or 6.28 s, the curve has gone through one-half cycle, or 3.14^{r} of an angle, as shown in Fig. 29-7.

In the equation $y = \boldsymbol{r} \sin t$, since $\omega t = t$, the angular velocity ω is 1 rad/s. Thus at the end of 2π s the curve has gone through one complete cycle, or $2\pi^{\text{r}}$ of angle.

Similarly, in the equation $y = \boldsymbol{r} \sin 2t$, the angular velocity ω is 2 rad/s. Then at the end of 2π s the curve has completed two cycles, or $4\pi^{\text{r}}$ of angle.

29-9 PERIOD

The time T required for a periodic function, or curve, to complete one cycle is called the *period*. Hence, if the frequency f is given by

$$f = \frac{\omega}{2\pi} \qquad \text{Hz}$$

it follows that

$$T = \frac{2\pi}{\omega} = f^{-1} \qquad \text{s} \tag{4}$$

For example, if a curve repeats itself 60 times in 1 s, it has a frequency of 60 Hz and a period of

$$T = \frac{1}{60} = 0.0167 \text{ s}$$

Similarly, in Fig. 29-7, the curve represented by $y = \boldsymbol{r} \sin \frac{1}{2}t$ has a frequency of

$$\frac{\omega}{2\pi} = \frac{0.5}{2\pi} = 0.0796 \text{ Hz}$$

and a period of 12.6 s. The curve of $y = \boldsymbol{r} \sin t$ has a frequency of

$$\frac{\omega}{2\pi} = \frac{1}{2\pi} = 0.159 \text{ Hz}$$

and a period of 6.28 s. The curve of $y = \boldsymbol{r} \sin 2t$ has a frequency of 0.318 Hz and a period of 3.14 s.

29-10 PHASE

In Fig. 29-8, two radius vectors are rotating

Fig. 29-7 $y = \boldsymbol{r} \sin t$———, $y = \boldsymbol{r} \sin 2t$– – – –, $y = \boldsymbol{r} \sin \frac{1}{2}t$——– – –——

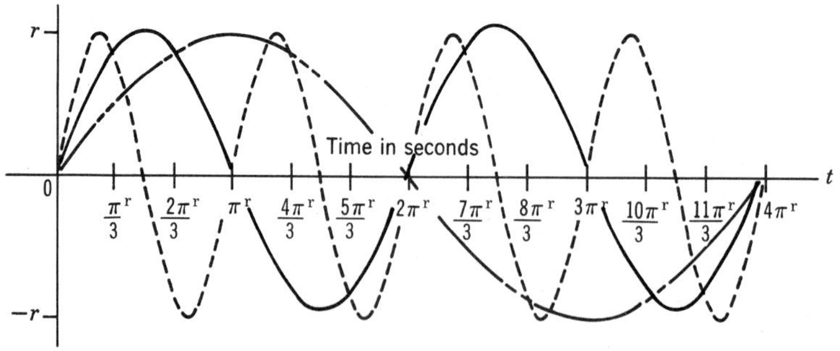

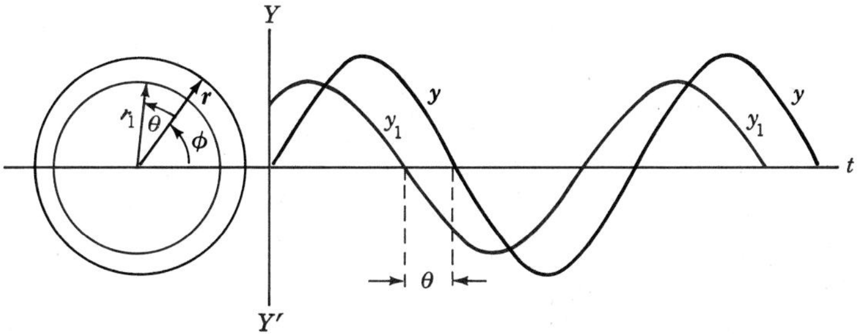

Fig. 29-8 $y = r \sin \omega t$, $y_1 = r_1 \sin (\omega t + \theta)$

about a point with equal angular velocities of ω and separated by the constant angle θ. That is, if r starts from the horizontal axis, then r_1 starts ahead of r by the angle θ and maintains this angular difference.

When $t = 0$, r starts from the horizontal axis to generate the curve $y = r \sin \omega t$. At the same time, r_1 is ahead of r by an angle θ; hence, r_1 generates the curve

$$y_1 = r_1 \sin (\omega t + \theta)$$

It will be noted that this *displaces* the y_1 curve along the horizontal by an angle θ as shown in the figure.

The angular difference θ between the two curves is called the *phase angle*, and since y_1 is *ahead* of y, we say that y_1 leads y. Thus, in the equation $y_1 = r_1 \sin (\omega t + \theta)$, θ is called the *angle of lead*. In Fig. 29-8, y_1 leads y by 30°; therefore, the equation for y_1 becomes

$$y_1 = r_1 \sin (\omega t + 30°)$$

In Fig. 29-9, the radius vectors r and r_1 are rotating about a point with equal angular velocities of ω, except that now r_1 is *behind* r by a constant angle θ. The phase angle between the two curves is θ, but in this case, y_1 lags y. Hence the equation for the curve generated by r_1 is

$$y_1 = r_1 \sin (\omega t - \theta)$$

In Fig. 29-9, the *angle of lag* is $\theta = 60°$; therefore, the equation for y_1 becomes

$$y_1 = r_1 \sin (\omega t - 60°)$$

29-11 SUMMARY

The general equation

$$y = r \sin (\omega t \pm \theta) \qquad (5)$$

Fig. 29-9 $y = r \sin \omega t$, $y_1 = r_1 \sin (\omega t - \theta)$

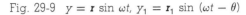

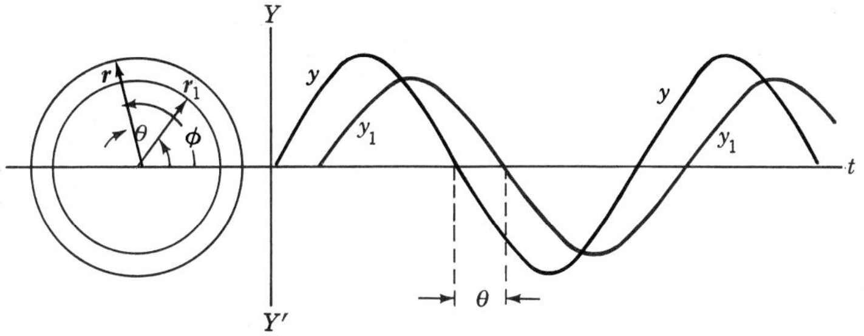

describes a periodic event, and its graph results in a periodic curve. By choosing the proper values for the three arbitrary constants r, ω, and θ, you can describe or plot any periodic sequence of events because a change in any one of these will change the curve accordingly. Hence,

1. If r is changed, the *amplitude* of the curve will be changed proportionally. For this reason, r is called the *amplitude factor.*
2. If ω is changed, the *frequency,* or period, of the curve will be changed. Thus, ω is called the *frequency factor.*
3. If θ is changed, the curve is moved along the time axis with no other change. Thus, if θ is made larger, the curve is displaced to the left and results in a leading phase angle. If θ is made smaller, the curve is moved to the right and results in a lagging phase angle. Hence the angle θ in the general equation is called the *phase angle* or the *angle of lead or lag.*

example 3 Discuss the equation

$$y = 147 \sin (377t + 30°)$$

solution Given $y = 147 \sin (377t + 30°)$. Comparing the given equation with the general equation, it is seen that $r = 147$, $\omega = 377$ rad/s, and $\theta = 30°$. Therefore, the curve represented by this equation is a sine curve with an amplitude of 147. The angular velocity is 377; hence, the frequency is

$$f = \frac{\omega}{2\pi} = \frac{377}{2\pi} = 60 \text{ Hz}$$

and the period is

$$T = f^{-1} = \frac{1}{60} = 0.0167$$

The curve has been displaced to the left 30°; that is, it leads the curve $y = r \sin 377t$ by a phase angle of 30°. Therefore, when $t = 0$, the curve begins at an angle of 30° with a value of

$$\begin{aligned} y &= 147 \sin (\omega t + 30°) \\ &= 147 \sin (0° + 30°) \\ &= 147 \times 0.5 = 73.5 \end{aligned}$$

PROBLEMS 29-3

In the following equations of periodic curves, specify (*a*) amplitude, (*b*) angular velocity, (c) frequency, (*d*) period, and (e) angle of lead or lag with respect to a curve of the same frequency but having no displacement angle.

1. $y = 100 \sin (2\pi t + 40°)$
2. $y = 157 \sin (377t - 12°)$
3. $i = 0.750 \sin (628t + 3°)$
4. $i = I_{max} \sin (31.4t - 20°)$
5. $e = E_{max} \sin (157t - 17°)$
6. $i_c = I_{c_{max}} \sin (1000\pi t + 37°)$

Plot the curves that represent the following motions:

7. $y = \sin 2\pi t$
8. $y = 10 \sin 10t$
9. $e = 141 \sin 120t$
10. $i = 0.5 \sin (120t + 30°)$
11. $i = 1.3 \sin (120t - 20°)$
12. $y = 16 \sin (377t + 10°)$
13. A radar antenna 60 cm long rotates in a horizontal plane at 20 rev/s in a counterclockwise direction, starting from east.

(a) Plot the curve that shows the projection of the antenna on a north-south centerline at any time.
(b) Write the equation for the curve.
(c) What is the distance of the end of the antenna from the east-west line at the end of 0.08 s?
(d) What is the distance of the end of the antenna from the north-south line at the end of 0.1 s?
(e) Through how many radians will the antenna turn in 0.25 s?

14. A radar scope scanning line rotates on the face of the oscilloscope just as a spoke on a wheel rotates with the wheel. If a scan line 175 mm long rotates in a positive direction at the rate of 12 sweeps/s, starting from a position 40° below the horizontal:
(a) Plot the curve that shows the projection of the line upon a vertical reference line at any time.
(b) Write the equation of the curve.
(c) What is the vertical projection of the line at the end of 0.0375 s?
(d) What is the horizontal projection of the line at the end of 0.833 s?
(e) Through how many radians will the line sweep in 2.5 s?

30

alternating currents—fundamental ideas

Thus far we have considered direct voltages and direct currents, that is, voltages that do not change in polarity and currents that do not change in their directions of flow.

In this chapter, you will begin the study of mathematics as applied to alternating currents. An *alternating current* is one that alternates, or changes its direction, periodically.

The fact that over 90% of the electric energy produced is generated in the form of alternating current makes this subject very important, for the operation of all radio and communication circuits is based on ac phenomena. The first requisite in the study of electronics engineering is a solid foundation in the principles of alternating currents.

30-1 GENERATION OF AN ALTERNATING ELECTROMOTIVE FORCE

A coil of wire that has its ends connected to slip rings and is rotating in a counterclockwise direction in a uniform magnetic field is shown in Fig. 30-1. That an alternating emf will be generated in the coil is apparent from a consideration of generated currents. For example, when the side of the coil *ab* moves from its present position away from the S pole, the emf generated in it will be directed from *b* to *a*; that is, *a* will be positive with respect to *b*. At the same time, the side of the coil *cd* is moving away from the N pole, thus cutting magnetic lines of force with a motion opposite to that of *ab*. Then the emf generated in *cd* will be directed from *c* to *d* and will add to the emf from *b* to *a* to send a current I_1 through the resistance R.

When the coil has rotated 90° from the position shown in Fig. 30-1, the plane of the coil is perpendicular to the magnetic field, and at this instant the sides of the coil are moving

Fig. 30-1 Representation of Elementary Alternator

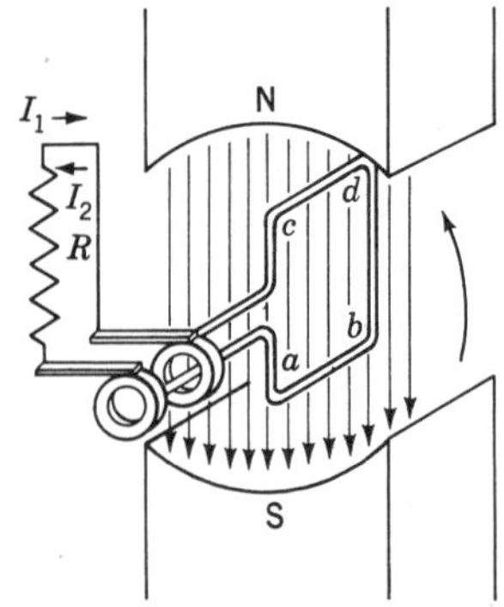

parallel to the magnetic field, thus cutting no lines of force. There is no emf generated at this instant.

As the side of the coil *ab* begins to move up toward the N pole, the emf generated in it will now be directed from *a* to *b*. Similarly, because the side of the coil *cd* is now moving down toward the S pole, the emf in *cd* will be directed from *d* to *c*. This reversal of the direction of generated emf is due to a change of direction of motion with respect to the direction of the lines of force. Therefore, the flow of current I_2 through R will be in the direction indicated by the arrow.

When the coil rotates so that the plane of the coil is again perpendicular to the lines of force (270° from the position shown in Fig. 30-1), no emf will be generated at that instant. Rotation beyond this position, however, causes an emf to be generated such that current flows in the original direction I_1. Such an emf, which periodically reverses its direction, is known as an *alternating electromotive force,* and the resulting current is known as an *alternating current.*

In some engineering textbooks the generation of an emf is explained as due to the change of magnetic flux through the rotating coil. In the final analysis, the results are the same. Here we are interested mainly in the behavior of the circuits connected to sources of alternating currents.

30-2 VARIATION OF AN ALTERNATING ELECTROMOTIVE FORCE

The first questions that come to mind are, "In what manner does an alternating emf vary? How can we represent that variation graphically?"

Figure 30-2 shows a cross section of the elementary alternator of Fig. 30-1. The circles represent either side of the rotating coil at successive instants during the rotation.

When a conductor passes through a magnetic field, there must be a component of its velocity at right angles to the lines of force in order to generate an emf. For example, a conductor must actually *cut* lines in order to develop an emf the amount of which will be proportional to the number of lines cut and the rate of cutting.

From studies of rotation and a consideration of Fig. 30-2, it is evident that the component of horizontal velocity of the rotating conductor is proportional to the sine of the angle of rotation. Because the horizontal velocity is perpendicular to the magnetic field, it is this component that develops an emf. For

Fig. 30-2 Generation of Voltage Sine Wave

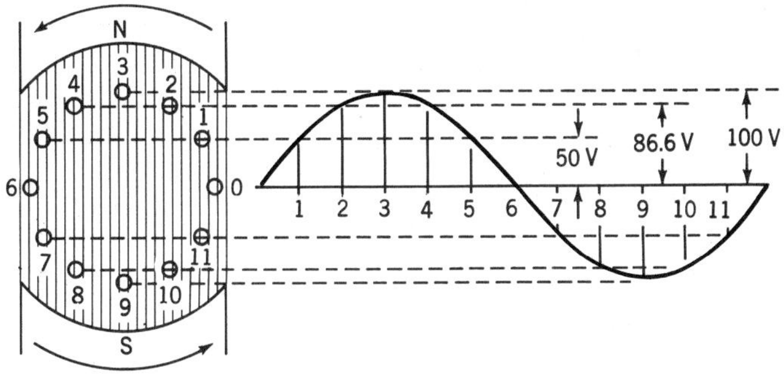

example, at position 0, where the angle of rotation is zero, the conductor is moving parallel to the field; hence, no voltage is generated. As the conductor rotates toward 90°, the component of horizontal velocity becomes greater, thus generating a higher voltage. Therefore, the sine curve of Fig. 30-2 is a graphical representation of the induced emf in a conductor rotating in a uniform magnetic field. The voltage starts from zero, increases in a positive direction to a maximum value (100 V in the figure) at 90°, decreases to zero at 180°, increases in the opposite or negative direction until it attains maximum negative value at 270°, and finally decreases to zero value again at 360°. It follows, then, that the induced emf can be completely described by the relation

$$e = \boldsymbol{E}_{max} \sin \theta \qquad V \qquad (1)$$

where e = instantaneous value of emf at any angle θ, V
$\boldsymbol{E}_{max}$ = maximum value of emf, V
θ = angular position of coil

30-3 VECTOR REPRESENTATION

Since the sine wave of emf is a periodic function, a simpler method of representing the relation of the emf induced in a coil to the angle of rotation is available. The rotating conductor can be replaced by a rotating radius vector whose length represents the magnitude of the maximum generated voltage $\boldsymbol{E}_{max}$. Then the instantaneous value for any position of the conductor can be represented by the vertical component of the vector (Sec. 28-4).

In Fig. 30-3, which is the vector diagram for the conductor at position 0 in Fig. 30-2, the vector $\boldsymbol{E}_{max}$ is at 0° position and therefore has no vertical component. Thus the value of the emf in this position is zero. Or, since

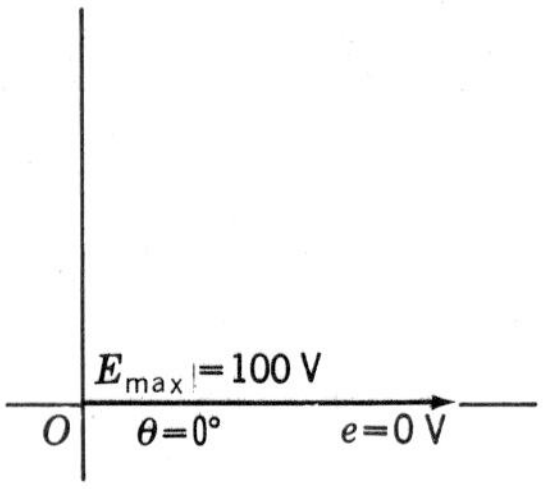

Fig. 30-3 e = 100 sin 0° = 0 V

$$e = \boldsymbol{E}_{max} \sin \theta$$

by substituting the values of $\boldsymbol{E}_{max}$ and θ,

$$e = 100 \sin 0° = 0$$

In Fig. 30-4, which is the vector diagram for the conductor at position 2 in Fig. 30-2, the coil has moved 60° from the zero position. The vector $\boldsymbol{E}_{max}$ is therefore at an angle of 60° from the reference axis, and the instantaneous value of the induced emf is represented by the vertical component of $\boldsymbol{E}_{max}$. Then, since

$$e = \boldsymbol{E}_{max} \sin \theta$$

by substituting the values of $\boldsymbol{E}_{max}$ and θ,

$$e = 100 \sin 60° = 86.6 \text{ V}$$

example 1 What is the instantaneous value of an alternating emf that has

Fig. 30-4 e = 100 sin 60° = 86.6 V
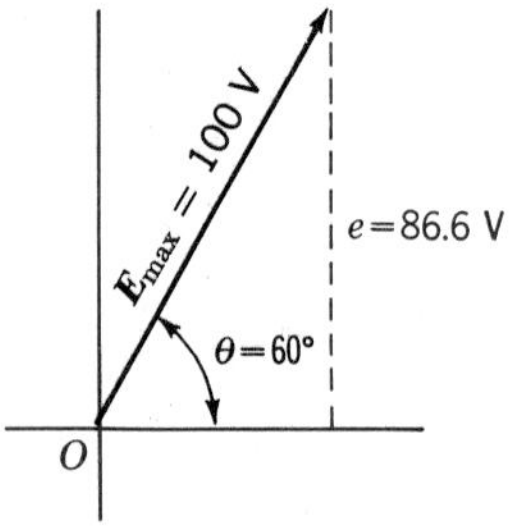

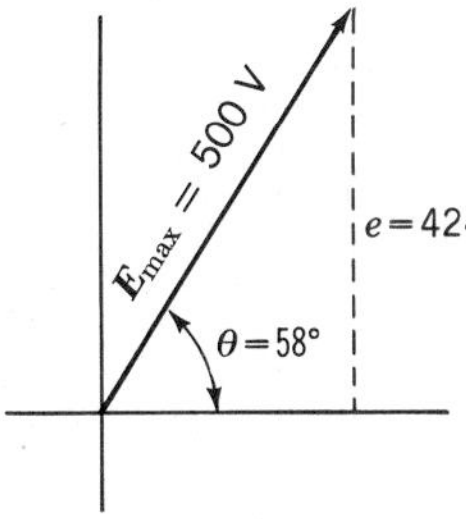

Fig. 30-5 $e = 500 \sin 58° = 424$ V

reached 58° of its cycle? The maximum value is 500 V.

solution Draw the vector diagram to scale as shown in Fig. 30-5. The instantaneous value is the vertical component of the vector $\boldsymbol{E}_{max}$.
Then, since

$$e = \boldsymbol{E}_{max} \sin \theta$$

by substituting the values of $\boldsymbol{E}_{max}$ and θ,

$$e = 500 \sin 58° = 424 \text{ V}$$

example 2 What is the instantaneous value of an alternating emf when it has reached 216° of its cycle? The maximum value is 163 V.

solution Draw the vector diagram to scale as shown in Fig. 30-6. The instantaneous value is the vertical component of the vector $\boldsymbol{E}_{max}$. Then, since

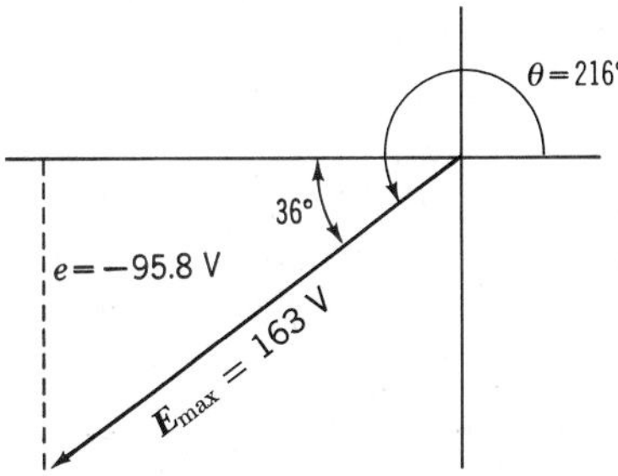

Fig. 30-6 $e = 163(-\sin 36°) = -95.8$ V

$$e = \boldsymbol{E}_{max} \sin \theta$$

by substituting the values of $\boldsymbol{E}_{max}$ and θ,

$$\begin{aligned} e = 163 \sin 216° &= 163[-\sin (216° - 180°)] \\ &= 163(-\sin 36°) = -95.8 \text{ V} \end{aligned}$$

A vector diagram drawn to scale should be made for every ac problem. This gives you a better insight into the functioning of alternating currents and at the same time serves as a good check on the mathematical solution.

Since the current in a circuit is proportional to the applied voltage, it follows that an alternating emf which varies periodically will produce a current of similar variation. Hence, the instantaneous current of a sine wave of alternating current is given by

$$i = \boldsymbol{I}_{max} \sin \theta \quad \text{A} \tag{2}$$

where i = instantaneous value of current, A
$\boldsymbol{I}_{max}$ = maximum value of current, A
θ = angular position of coil

PROBLEMS 30-1

1. An alternating current has a maximum value of 165 A. What are the instantaneous values of this current at the following points in its cycle: (*a*) 18°, (*b*) 67°, (*c*) 136°, (*d*) 242°, (*e*) 326°?
2. The instantaneous value of an alternating emf at 17° is 34.2 V. What is its maximum value?

3. The instantaneous value of an alternating emf at 334.4° is −190 V. What is its maximum value?
4. An alternating current has a maximum value of 750 mA. What are the instantaneous values of the current at the following points in its cycle: (*a*) 26°, (*b*) 341°, (*c*) 210°, (*d*) 297°, (*e*) 162°?
5. The instantaneous value of an alternating emf is 110 V at 71°. What will the value be at 232°?
6. The instantaneous value of an alternating emf at 289° is −22 V. What will the value be at 142°?
7. The instantaneous value of an alternating current at 99.9° is 3.2 A. What will the value be at 199.9°?
8. An alternating current has a maximum value of 365 mA. At what angles will it be 80% of its positive maximum value?
9. At what angles are the instantaneous values of an alternating current equal to 50% of the maximum negative value?
10. What is the instantaneous value of an alternating emf 110° after its maximum positive value of 165 V?

30-4 CYCLES, FREQUENCY, AND POLES

Each revolution of the coil in Fig. 30-1 results in one complete *cycle* which consists of one positive and one negative loop of the sine wave (Sec. 29-8). The number of cycles generated in 1 s is called the *frequency* of the alternating emf, and the *period* is the time required to complete one cycle. One half cycle is called an *alternation.* Thus, by a 60-Hz alternating current is meant that the current passes through 60 cycles per second, which results in a period of 0.0167 s. Also, a 60-Hz current completes 120 alternations per second.

Figure 30-7 represents a coil rotating in a four-pole machine. When one side of the coil has rotated from position 0 to position 4, it has passed under the influence of an N and an S pole, thus generating one complete sine wave, or electrical cycle. This corresponds to 2π electrical radians, or 360 electrical degrees, although the coil has rotated only 180 space degrees. Therefore, in one complete revolution the coil will generate two complete cycles, or 720 electrical degrees, so that for every *space degree* there result two *electrical time degrees.*

In any alternator the armature, or field, must move an angular distance equal to the angle formed by two consecutive like poles in order to complete one cycle. It is evident, then, that a two-pole machine must rotate at twice the speed of a four-pole machine to produce the same frequency. Therefore, to find the frequency of an alternator in cycles per second (hertz), *the number of pairs of poles is multiplied by the speed of the armature in revolutions per second.* That is,

Fig. 30-7 Elementary Four-Pole Alternator

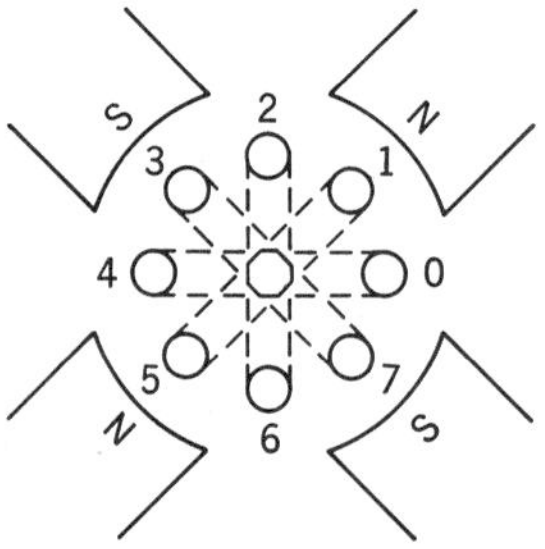

$$f = \frac{PS}{60} \quad \text{Hz} \qquad (3)$$

where f = frequency, Hz
P = number of pairs of poles
S = rotational speed of armature, or field, rev/min

example 3 What is the frequency of an alternator having four poles with a speed of 1800 rev/min?

solution $f = \dfrac{2 \times 1800}{60} = 60$ Hz

30-5 EQUATIONS OF VOLTAGES AND CURRENTS

Since each cycle consists of 360 electrical degrees, or 2π electrical radians, the variation of an alternating emf can be expressed in terms of time. Thus, a frequency of f Hz results in $2\pi f^{\mathrm{r}}/\mathrm{s}$, which is denoted by ω (Sec. 29-5). Hence, the instantaneous emf at any time t is given by the relation

$$e = \boldsymbol{E}_{max} \sin \omega t \quad \text{V} \qquad (4)$$

The instantaneous current is

$$i = \boldsymbol{I}_{max} \sin \omega t \quad \text{A} \qquad (5)$$

You should review Secs. 29-6 to 29-10 to ensure a complete understanding of the relations between the general equation for a periodic function and Eqs. (4) and (5). Thus, $\boldsymbol{E}_{max}$ and $\boldsymbol{I}_{max}$ are the amplitude factors of their respective equations, and ω is the frequency factor.

example 4 Write the equation of a 60-Hz alternating voltage that has a maximum value of 156 V.

solution The angular velocity ω is 2π times the frequency or

$$2\pi \times 60 = 377\ ^{\mathrm{r}}/\mathrm{s}$$

Substituting 156 V for $\boldsymbol{E}_{max}$ and 377 for ω in Eq. (4),

$$e = 156 \sin 377t \quad \text{V}$$

example 5 Write the equation of an RF current of 700 kHz that has a maximum value of 21.2 A.

solution
$\boldsymbol{I}_{max} = 21.2$ A and $f = 700$ kHz $= 7 \times 10^5$ Hz. Then

$$\omega = 2\pi f = 2\pi \times 7 \times 10^5$$
$$= 4.4 \times 10^6$$

Substituting these values in Eq. (5),

$$i = 21.2 \sin (4.4 \times 10^6)t \quad \text{A}$$

example 6 If the time $t = 0$ when the voltage of Example 4 is zero and increasing in a positive direction, what is the instantaneous value of the voltage at the end of 0.002 s?

solution Substituting 0.002 for t in the equation for the voltage,

$$e = 156 \sin (377 \times 0.002)$$
$$= 156 \sin 0.754^{\mathrm{r}} \quad \text{V}$$

where 0.754 is the time angle in *radians.* Then, since $1^{\mathrm{r}} \cong 57.3°$,

$$e = 156 \sin (0.754 \times 57.3°)$$
$$= 156 \sin 43.2°$$

Hence, $e = 107$ V

PROBLEMS 30-2

1. An alternator with 40 poles has a speed of 1200 rev/min and develops a maximum emf of 314 V.
   (*a*) What is the frequency of the alternating emf?
   (*b*) What is the period of the alternating emf?
   (*c*) Write the equation for the instantaneous emf at any time t.
2. An alternator with 8 poles has a speed of 3600 rev/min, and develops a maximum voltage of 120 V.
   (*a*) What is the frequency of the alternating emf?
   (*b*) Write the equation for the instantaneous value of the emf at any time t.
3. A 400-Hz generator which develops a maximum emf of 250 V has a speed of 1200 rev/min.
   (*a*) How many poles has it?
   (*b*) Write the equation of the voltage.
   (*c*) What is the value of the voltage when the time $t = 2$ ms?
4. An 800-Hz alternator generates a maximum of 163 V at 4000 rev/min.
   (*a*) How many poles has it?
   (*b*) Write the equation for the voltage.
   (*c*) What is the value of the emf when time $t = 500\ \mu s$?
5. At what speed must a 12-pole 60-Hz alternator be driven in order to develop its rated frequency?
6. The equation for a certain alternating current is $i = 84.6 \sin 377t$ mA. What is its frequency?
7. The equation for an alternating emf is $e = 0.05 \sin (3.14 \times 10^9)t$ V. What is the frequency of the emf?
8. The equation for an alternating current is

$$i = (2.75 \times 10^{-2}) \sin (2.7 \times 10^7)t \qquad \text{A}$$

   (*a*) What is the maximum instantaneous current?
   (*b*) What is the frequency?
9. A 500-MHz current has a maximum instantaneous value of 30 μA. Write the equation describing the current.
10. A broadcasting station operating at 1430 kHz develops a maximum potential of 0.362 mV across a listener's antenna. Write the equation for this emf.

30-6 AVERAGE VALUE OF CURRENT OR VOLTAGE

Since an alternating current or voltage is of sine-wave form, it follows that the average current or voltage of one cycle is zero owing to the reversal of direction each half cycle. The term *average value* is usually understood to mean the average value of one alternation without regard to positive or negative values. The average value of a sine wave, such as that shown in Fig. 30-2, can be computed to a fair degree of accuracy by taking the average of many instantaneous values between two consecutive zero points of the curve, the values chosen being separated by equal values of angle. Thus, the average value is equal to the average height of any voltage or current loop. The exact average value is $2 \div \pi \cong 0.637$ times the maximum value. Thus, if I_{av} and E_{av} denote the average values of alternating current and voltage, respectively, we obtain

$$I_{av} = \frac{2}{\pi}\boldsymbol{I}_{max} \cong 0.637\boldsymbol{I}_{max} \quad \text{A} \qquad (6)$$

and $$E_{av} = \frac{2}{\pi}\boldsymbol{E}_{max} \cong 0.637\boldsymbol{E}_{max} \quad \text{V} \qquad (7)$$

example 7 The maximum value of an alternating voltage is 622 V. What is the average value?

solution

$$E_{av} = 0.637\boldsymbol{E}_{max} = 0.637 \times 622$$
$$= 396 \text{ V}$$

30-7 EFFECTIVE VALUE OF CURRENT OR VOLTAGE

If a direct current of I A is caused to flow through a resistance of $R\ \Omega$, the resulting energy converted into heat equals I^2R W. We should not expect an alternating current with a maximum value of 1 A to produce as much heat as a direct current of 1 A, for the former does not maintain a constant value. Thus, the above ac ampere is not as effective as the dc ampere. The *effective value* of an alternating current is rated in terms of direct current; that is, an alternating current has an effective value of 1 A if, when it flows through a given resistance, it produces heat at the same rate as a dc ampere would.

The effective value of a sine wave of current can be computed to a fair degree of accuracy by taking equally spaced instantaneous values and extracting the square root of their average, or mean, squared values. For this reason, the effective value is often called the *root-mean-square* (rms) value. The exact effective value of an alternating current or voltage is $1/\sqrt{2} \cong 0.707$ times the maximum value. Thus, if I and E denote the effective values of current and voltage, respectively, we obtain

$$I = \frac{\boldsymbol{I}_{max}}{\sqrt{2}} \cong 0.707\boldsymbol{I}_{max} \quad \text{A} \qquad (8)$$

and $$E = \frac{\boldsymbol{E}_{max}}{\sqrt{2}} \cong 0.707\boldsymbol{E}_{max} \quad \text{V} \qquad (9)$$

It should be noted that all meters, unless marked to the contrary, read effective values of current and voltage.

example 8 The maximum value of an alternating voltage is 311 V. What is the effective value?

solution

$$E = 0.707\boldsymbol{E}_{max} = 0.707 \times 311$$
$$= 220 \text{ V}$$

example 9 An ac ammeter reads 15 A. What is the maximum value of the current?

solution 1

Since $$I = 0.707\boldsymbol{I}_{max}$$

then $$\boldsymbol{I}_{max} = \frac{I}{0.707}$$

Substituting 15 A for I,

$$\boldsymbol{I}_{max} = \frac{15}{0.707} = 21.2 \text{ A}$$

solution 2

Since $$I = \frac{\boldsymbol{I}_{max}}{\sqrt{2}}$$

then $I_{max} = I\sqrt{2} = 1.41I$

Substituting for I,

$$I_{max} = 1.41 \times 15 = 21.2 \text{ A}$$

Hence the maximum value of an alternating current or voltage is equal to 1.41 times the effective value.

PROBLEMS 30-3

1. What is the average value of an alternating emf whose maximum value is 77 V?
2. What is the maximum value of an alternating current whose average value is 56 mA?
3. The average value of an alternating emf is 10.5 V. What is the maximum value?
4. The maximum value of an alternating current is 173 μA. What is the average value?
5. The maximum value of an alternating emf is 180 V. What is the effective value?
6. An rms voltmeter indicates 117 V of alternating emf. What is the maximum value of the emf?
7. What is the effective value of an alternating current which has a maximum value of 30 A?
8. What is the effective value of an alternating emf which has an average value of 125 V?
9. What is the average value of an alternating current which has an effective value of 258 mA?
10. An rms ammeter indicates an alternating current reading of 33.8 A. What is the average value of the current?

30-8 PHASE RELATIONS—PHASE ANGLES

Nearly all ac circuits contain circuit elements, or components, that cause the voltage and current to pass through their corresponding zero values at different times. The effects of such conditions are given detailed consideration in the next chapter.

If an alternating voltage and the resulting alternating current of the same frequency pass through corresponding zero values at the same instant, they are said to be *in phase*.

If the current passes through a zero value before the corresponding zero value of the voltage, the current and voltage are *out of phase* and the current is said to *lead* the voltage.

Figure 30-8 illustrates a phasor diagram and the corresponding sine waves for a current of i A leading a voltage of e V by a *phase angle* of θ (Sec. 29-10). Hence, if the voltage is taken as reference, the general equation of the voltage is

$$e = E_{max} \sin \omega t \qquad \text{V} \tag{10}$$

and the current is given by

$$i = I_{max} \sin (\omega t + \theta) \qquad \text{A} \tag{11}$$

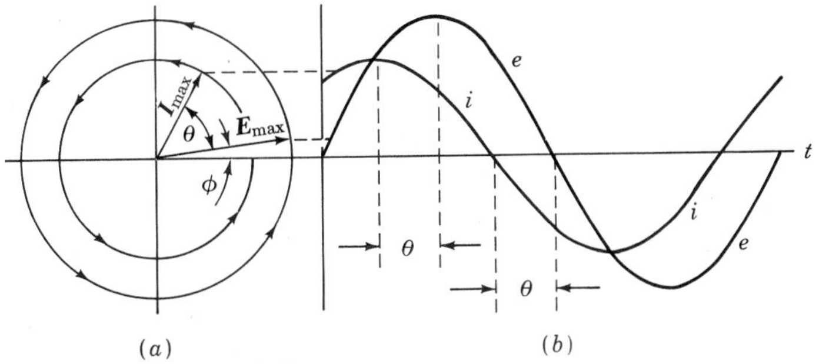

Fig. 30-8 Current i Leads Voltage e by Phase Angle θ

The instantaneous values of the voltage and current for any angle ϕ of the voltage are

$$e = \boldsymbol{E}_{max} \sin \phi \qquad \text{V} \qquad (12)$$

and

$$i = \boldsymbol{I}_{max} \sin (\phi + \theta) \qquad \text{A} \qquad (13)$$

example 10 In Fig. 30-8, the maximum values of the voltage and the current are 156 V and 113 A, respectively. The frequency is 60 Hz, and the current leads the voltage by 40°. (*a*) Write the equation for the voltage at any time *t*. (*b*) Write the equation for the current at any time *t*. (*c*) What is the instantaneous value of the current when the voltage has reached 10° of its cycle?

solution Given

$$\text{Maximum voltage} = \boldsymbol{E}_{max} = 156 \text{ V}$$
$$\text{Maximum current} = \boldsymbol{I}_{max} = 113 \text{ A}$$
$$\text{Frequency} = f = 60 \text{ Hz}$$
$$\text{Phase angle} = \theta = 40° \text{ lead}$$
$$\text{Voltage angle} = \phi = 10°$$

Draw a vector diagram as shown in Fig. 30-8*a*. (The circles are not necessary; they simply denote rotation of the vectors.)
(*a*) Substituting given values in Eq. (10),

$$e = 156 \sin 2\pi \times 60t$$

or

$$e = 156 \sin 377t \text{ V}$$

(*b*) Substituting given values in Eq. (11),

$$i = 113 \sin (377t + 40°) \text{ A}$$

note The quantity $377t$ is in *radians*.

(*c*) Substituting given values in Eq. (13),

$$i = 113 \sin (10° + 40°)$$

or

$$i = 113 \sin 50° = 86.6 \text{ A}$$

Figure 30-9 illustrates a vector diagram and the corresponding sine waves for a current of i A lagging a voltage of e V by a *phase angle* of θ. Therefore, if the voltage is taken as reference, the general equation of the voltage will be as given by Eq. (10) and the current will be

$$i = \boldsymbol{I}_{max} \sin (\omega t - \theta) \qquad \text{A} \qquad (14)$$

The instantaneous value of the current for any angle ϕ of the voltage is

$$i = \boldsymbol{I}_{max} \sin (\phi - \theta) \qquad \text{A} \qquad (15)$$

example 11 In Fig. 30-9, the maximum values of the voltage and the current are 170 V and 14.1 A, respectively. The frequency is 800 Hz, and the current lags the voltage by 40°. (*a*) Write the equation for the voltage at any time *t*. (*b*) Write the equation for the current at any time *t*. (*c*)

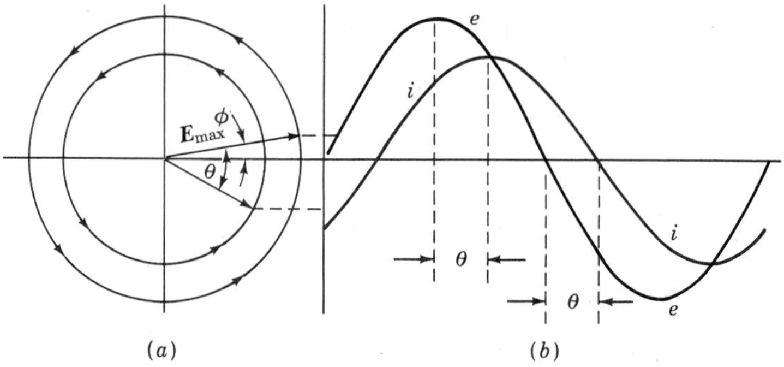

Fig. 30-9 Current i Lags Voltage e by Phase Angle θ

What is the instantaneous value of the current when the voltage has reached 10° of its cycle?

solution Given

Maximum voltage = $\boldsymbol{E}_{max}$ = 170 V
Maximum current = $\boldsymbol{I}_{max}$ = 14.1 A
Frequency = f = 800 Hz
Phase angle = θ = 40° lag
Voltage angle = ϕ = 10°

Draw a vector diagram as shown in Fig. 30-9a.

(*a*) Substituting given values in Eq. (10),

$$e = 170 \sin 2\pi \times 800t$$

or

$$e = 170 \sin 5030t \text{ V}$$

(*b*) Substituting given values in Eq. (14),

$$i = 14.1 \sin (5030t - 40°) \text{ A}$$

(*c*) Substituting given values in Eq. (15),

$$i = 14.1 \sin (10° - 40°)$$

or

$$i = 14.1 \sin (-30°) = -7.05 \text{ A}$$

example 12 In a certain ac circuit a current of 14 A lags a voltage of 220 V by an angle of 60°. What is the instantaneous value of the voltage when the current has completed 245° of its cycle?

note Unless otherwise specified, all voltages and currents are to be considered *effective* values.

solution Draw the vector diagram as shown in Fig. 30-10.

$$\boldsymbol{E}_{max} = \sqrt{2}E = \sqrt{2} \times 220 = 311 \text{ V}$$
$$\phi = 245° + \theta = 245° + 60°$$
$$= 305° = -55°$$

Then, substituting the values of $\boldsymbol{E}_{max}$ and θ in Eq. (12),

$$e = 311 \sin (-55°) = -255 \text{ V}$$

Fig. 30-10 Phasor Diagram of Example 12

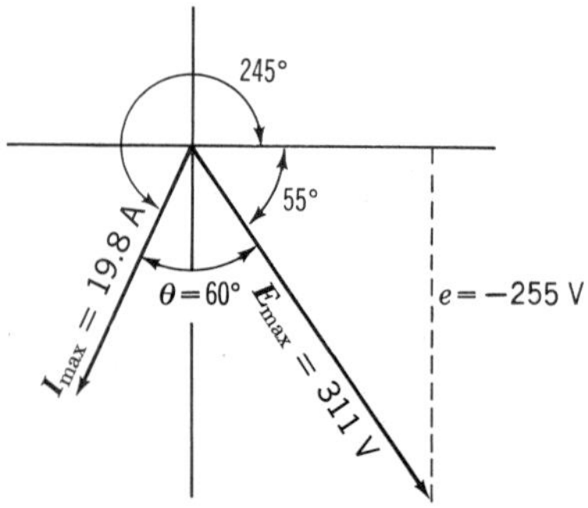

PROBLEMS 30-4

1. A 60-Hz alternator generates a maximum emf of 165 V and delivers a maximum current of 6.5 A. The current leads the voltage by an angle of 36°.
   (a) Write the equation for the current at any time *t*.
   (b) What is the instantaneous value of the current when the emf has completed 60° of its cycle?
2. A 25-Hz alternator generates 6.6 kV at 700 A. The current leads the voltage by an angle of 22°.
   (a) Write the equation for the current at any time *t*.
   (b) How much of the voltage cycle will have been completed the first time that the instantaneous current rises to 465A?
3. In the alternator of Prob. 1, what will be the instantaneous value of the current when the voltage has completed 200° of its cycle?
4. In the alternator of Prob. 2, what will be the instantaneous value of the current when the voltage has completed 350° of its cycle?
5. A 50-Hz alternator generates 2.3 kV with a current of 200 A. The phase angle is 25° lagging.
   (a) Write the equation for the current at any time *t*.
   (b) What is the instantaneous value of the current when the voltage has completed 192° of its cycle?
6. In the alternator of Prob. 5, what is the instantaneous value of the current when the voltage has completed 17° of its cycle?
7. A 60-Hz alternator generates a maximum of 170 V and delivers a maximum current of 42.4 A. If the instantaneous value of the current is 22.5 A when the instantaneous value of the emf is 112 V, what is the phase angle between the current and the emf?
8. In Prob. 7, what will be the instantaneous value of the emf when the instantaneous value of the current is −39.3 A for the first time?
9. A 400-Hz alternator develops 30 A at 230 V. If the instantaneous value of the emf is −85.8 V when the instantaneous value of the current is 23.5 A, what is the phase angle between current and emf?
10. (a) Write the equation for the current in Prob. 9.
    (b) In Prob. 9, what will be the instantaneous value of the current when the emf has reached its maximum value negatively?

31 phasor algebra

In Sec. 1-2 we commented briefly on the use of mathematics as a tool in electronics. One of the most valuable of all the mathematical tools, certainly the most valuable in the solution of ac circuits, is the operator j and complex numbers. The complex number operations to be developed in this chapter are so important in electronics that some (more expensive) electronic calculators have special function keys to simplify even further the mathematical processes involved.

31-1 PHASOR DEFINITIONS

Real and imaginary Complex numbers were introduced in Secs. 20-14 to 20-18, and, in keeping with traditional methods of notation, Secs. 20-13 and 20-14 used the expressions "real number" and "imaginary number." In the complex number $a + jb$, theoretical mathematicians refer to a as the *real part*, and to b as the *imaginary part* of the complex number.

Phase Since our development of the operator j did not depend upon any imaginary features, we now abandon the traditional definitions in favor of the more realistic expressions of electrical engineering. We will refer to a as the *in phase* portion of the complex number, and to b as the *out of phase* portion. These expressions are more in keeping with the correct understanding of the phase relationships of ac circuits which will be developed further in Chap. 32.

Rectangular form The form $a + jb$ is referred to as the *rectangular form* of the complex number. Later, in circuit analyses, we shall use the form $R + jX$.

Polar form When the rectangular components a and b of a complex number are resolved into a single magnitude $\boldsymbol{r}$ rotated through an angle θ from a reference axis, the resultant form $\boldsymbol{r}\underline{/\theta}$ is referred to as the *polar form*. Later, in circuit analyses, we shall use the form $\boldsymbol{Z}\underline{/\theta}$.

Any complex number is fully described by either its rectangular form or its polar form. You must be wholly satisfied with the following relationships, which are described in Fig. 31-1, before continuing the study of this chapter.

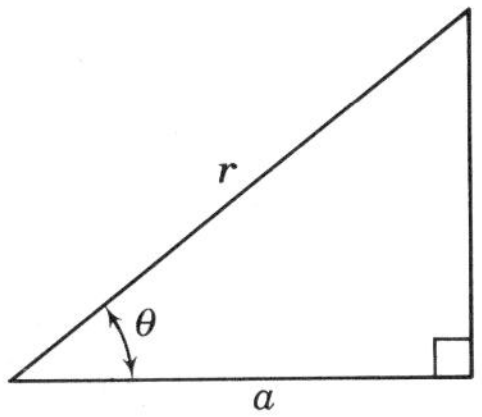

Fig. 31-1 Inter-relationships of rectangular and phasor forms of complex numbers.

$$a = r\cos\theta \qquad \text{and} \qquad b = r\sin\theta$$

$$\theta = \arctan\frac{b}{a} \qquad \text{and}$$

$$r = \frac{a}{\cos\theta} = \frac{b}{\sin\theta}$$

If

$$a + jb = r\underline{/\theta}$$

then

$$a = r\cos\theta$$

and

$$b = r\sin\theta$$

and

$$\theta = \arctan\frac{b}{a}$$

and

$$r = \frac{a}{\cos\theta} = \frac{b}{\sin\theta}$$

Chapter 28 developed the idea of *vectors,* and introduced the *phasor* as the two-dimensional electrical refinement of vectors to describe the polar or phasor relationships. Section 28-6 developed *phasor summation,* and you can see by comparing that section with Sec. 20-15, that a phasor relationship may be simply described by a complex number. *Phasor algebra* is merely the systematic analytical form of right triangle calculations developed in Chaps. 26 and 28.

31-2 ADDITION AND SUBTRACTION OF PHASORS IN RECTANGULAR FORM

As stated in Sec. 20-15, complex numbers, or phasors in rectangular form, can be added or subtracted by treating them as ordinary binomials.

example 1
Add 4.60 + j2.82 and 2.11 − j8.10.

solution

$$\begin{array}{r} 4.60 + j2.82 \\ 2.11 - j8.10 \\ \hline 6.71 - j5.28 \end{array}$$

Express the sum in polar form,

$$6.71 - j5.28 = 8.54\underline{/-38.2^\circ}$$

example 2 Subtract 3.7 + j4.62 from 14.6 − j8.84.

solution

$$\begin{array}{r} 14.6 - j8.84 \\ 3.7 + j4.62 \\ \hline 10.9 - j13.46 \end{array}$$

Express the result in polar form,

$$10.9 - j13.46 = 17.3\underline{/-51^\circ}$$

PROBLEMS 31-1

Perform the indicated operations and express the answers in both rectangular and polar forms. Check your results by graphical methods:

1. (8.3 − j11.3) + (12.4 + j22.6)
2. (18.4 + j25) + (81.2 − j110)
3. (400 + j298) + (700 + j102)
4. (16.95 − j17.8) + (−11.33 − j22.2)

5. $(115 + j925) + (-557 - j184)$
6. $(-488 - j603) + (172 + j168)$
7. $(23.8 - j44.5) - (12.6 - j8.1)$
8. $(8.37 - j3.4) - (-6.53 + j10.2)$
9. $(1100 - j200) - (-1400 - j600)$
10. $(75.3 - j38.7) - (137.4 + j47.1)$
11. $(32.6 + j3.4) - (22.6 - j5.6)$
12. $(-16.5 - j13.7) - (-16.5 + j86.3)$

31-3 MULTIPLICATION OF PHASORS IN RECTANGULAR FORM

Multiplication of complex numbers was explained in Sec. 20-16, where it was shown that phasors expressed in terms of their rectangular components are multiplied by treating them as ordinary binomials.

example 3 Multiply $8 + j5$ by $10 + j9$.

solution

$$\begin{array}{r} 8 + j5 \\ 10 + j9 \\ \hline 80 + j50 \qquad\qquad \\ + j72 + j^2 45 \\ \hline 80 + j122 + j^2 45 \end{array}$$

Since $j^2 = -1$, the product is

$$80 + j122 + (-1)45 = 80 + j122 - 45 = 35 + j122$$

Expressing the product in polar form,

$$35 + j122 = 127\underline{/74^\circ}$$

example 4 Multiply $80 + j39$ by $35 - j50$.

solution

$$\begin{array}{r} 80 + j39 \\ 35 - j50 \\ \hline 2800 + j1365 \qquad\qquad \\ - j4000 - j^2 1950 \\ \hline 2800 - j2635 - j^2 1950 \end{array}$$

Since $j^2 = -1$, the product is

$$2800 - j2635 - (-1)1950 = 2800 - j2635 + 1950 = 4750 - j2635$$

Expressing the product in polar form,

$$4750 - j2635 = 5430\underline{/-29^\circ}$$

31-4 DIVISION OF PHASORS IN RECTANGULAR FORM

As explained in Sec. 20-17, division of complex numbers, or phasors in rectangular form, is accomplished by rationalizing the denominator in order to obtain an in phase number for a divisor. Multiplying a complex number by its conjugate always results in a product that is a simple number not affected by the operator j.

example 5 Find the quotient of $\dfrac{50 + j35}{8 + j5}$.

solution Multiply both dividend and divisor (numerator and denominator) by the conjugate of the divisor, which is $8 - j5$. Thus,

$$\frac{50 + j35}{8 + j5} \cdot \frac{8 - j5}{8 - j5} = \frac{400 + j30 - j^2 175}{64 - j^2 25} = \frac{575 + j30}{89}$$

That is,

$$\frac{575 + j30}{89} = \frac{575}{89} + j\frac{30}{89} = 6.46 + j0.337$$

Express the quotient in polar form,

$$6.46 + j0.337 \cong 6.46\underline{/3.0^\circ}$$

example 6 Simplify $\dfrac{10}{3 + j4}$.

solution Multiply both numerator and denominator by the conjugate of the denominator, which is 3 − j4. Thus,

$$\frac{10}{3 + j4} \cdot \frac{3 - j4}{3 - j4} = \frac{10(3 - j4)}{9 - j^2 16}$$
$$= \frac{30 - j40}{25} = 1.2 - j1.6$$

Express the quotient in polar form,

$$1.2 - j1.6 = 2.0\underline{/-53.1^\circ}$$

PROBLEMS 31-2

Perform the indicated operations and express the answers in both rectangular and polar form:

1. (5 + j4)(2 − j6)
2. (12 + j14)(22 + j17)
3. (2.5 + j7.6)(3.8 − j1.5)
4. (470 − j35.0)(330 + j0.621)
5. (6.8 − j4.6)(5.6 − j7.2)
6. (2.7 − j9)(12 − j8)
7. (4 − j2) ÷ (3 + j5)
8. (7 − j5) ÷ (10 − j14)
9. (20 − j16) ÷ (3 + j5)
10. 1 ÷ (12 − j9)

31-5 ADDITION AND SUBTRACTION OF POLAR PHASORS

As explained in preceding sections, phasors expressed in polar form can be added or subtracted by graphical methods only if their directions are parallel.

In order to add or subtract them algebraically, phasors must be expressed in terms of their rectangular components.

example 7
Add $5.40\underline{/31.5^\circ}$ and $8.37\underline{/-75.4^\circ}$.

solution Converting the phasors into their rectangular components,

$$5.40\underline{/31.5^\circ} = 5.40(\cos 31.5^\circ + j \sin 31.5^\circ) = 4.60 + j2.82$$
$$8.37\underline{/-75.4^\circ} = 8.37(\cos 75.4^\circ - j \sin 75.4^\circ) = 2.11 - j8.10$$

Adding, Sum = 6.71 − j5.28

Expressing the sum in polar form,

$$6.71 - j5.28 = 8.54\underline{/-38.2^\circ}$$

Note that the phasors of this example are the same as those of Example 1 of Sec. 31-2.

example 8 Subtract $5.92\underline{/51.3^\circ}$ from $17.1\underline{/-31.2^\circ}$.

solution Converting the phasors into their rectangular components,

$$17.1\underline{/-31.2^\circ} = 17.1(\cos 31.2^\circ - j \sin 31.2^\circ)$$
$$= 14.6 - j8.86$$
$$5.92\underline{/51.3^\circ} = 5.92(\cos 51.3^\circ + j \sin 51.3^\circ)$$
$$= 3.7 + j4.62$$

Subtracting, Result = 10.9 − j13.48

Expressing the result in polar form,

$$10.9 - j13.48 = 17.3\underline{/-51^\circ}$$

Note that the phasors of this example are the same as those of Example 2 of Sec. 31-2.

PROBLEMS 31-3

Perform the indicated operations and express the results in both polar and rectangular form. Check results graphically.

1. $14\underline{/-53.7^\circ} + 25.8\underline{/61.2^\circ}$
2. $31\underline{/53.7^\circ} + 137\underline{/-53.7^\circ}$
3. $500\underline{/36.6^\circ} + 710\underline{/8.27^\circ}$
4. $24.6\underline{/-46.4^\circ} + 24.9\underline{/242.8^\circ}$
5. $933\underline{/82.9^\circ} + 590\underline{/198.3^\circ}$
6. $777\underline{/-129^\circ} + 241\underline{/44.3^\circ}$
7. $50.6\underline{/-61.9^\circ} - 15\underline{/-32.7^\circ}$
8. $9\underline{/-22.2^\circ} - 12.1\underline{/57.4^\circ}$
9. $1110\underline{/10.3^\circ} - 1510\underline{/203.4^\circ}$
10. $85\underline{/-27.15^\circ} - 145\underline{/18.91^\circ}$
11. $1000\underline{/-53.1^\circ} - 1500\underline{/-53.1^\circ}$
12. $10.64\underline{/-53^\circ} - 22.35\underline{/62.5^\circ}$

31-6 MULTIPLICATION OF POLAR PHASORS

In Example 3 of Sec. 31-3, it was shown that

$$(8 + j5)(10 + j9) = 127\underline{/74^\circ}$$

Now $\quad 8 + j5 = 9.44\underline{/32^\circ}$

and $\quad 10 + j9 = 13.45\underline{/42^\circ}$

Multiplying the magnitudes and adding the angles,

$$(9.44 \times 13.45)\underline{/32^\circ + 42^\circ} = 127\underline{/74^\circ}$$

which is the same product as that obtained by multiplying the phasors when expressed in terms of their rectangular components.

Similarly, in Example 4 of Sec. 31-3, it was shown that

$$(80 + j39)(35 - j50) = 5430\underline{/-29^\circ}$$

Now $\quad 80 + j39 = 89.0\underline{/26^\circ}$

and $\quad 35 - j50 = 61.0\underline{/-55^\circ}$

Multiplying the magnitudes and adding the angles,

$$(89 \times 61.0)\underline{/26^\circ + (-55^\circ)} = 5430\underline{/-29^\circ}$$

which is the same product as that obtained by multiplying the phasors when expressed in terms of their rectangular components.

From the foregoing, it is evident that the product of two polar phasors is found by multiplying their magnitudes and adding their angles algebraically.

31-7 DIVISION OF POLAR PHASORS

In Example 5 of Sec. 31-4, it was shown that

$$\frac{50 + j35}{8 + j5} = 6.46\underline{/3.0^\circ}$$

Now $\quad 50 + j35 = 61.0\underline{/35^\circ}$

and $\quad 8 + j5 = 9.44\underline{/32^\circ}$

Dividing the magnitudes and subtracting the angle of the divisor from the angle of the dividend,

$$\frac{61.0\underline{/35^\circ}}{9.44\underline{/32^\circ}} = \frac{61.0}{9.44}\underline{/35^\circ - 32^\circ} = 6.46\underline{/3.0^\circ}$$

which is the same quotient as that obtained by dividing the phasors when expressed in terms of their rectangular components.

Similarly, in Example 6 of Sec. 31-4, it was shown that

$$\frac{10}{3 + j4} = 2.0\underline{/-53.1^\circ}$$

Since 10 is a positive number, it is plotted on the 0° axis (Sec. 3-5) and expressed as

$$10\underline{/0^\circ}$$

Now $\quad 3 + j4 = 5\underline{/53.1^\circ}$

Dividing the magnitudes and subtracting the angle of the divisor from the angle of the dividend,

$$\frac{10\angle 0^\circ}{5\angle 53.1^\circ} = \frac{10}{5}\angle 0^\circ - 53.1^\circ$$
$$= 2.0\angle -53.1^\circ$$

which is the same quotient as that obtained by dividing the phasors when expressed in terms of their rectangular components.

From the foregoing, it is evident that the quotient of two polar phasors is found by dividing their magnitudes and subtracting the angle of the divisor from the angle of the dividend.

31-8 EXPONENTIAL FORM

In the preceding two sections it has been demonstrated that angles are added when phasors are multiplied and that angles are subtracted when one phasor is divided by another. These operations can be further justified from a consideration of the sine and cosine when expanded in series form.

By Maclaurin's theorem, a treatment of which is beyond the scope of this book, $\cos\theta$ and $\sin\theta$ can be expanded into series form as follows:

$$\cos\theta = 1 - \frac{\theta^2}{2!} + \frac{\theta^4}{4!} - \frac{\theta^6}{6!} + \cdots \tag{1}$$

$$\sin\theta = \theta - \frac{\theta^3}{3!} + \frac{\theta^5}{5!} - \frac{\theta^7}{7!} + \cdots \tag{2}$$

The symbol $n!$ denotes the product of 1, 2, 3, 4, . . . , n and is read "factorial n." Thus, 5! (factorial five) is $1 \times 2 \times 3 \times 4 \times 5$. Similarly, it can be shown that

$$\varepsilon^{j\theta} = 1 + j\theta - \frac{\theta^2}{2!} - j\frac{\theta^3}{3!} + \frac{\theta^4}{4!} + j\frac{\theta^5}{5!} - \frac{\theta^6}{6!} - j\frac{\theta^7}{7!} + \cdots \tag{3}$$

where ε is the base of the natural system of logarithms $\cong 2.718$. By collecting and factoring j terms, Eq. (3) can be written

$$\varepsilon^{j\theta} = \left(1 - \frac{\theta^2}{2!} + \frac{\theta^4}{4!} - \frac{\theta^6}{6!} + \cdots\right) + j\left(\theta - \frac{\theta^3}{3!} + \frac{\theta^5}{5!} - \frac{\theta^7}{7!} + \cdots\right) \tag{4}$$

Note that the first term of the right member of Eq. (4) is $\cos\theta$ as given in Eq. (1) and that the second term in the right member of Eq. (4) is $j\sin\theta$. Therefore,

$$\varepsilon^{j\theta} = \cos\theta + j\sin\theta \tag{5}$$

This expression, $\cos\theta + j\sin\theta$, is often referred to as cis θ, and some texts will actually refer to the *cis function*. You should bear in mind that *cis* is simply an abbreviation for cos + j sin.

Since a phasor, such as $Z\angle\theta$, can be expressed in terms of its rectangular components by the relation

$$Z\angle\theta = Z(\cos\theta + j\sin\theta) \tag{6}$$

it follows from Eqs. (5) and (6) that

$$Z\angle\theta = Z\varepsilon^{j\theta} \tag{7}$$

Similarly, it can be shown that

$$Z\angle-\theta = Z\varepsilon^{-j\theta} \tag{8}$$

Equations (7) and (8) show that the angles of phasors can be treated as exponents.

Two vectors $Z_1\angle\theta$ and $Z_2\angle\phi$ are multiplied by multiplying the magnitudes of the phasors and adding their angles algebraically. That is,

$$(Z_1\angle\theta)(Z_2\angle\phi) = Z_1Z_2\angle\theta + \phi$$

Also, $$\frac{Z_1\angle\theta}{Z_2\angle\phi} = \frac{Z_1}{Z_2}\angle\theta - \phi$$

and $$\frac{Z_a\angle\theta}{Z_b\angle -\phi} = \frac{Z_a}{Z_b}\angle\theta + \phi$$

example 9

Multiply $Z_1 = 8.4\angle 15°$ by $Z_2 = 10.5\angle 20°$.

solution

$$\begin{aligned} Z_1Z_2 &= 8.4 \times 10.5\angle 15° + 20° \\ &= 88.2\angle 35° \end{aligned}$$

example 10

Multiply $Z_a = 164\angle -39°$ by $Z_b = 2.2\angle -26°$.

solution

$$\begin{aligned} Z_aZ_b &= 164 \times 2.2\angle -39° + (-26°) \\ &= 361\angle -65° \end{aligned}$$

example 11

Divide $Z_1 = 54.2\angle 47°$ by $Z_2 = 18\angle 16°$.

solution

$$\frac{Z_1}{Z_2} = \frac{54.2}{18}\angle 47° - 16° = 3.01\angle 31°$$

example 12

Divide $Z_a = 886\angle 18°$ by $Z_b = 31.2\angle -50°$

solution

$$\begin{aligned} \frac{Z_a}{Z_b} &= \frac{886}{31.2}\angle 18° - (-50°) \\ &= 28.4\angle 68° \end{aligned}$$

31-9 POWERS AND ROOTS OF POLAR PHASORS

In addition to following the laws of exponents for multiplication and division, phasor angles can be used as any other exponents are used when powers or roots of phasors are desired. For example, to square a phasor, the magnitude is squared and the angle is multiplied by 2. Similarly, the root of a phasor is found by extracting the root of the magnitude and dividing the angle by the index of the root.

example 13

Find the square of $Z_1 = 14\angle 18°$.

solution

$$\begin{aligned} Z_1{}^2 &= (14\angle 18°)^2 = 14^2\angle 18° \times 2 \\ &= 196\angle 36° \end{aligned}$$

example 14

Find the square root of $Z_a = 625\angle 60°$.

solution

$$\begin{aligned} \sqrt{Z_a} &= \sqrt{625\angle 60°} \\ &= \sqrt{625}\angle 60° \div 2 \\ &= \pm 25\angle 30° \end{aligned}$$

Our treatment of this subject at this time is necessarily limited to the features which are of immediate use to us in our present studies. You will find in advanced work in mathematics that DeMoivre's theorem proves that there are as many answers to a root problem as there are roots to be taken: the third root of a phasor has three answers, each of the same magnitude but at a different angle. For our immediate purposes, however, Examples 13 and 14 show the basic operations.

PROBLEMS 31-4

Perform the indicated operations and express the results in both polar and rectangular form:

1. $5\angle 53.1^\circ \times 6.7\angle -63.4^\circ$
2. $21.4\angle 52.6^\circ \times 25.5\angle 25.6^\circ$
3. $(9.9\angle 69.9^\circ)(8.8\angle 82.2^\circ)$
4. $(183.3\angle -11^\circ)(3.26\angle 11^\circ)$
5. $(8.24\angle -34^\circ)(9.07\angle -52.6^\circ)$
6. $(9.5\angle -71.6^\circ)(8.26\angle -7.6^\circ)$
7. $10\angle 53.2^\circ \div 5\angle 36.8^\circ$
8. $92.3\angle -12.5^\circ \div 81\angle -64.6^\circ$
9. $3.86\angle -79.57^\circ \div 13.9\angle 69^\circ$
10. $1\angle 0^\circ \div 20\angle -36.8^\circ$
11. $\dfrac{66.8\angle 13^\circ}{4.73\angle 24^\circ}$
12. $\dfrac{1.87\angle -180^\circ}{3.54\angle -180^\circ}$

Perform the indicated operations:

13. $\sqrt{144\angle 30^\circ}$
14. $\sqrt{1024\angle -17^\circ}$
15. $(1.7\angle 22^\circ)^2$
16. $(0.31\angle -60^\circ)^2$
17. $\sqrt[3]{64\angle 270^\circ}$
18. $\sqrt[3]{1728\angle -21.9^\circ}$
19. $(3\angle 11^\circ)^3$
20. $(2\angle -16^\circ)^5$

32

alternating currents—series circuits

Because of the phenomena that occur in them, ac circuits make a very interesting subject for study. Unlike circuits that carry direct currents, in ac circuits the product of the voltage and current is seldom equal to the reading of a wattmeter connected in the circuit, the current may lag or lead the voltage, or the potential difference across an inductance or capacitance may be several times the supply voltage. This chapter deals with the computation of such effects in series circuits.

32-1 DEFINITIONS

In Chap. 9 we investigated *resistance* and defined it as the amount of opposition to current flow within a conductor. It may be helpful to think of resistance as the electrical phenomenon which always tends to oppose the flow of electric current and which always converts some of the energy of the current electricity into heat energy. This heat energy is dissipated, usually by radiation, and is *lost* so far as the circuit is concerned. In some cases, of course, the purpose of the circuit is to provide a conversion of electric energy into heat energy. This heat energy is then radiated away from the circuit, and it represents lost energy so far as the circuit is concerned.

In this chapter, we will also investigate relationships which are involved when alternating current flows under the influence of alternating emf's because when inductance and/or capacitance is involved in the circuit, we must abandon Ohm's law as a specific method of computation.

Inductance is the electrical phenomenon which always tends to oppose a change in electric current and which always converts some of the energy of current electricity into stored electromagnetic energy. This electromagnetic energy is stored by the inductance when the current is increasing, and it is released into the circuit when the current is decreasing. It is found that the current flow through an inductance lags the applied emf by 90 electrical degrees.

Capacitance is the electrical phenomenon which always tends to oppose a change in voltage and which converts some of the energy of current electricity into stored electrostatic energy. This electrostatic energy is stored by the capacitance as an electric charge on the plates of a capacitor when the applied emf is increasing, and it is released into the circuit when the applied emf is decreasing. It is found that the voltage across a capacitor lags the current flow "through" the capacitor by 90 electrical degrees.

It is the 90° phase angles between voltage and current in ac circuits containing inductance and capacitance, together with their associated resistances, that really bring the trigonometric functions into play. You should make a special effort to resolve any difficulties which may still exist in your ability to solve right triangles by trigonometry (Chap. 26) and the j operator (Chaps. 20 and 31), and you should also ensure that you are fully conversant with the S, T, and ST or SRT scales of your slide rule so that you can attack this chapter with confidence. Alternatively, you should be confident in the use of the trigonometric operations on your calculator.

32-2 THE RESISTIVE CIRCUIT

Figure 32-1 represents a 60-Hz alternator supplying 220 V to two resistances connected in series.

This circuit contains resistance only; therefore, Ohm's law applies in every respect. The internal resistance of the alternator and the resistance of the connecting wires being neglected, the current through the circuit is given by the familiar relation

$$I = \frac{E}{R_t} = \frac{E}{R_1 + R_2} = \frac{220}{30 + 25} = \frac{220}{55}$$
$$= 4 \text{ A}$$

Again, as with direct currents, the voltage drops, or potential differences, across the resistances are

$$E_1 = IR_1 = 4 \times 30 = 120 \text{ V}$$
$$E_2 = IR_2 = 4 \times 25 = 100 \text{ V}$$
$$\text{Applied voltage} = 220 \text{ V}$$

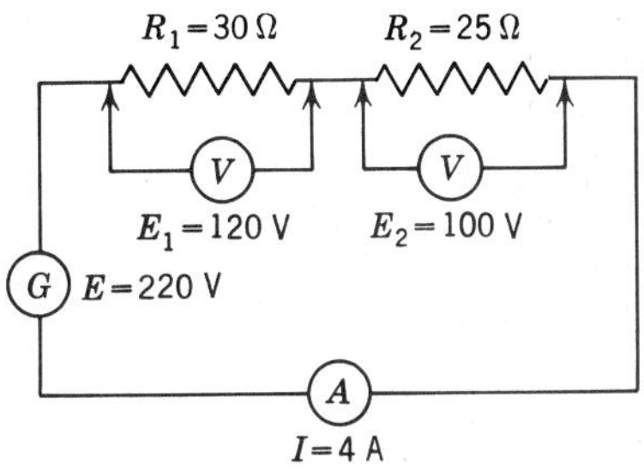

Fig. 32-1 Alternator Supplying Resistive Circuit

In an ac circuit containing only resistance, the voltage and current are in phase; that is, the voltage and current pass through corresponding parts of their cycles at the same instant.

From the above it follows that if

$$e = \boldsymbol{E}_{max} \sin \omega t = 311 \sin 377t \text{ V}$$

is the equation for the alternator voltage of Fig. 32-1, then the current through the circuit is

$$i = \boldsymbol{I}_{max} \sin (\omega t + \theta) = \boldsymbol{I}_{max} \sin (\omega t + 0°)$$
$$= 5.66 \sin 377t \text{ A}$$

Figure 32-2 is the phasor diagram for the circuit of Fig. 32-1. It will be noted that the voltage phasor and the current phasor coincide. This is as anticipated from the equations for the voltage and current, for they differ only in amplitude factors; the frequency factors are equal, and the phase angle is 0° (Secs. 29-7 to 29-9).

Fig. 32-2 Phasor Diagram for Circuit of Fig. 32-1

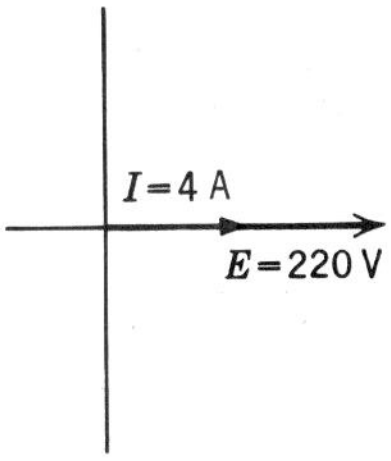

It is evident that Ohm's law says nothing about maximum, average, or effective values of current and voltage. Any of these values can be used; that is, maximum voltage can be used to find maximum current, average voltage can be used to find average current, etc. Naturally, maximum voltage is not used to find effective current unless the proper conversion constant is introduced into the equation. As previously stated, all voltage and current values here are to be considered as effective values unless otherwise specified (Sec. 30-7).

32-3 POWER IN THE RESISTIVE CIRCUIT

In dc circuits the power is equal to the product of the voltage and the current (Sec. 8-5). This is true of ac circuits for *instantaneous values* of voltage and current. That is, the *instantaneous power* is

$$p = ei \qquad \text{VA} \tag{1}$$

and is measured in *voltamperes* or *kilovoltamperes*, abbreviated VA and kVA (or sometimes V · A and kV · A), respectively.[1]

When a sine wave of voltage is impressed across a resistance, the relations among voltage, current, and power are as shown in Fig. 32-3. The voltage existing across the resistance is in phase with the current flowing through it. The power delivered to the resistance at any instant is represented by the height of the power curve, which is the product of the instantaneous values of voltage and current at that instant. The shaded area under the power curve represents the total power delivered to the circuit during one complete cycle of voltage. It will be noted that the power curve is of sine-wave form and has a frequency twice that of the voltage. Also, the power curve lies entirely above the x axis; there are no negative values of power.

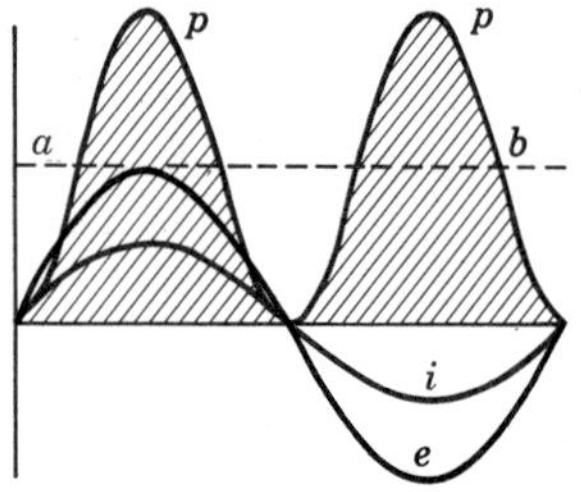

Fig. 32-3 Power Curves for Circuit Containing Only Resistance

The maximum height of the power curve is the product of the maximum values of voltage and current. Stated as an equation,

$$\boldsymbol{P}_{\max} = \boldsymbol{E}_{\max}\boldsymbol{I}_{\max} \tag{2}$$

The average power delivered to a resistance load is represented by the height of the line *ab* in Fig. 32-3, which is half the maximum height of the power curve, or its average height. Then, since

$$\text{Average power} = P = \tfrac{1}{2}\boldsymbol{P}_{\max}$$

by dividing both members of Eq. (2) by 2 we obtain

$$\tfrac{1}{2}\boldsymbol{P}_{\max} = \tfrac{1}{2}\boldsymbol{E}_{\max}\boldsymbol{I}_{\max}$$

Substituting for the value of $\frac{1}{2}\boldsymbol{P}_{\max}$ and factoring the denominator of the right member,

$$P = \frac{\boldsymbol{E}_{\max}\boldsymbol{I}_{\max}}{\sqrt{2}\sqrt{2}}$$

Substituting for the values in the right member (Sec. 30-7),

$$P = EI \qquad \text{W} \tag{3}$$

[1] The dot is preferred in general physics relationships, but it is customarily omitted in electricity and electronics use.

Hence, the alternating power consumed by a resistance load is equal to the product of the effective values of voltage and current. As in dc circuits, alternating power is measured in watts and kilowatts.

example 1 What is the power expended in the resistances of Fig. 32-1?

solution

Voltage across $R_1 = E_1 = 120$ V
Voltage across $R_2 = E_2 = 100$ V
Current through circuit $= I = 4$ V
Power expended in $R_1 = P_1 = E_1I = 120 \times 4 = 480$ W
Power expended in $R_2 = P_2 = E_2I = 100 \times 4 = \underline{400}$ W
Total $= 880$ W

Also, the total power is

$$P_t = EI = 220 \times 4 = 880 \text{ W}$$

Because $P = EI$, the usual Ohm's law relations hold for resistances in ac circuits. Hence,

$$P = I^2R \quad \text{W} \tag{4}$$

and $$P = \frac{E^2}{R} \quad \text{W} \tag{5}$$

Thus, the power consumed by R_1 of Fig. 32-1 can be computed by using Eq. (4) or (5). Hence,

$$P_1 = I^2R_1 = 4^2 \times 30 = 480 \text{ W}$$

or $$P_1 = \frac{E_1^2}{R_1} = \frac{120^2}{30} = 480 \text{ W}$$

PROBLEMS 32-1

1. A 400-Hz alternator supplies 88 V across a combination of three series resistors of 150, 67, and 22 Ω.
   (*a*) How much current flows in the circuit?
   (*b*) Write the equation for the alternator voltage at any time *t*.
   (*c*) Write the equation for the circuit current at any time *t*.
   (*d*) What is the voltage measured across the 67-Ω resistor?
   (*e*) How much power is dissipated by the 22-Ω resistor?
   (*f*) What is the instantaneous value of the current when the instantaneous emf is 26 V?
2. Given the circuit of Fig. 32-4:
   (*a*) Write the equation for the emf of the alternator at any time *t*.
   (*b*) Write the equation for the total current of the circuit.

Fig. 32-4 Circuit of Probs. 2 and 3

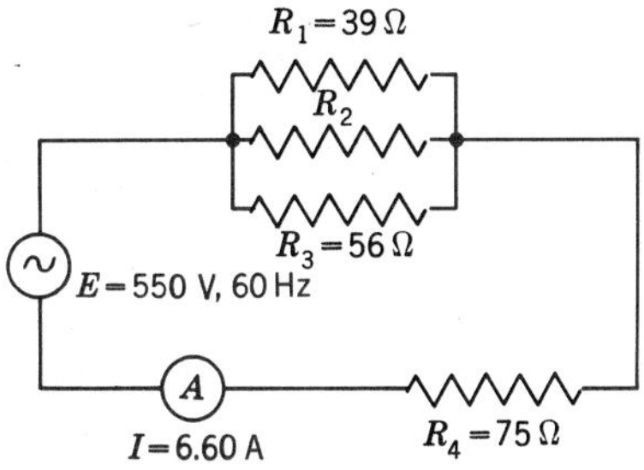

(c) What is the voltage across R_3?
(d) How much power is dissipated in R_2?
(e) How much current flows through R_1?
(f) What is the instantaneous value of the total current when the instantaneous alternator emf is 36.5 V?

3. In the circuit of Fig. 32-4, what is the instantaneous value of the voltage across R_2 when the instantaneous current through R_4 is 2.75 A?

4. A 10-kHz signal generator is connected to a 600-Ω resistive load. A milliwattmeter indicates that the resistor is dissipating 800 mW. What is the maximum instantaneous voltage developed at the generator terminals?

5. What is the equation of the current in Prob. 4?

32-4 THE INDUCTIVE CIRCUIT

A circuit, or an inductance coil, has the property of inductance when there is set up in it an emf due to a *change* of current through it. Thus, a circuit has an inductance of 1 H when a change of current of 1 A/s induces an emf of 1 V. Expressed as an equation,

$$E_{av} = L\frac{I}{t} \quad \text{V} \tag{6}$$

where E_{av} is the average voltage induced in a circuit of L H by a *change* of current of I A in t s.

An alternating current of $\boldsymbol{I}_{max}$ A makes *four changes* during each cycle. These are

1. From zero to maximum positive value
2. From maximum positive value to zero
3. From zero to maximum negative value
4. From maximum negative value to zero

The time required for one complete cycle of alternating current is $T = f^{-1}$ s (Sec. 29-9), and each of the above changes occurs in one-quarter of the time required for the completion of each cycle. Then the time for each change is $(4f)^{-1}$ s. Substituting this value of t, and $\boldsymbol{I}_{max}$ for I, in Eq. (6), we have

$$E_{av} = L\frac{\boldsymbol{I}_{max}}{(4f)^{-1}} = 4fL\boldsymbol{I}_{max} \tag{7}$$

Equation (7) is cumbersome if used in its present form, for it contains an average-voltage term and a maximum-current term. The equation can be expressed in terms of the relation between average and maximum values as given in Sec. 30-6:

$$E_{av} = \frac{2}{\pi}\boldsymbol{E}_{max}$$

Substituting in Eq. (7) for this value of E_{av}, we have

$$\frac{2}{\pi}\boldsymbol{E}_{max} = 4fL\boldsymbol{I}_{max}$$

which becomes

$$\boldsymbol{E}_{max} = 2\pi fL\boldsymbol{I}_{max} \tag{8}$$

Because both voltage and current in Eq. (8) are now in terms of maximum values, effective values can be used. Thus,

$$E = 2\pi fLI \quad \text{V} \tag{9}$$

The factors $2\pi fL$ in Eqs. (8) and (9) represent a reaction due to the frequency of the alternating current and the amount of inductance contained in the circuit. Hence, the alternating voltage E required to cause a current of I A with a frequency of f Hz to flow through an inductance of L H is given by Eq. (9). That is, the voltage must overcome the reaction $2\pi fL$, which is called the *inductive reactance*. From Eq. (9) the inductive reactance, which is denoted by X_L and expressed in ohms, is given by

$$\frac{E}{I} = 2\pi fL$$

or
$$X_L = 2\pi fL = \omega L \qquad \Omega \tag{10}$$

where f = frequency, Hz
L = inductance, H

Note the similarity of the relations between voltage and current for inductive reactance and resistance. Both inductive reactance and resistance offer an opposition to a flow of alternating current, both are expressed in ohms, and both are equal to the voltage divided by the current. Here the similarity ends; there is no inductive reactance to steady-state direct currents because there is no *change* in current, and, as explained later, inductive reactances consume no alternating power.

Figure 32-5 represents a 60-Hz alternator delivering 220 V to a coil having an inductance of 0.165 H. The opposition, or inductive reactance, to the flow of current is

$$X_L = 2\pi fL = 2\pi \times 60 \times 0.165 = 62.2\ \Omega$$

Although it is impossible to construct an inductance containing no resistance, to simplify basic considerations we shall consider the coil of Fig. 32-5 as being an inductance with negligible resistance. (The effects of inductance and resistance acting together are discussed in Sec. 32-8.) The current in the circuit due to the action of voltage and inductive reactance is

$$I = \frac{E_L}{X_L} = \frac{220}{62.2} = 3.54\ \text{A}$$

Fig. 32-5 $E_L = 220$ V, $L = 0.165$ H

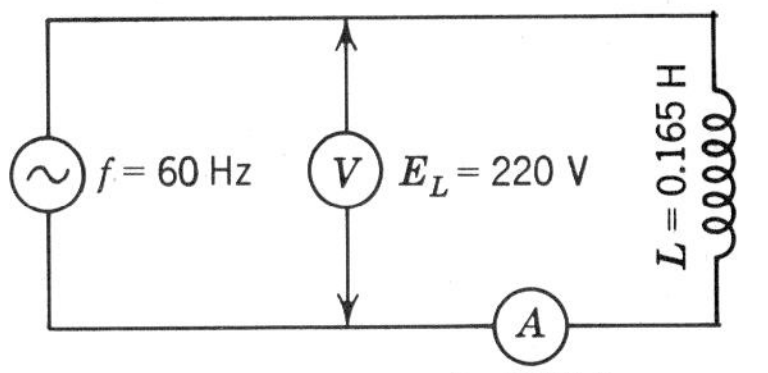

example 2 What is the inductive reactance of an inductance of 17 μH at a frequency of 2500 kHz?

solution

$$f = 2500\ \text{kHz} = 2.5 \times 10^6\ \text{Hz}$$
$$L = 17\ \mu\text{H} = 1.7 \times 10^{-5}\ \text{H}$$
$$X_L = 2\pi fL = 2\pi \times 2.5 \times 10^6 \times 1.7 \times 10^{-5}$$
$$= 2\pi \times 1.7 \times 2.5 \times 10 = 267\ \Omega$$

example 3 An inductor is connected to 115 V, 60 Hz. An ammeter connected in series with the coil reads 0.714 A. On the assumption that the coil contains negligible resistance, what is its inductance?

solution

$$E_L = 115\ \text{V}$$
$$f = 60\ \text{Hz}$$
$$I = 0.714\ \text{A}$$
$$X_L = \frac{E_L}{I} = \frac{115}{0.714} = 161\ \Omega$$

Since

$$X_L = 2\pi fL$$

then

$$L = \frac{X_L}{2\pi f} = \frac{161}{2\pi \times 60} = 0.427\ \text{H}$$

In a circuit containing inductance, a change of current induces an emf of such polarity that it always opposes the change of current. Because an alternating current is

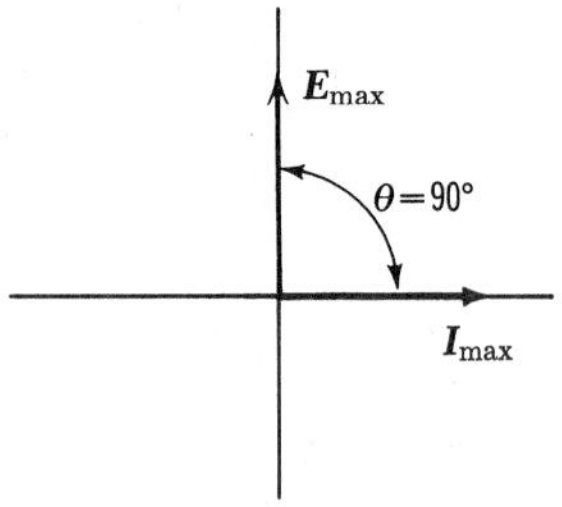

Fig. 32-6 Current Lags Voltage by 90°

constantly changing, in an inductive circuit there is always present a reaction that opposes this change. The net effect of this, in a *purely inductive circuit,* is to cause the *current to lag the voltage by* 90°. This is illustrated by the phasor diagram of Fig. 32-6, which shows the voltage of the circuit of Fig. 32-5 to be at maximum positive value when the current is passing through zero.

The instantaneous voltage across the inductance is given by

$$e = \boldsymbol{E}_{max} \sin \omega t \qquad \text{V}$$

or

$$e = 311 \sin 377t \qquad \text{V}$$

Since the current lags the voltage by a phase angle θ of 90°, the equation for the current through the inductance is

$$i = \boldsymbol{I}_{max} \sin (\omega t - \theta) \qquad \text{A} \qquad (11)$$

or

$$i = 5 \sin (377t - 90°) \qquad \text{A} \qquad (12)$$

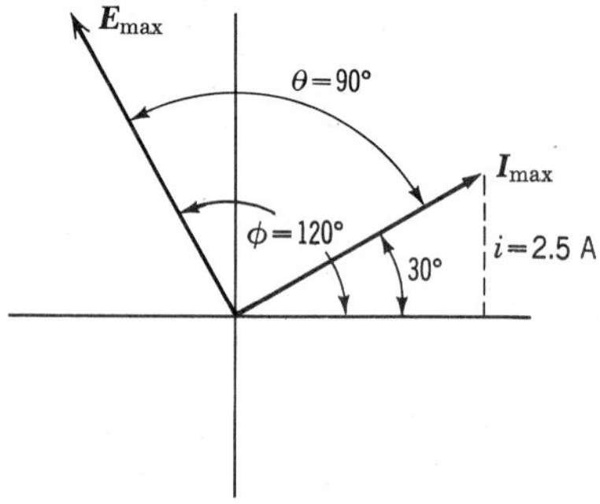

Fig. 32-7 Phasor Diagram of Example 4

If the voltage has completed $\phi°$ of its cycle, the instantaneous current is

$$i = 5 \sin (\phi - 90°) \qquad \text{A} \qquad (13)$$

example 4 What is the instantaneous value of the current in Fig. 32-5 when the voltage has completed 120° of its cycle?

solution Draw a phasor diagram of the current and voltage relations as shown in Fig. 32-7. The instantaneous value of the current is found from Eq. (13) and is

$$\begin{aligned} i &= \boldsymbol{I}_{max} \sin (\phi - 90°) \\ &= 5 \sin (120° - 90°) \\ &= 5 \sin 30° \\ &= 2.5 \text{ A} \end{aligned}$$

PROBLEMS 32-2

1. What is the reactance of a 15-mH coil at 60 Hz?
2. What is the reactance of a 15-mH coil at 1 kHz?
3. What is the reactance of a 15-mH coil at 1 MHz?
4. What is the inductance of a coil that exhibits a reactance of 754 Ω at a frequency of 400 Hz?
5. A tuning coil in a radio transmitter has an inductance of 270 μH. What is its reactance at a frequency of 1.5 MHz?
6. At what frequency will a television set coil with an inductance of 3.25 μH offer a reactance of 3740 Ω?

7. Assuming negligible resistance, what would be the current flow through an inductance of 0.067 H at a voltage of 100 V, 800 Hz?
8. What would be the equation of the current in Prob. 7?
9. A current of 379 μA at 2.5 V flows through a 5.25-μH coil. Assuming negligible resistance, what is the frequency of the applied emf?
10. An emf described by the equation $e = 311 \sin 314t$ V is applied to an inductor of 1.65 H. What is the equation of the current flow, assuming negligible resistance?
11. What is the instantaneous value of the current in Prob. 10 when the emf has completed 45° of its cycle?
12. What is the instantaneous value of the applied voltage in Prob. 10 when the current has completed 210° of its cycle?
13. What happens to the inductive reactance of a circuit when the inductance is fixed but the frequency of the applied emf is (*a*) doubled, (*b*) tripled, (*c*) halved?
14. What happens to the inductive reactance of a circuit when the frequency of the applied emf is held constant and the inductance is varied?

32-5 THE CAPACITIVE CIRCUIT

A capacitance is formed between two conductors when there is an insulating material between them. A circuit, or a capacitor, is said to have a capacitance of one farad when a *change* of one volt per second produces a current of one ampere. Expressed as an equation,

$$I_{av} = C\frac{E}{t} \quad \text{A} \qquad (14)$$

where I_{av} is the average current in amperes that is caused to flow through a capacitance of C F by a *change* of E V in t s.

In all probability the above definition does not clearly indicate to you *how much* electricity, or charge, a given capacitor will contain. Perhaps a more understandable definition is that a circuit, or a capacitor, has a capacitance of one farad when a difference of potential of one volt will produce on it one coulomb of charge. Expressed as an equation,

$$Q = CE \quad \text{C} \qquad (15)$$

where Q is the charge in coulombs placed on a capacitor of C F by a difference of potential of E V across the capacitor.

It was shown in Sec. 32-4 that the time t required for one change of an alternating emf was $(4f)^{-1}$ s. Thus, if an alternating emf of $\boldsymbol{E}_{max}$ V at a frequency of f Hz is impressed across a capacitor of C F, by substituting the above value of t, and $\boldsymbol{E}_{max}$ for E, in Eq. (14),

$$I_{av} = C\frac{\boldsymbol{E}_{max}}{(4f)^{-1}} = 4fC\boldsymbol{E}_{max} \quad \text{A} \qquad (16)$$

Again, as in Eq. (7), the above equation contains an average term and a maximum term. As given in Sec. 30-6,

$$I_{av} = \frac{2}{\pi}\boldsymbol{I}_{max} \quad \text{A}$$

Substituting in Eq. (16) for this value of I_{av}, we have

$$\frac{2}{\pi}\boldsymbol{I}_{max} = 4fC\boldsymbol{E}_{max}$$

which becomes

$$I_{max} = 2\pi f C E_{max} \qquad \text{A} \qquad (17)$$

Because both voltage and current in Eq. (17) are now in terms of maximum values, effective values can be used. Thus,

$$I = 2\pi f C E \qquad \text{A} \qquad (18)$$

The factors $2\pi fC$ represent a reaction due to the frequency of the alternating emf and the amount of capacitance; hence, it is evident that the amount of current in a purely capacitive circuit depends upon these factors. As in the case of resistive circuits and inductive circuits, the opposition to the flow of current is obtained by dividing the voltage by the current. Then, from Eq. (18),

$$\frac{E}{I} = \frac{1}{2\pi fC} \qquad \Omega \qquad (19)$$

The right member of Eq. (19), which represents the opposition to a flow of alternating current in a purely capacitive circuit, is called the *capacitive reactance*. It is denoted by X_C and expressed in ohms. Thus,

$$X_C = \frac{1}{2\pi fC} = \frac{1}{\omega C} \qquad \Omega \qquad (20)$$

where f = frequency, Hz
C = capacitance, F

Figure 32-8 represents a 60-Hz alternator delivering 220 V to a capacitor having a capacitance of 14.5 μF. The opposition, or capacitive reactance, to the flow of current is

$$X_C = \frac{1}{2\pi fC} = \frac{1}{2\pi \times 60 \times 14.5 \times 10^{-6}}$$
$$= \frac{10^4}{2\pi \times 6 \times 1.45} = 183\ \Omega$$

Fig. 32-8 $E_C = 220$ V, $C = 14.5\ \mu$F

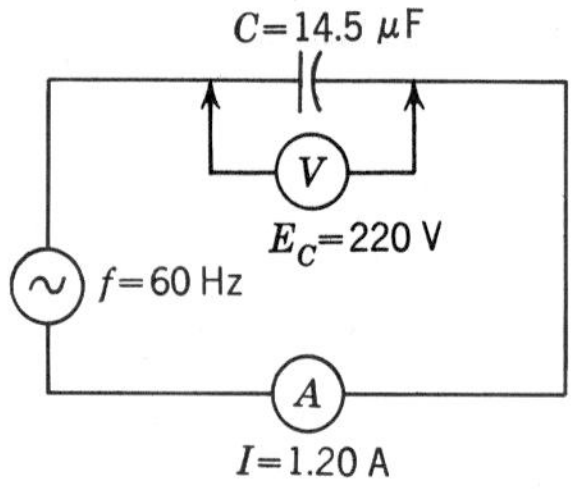

Neglecting the resistance of the connecting leads and the extremely small losses at low frequencies in a well-constructed capacitor, the current in the circuit due to the action of the voltage and capacitive reactance is

$$I = \frac{E_C}{X_C} = \frac{220}{183} = 1.20\ \text{A}$$

example 5 What is the capacitive reactance of a 350-pF capacitor at a frequency of 1200 kHz?

solution

$$f = 1200\ \text{kHz} = 1.2 \times 10^6\ \text{Hz}$$
$$C = 350\ \text{pF} = 3.5 \times 10^{-10}\ \text{F}$$
$$X_C = \frac{1}{2\pi fC}$$
$$= \frac{1}{2\pi \times 1.2 \times 10^6 \times 3.5 \times 10^{-10}}$$
$$= \frac{10^4}{2\pi \times 1.2 \times 3.5} = 379\ \Omega$$

example 6 A capacitor is connected across 110 V, 60 Hz. A milliammeter connected in series with the capacitor reads 350 mA. What is the capacitance of the capacitor?

solution

$$E_C = 110\ \text{V}$$
$$f = 60\ \text{Hz}$$
$$I = 350\ \text{mA} = 0.350\ \text{A}$$
$$X_C = \frac{E_C}{I} = \frac{110}{0.35} = 314\ \Omega$$

since $X_C = \dfrac{1}{2\pi fC}$

then
$$C = \frac{1}{2\pi f X_C} = \frac{1}{2\pi \times 60 \times 314}$$
$$= \frac{10^{-3}}{2\pi \times 6 \times 3.14}$$
$$= 8.44 \times 10^{-6}\ \text{F} = 8.44\ \mu\text{F}$$

Because current flows in a capacitor only when the voltage across the capacitor is changing, it is evident that, when an alternating voltage is impressed, current is flowing at all times because the potential difference across the capacitor is constantly changing. Furthermore, the greatest amount of current will flow when the voltage is changing most rapidly, and this occurs when the voltage passes through zero value. This property, in conjunction with the effects of the counter emf, *causes the current to lead the voltage by 90° in a purely capacitive circuit.* This is illustrated by the vector diagram of Fig. 32-9, which shows the current through the circuit of Fig. 32-8 to be at maximum positive value when the voltage is passing through zero.

The instantaneous voltage across the capacitor is given by

$$e = \boldsymbol{E}_{max} \sin \omega t \quad \text{V} \tag{21}$$

or

$$e = 311 \sin 377t \text{ V} \tag{22}$$

Therefore, the equation for the current is

Fig. 32-9 Current Leads Voltage by 90°

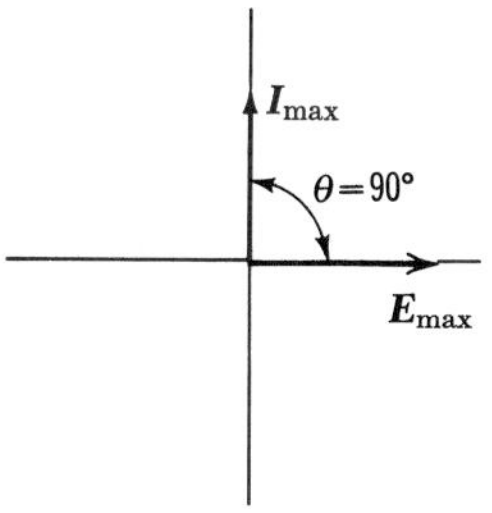

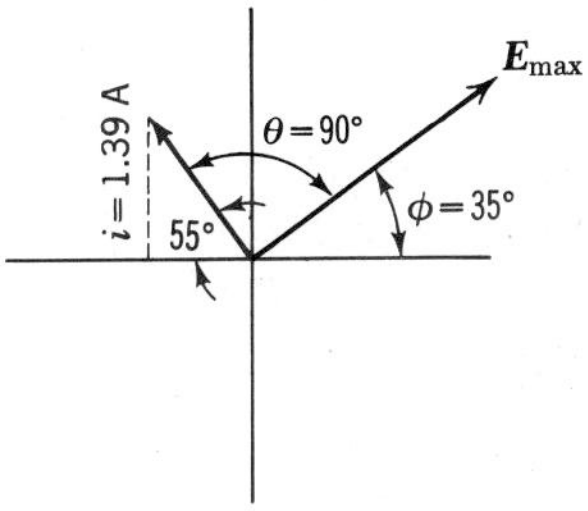

Fig. 32-10 Phasor Diagram for Example 7

$$i = \boldsymbol{I}_{max} \sin (377t + \theta) \quad \text{A} \tag{23}$$

or

$$i = 1.70 \sin (377t + 90°) \text{ A} \tag{24}$$

If the voltage has completed $\phi°$ of its cycle, the instantaneous current is

$$i = \boldsymbol{I}_{max} \sin (\phi + 90°) \quad \text{A} \tag{25}$$

example 7 What is the instantaneous value of the current in Fig. 32-8 when the voltage has completed 35° of its cycle?

solution Draw a phasor diagram of the current and voltage relations as shown in Fig. 32-10. The instantaneous value of the current is found from Eq. (25) and is

$$i = \boldsymbol{I}_{max} \sin (\phi + 90°)$$
$$= 1.70 \sin (35° + 90°)$$
$$= 1.70 \sin 125° = 1.39 \text{ A}$$

32-6 CAPACITORS IN SERIES

Figure 32-11 represents two capacitors C_1 and C_2 connected in series with a voltage

Fig. 32-11 Capacitors C_1 and C_2 Connected in Series

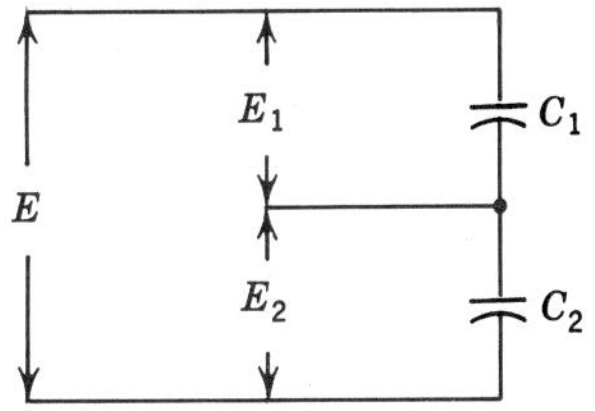

E across the combination. Because the capacitors are in series, the same quantity of electricity must be sent into each of them. Then, if E_1 and E_2 represent the potential differences across C_1 and C_2, respectively, Q represents the quantity of electricity in each capacitor and C_t is the capacitance of the combination. Hence,

$$E = \frac{Q}{C_t}$$

$$E_1 = \frac{Q}{C_1}$$

and $$E_2 = \frac{Q}{C_2}$$

Since $$E = E_1 + E_2 \qquad (26)$$

by substituting the values for all voltages into Eq. (26),

$$\frac{Q}{C_t} = \frac{Q}{C_1} + \frac{Q}{C_2}$$

or $$\frac{1}{C_t} = \frac{1}{C_1} + \frac{1}{C_2} \qquad (27)$$

Equation (27) resolves into

$$C_t = \frac{C_1 C_2}{C_1 + C_2} \qquad (28)$$

The above illustrates the fact that capacitors in series combine like resistances in parallel; that is, the reciprocal of the combined capacitance of capacitors in series is equal to the sum of the reciprocals of the capacitances of the individual capacitors.

example 8 What is the capacitance of a 6-μF capacitor in series with a capacitor of 4 μF?

solution $C_t = \dfrac{6 \times 4}{6 + 4} = 2.4\ \mu\text{F}$

PROBLEMS 32-3

1. What is the capacitive reactance of a 22-μF capacitor at a frequency of 400 Hz?
2. What is the capacitive reactance of a 22-μF capacitor at a frequency of 1 kHz?
3. What is the capacitive reactance of a 22-μF capacitor at a frequency of 100 kHz?
4. What is the reactance of a 50-pF capacitor at a frequency of 12 GHz?
5. A filter capacitor in a radio receiver has a capacitance of 0.0016 μF. What is its reactance at a frequency of 720 kHz?
6. What is the reactance of the capacitor of Prob. 5 if the frequency is increased to 1320 kHz?
7. How much current will flow in a capacitor of 6.3 pF when 475 V at 1 kHz is impressed across the capacitor, neglecting resistance?
8. What will be the current in the capacitor of Prob. 7 if the frequency is increased to 12 kHz?
9. When a 120-V 800-Hz emf is impressed across a capacitor, the current flow is 2.41 A. What is the capacitance?
10. A current of 452 mA flows through a 5-μF capacitor when the frequency of the applied emf is 60 Hz. What is the voltage?
11. What is the equation for the current in Prob. 10?

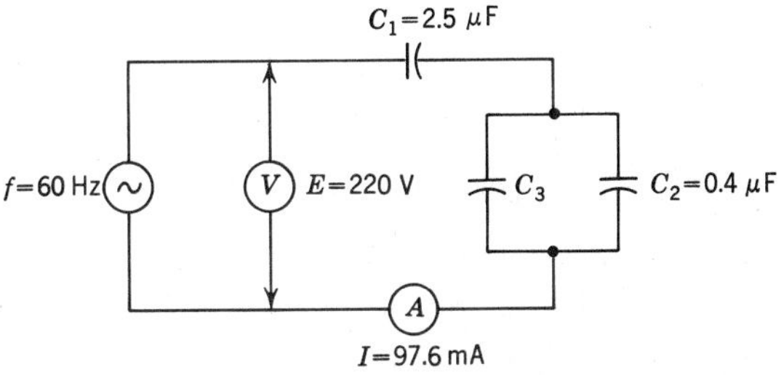

Fig. 32-12 Circuit of Prob. 18

12. What is the instantaneous value of the current in Prob. 10 when the emf has completed 230° of its cycle?

13. What is the resulting capacitance when a 220-pF capacitor is connected in series with a 500-pF capacitor?

14. Two capacitors, 20 and 200 pF, are connected in series. What is the resultant capacitance?

15. If an emf of 80 V at 15 kHz is impressed across the series circuit of Prob. 14, what will be the resultant current flow, neglecting resistance?

16. What happens to the capacitive reactance of a circuit when the capacitance is fixed but the frequency of the applied emf is (*a*) doubled, (*b*) tripled, (*c*) halved?

17. What happens to the capacitive reactance of a circuit when the frequency of the applied emf is held constant and the capacitance is varied?

18. Neglecting the resistance of the connecting wires in Fig. 32-12:
(*a*) Write the equation for the emf of the alternator.
(*b*) Write the equation for the circuit current.
(*c*) What is the voltage across C_1?
(*d*) What is the voltage across C_2?

32-7 POWER IN CIRCUITS CONTAINING ONLY INDUCTANCE OR CAPACITANCE

Figure 32-13 illustrates the voltage, current, and power relations when a sine wave of emf is impressed across an inductor whose resistance is negligible.

When the current is increasing from zero to maximum positive value, during the time interval from 1 to 2, power is being taken from the source of emf and is being stored in the magnetic field about the coil. As the current through the inductor decreases from maxi-

Fig. 32-13 Voltage, Current, and Power in an Inductive Circuit

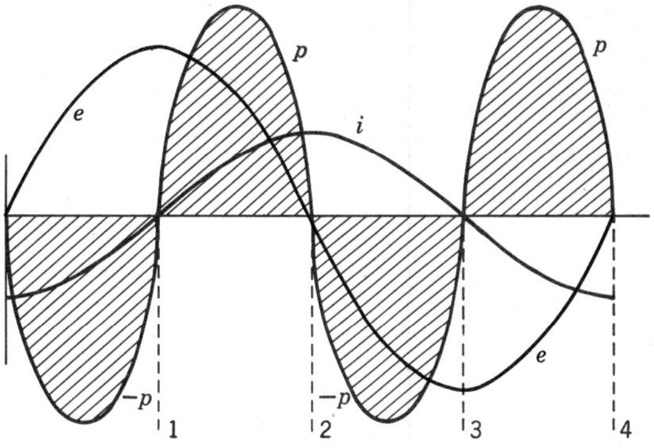

mum positive value to zero, during the time from 2 to 3, the magnetic field is collapsing, thus returning its power to the circuit. Thus, during the intervals from 1 to 2 and from 3 to 4, the inductor is taking power from the source that is represented by the *positive* power in the figure. During the intervals from 0 to 1 and 2 to 3, the inductor is returning power to the source that is represented by the *negative* power in the figure. As previously stated, the instantaneous power is equal to the product of the voltage and current; it is positive when the voltage and current are of like sign and negative when of unlike sign. Note that between points 3 and 4, although both the voltage and the current are negative, the power is positive.

When an alternating emf is impressed across a capacitor, power is taken from the source and stored in the capacitor as the voltage increases from zero to maximum positive value. As the voltage decreases from maximum positive value to zero, the capacitor discharges and returns power to the source. As in the case of the inductor, half of the power loops are positive and half are negative; therefore, no power is expended in either circuit, for the power alternately flows to and from the source. This power is called *reactive* or *apparent power* and is given by the relation

$$P = EI \qquad \text{VA}$$

32-8 RESISTANCE AND INDUCTANCE IN SERIES

It has been explained that in a circuit containing only resistance the voltage applied across the resistance and the current through the resistance are in phase and that in a circuit containing only reactance the voltage and current are 90° out of phase. However, circuits encountered in practice contain both resistance and reactance. Such

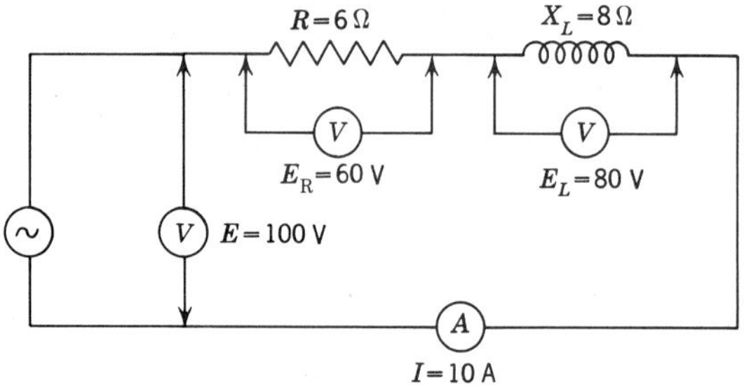

Fig. 32-14 Series Circuit Containing Resistance and Inductance

a condition is shown in Fig. 32-14, where an alternating emf of 100 V is impressed across a combination of 6 Ω resistance in series with 8 Ω inductive reactance.

As with dc circuits, the sum of the voltage drops around the circuit comprising the load must equal the applied emf. In the consideration of resistance and reactance, however, we are dealing with voltages that can no longer be added or subtracted arithmetically. That is because the voltage drop across the resistance is in phase with the current and the voltage drop across the inductive reactance is 90° ahead of the current.

Because the current is the same in all parts of a series circuit, we can use it as a reference and plot the voltage across the resistance and that across the inductive reactance as shown in Fig. 32-15. The resultant of these two voltages, which can be treated as rectangular components (Sec. 28-4), must be equal to the applied emf. Hence, if IR and IX_L are the potential differences across the resistance and inductive reactance, respectively,

$$E = \sqrt{(IR)^2 + (IX_L)^2} \qquad \text{V} \tag{29}$$

or

$$E = \sqrt{60^2 + 80^2} = 100 \text{ V}$$

The phase angle θ between voltage and current can be found by using any of the trigonometric functions. For example,

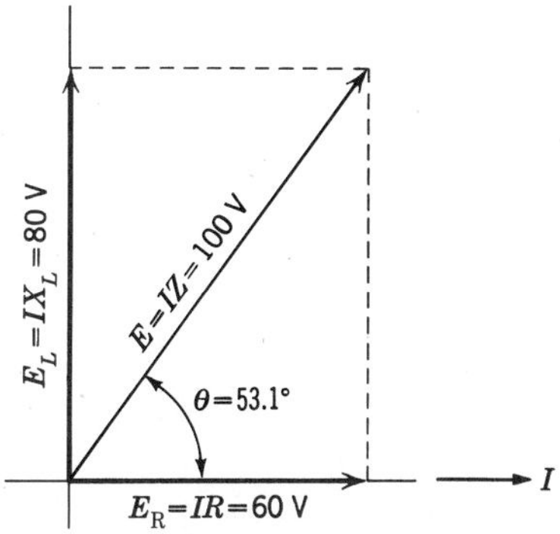

Fig. 32-15 Phasor Diagram for Circuit of Fig. 32-14

$$\tan\theta = \frac{IX_L}{IR} = \frac{80}{60} = 1.33$$

$$\therefore \theta = 53.1°$$

and it is apparent from the phasor diagram that the current through the circuit lags the applied voltage by this amount.

Although the foregoing demonstrates that the *phasor summation* of the voltage across the resistance and the voltage across the reactance is equal to the applied emf, no relation between applied voltage and circuit current has been given as yet.

Since

$$E = \sqrt{(IR)^2 + (IX_L)^2}$$

then

$$E = \sqrt{I^2R^2 + I^2X_L^2}$$

Factoring,

$$E = \sqrt{I^2(R^2 + X_L^2)}$$

Hence,

$$E = I\sqrt{R^2 + X_L^2} \quad \text{V} \qquad (30)$$

As previously stated, the applied voltage divided by the current results in a quotient that represents the opposition offered to the flow of current. Hence, from Eq. (30),

$$\frac{E}{I} = \sqrt{R^2 + X_L^2} \qquad (31)$$

The expression $\sqrt{R^2 + X_L^2}$ is called the *impedance* of the circuit. It is denoted by Z and measured in ohms. Therefore

$$Z = \sqrt{R^2 + X_L^2} \quad \Omega \qquad (32)$$

Applying Eq. (32) to the circuit of Fig. 32-14,

$$Z = \sqrt{6^2 + 8^2} = 10\ \Omega$$

and

$$I = \frac{E}{Z} = 10\ \text{A}$$

From Eq. (31), Eq. (32) can be written

$$E = IZ = I\sqrt{R^2 + X_L^2}$$

The foregoing illustrates that the factor I is common to all expressions, which is the same as saying that the current is the same in all parts of the circuit. Because this condition exists, it is permissible to plot the resistance and reactance as rectangular components as shown in Fig. 32-16; hence, the impedance of a series circuit is simply the phasor sum of the resistance and reactance. The various methods used in solving for the impedance are the same as those given for phasor summation of rectangular components in Example 4 of Sec. 28-4. Note that the values are identical.

Fig. 32-16 *Z* Can Be Plotted as Phasor Sum of *R* and X_L

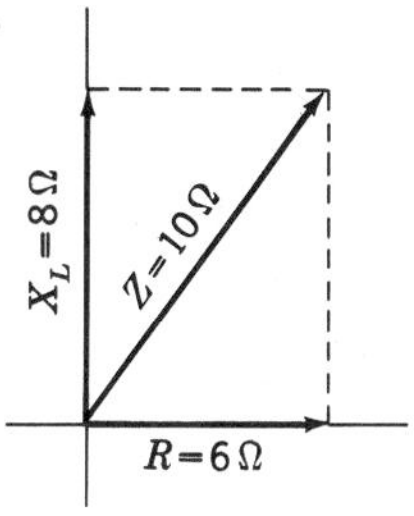

example 9 A circuit consisting of 120 Ω resistance in series with an inductance of 0.35 H is connected across a 440-V 60-Hz alternator. Determine (*a*) the phase angle between voltage and current, (*b*) the impedance of the circuit, and (*c*) the current through the circuit.

solution (*a*) Drawing and labeling the circuit is left to you. The inductive reactance is

$$X_L = 2\pi fL = 2\pi \times 60 \times 0.35$$
$$= 132\ \Omega$$

Draw the phasor impedance diagram as shown in Fig. 32-17. Then, since

$$\tan\theta = \frac{X_L}{R} = \frac{132}{120} = 1.10$$
$$\therefore \theta = 47.7°$$

Note that the phase angle denotes the position of the applied voltage with respect to the current, which is taken as a reference. Thus an inductive series circuit always has a "lagging" phase angle which is a *positive angle* when resistance, reactance, and impedance are plotted vectorially.

Fig. 32-17 Impedance Phasor Diagram for Circuit of Example 9

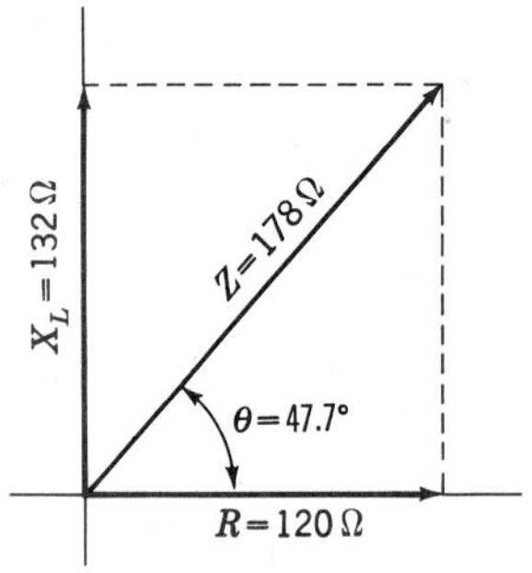

(*b*) $$Z = \frac{R}{\cos\theta} = \frac{120}{\cos 47.7°} = 178\ \Omega$$

or $$Z = \frac{X_L}{\sin\theta} = \frac{132}{\sin 47.7°} = 178\ \Omega$$

(*c*) $$I = \frac{E}{Z} = \frac{440}{178} = 2.47\ \text{A}$$

32-9 RESISTANCE AND CAPACITANCE IN SERIES

Figure 32-18 represents a circuit in which an alternating emf of 100 V is applied across a combination of 6 Ω resistance in series with 8 Ω capacitive reactance. Note the similarity between the circuits of Figs. 32-14 and 32-18. Both have the same values of resistance and absolute values of reactance. However, in the circuit of Fig. 32-18 the voltage drop across the capacitance reactance is 90° behind the current. Again using the current as a reference, because it is the same in all parts of the circuit, the voltage across the resistance and the voltage across the capacitive reactance are plotted as shown in Fig. 32-19 and treated as rectangular components of the applied emf. The impedance of the circuit is found in the same manner as that of the inductive circuit, that is, by phasor summation of the rectangular com-

Fig. 32-18 Series Circuit Consisting of Resistance and Capacitance

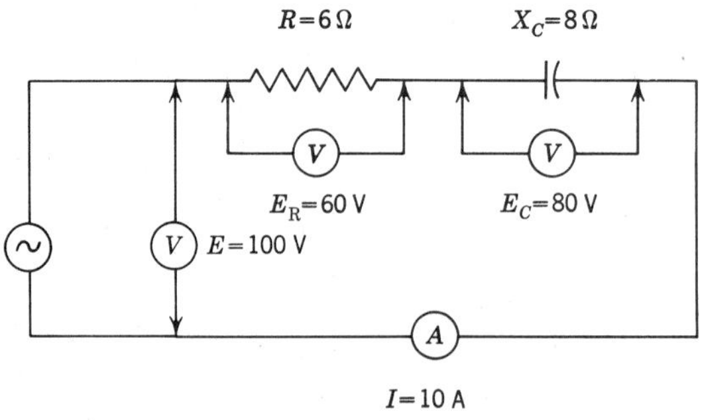

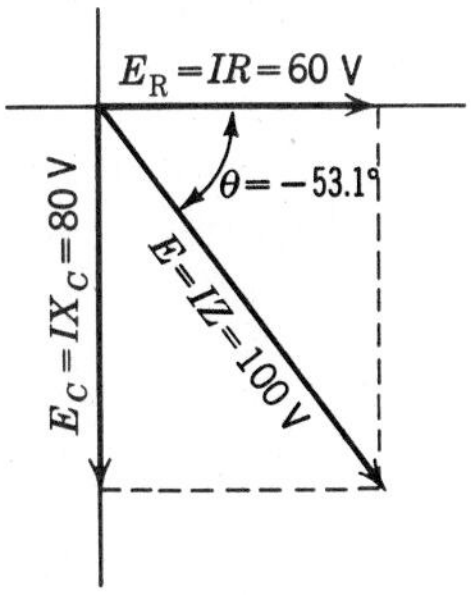

Fig. 32-19 Phasor Diagram for Circuit of Fig. 32-18

ponents. The phase angle is found by the same method.

$$\tan\theta = \frac{X_C}{R} = \frac{8}{6} = 1.33$$

$$\therefore \theta = -53.1°$$

In the capacitive circuit the current leads the voltage, and we prefix the impedance angle with a minus sign because of its position (Sec. 23-2).

example 10 A circuit consisting of 175 Ω resistance in series with a capacitor of 5.0 μF is connected across a source of 150 V, 120 Hz. Determine (*a*) the phase angle between voltage and current, (*b*) the impedance of the circuit, and (*c*) the current through the circuit.

solution

(*a*) Drawing and labeling the circuit is left to you. The capacitive reactance is

$$X_C = \frac{1}{2\pi fC}$$

$$= \frac{1}{2\pi \times 120 \times 5 \times 10^{-6}}$$

$$= \frac{10^4}{2\pi \times 1.2 \times 5} = 265\ \Omega$$

Draw the impedance diagram as shown in Fig. 32-20. Then, since

$$\tan\theta = \frac{X_C}{R} = \frac{265}{175} = 1.51$$

$$\therefore \theta = -56.6°$$

Thus the current is leading the voltage by 56.6°, as shown by the impedance phasor diagram.

(*b*) $$Z = \frac{R}{\cos\theta} = \frac{175}{\cos 56.6°} = 318\ \Omega$$

or $$Z = \frac{X_C}{\sin\theta} = \frac{265}{\sin 56.6°} = 318\ \Omega$$

(*c*) $$I = \frac{E}{Z} = \frac{150}{318} = 0.472\ \text{A}$$

Fig. 32-20 Impedance Phasor Diagram for Example 10

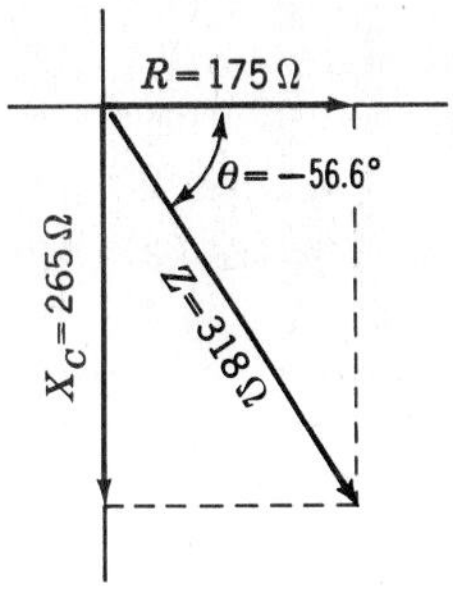

PROBLEMS 32-4

1. A series circuit consists of a 1.5-H inductor which has a resistance of 35 Ω. It is supplied with 220 V, 60 Hz. Find

(*a*) The inductive reactance.
(*b*) The impedance of the coil.
(*c*) The current flowing through the coil.
(*d*) The equation of the current.

(e) The voltage across the resistance of the coil.
(f) The voltage across the inductance of the coil.
(g) Why e + f does not equal 220 V.

2. A 500-V 8-MHz source is connected to a series circuit consisting of a 3.3-kΩ resistor and a 500-μH inductor of negligible resistance. Find
(a) The inductive reactance of the inductor.
(b) The impedance of the circuit.
(c) The current flowing through the circuit.
(d) The phase angle of the current.
(e) The voltage across the resistor.
(f) The voltage across the inductor.

3. In the circuit of Prob. 2, the applied emf is held constant while the frequency is decreased.
(a) Why will this cause the current to rise?
(b) When the current is twice that found in Prob. 2, find the impedance, the frequency, and the phase angle.

4. A 25-mH choke has a measured resistance of 40 Ω at 400 Hz. This choke is connected across 48 V at 400 Hz. Find (a) the impedance of the choke and (b) the current flow.

5. A 120-V 60-Hz source energizes a series circuit consisting of a 330-Ω resistor and a 22-μF capacitor. Find:
(a) The capacitive reactance.
(b) The impedance of the circuit.
(c) The current flow through the circuit.
(d) The voltage across the resistor.
(e) The voltage across the capacitor.

6. If the frequency of the 120-V source in Prob. 5 is doubled, what will be the current flow through the circuit?

7. What will be the impedance of the circuit of Prob. 5 if a 150-μF capacitor is connected in series with the original circuit?

8. What will be the current flow in the circuit of Prob. 5 if a 6.7-kΩ resistor is connected in series with the original circuit?

9. A series circuit consisting of a 1-kΩ resistor and a 150-pF capacitor is connected across 600 V at 4.3 MHz. Find:
(a) The impedance of the circuit.
(b) The current flowing through the circuit.
(c) The voltage across the resistor.
(d) The voltage across the capacitor.

10. In the circuit of Prob. 9, a 50-pF capacitor is connected in series with the original capacitor. Find:
(a) The current flow through the new circuit.
(b) The voltage across the resistor.
(c) The voltage across the 150-pF capacitor.
(d) The voltage across the 50-pF capacitor.

32-10 RESISTANCE, INDUCTANCE, AND CAPACITANCE IN SERIES

It has been shown that inductive reactance causes the current to lag the voltage and that capacitive reactance causes the current to lead the voltage; hence, these two reactions are exactly opposite in effect. Figure 32-21 represents a series circuit consisting of resistance, inductance, and capacitance connected across an alternator that supplies 220 V, 60 Hz. Now

$$\omega = 2\pi f = 2\pi \times 60 = 377$$
$$\therefore X_L = \omega L = 377 \times 0.35 = 132\ \Omega$$

and
$$X_C = \frac{1}{\omega C} = \frac{1}{377 \times 13 \times 10^{-6}}$$
$$= \frac{10^3}{3.77 \times 1.3} = 204\ \Omega$$

Figure 32-22 is an impedance phasor diagram of the conditions existing in the circuit. Since X_L and X_C are oppositely directed phasors, it is evident that the resultant reactance will have a magnitude equal to their algebraic sum and will be in the direction of the greater. Therefore, the net reactance of the circuit is a capacitive reactance of 72 Ω as illustrated in Fig. 32-22. Thus the entire circuit could be replaced by an equivalent series circuit consisting of 100 Ω resistance and 72 Ω capacitive reactance, provided that the frequency of the alternator remained constant.

Fig. 32-21 Series Circuit Consisting of R, L, and C

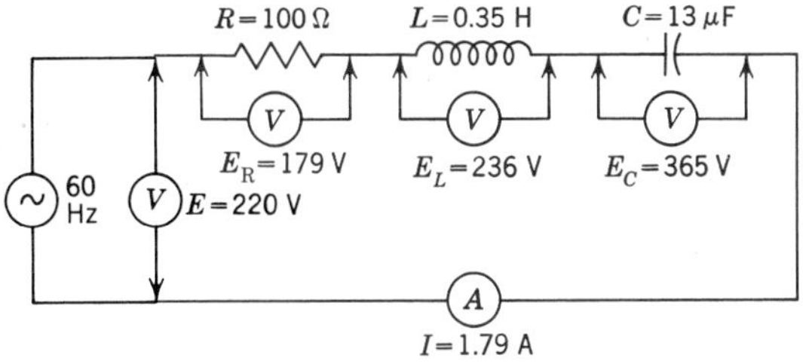

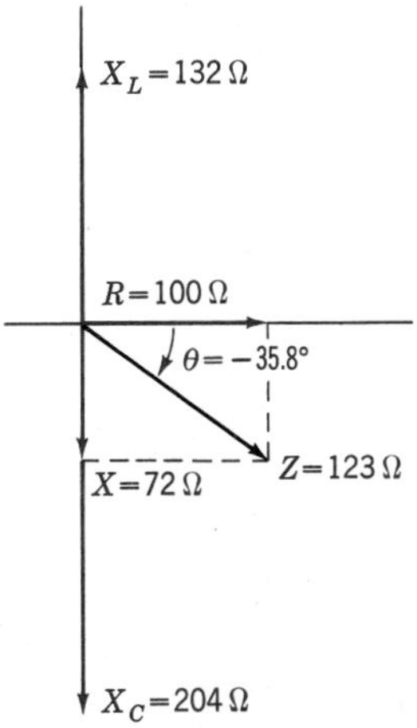

Fig. 32-22 Impedance Phasor Diagram for Circuit of Fig. 32-21

The impedance, current, and potential differences are found by the usual methods.

$$\tan\theta = \frac{X_C}{R} = \frac{72}{100} = 0.72$$
$$\therefore \theta = -35.8°$$
$$Z = \frac{X}{\sin\theta} = \frac{72}{\sin 35.8°} = 123\ \Omega$$
$$I = \frac{E}{Z} = \frac{220}{123} = 1.79\ \text{A}$$
$$E_R = IR = 1.79 \times 100 = 179\ \text{V}$$
$$E_L = IX_L = 1.79 \times 132 = 236\ \text{V}$$
$$E_C = IX_C = 1.79 \times 204 = 365\ \text{V}$$

Note that the potential difference across the reactances is greater than the emf impressed across the entire circuit. This is reasonable, for the applied emf is across the impedance of the circuit, which is a smaller value, in ohms, than the reactances. Because the current is common to all circuit components, it follows that the greatest potential difference will exist across the component offering the greatest opposition.

32-11 POWER IN A SERIES CIRCUIT

It has been shown that, in a circuit consisting of resistance only, no power is returned to the source of emf. Also, it has been shown that a circuit containing reactance alone consumes no power; that is, a reactance alternately receives and returns all power to the source. It is evident, therefore, that in a circuit containing both resistance and reactance there must be some power expended in the resistance and also some returned to the source by the reactance. Figure 32-23 represents the relation among voltage, current, and power in the circuit of Fig. 32-21.

As previously stated, the instantaneous power in the circuit is equal to the product of the applied voltage and the current through the circuit. When the voltage and current are of the same sign, they are acting together and taking power from the source. When their signs are unlike, they are operating in opposite directions and power is returned to the source. The *apparent power* is

$$P_a = EI \qquad \text{VA} \tag{33}$$

and the actual power taken by the circuit, which is called the *true power* or *active power*, is

$$P = I^2R \qquad \text{W} \tag{34}$$

or

$$P = E_RI \qquad \text{W} \tag{35}$$

Fig. 32-23 Voltage, Current, and Power Relations for Circuit of Fig. 32-21

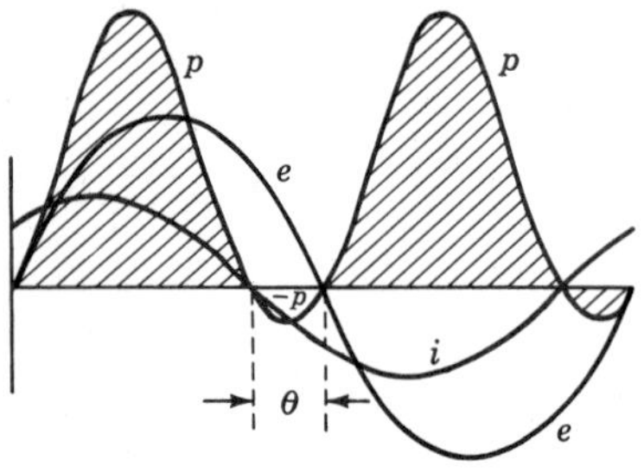

where E_R is the potential difference across the resistance of the circuit.

The *power factor* (PF) of a circuit is the ratio of the true power to the apparent power. That is,

$$\text{PF} = \frac{P}{P_a} \tag{36}$$

Substituting the value of P from Eq. (34) and that of P_a in Eq. (33),

$$\text{PF} = \frac{I^2R}{EI} = \frac{IR}{E}$$

Then, since

$$E = IZ$$

$$\text{PF} = \frac{IR}{IZ}$$

or

$$\text{PF} = \frac{R}{Z} \tag{37}$$

Hence, the power factor of a series circuit can be obtained by dividing the resistance of a circuit by its impedance. The power factor is often expressed in terms of the angle of lead or lag. From preceding vector diagrams, it is evident that

$$\frac{R}{Z} = \cos\theta$$

$$\therefore \text{PF} = \cos\theta \tag{38}$$

From Eq. (36),

$$P = (P_a)(\text{PF})$$

Substituting for P_a,

$$P = (EI)(\text{PF})$$

Substituting for the PF,

$$P = EI\cos\theta \tag{39}$$

From the foregoing it is seen that the power expended in a circuit can be obtained by utilizing different relations. For example, in the circuit of Fig. 32-21,

$$P = I^2R = 1.79^2 \times 100 = 320 \text{ W}$$
$$P = E_RI = 179 \times 1.79 = 320 \text{ W}$$

and

$$\begin{aligned} P &= EI \cos \theta \\ &= 220 \times 1.79 \times \cos 35.8° \\ &= 320 \text{ W} \end{aligned}$$

The power factor of a circuit can be expressed as a decimal or as a percent. Thus the power factor of this circuit is

$$\cos \theta = \cos 35.8° = 0.812$$

Expressed as percent,

$$\text{PF} = 100 \cos 35.8° = 81.2\%$$

32-12 A SIMPLIFIED SLIDE-RULE SOLUTION

There is a method of computing the impedance of series circuits which is convenient to slide rule operators; it employs the trigonometric relationships of a right triangle. The same relationships may be applied to calculators.

Given the resistance R and the reactance X, draw these as perpendicular sides of a right triangle, Fig. 32-24. The impedance is found as follows: First, divide the reactance by the resistance, using scales C and D, so that the phase angle θ may be read directly from the T scale. Second, divide the reactance by sin θ, using the D and S scales, and read impedance Z directly from the D scale. Immediately test this answer by dividing the resistance by cos θ, again using the D and S scales. If the two results for Z do not agree, recalculate θ in the first step. If the two values of Z agree, then your solution is correct. When dealing with very large angles, which do not lend themselves to accurate reading or interpolation at the top of the S scale, carefully use the cosine relationship alone.

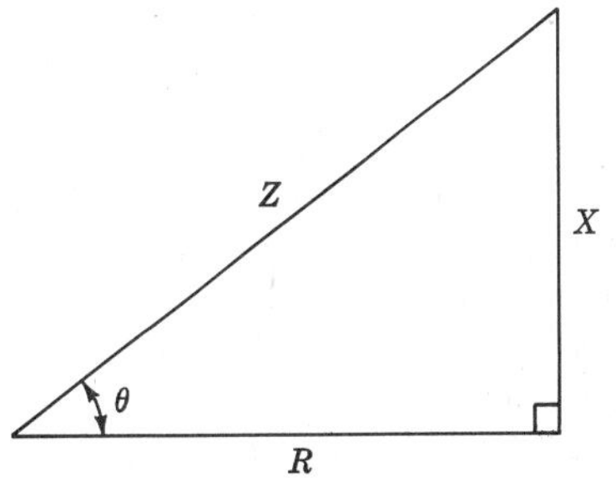

Fig. 32-24 Slide-Rule Solution of Series Circuits

In summary:

$$\theta = \arctan \frac{X}{R} \qquad Z = \frac{X}{\sin \theta} = \frac{R}{\cos \theta}$$

32-13 NOTATION FOR SERIES CIRCUITS

In Sec. 3-5 it was shown that positive and negative "real" numbers could be represented graphically by plotting them along a horizontal line. The positive numbers were plotted to the right of zero, and the negative numbers were plotted to the left. This idea was expanded in Sec. 16-3, where the original horizontal line was made the x axis for a system of rectangular coordinates.

In Sec. 20-12 the system of representation was extended to include what we referred to as "imaginary" numbers by agreeing to plot them along the y axis, the letter j being used as a symbol of 90° operation. Thus, when some number is prefixed with j, it means that the vector which the number represents is to be rotated through an angle of 90°. The rotation is positive, or in a counterclockwise direction, when the sign of j is positive and negative, or in a clockwise direction, when the sign of j is negative. In Sec. 31-1 we saw that "real" is more properly called "in phase," and "imaginary" is more properly called "out of phase."

From the foregoing, it is evident that resistance, when plotted on an impedance phasor diagram, is considered as an in phase number because it is plotted along the x axis. In this instance the term *real* may well define resistance, for it is only the opposition to the flow of current that consumes power.

Since reactances are displaced 90° from resistance in an impedance phasor diagram, it follows that inductive reactance can be prefixed with a plus j and capacitive reactance with a minus j. Thus, an inductive reactance of 75 Ω would be written j75 Ω and plotted on the positive y axis; a capacitive reactance of 86 Ω would be written −j86 Ω and plotted on the negative y axis.

In Sec. 31-1 we saw that a vector can be completely described in terms of either its polar or its rectangular components. For example, the circuit of Fig. 32-14 can be described as consisting of an impedance of 10 Ω at an angle of 53.1°, which would be written

$$Z = 10\angle 53.1^\circ\ \Omega$$

where the angle sign is included for emphasis and the number of degrees denotes the angle that the vector makes with the positive x axis. This is known as *polar form*. Since this impedance is made up of 6 Ω of resistance and 8 Ω of inductive reactance, we can write

$$Z = R + jX_L = 6 + j8\ \Omega$$

This is known as *rectangular form*.

The rectangular form is a very convenient method of notation. For example, instead of writing "A series circuit of 4 Ω resistance and 3 Ω capacitive reactance," we can write "A series circuit of 4 − j3 Ω." Figure 32-25 shows the various types of series circuits with their proper impedance phasor diagram and corresponding notation.

Note that the sign of the phase angle is the same as that of j in the rectangular form. 4 − j3 converts to a polar form with a negative angle $5\angle -36.9^\circ$. It must be understood that neither the rectangular form nor the polar form is a method for solving series circuits. These forms are simply convenient forms of notation that completely describe circuit conditions from both electrical and mathematical viewpoints.

In converting from rectangular to polar form, use the usual methods of solution of right triangles developed in Chap. 26.

example 11 Find the phasor impedance of the following series circuit: 250 − j100 Ω.

solution Given

$$Z = R - jX = 250 - j100\ \Omega$$

$$\tan\theta = \frac{X}{R} = \frac{100}{250} = 0.400$$

$$\therefore \theta = -21.8^\circ$$

$$Z = \frac{X}{\sin\theta} = \frac{100}{\sin 21.8^\circ} = 269\ \Omega$$

or

$$Z = \frac{R}{\cos\theta} = \frac{250}{\cos 21.8^\circ} = 269\ \Omega$$

Hence, $Z = 269\angle -21.8^\circ\ \Omega$

Converting from rectangular form to polar form, which is simply phasor summation of rectangular components, can be completed with a total of three movements on many slide rules.

Converting from polar form, in which the magnitude and angle are given, to rectan-

Circuit	Impedance phasor	Z Rectangular form	Z Polar form
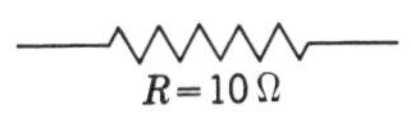 $R=10\,\Omega$	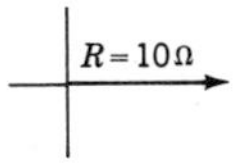 $R=10\,\Omega$	$Z=10+j0\,\Omega$	$Z=10\angle 0°\,\Omega$
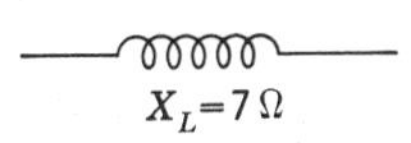 $X_L=7\,\Omega$	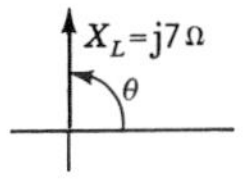 $X_L=j7\,\Omega$, θ	$Z=0+j7\,\Omega$	$Z=7\angle 90°\,\Omega$
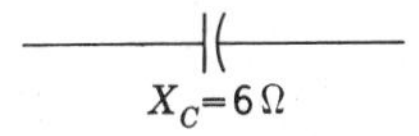 $X_C=6\,\Omega$	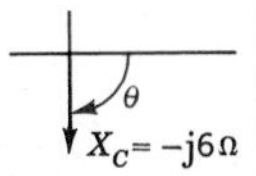 θ, $X_C=-j6\,\Omega$	$Z=0-j6\,\Omega$	$Z=6\angle -90°\,\Omega$
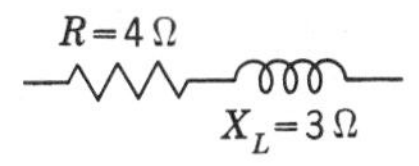 $R=4\,\Omega$, $X_L=3\,\Omega$	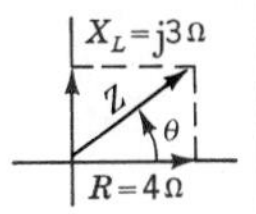 $X_L=j3\,\Omega$, Z, θ, $R=4\,\Omega$	$Z=4+j3\,\Omega$	$Z=5\angle 36.9°\,\Omega$
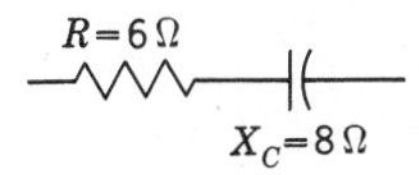 $R=6\,\Omega$, $X_C=8\,\Omega$	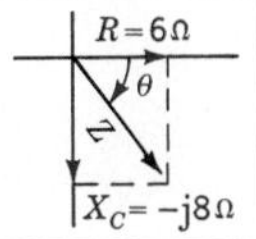 $R=6\,\Omega$, θ, Z, $X_C=-j8\,\Omega$	$Z=6-j8\,\Omega$	$Z=10\angle -53.1°\,\Omega$
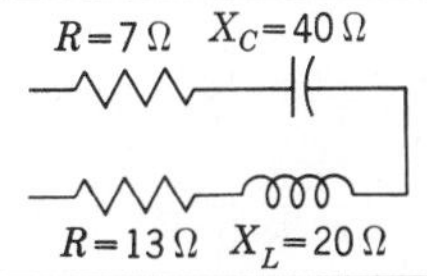 $R=7\,\Omega$ $X_C=40\,\Omega$, $R=13\,\Omega$ $X_L=20\,\Omega$	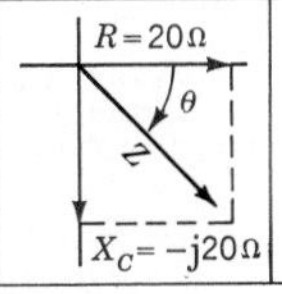 $R=20\,\Omega$, θ, Z, $X_C=-j20\,\Omega$	$Z=20-j20\,\Omega$	$Z=28.2\angle -45°\,\Omega$

Fig. 32-25 Phasor Notation for Series Circuits

gular form is simplified by making use of the trigonometric functions. Since

$$R = Z\cos\theta \qquad \Omega$$
$$X = Z\sin\theta \qquad \Omega$$

and

$$Z = R \pm jX \tag{40}$$

by substitution,

$$Z = Z\cos\theta + jZ\sin\theta \tag{41}$$

Factoring,

$$Z = Z(\cos\theta + j\sin\theta) \qquad \Omega \tag{42}$$

The ± sign is omitted in Eqs. (41) and (42) because, if the proper angles are used (positive or negative), the respective sine values will determine the proper sign of the reactance component.

example 12 A series circuit has an impedance of 269 Ω with a leading power factor of 0.928. What are the reactance and resistance of the circuit?

solution Given $Z = 269\ \Omega$ and PF = 0.928. The power factor, when expressed as a decimal, is equal to the cosine of the phase angle. Hence,

if $\qquad 0.928 = \cos\theta$

then $\qquad \theta = -21.8°$

The angle was given the minus sign because a "leading power factor" means the current leads the voltage. Therefore,

$$Z = 269\underline{/-21.8°}\ \Omega$$

Substituting these values in Eq. (41),

$$Z = 269 \cos 21.8° - j269 \sin 21.8°$$
$$= 250 - j100\ \Omega$$

32-14 THE GENERAL SERIES CIRCUIT

In a series circuit consisting of several resistances and reactances, the total resistance of the circuit is the sum of all the series resistances and the total reactance is the algebraic sum of the series reactances. That is, the total resistance is

$$R_t = R_1 + R_2 + R_3 + \cdots$$

and the reactance of the circuit is

$$X = j(\omega L_1 + \omega L_2 + \omega L_3 + \cdots) - j\left(\frac{1}{\omega C_1} + \frac{1}{\omega C_2} + \frac{1}{\omega C_3} + \cdots\right)$$

Hence, the impedance is

$$Z = R_t \pm jX \qquad \Omega$$

As an alternate method, such a circuit can always be reduced to an equivalent series circuit by combining inductances and capacitances before computing reactances. Thus, the total inductance is

$$L_t = L_1 + L_2 + L_3 + \cdots$$

and the capacitance of the circuit is obtained from

$$\frac{1}{C_t} = \frac{1}{C_1} + \frac{1}{C_2} + \frac{1}{C_3} + \cdots$$

However, when voltage drops across individual reactances are desired, it is best to find the equivalent circuit by combining reactances.

example 13 Given the circuit of Fig. 32-26, which is supplied by 220 V, 60 Hz. Find the (*a*) equivalent series circuit, (*b*) impedance of the circuit, (*c*) current, (*d*) power factor, (*e*) power expended in the circuit, (*f*) apparent power, (*g*) voltage drop across C_1, and (*h*) power expended in R_2.

solution

(*a*) $R_t = R_1 + R_2 + R_3 = 35 + 10 + 30$
$= 75\ \Omega$
$\omega = 2\pi f = 2\pi \times 60 = 377$
$L_t = L_1 + L_2 = 0.62 + 0.34 = 0.96\ \text{H}$
$X_L = \omega L = 377 \times 0.96 = 362\ \Omega$

$$X_{C_1} = \frac{1}{\omega C_1} = \frac{1}{377 \times 30 \times 10^{-6}} = \frac{10^3}{3.77 \times 3} = 88.4\ \Omega$$

$$X_{C_2} = \frac{1}{\omega C_2} = \frac{1}{377 \times 20 \times 10^{-6}} = \frac{10^3}{3.77 \times 2} = 132.6\ \Omega$$

$X_C = 88.4 + 132.6 = 221\ \Omega$
$X = X_L - X_C = 362 - 221 = 141\ \Omega$

The equivalent series circuit consists of a resistance of 75 Ω and an inductive react-

Fig. 32-26 Series Circuit of Example 13

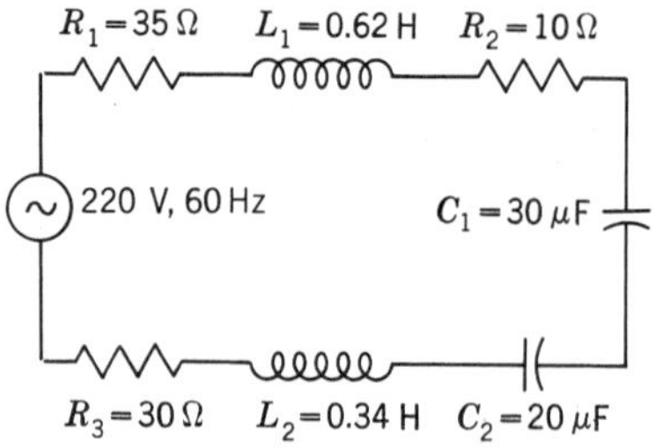

ance of 141 Ω. That is,

$$Z = 75 + j141\ \Omega$$

The impedance phasor diagram for the equivalent circuit is shown in Fig. 32-27.

(b) $\tan\theta = \dfrac{X}{R_t} = \dfrac{141}{75} = 1.88$

$\therefore \theta = 62°$

$$Z = \frac{R}{\cos\theta} = \frac{75}{\cos 62°} = 160\ \Omega$$

Hence, $Z = 160\underline{/62°}\ \Omega$

(c) $I = \dfrac{E}{Z} = \dfrac{220}{160} = 1.38\ \text{A}$

(d) $\text{PF} = \cos\theta = \cos 62° = 0.470$

Expressed as a percent, PF = 47.0%

(e) $P = EI\cos\theta = 220 \times 1.38 \times \cos 62° = 143\ \text{W}$
or $P = I^2R = 1.38^2 \times 75 = 143\ \text{W}$
(f) $P_a = EI = 220 \times 1.38 = 304\ \text{VA}$
(g) $E_{C_1} = IX_{C_1} = 1.38 \times 88.4 = 122\ \text{V}$
(h) $P_{R_2} = I^2R_2 = 1.38^2 \times 10 = 19\ \text{W}$

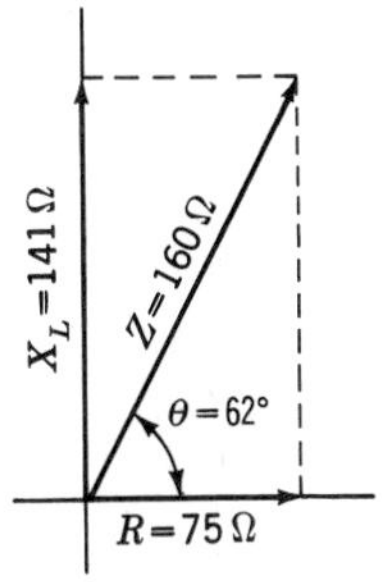

Fig. 32-27 Impedance Phasor Diagram for Circuit of Fig. 32-26

You will find it convenient to compute the value of the angular velocity $\omega = 2\pi f$ for all ac problems, for this factor is common to all reactance equations.

As with all electric circuit problems, a neat diagram of the circuit should be made, with all known circuit components, voltages, and currents clearly marked. In addition, a phasor or impedance diagram should be drawn to scale in order to check the mathematical solution.

PROBLEMS 32-5

Given the circuit of Fig. 32-28, with values as listed in Table 32-1. Draw an impedance phasor diagram for each circuit and find (a) the impedance of the circuit, (b) the current flowing through the circuit, (c) the equation of the current, (d) the PF of the circuit, and (e) the power expended in the circuit.

Table 32-1 PROBLEMS 1 TO 10

Problems	*E, V*	*f*	*R*	*L*	*C*
1	220	60 Hz	200 Ω	2 H	10 μF
2	450	1 kHz	67 Ω	5 mH	50 μF
3	110	50 Hz	2 kΩ	5.6 H	2.2 μF
4	850	400 Hz	500 Ω	2.5 H	100 μF
5	1200	5 MHz	220 Ω	67 μH	20 pF
6	1000	8 GHz	330 Ω	0.08 μH	0.005 pF
7	117	60 Hz	15 Ω	4.5 mH	2500 μF
8	2	10 kHz	27 Ω	3.5 μH	1.5 μF
9	1760	2.5 MHz	500 Ω	12.5 μH	850 pF
10	110	60 Hz	50 Ω	300 mH	22.0 μF

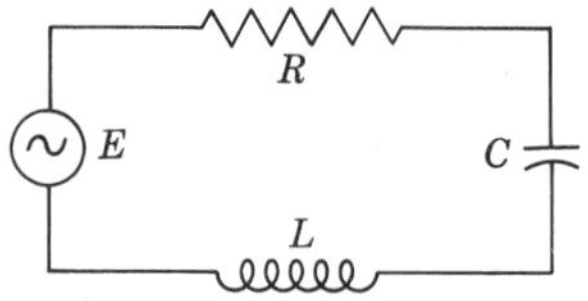

Fig. 32-28 Circuit for Probs. 1 to 10

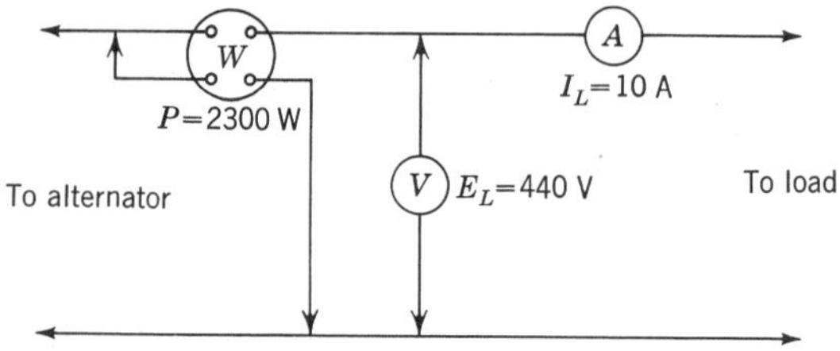

Fig. 32-29 Circuit of Prob. 14

11. A choke coil, when connected across a 230-V dc source, draws 1.15 A. When connected across 230 V, 60 Hz, the current is 665 mA.
(*a*) What is the resistance of the coil?
(*b*) What is its inductive reactance?
(*c*) What is the inductance?

12. Assuming that the resistance of the coil in Prob. 11 is unchanged, how much power would it draw when connected across 230 V, 400 Hz?

13. The following 60-Hz impedances are connected in series:

$$Z_1 = 30 - j40\ \Omega \qquad Z_2 = 5 + j12\ \Omega$$
$$Z_3 = 8 - j6\ \Omega \qquad Z_4 = 4 + j4\ \Omega$$

(*a*) What is the resultant impedance of the circuit?
(*b*) What value of pure reactance must be added in series to make the PF of the circuit 80% leading?

14. The meters represented in Fig. 32-29 are connected such a short distance from an inductive load that line drop from meters to load is negligible. What is the equivalent series circuit of the load?

15. A single-phase induction motor, with 440 V across its input terminals, delivers 10.8 mechanical horsepower at an efficiency of 90% and a PF of 86.6%.
(*a*) What is the line current?
(*b*) How much power is taken by the motor?

16. Given any series circuit, for example, 110 V at 60 Hz applied across 3 + j4 Ω. On the same set of axes and to the same scale, plot instantaneous values of the applied emf *e*, the potential difference across the resistance *R*, and the potential difference across the reactance *X*. What is your conclusion?

32-15 SERIES RESONANCE

It has been shown that the inductive reactance of a circuit varies directly as the frequency and that the capacitive reactance varies inversely as the frequency. That is, the inductive reactance will increase and the capacitive reactance will decrease as the frequency is increased, and vice versa. Then, for any value of inductance and capacitance in a circuit, there is a frequency at which the inductive reactance and the capacitive reactance are equal. This is called the *resonant frequency* of the circuit. Since, in a series circuit.

$$Z = R + j\left(\omega L - \frac{1}{\omega C}\right) \qquad \Omega$$

at resonance,

$$\omega L = \frac{1}{\omega C} \qquad (43)$$

Hence, $\qquad Z = R$

Therefore, at the resonant frequency of a series circuit, the resistance is the only circuit component that limits the flow of current, for the net reactance of the circuit is zero. Thus the current is in phase with the applied voltage, which results in a circuit power factor of 100%.

example 14 There is impressed 10 V at a frequency of 1 MHz across a circuit consisting of a coil of 92.2 μH in series with a capacitance of 275 pF. The effective resistance of the coil at this frequency is 10 Ω, and both the resistance of the connecting wires and the capacitance are negligible. (*a*) What is the impedance of the circuit? (*b*) How much current flows through the circuit? (c) What are the voltages across the reactances?

solution The resistance of the coil is treated as being in series with its inductive reactance.

(*a*)
$$\omega = 2\pi f = 6.28 \times 10^6$$
$$X_L = \omega L = 6.28 \times 10^6 \times 92.2 \times 10^{-6} = 6.28 \times 92.2 = 579\ \Omega$$
$$X_C = \frac{1}{\omega C} = \frac{1}{6.28 \times 10^6 \times 275 \times 10^{-12}} = \frac{10^4}{6.28 \times 2.75} = 579\ \Omega$$

Since $\quad X_L = X_C$
then $\quad Z = R = 10\ \Omega$

(*b*)
$$I = \frac{E}{Z} = \frac{10}{10} = 1\text{ A}$$

(c)
$$E_C = IX_C = 1 \times 579 = 579\text{ V}$$
$$E_L = IX_L = 1 \times 579 = 579\text{ V}$$

Note that the voltages across the inductance and capacitance are much greater than the applied voltage.

The *quality* or *merit* of an inductance, denoted by Q, is defined as the ratio of its inductive reactance to its resistance at a given frequency. Thus,

$$Q = \frac{\omega L}{R} \qquad (44)$$

Then, at resonance,

$$E_C = E_L = I\omega L \qquad \text{V}$$

Substituting for I,

$$E_C = E_L = \frac{E\omega L}{R} \qquad \text{V}$$

Substituting for $\frac{\omega L}{R}$,

$$E_C = E_L = EQ \qquad \text{V} \qquad (45)$$

Because the average radio circuit has purposely been designed for high Q values, it is seen that very high voltages can be developed in resonant series circuits.

32-16 RESONANT FREQUENCY

The resonant frequency of a circuit can be determined by rewriting Eq. (43). Thus,

$$2\pi f L = \frac{1}{2\pi f C}$$

$$\therefore f = \frac{1}{2\pi\sqrt{LC}} \qquad \text{Hz} \qquad (46)$$

where f, L, and C are in the usual units, hertz, henrys, and farads, respectively.

example 15 A series circuit consists of an inductance of 500 μH and a capacitor of 400 pF. What is the resonant frequency of the circuit?

solution

$$L = 500\ \mu\text{H} = 5 \times 10^{-4}\ \text{H}$$
$$C = 400\ \text{pF} = 4 \times 10^{-10}\ \text{F}$$
$$f = \frac{1}{2\pi\sqrt{LC}}$$
$$= \frac{1}{2\pi\sqrt{5 \times 10^{-4} \times 4 \times 10^{-10}}}$$
$$= \frac{10^7}{2\pi\sqrt{20}}$$
$$= 356\ 000\ \text{Hz}$$

or $$f = 356\ \text{kHz}$$

From Eq. (46) it is evident that the resonant frequency of a series circuit depends *only* upon the LC product. This means there are an infinite number of combinations of L and C that will resonate to a particular frequency.

example 16 How much capacitance is required to obtain resonance at 1500 kHz with an inductance of 45 μH?

solution

$$f = 1500\ \text{kHz} = 1.5 \times 10^6\ \text{Hz}$$
$$L = 45\ \mu\text{H} = 4.5 \times 10^{-5}\ \text{H}$$
$$\omega = 2\pi f = 2\pi \times 1.5 \times 10^6 = 9.42 \times 10^6$$

From Eq. (46),

$$C = \frac{1}{(2\pi f)^2 L} = \frac{1}{\omega^2 L}$$
$$\therefore C = \frac{1}{(9.42 \times 10^6)^2 \times 4.5 \times 10^{-5}}$$
$$= 250\ \text{pF}$$

PROBLEMS 32-6

1. 100 V 10 kHz is impressed across a series circuit consisting of a 220-pF capacitor of negligible resistance and an 800-mH coil with effective resistance of 125 Ω.
   (*a*) How much current flows through the circuit?
   (*b*) How much power does the circuit absorb from the source?
   (*c*) What are the voltages across the capacitor and the coil?

2. What is the Q of the coil in Prob. 1?

3. At what frequency would the circuit of Prob. 1 be resonant?

4. What type and value of "pure reactance" must be added to the circuit of Prob. 1 to make it resonant at 10 kHz?

5. A tuning capacitor is continuously variable between 20 pF and 350 pF.
   (*a*) What inductance must be connected in series with it to provide a lowest resonant frequency of 550 kHz?
   (*b*) What will then be the highest resonant frequency?

6. What is the equivalent circuit of a series circuit when operating at (*a*) resonant frequency, (*b*) at a frequency less than resonant frequency, and (*c*) at a frequency higher than resonant frequency?

33

alternating currents—parallel circuits

Parallel circuits are the most commonly encountered circuits in use. The average distribution circuit has many types of loads all connected in parallel with each other: lighting circuits, motors, transformers for various uses, etc. The same is true of electronic circuits, which range from the most simple parallel circuits to complex networks.

This chapter deals with the solutions of parallel circuits. Such solutions may reduce a parallel circuit to an equivalent series circuit that, when connected to the same source of emf as the given parallel circuit, would result in the same line current and phase angle; that is, the alternator would "see" the same load.

33-1 RESISTANCES IN PARALLEL

It was explained in Secs. 32-1 and 32-2 that, in an ac circuit containing resistance only, the voltage, current, and power relations were the same as in dc circuits. However, in order to build a foundation from which all parallel circuits can be analyzed, the case of paralleled resistances must be considered from a phasor viewpoint.

Figure 33-1 represents a 60-Hz 220-V alternator connected to three resistances in parallel.

Neglecting the internal resistance of the alternator and the resistance of the connecting wires, the emf of the alternator is impressed across each of the three resistances. If I_1, I_2, and I_3 represent the currents flowing through R_1, R_2, and R_3, respectively, then by Ohm's law,

$$I_1 = 2.5 \text{ A}$$
$$I_2 = 0.5 \text{ A}$$
$$I_3 = 2.0 \text{ A}$$

Since all currents are in phase, the total current flowing in the line, or external circuit, will be equal to the sum of the branch currents, or 5.0 A. The phasor diagram for the

Fig. 33-1 Alternator Connected to Three Resistors in Parallel

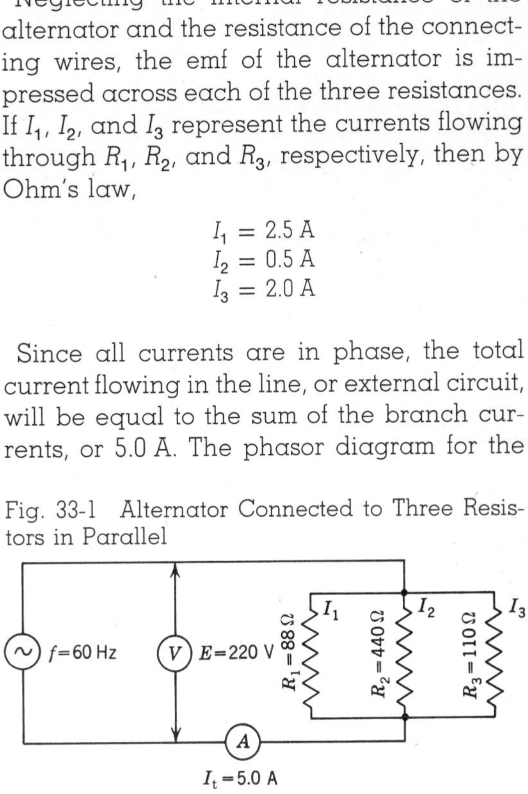

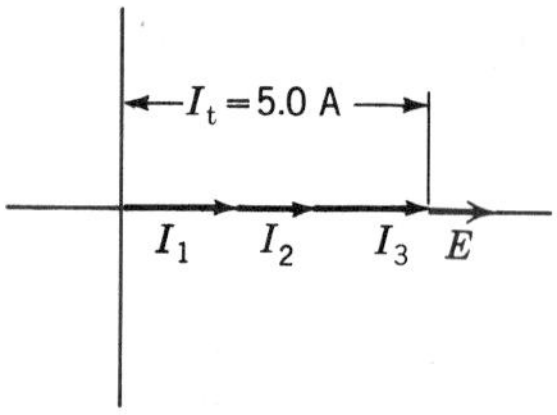

Fig. 33-2 Phasor Diagram for the Circuit of Fig. 33-1

three currents is shown in Fig. 33-2. All currents are plotted in phase with the applied emf, which is used as a reference phasor because the voltage is common to all resistances. Then, using rectangular phasor notation,

$$\begin{aligned} I_1 &= 2.5 + j0 \text{ A} \\ I_2 &= 0.5 + j0 \text{ A} \\ I_3 &= 2.0 + j0 \text{ A} \\ \hline I_t &= 5.0 + j0 \text{ A} = 5.0\angle 0^\circ \text{ A} \end{aligned}$$

As with all other circuits, the equivalent series impedance, which in this case is a pure resistance, is found by dividing the voltage across the circuit by the total current. That is,

$$Z = \frac{E}{I_t} = \frac{220}{5} = 44\ \Omega = 44\angle 0^\circ\ \Omega$$

33-2 CAPACITORS IN PARALLEL

Figure 33-3 represents two capacitors C_1 and C_2 connected in parallel across a voltage E. The quantity of charge in capacitor C_1 will be

Fig. 33-3 Capacitors C_1 and C_2 Connected in Parallel

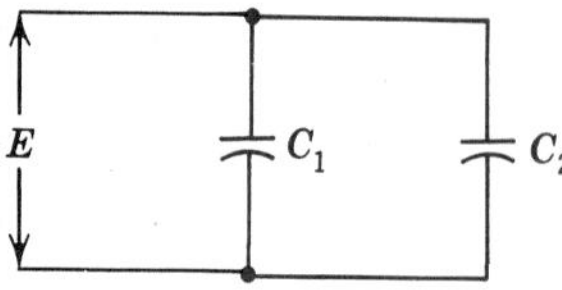

$$Q_1 = C_1 E \tag{1}$$

and that in capacitor C_2 will be

$$Q_2 = C_2 E \tag{2}$$

Since the total quantity in both capacitors is $Q_1 + Q_2$, then

$$Q_1 + Q_2 = C_p E \tag{3}$$

where C_p is the total capacitance of the combination. Adding Eqs. (1) and (2),

$$Q_1 + Q_2 = C_1 E + C_2 E$$

or

$$Q_1 + Q_2 = (C_1 + C_2)E$$

Substituting the value of $Q_1 + Q_2$ from Eq. (3),

$$C_p E = (C_1 + C_2)E$$

which results in

$$C_p = C_1 + C_2 \tag{4}$$

From the foregoing, it is apparent that capacitors in parallel combine like resistances in series; that is, the capacitance of paralleled capacitors is equal to the sum of the individual capacitances.

example 1 What is the capacitance of a 6-μF capacitor in parallel with a capacitor of 4 μF?

solution $C_p = 6 + 4 = 10\ \mu\text{F}$

33-3 INDUCTANCE AND CAPACITANCE IN PARALLEL

When a purely inductive reactance and a capacitive reactance are connected in parallel, as shown in Fig. 33-4, the currents flow-

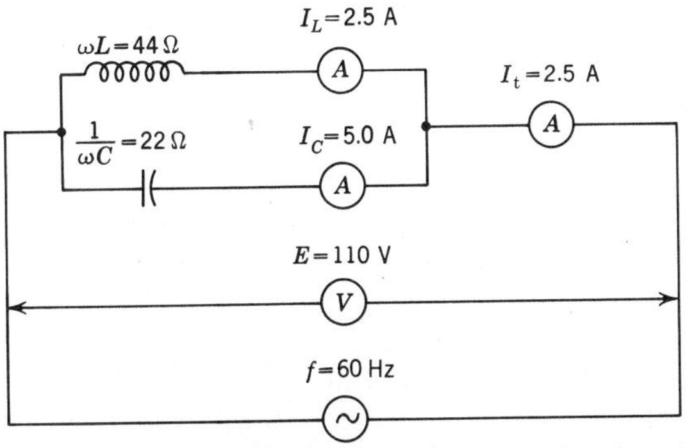

Fig. 33-4 X_L and X_C Connected in Parallel

ing through these reactances differ in phase by 180°.

The current flowing through the inductor is

$$I_L = \frac{E}{X_L} = \frac{E}{\omega L} = \frac{110}{44} = 2.5\ \text{A}$$

and that through the capacitor is

$$I_C = \frac{E}{X_C} = \omega CE = \frac{110}{22} = 5.0\ \text{A}$$

In series circuits, the current was used as the reference phasor because the current is the same in all parts of the circuit. In parallel circuits there are different values of currents in various parts of a circuit; therefore, the current cannot be used as the reference phasor.

Since the same voltage exists across two or more parallel branches, the applied voltage can be used as the reference phasor as shown in Fig. 33-5.

Note that the current I_L through the inductor is plotted as *lagging* the alternator voltage by 90° and the current I_C through the capacitor is *leading* the voltage by 90°. The total line current I_t, which is the phasor sum of the branch currents, is leading the applied voltage by 90°. That is using rectangular phasor notation,

$$\begin{aligned} I_L &= 0 - j2.5\ \text{A} \\ I_C &= 0 + j5.0\ \text{A} \\ \hline I_t &= 0 + j2.5\ \text{A} = 2.5\underline{/90^\circ}\ \text{A} \end{aligned}$$

Since the line current leads the alternator voltage by 90°, the equivalent series circuit consists of a capacitive reactance of

$$\frac{E}{I_t} = \frac{110}{2.5} = 44\ \Omega$$

That is, the parallel circuit could be replaced with a 60.3-μF capacitor which would result in a current of 2.5 A leading the voltage by 90°; in other words, the alternator would not sense the difference.

Note the difference between reactances in series and reactances in parallel. In a series circuit the *greatest* reactance of the circuit results in the equivalent series circuit containing the same kind of reactance. For this reason, it is said that reactances, or voltages across reactances, are the controlling factors of series circuits. In a parallel circuit the *least* reactance of the circuit, which passes the greatest current, results in the equivalent series circuit containing the same kind of reactance. For this reason, it is said that

Fig. 33-5 Phasor Diagram for the Circuit of Fig. 33-4

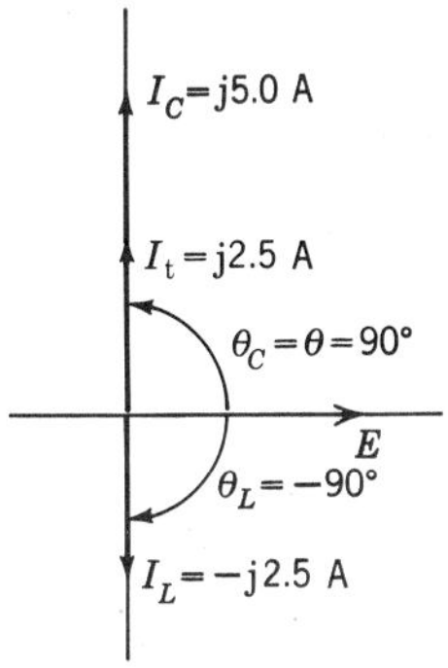

currents are the controlling factors of parallel circuits.

33-4 ASSUMED VOLTAGES

The solutions of the great majority of parallel circuits are facilitated by assuming a voltage to exist across a parallel combination. The current through each branch, due to the assumed voltage, is then added vectorially to obtain the total current. The assumed voltage is then divided by the total current, the quotient being the joint impedance of the parallel branches.

The assumed voltage should always be some power of 10 in order that you can make full use of the reciprocal scales and reciprocal relations on your slide rule.

In order to avoid small decimal quantities the assumed voltage should be greater than the largest impedance of any parallel branch.

example 2 Given the circuit of Fig. 33-6. What are the impedance and the power factor of the circuit at a frequency of 2.5 MHz?

solution C_1 and C_2 are in parallel; hence, the total capacitance is

$$C_p = C_1 + C_2 = 200 + 125 = 325 \text{ pF}$$

Fig. 33-6 Circuit of Example 2

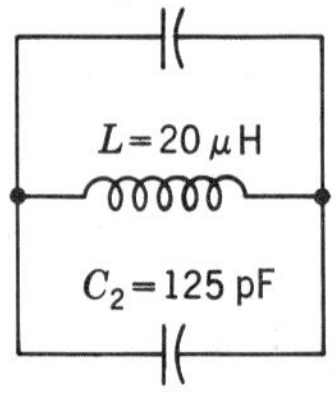

This simplifies the circuit to a capacitor C of 325 pF in parallel with an inductance L of 20 μH.

$$\omega = 2\pi f = 2\pi \times 2.5 \times 10^6 = 1.57 \times 10^7$$

$$X_L = \omega L = 1.57 \times 10^7 \times 2 \times 10^{-5} = 314\ \Omega$$

$$X_C = \frac{1}{\omega C} = \frac{1}{1.57 \times 10^7 \times 325 \times 10^{-12}}$$

$$= \frac{10^3}{1.57 \times 3.25} = 196\ \Omega$$

Assume 1000 V across the parallel branch. Then the current through the capacitors is

$$I_C = \frac{E_a}{X_C} = \frac{1000}{196} = 5.10 \text{ A}$$

and the current through the inductance is

$$I_L = \frac{E_a}{X_L} = \frac{1000}{314} = 3.18 \text{ A}$$

Since I_C leads the assumed voltage by 90° and I_L lags the assumed voltage by 90°, they are plotted with the assumed voltage as reference phasor as shown in Fig. 33-7. Then the total current I_t that would flow because of assumed voltage would be the phasor summation of I_C and I_L. Performing phasor summation:

$$I_C = 0 + j5.10 \text{ A}$$

$$\underline{I_L = 0 - j3.18 \text{ A}}$$

$$I_t = 0 + j1.92 \text{ A} = 1.92\angle 90^\circ \text{ A}$$

Again, since the total current leads the voltage by 90°, the equivalent series circuit consists of a capacitor whose capacitive reactance is

$$\frac{E_a}{I_t} = \frac{1000}{1.92} = 521\ \Omega$$

Since $\theta = 90°$,

$$\text{PF} = \cos\theta = 0$$

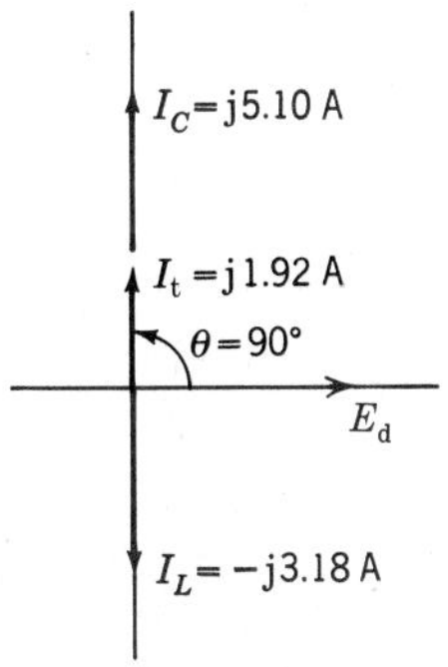

Fig. 33-7 Phasor Diagram for Circuit of Example 2

You should solve the circuit of Fig. 33-6 with different values of assumed voltages.

33-5 RESISTANCE AND INDUCTANCE IN PARALLEL

When a resistance and an inductive reactance are connected in parallel, as represented in Fig. 33-8, the currents that flow differ in phase by 90°. The current flowing through the resistance is

$$I_R = \frac{E}{R} = \frac{120}{20} = 6.0 \text{ A}$$

and that through the inductance is

$$I_L = \frac{E}{\omega L} = \frac{120}{15} = 8.0 \text{ A}$$

Since the current through the resistance is in phase with the applied voltage and the current through the inductance lags the applied voltage by 90°, I_R and I_L are plotted with the applied emf as reference phasor as shown in Fig. 33-9. Then the total current I_t, or line current, is the phasor sum of I_R and I_L. Performing phasor summation,

$$\begin{aligned} I_R &= 6.0 + j0 \text{ A} \\ I_L &= 0 - j8.0 \text{ A} \\ \hline I_t &= 6.0 - j8.0 \text{ A} \end{aligned}$$

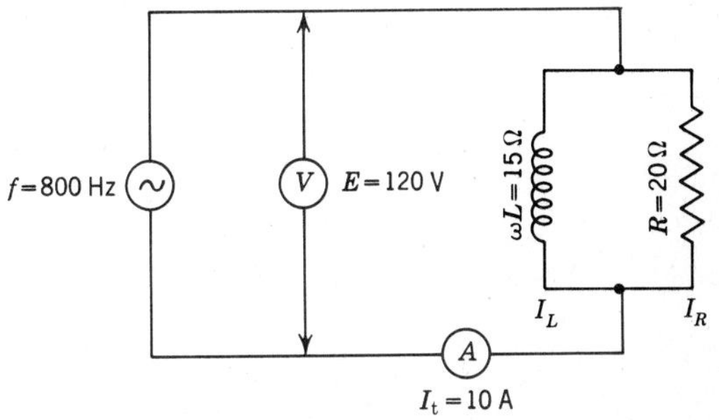

Fig. 33-8 R and X_L in Parallel

Hence, the total current, which consists of an in phase component of 6.0 A and a 90° lagging component of 8.0 A, is expressed in terms of its rectangular components. The magnitude and phase angle are then found by the usual trigonometric methods. Thus,

$$I_t = 10\angle{-53.1°} \text{ A}$$

The power factor of the circuit is

$$\begin{aligned} \text{PF} &= \cos\theta = \cos(-53.1°) \\ &= 0.60 \text{ lagging} \end{aligned}$$

The power expended in the circuit is

$$\begin{aligned} P &= EI\cos\theta = 120 \times 10 \times 0.60 \\ &= 720 \text{ W} \end{aligned}$$

or

$$P = I_R^2 R = 6^2 \times 20 = 720 \text{ W}$$

Fig. 33-9 Phasor Diagram for Circuit of Fig. 33-8

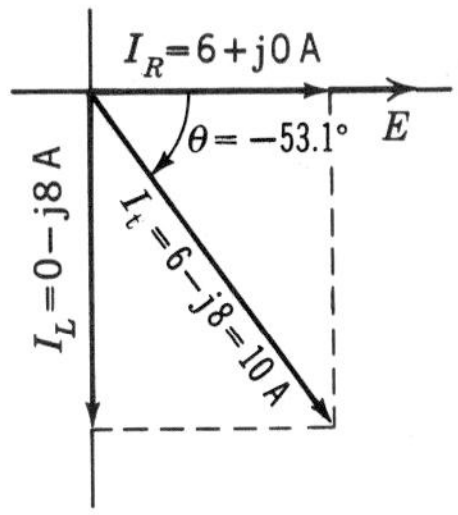

The equivalent impedance, or total impedance, of the circuit is

$$Z_t = \frac{E}{I_t} = \frac{120}{10} = 12\ \Omega$$

Since the entire circuit has a lagging PF of 0.60, it follows that the equivalent series circuit consists of a resistance and an inductive reactance in series, the phasor sum of which is 12 Ω at a phase angle θ such that $\cos\theta = 0.60$. Therefore, $\theta = 53.1°$, and

$$\begin{aligned} Z_t &= 12\underline{/53.1°}\ \Omega \\ &= 12\,(\cos 53.1° + \mathrm{j}\sin 53.1°) \\ &= 7.2 + \mathrm{j}9.6\ \Omega \end{aligned}$$

From the foregoing, it is evident that the parallel circuit of Fig. 33-8 could be replaced by a series circuit of 7.2 Ω resistance and 9.6 Ω inductive reactance and that the alternator would be working under exactly the same load conditions as before.

In order to justify such solutions, solve for the equivalent impedance of the circuit of Fig. 33-8 by using an assumed voltage and then using the *actual* voltage to obtain the power.

33-6 RESISTANCE AND CAPACITANCE IN PARALLEL

When resistance and capacitive reactance are connected in parallel, as represented in Fig. 33-10, the current through the resistance is in phase with the voltage across the parallel combination, and the current through the capacitive reactance leads this voltage by 90°.

Fig. 33-10 R and X_C in Parallel

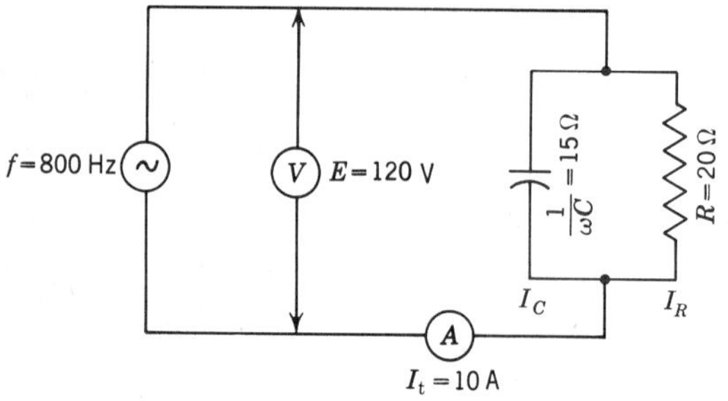

The circuit of Fig. 33-10 is similar to that of Fig. 33-8 except that Fig. 33-10 contains a capacitive reactance of 15 Ω in place of the inductive reactance of 15 Ω. The phasor diagram of currents is shown in Fig. 33-11, and it is evident that the total current is

$$I_t = 6.0 + \mathrm{j}8.0\ \mathrm{A} = 10\underline{/53.1°}\ \mathrm{A}$$

The power factor of the circuit is

$$\begin{aligned} \mathrm{PF} &= \cos\theta = \cos 53.1° \\ &= 0.60\ \text{leading} \end{aligned}$$

Similarly, the total impedance of the circuit is 12 Ω; and since the circuit has a leading PF of 0.60, it follows that the equivalent series circuit consists of resistance and capacitive reactance in series the phasor sum of which is 12 Ω at a phase angle θ such that $\cos\theta = 0.60$. Therefore,

$$\theta = -53.1°$$

and

$$\begin{aligned} Z_t &= 12\underline{/-53.1°}\ \Omega \\ &= 7.2 - \mathrm{j}9.6\ \Omega \end{aligned}$$

If the parallel circuit of Fig. 33-10 were replaced by a series circuit of 7.2 Ω resistance and 9.6 Ω capacitive reactance, the alterna-

Fig. 33-11 Phasor Diagram for Circuit of Fig. 33-10

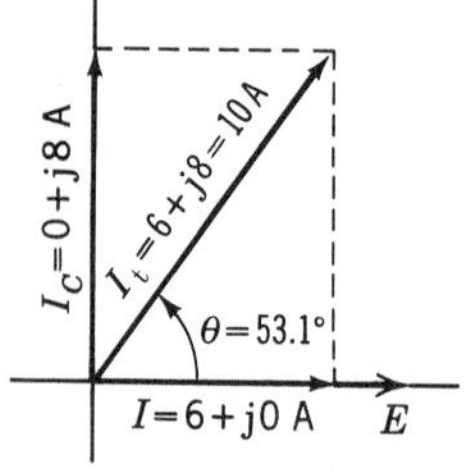

tor would be working under exactly the same load conditions as before.

33-7 RESISTANCE, INDUCTANCE, AND CAPACITANCE IN PARALLEL

When resistance, inductive reactance, and capacitive reactance are connected in parallel, as represented in Fig. 33-12, the line current is the phasor sum of the several currents.

The currents through the branches are

$$I_R = \tfrac{220}{40} = 5.5 \text{ A}$$
$$I_L = \tfrac{220}{10} = 22 \text{ A}$$
$$I_C = \tfrac{220}{18} = 12.2 \text{ A}$$

Performing phasor summation of these currents as shown in Fig. 33-13,

$$\begin{aligned} I_R &= 5.5 + j0 \quad \text{A} \\ I_L &= 0 - j22 \quad \text{A} \\ I_C &= 0 + j12.2 \text{ A} \\ \hline I_t &= 5.5 - j9.8 \text{ A} = 11.2\angle{-60.7^\circ} \text{ A} \\ \text{PF} &= \cos(-60.7^\circ) = 0.489 \text{ lagging} \end{aligned}$$

The total impedance is

$$Z_t = \frac{E}{I_t} = \frac{220}{11.2} = 19.6\ \Omega$$

Since the circuit has a lagging PF of 0.489, the equivalent series circuit consists of a re-

Fig. 33-12 *L*, *C*, and *R* in Parallel

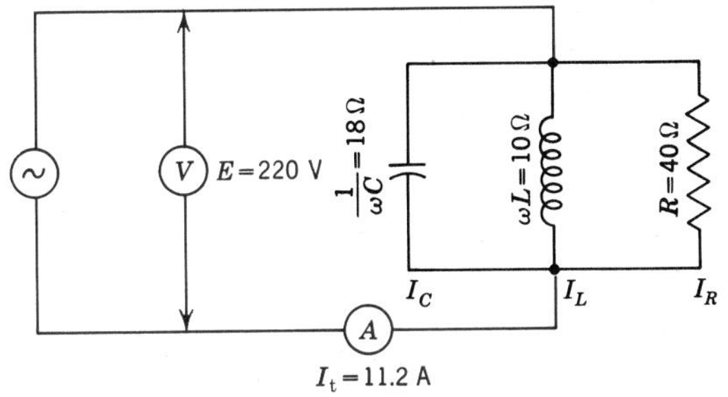

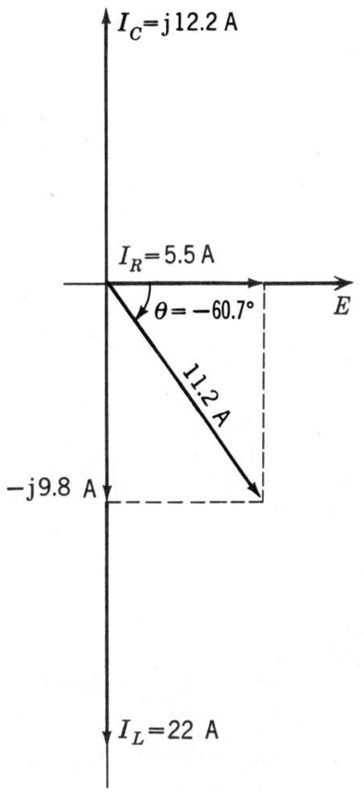

Fig. 33-13 Phasor Diagram for Circuit of Fig. 33-12

sistance and an inductive reactance. The phasor sum of these must be 19.6 Ω at a phase angle θ such that $\cos\theta = 0.489$. Therefore, $\theta = 60.7^\circ$ and

$$Z_t = 19.6\angle{60.7^\circ}\ \Omega = 9.59 + j17.1\ \Omega$$

which are the values which constitute the equivalent series circuit.

example 3 Given the circuit represented in Fig. 33-14. Solve for the equivalent series circuit at a frequency of 5 MHz.

Fig. 33-14 Circuit of Example 3

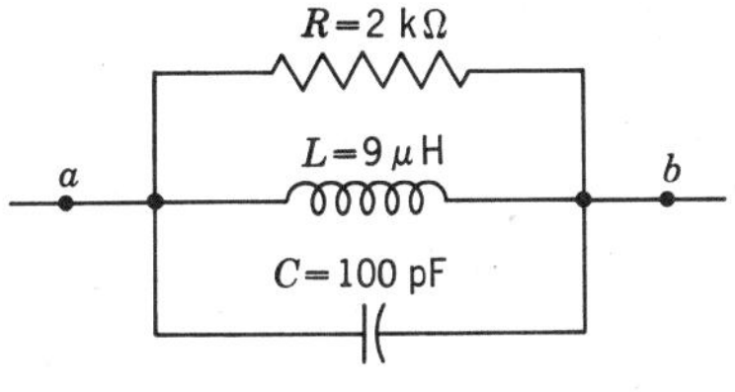

solution

$$f = 5\text{ MHz} = 5 \times 10^6\text{ Hz}$$
$$L = 9\ \mu\text{H} = 9 \times 10^{-6}\text{ H}$$
$$C = 100\text{ pF} = 10^{-10}\text{ F}$$
$$\omega = 2\pi f = 2\pi \times 5 \times 10^6 = 3.14 \times 10^7$$
$$X_L = \omega L = 3.14 \times 10^7 \times 9 \times 10^{-6} = 283\ \Omega$$
$$X_C = \frac{1}{\omega C} = \frac{1}{3.14 \times 10^7 \times 10^{-10}} = \frac{10^3}{3.14} = 318\ \Omega$$

Assume $E_a = 1000$ V applied between a and b.

$$I_R = \frac{E_a}{R} = \frac{1000}{2000} = 0.50\text{ A}$$
$$I_L = \frac{E_a}{X_L} = \frac{1000}{283} = 3.54\text{ A}$$
$$I_C = \frac{E_a}{X_C} = \frac{1000}{318} = 3.14\text{ A}$$

The total current I_t is the phasor sum of the three branch currents as represented in the phasor diagram of Fig. 33-15. Adding vectorially,

$$\begin{aligned} I_R &= 0.50 + j0 \\ I_L &= 0 \quad - j3.54\text{ A} \\ I_C &= 0 \quad + j3.14\text{ A} \\ \hline I_t &= 0.50 - j0.40\text{ A} = 0.640\underline{/-38.7^\circ}\text{ A} \\ \text{PF} &= \cos(-38.7^\circ) = 0.78\text{ lagging} \end{aligned}$$

The total impedance Z_t, which is the impedance between points a and b, is

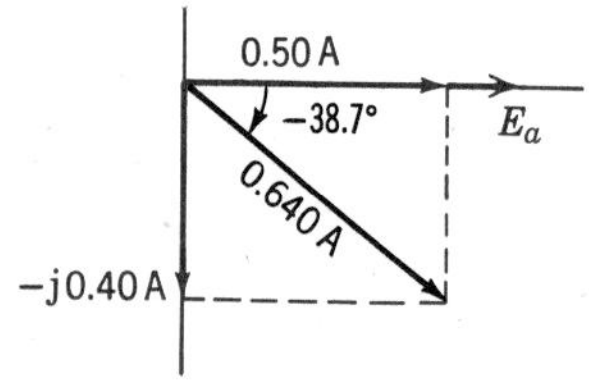

Fig. 33-15 Phasor Diagram of Circuit of Fig. 33-14

$$Z_t = Z_{ab} = \frac{E_a}{I_t} = \frac{1000}{0.64} = 1560\ \Omega$$

Since the current is lagging the voltage, the equivalent series circuit consists of a resistance and an inductive reactance. The phasor sum of these is 1560 Ω at a phase angle θ such that $\cos\theta = 0.78$. Therefore, $\theta = 38.7^\circ$ and

$$Z_t = 1560\underline{/38.7^\circ}\ \Omega = 1220 + j976\ \Omega$$

That is, the equivalent series circuit is a resistance of $R = 1220\ \Omega$ and an inductive reactance of $\omega L = 976\ \Omega$. Since

$$\omega L = 976\ \Omega$$

then

$$L = \frac{976}{\omega} = \frac{976}{3.14 \times 10^7} = 31.1\ \mu\text{H}$$

which results in the equivalent circuit as represented in Fig. 33-16 with the impedance phasor diagram of Fig. 33-17.

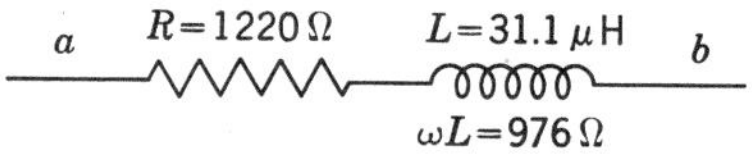

Fig. 33-16 Equivalent Series Circuit of Example 3

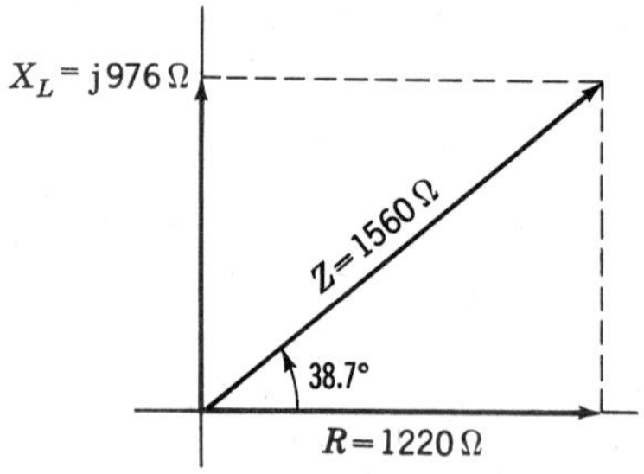

Fig. 33-17 Impedance Phasor Diagram for Equivalent Series Circuit

PROBLEMS 33-1

1. What is the resulting capacitance when a 500-pF capacitor is connected in parallel with a 220-pF capacitor?
2. Two capacitors, 50 and 500 pF, are connected in parallel. A current of 200 mA, 2.7 GHz, flows through the 500-pF capacitor. How much current flows through the 50-pF capacitor?
3. Neglecting the resistance of the connecting wires in Fig. 32-12,
   (a) Write the equation for the emf of the alternator.
   (b) Write the equation for the circuit current.
   (c) What is the voltage across C_1?
   (d) What is the capacitance of C_3?
   (e) How much current flows through C_2?
4. In Fig. 33-18, $R = 200\ \Omega$, $L = 2$ H, $C = 5\ \mu$F, $E = 220$ V, and $f = 60$ Hz.

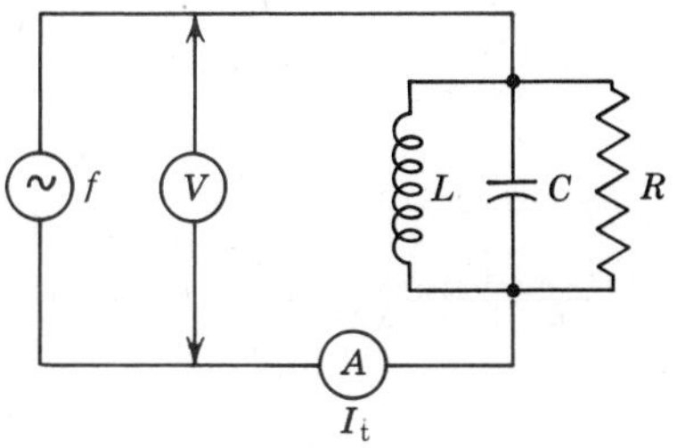

Fig. 33-18 Circuit for Probs. 4 to 6

   (a) What is the ammeter reading?
   (b) How much power is expended in the circuit?
   (c) What is the equivalent series circuit?
   (d) What is the power factor?
   (e) What is the equation of the current flowing through the ammeter?
5. Using the other values of Prob. 4, what must be the value of the inductance in the circuit in order to obtain a PF of (a) 0.8 lagging and (b) 1.0?
6. In Fig. 33-18, $R = 500\ \Omega$, $L = 6$ nH, $C = 0.02$ pF, $E = 1$ kV, and $f = 8$ GHz.
   (a) What is the reading of the ammeter?
   (b) What parallel capacitance must be added to the circuit in order to achieve unity PF?

33-8 PHASOR IMPEDANCES IN PARALLEL

It was shown in Sec. 13-2 that the reciprocal of the equivalent resistance R_p of several resistances in parallel is expressed by the relation

$$\frac{1}{R_p} = \frac{1}{R_1} + \frac{1}{R_2} + \frac{1}{R_3} + \frac{1}{R_4} + \cdots$$

and that when two resistances R_1 and R_2 are connected in parallel, the equivalent resistance is

$$R_p = \frac{R_1 R_2}{R_1 + R_2}$$

An analogous condition exists when two or more impedances are connected in parallel.

By following the line of reasoning used for resistances in parallel, the reciprocal of the equivalent impedance of several impedances in parallel is found to be

$$\frac{1}{Z_p} = \frac{1}{Z_1} + \frac{1}{Z_2} + \frac{1}{Z_3} + \frac{1}{Z_4} + \cdots \qquad (5)$$

Similarly, the equivalent impedance Z_p of two impedances Z_1 and Z_2 connected in parallel is

$$Z_p = \frac{Z_1 Z_2}{Z_1 + Z_2} \qquad (6)$$

Note that the impedances of Eqs. (5) and (6) are in polar form.

example 4 Find the equivalent impedance of the circuit of Fig. 33-19.

solution First express the given impedances in both rectangular and polar forms.

$$Z_1 = 75 - j30 = 80.8\angle -21.8^\circ\ \Omega$$
$$Z_2 = 35 + j50 = 61.0\angle 55^\circ\ \Omega$$

As pointed out in Sec. 31-5, phasors in polar form cannot be added algebraically; they must be added in terms of their rectangular components. Therefore, when the given impedance values are substituted in Eq. (6), the impedances in the denominator must be in rectangular form so that the indicated addition can be carried out. Substituting,

$$Z_p = \frac{(80.8\angle -21.8^\circ)(61.0\angle 55^\circ)}{(75 - j30) + (35 + j50)} = \frac{4930\angle 33.2^\circ}{110 + j20}$$

Because the denominator is in rectangular form and the numerator is in polar form, the denominator must be converted to polar form so the indicated division can be completed. Thus, performing phasor summation of the terms of the denominator,

$$Z_p = \frac{4930\angle 33.2^\circ}{112\angle 10.3^\circ} = \frac{4930}{112}\angle 33.2^\circ - 10.3^\circ = 44\angle 22.9^\circ\ \Omega$$

Fig. 33-19 Circuit of Example 4

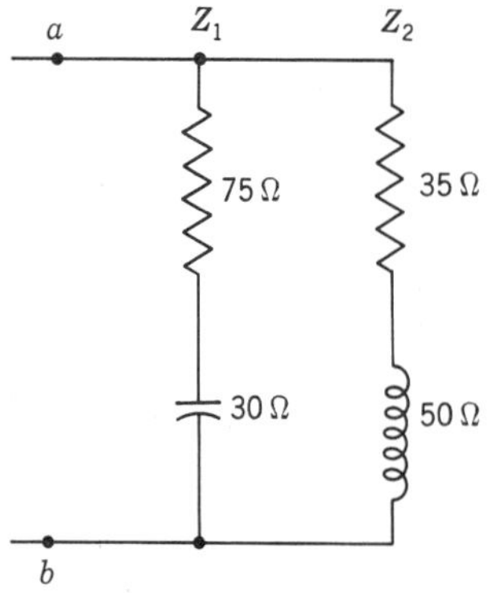

example 5 Find the equivalent impedance of the circuit of Fig. 33-20.

solution Expressing the impedance in rectangular and polar form,

$$Z_1 = 80 + j26 = 84.1\angle 18^\circ\ \Omega$$
$$Z_2 = 0 - j100 = 100\angle -90^\circ\ \Omega$$

Substituting these values in Eq. (6),

$$Z_p = \frac{(84.1\angle 18^\circ)(100\angle -90^\circ)}{(80 + j26) + (0 - j100)} = \frac{8410\angle -72^\circ}{80 - j74}$$

Performing the phasor summation in the

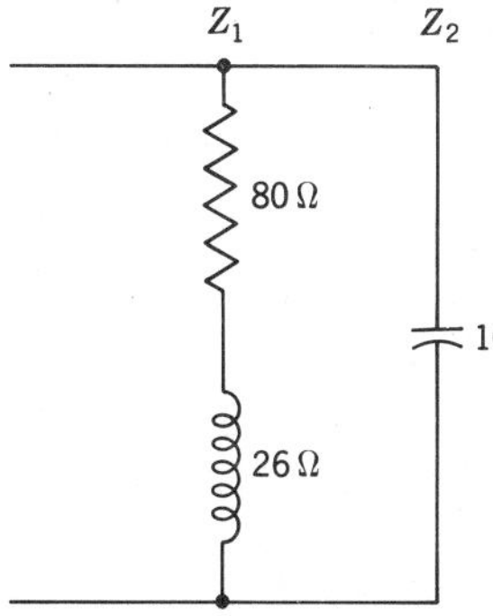

Fig. 33-20 Circuit of Example 5

denominator,

$$Z_p = \frac{8410\angle{-72^\circ}}{109\angle{-42.8^\circ}}$$

$$\therefore Z_p = 77.2\angle{-29.2^\circ}\ \Omega$$

The equivalent series circuit is found by the usual method of converting from rectangular form to polar form, namely,

$$\begin{aligned} 77.2\angle{-29.2^\circ} &= 77.2(\cos 29.2^\circ - j \sin 29.2^\circ) \\ &= 77.2 \cos 29.2^\circ - j77.2 \sin 29.2^\circ \\ &= 67.4 - j37.7\ \Omega \end{aligned}$$

33-9 SERIES-PARALLEL CIRCUITS

An equation for the equivalent impedance of a series-parallel circuit is obtained in the same manner as the equation for the equivalent resistance of a combination of resistances in series and parallel as outlined in Sec. 13-3. For example, in the circuit represented in Fig. 33-21, the total impedance is

$$Z_t = Z_s + \frac{Z_1 Z_2}{Z_1 + Z_2} \qquad (7)$$

example 6 In the circuit of Fig. 33-21, $Z_s = 12.4 + j25.6\ \Omega$, $Z_1 = 45 + j12.9\ \Omega$, and $Z_2 = 35 - j75\ \Omega$. Determine the equivalent impedance of the circuit.

solution Since Z_1 and Z_2 must be multiplied, it is necessary to express them in polar form.

$$Z_1 = 45 + j12.9 = 46.8\angle{16^\circ}\ \Omega$$

and $$Z_2 = 35 - j75 = 82.8\angle{-65^\circ}\ \Omega$$

Substituting the values in Eq. (7),

$$Z_t = (12.4 + j25.6) + \frac{(46.8\angle{16^\circ})(82.8\angle{-65^\circ})}{(45 + j12.9) + (35 - j75)}\ \Omega$$

The solution is completed in the usual manner and results in

$$Z_t = 53.2\angle{20^\circ}\ \Omega$$

From the foregoing examples, it is evident that an equation for the impedance of a network is expressed exactly as in direct-current problems, impedances in polar form being substituted for the resistances.

Fig. 33-21 Series-Parallel Circuit of Example 6

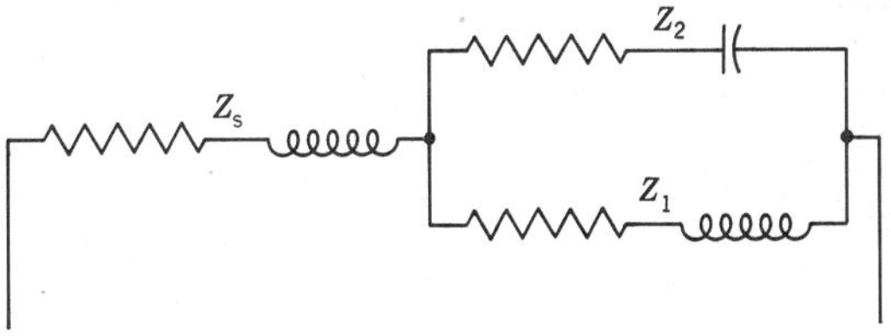

PROBLEMS 33-2

1. What is the equivalent impedance of two impedances $Z_1 = 151\angle{4.07^\circ}\ \Omega$ and $Z_2 = 50\angle{53.1^\circ}\ \Omega$ connected in parallel?
2. What is the equivalent impedance of two impedances $Z_a = 148.5\angle{42.2^\circ}\ \Omega$ and $Z_b = 145\angle{-12.7^\circ}\ \Omega$ connected in parallel?

3. What is the equivalent impedance of two impedances $Z_1 = 73.8 - j34.4\ \Omega$ and $Z_2 = 30 + j40\ \Omega$ connected in parallel?
4. What is the equivalent impedance of two impedances $Z_a = 276 - j180\ \Omega$ and $Z_b = 117 - j18.6\ \Omega$ connected in parallel?
5. What is the equivalent impedance of two impedances $Z_x = 60.5\angle 20^\circ\ \Omega$ and $Z_y = 100 + j0\ \Omega$ connected in parallel?
6. What is the equivalent impedance of two impedances $Z_1 = 355\angle 12^\circ\ \Omega$ and $Z_2 = 0 - j100\ \Omega$ connected in parallel?
7. What is the equivalent impedance of two impedances $Z_z = 251\angle -3^\circ\ \Omega$ and $Z_L = 0 + j70\ \Omega$ connected in parallel?
8. The joint impedance of two parallel impedances is $53.5\angle -42.4^\circ\ \Omega$. One of the impedances is $168\angle 27^\circ\ \Omega$. What is the other?
9. What impedance must be connected in parallel with $64.9 + j45.4\ \Omega$ to produce $43.7 + j155.5\ \Omega$?
10. In Fig. 33-22, $Z_s = 9.4 + j6.6\ \Omega$, $Z_1 = 78.5 - j35\ \Omega$, and $Z_2 = 33.6 + j48\ \Omega$. What is the single equivalent impedance Z_t?
11. In Fig. 33-22, $Z_s = 111.5\angle 21^\circ\ \Omega$, $Z_1 = 27.7 - j50\ \Omega$, and $Z_2 = 150 + j76.2\ \Omega$. What is Z_t?
12. In Fig. 33-22, $Z_s = 5 + j3.9\ \Omega$, $Z_1 = 57.2\angle -61^\circ\ \Omega$, and $Z_2 = 168\angle 27^\circ\ \Omega$. What is Z_t?
13. The primary current I_p of a coupled circuit is expressed by the equation

$$I_p = \frac{E}{Z_p + \dfrac{(\omega M)^2}{Z_s}} \quad \text{A}$$

Compute the value of I_p when $E = 110\angle 0^\circ$ V, $Z_p = 12 + j40\ \Omega$, $Z_s = 18 + j50\ \Omega$, and $\omega M (= 2\pi f \times \text{mutual inductance}) = 15$.
14. The secondary current I_s of a coupled circuit is expressed by the equation

$$I_s = \frac{-j\omega ME}{Z_p Z_s + (\omega M)^2} \quad \text{A}$$

Fig. 33-22 Circuit for Probs. 10, 11, and 12

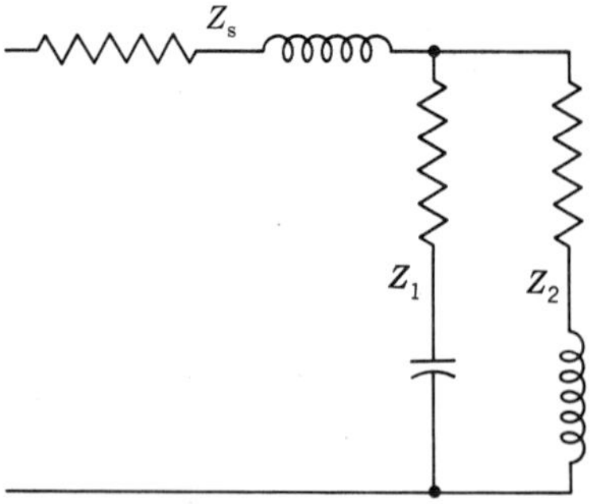

Compute the value of I_s if $\omega M = 15$, $E = 20$ V, $Z_p = 6 + j8\ \Omega$, and $Z_s = 20 + j12\ \Omega$.

33-10 PARALLEL RESONANCE

Communication circuits and electronic networks contain resonant parallel circuits. Figure 33-23 represents a typical parallel circuit consisting of an inductor and capacitor in parallel. The resistance of the capacitor, which is very small, can be neglected, and the resistance R represents the effective resistance of the inductor.

At low frequencies the inductive reactance is a low value whereas the capacitive reactance is high. Hence, a large current flows through the inductive branch and a small current flows through the capacitive branch The phasor sum of these currents causes a large lagging line current which, in effect, results in an equivalent series circuit of low impedance consisting of resistance and inductive reactance. At high frequencies the inductive reactance is large and the capacitive reactance is small. This results in a large leading line current with an attendant equivalent series circuit of low impedance consisting of resistance and capacitive reactance.

There is one frequency, between those mentioned above, at which the lagging component of current through the inductive branch is equal to the leading current through the capacitive branch. This condition results in a small line current that is in phase with the voltage across the parallel circuit and therefore an impedance that is equivalent to a very high resistance.

Fig. 33-23 Parallel LC Circuit. R Represents Effective Resistance of L

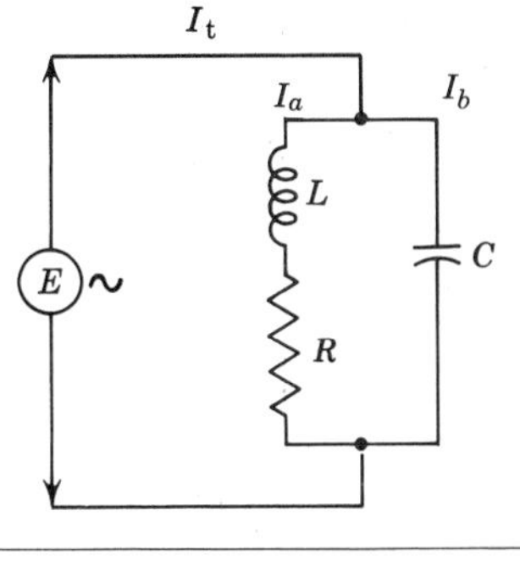

The resonant frequency of a parallel circuit is often a source of confusion to the student studying parallel resonance for the first time. The reason for this is that different definitions for the resonant frequency are encountered in various texts. Thus, the resonant frequency of a parallel circuit can be defined by any one of the following as:

1. The frequency at which the parallel circuit acts as a pure resistance.
2. The frequency at which the line current becomes minimum.
3. The frequency at which the inductive reactance equals the capacitive reactance. This is the same definition as that for the resonant frequency of a series circuit. That is,

$$\omega L = \frac{1}{\omega C}$$

or

$$f_r = \frac{1}{2\pi\sqrt{LC}} \qquad (8)$$

A little consideration of these definitions will convince you that, in high-Q circuits, the three resonant frequencies differ by an amount so small as to be negligible.

In the circuit of Fig. 33-23,

$$I_b = \frac{E}{\dfrac{1}{\omega C}} = \omega CE$$

Also,

$$I_a = \frac{E}{R + j\omega L}$$

Rationalizing (Sec. 20-17),

$$I_a = \frac{E}{R + j\omega L} \cdot \frac{R - j\omega L}{R - j\omega L} = \frac{E(R - j\omega L)}{R^2 + (\omega L)^2}$$

$$= \frac{ER}{R^2 + (\omega L)^2} - j\frac{\omega LE}{R^2 + (\omega L)^2}$$

In order to satisfy the first definition for resonant frequency, the line current must be in phase with the applied voltage; that is, the out-of-phase, or quadrature, component of the current through the inductive branch must be equal to the current through the capacitive branch. Thus,

$$\frac{\omega LE}{R^2 + (\omega L)^2} = \omega CE$$

D: ωE,

$$\frac{L}{R^2 + (\omega L)^2} = C$$

M: $[R^2 + (\omega L)^2]$,

$$L = [R^2 + (\omega L)^2]C \qquad (9)$$

or

$$\frac{L}{C} - R^2 = (\omega L)^2$$

Hence,

$$\omega = \frac{\sqrt{\frac{L}{C} - R^2}}{L}$$

$$= \sqrt{\frac{1}{LC} - \frac{R^2}{L^2}}$$

Substituting $2\pi f$ for ω,

$$2\pi f = \sqrt{\frac{1}{LC} - \frac{R^2}{L^2}}$$

Thus, the resonant frequency is

$$f = \frac{1}{2\pi}\sqrt{\frac{1}{LC} - \frac{R^2}{L^2}} \qquad (10)$$

If the Q of the inductance is at all large, then $\omega L \gg R$, which, for all practical purposes, makes the term $\frac{R^2}{L^2}$ in Eq. (10) of such low value that it can be neglected, and Eq. (10) is thus reduced to Eq. (8).

Work out several examples with different circuit values, and compare the resonant frequencies obtained from the formulas. In this connection, it is left to you as an exercise to show that in a parallel-resonant circuit, as represented in Fig. 33-23, the line current and applied voltage will be in phase (unity power factor) when

$$R^2 = X_L(X_C - X_L) \qquad (11)$$

33-11 IMPEDANCE OF PARALLEL-RESONANT CIRCUITS

When a parallel circuit is operating at the frequency at which the circuit acts as a pure resistance, the circuit has unity PF and the line current I_t (Fig. 33-23) consists of the in-phase component of I_a. That is,

$$I_t = \frac{ER}{R^2 + (\omega L)^2} \quad \text{A} \qquad (12)$$

Then, since $\qquad Z_t = \frac{E}{I_t} \quad \Omega$

substituting in Eq. (12) for I_t,

$$\frac{E}{Z_t} = \frac{ER}{R^2 + (\omega L)^2}$$

Hence,

$$Z_t = \frac{R^2 + (\omega L)^2}{R} \quad \Omega \qquad (13)$$

From Eq. (9),

$$R^2 + (\omega L)^2 = \frac{L}{C}$$

Substituting this value in Eq. (13),

$$Z_t = \frac{L}{CR} \quad \Omega \qquad (14)$$

example 7 In the circuit of Fig. 33-23, let $L = 203\ \mu\text{H}$, $C = 500\ \text{pF}$, and $R = 6.7\ \Omega$. (*a*) What is the resonant frequency of the circuit? (*b*) What is the impedance of the circuit at resonance?

solution

(*a*)
$$f = \frac{1}{2\pi\sqrt{LC}}$$
$$= \frac{1}{2\pi\sqrt{2.03 \times 10^{-4} \times 5 \times 10^{-10}}}$$
$$= 500\ \text{kHz}$$

(*b*)
$$Z_t = \frac{L}{CR}$$
$$= \frac{203 \times 10^{-6}}{500 \times 10^{-12} \times 6.7}$$
$$= \frac{203}{5 \times 6.7} \times 10^4$$
$$= 60.6\ \text{k}\Omega$$

If the value of C is unknown, Eq. (14) can be used in different form. Thus, by multiplying both numerator and denominator by ω,

$$Z_t = \frac{\omega L}{\omega CR} = \frac{1}{\omega C}\frac{\omega L}{R}$$

Since at resonance,

$$\omega L = \frac{1}{\omega C}$$

then
$$Z_t = \frac{(\omega L)^2}{R} \quad \Omega \qquad (15)$$

Moreover, since

$$Q = \frac{\omega L}{R}$$

substituting in Eq. (15),

$$Z_t = \omega L Q \quad \Omega \qquad (16)$$

example 8 In the circuit of Fig. 33-23, let $L = 70.4\ \mu\text{H}$ and $R = 5.31\ \Omega$. If the resonant frequency of the circuit is 1.2 MHz, determine (*a*) the impedance of the circuit at resonance and (*b*) the capacitance of the capacitor.

solution

$$f = 1.2\ \text{MHz} = 1.2 \times 10^6\ \text{Hz}$$
$$\omega = 2\pi f = 2\pi \times 1.2 \times 10^6 = 7.54 \times 10^6$$

(*a*)
$$Z_t = \frac{(\omega L)^2}{R} = \frac{(7.54 \times 10^6 \times 70.4 \times 10^{-6})^2}{5.31}$$
$$= 53.1\ \text{k}\Omega$$

(*b*) Since, at resonance, $\omega L = \dfrac{1}{\omega C}$ and

$\omega L = 531\ \Omega$,

then
$$\frac{1}{\omega C} = 531\ \Omega$$

Hence,
$$C = \frac{1}{531\omega} = 250\ \text{pF}$$

What is the Q of this circuit?

PROBLEMS 33-3

1. An inductor of $16\ \mu\text{H}$ and a capacitor of 50 pF are connected in parallel as shown in Fig. 33-23. If the effective resistance of the coil is $22\ \Omega$, find:
   (*a*) The resonant frequency of the circuit according to definition 1 (Sec. 33-10).
   (*b*) The resonant frequency according to definition 3.
   (*c*) The Q of the coil by using the frequency of part (*b*).
2. Repeat Prob. 1 for an effective resistance of the coil of $44\ \Omega$.

3. An inductor of 10 mH with a Q of 800 is connected in parallel with a 200-pF capacitor.
   (*a*) What is the resonant frequency of the circuit?
   (*b*) What is the impedance of the circuit at resonance?
   (*c*) What is the effective resistance of the inductor?
4. If the circuit of Prob. 3 is energized with 600 V at the resonant frequency, how much power will it absorb?
5. A coil with a Q of 71.6 is connected in parallel with a capacitor, and this circuit resonates at 356 kHz. The impedance at resonance is found to be 64 kΩ. What is the value of the capacitor?
6. An inductor is connected in parallel with a 254-pF capacitor, and the circuit is found to resonate at 999 kHz. A circuit magnification meter indicates that the Q of the inductor is 90.
   (*a*) What is the value of the inductance?
   (*b*) What is the effective resistance of the inductor?
   (*c*) What is the impedance of the circuit at resonance?
7. If the circuit of Prob. 6 is connected to 20 V at the resonant frequency, how much power will it absorb?
8. If the circuit of Prob. 6 is connected to a 20-V source at 499 kHz, (*a*) how much power will it absorb and (*b*) what will be the PF of the circuit?
9. If the circuit of Prob. 6 is connected to a 20-V source at 1499 kHz, what will be the PF?
10. An inductor with a measured Q of 100 resonates with a capacitor at 7.496 MHz with an impedance of 65.9 kΩ. What is the value of the inductance?
11. What is the capacitance of the test capacitor in Prob. 10?
12. 18.9 mA is the total current drain when a capacitor is in resonance with an inductor at 1.5 MHz and the parallel circuit is energized with a 1-kV source. The Q of the inductor is measured at 99.7. What is the value of the capacitor?

33-12 EQUIVALENT Y AND Δ CIRCUITS

When networks contain complex impedances, the equations for converting from a Δ network to an equivalent Y network, or vice versa, are derived by methods identical with those of Sec. 22-7. Thus, in Fig. 33-24, each equivalent Y impedance is equal to the product of the two *adjacent* Δ impedances divided by the summation of the Δ impedances, or

Fig. 33-24 Equivalent Y and Δ Impedances

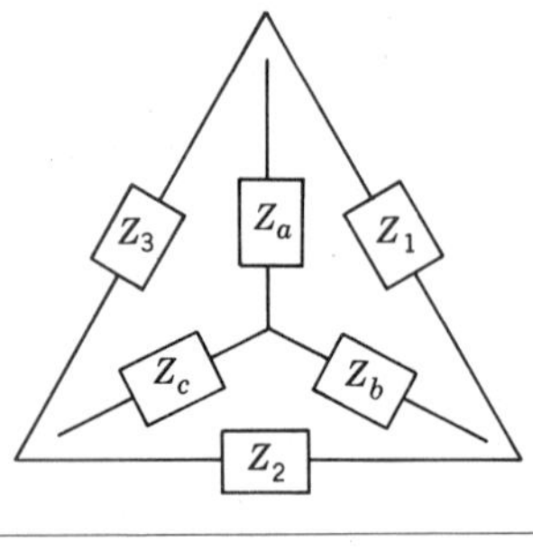

$$Z_a = \frac{Z_1 Z_3}{\Sigma Z_\Delta} \tag{17}$$

$$Z_b = \frac{Z_1 Z_2}{\Sigma Z_\Delta} \tag{18}$$

and
$$Z_c = \frac{Z_2 Z_3}{\Sigma Z_\Delta} \tag{19}$$

where
$$\Sigma Z_\Delta = Z_1 + Z_2 + Z_3$$

and all impedances are expressed in polar form.

Similarly, each equivalent Δ impedance is equal to the summation of the Y impedances divided by the *opposite* Y impedance. Thus,

$$Z_1 = \frac{\Sigma Z_Y}{Z_c} \tag{20}$$

$$Z_2 = \frac{\Sigma Z_Y}{Z_a} \tag{21}$$

and
$$Z_3 = \frac{\Sigma Z_Y}{Z_b} \tag{22}$$

where
$$\Sigma Z_Y = Z_a Z_b + Z_b Z_c + Z_a Z_c$$

and all impedances are expressed in polar form.

example 9 In Fig. 33-24,

$$Z_1 = 7.07 + j7.07\ \Omega$$
$$Z_2 = 4 + j3\ \Omega$$
and
$$Z_3 = 6 - j8\ \Omega$$

What are the values of the equivalent Y circuit?

solution Express all impedances in both rectangular and polar forms.

$$Z_1 = 7.07 + j7.07 = 10\angle 45^\circ\ \Omega$$
$$Z_2 = 4 + j3 = 5\angle 36.9^\circ\ \Omega$$
$$Z_3 = 6 - j8 = 10\angle -53.1^\circ\ \Omega$$
$$\Sigma Z_\Delta = (7.07 + j7.07) + (4 + j3) + (6 - j8)$$
$$= 17.2\angle 6.91^\circ\ \Omega$$

Substituting in Eq. (17),

$$Z_a = \frac{(10\angle 45^\circ)(10\angle -53.1^\circ)}{17.2\angle 6.91^\circ}$$
$$= 5.62 - j1.51\ \Omega$$

Substituting in Eq. (18),

$$Z_b = \frac{(10\angle 45^\circ)(5\angle 36.9^\circ)}{17.2\angle 6.91^\circ}$$
$$= 0.752 + j2.81\ \Omega$$

Substituting in Eq. (19),

$$Z_c = \frac{(5\angle 36.9^\circ)(10\angle -53.1^\circ)}{17.2\angle 6.91^\circ}$$
$$= 2.67 - j1.14\ \Omega$$

The solution can be checked by converting the above Y-network equivalents back to the original Δ by using Eqs. (20), (21), and (22).

example 10 Determine the equivalent impedance between points *a* and *c* in Fig. 33-25.

solution Convert one of the Δ circuits of Fig. 33-25 to its equivalent Y circuit. Thus, for the delta *abd*,

Fig. 33-25 Circuit of Example 19

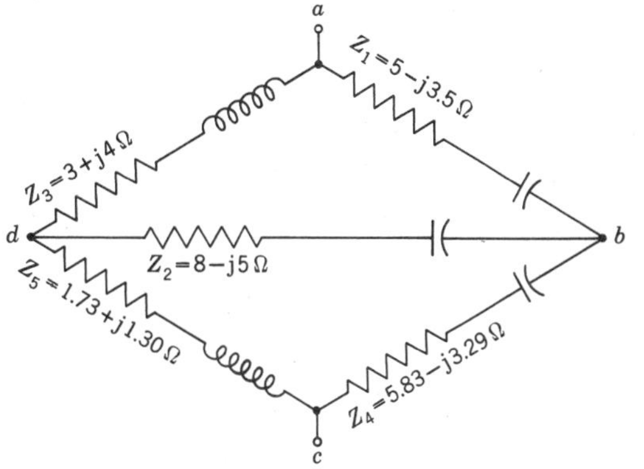

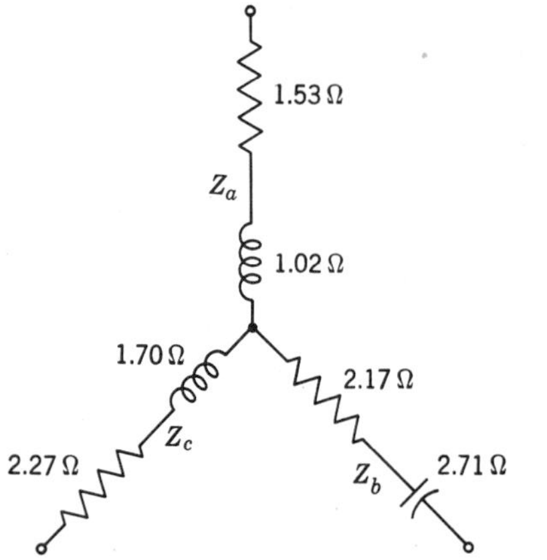

Fig. 33-26 Equivalent Y Impedances for Circuit of Fig. 33-25

$$Z_1 = 5 - j3.5 = 6.1\underline{/-35^\circ}\ \Omega$$
$$Z_2 = 8 - j5 = 9.44\underline{/-32^\circ}\ \Omega$$
$$Z_3 = 3 + j4 = 5\underline{/53.1^\circ}\ \Omega$$
$$\Sigma Z_\Delta = (5 - j3.5) + (8 - j5) + (3 + j4) = 16.6\underline{/-15.7^\circ}\ \Omega$$

Substituting in Eq. (17),

$$Z_a = \frac{(6.1\underline{/-35^\circ})(5\underline{/53.1^\circ})}{16.6\underline{/-15.7^\circ}} = 1.84\underline{/33.8^\circ} = 1.53 + j1.02\ \Omega$$

Substituting in Eq. (18),

$$Z_b = \frac{(6.1\underline{/-35^\circ})(9.44\underline{/-32^\circ})}{16.6\underline{/-15.7^\circ}} = 3.47\underline{/-51.3^\circ} = 2.17 - j2.71\ \Omega$$

Substituting in Eq. (19),

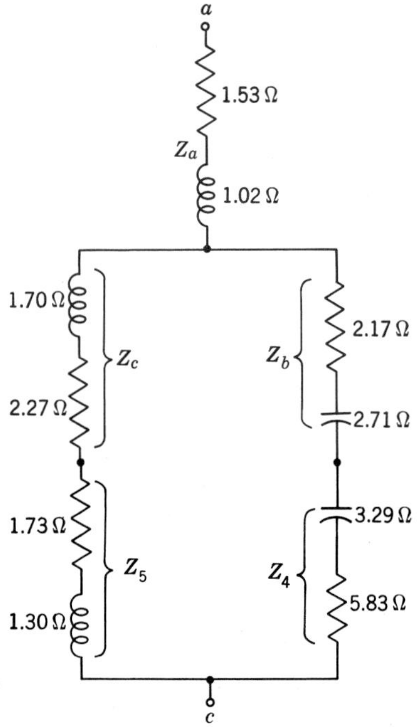

Fig. 33-27 Equivalent Y Impedances Connected to Remainder of Circuit of Fig. 33-25

$$Z_c = \frac{(9.44\underline{/-32^\circ})(5\underline{/53.1^\circ})}{16.6\underline{/-15.7^\circ}} = 2.84\underline{/36.8^\circ} = 2.27 + j1.70\ \Omega$$

The equivalent Y impedances are shown in Fig. 33-26.

The equivalent Y impedances are connected to the remainder of the circuit as shown in Fig. 33-27 and solved as an ordinary series-parallel circuit. See Eq. Fig. 33-1.

Eq. Fig. 33-1

$$Z_{ac} = Z_a + \frac{(Z_c + Z_5)(Z_b + Z_4)}{Z_c + Z_5 + Z_b + Z_4}$$
$$= 1.53 + j1.02 + \frac{[(2.27 + j1.70) + (1.73 + j1.30)][(2.17 - j2.71) + (5.83 - j3.29)]}{(2.27 + j1.70) + (1.73 + j1.30) + (2.17 - j2.71) + (5.83 - j3.29)}$$
$$= 5.45 + j2.0\ \Omega$$

As we saw in Sec. 22-7, the Δ network is more generally referred to in electronics as a π network and the Y or star network is often known as the T network. In the problems which follow, the two sets of expressions are used interchangeably.

PROBLEMS 33-4

1. In the circuit of Fig. 33-24, $Z_1 = 20 + j30\ \Omega$, $Z_2 = 25 + j50\ \Omega$, $Z_3 = 30 - j10\ \Omega$. Find the impedances of the equivalent Y circuit.
2. In the circuit of Fig. 33-24, $Z_1 = 3 + j4\ \Omega$, $Z_2 = 12 + j5\ \Omega$, $Z_3 = 8 - j6\ \Omega$. Find the equivalent Y-circuit values.
3. In the circuit of Fig. 33-24,

$$Z_a = 46.4\angle 75.55^\circ\ \Omega$$
$$Z_b = 43.8\angle -45.45^\circ\ \Omega$$
$$Z_c = 56.4\angle -37.45^\circ\ \Omega$$

Find the impedances of the equivalent π circuit.

4. In the circuit of Fig. 33-24, $Z_a = 50.9\angle 86.8^\circ\ \Omega$, $Z_b = 62.7\angle -20.2^\circ\ \Omega$, and $Z_c = 44.5\angle 8.8^\circ\ \Omega$. Find the equivalent Δ-circuit values.
5. In the circuit of Fig. 33-28, $Z_1 = 78\angle 22.6^\circ\ \Omega$, $Z_2 = 80\angle -53.1^\circ\ \Omega$, $Z_3 = 50\angle 45^\circ\ \Omega$, $Z_4 = 39\angle -67.4^\circ\ \Omega$, and $Z_5 = 100\angle 36.9^\circ\ \Omega$. Find Z_{ab}.
6. In Prob. 5, if $E = 100\angle 0^\circ$ V, find the current flow through impedance Z_4.
7. In the circuit of Fig. 33-28, $Z_1 = 102 + j190\ \Omega$, $Z_2 = 134 - j33\ \Omega$, $Z_3 = 380 - j210\ \Omega$, $Z_4 = 30 - j40\ \Omega$, and $Z_5 = 80 - j60\ \Omega$. What is the equivalent impedance Z_{ab}?
8. In Prob. 7, if $E = 440$ V, how much current flows through Z_5?
9. In Prob. 7, if $E = 200$ V, how much power is expended in Z_4?
10. In Prob. 7, if $E = 200$ V, how much current flows through Z_2?
11. In Fig. 33-28, $Z_1 = 90 - j120\ \Omega$, $Z_2 = 115 - j18\ \Omega$, $Z_3 = 168 - j58\ \Omega$, $Z_4 = 50 + j0\ \Omega$, and $Z_5 = 0 + j25\ \Omega$. Determine the equivalent impedance Z_{ab}.
12. In Prob. 11, if $E = 100$ V, how much current flows through Z_5?
13. In Prob. 11, if $E = 100$ V, how much power is expended in Z_1?
14. In Fig. 33-29, $Z_1 = 3 + j4\ \Omega$, $Z_2 = 37\angle 77.5^\circ\ \Omega$, $Z_3 = 40\angle -80^\circ\ \Omega$,

Fig. 33-28 Circuit for Probs. 5 to 13

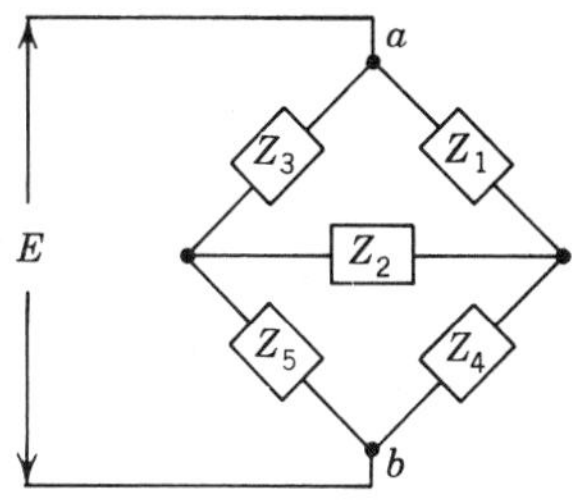

Fig. 33-29 Circuit for Prob. 14

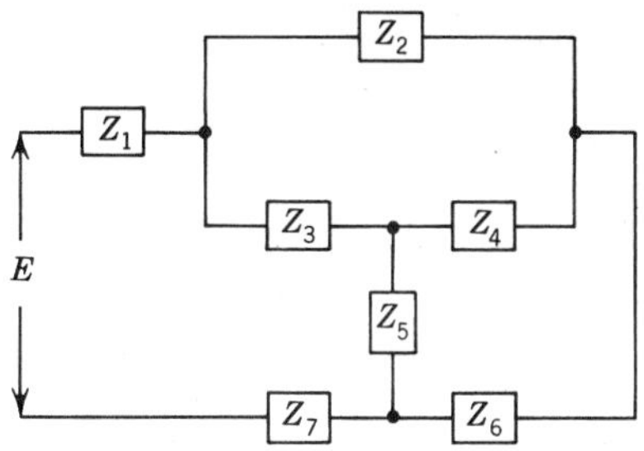

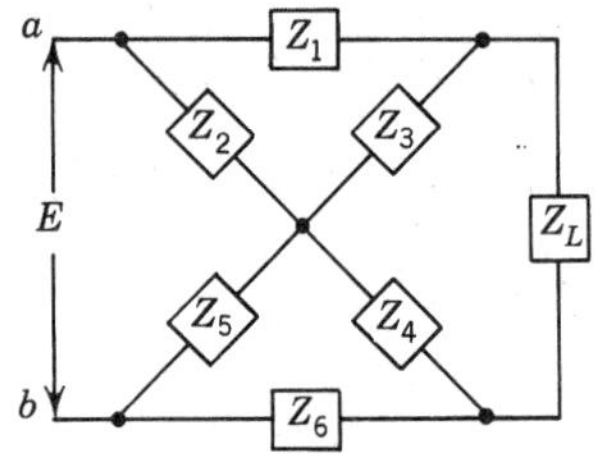

Fig. 33-30 Circuit for Probs. 15 to 18

$Z_4 = 64 - j50\ \Omega$, $Z_5 = 15 + j85\ \Omega$, $Z_6 = 40 - j36\ \Omega$, $Z_7 = 10\angle -53.1°\ \Omega$, and $E = 120$ V. How much current flows through Z_7?

15. In Fig. 33-30, $Z_1 = 254\angle 88.6°\ \Omega$, $Z_2 = 306\angle 86.1°\ \Omega$, $Z_3 = 437\angle -73.6°\ \Omega$, $Z_4 = 177\angle -87°\ \Omega$, $Z_5 = 288\angle 87.5°\ \Omega$, $Z_6 = 250\angle 89.1°\ \Omega$, and $Z_L = 680\angle 0°\ \Omega$. Determine the equivalent impedance Z_{ab}.

16. In Prob. 15, if $E = 475$ V, how much current flows through the load impedance Z_L?

17. In Fig. 33-30, $Z_1 = 63 + j5\ \Omega$, $Z_2 = 12 + j60\ \Omega$, $Z_3 = 20 + j90\ \Omega$, $Z_4 = 18 + j86\ \Omega$, $Z_5 = 8 + j52\ \Omega$, $Z_6 = 47 + j2\ \Omega$, $Z_L = 600 + j0\ \Omega$. Determine the equivalent impedance Z_{ab}.

18. In Prob. 17, if $E = 135$ V, how much power is dissipated in the load impedance Z_L?

34

logarithms

In problems pertaining to engineering, there often occurs the need for numerical computations involving multiplication, division, powers, or roots. Some of these problems can be solved more readily by the use of logarithms than by ordinary arithmetical processes.

The credit for the invention of logarithms is chiefly due to John Napier, whose tables appeared in 1614. This was an extremely important event in the development of mathematics; for by the use of logarithms:

1. Multiplication is reduced to addition.
2. Division is reduced to subtraction.
3. Raising to a power is reduced to one multiplication.
4. Extracting a root is reduced to one division.

In some phases of engineering, computation by logarithms is utilized to a great extent because of the high degree of accuracy desired and the amount of labor that is thereby saved. Because the slide rule is convenient and because slide-rule results meet the ordinary demands for accuracy in problems relating to electronics, it is not necessary to make wide use of logarithms for computations in the general field. However, it is essential that the electrical engineer and, more particularly, the electronics engineer have a thorough understanding of logarithmic processes.

34-1 DEFINITION

The *logarithm* of a quantity is the exponent of the power to which a given number, called the *base*, must be raised in order to equal the quantity.

example 1 Since $10^3 = 1000$, then $3 =$ logarithm of 1000 to the base 10.

example 2 Since $2^3 = 8$, then $3 =$ logarithm of 8 to the base 2.

example 3 Since $a^x = b$, then $x =$ logarithm of b to the base a.

34-2 NOTATION

If

$$b^x = N \tag{1}$$

then x is the logarithm of N to the base b. It may be helpful to mentally translate this expression to "x is the power to which b must be raised to obtain N." This statement is abbreviated by writing

$$x = \log_b N \tag{2}$$

It is evident that Eqs. (1) and (2) mean the same thing and are simply different methods of expressing the same relation among b, x, and N. Equation (1) is called the *exponential form*, and Eq. (2) is called the *logarithmic form*.

As an aid in remembering that *a logarithm is an exponent*, Eq. (1) can be written in the form

$$(\text{Base})^{\log} = \text{number}$$

The following example illustrates relations between exponential and logarithmic forms.

example 4

Exponential Notation	*Logarithmic Notation*
$2^4 = 16$	$4 = \log_2 16$
$3^5 = 243$	$5 = \log_3 243$
$25^{0.5} = 5$	$0.5 = \log_{25} 5$
$10^2 = 100$	$2 = \log_{10} 100$
$10^4 = 10\,000$	$4 = \log_{10} 10\,000$
$a^b = c$	$b = \log_a c$
$\varepsilon^x = y$	$x = \log_\varepsilon y$

From the foregoing examples, it is apparent that any positive number, other than 1, can be selected as a base for a system of logarithms. Because 1 raised to any power is 1, it cannot be used as a base.

Based on the definitions in Eqs. (1) and (2), you should satisfy yourself with the correctness of the following statement:

$$\log_a a^b = b$$

PROBLEMS 34-1

Express the following equations in logarithmic form:

1. $10^2 = 100$
2. $10^3 = 1000$
3. $7^2 = 49$
4. $4^3 = 64$
5. $4^{0.5} = 2$
6. $\varepsilon^1 = \varepsilon$
7. $a^1 = a$
8. $10^1 = 10$
9. $a^0 = 1$
10. $1 = 10^0$

Express the following equations in exponential form:

11. $3 = \log_{10} 1000$
12. $5 = \log_{10} 100\,000$
13. $2 = \log_5 25$
14. $3 = \log_4 64$
15. $0 = \log_6 1$
16. $0 = \log_a 1$
17. $4 = \log_5 625$
18. $0.5 = \log_9 3$
19. $s = \log_r t$
20. $2x = \log_3 M$

Find the value of x:

21. $3^x = 9$
22. $2^x = 16$
23. $10^x = 1\,000\,000$
24. $x = \log_2 32$
25. $4^x = 2$
26. $\log_8 x = 3$
27. Show that $\log_{10} 100 = \log_{10} 100\,000 - \log_{10} 1000$.
28. Show that $\log_p p = 1$.

29. What are the logarithms to the base 2 of 2, 4, 8, 16, 32, 64, 128, 256, and 512?

30. What are the logarithms to the base 3 of 3, 9, 27, 81, 243, 729, and 2187?

34-3 LOGARITHM OF A PRODUCT

The logarithm of a product is equal to the sum of the logarithms of the factors.

Consider the two factors M and N, and let x and y be their respective logarithms to the base a; then,

$$x = \log_a M \quad (3)$$

and

$$y = \log_a N \quad (4)$$

Writing Eq. (3) in exponential form,

$$a^x = M \quad (5)$$

Writing Eq. (4) in exponential form,

$$a^y = N \quad (6)$$

Then

$$M \cdot N = a^x \cdot a^y = a^{x+y}$$

$$\therefore \log_a (M \cdot N) = x + y = \log_a M + \log_a N$$

example 5

$$2 = \log_{10} 100 \qquad \text{or} \qquad 10^2 = 100$$

$$4 = \log_{10} 10\,000 \qquad \text{or} \qquad 10^4 = 10\,000$$

Then

$$100 \times 10\,000 = 10^2 \cdot 10^4$$

$$= 10^{2+4} = 10^6$$

$$\therefore \log_{10} (100 \times 10\,000) = 2 + 4$$

$$= \log_{10} 100 + \log_{10} 10\,000$$

The above proposition is also true for the product of more than two factors. Thus, by successive applications of the proof, it can be shown that

$$\log_a (A \cdot B \cdot C \cdot D) = \log_a A + \log_a B + \log_a C + \log_a D$$

34-4 LOGARITHM OF A QUOTIENT

The logarithm of the quotient of two numbers is equal to the logarithm of the dividend minus the logarithm of the divisor.

As in Sec. 34-3, let

$$x = \log_a M \quad (3)$$

and

$$y = \log_a N \quad (4)$$

Writing Eq. (3) in exponential form,

$$a^x = M \quad (5)$$

Writing Eq. (4) in exponential form,

$$a^y = N \quad (6)$$

Dividing Eq. (5) by Eq. (6),

$$\frac{a^x}{a^y} = \frac{M}{N}$$

That is,

$$a^{x-y} = \frac{M}{N} \quad (7)$$

Writing Eq. (7) in logarithmic form,

$$x - y = \log_a \frac{M}{N} \quad (8)$$

Substituting in Eq. (8) for the values of x and y,

$$\log_a M - \log_a N = \log_a \frac{M}{N}$$

example 6

$$2 = \log_{10} 100 \quad \text{or} \quad 10^2 = 100$$
$$4 = \log_{10} 10\,000 \quad \text{or} \quad 10^4 = 10\,000$$

Then $\dfrac{10\,000}{100} = \dfrac{10^4}{10^2} = 10^{4-2} = 10^2$

$$\therefore \log_{10} \frac{10\,000}{100} = 4 - 2 = \log_{10} 10\,000 - \log_{10} 100$$

34-5 LOGARITHM OF A POWER

The logarithm of a power of a number equals the logarithm of the number multiplied by the exponent of the power.

Again, let $\quad x = \log_a M \qquad (3)$

Then $\quad M = a^x \qquad (9)$

Raising both sides of Eq. (9) to the nth power,

$$M^n = a^{nx} \qquad (10)$$

Writing Eq. (10) in logarithmic form,

$$\log_a M^n = nx \qquad (11)$$

Substituting in Eq. (11) for the value of x,

$$\log_a M^n = n \log_a M$$

example 7

$$2 = \log_{10} 100 \quad \text{or} \quad 100 = 10^2$$

Since $\quad (10^2)^2 = 10^{2 \cdot 2} = 10^4 = 10\,000$

then $\quad \log_{10} 10\,000 = 4$

$$\therefore \log_{10} 100^2 = 2 \log_{10} 100 = 2 \cdot 2 = 4$$

34-6 LOGARITHM OF A ROOT

The logarithm of a root of a number is equal to the logarithm of the number divided by the index of the root.

Again, let $\quad x = \log_a M \qquad (3)$

Then $\quad M = a^x \qquad (9)$

Extracting the nth root of both sides of Eq. (9),

$$M^{1/n} = a^{x/n} \qquad (12)$$

Writing Eq. (12) in logarithmic form,

$$\log_a M^{1/n} = \frac{x}{n} \qquad (13)$$

Substituting in Eq. (13) for the value of x,

$$\log_a M^{1/n} = \frac{\log_a M}{n}$$

example 8

$$4 = \log_{10} 10\,000 \quad \text{or} \quad 10\,000 = 10^4$$

Since $\quad \sqrt{10\,000} = \sqrt{10^4} = 10^{4/2} = 10^2 = 100$

then $\quad \log_{10} \sqrt{10\,000} = \dfrac{\log_{10} 10\,000}{2} = \dfrac{4}{2} = 2$

34-7 SUMMARY

It is evident that if the logarithms of numbers are used for computations instead of the numbers themselves, then *multiplication, division, raising to powers,* and *extracting roots* are replaced by *addition, subtraction, multiplication,* and *division,* respectively. Because you are familiar with the laws of exponents, especially as applied to the powers of 10, the foregoing operations with logarithms involve no new ideas. The sole idea behind logarithms is that every positive number can be expressed as a power of some base. That is,

$$\text{Any positive number} = (\text{base})^{\log}$$

34-8 THE COMMON SYSTEM OF LOGARITHMS

Since 10 is the base of our number systems, both integral and decimal, the base 10 has been chosen for a system of logarithms. This system is called the *common system* or *Briggs's system*. The natural system, of which the base to five decimal places is 2.718 28, will be discussed later.

Hereafter, when no other base is stated, the base will be 10. For example, $\log_{10} 625$ will be written log 625, the base 10 being understood.

34-9 THE NATURAL SYSTEM OF LOGARITHMS

In the number system there exist certain special numbers whose value is not absolutely determined, but which are themselves extremely valuable to us. You are already familiar with π, which has a value of approximately $\frac{22}{7}$.

Another useful number is ε, which has a value of approximately 2.718 28. This unusual number turns out to be extremely valuable when used as a base for logarithms. Because it can be shown to be related to *natural* events, like the decay of charge on a capacitor which is discharged through a resistor or the decay of current when the magnetic field about an inductance collapses, it is called the base of the *natural logarithms*. Tables of natural logarithms, or logarithms to the base ε, are to be found in many published books of tables. In Sec. 34-25 we will see how to change logarithms to the base 10 into logarithms to the base ε or to other bases.

The notation for logarithms to the base ε is shown variously as $\log_\varepsilon$ or ln (pronounced "lon").

34-10 DEVELOPING A TABLE OF LOGARITHMS

Table 34-1 illustrates the connection between the power of 10 and the logarithms of certain numbers.

Table 34-1

Exponential Form	*Logarithmic Form*
$10^4 = 10\,000$	$\log 10\,000 = 4$
$10^3 = 1\,000$	$\log 1\,000 = 3$
$10^2 = 100$	$\log 100 = 2$
$10^1 = 10$	$\log 10 = 1$
$10^0 = 1$	$\log 1 = 0$
$10^{-1} = 0.1$	$\log 0.1 = -1$
$10^{-2} = 0.01$	$\log 0.01 = -2$
$10^{-3} = 0.001$	$\log 0.001 = -3$
$10^{-4} = 0.0001$	$\log 0.0001 = -4$

Inspection of Table 34-1 shows that only powers of 10 have integers for logarithms. Also, it is evident that the logarithm of any number between 10 and 100, for example, is between 1 and 2; that is, it is 1 plus a decimal. Similarly, the logarithm of any number between 100 and 1000 is between 2 and 3, and so on. Therefore, to represent all numbers, it is necessary for us to develop the fractional powers which represent numbers between 1 and 10. Then, by using powers of 10 to convert any number to a number between 1 and 10 times the appropriate power of 10 (Chap. 6), we may use our new fractional powers of 10 to find the logarithm of any number instead of just integral powers of 10.

In Sec. 20-4 we saw that $a^{1/2} = \sqrt{a}$. Accordingly, we can see that

$$10^{0.5} = 10^{1/2} = \sqrt{10} = 3.162\,277\,66$$

which gives us the first intermediate step in our table of logarithms between 1 and 10:

$$\log_{10} 3.162\,277\,66 = 0.5$$

Similarly,

$$10^{0.25} = (10^{0.5})^{0.5} = \sqrt{3.162\,277\,66} = 1.778\,279$$

or $\quad \log_{10} 1.778\,279 = 0.25$

By repeating the square roots time after time, we can obtain

$$\log_{10} 1.334 = 0.125 \qquad \text{etc.}$$

Then, by applying the laws of exponents developed in Sec. 4-3 and summarized in Sec. 20-1, we can determine that

$$3.162\,277\,66 \times 1.778\,279 = 10^{0.5} \times 10^{0.25}$$
$$= 10^{0.75} = 5.623\,41$$

or

$$\log 5.623\,41 = 0.75$$

Repeated applications of this method gives us such additional logarithms as

$$\log 4.2173 = 0.625$$

and

$$\log 2.37 = 0.375$$

You should use the values now developed to prove that $10^{0.75} \times 10^{0.25} = 10$, as a check on our method.

These various values can be plotted on a graph, as in Fig. 34-1, and the more convenient logarithms can be picked off the curve, or other more sophisticated methods of higher mathematics may be applied to yield Table 34-2 of logarithms of numbers between 1 and 10:

Table 34-2 LOGARITHMS OF NUMBERS BETWEEN 1 AND 10

Number	*Logarithm*
1	0.000 00
2	0.301 03
3	0.477 12
4	0.602 06
5	0.698 97
6	0.778 15
7	0.845 10
8	0.903 09
9	0.954 24
10	1.000 00

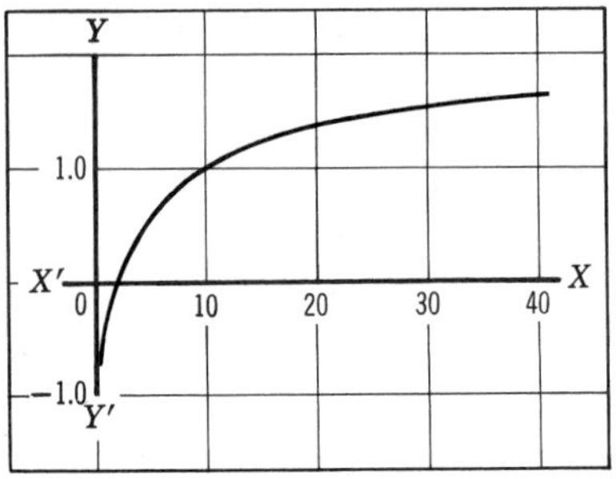

Fig. 34-1 Graph of the Equation $y = \log_{10}x$

Since we convert every number to its equivalent number between 1 and 10 times the appropriate power of 10, every logarithm we will ever look up will be a decimal fraction. Because of this universality of decimals as logarithms, almost every table of logarithms published omits the decimal point: log 2 will appear as simply 301 03 instead of 0.301 03.

From the foregoing discussion it will be evident that every logarithm has two parts: a decimal part which we read from the table of logarithms and an integer which we must provide each time from our knowledge of powers of 10.

example 9 Determine the logarithm of 200.

solution First, rewrite the number in standard form:

$$200 = 2.00 \times 10^2$$

Since $\log 2.00 = 0.301\,03$, this number could be written

$$2.00 \times 10^2 = 10^{0.301\,03} \times 10^2 = 10^{2.301\,03}$$

This power to which 10 is raised to be equal to 200 is the logarithm of 200. In other words,

$$\log 200 = 2.301\,03$$

This logarithm is made up of two parts: the decimal part from the table and the integer part which we developed from the "power of 10." Similarly, log 2000 = 3.301 03.

In the same manner, referring to Table 34-1, it follows that the logarithm of a number between 0.1 and 0.01 will be −2 followed by a decimal number from the table and the logarithm of a number between 0.001 and 0.0001 will be −4 followed by a decimal. Note carefully that the decimal number taken from the table is always positive, so that the logarithm of a number between 0.001 and 0.0001, which we have seen will be −4 followed by a *positive* decimal, may be written as −3 followed by a *negative* decimal.

The integral part of the logarithm, which we provide by ourselves, is called the *characteristic*, and it may be positive, negative, or zero. The fractional part, which is taken from the table, is called the *mantissa*, and it is always *positive*.

34-11 THE CHARACTERISTIC

The use of the base 10 makes it possible to simplify computation by means of logarithms, and to express logarithms in a compact tabular form. Any number, however large or small, may be expressed as a number between 1 and 10, multiplied by the appropriate whole-number power of 10. In Sec. 34-10 we developed a rudimentary table of logarithms of numbers between 1 and 10 and saw that these powers of 10 (logarithms) must be numbers between 0 and 1, usually expressed as decimal fractions, and called *mantissas*.

Let us now turn our attention to logarithms of numbers greater than 10. In Example 9 we converted 200 into 2.00×10^2 and saw that the logarithm of 200 is 2.301 03. The mantissa 0.301 03 is taken from the table to represent the 2—a quantity between 1 and 10. The whole number portion 2, the *characteristic*, is simply the whole-number power of 10 obtained when 200 is written 2.00×10^2.

example 10 What is the logarithm of 50 000?

solution 1

$$\begin{aligned} 50\,000 &= 5.00 \times 10^4 \\ &= 10^{0.698\,97} \times 10^4 \\ &= 10^{4.698\,97} \\ \log 50\,000 &= 4.698\,97 \end{aligned}$$

solution 2 Without actually rewriting 50 000 as 5×10^4, mentally determine the power of 10 necessary to do so. Move the decimal point four places to the right, and record 4. (with the decimal point) as the characteristic. Then determine the mantissa from the table: 0.698 97.

$$\log 50\,000 = 4.698\,97$$

When you are required to determine the log of a number less than 1, follow the same procedure exactly, finding the *negative* whole number power of ten.

example 11 What is the logarithm of 0.005 00?

solution 1

$$\begin{aligned} 0.005\,00 &= 5.00 \times 10^{-3} \\ &= 10^{0.698\,97} \times 10^{-3} \end{aligned}$$

There are several alternative methods of showing the logarithm of a fraction, each being useful for a different class of problem:

solution 2 Directly combine the two parts of the logarithm

$$10^{0.698\,97} \times 10^{-3} = 10^{-2.301\,03}$$

here, $-3.000\,00 + 0.698\,97 = -2.301\,03$, a totally negative number.

$$\log 0.005\,00 = -2.301\,03$$

This *negative logarithm* is the form provided by the better pocket calculators and by such slide rules as the K&E Decilon. It cannot be taken directly from most of the commonly available tables.

solution 3 Combine the two parts, but clearly identify the *negative characteristic* and the *positive mantissa*. This is done by rewriting the -3 characteristic as $\bar{3}$. Pronounced "bar three," it signifies that the characteristic alone is negative when it is combined with the positive mantissa:

$$\log 0.005\,00 = \bar{3}.698\,97$$

solution 4 Present the two parts in a form that keeps them separately identifiable:

$$\log 0.005\,00 = 0.698\,97 - 3$$
$$\text{or} \quad = 7.698\,97 - 10$$
$$\text{or} \quad = 3.698\,97 - 6$$

or any other combination of numbers that has a net positive mantissa 0.698 97 and a net negative characteristic equal to -3 ($7 - 10 = -3$, etc.).

Table 34-3

Number	*Standard Notation*	*Characteristic*
682	6.82×10^2	2
3765	3.765×10^3	3
14	1.4×10^1	1
1	1×10^0	0
0.004 25	4.25×10^{-3}	-3 or $7 - 10$
0.1	1×10^{-1}	-1 or $9 - 10$
0.000 072	7.2×10^{-5}	-5 or $5 - 10$

Rule

When any given number is expressed as a number between 1 and 10, multiplied by the appropriate whole-number power of 10, that whole-number power is the characteristic of the logarithm of the given number.

The foregoing is illustrated in Table 34-3.

34-12 THE MANTISSA

Note that all numbers whose logarithms are given below have the same significant figures. These logarithms were obtained by first finding log 2.207 from a table, as will be discussed later. The remaining logarithms were then obtained by applying the properties of logarithms as stated in Secs. 34-3 and 34-4. (See Eq. Fig. 34-1.)

Eq. Fig. 34-1

$$\log 2207 = \log 1000(2.207) = \log 1000 + \log 2.207 = 3 + 0.343\,80$$
$$\log 220.7 = \log 100(2.207) = \log 100 + \log 2.207 = 2 + 0.343\,80$$
$$\log 22.07 = \log 10(2.207) = \log 10 + \log 2.207 = 1 + 0.343\,80$$
$$\log 2.207 = \log 1(2.207) = \log 1 + \log 2.207 = 0 + 0.343\,80$$
$$\log 0.2207 = \log \frac{2.207}{10} = \log 2.207 - \log 10 = -1 + 0.343\,80$$
$$\log 0.022\,07 = \log \frac{2.207}{100} = \log 2.207 - \log 100 = -2 + 0.343\,80$$

From the above examples, it is apparent that the mantissa is not affected by a shift of the decimal point. That is, *the mantissa of the logarithm of a number depends only on the sequence of the significant figures in the number.* Because of this, 10 is ideally suited as a base for a system of logarithms to be used for computation.

PROBLEMS 34-2

Write the characteristics of the logarithms of the following numbers:

1. 37	**2.** 226	**3.** 688	**4.** 20.6
5. 7.27	**6.** 72.7	**7.** 727	**8.** 0.727
9. 0.000 727	**10.** 958 16	**11.** 95.816	**12.** 0.095 816
13. 1002	**14.** 10.02	**15.** 0.000 100 2	**16.** 1 002 000
17. 0.004	**18.** 2.65×10^6	**19.** 3.3×10^3	**20.** 8×10^{-12}

Find the value of each of the following expressions:

21. $\log 100 + \log 0.001$ **22.** $\log \sqrt{100}$ **23.** $\log \sqrt{\dfrac{1000}{10}}$

24. $\log \sqrt{1000} - \log \sqrt{100}$ **25.** $\log \sqrt{0.001}$

Write the following expressions in expanded form:

26. $\log \dfrac{278 \times 9.36}{81.1}$

SOLUTION: $\log \dfrac{278 \times 9.36}{81.1} = \log 278 + \log 9.36 - \log 81.1$

27. $\log \dfrac{6792 \times 20.9}{176}$ **28.** $\log \dfrac{3.66 \times (4.71 \times 10^2)}{3.42 \times 7280}$

29. $\log \sqrt{\dfrac{512 \times 0.36}{2\pi \times 177}}$ **30.** $\log \sqrt[5]{32\,000 \times 286 \times 159}$

31. $\log \left(\dfrac{159 \times 0.837}{82.2}\right)^3$ **32.** $\log \dfrac{pq^2r}{wy}$

Given $\log 27.36 = 1.437\ 12$, write the logarithms of the following numbers:

33. 2.736	**34.** 2736	**35.** 0.027 36
36. 0.000 273 6	**37.** 27 360	**38.** 2736×10^{-4}
39. 27.36×10^6	**40.** $0.002\ 736 \times 10^{-3}$	**41.** 27.36×10^{-12}

Given log 7.57 = 0.879 10, find the numbers that correspond to the following logarithms:

42. 1.879 10 **43.** 3.879 10 **44.** 5.879 10 − 10
45. 6.879 10 **46.** 9.879 10 − 10 **47.** 3.879 10 − 10
48. 2.879 10 **49.** 10.879 10 **50.** 2.879 10 − 10

34-13 TABLES OF LOGARITHMS

Because the characteristic of the logarithm of any number is obtainable by inspection, it is necessary to tabulate only the mantissas of the logarithms of numbers. Though mantissas can be computed by use of advanced mathematics, for convenience the mantissas of the logarithms to a number of significant figures have been computed and arranged in tables. Table 7 in the Appendix is a five-place table of logarithms; that is, the mantissas therein have been computed and rounded off to four decimal places.

In order for you to learn how to use tables of logarithms, Table 7 is used in the following sections and examples. In addition, inside the front cover of this book is a three-place table of mantissas. You will find that this table will serve most of your needs when working with logarithms related to electronic applications.

34-14 TO FIND THE LOGARITHM OF A GIVEN NUMBER

Table 34-4 is a portion of Table 7 in the Appendix.

Examination of the table shows that the first column has N at the top. N is an abbreviation for "number." The other columns are labeled 0, 1, 2, 3, 4, . . . , 9. Therefore, any number consisting of three significant figures has its first two figures in the N column and its third figure in another column. This will be illustrated in the following examples.

When finding the logarithm of a number, always write the characteristic at once, before looking for the mantissa.

example 12 Find the log 40.

solution $40 = 4 \times 10^1$; therefore, the characteristic is 1.

Since 40 has no third significant figure other than zero, the mantissa of 40 is found at the right of 40 in the N column, in the column headed 0. It is 0.602 06.

$$\therefore \log 40 = 1.602\,06$$

example 13 Find log 416.

solution $416 = 4.16 \times 10^2$; therefore, the characteristic is 2.

The first two digits of 416 are found in the

Table 34-4

N	0	1	2	3	4	5	6	7	8	9
40	60206	60314	60423	60531	60638	60746	60853	60959	61066	61172
41	61278	61384	61490	61595	61700	61805	61909	62014	62118	62221
42	62325	62428	62531	62634	62737	62839	62941	63043	63144	63246
43	63347	63448	63548	63649	63749	63849	63949	64048	64147	64246

N column and the third digit in the column headed 6. Then the mantissa is read in the row containing 41 and in the column headed 6. It is 0.619 09.

$$\therefore \log 416 = 2.619\,09$$

Similarly, $\log 4.16 = 0.619\,09$

$$\log 41.6 = 1.619\,09$$
$$\log 4160 = 3.619\,09$$
$$\log 0.004\,16 = 7.619\,09 - 10, \text{ etc.}$$

That is, the mantissa of any number having 416 as significant figures is 0.619 09.

example 14 Find log 4347.

solution 1 $4347 = 4.347 \times 10^3$; therefore, the characteristic is 3.

Since 4347 is between 4340 and 4350, its mantissa must be between the mantissas of 4340 and 4350.

Mantissa of 4350 =	0.638 49
Mantissa of 4340 =	0.637 49
Difference =	0.001 00

The *tabular difference* between these mantissas is 0.001 00, and it is apparent that an *increase* of 10 in the number causes the mantissa to *increase* by 0.001 00. Therefore, an increase of 7 in the number will increase the mantissa 0.7 as much. Hence the increase in the mantissa will be $0.001\,00 \times 0.7 = 0.000\,70$, and the mantissa of 4347 will be

$$.637\,49 + 0.000\,70 = 0.638\,19$$
$$\therefore \log 4347 = 3.638\,19$$

Similarly, $\log 43.47 = 1.638\,19$

$$\log 4.347 = 0.638\,19$$
$$\log 434\,700 = 5.638\,19$$
$$\log 0.000\,434\,7 = 6.638\,19 - 10, \text{ etc.}$$

That is, the mantissa of any number having 4347 as significant figures is 0.638 19.

The foregoing process of finding the mantissa, called *interpolation*, is based on the assumption that the increase in the logarithm is proportional to the increase in the number.

solution 2 The logarithmic table (Table 7) contains, to the right of the main columns, additional columns headed "Proportional Parts." In these portions of the tables the arithmetic (subtraction and subsequent multiplication in solution 1 of Example 14) has been done for you. The small numbers of the proportional parts combined with the five-figure numbers of the main tables provide the complete logarithmic value.

Find log 4347.

From Table 7, log 434	= 3.637 49
From the same horizontal line, add the proportional part for 7	= 70
Adding, log 4347	= 3.638 19

Further details relating to the use of the proportional parts, especially as they relate to repeated interpolation, will be found in the introductory note headed "Interpolation of Mathematical Tables" immediately preceding Table 9 in the Appendix.

example 15 Find log 0.000 042 735.

solution $0.000\,042\,735 = 4.2735 \times 10^{-5}$; therefore the characteristic is -5, $\bar{5}$, or $5 - 10$.

From Table 7, log 427	= $\bar{5}.630\,43$
Adding the proportional part for 3:	= 31
Adding the proportional part for 5:	= 5 | 1
Adding, log 0.000 042 735	= $\bar{5}.630\,79$ | 1

This mantissa, as written above, is another example of how the retention of decimals might easily give a false impression of accuracy. The table from which the mantissa is taken is correct to five significant figures. Therefore, any mantissa found by interpolation from such a table cannot be correct beyond five significant figures. Hence, it is correct to write

$$\log 0.000\,042\,735 = 5.63079 - 10$$

Summarizing, we have the following:

Rule

To find the logarithm of a number containing three significant figures:

1. Determine the characteristic.
2. Locate the first two significant figures in the column headed N.
3. In the same row and in the column headed by the third significant figure, find the required mantissa.

Rule

To find the logarithm of a number containing more than three significant figures:

1. Determine the characteristic.
2. Find the mantissa for the first three significant figures of the number.
3. Find the next higher mantissa, and take the tabular difference of the two mantissas.
4. Add to the lesser mantissa the product of the tabular difference and the remaining figures of the number considered as a decimal.

note To students who use slide-rule ln scales or engineering-type calculators:
Following the steps of Example 13, we can see that the logarithm of 0.004 27 consists of two parts, 7.6304 and -10. The single number combination of these two parts is $-10.000\,00 + 7.6304 = -2.3696$.

Alternatively, log 0.004 27 may be written $\overline{3}.6304$. Again, the logarithm consists of two parts, -3 and $+0.6304$. These also combine to yield -2.3696.

Punching 0.004 27 into a calculator, such as the Hewlett-Packard engineering model, gives a logarithm reading of $-2.369\,572\,125$, which rounds off to the value above.

Similarly, with the C index of your slide rule set opposite 10 on the black ln 3 scale, and the hairline at 0.004 27 on the red ln 3 scale, we read 2.37 on the C scale. The special notation for the red ln 3 scale indicates that the value on the C scale must be between -1 and -10. Therefore we understand that $\log 0.004\,27 \simeq -2.37$.

These negative logarithms are perfectly legitimate. But if further work is to be done using the tables, it may be necessary to convert them into logarithms with positive mantissas with negative (barred) characteristics.

PROBLEMS 34-3

Find the logarithms of the following numbers:

1. 7	**2.** 700	**3.** 70
4. 263	**5.** 721	**6.** 438
7. 103	**8.** 400	**9.** 382 000
10. 0.000 028 8	**11.** 9264	**12.** 5 989 000
13. 0.1101	**14.** 281 300	**15.** 252.66
16. 989 900	**17.** 3.142×10^{-6}	**18.** 202.8×10^{7}
19. 6.28	**20.** 3.1416	**21.** 2.7183

22. 159.1	**23.** 0.000 471	**24.** 864 000
25. 69 990	**26.** 2 003 000	**27.** 2.003×10^6
28. 0.000 03	**29.** 5×10^{-12}	**30.** 84.37×10^{-5}

34-15 TO FIND THE NUMBER CORRESPONDING TO A GIVEN LOGARITHM

The number corresponding to a given logarithm is called the *antilogarithm* and is written "antilog." For example, if log 692 = 2.840 11, then the number corresponding to the logarithm 2.840 11 is 692. That is,

$$\text{antilog } 2.840\,11 = 692$$

To find the antilog of a given logarithm, we reverse the process of finding the logarithm when the number is given.

example 16 Find the number whose logarithm is 3.910 09.

solution 1 The characteristic tells us only the position of the decimal point. Therefore, to find the significant figures of the number (antilog), the mantissa must be found in Table 7 in the Appendix. To the left of the mantissa 0.910 09, in column N, find the first two significant figures of the number, which are 81, and at the head of the column of the mantissa, find the third significant figure, which is 3. Hence, the number has the significant figures 813. The position of the decimal point is fixed by the characteristic, and because the characteristic is 3, there must be four figures to the left of the decimal point.

Thus, $\quad \text{antilog } 3.910\,09 = 8130$

Similarly,

$$\text{antilog } 0.910\,09 = 8.13$$
$$\text{antilog } 7.910\,09 - 10 = 0.008\,13$$
$$\text{antilog } 6.910\,09 = 8.13 \times 10^6, \text{ etc.}$$

A change in the characteristic changes only the position of the decimal point.

Note how an electronic calculator performing this operation in "scientific notation" reads out the antilogarithm as a number between 1 and 10, multiplied by the appropriate power of 10.

solution 2 Table 8, headed "Common Antilogarithms," gives the antilogarithms to the base 10. Note that it is shaded to distinguish it from Table 7. In the left-hand column, find 91, and on the 91 line, from the 0 column, record 8.12831. Then add the proportional parts:

Antilog 910:	=	8.12831
Add for 0:	=	000
Add for 9:	=	170 3
Add, and insert power:		8.13001 3 $\times 10^3$

PROBLEMS 34-4

Find the antilogarithms of the following logarithms:

1. 0.477 12	**2.** 2.477 12	**3.** 1.477 12
4. 2.551 45	**5.** 2.807 54	**6.** 2.873 32
7. 2.004 32	**8.** 2.698 97	**9.** 5.383 82
10. 6.698 10 − 10	**11.** 3.925 57	**12.** 6.669 50

13. 9.990 87 − 10
14. 5.151 37
15. 2.534 699
16. 5.391 464
17. 5.174 060 − 10
18. 9.847 14
19. 0.797 96
20. 0.497 206
21. 0.434 249
22. 2.576 226
23. 6.9921 − 10
24. 5.872 74
25. 4.902 997 7
26. 6.751 664
27. 3.237 544
28. 5.778 15 − 10
29. 8.903 09 − 20
30. 6.539 45 − 10

34-16 ADDITION AND SUBTRACTION OF LOGARITHMS

Since the mantissa of a logarithm is always positive, care must be exercised in adding or subtracting logarithms.

Adding logarithms with positive characteristics is the same as adding arithmetical numbers.

example 17 Add the logarithms 2.764 21 and 4.304 64.

solution

$$\begin{array}{r} 2.764\ 21 \\ \underline{4.304\ 64} \\ 7.068\ 85 \end{array}$$

When adding logarithms with negative characteristics, you must bear in mind that the mantissas are always positive.

example 18 Add the logarithms $\overline{4}.326\ 52$ and 6.284 37.

solution The mantissas are added as positive numbers, and the characteristics are added algebraically:

$$\begin{array}{rr} & \overline{4}.326\ 52 \\ & \underline{6.284\ 37} \\ \text{Sum} = & 2.610\ 89 \end{array}$$

example 19 Add the logarithms $\overline{4}.328\ 30$, $\overline{3}.764\ 22$, and $\overline{1}.104\ 82$.

solution

$$\begin{array}{rr} & \overline{4}.328\ 30 \\ & \overline{3}.764\ 22 \\ & \underline{\overline{1}.104\ 82} \\ \text{Sum} = & \overline{7}.197\ 34 \end{array}$$

In Example 19 the sum of the *mantissas* is 1.197 34, and the 1 must be carried over for addition with the characteristics. Since the 1 from the mantissa sum is positive and the characteristics are negative, the two are added algebraically to obtain −7.

example 20 Subtract the logarithm 6.986 02 from the logarithm 4.107 37.

solution

$$\begin{array}{rr} & 4.107\ 37 \\ & \underline{6.986\ 02} \\ \text{Remainder} = & \overline{3}.121\ 35 \end{array}$$

example 21 Subtract the logarithm $\overline{5}.785\ 67$ from the logarithm $\overline{2}.672\ 58$.

solution

$$\begin{array}{rr} & \overline{2}.672\ 58 \\ & \underline{\overline{5}.785\ 67} \\ \text{Remainder} = & 2.886\ 91 \end{array}$$

In Example 21, in order to subtract the mantissas, it was necessary to subtract 7 from 6. The 6 was made to be 16 (1.6). To compensate for this *increase* in the value of the mantissa, the characteristic is changed from −2 to −3. In effect, *borrowing* 1 from the −2 characteristic makes it −3.

Another method of handling logarithms whose characteristics are negative is to express them as logarithms with a positive characteristic, and write the proper multiple of negative 10 after the mantissa.

example 22 Add the logarithms $\bar{4}.326\ 52$ and 6.284 37.

solution

$$\bar{4}.326\ 52 = 6.326\ 52 - 10$$

$$\begin{array}{rl} & 6.326\ 52 - 10 \\ & \underline{6.284\ 37} \\ \text{Sum} = & 12.610\ 89 - 10 = 2.610\ 89 \end{array}$$

Note that this is the same as Example 18. If −10, −20, −30, −40, etc., appear in the sum after the mantissa and the characteristic is greater than 9, subtract from both characteristic and mantissa a multiple of 10 that will make the characteristic less than 10.

example 23 Add the logarithms $\bar{4}.328\ 30$, $\bar{3}.764\ 22$, and $\bar{1}.104\ 82$.

solution

$$\begin{array}{rr} & 6.328\ 30 - 10 \\ & 7.764\ 22 - 10 \\ & \underline{9.104\ 82 - 10} \\ & 23.197\ 34 - 30 \\ \text{Sum} = & 3.197\ 34 - 10 \end{array}$$

Note that this is the same as Example 19. When a larger logarithm is subtracted from a smaller, the characteristic of the smaller should be increased by 10 and −10 should be written after the mantissa to preserve equality.

example 24 Subtract the logarithm 6.986 02 from the logarithm 4.107 37.

solution

$$\begin{array}{rr} 4.107\ 37 = & 14.107\ 37 - 10 \\ & \underline{6.986\ 02 } \\ \text{Remainder} = & 7.121\ 35 - 10 \end{array}$$

Also, when a negative logarithm is subtracted from a positive logarithm, the characteristic of the minuend should be made positive by adding to it the proper multiple of 10 and writing that multiple negative after the mantissa in order to preserve equality.

example 25 Subtract the logarithm 5.785 63 − 10 from the logarithm 1.672 57.

solution Adding 10 to the characteristic,

$$\begin{array}{rr} 1.672\ 57 = & 11.672\ 57 - 10 \\ & \underline{5.785\ 63 - 10} \\ \text{Remainder} = & 5.886\ 94 \end{array}$$

example 26 Subtract the logarithm 8.675 43 − 20 from the logarithm 2.462 58.

solution Adding 20 to the characteristic,

$$\begin{array}{rr} 2.462\ 58 = & 22.462\ 58 - 20 \\ & \underline{8.675\ 43 - 20} \\ \text{Remainder} = & 13.787\ 15 \end{array}$$

PROBLEMS 34-5

Add the following logarithms:

1. 2.824 12 + 3.127 37

2. 6.203 81 + 1.536 90

3. $\bar{6}.232\ 86 + \bar{4}.170\ 33$

4. 8.203 65 − 10 + 1.927 32

5. $\bar{3}.464\ 80 + \bar{2}.808\ 86$

6. 9.352 82 − 10 + 5.865 33 − 10

Perform the indicated subtractions:

7. $3.258\ 79 - 0.699\ 01$

8. $0.434\ 36 - \bar{3}.572\ 82$

9. $\bar{2}.628\ 58 - \bar{4}.280\ 71$

10. $\bar{4}.392\ 64 - 2.610\ 24$

11. $3.293\ 78 - (9.437\ 87 - 10)$

12. $9.538\ 65 - 10 - (9.749\ 32 - 10)$

34-17 MULTIPLICATION WITH LOGARITHMS

It was shown in Sec. 34-3 that the logarithm of a product is equal to the sum of the logarithms of the factors. This property, with the aid of the tables, is of value in multiplication.

example 27 Find the product of 2.79×684.

solution Let p = the desired product; then

$$p = 2.79 \times 684 \tag{14}$$

Taking the logarithms of both members of Eq. (14),

$$\log p = \log 2.79 + \log 684$$

Looking up the logarithms, tabulating them, and adding them,

$$\begin{aligned} \log 2.79 &= 0.445\ 60 \\ \log 684 &= 2.835\ 06 \\ \hline \log p &= 3.280\ 66 \end{aligned}$$

Interpolating to find the value of p,

Antilog 280	=	1.90546
Add for 6	=	266
Add for 6	=	26|6
Antilog 3.280 66	=	1.90838|6 $\times 10^3$

There is no need to express the result of the above interpolation beyond five significant figures. It is correct to report $p = 1.9084 \times 10^3$.

example 28 Given $X_L = 2\pi fL$. Find the value of X_L when $f = 10\ 600\ 000$ and $L = 0.000\ 025\ 1$. Use $2\pi = 6.28$.

solution $X_L = 6.28 \times 10\ 600\ 000 \times 0.000\ 025\ 1$

Taking logarithms,

$$\log X_L = \log 6.28 + \log 10\ 600\ 000 + \log 0.000\ 025\ 1$$

Tabulating,

$$\begin{aligned} \log 6.28 &= 0.797\ 96 \\ \log 10\ 600\ 000 &= 7.025\ 31 \\ \log 0.000\ 025\ 1 &= 5.399\ 67 - 10 \\ \hline \log X_L &= 13.222\ 94 - 10 = 3.222\ 94 \end{aligned}$$

By interpolation,

$$X_L = 1.6709 \times 10^3$$

In using logarithms, a form should be written out for all the work before beginning any computations. The form should provide places for all logarithms as taken from Table 7 and for other work necessary to complete the problem.

34-18 COMPUTATION WITH NEGATIVE NUMBERS

Because a negative number has an imaginary logarithm, the logarithms of negative numbers cannot be used in computation. However, the numerical results of multiplications and divisions are the same regardless of the algebraic signs of the factors. Therefore, to make computations involving negative numbers, first determine whether the final result will be positive or negative. Then find the numerical value of the expression by logarithms, considering all numbers as positive, and affix the proper sign to the result.

PROBLEMS 34-6

Compute by logarithms:

1. 8×32
2. 47×5
3. 5×50
4. 0.6×24
5. $3 \times 18 \times 0.7$
6. $12 \times (-16)$
7. $(-95) \times 2.6$
8. $0.007 \times (-22)$
9. 296×8.02
10. $0.425 \times (-0.0036)$
11. 37.7×266
12. $3250 \times (-2.03)$
13. $5.243 \times (-0.1872)$
14. $3 \times 6 \times 47$
15. $2.84 \times 72.4 \times 369$
16. $6.01 \times 444 \times 0.009\,13$
17. $(-0.003\,96) \times 500 \times 681$
18. $14.83 \times (-2.222) \times 0.1123$
19. $242.6 \times 471.8 \times 0.000\,082\,17$
20. $(-4627) \times 9126 \times (-7336)$

34-19 DIVISION BY LOGARITHMS

It was shown in Sec. 34-4 that the logarithm of the quotient of two numbers is equal to the logarithm of the dividend minus the logarithm of the divisor. This property allows division by the use of logarithms.

example 29 Find the value of $\frac{948}{237}$, by using logarithms.

solution Let q = quotient.

Then $$q = \frac{948}{237}$$

Taking logarithms, $\log q = \log 948 - \log 237$

Tabulating, $$\log 948 = 2.976\,81$$ $$\log 237 = 2.374\,75$$

Subtracting, $$\log q = 0.602\,06$$

Taking antilogs, $$q = 4$$

example 30 Find the value of $\frac{-24.68}{682\,700}$ by using logarithms.

solution By inspection the quotient will be negative. Let

$$q = \text{quotient}$$

Then $$q = \frac{-24.68}{682\,700}$$

Taking logarithms,

$$\log q = \log 24.68 - \log 682\,700$$

Interpolating and tabulating,

$$\log 24.68 = 11.392\,35 - 10$$
$$\log 682\,700 = 5.834\,23$$

Subtracting, $\log q = 5.558\,12 - 10$

Taking antilogs, and inserting minus sign,

$$q = -3.615 \times 10^{-5}$$

note log 24.68 = 1.392 35, but 10 was added to the characteristic and subtracted after the mantissa in order to facilitate the subtraction of a larger logarithm, as explained in Sec. 34-16.

PROBLEMS 34-7

Compute by logarithms:

1. $\dfrac{12}{4}$ 2. $\dfrac{81}{9}$ 3. $\dfrac{340}{17}$ 4. $\dfrac{1920}{-6.4}$

5. $\dfrac{0.245}{-0.000\,35}$ 6. $\dfrac{426}{-1137}$ 7. $\dfrac{-2325}{4.023}$ 8. $\dfrac{0.000\,517\,9}{-3.648}$

9. $\dfrac{3906}{0.000\,800\,2}$ 10. $\dfrac{-25.83}{-0.003\,142}$

34-20 COMBINED MULTIPLICATION AND DIVISION

When it is necessary to perform several steps of multiplication and division in one problem, make up a skeleton form indicating the logarithmic operations required. Insert the characteristics as you make up the skeleton. Then, turning to the tables, insert the logarithms of all the factors, first of the numerator, and then of the denominator. The systematic tabulation of the operations will make it easier to keep in touch with the steps to be followed.

example 31 Evaluate, by means of logarithms

$$N = \frac{(14.63)^2}{0.003\,62 \times 8767}$$

solution First, prepare the skeleton form of the logarithmic operations:

$$\begin{array}{rll}
\log 14.63 = & 1. & \\
& \times 2 & \\
\hline
\log \text{numerator} = & & = \\
\log 0.003\,62 = & \bar{3}. & \\
\log 8767 = & 3. & + \\
\hline
\log \text{denominator} = & & = \underline{\qquad\qquad} \; - \\
\log N & & = \\
N & & =
\end{array}$$

Then complete the tabulation, referring to the tables, and perform the calculations:

$$\begin{array}{rll}
\log 14.63 = & 1.165\,23 & \\
& \times 2 & \\
\hline
\log \text{numerator} = & 2.330\,46 & = 2.330\,46 \\
\log 0.003\,62 = & \bar{3}.558\,71 & \\
\log 8767 = & 3.942\,85 & \\
\hline
\log \text{denominator} = & 1.501\,56 & = \underline{1.501\,56} \; - \\
\log N & & = 0.828\,90 \\
N & & = 6.7436
\end{array}$$

example 32 Evaluate, by means of logarithms

$$\phi = \frac{(64.28)(0.009\,73)}{(4006)(0.051\,34)(0.002\,085)}$$

solution Make up the skeleton of the operations:

$$\begin{aligned}
\log 64.28 &= 1. \\
\log 0.009\,73 &= \overline{3}. \qquad\qquad + \\
\hline
\text{log numerator} &= \qquad\qquad = \\
\log 4006 &= 3. \\
\log 0.051\,34 &= \overline{2}. \\
\log 0.002\,085 &= \overline{3}. \qquad\qquad + \\
\hline
\text{log denominator} &= \qquad\qquad = \qquad\qquad - \\
\log \phi &= \\
\phi &=
\end{aligned}$$

Then, from the tables, complete the tabulation and perform the calculations:

$$\begin{aligned}
\log 64.28 &= 1.808\,08 \\
\log 0.009\,73 &= \overline{3}.988\,11 + \\
\text{log numerator} &= \overline{1}.796\,19 = \overline{1}.796\,19 \\
\log 4006 &= 3.602\,70 \\
\log 0.051\,34 &= \overline{2}.710\,46 \\
\log 0.002\,085 &= \overline{3}.319\,16 + \\
\text{log denominator} &= \overline{1}.632\,32 = \overline{1}.632\,32 - \\
\log \phi &= 0.163\,87 \\
\phi &= 1.458\,4
\end{aligned}$$

PROBLEMS 34-8

Use logarithms to compute the results of the following:

1. $\dfrac{2.4 \times 3.5}{1.7}$

2. $\dfrac{5.6 \times 8.9}{4.7 \times 9.3}$

3. $\dfrac{22.1 \times 1.08}{12.65 \times 0.78}$

4. $\dfrac{86.3 \times 0.0297}{0.0379}$

5. $\dfrac{-0.536}{734.4 \times 0.005\,83}$

6. $\dfrac{2.006}{3.142 \times 0.833}$

7. $\dfrac{0.000\,009\,207}{4.98 \times 0.000\,000\,707}$

8. $\dfrac{1}{6.28 \times 427\,000\,000 \times 0.000\,050}$

9. $\dfrac{1}{4.73 \times 5222 \times 0.000\,680\,7}$

10. $\dfrac{6.28 \times 0.000\,159 \times 326}{0.003\,68 \times 436 \times 0.0278}$

34-21 RAISING TO A POWER BY LOGARITHMS

It was shown in Sec. 34-5 that the logarithm of a power of a number is equal to the logarithm of the number multiplied by the exponent of the power.

example 33 Find by logarithms the value of 12^3.

solution

$$\begin{aligned}
\log 12^3 &= 3 \log 12 \\
\log 12 &= 1.079\,18 \\
\textbf{M: } 3 \qquad & \qquad\quad 3 \\
\hline
& 3.237\,54 = \log 1728 \\
\therefore 12^3 &= 1728
\end{aligned}$$

example 34 Find by logarithms the value of $0.056\,3^5$.

solution

$$\log 0.056\,3^5 = 5 \log 0.056\,3$$
$$\log 0.056\,3 = 8.750\,51 - 10$$
$$5$$

M: 5

$$5 \log 0.0563 = 43.752\,55 - 50$$
$$= 3.752\,55 - 10$$
$$\text{antilog } 3.7525 - 10 = 5.6566 \times 10^{-7}$$
$$\therefore 0.0563^5 = 5.6566 \times 10^{-7}$$

example 35 Find by logarithms the value of 5^{-3}.

solution By the laws of exponents,

$$5^{-3} = \frac{1}{5^3}$$

Then

$$\log 5^{-3} = \log 1 - \log 5^3$$
$$= \log 1 - 3 \log 5$$

Multiplying,

$$\log 5 = 0.698\,97$$
$$3$$
$$3 \log 5 = 2.096\,91$$

$$\log 1 = 10.000\,00 - 10$$
$$3 \log 5 = 2.096\,91$$
$$\log 5^{-3} = 7.903\,09 - 10$$
$$\text{antilog } 7.903\,09 - 10 = 0.008$$
$$\therefore 5^{-3} = 0.008$$

34-22 EXTRACTING ROOTS BY LOGARITHMS

It was shown in Sec. 34-6 that the logarithm of a root of a number is equal to the logarithm of the number divided by the index of the root.

example 36 Find by logarithms the value of $\sqrt[3]{815}$.

solution By the laws of exponents,

$$\sqrt[3]{815} = 815^{\frac{1}{3}}$$

Then

$$\log 815^{\frac{1}{3}} = \tfrac{1}{3} \log 815$$
$$\log 815 = 2.911\,15$$
$$\tfrac{1}{3} \log 815 = \frac{2.911\,15}{3} = 0.970\,38$$
$$\text{antilog } 0.970\,38 = 9.3406$$
$$\therefore \sqrt[3]{815} = 9.34 \text{ to three significant figures}$$

example 37 Find by logarithms the value of $\sqrt[4]{0.009\,55}$.

solution

$$\sqrt[4]{0.009\,55} = 0.009\,55^{\frac{1}{4}}$$

Then

$$\log 0.009\,55^{\frac{1}{4}} = \tfrac{1}{4} \log 0.009\,55$$
$$\log 0.009\,55 = 7.980\,00 - 10$$
$$\tfrac{1}{4} \log 0.009\,55 = 1.995\,00 - 2.5$$

This result, though correct, is not in the standard form for a negative characteristic. This inconvenience can be obviated by writing the logarithm in such a manner that the negative part when divided results in a quotient of -10. Thus,

$$\log 0.009\,55 = 7.980\,00 - 10$$

would be written

$$\log 0.009\,55 = 37.980\,00 - 40$$

Since it is necessary to divide the logarithm by 4 in order to obtain the fourth root, 30 was subtracted from the negative part to make it exactly divisible by 4. Therefore, to preserve equality, it was necessary to add 30 to the positive part. Then

$$\log \sqrt[4]{0.009\,55} = \frac{37.980\,00 - 40}{4}$$
$$= 9.495\,00 - 10$$
$$\text{antilog } 9.495\,00 - 10 = 0.312\,61$$
$$\therefore \sqrt[4]{0.009\,55} = 0.312\,61$$

34-23 FRACTIONAL EXPONENTS

Computations involving fractional exponents are made by combining the operations of raising to powers and extracting roots.

example 38 Find by logarithms the value of $\sqrt[4]{0.0542^3}$.

solution

$$\sqrt[4]{0.0542^3} = 0.0542^{\frac{3}{4}}$$

Then

$$\log 0.0542^{\frac{3}{4}} = \tfrac{3}{4}\log 0.0542$$

$$\log 0.0542 = 8.734\,00 - 10$$

$$3\log 0.0542 = 26.202\,00 - 30$$

Adding 10 to the characteristic and subtracting 10 from the negative part in order to make it evenly divisible by 4,

$$3\log 0.0542 = 36.202\,00 - 40$$

$$\tfrac{3}{4}\log 0.0542 = \frac{36.202\,00 - 40}{4}$$

$$= 9.050\,50 - 10$$

$$\text{antilog } 9.050\,50 - 10 = 0.112\,33$$

$$\therefore \sqrt[4]{0.0542^3} = 0.112$$

Instead of adding 10 to the characteristic, as above, it would also have been correct to subtract 10 from the characteristic and add 10 to the mantissa, and thus obtain 16.202 00 − 20. It is immaterial what numbers are added and subtracted as long as the resulting negative characteristic portion will yield an integral quotient.

PROBLEMS 34-9

Use logarithms to compute the results of the following:

1. 12.8^2
2. 82.3^5
3. 0.0176^4
4. 0.463^6
5. $\sqrt[3]{180}$
6. $\sqrt[4]{782}$
7. $1237^{\frac{1}{4}}$
8. $0.643^{\frac{1}{5}}$
9. $0.862^{\frac{1}{2}}$
10. $\sqrt[6]{4258}$
11. $127^{\frac{2}{3}}$
12. $\sqrt[3]{2.61^4}$
13. $30.6^{\frac{3}{2}}$
14. $164^{\frac{2}{3}}$
15. $\sqrt{\dfrac{196 \times 0.083}{12.1}}$
16. $\sqrt[3]{\dfrac{(-0.436) \times 30.8}{0.0287}}$
17. $\left(\dfrac{224}{363}\right)^{\frac{3}{2}}$
18. $\left(\dfrac{9764}{238.3}\right)^{1.5}$
19. $\sqrt[6]{0.000\,028\,6} \times \sqrt[4]{629}$
20. $\left(\dfrac{0.000\,000\,587}{0.000\,001\,72}\right)^{\frac{5}{2}}$

34-24 PRECAUTIONS TO BE OBSERVED

We have now investigated the common operations involving the use of logarithms in performing mathematical computations, and we know that to use logarithms to perform multiplications, we add logarithms; to perform division, we subtract logarithms; to raise to a power, we multiply the logarithms by the power; and to extract a root, we divide the logarithms by the root.

There are times, however, when a problem introduces the *use of logarithms*, apart from the employment of logarithms in computing an arithmetical solution. Consider carefully the following examples:

example 39 Compute, by means of logarithms, 125 × 13.6.

solution This is a standard multiplication problem of the type which we successfully mastered in Problems 34-6. We find the logarithm of each number, add the logarithms, take the antilogarithm of the sum to determine the value:

$$\begin{aligned} \log 125 &= 2.096\,91 \\ \log 13.6 &= \underline{1.133\,54} \\ \log \text{answer} &= 3.230\,45 \\ \text{Answer} &= \text{antilog } 3.230\,45 = 1.7 \times 10^3 \end{aligned}$$

example 40 Compute (log 125)(log 13.6).

solution This problem calls for us to multiply the logarithm of 125, whatever that may be, by the logarithm of 13.6, whatever that may be. We can determine what these logarithms are and rewrite the problem:

$$(\log 125)(\log 13.6) = (2.096\,91)(1.133\,54)$$

In other words, we have replaced the log expressions in the problems with the numbers which *are* the logarithms as called for. Then, having made this substitution, we perform the actual required operation, that is, multiply 2.096 91 by 1.133 54, to obtain 2.38.

Note carefully that this problem did not introduce the addition of logarithms and the taking of antilogarithms in order to arrive at an answer. It *may* have suited our convenience to perform the necessary multiplication by means of logarithms, but that would introduce an additional problem.

example 41 Compute, by means of logarithms,

$$(\log 125)(\log 13.6)$$

solution As in Example 40, first rewrite the problem:

$$(\log 125)(\log 13.6) = (2.096\,91)(1.133\,54)$$

To perform this multiplication operation by means of logarithms, we follow the usual procedures of interpolation, addition of logarithms, and subsequent antilogarithm:

$$\begin{aligned} \log 2.096\,91 &= 0.321\,633\,1 \\ \log 1.133\,54 &= \underline{0.054\,448\,4} \\ \log \text{answer} &= 0.376\,081\,5 \\ \text{Answer} &= \text{antilog } 0.376\,081\,5 \\ &= 2.377\,284\,13 = 2.38 \end{aligned}$$

In Example 41, because the problem called for a logarithmic performance of arithmetic, we performed logarithmic calculations. In Example 40 we arrived at the same value by other methods, despite the fact that logarithms *appeared* in the problem.

It is essential that you be aware at all times of the difference between performing operations by means of logarithms and performing operations which somehow involve the logarithms of numbers. This difference will appear in several of the problems of Chap. 35, and Problems 34-10 are included at this point to give you practice in recognizing the different types of problems which may arise.

PROBLEMS 34-10

Evaluate the following:

1. $\log 37.2 + \log 9.83$ **2.** $\log 16.3 - \log 7.03$ **3.** $\log 3.68 - \log 5.66$

4. $(\log 87.2)(\log 15.7)$ **5.** $\dfrac{\log 265}{\log 17.6}$ **6.** $\dfrac{\log 20.3}{\log 65.2}$

7. $(\log 3.97)\left(\dfrac{\log 16.3}{\log 8.6}\right)$ **8.** $(\log 224)^2$ **9.** $\log 224^2$

10. $\dfrac{\log 0.987}{\log 3.5}$

11 to **17.** Evaluate Probs. 4 to 10 by using logarithms for all calculations.

34-25 CHANGE OF BASE

In Problems 34-1 we found logarithms of numbers to many bases besides 10, and it is often convenient for us to be able to find the logarithms of numbers to certain bases other than 10 without developing a set of tables for other bases. An interesting development shows us how this may be achieved.

$$N = a^x \qquad (15)$$

which we may rewrite

$$x = \log_a N \qquad (16)$$

Taking logarithms of both sides of Eq. (15) to the base b:

$$\log_b N = \log_b a^x \qquad (17)$$

Substituting Eq. (11) into Eq. (17),

$$\log_b N = x \log_b a \qquad (18)$$

Substituting Eq. (16) into Eq. (18),

$$\log_b N = \log_a N \cdot \log_b a \qquad (19)$$

Since it can be shown that

$$\log_b a = \frac{1}{\log_a b} \qquad (20)$$

Equation (19) may be written in the form

$$\log_b N = \frac{\log_a N}{\log_a b} \qquad (21)$$

If, then, we have a table of logarithms to the base 10 and find it necessary to produce the logarithm of any number to any other base b, we simply divide the logarithm to the base 10 of the given number by the logarithm to the base 10 of the other base number b:

$$\log_b N = \frac{\log_{10} N}{\log_{10} b} \qquad (22)$$

We are primarily concerned with the natural system of logarithms, which has for its base the number $\varepsilon = 2.718\,28\ldots$ (Sec. 34-9). Many relationships in electronics as well as other branches of science involve logarithms to this base.

Although we will develop and use logarithms to the base ε in Sec. 34-26, you will often find that only tables of logarithms to the base 10 are immediately available. Using the relationship expressed in formulas (19) and (21), you will be able to perform the necessary operations.

$$\log_\varepsilon N = 2.302\,59 \log_{10} N \qquad (23)$$
$$\log_{10} N = 0.434\,29 \log_\varepsilon N \qquad (24)$$

example 42

$$\begin{aligned}\log_\varepsilon 1000 &= 2.302\,59 \log_{10} 1000\\ &= 2.305\,59 \times 3\\ &= 6.907\,77\end{aligned}$$

example 43

$$\begin{aligned}\log_{10} 100 &= 0.434\,29 \log_\varepsilon 100\\ &= 0.434\,29 \times 4.6052\\ &= 2.0000\end{aligned}$$

example 44 Given $x = \log_\varepsilon 48$. Solve for x.

solution

$$\begin{aligned} \log_\varepsilon 48 &= 2.302\,59 \log_{10} 48 \\ &= 2.302\,59 \times 1.681\,24 \\ x &= 3.8712 \end{aligned}$$

34-26 NATURAL LOGARITHMS

Because so many calculations in electronics do involve logarithms to the base ε, Table 12 is included in the Appendix to meet such requirements. A few special notes will simplify your use of these natural logarithms:

1. The laws of logarithms [Eqs. (6), (8), (11), and (13)] apply to any logarithmic system, regardless of the base used. Therefore, natural logarithms may be used instead of common logarithms, if you prefer, for any problem involving multiplication, division, raising to powers, or extracting roots.
2. For convenience, we often replace the notation $\log_\varepsilon$ with the special symbol ln, pronounced lon.
3. Since $\ln \varepsilon = 1$, the characteristics in natural logarithms do not represent powers of ten. (They represent powers of ε.) Accordingly, Table 12 gives the *entire* logarithm of a number, and not just its mantissa.

example 45

$$\begin{aligned} \ln 2.70 &= 0.993\,25 \\ \ln 2.72 &= 1.000\,63 \\ \ln 5.05 &= 1.619\,39 \\ \ln 7.38 &= 1.998\,77 \\ \ln 7.39 &= 2.000\,13 \end{aligned}$$

4. Because the characteristic represents a power of ε, it is necessary to build up natural logarithms of very large and very small numbers, using Eq. (6) for the purpose.

example 46 Using Table 12, find the natural logarithm of 127.4.

solution Using the law expressed in Eq. (6),

$$\begin{aligned} \ln 127.4 &= \ln (1.274 \times 10^2) \\ &= \ln 1.274 + \ln 10^2 \end{aligned}$$

From the main table,

$$\ln 1.274 = 0.242\,22$$

From the table at the bottom of the main table,

$$\begin{aligned} \ln 10^2 &= \underline{4.605\,17} \\ \text{Adding, } \ln 127.4 &= 4.847\,39 \end{aligned}$$

example 47 Find ln 0.001 274.

solution

$$\begin{aligned} \ln 0.001\,274 &= \ln (1.274 \times 10^{-3}) \\ &= \ln 1.274 + \ln 10^{-3} \\ \ln 1.274 &= 0.242\,22 \\ \ln 10^{-3} &= \underline{\overline{7}.092\,24} \\ \ln 0.001\,274 &= \overline{7}.334\,46 \end{aligned}$$

(Users of slide rules with LL or lon scales, or of calculators, will observe that the value given for ln 0.001 274 is −6.665 59, which, of course, is the combination of −7.000 00 + 0.334 406.)

5. To interpolate, we turn to the proportional parts portion of the table, using successive tenths of the interpolation quantities after the first step.

example 48 Find ln 5.273 628.

solution From the main table,

ln 5.27	= 1.662 03
From the proportional parts, add for 3:	57
Note that this first interpolation is added completely.	
From the same line, the addition for 6 would be 114. Since this is the *second* interpolation, add:	11 4
The interpolation for 2 would be 38. Since this is the *third* interpolation, add:	3 8
The interpolation for 8 would be 152. Add:	1 52
Adding, ln 5.273 628	= 1.662 719 32

(A student has suggested that each stage of interpolation after the first "sticks out one more place.")

Bear in mind always that the entries in Table 12, as in all such tables, have been rounded off, and so, of course, have the proportional parts. Accordingly, confidence must not be placed in the extreme number of decimal places of such an interpolated logarithm. Instead, the resultant figure should itself be rounded off to the number of places in the original table. We would say here ln 5.273 628 = 1.662 72.

6. To find natural antilogarithms, we cannot use the characteristic as a power of 10, but must step through the procedure:

example 49 Find x = antiln 8.952 87.

solution Since this logarithm exceeds the values in the main table, we first use the table values to reduce it:

given	8.952 87
	6.907 76 = ln 10^3
	2.045 11

From the main table, we find

2.045 11 = ln 7.73

Accordingly,

antiln 8.95 287 = 7.73 × 10^3.

example 50 Find x = antiln $\bar{2}$.614 715.

solution The negative characteristic indicates that the decimal number is quite small, involving a negative power of 10.

given	$\bar{2}$.614 715

From the table

$\bar{3}$.697 41 = ln 10^{-1}

0.917 305

From the main table,

0.916 29 = ln 2.50

1015

For the first interpolation, the 101 is the maximum amount, and we use

78 : 2

235

For the second interpolation,

235: 6

Accordingly,

antiln $\bar{2}$.614 716 = 2.5026 × 10^{-1}

34-27 GRAPH OF $y = \log_{10} x$

The graph of $y = \log_{10} x$ is shown in Fig. 34-1. A study of this graph shows the following:

1. A negative number has no real logarithm.
2. The logarithm of a positive number less than 1 (a decimal between 0 and 1) is negative.
3. The logarithm of 1 is zero.
4. The logarithm of a positive number greater than 1 is positive.
5. As the number approaches zero, its logarithm decreases without limit.
6. As the number increases indefinitely, its logarithm increases without limit.

Is the method of interpolation that treats a short distance on the logarithmic curve as a straight line sufficiently accurate for computation?

34-28 LOGARITHMIC EQUATIONS

An equation in which there appears the logarithm of some expression involving the unknown quantity is called a *logarithmic equation.*

Logarithmic equations have wide application in electric circuit analysis. In addition, the communications engineer uses them in computations involving decibels and transmission line characteristics.

example 51 Solve the equation $4 \log x + 3.796\,00 = 4.699\,09 + \log x$.

solution

Given

$$4 \log x + 3.796\,00 = 4.699\,09 + \log x$$

Transposing,

$$4 \log x - \log x = 4.699\,09 - 3.796\,00$$

Collecting terms,

$$3 \log x = 0.903\,09$$

D: 3,

$$\log x = 0.301\,03$$

From tables or slide rule,

$$x = 2$$

In solving logarithmic equations, the logarithm of the unknown, as $\log x$ in Example 51, is considered as any other literal coefficient. That is, in general, the rules for solving ordinary algebraic equations apply to logarithmic equations.

A common error made by students in solving logarithmic equations is confusing coefficients of logarithms with coefficients of the unknown. For example,

$$3 \log x \neq \log 3x$$

because the left member denotes the product of 3 times the logarithm of x, whereas the right member denotes the logarithm of the quantity 3 times x, that is, $\log (3x)$.

example 52 Given $500 = 276 \log \dfrac{d}{0.05}$. Solve for d.

solution 1

Given $\quad 500 = 276 \log \dfrac{d}{0.05}$

Then $\quad 500 = 276 (\log d - \log 0.05)$

D: 276, $\quad 1.81 = \log d - \log 0.05$

Transposing,

$$\log d = 1.81 + \log 0.05$$

Substituting $8.698\,97 - 10$ for $\log 0.05$,

$$\log d = 1.81 + 8.698\,97 - 10$$

Collecting terms,

$$\log d = 0.508\,97$$

From tables or slide rule,

$$d = 3.23$$

solution 2

Given $\quad 500 = 276 \log \dfrac{d}{0.05}$

D: 276, $\quad 1.81 = \log \dfrac{d}{0.05}$

Taking antilogs of both members,

$$64.57 = \frac{d}{0.05}$$

Solving for d, $\quad d = 3.23$

34-29 EXPONENTIAL EQUATIONS

An equation in which the unknown appears in an exponent is called an *exponential equation*. In the equation

$$x^3 = 125$$

it is necessary to find some value of x that, when cubed, will equal 125. In this equation *the exponent is a constant.*

In the *exponential equation*

$$5^x = 125$$

the situation is different. *The unknown appears as an exponent,* and it is now necessary to find to what power 5 must be raised to obtain 125.

Some exponential equations can be solved by inspection. For example, the value of x in the foregoing equation is 3. In general, taking the logarithms of both sides of an exponential equation will result in a logarithmic equation that can be solved by the usual methods.

example 53 Given $4^x = 256$. Solve for x.

solution

Given $\quad 4^x = 256$

Taking the logarithms of both members,

$$\log 4^x = \log 256$$

or $\quad x \log 4 = \log 256$

D: log 4, $\quad x = \dfrac{\log 256}{\log 4}$

From tables or slide rule,

$$x = \frac{2.408\,24}{0.602\,06} = 4$$

check $\quad 4^4 = 256$

example 54 Given $5^{x-3} = 52$. Solve for x.

solution

Given $\quad 5^{x-3} = 52$

Taking the logarithms of both members,

$$\log 5^{x-3} = \log 52$$

or $\quad (x - 3) \log 5 = \log 52$

D: log 5, $\quad x - 3 = \dfrac{\log 52}{\log 5}$

From tables or slide rule,

$$x - 3 = \frac{1.716\,00}{0.698\,97}$$

A: 3, $\quad x = \dfrac{1.716\,00}{0.698\,97} + 3$

or $\quad x = 5.455$

How would you check this solution?

PROBLEMS 34-11

Solve the following equations:

1. $x = \log_\varepsilon 226$ **2.** $x = \log_\varepsilon 4.38$ **3.** $\log x + 3 \log x = 6$

4. $\log x + \log 6x = 8.5$ **5.** $\log 5x + 2 \log x = 6.88$ **6.** $\log \frac{P}{3} = 0.573$

(**hint** $\log 6x = \log 6 + \log x$)

7. $\log \frac{P_1}{14} = 2.86$ **8.** $\log \frac{12}{E} = 3$ **9.** $\log x^2 - \log x = 6.75$

10. $x^4 = 462$ **11.** $4^x = 167$ **12.** $5^x = 37.3$

13. $2^m = 0.88$ **14.** $3^{q-3} = 14$ **15.** $4^{3x} = 14$

16. $M^{2.3} = 25$ **17.** $x = \log_6 1296$ **18.** $x = \log_3 2187$

19. If $10 \log L_2 = \frac{3}{2}(10 \log L_1)$, solve for L_1.

20. If $20 \log \dfrac{2Z_1}{2Z_1 - Z_a} = 20 \log \dfrac{-Z_b}{-Z_b + \dfrac{Z_1}{2}}$, solve for Z_1 in terms of Z_a and Z_b.

21. If $V_g = \dfrac{2.3T}{11\,600} \log \dfrac{I_o}{I_g}$, solve for I_o.

22. If $i = \dfrac{V}{L} t\varepsilon^{S_c t}$, solve for S_c.

23. If $i_c = \dfrac{E}{R}\varepsilon^{\frac{-t}{RC}}$, solve for (a) E, (b) C, (c) t.

24. If $I_k = AT^2\varepsilon^{\frac{-B}{T}}$, solve for (a) A, (b) B.

25. If $i_L = \dfrac{E}{R}(1 - \varepsilon^{-\frac{Rt}{L}})$, solve for (a) E, (b) L, (c) t.

26. If $q = CE(1 - \varepsilon^{-\frac{t}{RC}})$, solve for (a) E, (b) R, (c) t.

27. If $I_p + I_g = K\left(E + \dfrac{E_p}{\mu}\right)^{\frac{3}{2}}$, solve for (a) E, (b) E_p, (c) μ.

28. In an inductive circuit, the equation for the growth of current is given by

$$i = \frac{E}{R}(1 - \varepsilon^{-\frac{Rt}{L}}) \quad \text{A} \tag{25}$$

where i = current, A
t = any elapsed time after switch is closed, s
E = constant impressed voltage, V
L = inductance of the circuit, H
R = circuit resistance, Ω
ε = base of natural system of logarithms

A circuit of 0.75-H inductance and 15-Ω resistance is connected across a

12-V battery. What is the value of the current at the end of 0.06 s after the circuit is closed?

SOLUTION: The circuit is shown in Fig. 34-2.

Given $$i = \frac{E}{R}(1 - \varepsilon^{-\frac{Rt}{L}})$$

Substituting the known values,

$$i = \frac{12}{15}(1 - \varepsilon^{-\frac{15 \times 0.06}{0.75}})$$

$$i = 0.8(1 - \varepsilon^{-1.2})$$

Multiplying, $$i = 0.8 - 0.8\varepsilon^{-1.2}$$

or $$i = 0.8 - \frac{0.8}{\varepsilon^{1.2}} \qquad (26)$$

Now evaluate $\varepsilon^{1.2}$, $\log_{10} \varepsilon^{1.2} = 1.2 \log_{10} \varepsilon = 1.2 \times 0.434\,29$
$= 0.521\,15$

Taking antilogs, $\varepsilon^{1.2} = 3.32$

Substituting the value of $\varepsilon^{1.2}$ in Eq. (26),

$$i = 0.8 - \frac{0.8}{3.32} = 0.559 \text{ A}$$

The growth of the current in the circuit of Fig. 34-2 is shown graphically in Fig. 34-3.

29. The inductance of the circuit in Fig. 34-2 is halved, and the resistance is thus reduced to 0.71 times its original value. If other circuit values remain the same, what will be the value of the current 0.08 s after the switch is closed?

30. Using the circuit values for the circuit of Fig. 34-2, what will be the value of the current (*a*) 0.005 s after the switch is closed and (*b*) 0.5 s after the switch is closed?

Fig. 34-2 Circuit for Probs. 28 to 31

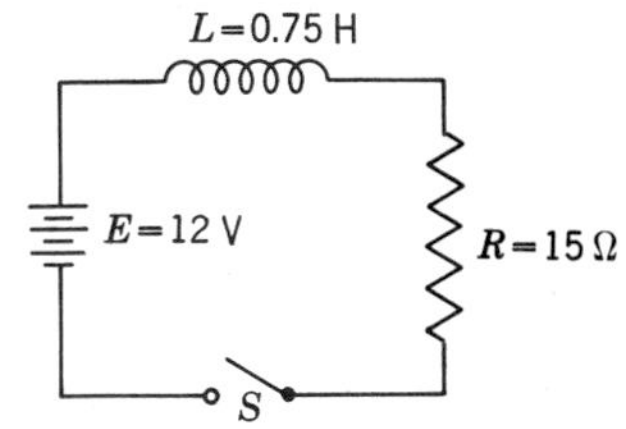

Fig. 34-3 Graph of Current in *RL* Circuit of Prob. 28

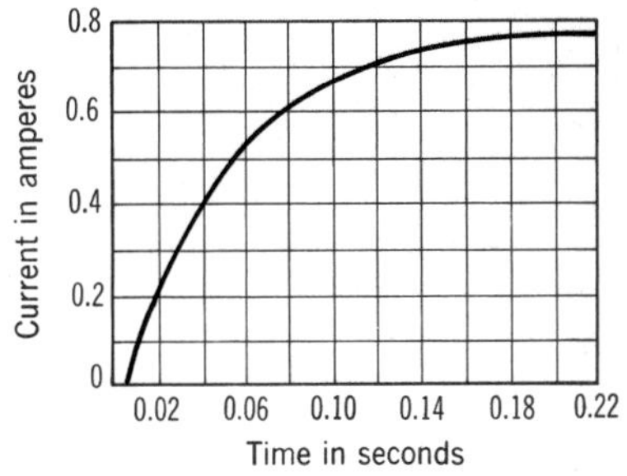

31. In the circuit of Fig. 34-2, after the switch is closed, how long will it take the current to reach 50% of its maximum value?

32. If $\frac{L}{R}$ is substituted for t in the equation

$$i = \frac{E}{R}(1 - \varepsilon^{-\frac{Rt}{L}})$$

show that the value of the current will be 63.2% of its steady-state value. The numerical value of L/R in seconds is known as the *time constant* of the inductive circuit. It is useful in determining the rapidity with which current rises or falls in one inductive circuit in comparison with others.

33. A 220-V generator shunt field has an inductance of 12 H and a resistance of 80 Ω. How long after the line voltage is applied does it take for the current to reach 75% of its maximum value?

34. A relay of 1.2 H inductance and 500 Ω resistance is to be used for keying a radio transmitter. The relay is to be operated from a 110-V line, and 0.175 A is required to close the contacts. How many words per minute will the relay carry if each word is considered as five letters of five impulses per letter? The time of opening of the contacts is the same as the time required to close them.

hint $0.175 = \frac{110}{500}(1 - \varepsilon^{-\frac{500t}{1.2}})$. t is the time required to close the relay.

35. How many words per minute would the relay of Prob. 34 carry if 50 Ω resistance were connected in series with it? The line voltage remains at 110 V.

36. In a capacitive circuit the equation for the current is given by

$$i = \frac{E}{R}\varepsilon^{-\frac{t}{RC}} \quad \text{A} \tag{27}$$

where i = current, A
t = any elapsed time after switch is closed, s
E = impressed voltage, V
C = capacitance of the circuit, F
R = circuit resistance, Ω
ε = base of natural system of logarithms

A capacitance of 500 μF in series with 1 kΩ is connected across a 50-V generator.

(a) What is the value of the current at the instant the switch is closed?

hint $t = 0$.

(*b*) What is the value of the current 0.02 s after the switch is closed? The circuit is shown in Fig. 34-4.

37. In the circuit of Fig. 34-4, how long after the switch is closed will the current have decayed to 30% of its initial value if $E = 110$ V, $R = 500\ \Omega$, $C = 20\ \mu$F, and $i = \dfrac{0.3E}{R}$. $t = ?$

SOLUTION: $i = \dfrac{0.3E}{R} = \dfrac{0.3 \times 110}{500} = 0.066$ A

Substituting in Eq. (27), $0.066 = \dfrac{110}{500}\varepsilon^{-\frac{t}{500 \times 20 \times 10^{-6}}}$

Simplifying, $0.066 = 0.22\,\varepsilon^{-\frac{t}{10^{-2}}}$

or $0.066 = 0.22\,\varepsilon^{-100t}$

D: 0.22, $0.3 = \varepsilon^{-100t}$

By the law of exponents, $0.3 = \dfrac{1}{\varepsilon^{100t}}$

M: ε^{100t} $0.3\,\varepsilon^{100t} = 1$

D: 0.3, $\varepsilon^{100t} = 3.33$

Taking logarithms, $\log_{10} \varepsilon^{100t} = \log_{10} 3.33$

That is, $100t \log_{10} \varepsilon = \log_{10} 3.33$

Then $100t \times 0.4343 = 0.5224$

or $43.43t = 0.5224$

$\therefore t = 0.012$ s

The decay of the current in the circuit of Fig. 34-4 is shown graphically in Fig. 34-5.

38. A 20-μF capacitor in series with a resistance of 680 Ω is connected across a 110-V source.

(*a*) What is the initial value of the current?

Fig. 34-5 Graph of Current in *RC* Circuit of Prob. 37

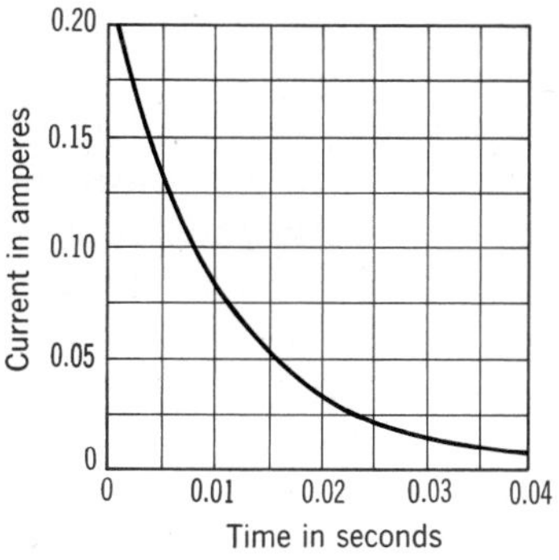

Fig. 34-4 Circuit of Probs. 36 and 37

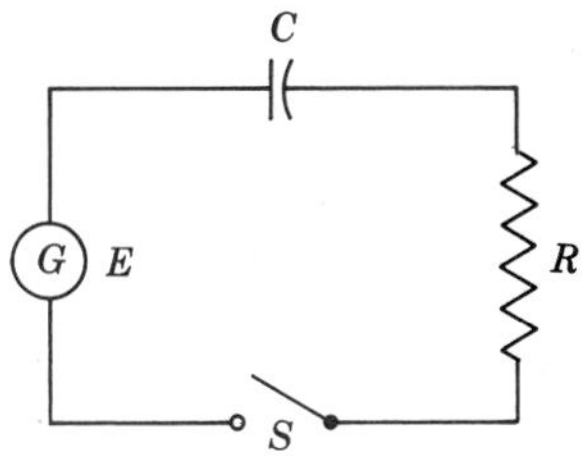

(*b*) How long after the switch is closed will the current have decayed to 36.8% of its initial value?

(*c*) Is the time obtained in (*b*) equal to *CR* s? The product of *CR*, in seconds, is the time constant of a capacitive circuit.

39. The quantity of charge on a capacitor is given by

$$q = CE(1 - \varepsilon^{-\frac{t}{CR}})\ \text{C} \qquad (28)$$

where q is the quantity of electricity in coulombs.

(*a*) Calculate the charge q in coulombs on a capacitor of 50 μF in series with a resistance of 3.3 kΩ, 0.008 s after being connected across a 70-V source.

(*b*) What is the voltage across the capacitor at the end of 0.02 s?

40. A key-click filter consisting of a 2-μF capacitor in series with a resistance is connected across the keying contacts of a transmitter. If the average time of impulse is 0.004 s, calculate the value of the series resistance required in order that the capacitor can discharge 90% in this time.

hint Under steady-state conditions, $q = CE$. Then

$$0.9CE = CE(1 - \varepsilon^{-\frac{t}{RC}})$$

41. The emission current in amperes of a heated filament is given by

$$I = AT^2 \varepsilon^{-\frac{B}{T}} \quad \text{A} \qquad (29)$$

For a tungsten filament, $A = 60$ and $B = 52\,400$. Find the current of such a filament at a temperature $T = 2500$ K.

42. An important triode formula is

$$I_p + I_g = K\left(E_g + \frac{E_p}{\mu}\right)^{\frac{3}{2}} \quad \text{A} \qquad (30)$$

where I_p = plate current, A
I_g = grid current, A
E_g = grid voltage, V
E_p = plate voltage, V
μ = amplification factor

Calculate $I_p + I_g$ if $K = 0.0005$, $E_g = 6$ V, $E_p = 270$ V, and $\mu = 15$.

43. The diameter of No. 0000 wire is 11.68 mm, and that of No. 36 is 0.127 mm. There are 38 wire sizes between No. 0000 and No. 36; therefore, the ratio between cross-sectional areas of successive sizes is the thirty-ninth root of

the ratio of the area of No. 0000 wire to that of No. 36 wire, or $\sqrt[39]{\frac{11.68^2}{0.127^2}}$.
Compute the value of this ratio. Because this ratio is nearly equal to $\sqrt[3]{2}$, we can use the approximation that the cross-sectional area of a wire doubles for every decrease of three sizes, as explained in Sec. 9-4. Calculate the percent error introduced by using $\sqrt[3]{2}$.

35

applications of logarithms

We have seen that logarithms can be extremely useful in the performance of arithmetic operations. Multiplication, division, raising to powers, and extracting roots are all important applications of logarithms which will be explored further in this chapter.

Similarly, proficiency in the use of logarithmic equations is an essential part of the electronics technician's mathematical toolbox. The broad application of these equations to computers, power measurement, amplification, attenuators, and transmission lines all testify to their importance.

In this chapter, we will see how logarithmic calculations are applied to the fields mentioned above and we will investigate briefly two extremely important applications of logarithms to our everyday work in electronics—the slide rule and preferred values.

35-1 THE SLIDE RULE

In Sec. 6-1 we introduce the idea of the slide rule as a mechanical analog computer. That is so because the distances on the slide rule are analogous to the numerical values which they bear. Let us examine a simple slide rule, Fig. 35-1. You may wish to follow the development by making a simple cardboard rule in order to guarantee your understanding of the construction and background knowledge of the use of the slide rule.

Figure 35-1*a* shows a 10-in (250-mm) line on which our slide rule will be developed. In (*b*), we label the left-hand end of the line

Fig. 35-1(a) Ten-Inch Base Line for Development of Slide Rule

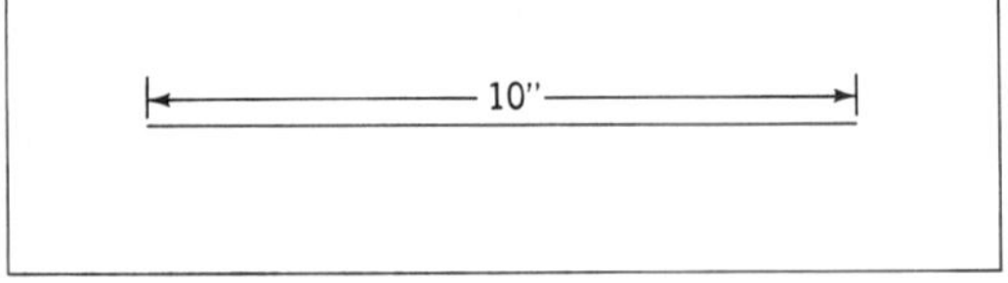

(*a*)

Fig. 35-1(b) $\log_{10} 1 = 0$. Zero Distance along Base Line Represents $\log_{10} 1$

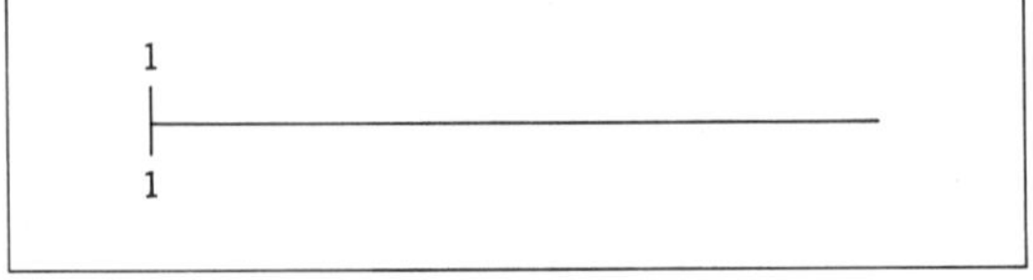

(*b*)

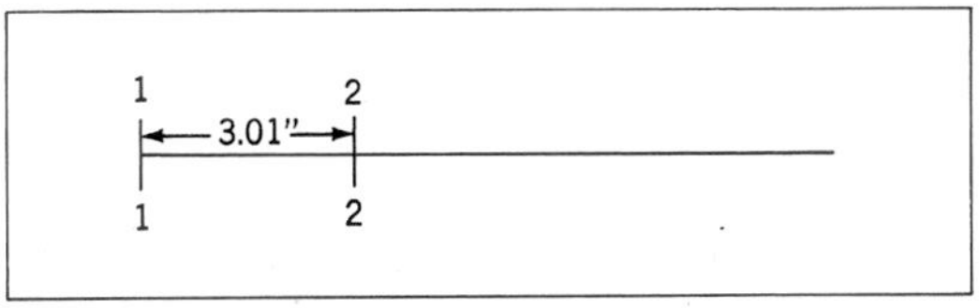

(c)

Fig. 35-1(c) 3.01 in along Base Line Represents $\log_{10} 2$

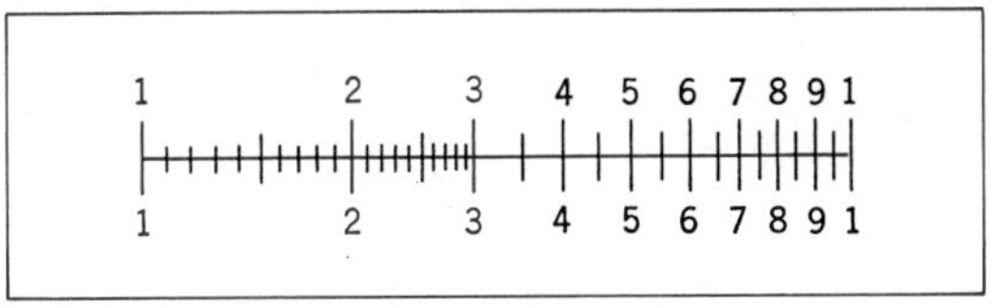

(e)

Fig. 35-1(e) Entire Base Line Divided Logarithmically

1, since $\log_{10} 1 = 0$. That is, zero distance from the left end of the line represents the numerical value 1.

In (c) we have added the 2 marker at a point 3.01 in (75 mm) from the left end of the line. $\log_{10} 2 = 0.301\,03$; so we make our mark at (0.301)(10 in) = 3.01 in (75 mm). In part (d), we have added the 3 marker at (0.4771)(10 in) = 4.771 in (119 mm) from the left end of the line, and in (e) we have shown the completed rule, with the distances from the left end representing the values of the numbers marked on the scale.

In Fig. 35-1*f*, we see the scale cut down the dividing line and the two parts labeled C and D, the common names given to the two simplest scales on any slide rule. Nearly

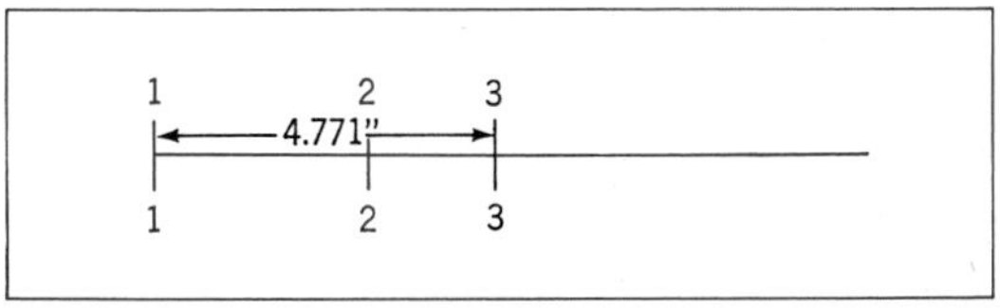

(d)

Fig. 35-1(d) 4.771 in along Base Line Represents $\log_{10} 3$

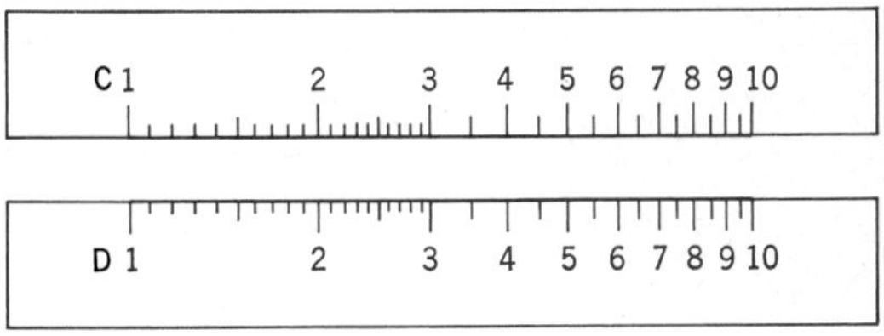

(f)

Fig. 35-1(f) Home-Made Slide Rule Divided into C and D Scales

every other scale on almost all slide rules is related to the D scale.

35-2 MULTIPLICATION AND DIVISION ON THE SLIDE RULE

example 1 Multiply 3×2 on the slide rule.

solution To multiply 3×2 by means of logarithms, we add the logs of the numbers involved and then take the antilog of the sum. To perform the same multiplication on the slide rule, we add logarithmic distances. Starting with the index 1 on the D scale, we proceed a distance equal to the logarithm of 2 (Fig. 35-2*a*). This takes us to the point we previously identified as 2. To this dis-

Fig. 35-2(a) Commencing to Multiply 2 and 3, Set a Distance on the D Scale Equivalent to the Logarithm of 2

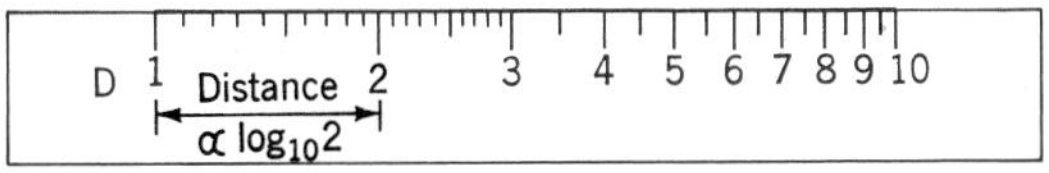

(a)

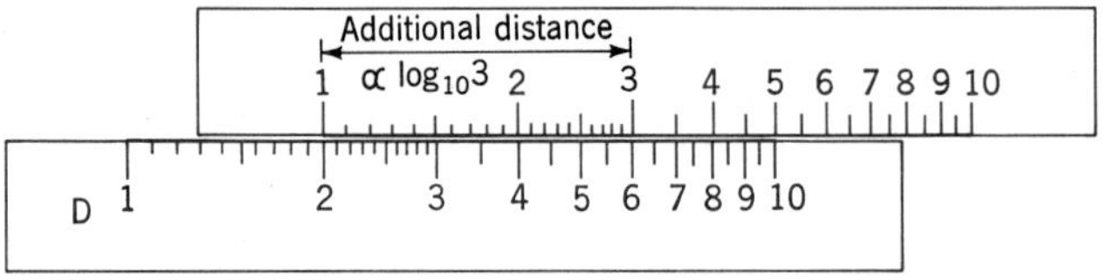

(b)

Fig. 35-2(b) To the Logarithm of 2 Add the Logarithm of 3

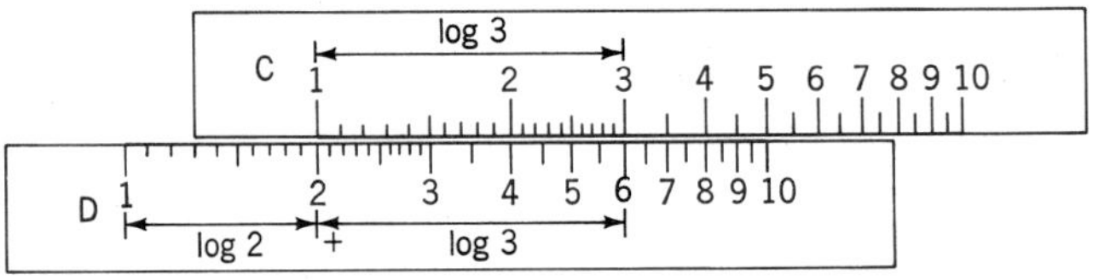

(c)

Fig. 35-2(c) log 2 + log 3 = log 6

tance we want to add a distance equivalent to log 3. This is easily done by setting the index of the C scale opposite D2 to mark the place. Then, following up the C scale a distance equivalent to log 3 takes us to the point on the C scale previously established as 3. Accordingly, we have gone distances from the D index equal to log 2 + log 3. The antilog of this sum is 6, and on the D scale opposite C3 we find 6—the slide rule has performed the operations of finding logarithms, adding them, and taking the antilog of the sum.

example 2 Divide 9 by 2 on the slide rule.

solution By logarithmic computation, we subtract log 2 from log 9 and take the antilog of the difference. On the slide rule, our starting point is 9 on the D scale, that is, a distance from the D index equal to log 9 (Fig. 35-3*a*). From this starting point we will subtract a distance equivalent to log 2. We do this by setting C2 opposite D9 (Fig. 35-3*b*), and moving down the C scale to its index, opposite which we read 4.5, the antilog of the difference of log 9 and log 2. Again, the slide rule automatically takes logs, subtracts them, and provides the antilog (Fig. 35-3*c*).

Now, of course, the great significance of Chap. 6 appears: the slide rule handles only the *mantissas* of the logarithms; you must provide the *characteristics* yourself. The simplest way to approximate an answer, and to guarantee the accuracy of the calculation, is to use the method of powers of 10.

Fig. 35-3(a) Commencing to Divide 9 by 2, Set a Distance on the D Scale Equivalent to the Logarithm of 9

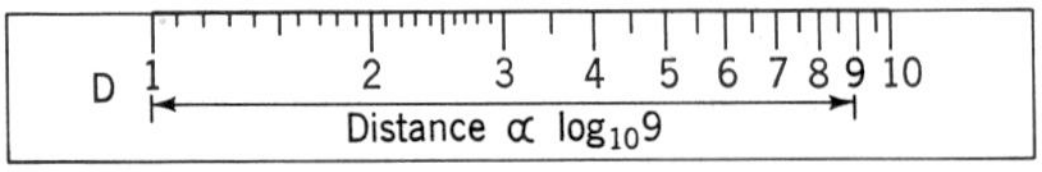

(a)

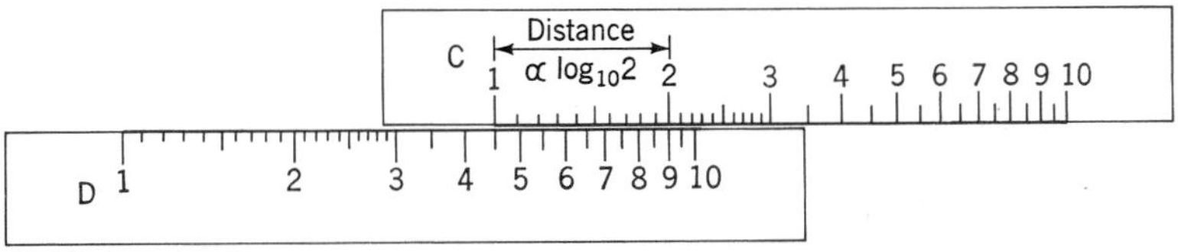

(b)

Fig. 35-3(b) From the Logarithm of 9 Subtract the Logarithm of 2

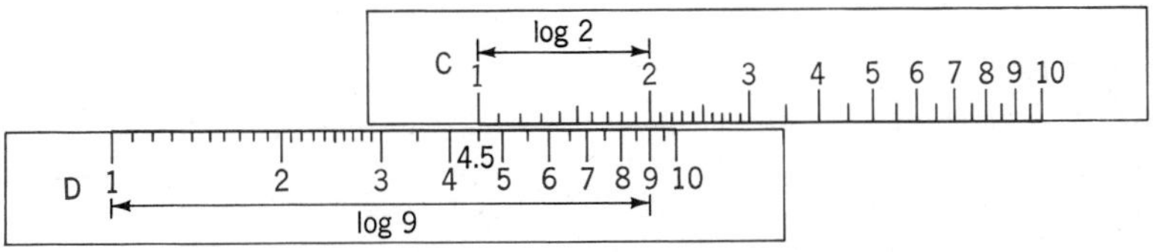

(c)

Fig. 35-3(c) log 9 − log 2 = log 4.5

35-3 OTHER SLIDE-RULE CALCULATIONS

When you buy a slide rule of reasonable quality, you should receive an instruction booklet showing in varying detail the different operations possible with the particular rule. You should review these operations in the light of the foregoing and the notes immediately following.

SQUARES AND SQUARE ROOTS By making our scale on a 5-in line instead of a 10-in one, we achieve a set of A and B scales for use in squaring numbers and extracting square roots. Every number on the A scale represents the square of the opposite number on the D scale. Of course, we need two 5-in A scales to cover the D scale. Then, any number on the D scale represents the square root of the opposite number on the A scale. Some rules you may meet will have the A scale divided from 1 to 10 and the divisions from 1 to 10 repeated. Others will have their A scale divided from 1 to 10 to 100, which helps you remember how to find the roots of large numbers.

example 3 Find the square root of 64.

solution To find the root of 64, which consists of two (an even number of) digits, we use the "upper half" of the A scale in order to read 8 on the D scale. If we had selected 64, without reference to a decimal point, on the lower half, we would have read 2.53 on the D scale—obviously wrong.

Rule

To read square roots of numbers, use the lower half of the A scale for numbers with an odd number of digits (in front of the decimal point) and the upper half of the A scale for numbers with an even number of digits.

If your slide rule has a K scale, note that it has three complete cycles, 1 to 10, 10 to 100, and 100 to 1000, to give the cube of numbers on the D scale between 1 and 10 or to find the cube root of K numbers on the D scale.

TRIGONOMETRIC FUNCTIONS An extremely useful scale is the S (sine) scale, especially if you have it related to the D scale. This scale is divided according to the logarithms of $\sin\theta$ in the left-to-right direction. You should prove from your knowledge of trigonometry that reading the S scale from right to left will give the cosines on the D scale. Some rules have cosine angles on the S scale marked in red.

Many rules make use of the approximate equality of $\sin\theta$ and $\tan\theta$ investigated in Sec. 25-4 for angles up to approximately 5.73° by providing a separate ST or SRT scale for computations involving these small angles. Remember that the D scale is multiplied by 10^{-2} with the ST and 10^{-1} with the S scales.

Similar to the S scale is the T scale, which gives tangents and cotangents of angles from 5.73° left to right and from 45° right to left.

LOGARITHMS Many rules include an L (logarithm) scale, which gives the mantissas of numbers on the D scale. Again, as with tables, you must provide the characteristics. Compare for accuracy the logarithms read from your L scale with those in the three-place table inside the front cover.

INVERTED SCALES With some practice, you will find the CI (C inverted) scale very time saving when several steps of multiplication and division are to be performed. Recall that multiplying by $\frac{1}{2}$ is equivalent to dividing by 2. A multiplication operation with the CI is equivalent to a C division, and vice versa. This scale acts like the colog of D, and the CIF has the same relationship to the CF and DF scales.

FOLDED SCALES Many operations in electronics calculations involve multiplication by π. These operations may be performed automatically by moving from D to DF, which is simply a D scale "folded" at π. Thus, any number on DF represents π times the opposite number on D.

PAIRED SCALES For greater ease in performing operations, several of the slide rule scales are duplicated. We have already discovered that C is identical with D. Similarly, B is identical with A, and CF is identical with DF. These paired scales enable us to continue a series of calculations when squaring, extracting roots, multiple operations, or multiplication or division by π takes us from one pair of scales to another.

SPECIALTY SCALES Several slide rules offer a family of scales identified as LL or Ln. These scales are related to D, and they are the so-called log-log or lon scales. They enable us to find any power or root of any number within the limits of the rule. They can be extremely helpful in the solution of such problems as 1 to 18 of Problems 34-11.

Also available on some rules are the hyperbolic functions, identified as Th and Sh scales. These are useful in the solution of transmission line problems.

You should always carefully check an unfamiliar slide rule by confirming the interrelationships of the various scales. Since it is easy to memorize some common relationships (e.g., $\sin 30° = 0.5$, $\tan 60° = \sqrt{3}$, and $\log 2 = 0.301$), it is relatively simple to determine whether your S scale is related to D or CI, or whether LLOO is tied to D or A.

PROBLEMS 35-1

Rework the problems of Chaps. 6, 26, 27, and 34 by slide rule according to the limitations of your particular rule.

35-4 PREFERRED VALUES

In the determination of the values of resistors, capacitors, and inductors which may be required in a circuit, such as those calculated in Sec. 15-2, we often find that the values available off the shelf are not identical with our calculated values. We may desire to have a 620-Ω resistor, and the lab assistant says, "Use a 560- or a 680-Ω. Either will be close enough." How can the lab assistant say so, unhesitatingly? How does the lab assistant know? In other words, how do we arrive at *preferred values*?

Under the prompting of the industry as a whole, the Electrical Industries Association has established lists of suggested figures for the guidance of manufacturers and technicians. Several series of values are normally listed, depending on the quality of service required. Most commonly used are the *R*6 and *R*12 series, which list the 6 and 12 values that cover all the requirements for 20% and 10% tolerances, respectively. Becoming more and more called upon is the *R*24 series, which gives values for 5% tolerance. Naturally, the price of the more exact values is considerably higher than the price of the other values, and the *R*6 and *R*12 values meet the demands of ordinary service quite satisfactorily.

Each of the series is developed from a logarithmic progression based on an appropriate root of 10. To develop the *R*6 series we take the $\frac{1}{6}$, $\frac{2}{6}$, $\frac{3}{6}$, $\frac{4}{6}$, $\frac{5}{6}$, and $\frac{6}{6}$ roots of 10, in order. Table 35-1 shows the development of the *R*6 series of preferred values.

You should confirm, by using logarithms, that $10^{\frac{4}{6}} \cong 4.642$. Now, the calculated values may be rounded off to easy-to-remember two-significant-figure numbers in order to arrive at the preferred values. Naturally, all these values may be multiplied by any power of 10, so that memorizing six numbers is all that is needed to cover the entire range of 20% values. The maximum error of ±20% has been arrived at by choosing desired values midway between the two preferred values and determining the percentage error. If we required a 4-kΩ resistor, then choosing either 3.3-kΩ or 4.7-kΩ will not introduce more than a 20% error. Obviously, then, any value closer to a preferred value than one midway between the two must be closer than 20% tolerance. The advantages to manufacturers, sales agencies, and technicians will be obvious at once.

When greater accuracy (less tolerance) is required, we may use the *R*12 series for ±10% or even the *R*24 series for ±5% values. Naturally, the 5% shows all the values in the 10% and 20% series plus intermediate values to round out the series.

Table 35-1 *R*6 SERIES OF PREFERRED VALUES

x	$10^{\frac{x}{6}}$	*Preferred Value*	*Difference*	*Percent Difference*	*Max % Error*
0	1.000	1.0	0.5	50	±20
1	1.468	1.5	0.7	46	18.9
2	2.155	2.2	1.1	50	20
3	3.162	3.3	1.4	42.5	17.5
4	4.642	4.7	2.1	44.6	18.3
5	6.813	6.8	3.2	47	19.1
6	10.	10	5.0	50	20
		15			

PROBLEMS 35-2

1. Using successive twelfth roots of 10, by logarithms, list the preferred values and the maximum percentage errors for the *R*12 series of preferred values.
2. Using successive twenty-fourth roots of 10, by logarithms, list the preferred values and the maximum percentage errors for the *R*24 series of preferred values.
3. The standard published values of capacitors made by a prominent manufacturer follows the *R*10 series. Using successive tenth roots of 10, determine the nominal value of electrolytic capacitors available from this manufacturer, between 100 and 1000 pF. What will be the probable published tolerance?
4. The permeability ratings of a popular line of potentiometer cores follows the *R*5 series. Using successive fifth roots of 10, develop the nominal values between 1 and 100 mH. What will be the probable published tolerance?
5. If your slide rule has a set of Ln scales, confirm your logarithmic calculations in Probs. 1 to 4.

35-5 POWER RATIOS—THE DECIBEL

The Weber-Fechner law states that "the minimum change in stimulus necessary to produce a perceptible change in response is proportional to the stimulus already existing." With respect to our sense of hearing, this means that the ear considers as equal changes of sound intensity those changes which are in the same *ratio*.

The above is more easily understood from a consideration of sound intensities. Any volume of sound must be changed approximately 25% before the ear notes a change in volume. If the volume is increased by this amount, in order for the ear to detect another increase in volume, the new value must be increased by an additional 25%. For example, the output of an amplifier delivering 16 W would have to be increased to a new output of 20 W in order for the ear to discern the increase in volume. Then, in order for the ear to detect an additional increase in volume, the output would have to be increased 25% of 20 W to a new output of 25 W.

From the foregoing it is apparent that a *change* of volume, for example, from 10 to 20 mW (a 10-mW change), would seem the same as the *change* from 100 to 200mW (a 100-mW change) because $\frac{20}{10} = \frac{200}{100}$. Since these changes in hearing response are equally spaced on a logarithmic scale, it follows that the ear responds logarithmically to variations in sound intensity. Therefore, any unit used for expressing power gains or losses in communication circuits must, in order to be practical, vary logarithmically.

One of the earliest of such units was the international transmission unit, the *bel* (B), so called to honor the inventor of the telephone, Alexander Graham Bell. The definition of the bel is

$$\text{bel} = \log_{10} \frac{P_2}{P_1}$$

where P_1 is the initial, or reference, power and P_2 is the final, or referred, power.

In normal practice, the number of bels is quite small, invariably a decimal number, and a derived unit, the *decibel* is used as the practical indicator of power ratio. The abbreviation for decibel is dB. A difference of 1 dB between two sound intensities is just discernible to the ear. Since deci means one-tenth, a decibel is one-tenth the size of a bel, and

$$\text{Number of decibels} = \text{dB} = 10 \log \frac{P_2}{P_1} \quad (1)$$

You should refer to the second paragraph of this section and prove that the difference between two discernible sound intensities is actually 0.969 dB.

example 4 A power of 10 mW is required to drive an AF amplifier. The output of the amplifier is 120 mW. What is the gain, expressed in decibels?

solution $P_1 = 10$ mW, and $P_2 = 120$ mW. dB = ? Substituting in Eq. (1),

$$\text{dB} = 10 \log \frac{120}{10} = 10 \log 12$$
$$= 10.8 \text{ dB gain}$$

example 5 A network has a loss of 16 dB. What power ratio corresponds to this loss?

solution

Given $$\text{dB} = 10 \log \frac{P_2}{P_1} \quad (1)$$

Substituting 16 for dB,

$$16 = 10 \log \frac{P_2}{P_1}$$

D: 10, $$1.6 = \log \frac{P_2}{P_1}$$

Taking antilogs of both members,

$$39.8 = \frac{P_2}{P_1}$$

Thus, a loss of 16 dB corresponds to a power ratio 39.8 : 1.

Because dB is 10 times the log of the power ratio, it is evident that power ratios of 10 = 10 dB, 100 = 20 dB, 1000 = 30 dB, etc. Therefore, it could have been determined by inspection that the 16-dB loss in the preceding example represented a power ratio somewhere between 10 and 100. This is evident by the figure 1 of 16 dB. The second digit 6 of 16 dB is ten times the logarithm of 3.98; hence, 16 dB represents a power ratio of 39.8.

A loss in decibels is customarily denoted by the minus sign. Thus, a loss of 16 dB is written −16 dB.

example 6 A certain radio receiver utilizes a type 6F6 vacuum tube as a final audio stage that delivers 4500 mW to the loudspeaker. The owner is considering modifying the circuit in order to substitute a type 6L6 tube for the 6F6. The 6L6 tube will deliver 6500 mW to the speaker. Is the gain in power sufficient to warrant the expense of making this change?

solution By a change to the 6L6 tube the power output is increased by a ratio of 1.44, nearly 1.5 times. Those not familiar with the use of the decibel would probably think that an increase in power of almost 45% would justify the change. However, when the power ratio is expressed in terms of decibels, it is evident that, as far as the ear is concerned, very little is gained. Substituting in Eq. (1),

$$\text{dB} = 10 \log \frac{6500}{4500} = 10 \times 0.1596 = 1.6$$

Such an increase in power would hardly be worth the owner's effort.

Expressing the gain or loss of various circuits or apparatus in decibels obviates the necessity of computing gains or losses by multiplication and division. Because the decibel is a logarithmic unit, the total gain of a circuit is found by adding the individual decibel gains and losses of the various circuit components.

example 7 A dynamic microphone with an output of −85 dB is connected to a preamplifier with a gain of 60 dB. The output of the preamplifier is connected through an attenuation pad with a loss of 10 dB to a final amplifier with a gain of 90 dB. What is the total gain?

solution In this example, all decibel values have been taken from a common reference level. Because the microphone is 85 dB below reference level, the preamplifier brings the level up to $-85 + 60 = -25$ dB. The attenuation pad then reduces the level to $-25 - 10 = -35$ dB. Finally, the final amplifier causes a net gain of $-35 + 90 = 55$ dB gain. Hence, it is apparent that the overall gain in any system is simply the algebraic sum of the decibel gains or losses of the associated circuit components. Thus, $-85 + 60 - 10 + 90 = 55$ dB gain.

35-6 POWER REFERENCE LEVELS

It is essential that you remember that the decibel is not an absolute quantity, but represents merely a change in power relative to the power at some different time or place. It is meaningless to say that a given amplifier has an output of so many dB unless that output is referred to a specific power level. If we know what the output power is, then the *ratio* of the output power to that specific input power may be expressed in dB.

Several reference levels ("zero-reference" or "zero-dB") have been developed within the industry. Some of these have already been dropped generally; some are used in isolated communities or within individual companies; others are in general use throughout the entire electronics industry. Some of the more common levels are discussed below.

dBm The most common reference level used in the telephone industry is one milliwatt. And since many radio and television programs are carried between studio and transmitter by telephone systems, we should be able to understand telephone transmission engineers when they talk about relative powers. The rather widespread use of the expression "decibels above or below one milliwatt" is usually abbreviated ±dBm. Signal power in communications systems is almost always being amplified (multiplication) or attenuated (division). It is far more convenient to add or subtract dB than to calculate the power in milliwatts or watts by long processes of multiplication or division. Thus, when a telephone engineer speaks of a power level of 25 dBm, the listeners can readily understand that, if $P_1 = 1$ mW, P_2 is 25 dB higher.

example 8 What is the output power represented by a level of 25 dBm?

solution dBm means "decibels referred to a reference power level of 1 mW"; that is, $P_1 = 1$ mW. Then, an amplification of 25 dB means:

$$25 = 10 \log_{10} \frac{P_2}{1 \text{ mW}}$$

$$\log P_2 = 2.5$$

$$P_2 = 316.23 \text{ mW}$$

Because circuits do not amplify or attenuate all frequencies by the same amount, the industry often reserves the term dBm for an input signal of a single-frequency (pure) sine wave (often 400 Hz or 1 kHz). However, dBm is often applied to more complex waveforms because of the convenience of calculations.

6 mW Several radio receiver and audio amplifier manufacturers use 0.006 W (6 mW) as their reference, or zero-dB, level.

example 9 How much power is represented by a gain of 23 dB if zero level is 6 mW?

solution 1 Substituting 23 for dB and 6 for P_1 in Eq. (1),

$$23 = 10 \log \frac{P_2}{6}$$

D: 10, $$2.3 = \log \frac{P_2}{6}$$

Taking antilogs of both members,

$$199.5 = \frac{P_2}{6}$$

$$\therefore P_2 = 1197 \text{ mW}$$

check

$$23 = 10 \log \frac{1197}{6}$$

$$23 = 10 \log 199.5$$

$$23 = 10 \times 2.3$$

solution 2

$$2.3 = \log \frac{P_2}{6}$$

or $$2.3 = \log P_2 - \log 6$$

Transposing,

$$\log P_2 = 2.3 + \log 6$$

Substituting the value of log 6,

$$\log P_2 = 2.3 + 0.778$$

$$\log P_2 = 3.078$$

Taking antilogs,

$$P_2 = 1197 \text{ mW}$$

example 10 How much power is represented by −64 dB if zero level is 6 mW?

solution 1 Substituting −64 for dB and 6 for P_1 in Eq. (1),

$$-64 = 10 \log \frac{P_2}{6}$$

D: 10, $$-6.4 = \log \frac{P_2}{6}$$

The left member of the above equation is a logarithm with a negative mantissa because the entire number 6.4 is negative. Hence, to express this logarithm with a positive mantissa the equation is written

$$3.6 - 10 = \log \frac{P_2}{6}$$

Taking antilogs of both members,

$$3.98 \times 10^{-7} = \frac{P_2}{6}$$

$$\therefore P_2 = 2.39 \times 10^{-6} \text{ mW}$$

check

$$-64 = 10 \log \frac{2.39 \times 10^{-6}}{6}$$

$$= 10 \log 3.98 \times 10^{-7}$$

$$-64 = 10(3.6 - 10)$$

$$-64 = -64$$

solution 2

$$-6.4 = \log \frac{P_2}{6}$$

Then $$-6.4 = \log P_2 - \log 6$$

Transposing, $$\log P_2 = \log 6 - 6.4$$

Substituting the value of log 6,

$$\log P_2 = 0.778 - 6.4$$

$$= (10.78 - 10) - 6.4$$

$$\therefore P_2 = 2.39 \times 10^{-6} \text{ mW}$$

If the larger power is always placed in the numerator of the power ratio, the quotient will always be greater than 1; therefore, the characteristic of the logarithm of the ratio will always be zero or a positive value. In this manner the use of a negative characteristic is avoided. As an illustration, from Example 10,

$$-6.4 = \log \frac{P_2}{6}$$

which is the same as $6.4 = \log 6 - \log P_2$

Hence, $6.4 = \log \frac{6}{P_2}$

It is always apparent whether there is a gain or a loss in decibels; therefore, the proper sign can be affixed after working the problem.

VU The volume unit, abbreviated VU, is used in broadcasting, and it is based on the amplitude of the program frequencies throughout the system. The standard volume indicator (VU meter) is calibrated in decibels with zero level corresponding to 1 mW of power in a 600-Ω line under steady-state conditions, usually at a frequency between 35 Hz and 10 kHz. Owing to the ballistic characteristics of the instrument, the scale markings are referred to as volume units and correspond to dBm only in the case of steady-state sine-wave signals.

dBRN AND dBA The signal-to-noise ratio is very important in most electronic amplifiers and communications circuits. When engineers establish a reference noise level, then the signal power may be expressed as being so many dB above this arbitrary reference level. The expression "decibels referred to an arbitrary reference noise level" is abbreviated dBRN. Often this reference noise level is set at −90 dBm. You should confirm that this represents 1 pW of power.

Then, when an original established reference noise level is adjusted to some new level, as it sometimes is in the telephone industry, the abbreviation dBA indicates "decibels referred to some adjusted reference noise level."

dBRAP A sound may be heard by "the average human ear" (whatever that is) if it has a sound power of 10^{-16} W or more. This minimum power represents the threshold of hearing, and it is called reference acoustical power. Any noise or signal of any kind must be above this power to be heard, and it may then be compared to this minimum power. Thus, dBRAP means a power ratio in dB when $P_1 = 10^{-16}$ W. Sound engineers often call the number of dBRAP by the name *phons*.

OTHER SPECIALIZED TERMS Other reference levels, used in more specialized fields are:

- dBW — dB referred to 1 W as zero-dB reference level.
- dBk — dB referred to 1 kW as reference level.
- dBV — dB referred to 1 V as zero reference signal level.

These, and many other zero reference levels, need introduce no great problem to you. It is only necessary to remember that dB represents a power ratio which must be referred to some original or arbitrary reference level.

35-7 CURRENT AND VOLTAGE RATIOS

Fundamentally, the decibel is a measure of the ratio of two powers. However, voltage ratios and current ratios can be utilized for computing the decibel gain or loss provided that the input and output impedances are taken into account.

In the following derivations, P_1 and P_2 will represent the power input and power output, respectively, and R_1 and R_2 will represent the input and output impedances, respectively. Then

$$P_1 = \frac{E_1^2}{R_1} \quad \text{and} \quad P_2 = \frac{E_2^2}{R_2}$$

Since $\quad \text{dB} = 10 \log \frac{P_2}{P_1}$

substituting for P_1 and P_2,

$$\text{dB} = 10 \log \frac{\dfrac{E_2^2}{R_2}}{\dfrac{E_1^2}{R_1}}$$

$$\therefore \text{dB} = 10 \log \frac{E_2^2 R_1}{E_1^2 R_2}$$

$$= 10 \log \left(\frac{E_2}{E_1}\right)^2 \frac{R_1}{R_2}$$

$$= 10 \log \left(\frac{E_2}{E_1}\right)^2 + 10 \log \frac{R_1}{R_2}$$

$$= 20 \log \frac{E_2}{E_1} + 10 \log \frac{R_1}{R_2} \qquad (2)$$

$$= 20 \log \frac{E_2 \sqrt{R_1}}{E_1 \sqrt{R_2}} \qquad (3)$$

Similarly,

$$P_1 = I_1^2 R_1 \quad \text{and} \quad P_2 = I_2^2 R_2$$

Then, since $\quad \text{dB} = 10 \log \frac{P_2}{P_1}$

by substituting for P_1 and P_2,

$$\text{dB} = 10 \log \frac{I_2^2 R_2}{I_1^2 R_1}$$

$$= 20 \log \frac{I_2}{I_1} + 10 \log \frac{R_2}{R_1} \qquad (4)$$

$$= 20 \log \frac{I_2 \sqrt{R_2}}{I_1 \sqrt{R_1}} \qquad (5)$$

If, in both the above cases, the impedances R_1 and R_2 are *equal*, they will cancel and the following formulas will result:

$$\text{Number of dB} = 20 \log \frac{E_2}{E_1} \qquad (6)$$

and

$$\text{Number of dB} = 20 \log \frac{I_2}{I_1} \qquad (7)$$

It is evident that voltage or current ratios can be translated into decibels *only* when the impedances across which the voltages exist or into which the currents flow are taken into account.

example 11 An amplifier has an input resistance of 200 Ω and an output resistance of 6400 Ω. When 0.5 V is applied across the input, a voltage of 400 V appears across the output. (*a*) What is the power output of the amplifier? (*b*) What is the gain in decibels?

solution

(*a*) $\quad$ Power output $= P_o = \dfrac{E_o^2}{R_o}$

$$= \frac{400^2}{6400}$$

$$= 25 \text{ W}$$

(*b*) $\quad$ Power input $= P_i = \dfrac{E_i^2}{R_i}$

$$= \frac{0.5^2}{200}$$

$$= 1.25 \times 10^{-3} \text{ W}$$

Power gain $= 10 \log \dfrac{P_o}{P_i}$

$$= 10 \log \frac{25}{1.25 \times 10^{-3}}$$

$$= 43 \text{ dB}$$

Check the solution by substituting the values of the voltages and resistances in Eq. (3).

$$\text{dB} = 20 \log \frac{E_o}{E_i}\sqrt{\frac{R_i}{R_o}}$$

$$= 20 \log \frac{400}{0.5}\sqrt{\frac{200}{6400}}$$

$$= 43$$

35-8 THE MERIT, OR GAIN, OF AN ANTENNA

The merit of an antenna, especially one designed for directive transmission or reception, is usually expressed in terms of antenna *gain*. The gain is generally taken as the ratio of the power that must be supplied some standard-comparison antenna to the power that must be supplied the antenna under test in order to produce the same field strengths in the desired direction at the receiving antenna. Similarly, the gain of one antenna over another could be taken as the ratio of their respective radiated fields.

The "effective radiated power" of an antenna is the product of the antenna power and the antenna power gain.

example 12 One kilowatt is supplied to a rhombic antenna, which results in a field strength of 20 μV/m at the receiving station. In order to produce the same field strength at the receiving station, a half-wave antenna, properly oriented and located near the rhombic, must be supplied with 16.6 kW. What is the gain of the rhombic?

solution Because the same antenna is used for reception, both transmitting antennas deliver the same power to the receiver. Hence,

$$\text{dB} = 10 \log \frac{P_2}{P_1} = 10 \log \frac{16.6}{1} = 12.2$$

PROBLEMS 35-3

1. How many decibels correspond to a power ratio of (*a*) 20, (*b*) 25, (*c*) 62.5, (*d*) $\frac{1}{177}$?
2. Referred to equal impedances, how many decibels correspond to a voltage ratio of (*a*) 42, (*b*) 100, (*c*) $\frac{1}{130}$, (*d*) $\frac{7}{180}$?
3. If 0 dB is taken as 6 mW, how much voltage across a 90-Ω load does this represent?
4. If 0 dB is taken as 6 mW, how much voltage across a 600-Ω load does this represent?
5. What is the voltage across a 600-Ω line at zero dBm?
6. What is the voltage across a 600-Ω line at 10 dBm?
7. If reference level is taken as 12.5 mW, how much voltage across a 300-Ω load does this represent?
8. If reference level is taken as 12.5 mW, how much voltage across a 600-Ω load does this represent? How much current flows through the load?
9. If 0 dB is 6 mW, compute the power in milliwatts, and the voltage across a 600-Ω load for the following output power meter readings: (*a*) 3 dB, (*b*) 10 dB, (*c*) −10 dB, (*d*) −80 dB.
10. If 0 dB is 1 mW, compute the power in milliwatts and the voltage across

a 600-Ω load for the following output meter readings: (a) 5 dB, (b) 10 dB, (c) 20 dB, (d) −10 dB.

11. An amplifier is rated as having a 90-dB gain. What power ratio does this represent?

12. The amplifier of Prob. 11 has equal input and output impedances. What is the ratio of the output current to the input current?

13. An amplifier has a gain of 60 dBm. If the input power is 1 mW, what is the output power?

14. If a high-selectivity tuned circuit has a very high Q, spurious signals which are 10% lower or higher in frequency will be attenuated at least 50 dB. What power ratio is represented by this level?

15. The manufacturer of a high-fidelity 100-W power amplifier claims that hum and noise in the amplifier is 90 dB below full power output. How much hum and noise power does this represent?

16. In the amplifier of Prob. 14, what will be the dB level of noise to signal when the amplifier is producing 3 W of output power?

17. A network has a loss of 80 dB. What power ratio corresponds to this loss?

18. If the network in Prob. 17 has equal input and output impedances, what is the ratio of the output voltage to the input voltage?

19. In single-sideband operation, the signals appearing in the unwanted set of sidebands should be attenuated by at least 30 dB. What is the ratio of output powers of the desired signal to the unwanted signal?

20. The noise level of a certain telephone line used for wired music programs is 60 dB down from the program level of 12.5 mW. How much noise power is represented by this level?

21. A certain crystal microphone is rated at −80 dB. There is on hand a final AF amplifier rated at 60 dB. How much gain must be provided by a preamplifier in order to drive the final amplifier to full output if an attenuator pad between the microphone and preamplifier has a loss of 20 dB? (All dB ratings are taken from the same reference.)

22. The output of a 200-Ω dynamic microphone is rated at −81.5 dB from a reference level of 6 mW. This microphone is to be used with an amplifier which is to have a power output of 25 W. What gain must be provided between the microphone and the amplifier output?

23. If the amplifier of Prob. 21 has an output impedance of 2.7 kΩ, what is the overall voltage ratio from microphone output to amplifier output?

24. What is the equivalent power amplification in the amplifier of Prob. 23?

25. It is desired to use the amplifier of Prob. 21 with a phonograph pickup which is rated at −20 dBm. To keep from overloading the amplifier, how much loss must be introduced between pickup and input?

26. An amplifier has a normal output of 30 W. A selector switch is arranged to reduce the output in 5-dB steps. What power outputs correspond to reduction in output of 5, 10, 15, 20, 25, and 30 dB?

27. An amplifier is operating at 37 dBm with a gain of 50 dB. The input resistance of the amplifier is 22 kΩ. What is the input voltage to the amplifier?

28. A type 2N45 transistor has the following ratings when used as a class A power amplifier:

• Collector voltage, V	−20
• Emitter current, mA	5
• Input impedance, Ω	10
• Source impedance, Ω	50
• Load impedance, Ω	4500
• Power output, mW	45
• Power gain, dB	23

What is the power input?

29. An amplifier has an input impedance of 600 Ω and an output impedance of 6000 Ω. The power output is 30 W when 1.9 V is applied across the input.
(*a*) What is the voltage gain of the amplifier?
(*b*) What is the power gain in decibels?
(*c*) What is the power input?

30. An amplifier has an input impedance of 500 Ω and an output impedance of 4500 Ω. When 0.10 V is applied across the input, a voltage of 350 V appears across the output.
(*a*) What is the power output of the amplifier?
(*b*) What is the power gain in decibels?
(*c*) What is the voltage gain of the amplifier?

31. A dynamic microphone with an output level of −72 dB is connected to a speech amplifier consisting of three voltage amplifier stages. The first voltage amplifier stage has a voltage gain of 100, and the second has a voltage gain of 9. The interstage transformer between the second and third voltage amplifier stages has a step-up ratio of 3:1, and the third stage has a voltage gain of 8. The driver stage and modulator have a gain of 23 dB. If zero power level is 6 mW, what is the output power of the modulator? (***hint*** transformers do not introduce *power changes*.)

Microphone ▶	Amplifier ▶	Amplifier ▶	Transformer ▶	Amplifier ▶	Modulator
−72 dB	100	9	3	8	23 dB

32. How many decibels gain is necessary to produce a 60-μW signal in 600-Ω telephones if the received signal supplies 9 μV to the 80-Ω line that feeds the receiver?

33. In the receiver of Prob. 32, if the overall gain is increased to 96 dB, what received signal will produce the 60-μW signal in the telephones?

34. The voltage across the 600-Ω telephones is adjusted to 1.73 V. When the AF filter is cut in, the voltage is reduced to 1.44 V. What is the "insertion loss" of the filter?

35. The input power to a 50-km line is 10 mW, and 40 μW is delivered at the end of the line. What is the attenuation in decibels per kilometer?

36. It is desired to raise the power level at the end of the line of Prob. 35 to that of the original output. What is the voltage gain of the required amplifier?

37. In Prob. 35, what is the ratio of input power to output power?

38. One of the original attenuation units was the *neper*, which is given by

$$\text{Number of nepers} = \log_\varepsilon \frac{I_1}{I_2}$$

Since

$$\text{Number of dB} = 20 \log_{10} \frac{I_1}{I_2}$$

what is the relation between nepers and decibels for equal impedances?

hint $\log_\varepsilon \frac{I_1}{I_2} = 2.30 \log_{10} \frac{I_1}{I_2}$

39. A television transmitting antenna has a power gain of 8.6 dB. If the power input to the antenna is 15 kW, what is the effective radiated power?

40. Five hundred watts is supplied to a directive antenna, which results in a field strength of 5 μV/m at a receiving station. In order to produce the same field strength at the same receiving station, the standard-comparison antenna must be supplied with 8 kW. What is the decibel gain of the directive antenna?

41. A rhombic transmitting antenna produces a field strength of 98 μV/m at a receiving test station. The standard-comparison antenna delivers a field strength of 5 μV/m. What is the decibel gain of the rhombic antenna?

42. A broadcasting station is rated at 1 kW. If the received signals vary as the square root of the radiated power, how much gain in decibels would be apparent to a nearby listener if the broadcasting station doubled its power?

35-9 TRANSMISSION LINES

A transmission line is a device consisting of one or more electric conductors and designed for the purpose of transferring electric energy from one point to another. The transmission line has a wide variety of uses: in one form it can carry electric power to a city several miles distant from the power plant; in another form it can be used for carrying chain broadcast programs from one studio to several broadcast stations; and in still another form it can carry RF energy from a radio transmitter to an antenna or from an antenna to a radio receiver.

The most common types of transmission lines are:

1. The two-wire open-air line as shown in Fig. 35-4*a*. This line consists of two parallel conductors whose spacing is carefully held constant.
2. The concentric-conductor line, as illustrated in Fig. 35-4*b*, which consists of tubular conductors one inside the other.

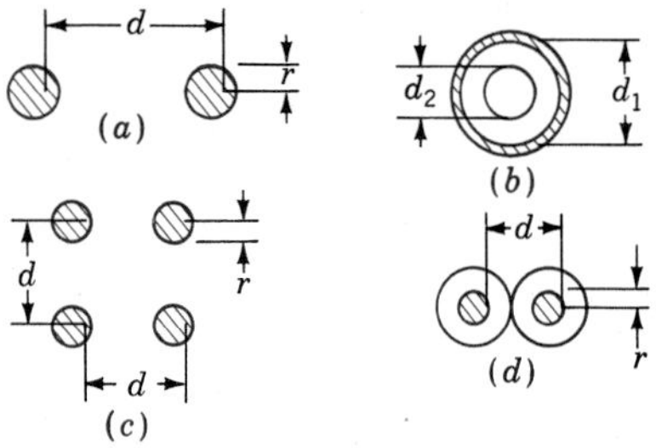

Fig. 35-4 Types of Transmission Lines

3. The four-wire open-air line as shown in Fig. 35-4c. In this type of line the diagonally opposite wires are connected to each other for effecting an electrical balance.
4. The twisted-pair line, as shown in Fig. 35-4*d*, which may consist of lamp cord, a telephone line, or other insulated conductors.

Any conductor has a definite amount of self-inductance, capacitance, and resistance per unit length. These properties account for the behavior of transmission lines in their various forms and uses.

The derivations of the transmission line equations that follow can be found in advanced engineering texts.

35-10 THE INDUCTANCE OF A LINE

The inductance of a two-wire open-air line is given by the equation

$$L = 0.621\, l\left(0.161 + 1.48 \log_{10} \frac{d}{r}\right) \times 10^{-3}\ \text{H} \quad (8)$$

where L = inductance of line and return, H
l = length of line, km
d = distance between conductor centers
r = radius of each wire, same units as d

example 13 What is the inductance of a line 145 km long consisting of No. 0000 copper wire spaced 1.5 m apart?

solution Diameter of No. 0000 copper wire = 11.68 mm; thus radius = 5.84 mm.

$$\frac{d}{r} = \frac{1500}{5.84} = 256.85$$

$$\log_{10} 256.85 = 2.409\ 68, \text{ say } 2.41$$

Then $L = 0.621 \times 145(0.161 + 1.48 \times 2.41)\ \text{mH}$
$= 336\ \text{mH}$

For radio frequencies, more accurate results are obtained by the approximate relation

$$L \cong 9.21 \times 10^{-9} \log_{10} \frac{d}{r} \quad \text{H/cm} \quad (9)$$

where L is the inductance in henrys per centimeter and d and r have the same values as in Eq. (8).

35-11 THE CAPACITANCE OF A LINE

The capacitance of a two-wire open-air line is

$$C = \frac{0.0121\, l}{\log \frac{d}{r}} \quad \mu\text{F} \quad (10)$$

where C = capacitance of line, μF
l = length of line, km
d = distance between wire centers
r = radius of wire, in same units as d

example 14 What is the capacitance per kilometer of a line consisting of No. 00 copper wire spaced 1.2 m apart?

solution Diameter of No. 00 copper wire = 9.266 mm; thus radius = 4.633 mm.

$$\frac{d}{r} = \frac{1200}{4.633} = 259$$

$$\log_{10} 259 = 2.413\,30 \qquad \text{say } 2.413$$

Then
$$C = \frac{0.0121}{2.413} \mu\text{F/km}$$
$$= 5.01 \text{ nF/km}$$

For radio frequencies, more accurate results are obtained by the equation

$$C \cong \frac{1}{9.21 \times 10^{-9}\, c^2 \log_{10} \dfrac{d}{r}} \quad \text{F/cm} \qquad (11)$$

where C is the capacitance in farads per centimeter, c is the velocity of light (3×10^{10} cm/s), and d and r have the same values as in Eq. (10).

The capacitance of submarine cables and of cables laid in metal sheaths is given by

$$C = \frac{0.0241Kl}{\log_{10} \dfrac{d_1}{d_2}} \quad \mu\text{F} \qquad (12)$$

where C = capacitance of line, μF
K = relative dielectric constant of insulation
l = length of line, km
d_1 = inside diameter of outer conductor
d_2 = outside diameter of inner conductor

example 15 A No. 14 copper wire is lead-sheathed. The wire is insulated with 3 mm gutta percha ($K = 4.1$). What is the capacitance of 1 km of this cable?

solution

$$d_2 = \text{diameter of No. 14} = 1.63 \text{ mm}$$
$$d_1 = 1.63 + 2(3) = 7.63 \text{ mm}$$
$$\log \frac{d_1}{d_2} = \log \frac{7.63}{1.63} = \log 4.687 = 0.6703$$
$$l = 1 \text{ km}$$

Then
$$C = \frac{0.0241Kl}{\log \dfrac{d_1}{d_2}} = \frac{0.0241 \times 4.1 \times 1}{0.6703} = 0.147\ \mu\text{F}$$

PROBLEMS 35-4

1. What is the inductance of a 120-km line consisting of two No. 00 wires spaced 1 m between centers?
2. What is the inductance of a 30-km line consisting of two No. 6 copper wires spaced 60 cm between centers?
3. A transmission line is 3.7 km long and consists of two No. 0 solid copper wires spaced 40 cm between centers. Determine (*a*) the inductance of the line and (*b*) the capacitance of the line.
4. If the spacing of the line of Prob. 3 were 1 m between centers, what would be (*a*) the inductance and (*b*) the capacitance?
5. A 40-km long two-wire line is to be constructed of No. 0 solid copper wire. What must be the minimum spacing between centers to keep the capacitance below 0.250 μF?
6. A 22-km two-wire line consisting of No. 00 solid copper wire is spaced

180 mm between wire centers. What is the capacitance of the line in microfarads per kilometer?

7. A lead-sheathed underground cable is to be constructed with solid copper wire covered with 12.5 mm of rubber insulation ($K = 4.3$). If the maximum capacitance per kilometer must be limited to 0.15 μF, $\pm$10%, what size conductor should be used?
8. A lead-sheathed cable consisting of No. 0 copper wire with 12.5 mm of rubber insulation ($K = 4.3$) is broken. A capacitance bridge measures 0.26 μF between the conductor and the sheath. How far out is the open circuit?
9. What is the capacitance per kilometer of the cable of Prob. 8?
10. The cable of Prob. 7 becomes open-circuited 5 km out. What reading will be given on a capacitance bridge?
11. The value of the current in a line at a point l km from the source of power is given by

$$i = I_0 \varepsilon^{-\kappa l}$$

where I_0 is the current at the source and κ is the attenuation constant. In a certain line, with $\kappa = 0.02$ dB/km, find the length of line where i is 10% of the original current I_0.

12. If the attenuation of a line is 0.012 dB/km, how far out from the power source will the current have decreased to 70.7% of its original value?
13. A two-wire open-air transmission line is used to couple a receiving antenna to the receiver. The line is 155 m long and consists of No. 10 wire spaced 15 cm between centers. Using Eqs. (9) and (11), find:
    (*a*) Inductance per centimeter of line
    (*b*) Capacitance per centimeter of line
    (*c*) Inductance of the entire line
    (*d*) Capacitance of the entire line
14. A two-wire open-air transmission line is used to couple a radio transmitter to an antenna. The line is 250 m long, and it consists of No. 14 wire spaced 14 cm between centers. Using Eqs. (9) and (11), find the (*a*) inductance of the line and (*b*) capacitance of the line.

35-12 CHARACTERISTIC IMPEDANCES OF RF TRANSMISSION LINES

The most important characteristic of a transmission line is the *characteristic impedance*, denoted by Z_0 and expressed in ohms. This impedance is often called *surge impedance*, *surge resistance*, or *iterative impedance*.

The value of the characteristic impedance is determined by the construction of the line, that is, by the size of the conductors and their spacing. At radio frequencies, the characteristic impedance can be considered to be a resistance the value of which is given by

$$Z_0 = \sqrt{\frac{L}{C}} \quad \Omega \tag{13}$$

where L and C are the inductance and capacitance, respectively, per unit length of line as given in Eqs. (9) and (11). The unit of length selected for L and C is immaterial as long as the *same* unit is used for both.

Substituting the values of L and C for a two-wire open-air transmission line in Eq. (13) results in

$$Z_0 = 276 \log_{10} \frac{d}{r} \quad \Omega \tag{14}$$

where d is the spacing between wire centers and r is the radius of the conductors *in the same units as d*. Note that the characteristic impedance is *not* a function of the length of the line.

Equation (14) is valid when you can conveniently neglect the capacitance of each line to ground. If you cannot, replace $\frac{d}{r}$ with $\frac{d}{D}$, where D is the diameter of the conductors.

example 16 A transmission line is made of No. 10 wire spaced 30 cm between centers. What is the characteristic impedance of the line?

solution $d = 30$ cm $= 300$ mm. Diameter of No. 10 wire $= 2.588$ mm; therefore $r = 1.294$ mm.

$$Z = 276 \log \frac{d}{r} = 276 \log \frac{300}{1.294}$$
$$= 276 \log 231.8 = 276 \times 2.365$$
$$= 653\ \Omega$$

The characteristic impedance of a concentric line is given by

$$Z_0 = 138 \log_{10} \frac{d_1}{d_2} \quad \Omega \tag{15}$$

where d_1 is the inside diameter of the outer conductor and d_2 is the outside diameter of the inner conductor.

example 17 The outer conductor of a concentric transmission line consists of copper tubing 1.6 mm thick with an outside diameter of 25 mm. The copper tubing which forms the inner conductor is 0.8 mm thick with an outside diameter of 6 mm. What is the characteristic impedance of the line?

solution

$$d_1 = 25 - (2 \times 1.6) = 21.8 \text{ mm}$$
$$d_2 = 6 \text{ mm}$$
$$Z_0 = 138 \log \frac{d_1}{d_2} = 138 \log \frac{21.8}{6}$$
$$= 138 \log 3.633$$
$$= 138 \times 0.5603 = 77.3\ \Omega$$

PROBLEMS 35-5

1. What is the characteristic impedance of a two-line open-air transmission line consisting of No. 10 wire spaced 150 mm between centers?
2. It is desired to use No. 14 wire to provide a transmission line with a characteristic impedance of approximately 500 Ω. What logical spacing between centers should be used?
3. If a 50 mm spacing is used for the line of Prob. 2, what percentage of error is introduced by assuming that the line does have a characteristic impedance of 500 Ω?

4. It is necessary to construct a 600-Ω transmission line to couple a radio transmitter to its antenna, and No. 10 wire is readily available. What should be the spacing between wire centers?
5. The impedance at the center of a half-wave antenna is approximately 74 Ω. For maximum power transfer between transmission line and antenna, the impedance of the line must match that of the antenna. Is it physically possible to construct an *open-wire* line with a characteristic impedance as low as 74 Ω?
6. Plot a graph of the characteristic impedance in ohms against the ratio $\frac{d}{r}$ for two-wire open-air transmission lines. Use values of $\frac{d}{r}$ between 1 and 150.
7. It is desired to construct a 600-Ω two-wire line at a certain radio station. In the stock room there are on hand a large number of 30-cm spreader insulators. That is, these spreaders will space the *wires* 30 cm. What size wire should be ordered to obtain as nearly as possible the desired impedance if the 30-cm spreaders are used?

   hint $d = 30 + 2r$.

8. What outside-diameter tubing should be used to construct a quarter-wave matching stub having an impedance of approximately 300 Ω if spreaders 40 mm long are used?
9. The outer conductor of a concentric transmission line is a copper pipe 5 mm thick with an outside diameter of 70 mm. The inner conductor is a copper rod 6 mm in diameter. What is the characteristic impedance of the line?
10. The inside diameter of the outer conductor of a coaxial line is 10 mm. The surge impedance is 90 Ω. What is the diameter of the inner conductor?
11. Plot a graph of the characteristic impedance in ohms against the ratio $\frac{d_1}{d_2}$ for concentric transmission lines. Use values of $\frac{d_1}{d_2}$ between 2 and 10.
12. A particular grade of twisted-pair transmission line, which has a surge impedance of 72 Ω, has a loss of 0.2 dB/m. For a 30-m length of line, determine (*a*) the total loss in decibels and (*b*) the efficiency of transmission.

   hint $\% \text{ efficiency} = \frac{\text{power output}}{\text{power input}} \times 100$

13. The twisted-pair line of Prob. 12 is replaced by a coaxial cable that has a loss of 0.01 dB/m. What is the new efficiency of transmission?
14. For a two-wire transmission line, the attenuation in decibels *per meter of wire* is given by the equation

$$\alpha = \frac{0.0157\, R_{ac}}{\log_{10} \frac{d}{r}} \qquad \text{dB/m} \tag{16}$$

where R_{ac} is the ac resistance of one meter of *wire*. One kilowatt of power, at a frequency of 16 MHz, is delivered to a 460-m line consisting of No. 8 wire spaced 30 cm between centers. The RF resistance of No. 8 wire is 49 times the dc resistance. (*a*) What is the line loss in decibels and (*b*) what is the efficiency of transmission?

15. If the spacing of the line in Prob. 14 should be changed to 20 cm between centers, (*a*) what will be the line loss in decibels, and (*b*) what is the efficiency of transmission?

16. For a concentric transmission line, the attenuation in decibels per meter of *line* is expressed by the relation

$$\alpha = \frac{38.33\, f(d_1 + d_2)10^{-6}}{d_1 d_2 \log_{10} \frac{d_1}{d_2}} \qquad \text{dB/m} \tag{17}$$

where d_1 and d_2 are in centimeters and have the same meaning as in Eq. (15), and f is the frequency in megahertz. A concentric line 400 m long consists of an outer conductor with an inside diameter of 3.2 cm and an inner conductor that is 0.8 cm in diameter. At a frequency of 27.8 MHz, (*a*) what is the line loss in dB, and (*b*) what is the efficiency of transmission?

17. The capacitance of a vertical antenna which is shorter than one-quarter wavelength at its operating frequency can be computed by the equation

$$C_a = \frac{55\,177l}{\left(\log_\varepsilon \frac{200l}{d} - 1\right)\left[1 - \left(\frac{fl}{75}\right)^2\right]} \qquad \text{pF} \tag{18}$$

where C_a = capacitance of antenna, pF
l = height of antenna, m
d = diameter of antenna conductor, cm
f = operating frequency, MHz

Determine the capacitance of a vertical antenna that is 85 m high and consists of 1.5-cm wire. The antenna is being operated on 214 kHz.

18. The RF resistance of a copper concentric transmission line can be computed by

$$R_{ac} = 8.33\, f\left(\frac{1}{d_1} + \frac{1}{d_2}\right) \times 10^{-3} \qquad \Omega/\text{m} \tag{19}$$

where f = frequency, MHz
d_1 = inside diameter of outer conductor, cm
d_2 = outside diameter of inner conductor, cm

What is the resistance of a concentric line 76 m long operating at 132 MHz if $d_1 = 3.8$ cm and $d_2 = 0.48$ cm?

19. If an antenna is matched to a coaxial transmission line, the percent efficiency is given by

$$\eta = \frac{100R_T}{Z_0 + R_T} \quad \% \tag{20}$$

where Z_0 = characteristic impedance of the concentric line
R_T = effective resistance of the line due to attenuation, obtainable from the line constants

$$R_T = Z_0(\varepsilon^{\frac{rl}{Z_0}} - 1) \quad \Omega \tag{21}$$

where r = RF resistance per meter of line as found in Eq. (19)
l = length of line, m

Find the efficiency of transmission of a matched concentric transmission line with a characteristic impedance of 300 Ω. The line is 24 m long, and it has an RF resistance of 0.72 Ω/m.

20. What is the efficiency of transmission of a matched concentric transmission line with a characteristic impedance of 90 Ω if the line is 335 m long and has an RF resistance of 0.33 Ω/m?

36

number systems for computers

Have you ever wondered about our numbering systems—the seldom-discussed "philosophy" of how we count? In this chapter, we shall explore the background of counting systems and apply the knowledge gained to the electronic computing field.

36-1 NUMBERS IN GENERAL

Recall from Sec. 6-17 how we referred to the problem of adding 5×10^3 to 3×10^2:

$$\begin{aligned} 5 \times 10^3 &= 5000 \\ 3 \times 10^2 &= \underline{\ \ 300} \\ 5 \times 10^3 + \overline{3 \times 10^2} &= \overline{5300} = 5.3 \times 10^3 \end{aligned}$$

In other words, a number like 5300 may be thought of as being made up of two separate parts, 5×10^3 and 3×10^2. Similarly, all the numbers in our decimal system may be broken down into different factors multiplied by suitable powers of 10. For example, 5328 may be thought of as—indeed, it really is

5000	or	5×10^3
300		3×10^2
20		2×10^1
8		8×10^0

and we could write 5328 in the form

$$5 \times 10^3 + 3 \times 10^2 + 2 \times 10^1 + 8 \times 10^0$$

In fact, the very way we place the digits in their appropriate places carries out the sense of powers of 10. In many elementary schools, students learn the "place names" of the digits in a long number like this:

```
7 654 321
| ||| ||└─units
| ||| |└──tens
| ||| └───hundreds
| ||└─────thousands
| |└──────ten thousands
| └───────hundred thousands
└─────────millions
```

and so on, and we would pronounce the whole number by using most of those place names: "seven million, six hundred fifty-four thousand, three hundred twenty-one."

36-2 BINARY NUMBERS

When we talk about decimal numbers, or the decimal number system, we mean we are counting in units of 10. That is, our numbering system has a *radix* of 10.

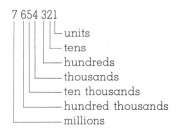

In the *binary* system, which is used extensively in digital computers, the radix is 2 and every number in the system represents an appropriate factor times the suitable power of 2:

	or, in binary
$0 = 0 \times 2^0$	$= 0_2$
$1 = 1 \times 2^0$	$= 1_2$
$2 = 1 \times 2^1 + 0 \times 2^0$	$= 10_2$
$3 = 1 \times 2^1 + 1 \times 2^0$	$= 11_2$
$4 = 1 \times 2^2 + 0 \times 2^1 + 0 \times 2^0$	$= 100_2$

Stop and be sure. 100_2, that is, one hundred in the binary numbering system, means: from the position of the digits,

$$1 \times 2^2 + 0 \times 2^1 + 0 \times 2^0$$

or

$$4 + 0 + 0 = 4$$

If you are sure, go on. If you are not sure, go back to the introduction and start the chapter again. When you are sure of the notion that a power of 2 must be connected with each digit in the binary number and the particular power depends upon the location of the digit in the number, go on and prove the following extension of the binary table:

$5 = 101_2$	$11 = 1011_2$
$6 = 110_2$	$12 = 1100_2$
$7 = 111_2$	$13 = 1101_2$
$8 = 1000_2$	$14 = 1110_2$
$9 = 1001_2$	$15 = 1111_2$
$10 = 1010_2$	$16 = 10000_2$

example 1 Write the decimal equivalent of the number 10011001_2.

solution Taking our cue from the position of the digits in the number and keeping track of the appropriate powers of 2, we convert each digit to its decimal equivalent, evaluating from the right:

$$\begin{aligned} 1 \times 2^0 &= 1_{10} \\ 0 \times 2^1 &= 0_{10} \\ 0 \times 2^2 &= 0_{10} \\ 1 \times 2^3 &= 8_{10} \\ 1 \times 2^4 &= 16_{10} \\ 0 \times 2^5 &= 0_{10} \\ 0 \times 2^6 &= 0_{10} \\ 1 \times 2^7 &= 128_{10} \\ \hline 10011001_2 &= 153_{10} \end{aligned}$$

You should use the subscripts to designate the *system* in which you are counting until you are satisfied with your confidence in intersystem conversions.

PROBLEMS 36-1

Write the following binary numbers in decimal form:

1. 000101	**2.** 001010	**3.** 000001	**4.** 001011	**5.** 000111
6. 100111	**7.** 101010	**8.** 110001	**9.** 100011	**10.** 111101

Now let us consider the reverse operation: converting a decimal number into its binary equivalent. Again we are looking for factors (either 1 or 0) times suitable powers of 2. The number 153, for instance, contains 128, which is 2^7. The remainder, $153 - 128 = 25$, contains 16, which is 2^4. The next remainder, $25 - 16 = 9$, contains 8, which is 2^3, and the last remainder, $9 - 8 = 1$, is 2^0. However, to write the complete binary equivalent, we must show the factors (zero) of 2^6, 2^5, 2^2, and 2^1.

$$153_{10} = 10011001_2$$

Obviously, it would be a tremendous help to know the whole 2^x table, and students anticipating advanced studies in computer designing, programming, or servicing will make these conversions by memorizing the 2^x table, say, to $2^{10} = 1024$. However, a ready mechanical method of arriving at the same binary number, without forgetting the missing powers of 2, is to convert the multiplication process into one of repeated division:

Radix Divisor	*Decimal Number to Be Converted*	*Remainder*
2	)153	1
2	)76	0
2	)38	0
2	)19	1
2	)9	1
2	)4	0
2	)2	0
	1 →	

Read Up

Writing the quotient and the remainders in order "backwards," we arrive at

$$153_{10} = 10011001_2$$

PROBLEMS 36-2

Convert the following decimal numbers into binary form:

1. 6 **2.** 12 **3.** 18 **4.** 23 **5.** 31
6. 88 **7.** 97 **8.** 126 **9.** 177 **10.** 361

36-3 OCTAL NUMBERS

Modern computers speak to us in binary numbers, but their internal workings are often in octal numbers, the computers translating their octal results into binary readouts. Octal numbers are based on a counting system whose radix is 8:

		Or
0_{10}	$= 0 \times 8^0$	$= 0_8$
1	$= 1 \times 8^0$	$= 1_8$
2	$= 2 \times 8^0$	$= 2_8$
3	$= 3 \times 8^0$	$= 3_8$
4	$= 4 \times 8^0$	$= 4_8$
5	$= 5 \times 8^0$	$= 5_8$
6	$= 6 \times 8^0$	$= 6_8$
7	$= 7 \times 8^0$	$= 7_8$
8	$= 1 \times 8^1 + 0 \times 8^0$	$= 10_8$
9	$= 1 \times 8^1 + 1 \times 8^0$	$= 11_8$
10	$= 1 \times 8^1 + 2 \times 8^0$	$= 12_8$
11	$= 1 \times 8^1 + 3 \times 8^0$	$= 13_8$
12	$= 1 \times 8^1 + 4 \times 8^0$	$= 14_8$
13	$= 1 \times 8^1 + 5 \times 8^0$	$= 15_8$
14	$= 1 \times 8^1 + 6 \times 8^0$	$= 16_8$
15	$= 1 \times 8^1 + 7 \times 8^0$	$= 17_8$
16	$= 2 \times 8^1 + 0 \times 8^0$	$= 20_8$

Just as the binary system uses digits up to, but not including, 2, so the octal system uses only digits below its radix, 8.

Write the following number in decimal form:

$$2731_8$$

As in the binary, we take our cue from the position of the digits in the number and introduce the appropriate powers of 8, reading from the right:

$$\begin{aligned} 1 \times 8^0 &= 1 \\ 3 \times 8^1 &= 24 \\ 7 \times 8^2 &= 448 \\ 2 \times 8^3 &= 1024 \\ \hline 2731_8 &= 1497_{10} \end{aligned}$$

PROBLEMS 36-3

Convert the following octal numbers into their decimal equivalents:

1. 00002	**2.** 00017	**3.** 00063	**4.** 00102	**5.** 00077
6. 00100	**7.** 01124	**8.** 01035	**9.** 06270	**10.** 22453

The conversion of decimal numbers to octal equivalents is achieved in the same fashion as in the binary, except that the divisor is the radix 8 instead of 2:

example 2

Radix Divisor	*Decimal Number to Be Converted*	*Remainder*
8	)1497	1
8	)187	3
8	)23	7
	2 →	Read up

$$1497_{10} = 2731_8$$

PROBLEMS 36-4

Convert the following decimal numbers to their octal equivalents:

1. 25	**2.** 37	**3.** 84	**4.** 127	**5.** 165
6. 477	**7.** 823	**8.** 1062	**9.** 3928	**10.** 5000

36-4 SYSTEMS WITH ANY RADIX

Just as we have developed binary numbers with radix 2 or octal numbers with radix 8, so we may develop any number system. Consider, for example, quinary numbers: the digits in a quinary number will consist of appropriate factors times suitable powers of 5. The factors may be 0, 1, 2, 3, and 4, but not 5 or higher.

$$22_5 = 2 \times 5^1 + 2 \times 5^0 = 12_{10}$$

PROBLEMS 36-5

Write the following decimal numbers in the systems of the indicated radices:

Number:	Radix:	Number:	Radix:
1. 9	3	**2.** 12	4
3. 27	5	**4.** 256	16
5. 256	4	**6.** 565	3
7. 1728	12	**8.** 1728	7
9. 5280	6	**10.** 672	5

Write the decimal equivalents of the numbers given:

11. 224_5	**12.** 163_7	**13.** 323_4	**14.** 201_3	**15.** 003_{12}
16. 0725_9	**17.** 0106_8	**18.** 2388_9	**19.** 51402_6	**20.** 73006_8

36-5 CONVERSION BETWEEN SYSTEMS

We have already seen how to convert from any numbering system to decimal and from decimal to any other. Thus, if we should be required to convert a number with any given radix a into a system with some other radix b, we could do so in two steps: (1) convert the given number into its decimal equivalent. (2) convert the decimal equivalent into the new system.

example 3 Convert 5134_6 into its binary equivalent.

solution In the first step, convert 5134_6 to 1138_{10}. In the second step, convert 1138_{10} to 10001110010_2.

Actually, most of the conversions which concern us are between the binary and the octal systems.

example 4 Convert 1772_8 into its binary equivalent.

solution

$$1772_8 = 1018_{10} = 1111111010_2 = 1\ 111\ 111\ 010_2$$

In the various numbering systems, no change of value is introduced if we add zeros to the *left* of a number, so that we may change the appearance of $1\ 111\ 111\ 010_2$ to $001\ 111\ 111\ 010_2$ without introducing any value change but yielding a number which consists of a quantity of groups of three binary digits. (BInary digiTS are often referred to as *bits*.) By good advance planning, (1) the octal numbering system uses ordinary arabic numerals up to 7, and (2) the largest binary number consisting of three digits is 7 ($= 111_2$). If we evaluate each digit in the octal number into its three-bit binary equivalent, we arrive at

1	7	7	2_8
001	111	111	010_2

Thus, $1772_8 = 001\ 111\ 111\ 010_2$.

example 5 Convert 5317_8 into its binary equivalent.

solution Replace each octal digit in turn with its binary three-bit equivalent:

$$5317_8 = 101\ 011\ 001\ 111_2$$

example 6 Convert 10110011001_2 to its octal equivalent.

solution From the right, mark off the given binary number into groups of three bits:

$$010\ 110\ 011\ 001$$

Replace each three-bit group with its regular decimal equivalent to arrive at the octal equivalent of the number:

$$010\ 110\ 011\ 001_2 = 2631_8$$

PROBLEMS 36-6

Convert the following octal numbers to their binary equivalents:

1. 361_8 **2.** 277_8 **3.** 532_8 **4.** 465_8 **5.** 106_8
6. 737_8 **7.** 5266_8 **8.** 4137_8 **9.** 7777_8 **10.** 1000_8

Convert the following binary numbers to their octal equivalent:

11. 000101_2 **12.** 011001_2 **13.** 11101_2
14. 0011_2 **15.** $110\ 111\ 101_2$ **16.** 100100100_2
17. 110011010_2 **18.** $001\ 001\ 111_2$ **19.** 10101010_2
20. 10111010_2

36-6 BINARY ADDITION

The addition of two quantities $a + b$, may, in binary devices, have only four possible values:

$$0 + 0 = 0 \qquad 0 + 1 = 1$$
$$1 + 0 = 1 \qquad 1 + 1 = 10$$

because of the dichotomous (two-state, on-off, open-closed, flipped-flopped, 1-0) nature of switching devices, and therefore the sum S of the addition $a + b$ will be limited to the four possible answers shown above. The first three forms present no difficulty, and we can add binary numbers which involve them very easily:

$$\begin{array}{r} 11001 \\ 00100 \\ \hline 11101 \end{array} \quad \text{or} \quad \begin{array}{r} 25 \\ 4 \\ \hline 29 \end{array}$$

But the addition of $1 + 1$ involves us in a two-part answer: 10. The 0 part of this answer is the *sum*, and the 1 part is the *carry*. This is similar to ordinary arithmetic. When the addition of two numbers requires it, say, $9 + 5$, we "put down 4 and carry 1."

example 7 Add 100110 and 110101.

solution Set the two numbers down in traditional addition form, one above the other. Addition of $0 + 0$, $0 + 1$, and $1 + 0$ involves nothing new. When adding $1 + 1$, put down 0 and carry 1 over to the next stage of addition:

$$\begin{array}{r} 1 \\ 100110 \\ 110101 \\ \hline 1011011 \end{array} \quad \text{or} \quad \begin{array}{r} 38 \\ 53 \\ \hline 91 \end{array}$$

PROBLEMS 36-7

Add the following binary numbers:

1. 010001	**2.** 100101	**3.** 1001101	**4.** 0110110	**5.** 100111
101000	010101	0100011	0100111	010101
6. 101111	**7.** 100011	**8.** 110010	**9.** 011010	**10.** 100101
010111	011110	011010	011010	111011

11 to **20.** Prove each of your answers by converting the individual parts into their decimal equivalents.

36-7 SUBTRACTION OF BINARY NUMBERS

Similarly to binary addition, binary subtraction is limited to four possibilities:

$$0 - 0 = 0$$
$$1 - 0 = 1$$
$$1 - 1 = 0$$
$$0 - 1 = 1$$

and carry 1 (or "borrow" 1)

When we are subtracting one ordinary number from another and come upon a step involving 5 − 8, we borrow 1 from the digit to the left of the 5, subtract 8 from 15, and obtain 7. Binary subtraction is no different.

example 8 Subtract 0110 from 1011.

solution Set the numbers in column form, the subtrahend below the minuend. When we must subtract 1 from 0, we borrow 1 from the number to the left of the 0 to make it 10. Then, 10 − 1 = 1:

$$\begin{array}{r} 1011 \\ -0110 \\ \hline 0101 \end{array} \qquad \begin{array}{r} 11 \\ -\ 6 \\ \hline 5 \end{array}$$

example 9

$$\begin{array}{r} 1000101 \\ -0110011 \\ \hline 0010010 \end{array}$$

You should convert these two binary numbers into their equivalent decimal numbers and test the solution.

PROBLEMS 36-8

Perform the following binary subtractions:

1. 010011	**2.** 011011	**3.** 001101	**4.** 110111
−001010	−010111	−000100	−011101
5. 111000	**6.** 110100	**7.** 110110	**8.** 100111
−010001	−101111	−011111	−100011

9. $\begin{array}{r} 111111 \\ -111010 \\ \hline \end{array}$ **10.** $\begin{array}{r} 100110 \\ -100101 \\ \hline \end{array}$

11 to **20.** Prove each answer by converting all parts of each problem into their equivalent decimal forms.

36-8 SUBTRACTION BY ADDING COMPLEMENTS

One of the oldest rules in subtraction is "change the sign and add." This policy makes binary subtraction extremely simple. Changing the sign of a binary number is like changing the condition of a switch. *On* becomes *off*, and *open* becomes *closed*. *Flipped* becomes *flopped*, 1 becomes 0, and 0 becomes 1.

example 10 Subtract 01101 from 11001 by means of complementation.

solution Rewrite the problem, changing the subtrahend to its 1's complement; then add:

$$\begin{array}{r} 11001 \\ -01101 \\ \hline \end{array} \text{ becomes } \begin{array}{r} 11001 \\ +10010 \\ \hline 101011 \end{array} \text{ that is, } \begin{array}{r} 25 \\ -13 \\ \hline 43 \end{array}$$

43?! Well, when the answer to such a process comes out with one more digit than the number of digits we had to start with, we transfer this extra digit as an "end-carry" and add it back in:

$$\begin{array}{r} 11001 \\ 10010 \\ \hline 101011 \\ + \llcorner\!\!\longrightarrow 1 \\ \hline 01100 \end{array} \quad \text{which is } 12_{10}$$

example 11 Perform the subtraction 11101101 − 01001011 by means of 1's complement.

solution Rewrite the subtrahend into its 1's complement and add. Bring down the extra 1, if any, and add it as an end-carry:

$$\begin{array}{r} 11111 \\ 11101101 \\ +10110100 \\ \hline 110100001 \\ \llcorner\!\!\longrightarrow 1 \\ \hline 10100010 \end{array} \qquad \begin{array}{r} \\ 237 \\ -75 \\ \hline \\ \\ 162 \end{array}$$

PROBLEMS 36-9

Perform the following subtractions by means of complementation:

1. 110010 − 100111 **2.** 101101 − 010010 **3.** 011001 − 001101
4. 001101 − 000110 **5.** 010101 − 001001 **6.** 101011 − 001010
7. 111101 − 110010 **8.** 101111 − 001100 **9.** 110010 − 001101
10. 001110 − 001001

11 to **20.** Prove each answer by converting all the parts into their decimal equivalents.

36-9 BINARY MULTIPLICATION

Since 1 times anything is the thing itself and 0 times anything is 0, binary multiplication is very easy.

example 12 Multiply 1101 by 100.

solution Set down the numbers as for ordinary multiplication and multiply in the usual way. Add the partial answer rows in binary form:

```
   1101      13
 × 100      × 4
 ------     ---
   0000
  0000
 1101
 ------
 110100      52
```

example 13 Multiply 10011 by 101.

solution As before, multiply by long multiplication methods. There is no need to write a complete line of 0's—just set down the right-hand 0 and shift the line for the following multiplier one step to the left:

```
  10011      19
 ×  101     × 5
 -------    ---
  10011
 100110
 -------
 1011111     95
```

PROBLEMS 36-10

Multiply:

1. 101111 by 10 **2.** 110011 by 11 **3.** 100101 by 101
4. 010111 by 100 **5.** 101001 by 111 **6.** 110011 by 110
7. 100111001 by 1001 **8.** 11001110 by 1101 **9.** 101001101 by 1001
10. 111001111 by 1011
11 to **20.** Prove each solution to Probs. 1 to 10 by converting all parts into their equivalent decimal forms.

36-10 BINARY DIVISION

Dividing by binary numbers is as easy as multiplying. Either the divisor is smaller than the dividend and the quotient is 1 or the divisor is larger than the dividend and the quotient is 0.

example 14 Divide 1000001 by 101.

solution Write the numbers as for ordinary long division. Will the three-bit divisor go into the first three bits of the dividend or not? If it will, put down a 1 as the first item in the quotient, and carry on. If it will not, bring down the next digit in the dividend, and put down a 0 as the first item of the quotient:

```
       01101        13
 101)1000001      5)65
     000
     ----
     1000
      101
     ----
      0110
       101
      ----
       0010
       0000
       ----
        0101
         101
        ----
          xx
```

PROBLEMS 36-11

Perform the following divisions:

1. 010101 by 111 **2.** 011110 by 101 **3.** 011011 by 011
4. 010100 by 100 **5.** 110000 by 1000 **6.** 001001011 by 1111
7. 001101100 by 1001 **8.** 101000100 by 10010 **9.** 101111100 by 10011
10. 1101010011 by 10111
11 to **20.** Prove each answer by converting all parts into their decimal equivalents.

In this book of basic mathematics for electronics, we will not attempt to go deeper into this fascinating subject of binary operations. If you become involved with digital devices, you will find other useful relationships in books which specialize in computer arithmetic. We trust that, at that time, this chapter will help you to relearn the subject.

37 boolean algebra

More and more, electronic devices are being put to work in computing machines and controlling machines. First, electronic tubes superseded relays, and then transistors took the place of tubes. Now, newer and more exotic devices are being added to the list of computer and control components.

And with these applications of electronic devices, there is a growing need for technologists to know at least something about the logic operations of computers. The subject, generally, is known as *Boolean algebra* in honor of George Boole (1815–1864), who developed the work upon which the subject is now based. It is also often referred to as propositional calculus, mathematical logic, and truth-functional logic.

Here we are going to explore the basic ideas of Boolean algebra to see how we can put *logic* to work for us in two ways: (1) to describe circuits mathematically, after they have been designed or assembled and (2) to design circuits mathematically before they are assembled. We are not going to do any work in the *philosophical* field, where logic and its algebra are extremely useful. Several excellent books have been written from that point of view, whereas there has been little introductory work from the point of view of switching or logic circuits.

37-1 THE SYMBOLS OF LOGIC CIRCUITRY

Different associations, different manufacturers, different authors, and different publishers all have their own ideas as to what symbols should be used in logic circuits. Table 37-1 shows the ANSI Y32.14 standard symbols which will be used in this book. However, you must be prepared to recognize others in textbooks, technical journals, trade magazines, and manufacturers' literature.

37-2 THE SYMBOLS OF MATHEMATICAL LOGIC

Just as the symbol for resistance appearing in circuit diagrams is replaced in the electronics mathematics by the symbol *R*, so the

Table 37-1

AND	NOT	OR	NOR	NAND

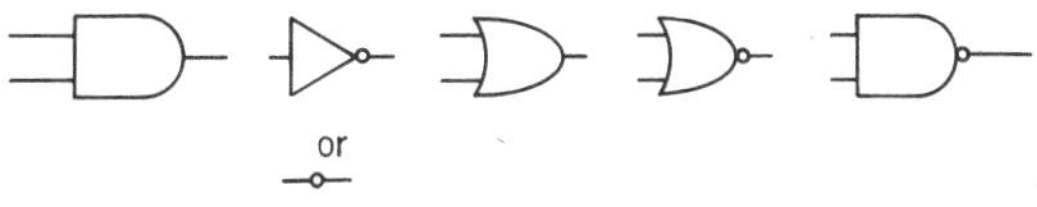

symbols of logic circuitry shown in Table 37-1 are replaced in the logic mathematics by their own special mathematical symbols, and these are shown in Table 37-2. Let us look further into the meanings of the circuit symbols and see what mathematical expressions are required.

AND The AND symbol means that an output signal will be produced by the particular device, regardless of the total amount of circuitry involved, only when both the a and b input signals are applied. Our mathematical counterpart must carry this meaning of AND.

OR The OR symbol means that an output will be produced by the device when either the a input or the b input signals are applied or when *both* are applied. Our mathematical replacement must give this meaning of "either . . . OR . . . , or both."

NOT The NOT (inverter) symbol means that either (1) there will *not* be an output when the input signal *is* applied, or (2) there *is* an output when the input signal *is not* applied. Our mathematical symbol must carry the meaning of "not," or "reversed."

Now we must develop mathematical operators, sometimes referred to as *truth functors*, which will simply and effectively describe these circuit requirements. Table 37-2 shows the variety of symbols used in the literature, and, again, the symbol at the head of each column is the one to be used throughout this book. As well as the appearance of the symbols and their general purpose, we must take particular pains to be able to pronounce them:

Table 37-2 LOGIC MATHEMATICAL SYMBOLS

	AND	*OR*	*NOT*
Symbols used in this text	· juxtaposition	+	—
Other symbols sometimes used	&	v	

AND $a \cdot b$ may be pronounced
- a and b
- both a and b
- the logical product of a and b
- a conjunct b
- the conjunction of a and b
- a in series with b
- if, and only if, a as well as b

OR $a + b$ may be pronounced
- a or b or both
- either a or b (or both)
- the inclusive OR of a and b
- the disjunction of a and b
- the alternation of a and b
- the logical sum of a and b
- a in parallel with b
- at least one of a and b
- if, and only if, a or b or both
- true if, and only if, a or b or both

NOT $\bar{a}$ may be pronounced
- not a
- the complement of a
- the inverse of a
- the negation of a
- the rejection of a

NOT	$\bar{a}$	it is false that a a is not assertable "not a" is true the valence of a is false

These pronunciations are the ones often met with in dealing with logic statements. Those appearing at the end of each group are the ones more usually found in philosophical statements, and they are included as a general-interest addition to our main study. At the same time, special symbols are often used for the *exclusive* OR operator, when we want to say "either a OR b, but not both together." Note that our definition of OR does not suit this requirement. However, we will say this in symbol form later *without* using any other special symbol.

AGGREGATE SYMBOLS: (), [] In addition to the operator symbols are the symbols of aggregation, already met with in Sec. 3-9. Everything inside an aggregate symbol is subject to the operator symbol which may be applied to the aggregate: $\overline{(a + b)}$ means "when input signal a or input signal b or both are applied, there will be no output signal." (Can you see that this could be said, "not a and not b"?)

TRUTH SYMBOLS: 1, 0 In addition to the operators and aggregates, we require "truth symbols" to say whether or not a signal is *true* or *false*, whether there is a signal or there is not a signal, whether a switch is closed or open. Sometimes the letters T and F are used for these designations, but more frequently 1 and 0 are used. (See how these *two possible states* lead us into applications of *binary* arithmetic.)

Thus, if switch . . . a . . . is closed, its value is 1. When switch . . . c . . . is open, its value is 0.

example 1 Express in logical mathematical symbols the statement "It is raining and the wind is blowing."

solution First of all, select identification symbols to stand for the two propositions which make up the statement, say r for "it is raining" and b for "the wind is blowing." Second, since these two propositions are connected, we must choose the operational symbol which will represent AND, using the · or mere juxtaposition of the identification symbols.

"It is raining and the wind is blowing" $= r \cdot b$
or $= rb$

example 2 Express in logical symbols the statement "Either switch p is open when switch q is closed or switch p is closed when switch q is open."

solution Select identification symbols:

p = switch p closed
$\bar{p}$ = switch p open
q = switch q closed
$\bar{q}$ = switch q open

Then select the operational symbols to represent the conditions:

1. The requirements of "either . . . OR . . ." are met by the use of + = OR.
2. The requirements of "when" = "at the same time" = AND is met with · or juxtaposition

"Either switch p is open when switch q is closed or switch p is closed when switch q is open" $= \bar{p}q + p\bar{q}$.

PROBLEMS 37-1

By using s to represent "We are going to school" and l to represent "We are learning something new," write in symbolic form the following statements:

1. We are going to school, and we are learning something new.
2. We are going to school, but we are not learning something new.
3. Either we are going to school or we are learning something new, or both.
4. We are not going to school, but we are learning something new.
5. When we are going to school, then we are learning something new.
6. We are not going to school, therefore, we are not learning anything new.
7. Either we are not going to school or we are learning something new, or both.
8. We are neither going to school nor learning something new.
9. We are (*a*) both going to school and learning something new or else (*b*) we are not going to school and we are not learning something new.
10. Either we are going to school or we are learning something new, but not both.

37-3 THE AXIOMATIC TAUTOLOGIES

In Sec. 5-2 we have already learned that an axiom is a statement which is so self-evident that it need not be formally proved. And a tautology is nothing more than a statement or equation which shows two different ways of saying the same thing. This is a specific mathematician's version of the dictionary definition. For example, $\sin\theta = \frac{\text{opp}}{\text{hyp}}\theta$ is a tautology. Sometimes it is convenient to use one relationship; sometimes the other.

While philosophical logic introduces many tautologies and develops them with great care, the following brief introduction will serve the purposes of most students working in this text. Some, who go on to computer or control engineering, will want to study further to broaden their scope in the subject.

T.1 $$a \cdot a = a$$

This is the *redundancy law of multiplication.* It means that whenever a circuit design calls for a contact on relay a to be closed and later calls for another contact on the same relay a to be closed in series with the first, we really need only a single contact on relay a.

T.2 $$a + a = a$$

This is the *redundancy law of addition.* It means that when a circuit calls for a contact on relay a to be closed and later for another contact on the same relay to be closed in parallel with the first, we need only a single contact on relay a.

These first two tautologies, or laws, really say, "Saying the same thing over and over again does not make it any more true."

T.3 $$a \cdot b = b \cdot a$$

This is the *commutative law of multiplication.* In the mathematics of logic, as in many other systems (but not all) it does not matter what the order of the multiplication is or, in switching algebra, what the physical order of the switches in series is.

T.4 $a + b = b + a$

This is the *addition law of commutation.* It does not matter whether a is in parallel with b or whether b is in parallel with a.

T.5 $(a \cdot b)c = a \cdot (b \cdot c)$

This is the *associative law of multiplication* and means, again, that the order of switches in series or the order of factors in multiplication does not matter.

T.6 $(a + b) + c = a + (b + c)$

The *associative law of addition,* which is applied in the same way as T.5 and in ordinary algebra.

T.7 $\bar{\bar{a}} = a$

This is the *law of double complementation,* and it means that an inverted inversion has the same effect as the original proposition. (A switch, which can only be open or closed, if changed in position twice, is back in its original position.)

note Ordinary English grammar does not follow this definition because we do not always understand that two negatives make a positive in an ordinary English statement.

T.8 $a + \bar{a} = 1$

This is the *first law of complementation.* Since the circuit will always give an output signal if one contact is normally closed and the other, in parallel, is normally open, a *true* indication will always appear.

T.9 $a \cdot \bar{a} = 0$

This is the *second law of complementation.* It is impossible to achieve an output signal with one contact open in series with another that is closed.

T.10 $a(b + \bar{b}) = a$

This tautology says that a contact a in series with a circuit that is always operating (T.8) will have the same effect as if that contact were alone.

T.11 $a + (b \cdot \bar{b}) = a$

Any contact a in parallel with a permanent open circuit (T.9) will have the same effect as a alone.

T.12 $\overline{a \cdot b} = \bar{a} + \bar{b}$

This is the first of De Morgan's *laws of negation.* Some serious thought, coupled with the work which will follow, will prove the truth of this and the next tautology.

T.13 $\overline{a + b} = \bar{a} \cdot \bar{b}$

The second of De Morgan's laws of negation.

Some additional tautologies will be found inside the back cover, and these will be referred to in the text below.

37-4 TRUTH TABLES

Analysis of circuits by mathematical logic may be carried out by purely algebraic means, using the tautologies, and this method will be investigated shortly. But another useful method of analyzing circuits is the method of *truth tables.* These are a fairly systematic mechanical method of examining the possible combinations of truths (or circuit conditions) existing in a particular problem. For instance, consider the circuit in Fig. 37-1,

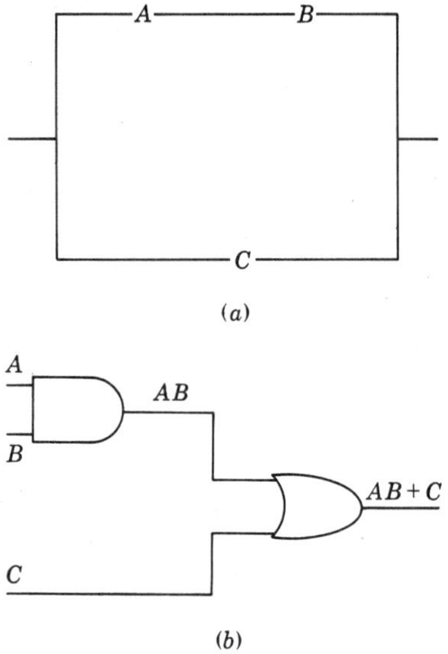

Fig. 37-1 Switching Circuit for $a \cdot b + c$ in Table 37-3

which may be described mathematically as $a \cdot b + c$. We may set up the truth table for this circuit to determine which combinations of closed (1) or open (0) conditions of the switches will produce an output signal.

The first step is to list the three possible contacts, a, b, and c, and the possibilities appearing in the formula. This step gives us the row of headings across the top of Table 37-3. Under these headings there will appear eight rows of data and calculations: 2^3, where the 2 represents the two possible states 1 or 0 and the 3 represents the three different switches, or contacts, a, b, and c. Note the mechanical method of establishing the possible combinations: each of the contacts will be open for half of the possibilities, and each will be closed for half. By making the first half of the eight possibilities for a 1 and the second half 0, then half of a's 1 conditions will see b 1 and half will see b 0, and so on.

Now, referring to the circuit, Fig. 37-1, and Table 37-3, check the circuit for each row of combinations:

- Combination 1. When switches a, b, and c are all closed (1), there is a complete circuit through the series leg (ab) and a complete circuit through the parallel switch c. Then the two closed parallel circuits will give a true (1) result, and there will be an output signal.
- Combination 2. When both a and b are closed, then even with c open, there will be an output signal and again the last, or *total circuit*, column reads 1.
- Combination 3. Here a and c are closed and b is open. Hence, even when the series leg is an open circuit (0), the closed switch c in parallel yields an output signal.
- Combination 4. When a is the only closed switch, then open b prevents a signal getting through the series leg and open c in parallel means that there will be *no* output

Table 37-3 TRUTH TABLE FOR $ab + c$

Combination	a	b	c	$a \cdot b$	$a \cdot b + c$
1	1	1	1	1	1
2	1	1	0	1	1
3	1	0	1	0	1
4	1	0	0	0	0
5	0	1	1	0	1
6	0	1	0	0	0
7	0	0	1	0	1
8	0	0	0	0	0

signal from the circuit. The final column reads 0.

- Combination 5. In combinations 5 through 8, since a is open, the condition of b has no effect, since the series leg is of necessity open. (See column ab.) Switch c, in parallel with this open circuit, determines that there will be an output signal when c is closed and no output signal when c is open.

You must satisfy yourself that there are no other possible switch combinations and that there will be a complete circuit, or an output signal, only for combinations 1, 2, 3, 5, and 7 and no output signal for combinations 4, 6, and 8. The formula for the circuit, $ab + c$, is sometimes said to be a tautology for the five closed combinations, although this is a loose use of the word.

PROBLEMS 37-2

Prepare the truth tables for the following expressions:

1. $ac + bc$
2. $c(a + b)$
3. $\overline{a + b + c}$
4. $\bar{a} \cdot \bar{b} \cdot \bar{c}$
5. $a + \bar{a} = 1$
6. $a(b + c) = ab + ac$
7. $a(a + b) = a$
8. $a + ab = a$
9. $p + \bar{p}q = p + q$
10. $\overline{a \cdot b} = \bar{a} \cdot \bar{b}$
11. $\overline{a + b} = \bar{a} \cdot \bar{b}$
12. $a + bc = (a + b)(a + c)$
13. $\overline{x + y} + \overline{x + z} = \overline{x + yz}$
14. $(p + q)(\bar{q} + r)(q + 1) = p\bar{q} + qr$
15. $(a + b)(\bar{a} + c)(b + c) = \bar{a}b + ac$
16. $(a + c)(a + d)(b + c)(b + d) = ab + cd$

37-5 PROPOSITIONAL INVESTIGATIONS

Sometimes it happens that a proposed circuit is described in Boolean algebra in a rather complicated manner and it is possible to use the tautologies in order to simplify it.

example 3 A designer asks for a circuit which will perform the following switching function:

$$\overline{a + b} + \overline{a + c}$$

Can we simplify the circuit requirements before drawing modules from stock and putting them together as requisitioned?

solution Choosing the appropriate tautologies (and here practice is the only cure), we alter the appearance of the original problem formula and see what might be done. (In the example, each step below has been identified with the number of the tautology applied. (See inside back cover.)

- Given $\overline{a + b} + \overline{a + c}$
- T.13 $\overline{a + b}$ may be written $\bar{a} \cdot \bar{b}$
- T.13 $\overline{a + c}$ may be written $\bar{a} \cdot \bar{c}$

and the formula becomes $\bar{a} \cdot \bar{b} + \bar{a} \cdot \bar{c}$

- T.14 $\bar{a}(\bar{b} + \bar{c})$
- T.12 $\bar{a}(\overline{b \cdot c})$
- T.13 $\overline{a + bc}$

Compare the original circuit, as requested, with the simplified version (Fig. 37-2a versus b). You should prepare a truth table for the two circuits, and prove that the two forms are tautological, that is, when one set of switches is true, then the other is also true for all possible identical combinations. Check also to satisfy yourself that there are no combinations other than 2^3.

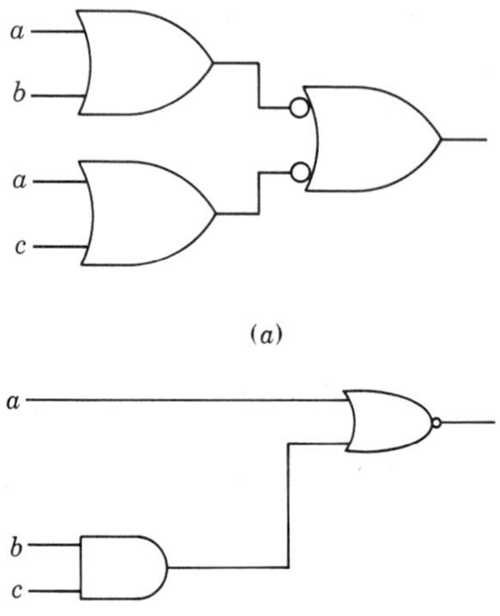

(a)

(b)

Fig. 37-2 Equivalent Switching Combinations of Example 3

PROBLEMS 37-3

Use truth tables to prove the following statements:

1. $\bar{a}b(a + b) = \bar{a}b$
2. $(a + b)(\bar{a} + c)(b + c) = \bar{a}b + ac$
3. $(\bar{a}b + a)(\bar{a}b + c) = (a + b)(\bar{a} + c)(b + c)$
4. $a(\bar{a} + b)(\bar{a} + b + c) = ab$
5. $abc(a + b + c) = abc(ab + bc + ac) + abc(abc + ab)$
6. $\bar{q}t + qt + \bar{q}\cdot\bar{t} = \bar{q}(qt) + \bar{q}\overline{(q\cdot t)}$
7. $st + vw = (s + v)(s + w)(t + v)(t + w)$
8. $ABC + A\bar{B}C + AB\bar{C} + A\bar{B}\bar{C} + \bar{A}BC + \bar{A}\bar{B}C + \bar{A}B\bar{C} = A + B + C$
9. $(\alpha + \beta)(\alpha + \gamma) = \alpha + \beta\gamma$
10. $\overline{(a\cdot b + bc + ac)} = \bar{a}\cdot\bar{b} + \bar{b}\cdot\bar{c} + \bar{a}\cdot\bar{c}$

37-6 SWITCHING NETWORKS

While actual switches may be adjusted so that some contacts *make* before others *break*, or vice versa, or some close or open in a special sequence, in general, every individual switch is either open or closed, off or on, flipped or flopped. This two-state condition lends itself to binary operation (1 or 0), and to Boolean analysis. When a switch is closed, it provides, theoretically, perfect permittance to a current flow, and when it is open, perfect hindrance. It is convenient to define Y_{pq} as the permittance of a circuit between the points p and q and Z_{pq} as the hindrance of the circuit between the same points. Obviously, $Y_{pq} = \bar{Z}_{pq}$.

example 4 Write the expressions for the permittance and the hindrance of the circuit of Fig. 37-3.

solution To write the expression for the permittance of the circuit Y_{lm} we agree that

$$Y_{lm} = Y_a(Y_b + Y_cY_d)$$

where Y_a is the permittance of switch a, and so on. We may write this simply as

$$Y_{lm} = a(b + cd)$$

and we understand that the letter designation for a switch without an overbar indicates that the switch is closed, that is, offers perfect permittance. Studying the circuit, you can see that when contact a is closed and then either b or c and d in series is closed, the circuit will offer permittance—there will be an output signal.

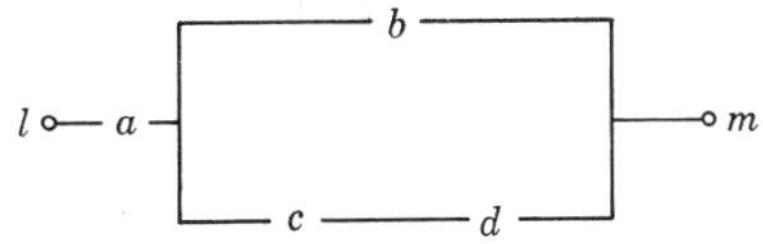

Fig. 37-3 Switching Circuit of Example 4

Similarly, the hindrance of a contact, that is, an open switch, is indicated by the letter designation with an overbar, so that Z_{lm} must be written:

$$Z_{lm} = \bar{a} + (\bar{b})(\bar{c} + \bar{d})$$

When contact a is open, or else when both b is open and either c or d is open, then there will be no output signal—or perfect hindrance.

You should prepare a set of truth tables to show that $Y_{lm} = \overline{Z}_{lm}$.

PROBLEMS 37-4

1. Write the expressions for (a) the hindrance and (b) the permittance of the circuit of Fig. 37-4.
2. Write the expressions for (a) the hindrance and (b) the permittance of the circuit of Fig. 37-5.
3. Write the expressions for (a) the hindrance and (b) the permittance of the circuit of Fig. 37-6.
4. Write the expressions for (a) the hindrance and (b) the permittance of the circuit of Fig. 37-7.

Fig. 37-4 Switching Circuit for Prob. 1

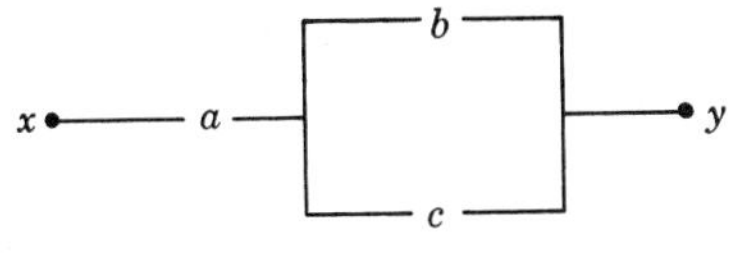

Fig. 37-5 Switching Circuit for Prob. 2

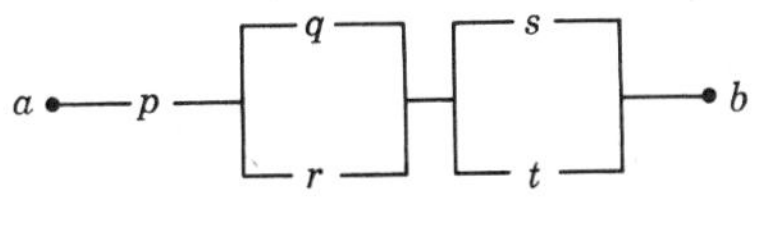

Fig. 37-6 Switching Circuit for Prob. 3

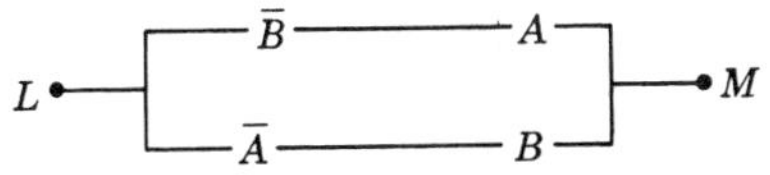

Fig. 37-7 Switching Circuit for Prob. 4

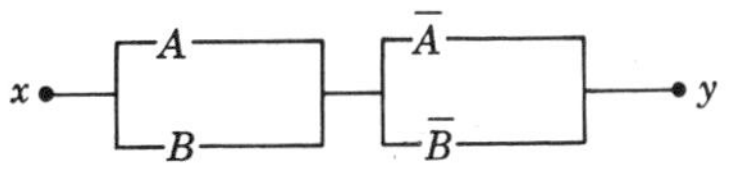

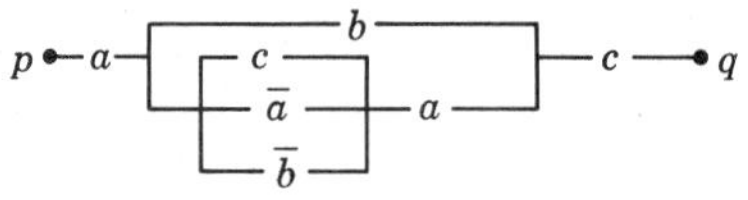

Fig. 37-8 Switching Circuit for Prob. 5

5. Write the expressions for (*a*) the hindrance and (*b*) the permittance of the circuit of Fig. 37-8.

Draw circuits for the following expressions:

6. $Y_{pq} = a(b + c)(ad)$

7. $Y_{lm} = xy(\overline{y}z + \overline{x})a$

8. $Y_{ab} = [\alpha(\beta + \overline{\gamma}) + \beta]\gamma$

9. $Z_{cd} = A[BC + C(\overline{A} + B)] + \overline{B} \cdot \overline{C}$

10. $Y_{pq} = \overline{A} \cdot \overline{B}(C + D)\overline{B} + \overline{D}$

Equivalent switching networks may be developed mathematically by using the tautologies of Boolean algebra, whereby somewhat complicated circuits may be reduced to circuits which will perform identical services with less hardware or, alternatively, to circuits which will perform identical services with readily available, although not simpler, hardware.

example 5 Given the switching network of Fig. 37-9, develop a simpler circuit which will provide an identical switching service.

Fig. 37-9 Switching Circuit of Example 5

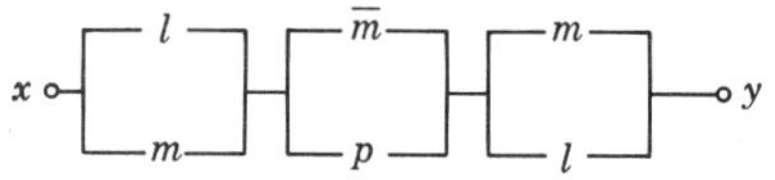

solution Write either the permittance or hindrance function of the circuit:

$$\begin{array}{ll} Y_{xy} = & (l + m)(\overline{m} + p)(m + l) \\ \text{T.4} & (l + m)(\overline{m} + p)(l + m) \\ \text{T.2} & (l + m)(\overline{m} + p) \end{array}$$

That is, the network of Fig. 37-9 may be replaced by that of Fig. 37-10. You should prepare a truth table to prove that the two circuits are tautological.

Fig. 37-10 Simpler Circuit Equivalent of Fig. 37-9

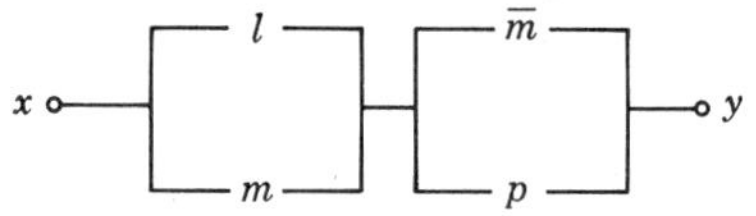

PROBLEMS 37-5

1. By using the appropriate tautologies, develop a simpler circuit to replace that of Fig. 37-11.

Fig. 37-11 Switching Circuit for Prob. 1

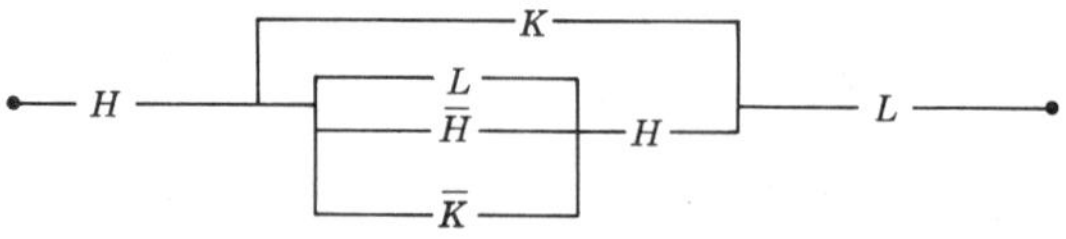

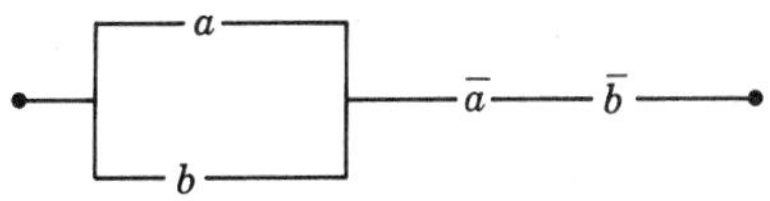

Fig. 37-12 Switching Circuit for Prob. 2

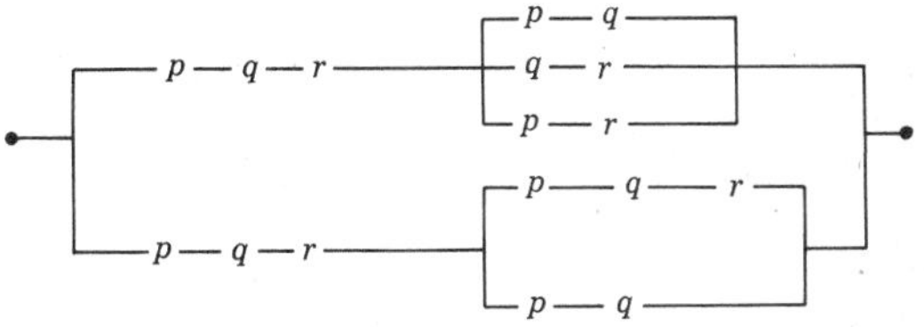

Fig. 37-13 Switching Circuit for Prob. 3

2. Develop a simpler circuit to replace that of Fig. 37-12.
3. Develop a simpler circuit to replace that of Fig. 37-13.
4. Develop a simpler circuit to replace that of Fig. 37-14.
5. Develop a simpler circuit to replace that of Fig. 37-15.
6 to **10.** Check each of your solutions above by means of truth tables.

Fig. 37-14 Switching Circuit for Prob. 4

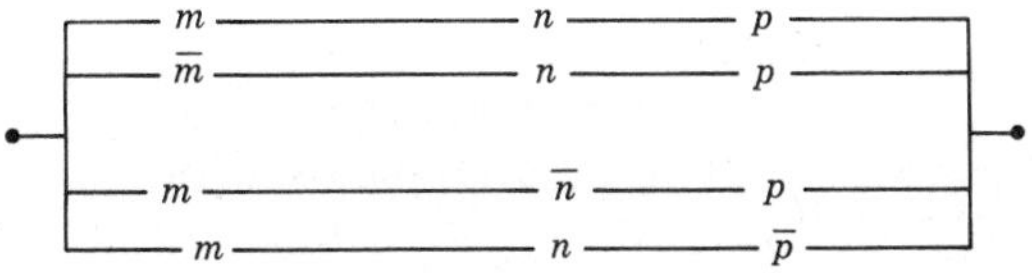

Fig. 37-15 Switching Circuit for Prob. 5

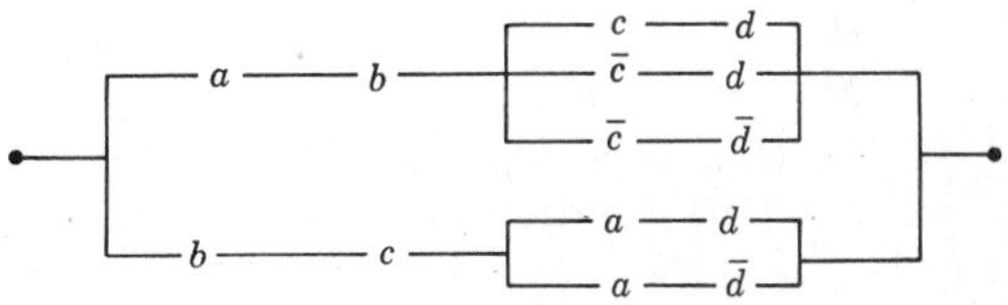

37-7 COMPUTER GATING APPLICATIONS

The standard computer gating symbols are shown in Table 37-1. These simple symbols (and the circuits for which they stand) may be combined into *adders* or *half-adders* or other more complex components. Let us look at a few of the simple tautologies as they would appear in gating configurations.

example 6 Tautology T.14 states that $a(b + c) = ab + ac$. The two circuit configurations are shown in Fig. 37-16.

investigation You should check the two parts of Fig. 37-16 and satisfy yourself that the two circuits do perform the same functions. Then, by preparing a truth table for the two statements, you will see that when $a(b + c)$ is 1, so also is $ab + ac$, and when $a(b + c)$ is 0, so also is $ab + ac$. Then, since the two forms have been proved by tracing and by truth table to be tautological, the end results of using one will be identical with those of using the other. There may be times when availability of circuit wiring boards or parts may make it more desirable

Fig. 37-16 Equivalent Circuits of inclusive OR

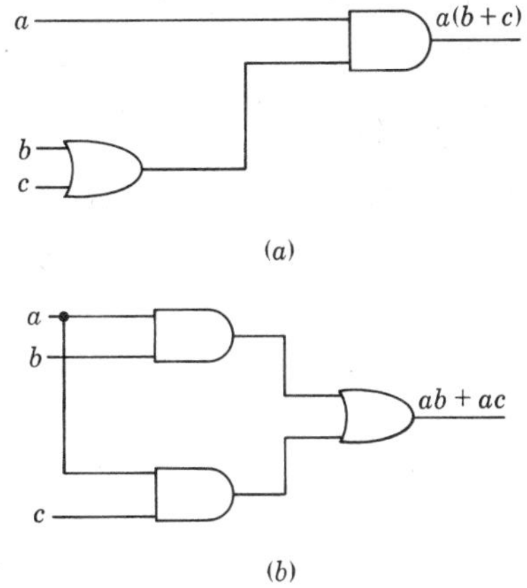

to use one circuit rather than the other, but the results will be the same regardless of the circuit configuration chosen.

You can see, then, that it may often be convenient to spend time exploring the possibilities mathematically, before even breadboarding a circuit, in order to reduce the total number of components or the number of different components required.

PROBLEMS 37-6

1. Write the output expression for the circuit of Fig. 37-17 and develop an alternate circuit. Test your answer by means of a truth table.
2. Write the output expression for the circuit of Fig. 37-18 and develop an alternate circuit. Test your answer by means of a truth table.
3. The *half-adder* circuit produces two outputs, a sum S and a carry C. The circuit is shown in Fig. 37-19. Show that the same result can be achieved by using three AND gates, one OR gate, and one INVERTER.
4. The classic *full adder*, shown in Fig. 37-20, involves the two quantities to be added (a and b) by a digital computer, plus the carry from the preceding step (c_p). The circuit requires eight AND gates, two OR gates, and nine INVERTERS. Show that the carry portion of the output may be simplified with a saving of one AND gate and three INVERTERS.

Fig. 37-17 Switching Circuit for Prob. 1

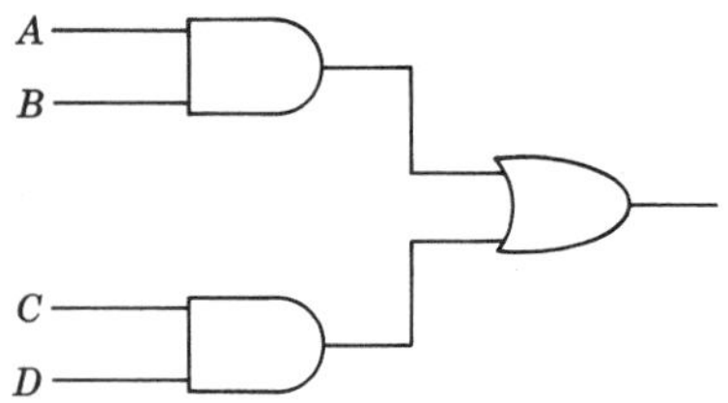

Fig. 37-18 Switching Circuit for Prob. 2

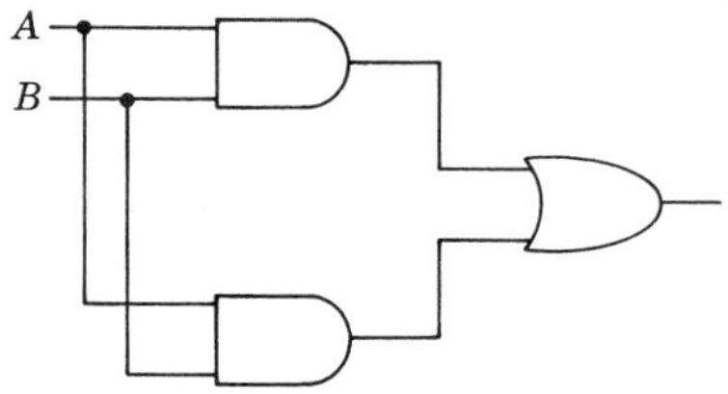

Fig. 37-19 Half-Adder Circuit of Prob. 3

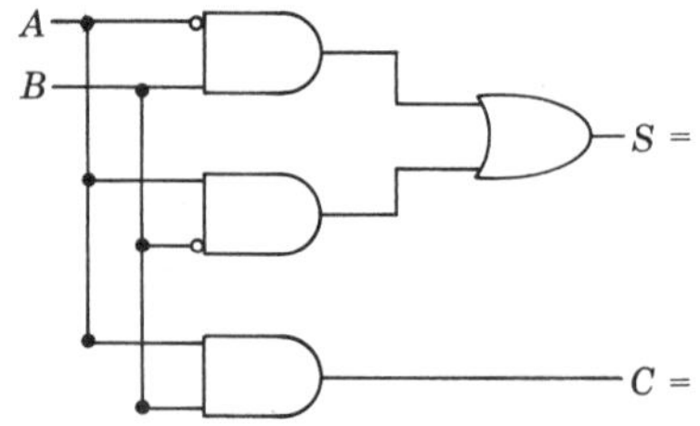

Fig. 37-20 Full-Adder Circuit of Prob. 4

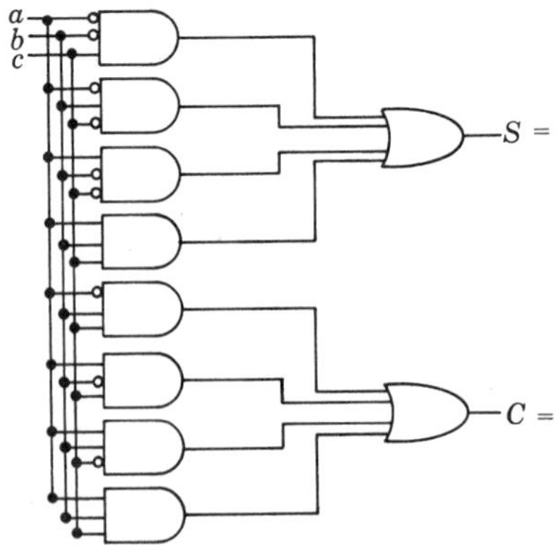

37-8 NOR COMBINATIONS

Over the last few years, many manufacturers have found it convenient for a number of reasons to build their logic circuits as multiples of a single type of gate. Often NOR gates (Fig. 37-21) are used because of the simplicity of circuit elements and design. In Problems 37-7 you will be asked to determine the gating equivalents of various combinations of NOR gates.

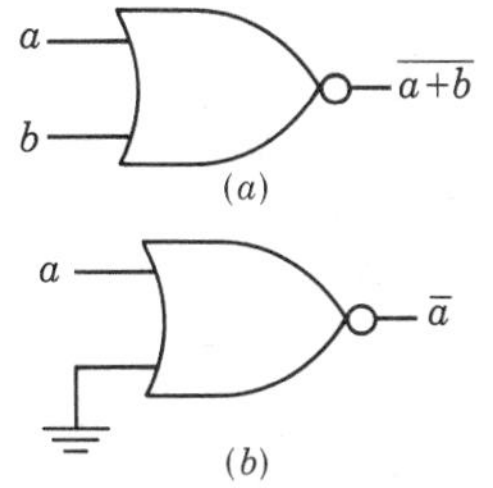

Fig. 37-21(a) The NOR Gate Delivers a Negation of the OR Combination: $\overline{a+b} = \overline{a} \cdot \overline{b}$. (b) When an Input is Grounded, It Represents a 0. Hence $\overline{a+0} = \overline{a} \cdot 1 = \overline{a}$, and We Have a NOT Gate

PROBLEMS 37-7

1. Write the output expression for the gating circuit of Fig. 37-22, and determine the simplest equivalent function.
2. Write the output expression for the gating circuit of Fig. 37-23, and determine the simplest equivalent function.
3. Write the output expression for the gating circuit of Fig. 37-24, and determine the simplest equivalent function.
4. Write the output expression for the gating circuit of Fig. 37-25, and determine the simplest equivalent function.

Fig. 37-22 NOR Gate Combination for Prob. 1

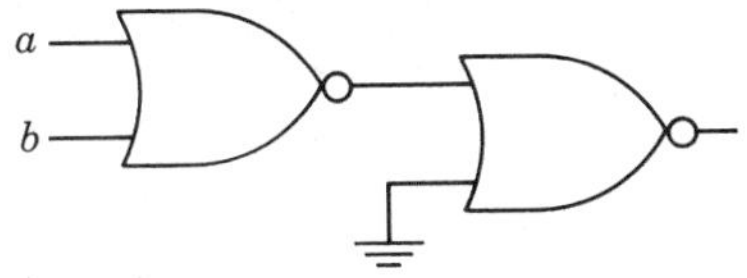

Fig. 37-23 NOR Gate Combination for Prob. 2

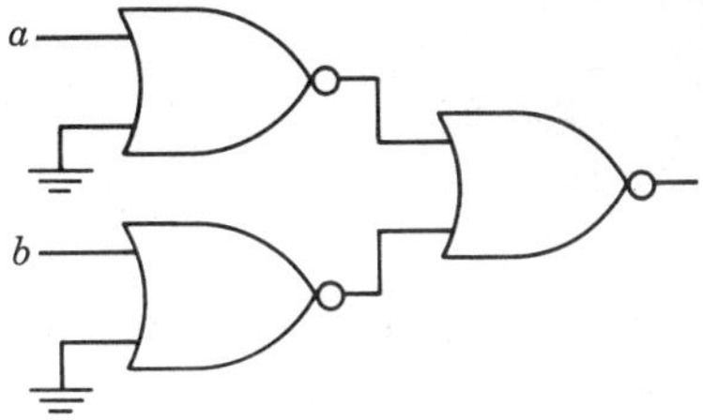

Fig. 37-24 NOR Gate Combination for Prob. 3

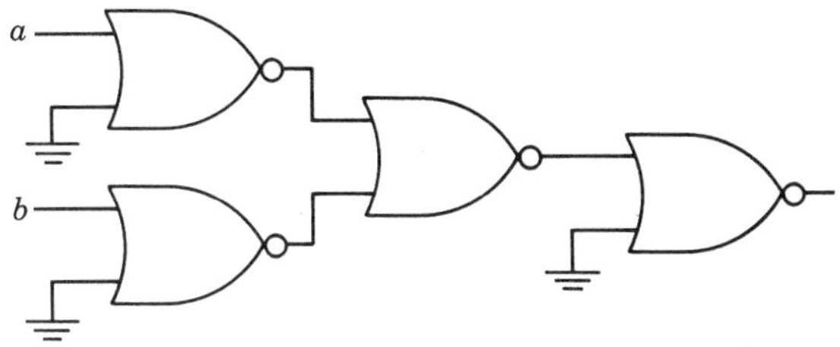

Fig. 37-25 NOR Gate Combination for Prob. 4

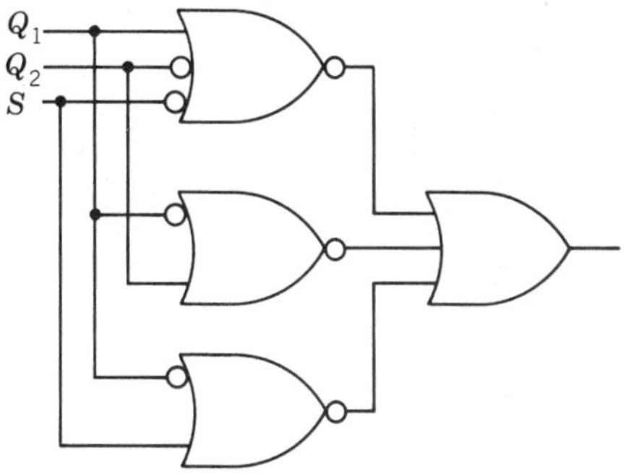

Table 1 MATHEMATICAL SYMBOLS

$\times$ or $\cdot$	multiplied by
$\div$ or :	divided by
$+$	positive, plus, add, OR
$-$	negative, minus, subtract
$\pm$	positive or negative, plus or minus
$\mp$	negative or positive, minus or plus
$=$ or $::$	equals
$\equiv$	identity
$\cong$	is approximately equal to
$\neq$	does not equal
$>$	is greater than
$\gg$	is much greater than
$<$	is less than
$\ll$	is much less than
$\geqq$	greater than or equal to
$\leqq$	less than or equal to
$\therefore$	therefore
$\angle$	angle
$\perp$	perpendicular to
$\parallel$	parallel to
$\|n\|$	absolute value of n
Δ	increment of
%	percent
$\propto$	is proportional to

Table 2 LETTER SYMBOLS

Term	*Symbol*	*Term*	*Symbol*
Altitude	a	Ohm	Ω
Area	A	Period (of time)	T
Base	B	Plate (anode)	P
Capacitance	C	Power	P
Cathode	K	Reactance	X
Collector	C	Resistance	R, r
Current	I,i	Resonant frequency	f_r
Diode	D	Rise time	t_r
Electromotive force	E, e	Speed of light	c
Emitter	E	Temperature	t
Frequency	f	Time	t
Grid	G	Transistor	Q
Impedance	Z	Tube (valve)	V
Inductance	L	Voltage	E, e, V, v
Length	l	Wavelength	λ
Number of turns	n	Width	w

Table 3 ABBREVIATIONS

Term	*Abbreviation*
Alternating current	ac
Ampere	A
Ampere-hour	Ah
Amplitude modulation	AM
Antilogarithm	antilog
Audio frequency	AF
Bel	B
Candela	cd
Centimeter	cm
Circular	cir
Clockwise	cw
Cologarithm	colog
Continuous wave	CW
Cosecant	csc
Cosine	cos
Cotangent	cot
Coulomb	C
Counterclockwise	ccw
Counter electromotive force	cemf
Cubic	$\ldots^3$
Cubic centimeter	cm^3
Cubic foot	ft^3
Cubic inch	in^3
Cubic meter	m^3
Cubic yard	yd^3
Cycles per second	Hz
Decibel	dB
Decibels referred to a level of one milliwatt	dBm
Degree (interval or change)	deg
Degrees Celsius	°C
Degrees Fahrenheit	°F
Degrees Kelvin	K
Diameter	diam
Direct current	dc
Dozen	spell
Efficiency	spell
Electromotive force	emf
Equation	Eq.
Farad	F
Foot, feet	ft
Feet per minute	ft/min
Feet per second	ft/s
Feet per second squared	ft/s^2
Figure	Fig.
Frequency	spell
Frequency modulation	FM
Giga (prefix, $= 1 \times 10^9$)	G
Gigacycles per second	GHz
Gigahertz	GHz
Gram	g
Henry	H
Hertz	Hz
High frequency	HF
Highest common factor	HCF
Hour	h
Hundred	spell, or $\times 10^2$
Inch	in
Inches per second	in/s
Intermediate frequency	IF
Kilo (prefix, $= 1 \times 10^3$)	k
Kilocycles per second	kHz
Kilogram	kg
Kilohertz	kHz
Kilohm	kΩ
Kilometer	km
Kilometers per hour	km/h
Kilovars	kvar
Kilovolt	kV
Kilovoltampere	kVA
Kilowatt	kW
Kilowatthour	kWh
Knot	kn
Logarithm (common, base 10)	log
Logarithm (any base)	$\log_a$
Logarithm (natural base ε)	$\log_\varepsilon$, ln
Low frequency	LF
Lowest common denominator	LCD
Lowest common multiple	LCM
Lumen	lm
Maximum	max
Mega (prefix, $= 1 \times 10^6$)	M
Megacycles per second	MHz
Megahertz	MHz
Megavolt	MV
Megawatt	MW
Megohm	MΩ
Meter	m
Meter-kilogram-second system	MKS
Meters per second	m/s
Mho	S
Micro (prefix, $= 1 \times 10^{-6}$)	μ

Table 3 ABBREVIATIONS *(Continued)*

Term	Abbreviation	Term	Abbreviation
Microampere	μA	Pico (prefix, $= 1 \times 10^{-12}$)	p
Microfarad	μF	Picoampere	pA
Microhenry	μH	Picofarad	pF
Micromho	μS	Picosecond	ps
Micromicro (prefix, $= 1 \times 10^{-12}$)	p	Picowatt	pW
Micromicrofarad	pF	Pound	lb
Microsecond	μs	Power factor	PF
Microsiemens	μS	Problem	Prob
Microvolt	μV	Radian	$\dots^r$
Microwatt	μW	Radians per second	^{r}/s
Mile	mi	Radio frequency	RF
Miles per hour	mi/h	Radius	r, R
Miles per minute	mi/min	Range (distance)	R
Miles per second	mi/s	Revolutions per minute	rev/min
Milli (prefix, $= 1 \times 10^{-3}$)	m	Revolutions per second	rev/s
Milliampere	mA	Root mean square	rms
Millihenry	mH	Secant	sec
Millimeter	mm	Second	s
Millisecond	ms	Siemens	S
Millivolt	mV	Sine	sin
Milliwatt	mW	Square centimeter	cm^2
Minimum	min	Square foot	ft^2
Minute	min	Square inch	in^2
Nano (prefix, $= 1 \times 10^{-9}$)	n	Square meter	m^2
Nanoampere	nA	Square yard	yd^2
Nanofarad	nF	Tangent	tan
Nanosecond	ns	Ultrahigh frequency	UHF
Nanowatt	nW	Var (reactive voltampere)	var
Neper	Np	Very high frequency	VHF
Number	No. or spell	Volt	V
Ohms	Ω	Voltampere	VA
Ohms per kilometer	Ω/km	Watt	W
Ounce	oz	Watthour	Wh
Peak-to-peak	p-p	Wattsecond	Ws
		Webers per square meter	Wb/m^2
		Yard	yd

Table 4 GREEK ALPHABET

Name	*Capital*	*Lower Case*	*Commonly Used To Designate*
Alpha	A	α	angles, area, coefficients
Beta	B	β	angles, flux density, coefficients
Gamma	Γ	γ	conductivity, specific gravity
Delta	Δ	δ	variation, density
Epsilon	E	ε	base of natural logarithms
Zeta	Z	ζ	impedance, coefficients, coordinates
Eta	H	η	hysteresis coefficient, efficiency
Theta	Θ	θ	temperature, phase angle
Iota	I	ι	
Kappa	K	κ	dielectric constant, susceptibility
Lambda	Λ	λ	wavelength
Mu	M	μ	micro, amplification factor, permeability
Nu	N	ν	reluctivity
Xi	Ξ	ξ	
Omicron	O	o	
Pi	Π	π	ratio of circumference to diameter = 3.1416
Rho	P	ρ	resistivity
Sigma	Σ	σ	summation
Tau	T	τ	time constant, time phase displacement
Upsilon	Υ	υ	
Phi	Φ	ϕ	magnetic flux, angles
Chi	X	χ	
Psi	Ψ	ψ	dielectric flux, phase difference
Omega	Ω	ω	capital, ohms; lower case, angular velocity

Table 5 STANDARD ANNEALED COPPER WIRE SOLID*
AMERICAN WIRE GAGE (BROWN AND SHARPE) (20°C)

Gage	*Diameter, mm*	*Cross Section, sq mm*	*Ohms per Kilometer*	*Meters per Ohm*	*Kilograms per Kilometer*
0000	11.68	107.2	0.160 8	6 219	953.2
000	10.40	85.01	0.202 8	4 931	755.8
00	9.266	67.43	0.255 7	3 911	599.5
0	8.252	53.49	0.322 3	3 102	475.5
1	7.348	42.41	0.406 5	2 460	377.0
2	6.543	33.62	0.512 8	1 950	298.9
3	5.827	26.67	0.646 6	1 547	237.1
4	5.189	21.15	0.815 2	1 227	188.0
5	4.620	16.77	1.028	972.4	149.0
6	4.115	13.30	1.297	771.3	118.2
7	3.665	10.55	1.634	612.0	93.80
8	3.264	8.367	2.061	485.3	74.38
9	2.906	6.631	2.600	384.6	58.95
10	2.588	5.261	3.277	305.2	46.77
11	2.30	4.17	4.14	242	37.1
12	2.05	3.31	5.21	192	29.4
13	1.83	2.63	6.56	152	23.4
14	1.63	2.08	8.28	121	18.5
15	1.45	1.65	10.4	95.8	14.7
16	1.29	1.31	13.2	75.8	11.6
17	1.15	1.04	16.6	60.3	9.24
18	1.02	0.823	21.0	47.7	7.32
19	0.912	.653	26.4	37.9	5.81
20	.813	.519	33.2	30.1	4.61
21	.724	.412	41.9	23.9	3.66
22	.643	.324	53.2	18.8	2.88
23	.574	.259	66.6	15.0	2.30
24	.511	.205	84.2	11.9	1.82
25	.455	.162	106	9.42	1.44
26	.404	.128	135	7.43	1.14
27	.361	.102	169	5.93	0.908
28	.320	.080 4	214	4.67	.715
29	.287	.064 7	266	3.75	.575
30	.254	.050 7	340	2.94	.450
31	.226	.040 1	430	2.33	.357
32	.203	.032 4	532	1.88	.288
33	.180	.025 5	675	1.48	.227
34	.160	.020 1	857	1.17	.179
35	.142	.015 9	1 090	0.922	.141

Table 5 *(Continued)*

Gage	*Diameter, mm*	*Cross Section, sq mm*	*Ohms per Kilometer*	*Meters per Ohm*	*Kilograms per Kilometer*
36	.127	.012 7	1 360	.735	.113
37	.114	.010 3	1 680	.595	.091 2
38	.102	.008 11	2 130	.470	.072 1
39	.089	.006 21	2 780	.360	.055 2
40	.079	.004 87	3 540	.282	.043 3
41	.071	.003 97	4 340	.230	.035 3
42	.064	.003 17	5 440	.184	.028 2
43	.056	.002 45	7 030	.142	.021 8
44	.051	.002 03	8 510	.118	.018 0
45	.0047	.001 57	11 000	.0910	.014 0
46	.0399	.001 25	13 800	.0724	.011 1
47	.0356	.000 993	17 400	.0576	.008 83
48	.0315	.000 779	22 100	.0452	.006 93
49	.0282	.000 624	27 600	.0362	.005 55
50	.0251	.000 497	34 700	.0288	.004 41
51	.0224	.000 392	43 900	.0228	.003 49
52	.0198	.000 308	55 900	.0179	.002 74
53	.0178	.000 248	69 400	.0144	.002 21
54	.0157	.000 195	88 500	.0113	.001 73
55	.0140	.000 153	112 000	.008 89	.001 36
56	.0124	.000 122	142 000	.007 06	.001 08

*Bureau of Standards Handbook 100, reproduced by permission.

Table 6 CONVERSION FACTORS*

Multiply	*By*	*To Obtain*
Avoirdupois ounces	28.35	Grams
Avoirdupois pounds	0.4536	Kilograms
British thermal units	1.054×10^3	Joules
Circular mils	5.067×10^{-4}	Square millimeters
Coulombs	6.242×10^{18}	Electric charges
Feet	0.3048	Meters
Gallons (imperial)	4.546	Liters
Gallons (U.S. dry)	4.405	Liters
Gallons (U.S. liquid)	3.785	Liters
Horsepower	0.746	Kilowatts
Inches	25.4	Millimeters
Kilograms	2.205	Avoirdupois pounds
Kilometers	3.28×10^3	Feet
Meters	3.28	Feet
Meters	39.37	Inches
Microns (micrometers)	1×10^{-6}	Meters
Mils	2.54×10^{-2}	Millimeters
Nautical miles	1.852×10^3	Meters
Statute miles	1.609	Kilometers
Yards	0.9144	Meters

*Selected from H. F. R. Adams, *SI Metric Units: An Introduction*, McGraw-Hill Ryerson Ltd., 1974, by permission.

INTERPOLATION OF MATHEMATICAL TABLES

The easiest and most accurate method of obtaining trigonometric or logarithmic values is by using a calculator with this facility. Values so obtained will be accurate to the capacity of the individual calculator (8, 10, or 12 places).

When comparing results from calculators, you must be aware that different calculator programs use different basic formulas for the computation of logarithms. The readouts of different model calculators should all agree with each other, and with interpolated tables, to at least four, and preferably five, places.

You can also obtain these values accurately from appropriate tables. However, tables typically show only selected values for x: $x = 1.10, 1.11, 1.12$, etc. To find a log or trig value for x when $x = 1.553$, you have to either interpolate for an estimated value or use the proportional parts table.

Both interpolation and proportional parts assume a straight-line curve between two points. This assumption is obviously incorrect for nonlinear curves, but it is acceptable for most applications if the interval is reasonably small.

Interpolation with this text's tables would be based on an interval of 0.001, while the interval for proportional parts is 0.1. Thus, manual interpolation, while more difficult, will be more accurate than proportional parts.

The logarithmic and trigonometric tables in this book have all been prepared with decimal proportional parts. This provision greatly simplifies the technique of interpolation, as can be seen from the examples below. Although you should use the method demonstrated, you must not be misled into thinking that the extra decimal places imply greater accuracy. The entries in the tables have all been rounded off, and so have the proportional parts. After following through the interpolation process, round off the result to the same number of places as the main body of the tables.

example 1 Find log 1.445 37.

solution

From Table 7,	log 1.44 =	0.158 36
From the same horizontal line, from the proportional parts for 5, add 152:		152
From the same horizontal line, from the proportional parts for 3, add one-tenth of 91:		9 1
Again, for 7, add one-hundredth of 213:		2 1 3
Adding, log 1.445 37	=	0.159 992 3

Rounding off, log 1.445 37 = 0.159 99

example 2 Find antilog 0.456 617.

solution Note that Table 8 has been screened to help you distinguish it from the logarithms Table 7.

From Table 8, antilog 0.456	=	2.857 59
From the same horizontal line, from the proportional parts column for 6, add 393:		3 93
For 1, add one-tenth of 65:		6 5
For 7, add one-hundredth of 459:		4 5 9
Antilog 0.456 617	=	2.861 6309

Rounding off, antilog 0.456 617 = 2.8616

example 3 When a zero is involved in the interpolation, allow for it: Find log 3.8207.

solution

From Table 7, log 3.82	=	0.582 06
Interpolation for 0 is 0:		00
Interpolation for 7: add one-tenth of 79:		7 9
log 3.8207	=	0.582 139

Rounding off, log 3.8207 = 0.582 14

example 4 Find ln 25.368.

solution First, in order to use Table 12, we must rewrite ln 25.368 as ln (2.5368×10^1), and rewrite it again as ln 2.5368 + ln 10^1. (See Chap. 34 for details.)
To find ln 2.5368, follow the usual rules for interpolation:

ln 2.53	=	0.928 22
Interpolation for 6: add completely 235:		235
Interpolation for 8: add one-tenth of 314:		31 4
ln 2.5368	=	0.930 88 4
From the table below Table 12, ln 10^1	=	2.302 59
Adding, ln 25.368	=	3.233 47 4

Rounding, ln 25.368 = 3.233 47

example 5 Find antiln $\overline{8}.909\,928$. Since there are no antiln tables, look for clues in the data. There are two clues here. The six decimal places indicate the need for interpolation. The $\overline{8}$ indicates a negative power of 10 in the solution. From the table for 10^{-n}, select the $\overline{10}$ characteristic:

We have	$\bar{8}.909\ 928$	
Subtracting,	$\overline{10}.789\ 66$:	$\times 10^{-4}$
	2.120 268	
From the body of Table 12,	2.119 86 :8.33	
Subtracting,	408	
From the same horizontal line, the nearest proportional part to 40 is	36 :0.003	
Subtracting,	48	
For the next interpolation, stepping one place to the right, find	48:0.0004	
antiln $\bar{8}.909\ 928$	$= 8.3334 \times 10^{-4}$	

example 6 Find sin 67.632°.

solution

From Table 9, sin 67.6°	= 0.924 55
On the same horizontal line, from the proportional parts column for 3, add 20:	20
For 2, add one-tenth of 13:	13
sin 67.632°	= 0.924 763

Rounding, sin 67.632° = 0.924 76

Interpolation for tangents is exactly the same as for sines, except use Table 11.

example 7 Find arcsin 0.928 382. The six decimal places reveal that an interpolation is needed.

solution

We have	0.928 382	
From the body of Table 9,	0.927 84	:sin 68.1°
	542	
The first interpolation in Example 6 shows that we are looking for a proportional part equal to or less than 54:	51 :	0.08°
	32	
For the second interpolation, look for a proportional part of 32 or less:	32:	0.005°
arcsin 0.928 382		= 68.185°

example 8 When interpolating cosines, the proportional parts must be *subtracted.* Note that the proportional parts portion of Table 10 has been screened to remind you of this.
Find cos 27.864°.

solution

From Table 10,		cos 27.8° = 0.884 58
Interpolation for 6: Subtract 48:	48	
Interpolation for 4, Subtract one-tenth of 32:	32	
Total to be subtracted:	512	512
Subtracting, cos 27.864°		= 0.884 068

Rounding off, cos 27.864° = 0.884 07

example 9 Find arccos 0.207 28.

solution

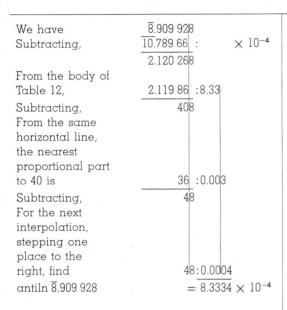

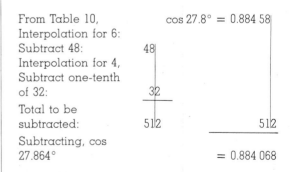

From the body of	0.207 91	= cos 78.0°
Table 10,	0.207 28	
Subtracting,	63	

From the same horizontal line, from the proportional parts column, subtract:	51	: *add*	0.03°
	120		
For the second interpolation, look for an interpolation near 120:	120	: *add*	0.007°
arccos 0.207 28		=	78.037°

Rounding off, arccos 0.207 28 = 78.04°

example 10 All the logarithms developed from Tables 7 and 12 *must* have a positive mantissa, although the characteristic may be positive or negative. You must bear in mind that the barred characteristic is a negative number, which *may* be combined with a positive mantissa to yield a net negative logarithm:

$$\log 0.005 = \bar{3}.698\,97$$
$$= -3.000\,00 + 0.698\,97$$
$$= -2.301\,03$$

An engineering calculator, such as those manufactured by Hewlett-Packard, gives a direct reading. log 0.005 = −2.301 029 996.

Similarly, a suitable slide rule, such as the K&E Decilon, gives log 0.005 = −2.3+.

These negative logarithms, consisting of both negative characteristic and negative mantissa, may be used directly in calculations. But if it is necessary to use antilog tables, you must first adjust the negative logarithm into the original form, with a negative characteristic and a positive mantissa. Then the positive mantissa may be "antilogged," and the negative characteristic will directly provide the power of 10.

Find antilog −2.30103

First, add and subtract a quantity larger than 2, performing the operation to yield a positive mantissa:

$$\begin{array}{l} +3.000\,00 - 3 \\ -2.301\,03 \\ \hline +0.698\,97 - 3 = \bar{3}.698\,97 \end{array}$$

Antilog $\bar{3}.698\,97 = 5 \times 10^{-3}$

Table 7 COMMON LOGARITHMS

	0.00	0.01	0.02	0.03	0.04	0.05	0.06	0.07	0.08	0.09
1.0	.00000	.00432	.00860	.01284	.01703	.02119	.02531	.02938	.03342	.03743
1.1	.04139	.04532	.04922	.05308	.05690	.06070	.06446	.06819	.07188	.07555
1.2	.07918	.08279	.08636	.08991	.09342	.09691	.10037	.10380	.10721	.11059
1.3	.11394	.11727	.12057	.12385	.12710	.13033	.13354	.13672	.13988	.14301
1.4	.14613	.14922	.15229	.15534	.15836	.16137	.16435	.16732	.17026	.17319
1.5	.17609	.17898	.18184	.18469	.18752	.19033	.19312	.19590	.19866	.20140
1.6	.20412	.20683	.20952	.21219	.21484	.21748	.22011	.22272	.22531	.22789
1.7	.23045	.23300	.23553	.23805	.24055	.24304	.24551	.24797	.25042	.25285
1.8	.25527	.25768	.26007	.26245	.26482	.26717	.26951	.27184	.27416	.27646
1.9	.27875	.28103	.28330	.28556	.28780	.29003	.29226	.29447	.29667	.29885
2.0	.30103	.30320	.30535	.30750	.30963	.31175	.31387	.31597	.31806	.32015
2.1	.32222	.32428	.32634	.32838	.33041	.33244	.33445	.33646	.33846	.34044
2.2	.34242	.34439	.34635	.34830	.35025	.35218	.35411	.35603	.35793	.35984
2.3	.36173	.36361	.36549	.36736	.36922	.37107	.37291	.37475	.37658	.37840
2.4	.38021	.38202	.38382	.38561	.38739	.38917	.39094	.39270	.39445	.39620
2.5	.39794	.39967	.40140	.40312	.40483	.40654	.40824	.40993	.41162	.41330
2.6	.41497	.41664	.41830	.41996	.42160	.42325	.42488	.42651	.42813	.42975
2.7	.43136	.43297	.43457	.43616	.43775	.43933	.44091	.44248	.44404	.44560
2.8	.44716	.44871	.45025	.45179	.45332	.45484	.45637	.45788	.45939	.46090
2.9	.46240	.46389	.46538	.46687	.46835	.46982	.47129	.47276	.47422	.47567
3.0	.47712	.47857	.48001	.48144	.48287	.48430	.48572	.48714	.48855	.48996
3.1	.49136	.49276	.49415	.49554	.49693	.49831	.49969	.50106	.50243	.50379
3.2	.50515	.50651	.50786	.50920	.51055	.51188	.51322	.51455	.51587	.51720
3.3	.51851	.51983	.52114	.52244	.52375	.52504	.52634	.52763	.52892	.53020
3.4	.53148	.53275	.53403	.53529	.53656	.53782	.53908	.54033	.54158	.54283
3.5	.54407	.54531	.54654	.54777	.54900	.55023	.55145	.55267	.55388	.55509
3.6	.55630	.55751	.55871	.55991	.56110	.56229	.56348	.56467	.56585	.56703
3.7	.56820	.56937	.57054	.57171	.57287	.57403	.57519	.57634	.57749	.57864
3.8	.57978	.58092	.58206	.58320	.58433	.58546	.58659	.58771	.58883	.58995
3.9	.59106	.59218	.59329	.59439	.59550	.59660	.59770	.59879	.59988	.60097
4.0	.60206	.60314	.60423	.60531	.60638	.60746	.60853	.60959	.61066	.61172
4.1	.61278	.61384	.61490	.61595	.61700	.61805	.61909	.62014	.62118	.62221
4.2	.62325	.62428	.62531	.62634	.62737	.62839	.62941	.63043	.63144	.63246
4.3	.63347	.63448	.63548	.63649	.63749	.63849	.63949	.64048	.64147	.64246
4.4	.64345	.64444	.64542	.64640	.64738	.64836	.64933	.65031	.65128	.65225
4.5	.65321	.65418	.65514	.65610	.65706	.65801	.65896	.65992	.66087	.66181
4.6	.66276	.66370	.66464	.66558	.66652	.66745	.66839	.66932	.67025	.67117
4.7	.67210	.67302	.67394	.67486	.67578	.67669	.67761	.67852	.67943	.68034
4.8	.68124	.68215	.68305	.68395	.68485	.68574	.68664	.68753	.68842	.68931
4.9	.69020	.69108	.69197	.69285	.69373	.69461	.69548	.69636	.69723	.69810

PROPORTIONAL PARTS

	1	2	3	4	5	6	7	8	9
1.0	42	85	127	170	212	254	297	339	381
	40	81	121	162	202	242	283	323	364
1.1	39	77	116	154	193	232	270	309	347
	37	74	111	148	185	222	259	296	333
1.2	35	71	106	142	177	213	248	284	319
	34	68	102	136	170	204	238	273	307
1.3	33	66	98	131	164	197	229	262	295
	32	63	95	126	158	190	221	253	284
1.4	30	61	91	122	152	183	213	244	274
	29	59	88	118	147	177	206	236	265
1.5	28	57	85	114	142	171	199	228	256
	28	55	83	110	138	165	193	221	248
1.6	27	53	80	107	134	160	187	214	241
	26	52	78	104	130	156	182	207	233
1.7	25	50	76	101	126	151	176	201	227
	24	49	73	98	122	147	171	196	220
1.8	24	48	71	95	119	143	167	190	214
	23	46	69	93	116	139	162	185	208
1.9	23	45	68	90	113	135	158	180	203
	22	44	66	88	110	132	154	176	198
2.0	21	42	64	85	106	127	148	170	191
2.1	20	40	61	81	101	121	141	162	182
2.2	19	39	58	77	97	116	135	154	174
2.3	18	37	55	74	92	111	129	148	166
2.4	18	35	53	71	89	106	124	142	160
2.5	17	34	51	68	85	102	119	136	153
2.6	16	33	49	66	82	98	115	131	148
2.7	16	32	47	63	79	95	111	126	142
2.8	15	30	46	61	76	91	107	122	137
2.9	15	29	44	59	74	88	103	118	133
3.0	14	28	43	57	71	85	100	114	128
3.1	14	28	41	55	69	83	97	110	124
3.2	13	27	40	53	67	80	94	107	120
3.3	13	26	39	52	65	78	91	104	117
3.4	13	25	38	50	63	76	88	101	113
3.5	12	24	37	49	61	73	86	98	110
3.6	12	24	36	48	59	71	83	95	107
3.7	12	23	35	46	58	69	81	93	104
3.8	11	23	34	45	56	68	79	90	102
3.9	11	22	33	44	55	66	77	88	99
4.0	11	21	32	43	54	64	75	86	97
4.1	10	21	31	42	52	63	73	84	94
4.2	10	20	31	41	51	61	72	82	92
4.3	10	20	30	40	50	60	70	80	90
4.4	10	20	29	39	49	59	68	78	88
4.5	10	19	29	38	48	57	67	76	86
4.6	9	19	28	37	47	56	65	75	84
4.7	9	18	27	37	46	55	64	73	82
4.8	9	18	27	36	45	54	63	72	81
4.9	9	18	26	35	44	53	61	70	79

Table 7 COMMON LOGARITHMS (continued)

	0.00	0.01	0.02	0.03	0.04	0.05	0.06	0.07	0.08	0.09	PROPORTIONAL PARTS 1	2	3	4	5	6	7	8	9
5.0	.69897	.69984	.70070	.70157	.70243	.70329	.70415	.70501	.70586	.70672	9	17	26	34	43	52	60	69	77
5.1	.70757	.70842	.70927	.71012	.71096	.71181	.71265	.71349	.71433	.71517	8	17	25	34	42	51	59	67	76
5.2	.71600	.71684	.71767	.71850	.71933	.72016	.72099	.72181	.72263	.72346	8	17	25	33	41	50	58	66	74
5.3	.72428	.72509	.72591	.72673	.72754	.72835	.72916	.72997	.73078	.73159	8	16	24	32	41	49	57	65	73
5.4	.73239	.73320	.73400	.73480	.73560	.73640	.73719	.73799	.73878	.73957	8	16	24	32	40	48	56	64	72
5.5	.74036	.74115	.74194	.74273	.74351	.74429	.74507	.74586	.74663	.74741	8	16	23	31	39	47	55	63	70
5.6	.74819	.74896	.74974	.75051	.75128	.75205	.75282	.75358	.75435	.75511	8	15	23	31	38	46	54	61	69
5.7	.75587	.75664	.75740	.75815	.75891	.75967	.76042	.76118	.76193	.76268	8	15	23	30	38	45	53	60	68
5.8	.76343	.76418	.76492	.76567	.76641	.76716	.76790	.76864	.76938	.77012	7	15	22	30	37	45	52	59	67
5.9	.77085	.77159	.77232	.77305	.77379	.77452	.77525	.77597	.77670	.77743	7	15	22	29	36	44	51	58	66
6.0	.77815	.77887	.77960	.78032	.78104	.78176	.78247	.78319	.78390	.78462	7	14	22	29	36	43	50	57	65
6.1	.78533	.78604	.78675	.78746	.78817	.78888	.78958	.79029	.79099	.79169	7	14	21	28	35	42	49	56	64
6.2	.79239	.79309	.79379	.79449	.79518	.79588	.79657	.79727	.79796	.79865	7	14	21	28	35	42	49	56	63
6.3	.79934	.80003	.80072	.80140	.80209	.80277	.80346	.80414	.80482	.80550	7	14	21	27	34	41	48	55	62
6.4	.80618	.80686	.80754	.80821	.80889	.80956	.81023	.81090	.81158	.81224	7	13	20	27	34	40	47	54	61
6.5	.81291	.81358	.81425	.81491	.81558	.81624	.81690	.81757	.81823	.81889	7	13	20	27	33	40	46	53	60
6.6	.81954	.82020	.82086	.82151	.82217	.82282	.82347	.82413	.82478	.82543	7	13	20	26	33	39	46	52	59
6.7	.82607	.82672	.82737	.82802	.82866	.82930	.82995	.83059	.83123	.83187	6	13	19	26	32	39	45	51	58
6.8	.83251	.83315	.83378	.83442	.83506	.83569	.83632	.83696	.83759	.83822	6	13	19	25	32	38	44	51	57
6.9	.83885	.83948	.84011	.84073	.84136	.84198	.84261	.84323	.84386	.84448	6	12	19	25	31	37	44	50	56
7.0	.84510	.84572	.84634	.84696	.84757	.84819	.84880	.84942	.85003	.85065	6	12	18	25	31	37	43	49	55
7.1	.85126	.85187	.85248	.85309	.85370	.85431	.85491	.85552	.85612	.85673	6	12	18	24	30	36	43	49	55
7.2	.85733	.85794	.85854	.85914	.85974	.86034	.86094	.86153	.86213	.86273	6	12	18	24	30	36	42	48	54
7.3	.86332	.86392	.86451	.86510	.86570	.86629	.86688	.86747	.86806	.86864	6	12	18	24	30	35	41	47	53
7.4	.86923	.86982	.87040	.87099	.87157	.87216	.87274	.87332	.87390	.87448	6	12	17	23	29	35	41	47	52
7.5	.87506	.87564	.87622	.87679	.87737	.87795	.87852	.87910	.87967	.88024	6	12	17	23	29	35	40	46	52
7.6	.88081	.88138	.88195	.88252	.88309	.88366	.88423	.88480	.88536	.88593	6	11	17	23	28	34	40	45	51
7.7	.88649	.88705	.88762	.88818	.88874	.88930	.88986	.89042	.89098	.89154	6	11	17	22	28	34	39	45	50
7.8	.89209	.89265	.89321	.89376	.89432	.89487	.89542	.89597	.89653	.89708	6	11	17	22	28	33	39	44	50
7.9	.89763	.89818	.89873	.89927	.89982	.90037	.90091	.90146	.90200	.90255	5	11	16	22	27	33	38	44	49
8.0	.90309	.90363	.90417	.90472	.90526	.90580	.90634	.90687	.90741	.90795	5	11	16	22	27	32	38	43	49
8.1	.90849	.90902	.90956	.91009	.91062	.91116	.91169	.91222	.91275	.91328	5	11	16	21	27	32	37	43	48
8.2	.91381	.91434	.91487	.91540	.91593	.91645	.91698	.91751	.91803	.91855	5	11	16	21	26	32	37	42	47
8.3	.91908	.91960	.92012	.92065	.92117	.92169	.92221	.92273	.92324	.92376	5	10	16	21	26	31	36	42	47
8.4	.92428	.92480	.92531	.92583	.92634	.92686	.92737	.92788	.92840	.92891	5	10	15	21	26	31	36	41	46
8.5	.92942	.92993	.93044	.93095	.93146	.93197	.93247	.93298	.93349	.93399	5	10	15	20	25	30	36	41	46
8.6	.93450	.93500	.93551	.93601	.93651	.93702	.93752	.93802	.93852	.93902	5	10	15	20	25	30	35	40	45
8.7	.93952	.94002	.94052	.94101	.94151	.94201	.94250	.94300	.94349	.94399	5	10	15	20	25	30	35	40	45
8.8	.94448	.94498	.94547	.94596	.94645	.94694	.94743	.94792	.94841	.94890	5	10	15	20	25	29	34	39	44
8.9	.94939	.94988	.95036	.95085	.95134	.95182	.95231	.95279	.95328	.95376	5	10	15	19	24	29	34	39	44
9.0	.95424	.95472	.95521	.95569	.95617	.95665	.95713	.95761	.95809	.95856	5	10	14	19	24	29	34	38	43
9.1	.95904	.95952	.95999	.96047	.96095	.96142	.96190	.96237	.96284	.96332	5	9	14	19	24	28	33	38	43
9.2	.96379	.96426	.96473	.96520	.96567	.96614	.96661	.96708	.96755	.96802	5	9	14	19	23	28	33	38	42
9.3	.96848	.96895	.96942	.96988	.97035	.97081	.97128	.97174	.97220	.97267	5	9	14	19	23	28	33	37	42
9.4	.97313	.97359	.97405	.97451	.97497	.97543	.97589	.97635	.97681	.97727	5	9	14	18	23	28	32	37	41
9.5	.97772	.97818	.97864	.97909	.97955	.98000	.98046	.98091	.98137	.98182	5	9	14	18	23	27	32	36	41
9.6	.98227	.98272	.98318	.98363	.98408	.98453	.98498	.98543	.98588	.98632	5	9	14	18	23	27	32	36	41
9.7	.98677	.98722	.98767	.98811	.98856	.98900	.98945	.98989	.99034	.99078	4	9	13	18	22	27	31	36	40
9.8	.99123	.99167	.99211	.99255	.99300	.99344	.99388	.99432	.99476	.99520	4	9	13	18	22	26	31	35	40
9.9	.99564	.99607	.99651	.99695	.99739	.99782	.99826	.99870	.99913	.99957	4	9	13	17	22	26	31	35	39

Table 8 COMMON ANTILOGARITHMS

	0.000	0.001	0.002	0.003	0.004	0.005	0.006	0.007	0.008	0.009	PROPORTIONAL PARTS 1	2	3	4	5	6	7	8	9
.00	1.00000	1.00231	1.00462	1.00693	1.00925	1.01158	1.01391	1.01625	1.01859	1.02094	23	46	69	93	116	139	163	186	209
.01	1.02329	1.02565	1.02802	1.03039	1.03276	1.03514	1.03753	1.03992	1.04232	1.04472	23	47	71	95	119	143	166	190	214
.02	1.04713	1.04954	1.05196	1.05439	1.05682	1.05925	1.06170	1.06414	1.06660	1.06905	24	48	73	97	121	146	170	195	219
.03	1.07152	1.07399	1.07647	1.07895	1.08143	1.08393	1.08643	1.08893	1.09144	1.09396	24	49	74	99	124	149	174	199	224
.04	1.09648	1.09901	1.10154	1.10408	1.10662	1.10917	1.11173	1.11429	1.11686	1.11944	25	51	76	102	127	153	178	204	229
.05	1.12202	1.12460	1.12720	1.12980	1.13240	1.13501	1.13763	1.14025	1.14288	1.14551	26	52	78	104	130	156	182	209	235
.06	1.14815	1.15080	1.15345	1.15611	1.15878	1.16145	1.16413	1.16681	1.16950	1.17220	26	53	80	106	133	160	187	213	240
.07	1.17490	1.17761	1.18032	1.18304	1.18577	1.18850	1.19124	1.19399	1.19674	1.19950	27	54	82	109	136	164	191	218	246
.08	1.20226	1.20504	1.20781	1.21060	1.21339	1.21619	1.21899	1.22180	1.22462	1.22744	28	56	84	112	140	168	196	224	252
.09	1.23027	1.23310	1.23595	1.23880	1.24165	1.24451	1.24738	1.25026	1.25314	1.25603	28	57	85	114	143	171	200	229	257
.10	1.25893	1.26183	1.26474	1.26765	1.27057	1.27350	1.27644	1.27938	1.28233	1.28529	29	58	87	117	146	175	205	234	263
.11	1.28825	1.29122	1.29420	1.29718	1.30017	1.30317	1.30617	1.30918	1.31220	1.31522	30	60	90	120	150	180	210	240	270
.12	1.31826	1.32130	1.32434	1.32739	1.33045	1.33352	1.33660	1.33968	1.34276	1.34586	30	61	92	122	153	184	214	245	276
.13	1.34896	1.35207	1.35519	1.35831	1.36144	1.36458	1.36773	1.37088	1.37404	1.37721	31	62	94	125	157	188	219	251	282
.14	1.38038	1.38357	1.38676	1.38995	1.39316	1.39637	1.39959	1.40281	1.40605	1.40929	32	64	96	128	160	192	225	257	289
.15	1.41254	1.41579	1.41906	1.42233	1.42561	1.42889	1.43219	1.43549	1.43880	1.44212	32	65	98	131	164	197	230	263	296
.16	1.44544	1.44877	1.45211	1.45546	1.45881	1.46218	1.46555	1.46893	1.47231	1.47571	33	67	101	134	168	202	235	269	303
.17	1.47911	1.48252	1.48594	1.48936	1.49279	1.49624	1.49968	1.50314	1.50661	1.51008	34	68	103	137	172	206	241	275	310
.18	1.51356	1.51705	1.52055	1.52405	1.52757	1.53109	1.53462	1.53815	1.54170	1.54525	35	70	105	141	176	211	246	282	317
.19	1.54882	1.55239	1.55597	1.55955	1.56315	1.56675	1.57036	1.57398	1.57761	1.58125	36	72	108	144	180	216	252	288	324
.20	1.58489	1.58855	1.59221	1.59588	1.59956	1.60325	1.60694	1.61065	1.61436	1.61808	36	73	110	147	184	221	258	295	332
.21	1.62181	1.62555	1.62930	1.63305	1.63682	1.64059	1.64437	1.64816	1.65196	1.65577	37	75	113	151	188	226	264	302	339
.22	1.65959	1.66341	1.66725	1.67109	1.67494	1.67880	1.68267	1.68655	1.69044	1.69434	38	77	115	154	193	231	270	309	347
.23	1.69824	1.70216	1.70608	1.71002	1.71396	1.71791	1.72187	1.72584	1.72982	1.73380	39	79	118	158	197	237	276	316	356
.24	1.73780	1.74181	1.74582	1.74985	1.75388	1.75792	1.76198	1.76604	1.77011	1.77419	40	80	121	161	202	242	283	323	364
.25	1.77828	1.78238	1.78649	1.79061	1.79473	1.79887	1.80302	1.80717	1.81134	1.81552	41	82	124	165	207	248	289	331	372
.26	1.81970	1.82390	1.82810	1.83231	1.83654	1.84077	1.84502	1.84927	1.85353	1.85780	42	84	127	169	211	254	296	339	381
.27	1.86209	1.86638	1.87068	1.87499	1.87932	1.88365	1.88799	1.89234	1.89671	1.90108	43	86	130	173	216	260	303	346	390
.28	1.90546	1.90985	1.91426	1.91867	1.92309	1.92752	1.93197	1.93642	1.94089	1.94536	44	88	133	177	221	266	310	355	399
.29	1.94984	1.95434	1.95884	1.96336	1.96789	1.97242	1.97697	1.98153	1.98609	1.99067	45	90	136	181	227	272	317	363	408
.30	1.99526	1.99986	2.00447	2.00909	2.01372	2.01837	2.02302	2.02768	2.03236	2.03704	46	92	139	185	232	278	325	371	418
.31	2.04174	2.04644	2.05116	2.05589	2.06063	2.06538	2.07014	2.07491	2.07970	2.08449	47	95	142	190	237	285	332	380	428
.32	2.08930	2.09411	2.09894	2.10378	2.10863	2.11349	2.11836	2.12324	2.12814	2.13304	48	97	145	194	243	291	340	389	437
.33	2.13796	2.14289	2.14783	2.15278	2.15774	2.16272	2.16770	2.17270	2.17771	2.18273	49	99	149	199	248	298	348	398	448
.34	2.18776	2.19280	2.19786	2.20293	2.20800	2.21309	2.21820	2.22331	2.22844	2.23357	50	101	152	203	254	305	356	407	458
.35	2.23872	2.24388	2.24905	2.25424	2.25944	2.26464	2.26986	2.27510	2.28034	2.28560	52	104	156	208	260	312	365	417	469
.36	2.29087	2.29615	2.30144	2.30675	2.31206	2.31739	2.32274	2.32809	2.33346	2.33884	53	106	160	213	266	320	373	426	480
.37	2.34423	2.34963	2.35505	2.36048	2.36592	2.37137	2.37684	2.38232	2.38781	2.39332	54	109	163	218	273	327	382	436	491
.38	2.39883	2.40436	2.40991	2.41546	2.42103	2.42661	2.43220	2.43781	2.44343	2.44906	55	111	167	223	279	335	391	447	502
.39	2.45471	2.46037	2.46604	2.47172	2.47742	2.48313	2.48886	2.49459	2.50035	2.50611	57	114	171	228	285	343	400	457	514
.40	2.51189	2.51768	2.52348	2.52930	2.53513	2.54097	2.54683	2.55270	2.55859	2.56448	58	117	175	234	292	351	409	468	526
.41	2.57040	2.57632	2.58226	2.58821	2.59418	2.60016	2.60615	2.61216	2.61818	2.62422	59	119	179	239	299	359	419	478	538
.42	2.63027	2.63633	2.64241	2.64850	2.65461	2.66073	2.66686	2.67301	2.67917	2.68534	61	122	183	245	306	367	428	490	551
.43	2.69153	2.69774	2.70396	2.71019	2.71644	2.72270	2.72898	2.73527	2.74157	2.74789	62	125	188	250	313	376	438	501	564
.44	2.75423	2.76058	2.76694	2.77332	2.77971	2.78612	2.79254	2.79898	2.80543	2.81190	64	128	192	256	320	384	449	513	577
.45	2.81838	2.82488	2.83139	2.83792	2.84446	2.85102	2.85759	2.86418	2.87078	2.87740	65	131	196	262	328	393	459	525	590
.46	2.88403	2.89068	2.89734	2.90402	2.91072	2.91743	2.92415	2.93089	2.93765	2.94442	67	134	201	268	335	403	470	537	604
.47	2.95121	2.95801	2.96483	2.97167	2.97852	2.98538	2.99226	2.99916	3.00608	3.01301	68	137	206	274	343	412	481	549	618
.48	3.01995	3.02691	3.03389	3.04089	3.04789	3.05492	3.06196	3.06902	3.07610	3.08319	70	140	211	281	351	422	492	562	633
.49	3.09030	3.09742	3.10456	3.11172	3.11889	3.12608	3.13329	3.14051	3.14775	3.15500	71	143	215	287	359	431	503	575	647

Table 8 COMMON ANTILOGARITHMS (continued)

	0.000	0.001	0.002	0.003	0.004	0.005	0.006	0.007	0.008	0.009	PROPORTIONAL PARTS 1	2	3	4	5	6	7	8	9
.50	3.16228	3.16957	3.17687	3.18420	3.19154	3.19890	3.20627	3.21366	3.22107	3.22849	73	147	220	294	368	441	515	589	662
.51	3.23594	3.24340	3.25087	3.25837	3.26588	3.27341	3.28095	3.28852	3.29610	3.30370	75	150	226	301	376	452	527	602	678
.52	3.31131	3.31894	3.32660	3.33426	3.34195	3.34965	3.35738	3.36512	3.37287	3.38065	77	154	231	308	385	462	539	617	694
.53	3.38844	3.39625	3.40408	3.41193	3.41979	3.42768	3.43558	3.44350	3.45144	3.45939	78	157	236	315	394	473	552	631	710
.54	3.46737	3.47536	3.48337	3.49140	3.49945	3.50752	3.51560	3.52371	3.53183	3.53997	80	161	242	323	403	484	565	646	726
.55	3.54813	3.55631	3.56451	3.57273	3.58096	3.58922	3.59749	3.60579	3.61410	3.62243	82	165	247	330	413	495	578	661	743
.56	3.63078	3.63915	3.64754	3.65595	3.66438	3.67282	3.68129	3.68978	3.69828	3.70681	84	169	253	338	422	507	592	676	761
.57	3.71535	3.72392	3.73250	3.74111	3.74973	3.75837	3.76704	3.77572	3.78443	3.79315	86	173	259	346	432	519	605	692	778
.58	3.80189	3.81066	3.81944	3.82825	3.83707	3.84592	3.85478	3.86367	3.87258	3.88150	88	177	265	354	442	531	619	708	797
.59	3.89045	3.89942	3.90841	3.91742	3.92645	3.93550	3.94457	3.95367	3.96278	3.97192	90	181	271	362	453	543	634	724	815
.60	3.98107	3.99025	3.99945	4.00867	4.01791	4.02717	4.03645	4.04576	4.05509	4.06443	92	185	278	370	463	556	649	741	834
.61	4.07380	4.08319	4.09261	4.10204	4.11150	4.12098	4.13048	4.14000	4.14954	4.15911	94	189	284	379	474	569	664	759	854
.62	4.16869	4.17830	4.18794	4.19759	4.20727	4.21697	4.22669	4.23643	4.24620	4.25598	97	194	291	388	485	582	679	776	873
.63	4.26580	4.27563	4.28549	4.29536	4.30527	4.31519	4.32514	4.33511	4.34510	4.35512	99	198	298	397	496	596	695	794	894
.64	4.36516	4.37522	4.38531	4.39542	4.40555	4.41570	4.42588	4.43609	4.44631	4.45656	101	203	305	406	508	610	711	813	915
.65	4.46684	4.47713	4.48745	4.49780	4.50817	4.51856	4.52898	4.53942	4.54988	4.56037	104	208	312	416	520	624	728	832	936
.66	4.57088	4.58142	4.59198	4.60257	4.61318	4.62381	4.63447	4.64515	4.65586	4.66659	106	212	319	425	532	638	745	851	958
.67	4.67735	4.68813	4.69894	4.70977	4.72063	4.73151	4.74242	4.75335	4.76431	4.77529	108	217	326	435	544	653	762	871	980
.68	4.78630	4.79733	4.80839	4.81948	4.83059	4.84172	4.85289	4.86407	4.87528	4.88652	111	222	334	445	557	668	780	891	1003
.69	4.89779	4.90908	4.92040	4.93174	4.94311	4.95450	4.96592	4.97737	4.98884	5.00035	114	228	342	456	570	684	798	912	1026
.70	5.01187	5.02343	5.03501	5.04661	5.05825	5.06991	5.08159	5.09331	5.10505	5.11682	116	233	350	466	583	700	817	933	1050
.71	5.12861	5.14044	5.15229	5.16416	5.17607	5.18800	5.19996	5.21195	5.22396	5.23600	119	238	358	477	597	716	836	955	1075
.72	5.24807	5.26017	5.27230	5.28445	5.29663	5.30884	5.32108	5.33335	5.34564	5.35797	122	244	366	488	611	733	855	977	1100
.73	5.37032	5.38270	5.39511	5.40754	5.42001	5.43250	5.44503	5.45758	5.47016	5.48277	125	250	375	500	625	750	875	1000	1125
.74	5.49541	5.50808	5.52077	5.53350	5.54626	5.55904	5.57186	5.58470	5.59758	5.61048	128	256	384	512	640	768	896	1024	1152
.75	5.62341	5.63638	5.64937	5.66239	5.67545	5.68853	5.70164	5.71479	5.72796	5.74116	130	261	392	523	654	785	916	1047	1178
.76	5.75440	5.76766	5.78096	5.79429	5.80764	5.82103	5.83445	5.84790	5.86138	5.87489	134	268	402	536	670	804	938	1072	1206
.77	5.88844	5.90201	5.91562	5.92925	5.94292	5.95662	5.97035	5.98412	5.99791	6.01174	137	274	411	548	685	822	960	1097	1234
.78	6.02560	6.03949	6.05341	6.06736	6.08135	6.09537	6.10942	6.12350	6.13762	6.15177	140	280	421	561	701	842	982	1122	1263
.79	6.16595	6.18016	6.19441	6.20869	6.22300	6.23735	6.25173	6.26614	6.28058	6.29506	143	287	430	574	718	861	1005	1148	1292
.80	6.30957	6.32412	6.33870	6.35331	6.36796	6.38263	6.39735	6.41210	6.42688	6.44169	146	293	440	587	734	881	1028	1175	1322
.81	6.45654	6.47143	6.48634	6.50130	6.51628	6.53131	6.54636	6.56145	6.57658	6.59174	150	300	451	601	751	902	1052	1203	1353
.82	6.60693	6.62217	6.63743	6.65273	6.66807	6.68344	6.69885	6.71429	6.72977	6.74528	153	307	461	615	769	923	1077	1231	1385
.83	6.76083	6.77642	6.79204	6.80769	6.82339	6.83912	6.85488	6.87068	6.88652	6.90240	157	314	472	629	787	944	1102	1259	1417
.84	6.91831	6.93426	6.95024	6.96627	6.98232	6.99842	7.01455	7.03072	7.04693	7.06318	161	322	483	644	805	966	1128	1289	1450
.85	7.07946	7.09578	7.11214	7.12853	7.14496	7.16143	7.17794	7.19449	7.21107	7.22770	164	329	494	659	824	989	1154	1319	1484
.86	7.24436	7.26106	7.27780	7.29458	7.31139	7.32825	7.34514	7.36207	7.37904	7.39605	168	337	506	674	843	1012	1181	1349	1518
.87	7.41310	7.43019	7.44732	7.46449	7.48170	7.49894	7.51623	7.53356	7.55092	7.56833	172	345	518	690	863	1036	1208	1381	1554
.88	7.58578	7.60326	7.62079	7.63836	7.65597	7.67361	7.69130	7.70903	7.72681	7.74462	176	353	530	706	883	1060	1236	1413	1590
.89	7.76247	7.78037	7.79830	7.81628	7.83430	7.85236	7.87046	7.88860	7.90679	7.92501	180	361	542	723	904	1084	1265	1446	1627
.90	7.94328	7.96159	7.97995	7.99834	8.01678	8.03526	8.05378	8.07235	8.09096	8.10961	185	370	555	740	925	1110	1295	1480	1665
.91	8.12831	8.14704	8.16582	8.18465	8.20352	8.22243	8.24138	8.26038	8.27942	8.29851	189	378	567	757	946	1135	1325	1514	1703
.92	8.31764	8.33681	8.35603	8.37529	8.39460	8.41395	8.43335	8.45279	8.47227	8.49180	193	387	581	774	968	1162	1356	1549	1743
.93	8.51138	8.53100	8.55067	8.57038	8.59014	8.60994	8.62979	8.64968	8.66962	8.68960	198	396	594	793	991	1189	1387	1586	1784
.94	8.70964	8.72971	8.74984	8.77001	8.79023	8.81049	8.83080	8.85116	8.87156	8.89201	202	405	608	811	1014	1217	1420	1622	1825
.95	8.91251	8.93305	8.95365	8.97429	8.99498	9.01571	9.03649	9.05733	9.07821	9.09913	207	415	622	830	1037	1245	1453	1660	1868
.96	9.12011	9.14113	9.16220	9.18333	9.20450	9.22571	9.24698	9.26830	9.28966	9.31108	212	424	637	849	1062	1274	1487	1699	1911
.97	9.33254	9.35406	9.37562	9.39723	9.41890	9.44061	9.46237	9.48418	9.50605	9.52796	217	434	652	869	1086	1304	1521	1739	1956
.98	9.54993	9.57194	9.59401	9.61612	9.63829	9.66051	9.68278	9.70510	9.72747	9.74990	222	444	667	889	1112	1334	1557	1779	2002
.99	9.77237	9.79490	9.81748	9.84011	9.86279	9.88553	9.90832	9.93116	9.95405	9.97700	227	455	682	910	1138	1365	1593	1821	2048

Table 9 NATURAL SINES

	0.0	0.1	0.2	0.3	0.4	0.5	0.6	0.7	0.8	0.9	PROPORTIONAL PARTS 1	2	3	4	5	6	7	8	9
0	0.00000	.00175	.00349	.00524	.00698	.00873	.01047	.01222	.01396	.01571	17	35	52	70	87	105	122	140	157
1	.01745	.01920	.02094	.02269	.02443	.02618	.02792	.02967	.03141	.03316	17	35	52	70	87	105	122	140	157
2	.03490	.03664	.03839	.04013	.04188	.04362	.04536	.04711	.04885	.05059	17	35	52	70	87	105	122	139	157
3	.05234	.05408	.05582	.05756	.05931	.06105	.06279	.06453	.06627	.06802	17	35	52	70	87	105	122	139	157
4	.06976	.07150	.07324	.07498	.07672	.07846	.08020	.08194	.08368	.08542	17	35	52	70	87	104	122	139	157
5	.08716	.08889	.09063	.09237	.09411	.09585	.09758	.09932	.10106	.10279	17	35	52	69	87	104	122	139	156
6	.15453	.10626	.10800	.10973	.11147	.11320	.11494	.11667	.11840	.12014	17	35	52	69	87	104	121	139	156
7	.12187	.12360	.12533	.12706	.12880	.13053	.13226	.13399	.13572	.13744	17	35	52	69	87	104	121	138	156
8	.13917	.14090	.14263	.14436	.14608	.14781	.14954	.15126	.15299	.15471	17	35	52	69	86	104	121	138	155
9	.15643	.15816	.15988	.16160	.16333	.16505	.16677	.16849	.17021	.17193	17	34	52	69	86	103	120	138	155
10	.17365	.17537	.17708	.17880	.18052	.18224	.18395	.18567	.18738	.18910	17	34	51	69	86	103	120	137	154
11	.19081	.19252	.19423	.19595	.19766	.19937	.20108	.20279	.20450	.20620	17	34	51	68	86	103	120	137	154
12	.20791	.20962	.21132	.21303	.21474	.21644	.21814	.21985	.22155	.22325	17	34	51	68	85	102	119	136	153
13	.22495	.22665	.22835	.23005	.23175	.23345	.23514	.23684	.23853	.24023	17	34	51	68	85	102	119	136	153
14	.24192	.24362	.24531	.24700	.24869	.25038	.25207	.25376	.25545	.25713	17	34	51	68	84	101	118	135	152
15	.25882	.26050	.26219	.26387	.26556	.26724	.26892	.27060	.27228	.27396	17	34	50	67	84	101	118	135	151
16	.27564	.27731	.27899	.28067	.28234	.28402	.28569	.28736	.28903	.29070	17	33	50	67	84	100	117	134	151
17	.29237	.29404	.29571	.29737	.29904	.30071	.30237	.30403	.30570	.30736	17	33	50	67	83	100	117	133	150
18	.30902	.31068	.31233	.31399	.31565	.31730	.31896	.32061	.32227	.32392	17	33	50	66	83	99	116	132	149
19	.32557	.32722	.32887	.33051	.33216	.33381	.33545	.33710	.33874	.34038	16	33	49	66	82	99	115	132	148
20	.34202	.34366	.34530	.34694	.34857	.35021	.35184	.35347	.35511	.35674	16	33	49	65	82	98	114	131	147
21	.35837	.36000	.36162	.36325	.36488	.36650	.36812	.36975	.37137	.37299	16	32	49	65	81	97	114	130	146
22	.37461	.37622	.37784	.37946	.38107	.38268	.38430	.38591	.38752	.38912	16	32	48	64	81	97	113	129	145
23	.39073	.39234	.39394	.39555	.39715	.39875	.40035	.40195	.40355	.40514	16	32	48	64	80	96	112	128	144
24	.40674	.40833	.40992	.41151	.41310	.41469	.41628	.41787	.41945	.42104	16	32	48	64	79	95	111	127	143
25	.42262	.42420	.42578	.42736	.42894	.43051	.43209	.43366	.43523	.43680	16	32	47	63	79	95	110	126	142
26	.43837	.43994	.44151	.44307	.44464	.44620	.44776	.44932	.45088	.45243	16	31	47	62	78	94	109	125	141
27	.45399	.45554	.45710	.45865	.46020	.46175	.46330	.46484	.46639	.46793	15	31	46	62	77	93	108	124	139
28	.46947	.47101	.47255	.47409	.47562	.47716	.47869	.48022	.48175	.48328	15	31	46	61	77	92	107	123	138
29	.48481	.48634	.48786	.48938	.49090	.49242	.49394	.49546	.49697	.49849	15	30	46	61	76	91	106	122	137
30	.50000	.50151	.50302	.50453	.50603	.50754	.50904	.51054	.51204	.51354	15	30	45	60	75	90	105	120	135
31	.51504	.51653	.51803	.51952	.52101	.52250	.52399	.52547	.52696	.52844	15	30	45	60	74	89	104	119	134
32	.52992	.53140	.53288	.53435	.53583	.53730	.53877	.54024	.54171	.54317	15	29	44	59	74	88	103	118	132
33	.54464	.54610	.54756	.54902	.55048	.55194	.55339	.55484	.55630	.55775	15	29	44	58	73	87	102	116	131
34	.55919	.56064	.56208	.56353	.56497	.56641	.56784	.56928	.57071	.57215	14	29	43	58	72	86	101	115	129
35	.57358	.57501	.57643	.57786	.57928	.58070	.58212	.58354	.58496	.58637	14	28	43	57	71	85	99	114	128
36	.58779	.58920	.59061	.59201	.59342	.59482	.59622	.59763	.59902	.60042	14	28	42	56	70	84	98	112	126
37	.60182	.60321	.60460	.60599	.60738	.60876	.61015	.61153	.61291	.61429	14	28	42	55	69	83	97	111	125
38	.61566	.61704	.61841	.61978	.62115	.62251	.62388	.62524	.62660	.62796	14	27	41	55	68	82	96	109	123
39	.62932	.63068	.63203	.63338	.63473	.63608	.63742	.63877	.64011	.64145	13	27	40	54	67	81	94	108	121
40	.64279	.64412	.64546	.64679	.64812	.64945	.65077	.65210	.65342	.65474	13	27	40	53	66	80	93	106	119
41	.65606	.65738	.65869	.66000	.66131	.66262	.66393	.66523	.66653	.66783	13	26	39	52	65	78	92	105	118
42	.66913	.67043	.67172	.67301	.67430	.67559	.67688	.67816	.67944	.68072	13	26	39	51	64	77	90	103	116
43	.68200	.68327	.68455	.68582	.68709	.68835	.68962	.69088	.69214	.69340	13	25	38	51	63	76	89	101	114
44	.69466	.69591	.69717	.69842	.69966	.70091	.70215	.70339	.70463	.70587	12	25	37	50	62	75	87	100	112

Table 9 NATURAL SINES (continued)

	0.0	0.1	0.2	0.3	0.4	0.5	0.6	0.7	0.8	0.9	PROPORTIONAL PARTS 1	2	3	4	5	6	7	8	9
45	.70711	.70834	.70957	.71080	.71203	.71325	.71447	.71569	.71691	.71813	12	24	37	49	61	73	86	98	110
46	.71934	.72055	.72176	.72297	.72417	.72537	.72657	.72777	.72897	.73016	12	24	36	48	60	72	84	96	108
47	.73135	.73254	.73373	.73491	.73610	.73728	.73846	.73963	.74080	.74198	12	24	35	47	59	71	83	94	106
48	.74314	.74431	.74548	.74664	.74780	.74896	.75011	.75126	.75241	.75356	12	23	35	46	58	69	81	93	104
49	.75471	.75585	.75700	.75813	.75927	.76041	.76154	.76267	.76380	.76492	11	23	34	45	57	68	79	91	102
50	.76604	.76717	.76828	.76940	.77051	.77162	.77273	.77384	.77494	.77605	11	22	33	44	56	67	78	89	100
51	.77715	.77824	.77934	.78043	.78152	.78261	.78369	.78478	.78586	.78694	11	22	33	43	54	65	76	87	98
52	.78801	.78908	.79016	.79122	.79229	.79335	.79441	.79547	.79653	.79758	11	21	32	42	53	64	74	85	96
53	.79864	.79968	.80073	.80178	.80282	.80386	.80489	.80593	.80696	.80799	10	21	31	42	52	62	73	83	93
54	.80902	.81004	.81106	.81208	.81310	.81412	.81513	.81614	.81714	.81815	10	20	30	41	51	61	71	81	91
55	.81915	.82015	.82115	.82214	.82314	.82413	.82511	.82610	.82708	.82806	10	20	30	40	49	59	69	79	89
56	.82904	.83001	.83098	.83195	.83292	.83389	.83485	.83581	.83676	.83772	10	19	29	39	48	58	67	77	87
57	.83867	.83962	.84057	.84151	.84245	.84339	.84433	.84526	.84619	.84712	9	19	28	38	47	56	66	75	84
58	.84805	.84897	.84989	.85081	.85173	.85264	.85355	.85446	.85536	.85627	9	18	27	36	46	55	64	73	82
59	.85717	.85806	.85896	.85985	.86074	.86163	.86251	.86340	.86427	.86515	9	18	27	35	44	53	62	71	80
60	.86603	.86690	.86777	.86863	.86949	.87036	.87121	.87207	.87292	.87377	9	17	26	34	43	52	60	69	77
61	.87462	.87546	.87631	.87715	.87798	.87882	.87965	.88048	.88130	.88213	8	17	25	33	42	50	58	67	75
62	.88295	.88377	.88458	.88539	.88620	.88701	.88782	.88862	.88942	.89021	8	16	24	32	40	48	56	64	73
63	.89101	.89180	.89259	.89337	.89415	.89493	.89571	.89649	.89726	.89803	8	16	23	31	39	47	55	62	70
64	.89879	.89956	.90032	.90108	.90183	.90259	.90334	.90408	.90483	.90557	8	15	23	30	38	45	53	60	68
65	.90631	.90704	.90778	.90851	.90924	.90996	.91068	.91140	.91212	.91283	7	14	22	29	36	43	51	58	65
66	.91355	.91425	.91496	.91566	.91636	.91706	.91775	.91845	.91914	.91982	7	14	21	28	35	42	49	56	63
67	.92050	.92119	.92186	.92254	.92321	.92388	.92455	.92521	.92587	.92653	7	13	20	27	33	40	47	53	60
68	.92718	.92784	.92849	.92913	.92978	.93042	.93106	.93169	.93232	.93295	6	13	19	26	32	38	45	51	58
69	.93358	.93420	.93483	.93544	.93606	.93667	.93728	.93789	.93849	.93909	6	12	18	24	31	37	43	49	55
70	.93969	.94029	.94088	.94147	.94206	.94264	.94322	.94380	.94438	.94495	6	12	17	23	29	35	41	47	52
71	.94552	.94609	.94665	.94721	.94777	.94832	.94888	.94943	.94997	.95052	6	11	17	22	28	33	39	44	50
72	.95106	.95159	.95213	.95266	.95319	.95372	.95424	.95476	.95528	.95579	5	10	16	21	26	31	37	42	47
73	.95630	.95681	.95732	.95782	.95832	.95882	.95931	.95981	.96029	.96078	5	10	15	20	25	30	35	40	45
74	.96126	.96174	.96222	.96269	.96316	.96363	.96410	.96456	.96502	.96547	5	9	14	19	23	28	33	37	42
75	.96593	.96638	.96682	.96727	.96771	.96815	.96858	.96902	.96945	.96987	4	9	13	17	22	26	31	35	39
76	.97030	.97072	.97113	.97155	.97196	.97237	.97278	.97318	.97358	.97398	4	8	12	16	20	24	29	33	37
77	.97437	.97476	.97515	.97553	.97592	.97630	.97667	.97705	.97742	.97778	4	8	11	15	19	23	26	30	34
78	.97815	.97851	.97887	.97922	.97958	.97992	.98027	.98061	.98096	.98129	3	7	10	14	17	21	24	28	31
79	.98163	.98196	.98229	.98261	.98294	.98325	.98357	.98389	.98420	.98450	3	6	10	13	16	19	22	25	29
80	.98481	.98511	.98541	.98570	.98600	.98629	.98657	.98686	.98714	.98741	3	6	9	12	14	17	20	23	26
81	.98769	.98796	.98823	.98849	.98876	.98902	.98927	.98953	.98978	.99002	3	5	8	10	13	15	18	21	23
82	.99027	.99051	.99075	.99098	.99122	.99144	.99167	.99189	.99211	.99233	2	5	7	9	11	14	16	18	21
83	.99255	.99276	.99297	.99317	.99337	.99357	.99377	.99396	.99415	.99434	2	4	6	8	10	12	14	16	18
84	.99452	.99470	.99488	.99506	.99523	.99540	.99556	.99572	.99588	.99604	2	3	5	7	8	10	12	13	15
85	.99619	.99635	.99649	.99664	.99678	.99692	.99705	.99719	.99731	.99744	1	3	4	5	7	8	10	11	12
86	.99756	.99768	.99780	.99792	.99803	.99813	.99824	.99834	.99844	.99854	1	2	3	4	5	6	7	9	10
87	.99863	.99872	.99881	.99889	.99897	.99905	.99912	.99919	.99926	.99933	1	2	2	3	4	5	5	6	7
88	.99939	.99945	.99951	.99956	.99961	.99966	.99970	.99974	.99978	.99982	0	1	1	2	2	3	3	4	4
89	.99985	.99988	.99990	.99993	.99995	.99996	.99998	.99999	.99999		0	0	0	1	1	1	1	1	1
90	1.00000																		

Table 10 NATURAL COSINES

	0.0	0.1	0.2	0.3	0.4	0.5	0.6	0.7	0.8	0.9	1	2	3	4	5	6	7	8	9
											PROPORTIONAL PARTS*								
0	1.00000	1.00000	.99999	.99999	.99998	.99996	.99995	.99993	.99990	.99988	0	0	0	1	1	1	1	1	1
1	.99985	.99982	.99978	.99974	.99970	.99966	.99961	.99956	.99951	.99945	0	1	1	2	2	3	3	4	4
2	.99939	.99933	.99926	.99919	.99912	.99905	.99897	.99889	.99881	.99872	1	2	2	3	4	5	5	6	7
3	.99863	.99854	.99844	.99834	.99824	.99813	.99803	.99792	.99780	.99768	1	2	3	4	5	6	7	9	10
4	.99756	.99744	.99731	.99719	.99705	.99692	.99678	.99664	.99649	.99635	1	3	4	5	7	8	10	11	12
5	.99619	.99604	.99588	.99572	.99556	.99540	.99523	.99506	.99488	.99470	2	3	5	7	8	10	12	13	15
6	.99452	.99434	.99415	.99396	.99377	.99357	.99337	.99317	.99297	.99276	2	4	6	8	10	12	14	16	18
7	.99255	.99233	.99211	.99189	.99167	.99144	.99122	.99098	.99075	.99051	2	5	7	9	11	14	16	18	21
8	.99027	.99002	.98978	.98953	.98927	.98902	.98876	.98849	.98823	.98796	3	5	8	10	13	15	18	21	23
9	.98769	.98741	.98714	.98686	.98657	.98629	.98600	.98570	.98541	.98511	3	6	9	12	14	17	20	23	26
10	.98481	.98450	.98420	.98389	.98357	.98325	.98294	.98261	.98229	.98196	3	6	10	13	16	19	22	25	29
11	.98163	.98129	.98096	.98061	.98027	.97992	.97958	.97922	.97887	.97851	3	7	10	14	17	21	24	28	31
12	.97815	.97778	.97742	.97705	.97667	.97630	.97592	.97553	.97515	.97476	4	8	11	15	19	23	26	30	34
13	.97437	.97398	.97358	.97318	.97278	.97237	.97196	.97155	.97113	.97072	4	8	12	16	20	24	29	33	37
14	.97030	.96987	.96945	.96902	.96858	.96815	.96771	.96727	.96682	.96638	4	9	13	17	22	26	31	35	39
15	.96593	.96547	.96502	.96456	.96410	.96363	.96316	.96269	.96222	.96174	5	9	14	19	23	28	33	37	42
16	.96126	.96078	.96029	.95981	.95931	.95882	.95832	.95782	.95732	.95681	5	10	15	20	25	30	35	40	45
17	.95630	.95579	.95528	.95476	.95424	.95372	.95319	.95266	.95213	.95159	5	10	16	21	26	31	37	42	47
18	.95106	.95052	.94997	.94943	.94888	.94832	.94777	.94721	.94665	.94609	6	11	17	22	28	33	39	44	50
19	.94552	.94495	.94438	.94380	.94322	.94264	.94206	.94147	.94088	.94029	6	12	17	23	29	35	41	47	52
20	.93969	.93909	.93849	.93789	.93728	.93667	.93606	.93544	.93483	.93420	6	12	18	24	31	37	43	49	55
21	.93358	.93295	.93232	.93169	.93106	.93042	.92978	.92913	.92849	.92784	6	13	19	26	32	38	45	51	58
22	.92718	.92653	.92587	.92521	.92455	.92388	.92321	.92254	.92186	.92119	7	13	20	27	33	40	47	53	60
23	.92050	.91982	.91914	.91845	.91775	.91706	.91636	.91566	.91496	.91425	7	14	21	28	35	42	49	56	63
24	.91355	.91283	.91212	.91140	.91068	.90996	.90924	.90851	.90778	.90704	7	14	22	29	36	43	51	58	65
25	.90631	.90557	.90483	.90408	.90334	.90259	.90183	.90108	.90032	.89956	8	15	23	30	38	45	53	60	68
26	.89879	.89803	.89726	.89649	.89571	.89493	.89415	.89337	.89259	.89180	8	16	23	31	39	47	55	62	70
27	.89101	.89021	.88942	.88862	.88782	.88701	.88620	.88539	.88458	.88377	8	16	24	32	40	48	56	64	73
28	.88295	.88213	.88130	.88048	.87965	.87882	.87798	.87715	.87631	.87546	8	17	25	33	42	50	58	67	75
29	.87462	.87377	.87292	.87207	.87121	.87036	.86949	.86863	.86777	.86690	9	17	26	34	43	52	60	69	77
30	.86603	.86515	.86427	.86340	.86251	.86163	.86074	.85985	.85896	.85806	9	18	27	35	44	53	62	71	80
31	.85717	.85627	.85536	.85446	.85355	.85264	.85173	.85081	.84989	.84897	9	18	27	36	46	55	64	73	82
32	.84805	.84712	.84619	.84526	.84433	.84339	.84245	.84151	.84057	.83962	9	19	28	38	47	56	66	75	84
33	.83867	.83772	.83676	.83581	.83485	.83389	.83292	.83195	.83098	.83001	10	19	29	39	48	58	67	77	87
34	.82904	.82806	.82708	.82610	.82511	.82413	.82314	.82214	.82115	.82015	10	20	30	40	49	59	69	79	89
35	.81915	.81815	.81714	.81614	.81513	.81412	.81310	.81208	.81106	.81004	10	20	30	41	51	61	71	81	91
36	.80902	.80799	.80696	.80593	.80489	.80386	.80282	.80178	.80073	.79968	10	21	31	42	52	62	73	83	93
37	.79864	.79758	.79653	.79547	.79441	.79335	.79229	.79122	.79016	.78908	11	21	32	42	53	64	74	85	92
38	.78801	.78694	.78586	.78478	.78369	.78261	.78152	.78043	.77934	.77824	11	22	33	43	54	65	76	87	98
39	.77715	.77605	.77494	.77384	.77273	.77162	.77051	.76940	.76828	.76717	11	22	33	44	56	67	78	89	100
40	.76604	.76492	.76380	.76267	.76154	.76041	.75927	.75813	.75700	.75585	11	23	34	45	57	68	79	91	102
41	.75471	.75356	.75241	.75126	.75011	.74896	.74780	.74664	.74548	.74431	12	23	35	46	58	69	81	93	104
42	.74314	.74198	.74080	.73963	.73846	.73728	.73610	.73491	.73373	.73254	12	24	35	47	59	71	83	94	106
43	.73135	.73016	.72897	.72777	.72657	.72537	.72417	.72297	.72176	.72055	12	24	36	48	60	72	84	96	108
44	.71934	.71813	.71691	.71569	.71447	.71325	.71203	.71080	.70957	.70834	12	24	37	49	61	73	86	98	110

* Subtract proportional parts.

Table 10 NATURAL COSINES (continued)

	0.0	0.1	0.2	0.3	0.4	0.5	0.6	0.7	0.8	0.9	PROPORTIONAL PARTS* 1	2	3	4	5	6	7	8	9
45	.70711	.70587	.70463	.70339	.70215	.70091	.69966	.69842	.69717	.69591	12	25	37	50	62	75	87	100	112
46	.69466	.69340	.69214	.69088	.68962	.68835	.68709	.68582	.68455	.68327	13	25	38	51	63	76	89	101	114
47	.68200	.68072	.67944	.67816	.67688	.67559	.67430	.67301	.67172	.67043	13	26	39	51	64	77	90	103	116
48	.66913	.66783	.66653	.66523	.66393	.66262	.66131	.66000	.65869	.65738	13	26	39	52	65	78	92	105	118
49	.65606	.65474	.65342	.65210	.65077	.64945	.64812	.64679	.64546	.64412	13	27	40	53	66	80	93	106	119
50	.64279	.64145	.64011	.63877	.63742	.63608	.63473	.63338	.63203	.63068	13	27	40	54	67	81	94	108	121
51	.62932	.62796	.62660	.62524	.62388	.62251	.62115	.61978	.61841	.61704	14	27	41	55	68	82	96	109	123
52	.61566	.61429	.61291	.61153	.61015	.60876	.60738	.60599	.60460	.60321	14	28	42	55	69	83	97	111	125
53	.60182	.60042	.59902	.59763	.59622	.59482	.59342	.59201	.59061	.58920	14	28	42	56	70	84	98	112	126
54	.58779	.58637	.58496	.58354	.58212	.58070	.57928	.57786	.57643	.57501	14	28	43	57	71	85	99	114	128
55	.57358	.57215	.57071	.56928	.56784	.56641	.56497	.56353	.56208	.56064	14	29	43	58	72	86	101	115	129
56	.55919	.55775	.55630	.55484	.55339	.55194	.55048	.54902	.54756	.54610	15	29	44	58	73	87	102	116	131
57	.54464	.54317	.54171	.54024	.53877	.53730	.53583	.53435	.53288	.53140	15	29	44	59	74	88	103	118	132
58	.52992	.52844	.52696	.52547	.52399	.52250	.52101	.51952	.51803	.51653	15	30	45	60	74	89	104	119	134
59	.51504	.51354	.51204	.51054	.50904	.50754	.50603	.50453	.50302	.50151	15	30	45	60	75	90	105	120	135
60	.50000	.49849	.49697	.49546	.49394	.49242	.49090	.48938	.48786	.48634	15	30	46	61	76	91	106	122	137
61	.48481	.48328	.48175	.48022	.47869	.47716	.47562	.47409	.47255	.47101	15	31	46	61	77	92	107	123	138
62	.46947	.46793	.46639	.46484	.46330	.46175	.46020	.45865	.45710	.45554	15	31	46	62	77	93	108	124	139
63	.45399	.45243	.45088	.44932	.44776	.44620	.44464	.44307	.44151	.43994	16	31	47	62	78	94	109	125	141
64	.43837	.43680	.43523	.43366	.43209	.43051	.42894	.42736	.42578	.42420	16	32	47	63	79	95	110	126	142
65	.42262	.42104	.41945	.41787	.41628	.41469	.41310	.41151	.40992	.40833	16	32	48	64	79	95	111	127	143
66	.40674	.40514	.40355	.40195	.40035	.39875	.39715	.39555	.39394	.39234	16	32	48	64	80	96	112	128	144
67	.39073	.38912	.38752	.38591	.38430	.38268	.38107	.37946	.37784	.37622	16	32	48	64	81	97	113	129	145
68	.37461	.37299	.37137	.36975	.36812	.36650	.36488	.36325	.36162	.36000	16	32	49	65	81	97	114	130	146
69	.35837	.35674	.35511	.35347	.35184	.35021	.34857	.34694	.34530	.34366	16	33	49	65	82	98	114	131	147
70	.34202	.34038	.33874	.33710	.33545	.33381	.33216	.33051	.32887	.32722	16	33	49	66	82	99	115	132	148
71	.32557	.32392	.32227	.32061	.31896	.31730	.31565	.31399	.31233	.31068	17	33	50	66	83	99	116	132	149
72	.30902	.30736	.30570	.30403	.30237	.30071	.29904	.29737	.29571	.29404	17	33	50	67	83	100	117	133	150
73	.29237	.29070	.28903	.28736	.28569	.28402	.28234	.28067	.27899	.27731	17	33	50	67	84	100	117	134	151
74	.27564	.27396	.27228	.27060	.26892	.26724	.26556	.26387	.26219	.26050	17	34	50	67	84	101	118	135	151
75	.25882	.25713	.25545	.25376	.25207	.25038	.24869	.24700	.24531	.24362	17	34	51	68	84	101	118	135	152
76	.24192	.24023	.23853	.23684	.23514	.23345	.23175	.23505	.22835	.22665	17	34	51	68	85	102	119	136	153
77	.22495	.22325	.22155	.21985	.21814	.21644	.21474	.21303	.21132	.20962	17	34	51	68	85	102	119	136	153
78	.20791	.20620	.20450	.20279	.20108	.19937	.19766	.19595	.19423	.19252	17	34	51	68	86	103	120	137	154
79	.19081	.18910	.18738	.18567	.18395	.18224	.18052	.17880	.17708	.17537	17	34	51	69	86	103	120	137	154
80	.17365	.17193	.17021	.16849	.16677	.16505	.16333	.16160	.15988	.15816	17	34	52	69	86	103	120	138	155
81	.15643	.15471	.15299	.15126	.14954	.14781	.14608	.14436	.14263	.14090	17	35	52	69	86	104	121	138	155
82	.13917	.13744	.13572	.13399	.13226	.13053	.12880	.12706	.12533	.12360	17	35	52	69	87	104	121	138	156
83	.12187	.12014	.11840	.11667	.11494	.11320	.11147	.10973	.10800	.10626	17	35	52	69	87	104	121	139	156
84	.10453	.10279	.10106	.09932	.09758	.09585	.09411	.09237	.09063	.08889	17	35	52	69	87	104	122	139	156
85	.08716	.08542	.08368	.08194	.08020	.07846	.07672	.07498	.07324	.07150	17	35	52	70	87	104	122	139	157
86	.06976	.06802	.06627	.06453	.06279	.06105	.05931	.05756	.05582	.05408	17	35	52	70	87	105	122	139	157
87	.05234	.05059	.04885	.04711	.04536	.04362	.04188	.04013	.03839	.03664	17	35	52	70	87	105	122	139	157
88	.03490	.03316	.03141	.02967	.02792	.02618	.02443	.02269	.02094	.01920	17	35	52	70	87	105	122	140	157
89	.01745	.01571	.01396	.01222	.01047	.00873	.00698	.00524	.00349	.00175	17	35	52	70	87	105	122	140	157
90	.00000																		

*Subtract proportional parts.

Table 11 NATURAL TANGENTS

	0.0	0.1	0.2	0.3	0.4	0.5	0.6	0.7	0.8	0.9	PROPORTIONAL PARTS*								
											1	2	3	4	5	6	7	8	9
0	0.00000	.00175	.00349	.00524	.00698	.00873	.01047	.01222	.01396	.01571	17	35	52	70	87	105	122	140	157
1	.01746	.01920	.02095	.02269	.02444	.02619	.02793	.02968	.03143	.03317	17	35	52	70	87	105	122	140	157
2	.03492	.03667	.03842	.04016	.04191	.04366	.04541	.04716	.04891	.05066	17	35	52	70	87	105	122	140	157
3	.05241	.05416	.05591	.05766	.05941	.06116	.06291	.06467	.06642	.06817	18	35	53	70	88	105	123	140	158
4	.06993	.07168	.07344	.07519	.07695	.07870	.08046	.08221	.08397	.08573	18	35	53	70	88	105	123	140	158
5	.08749	.08925	.09101	.09277	.09453	.09629	.09805	.09981	.10158	.10334	18	35	53	70	88	106	123	141	159
6	.10510	.10687	.10863	.11040	.11217	.11394	.11570	.11747	.11924	.12101	18	35	53	71	88	106	124	141	159
7	.12278	.12456	.12633	.12810	.12988	.13165	.13343	.13521	.13698	.13876	18	36	53	71	89	107	124	142	160
8	.14054	.14232	.14410	.14588	.14767	.14945	.15124	.15302	.15481	.15660	18	36	54	71	89	107	125	143	161
9	.15838	.16017	.16196	.16376	.16555	.16734	.16914	.17093	.17273	.17453	18	36	54	72	90	108	126	144	161
10	.17633	.17813	.17993	.18173	.18353	.18534	.18714	.18895	.19076	.19257	18	36	54	72	90	108	126	144	162
11	.19438	.19619	.19801	.19982	.20164	.20345	.20527	.20709	.20891	.21073	18	36	55	73	91	109	127	145	164
12	.21256	.21438	.21621	.21804	.21986	.22169	.22353	.22536	.22719	.22903	18	37	55	73	92	110	128	146	165
13	.23087	.23271	.23455	.23639	.23823	.24008	.24193	.24377	.24562	.24747	18	37	55	74	92	111	129	148	166
14	.24933	.25118	.25304	.25490	.25676	.25862	.26048	.26235	.26421	.26608	19	37	56	74	93	112	130	149	168
15	.26795	.26982	.27169	.27357	.27545	.27732	.27921	.28109	.28297	.28486	19	38	56	75	94	113	132	150	169
16	.28675	.28864	.29053	.29242	.29432	.29621	.29811	.30001	.30192	.30382	19	38	57	76	95	114	133	152	171
17	.30573	.30764	.30955	.31147	.31338	.31530	.31722	.31914	.32106	.32299	19	38	58	77	96	115	134	154	173
18	.32492	.32685	.32878	.33072	.33266	.33460	.33654	.33848	.34043	.34238	19	39	58	78	97	116	136	155	175
19	.34433	.34628	.34824	.35020	.35216	.35412	.35608	.35805	.36002	.36199	20	39	59	79	98	118	137	157	177
20	.36397	.36595	.36793	.36991	.37190	.37388	.37588	.37787	.37986	.38186	20	40	60	80	99	119	139	159	179
21	.38386	.38587	.38787	.38988	.39190	.39391	.39593	.39795	.39997	.40200	20	40	60	81	101	121	141	161	181
22	.40403	.40606	.40809	.41013	.41217	.41421	.41626	.41831	.42036	.42242	20	41	61	82	102	123	143	164	184
23	.42447	.42654	.42860	.43067	.43274	.43481	.43689	.43897	.44105	.44314	21	42	62	83	104	125	145	166	187
24	.44523	.44732	.44942	.45152	.45362	.45573	.45784	.45995	.46206	.46418	21	42	63	84	105	126	148	169	190
25	.46631	.46843	.47056	.47270	.47483	.47698	.47912	.48127	.48342	.48557	21	43	64	86	107	129	150	171	193
26	.48773	.48989	.49206	.49423	.49640	.49858	.50076	.50295	.50514	.50733	22	44	65	87	109	131	153	174	196
27	.50953	.51173	.51393	.51614	.51835	.52057	.52279	.52501	.52724	.52947	22	44	67	89	111	133	155	177	200
28	.53171	.53395	.53620	.53844	.54070	.54296	.54522	.54748	.54975	.55253	23	45	68	90	113	136	158	181	203
29	.55431	.55659	.55888	.56117	.56347	.56577	.56808	.57039	.57271	.57503	23	46	69	92	115	138	161	184	207
30	.57735	.57968	.58201	.58435	.58670	.58905	.59140	.59376	.59612	.59849	24	47	71	94	118	141	165	188	212
31	.60086	.60324	.60562	.60801	.61040	.61280	.61520	.61761	.62003	.62245	24	48	72	96	120	144	168	192	216
32	.62487	.62730	.62973	.63217	.63462	.63707	.63953	.64199	.64446	.64693	25	49	74	98	123	147	172	196	221
33	.64941	.65189	.65438	.65688	.65938	.66189	.66440	.66692	.66944	.67197	25	50	75	100	126	151	176	201	226
34	.67451	.67705	.67960	.68215	.68471	.68728	.68985	.69243	.69502	.69761	26	51	77	103	128	154	180	206	231
35	.70021	.70281	.70542	.70804	.71066	.71329	.71593	.71857	.72122	.72388	26	53	79	105	132	158	184	211	237
36	.72654	.72921	.73189	.73457	.73726	.73996	.74267	.74538	.74810	.75082	27	54	81	108	135	162	189	216	243
37	.75355	.75629	.75904	.76180	.76456	.76733	.77010	.77289	.77568	.77848	28	55	83	111	139	166	194	222	250
38	.78129	.78410	.78692	.78975	.79259	.79544	.79829	.80115	.80402	.80690	28	57	85	114	142	171	199	228	256
39	.80978	.81268	.81558	.81849	.82141	.82434	.82727	.83022	.83317	.83613	29	59	88	117	147	176	205	235	264
40	.83910	.84208	.87507	.84806	.85107	.85408	.85710	.86014	.86318	.86623	30	60	91	121	151	181	211	241	272
41	.86929	.87236	.87543	.87852	.88162	.88473	.88784	.89097	.89410	.89725	31	62	93	124	156	187	218	249	280
42	.90040	.90357	.90674	.90993	.91313	.91633	.91955	.92277	.92601	.92926	32	64	96	128	161	193	225	257	289
43	.93252	.93578	.93906	.94235	.94565	.94896	.95229	.95562	.95897	.96232	33	66	100	133	166	199	232	265	299
44	.96569	.96907	.97246	.97586	.97927	.98270	.98613	.98958	.99304	.99652	34	69	103	137	172	206	240	274	309

*Proportional parts not very reliable for angles greater than 66°.

Table 11 NATURAL TANGENTS (continued)

	0.0	0.1	0.2	0.3	0.4	0.5	0.6	0.7	0.8	0.9	PP 1	PP 2	PP 3	PP 4	PP 5	PP 6	PP 7	PP 8	PP 9
45	1.00000	1.00350	1.00701	1.01053	1.01406	1.01761	1.02117	1.02474	1.02832	1.03192	36	71	107	142	178	213	249	284	320
46	1.03553	1.03915	1.04279	1.04644	1.05010	1.05378	1.05747	1.06117	1.06489	1.06862	37	74	111	147	184	221	258	295	332
47	1.07237	1.07613	1.07990	1.08369	1.08749	1.09131	1.09514	1.09899	1.10285	1.10672	38	76	115	153	191	229	268	306	344
48	1.11061	1.11452	1.11844	1.12238	1.12633	1.13029	1.13428	1.13828	1.14229	1.14632	40	80	119	159	199	239	278	318	358
49	1.15037	1.15443	1.15851	1.16261	1.16672	1.17085	1.17500	1.17916	1.18334	1.18754	41	83	124	166	207	248	290	331	372
50	1.19175	1.19599	1.20024	1.20451	1.20879	1.21310	1.21742	1.22176	1.22612	1.23050	43	86	129	173	216	259	302	345	388
51	1.23490	1.23931	1.24375	1.24820	1.25268	1.25717	1.26169	1.26622	1.27077	1.27535	45	90	135	180	225	270	315	360	405
52	1.27994	1.28456	1.28919	1.29385	1.29853	1.30323	1.30795	1.31269	1.31745	1.32224	47	94	141	188	236	283	330	377	424
53	1.32704	1.33187	1.33673	1.34160	1.34650	1.35142	1.35637	1.36134	1.36633	1.37134	49	99	148	197	247	296	345	395	444
54	1.37638	1.38145	1.38653	1.39165	1.39679	1.40195	1.40714	1.41235	1.41759	1.42286	52	104	155	207	259	311	362	414	466
55	1.42815	1.43347	1.43881	1.44418	1.44958	1.45501	1.46046	1.46595	1.47146	1.47699	54	109	163	218	272	326	381	435	490
56	1.48256	1.48816	1.49378	1.49944	1.50512	1.51084	1.51658	1.52235	1.52816	1.53400	57	115	172	229	287	344	401	458	516
57	1.53986	1.54576	1.55170	1.55766	1.56366	1.56969	1.57575	1.58184	1.58797	1.59414	60	121	181	242	302	363	423	484	544
58	1.60033	1.60657	1.61283	1.61914	1.62578	1.63185	1.63826	1.64471	1.65120	1.65772	64	128	192	256	320	384	448	512	576
59	1.66428	1.67088	1.67752	1.68419	1.69091	1.69766	1.70446	1.71129	1.71817	1.72509	68	136	203	271	339	407	474	542	610
60	1.73205	1.73905	1.74610	1.75319	1.76032	1.76749	1.77471	1.78198	1.78929	1.79665	72	144	216	288	360	432	504	576	648
61	1.80405	1.81150	1.81899	1.82654	1.83413	1.84177	1.84946	1.85720	1.86499	1.87283	77	153	230	307	383	460	537	613	690
62	1.88073	1.88867	1.89667	1.90472	1.91282	1.92098	1.92920	1.93746	1.94579	1.95417	82	164	246	328	409	491	573	655	737
63	1.96261	1.97111	1.97966	1.98828	1.99695	2.00569	2.01449	2.02335	2.03227	2.04125	88	175	263	351	438	526	614	702	789
64	2.05030	2.05942	2.06860	2.07785	2.08716	2.09654	2.10600	2.11552	2.12511	2.13477	94	188	283	377	471	565	659	754	848
65	2.14451	2.15432	2.16420	2.17416	2.18419	2.19430	2.20449	2.21475	2.22510	2.23553	102	203	305	406	508	609	711	812	914
66	2.24604	2.25663	2.26730	2.27806	2.28891	2.29984	2.31086	2.32197	2.33317	2.34447	110	220	329	439	549	659	769	879	988
67	2.35585	2.36733	2.37891	2.39058	2.40235	2.41421	2.42618	2.43825	2.45043	2.46270									
68	2.47509	2.48758	2.50018	2.51289	2.52571	2.53865	2.55170	2.56487	2.57815	2.59156									
69	2.60509	2.61874	2.63252	2.64642	2.66046	2.67462	2.67892	2.70335	2.71792	2.73263									
70	2.74748	2.76247	2.77761	2.79289	2.80833	2.82391	2.83965	2.85555	2.87161	2.88783									
71	2.90421	2.92076	2.93748	2.95437	2.97144	2.98868	3.00611	3.02372	3.04152	3.05950									
72	3.07768	3.09606	3.11464	3.13341	3.15240	3.17159	3.19100	3.21063	3.23048	3.25055									
73	3.27085	3.29139	3.31216	3.33317	3.35443	3.37594	3.39771	3.41973	3.44202	3.46458									
74	3.48741	3.51053	3.53393	3.55761	3.58160	3.60588	3.63048	3.65538	3.68061	3.70616									
75	3.73205	3.75828	3.78485	3.81177	3.83906	3.86671	3.89474	3.92316	3.95196	3.98117									
76	4.01078	4.04081	4.07127	4.10216	4.13350	4.16530	4.19756	4.23030	4.26352	4.29724									
77	4.33148	4.36623	4.40152	4.43735	4.47374	4.51071	4.54826	4.58641	4.62518	4.66458									
78	4.70463	4.74534	4.78673	4.82882	4.87162	4.91516	4.95945	5.00451	5.05037	5.09704									
79	5.14455	5.19293	5.24218	5.29235	5.34345	5.39552	5.44857	5.50264	5.55777	5.61397									
80	5.67128	5.72974	5.78938	5.85024	5.91236	5.97576	6.04051	6.10664	6.17419	6.24321									
81	6.31375	6.38587	6.45961	6.53503	6.61219	6.69116	6.77199	6.85475	6.93952	7.02637									
82	7.11537	7.20661	7.30018	7.39616	7.49465	7.59575	7.69957	7.80622	7.91582	8.02848									
83	8.14435	8.26355	8.38625	8.51259	8.64275	8.77689	8.91520	9.05789	9.20516	9.35724									
84	9.51436	9.67680	9.84482	10.019	10.199	10.385	10.579	10.780	10.988	11.205									
85	11.430	11.664	11.909	12.163	12.429	12.706	12.996	13.300	13.617	13.951									
86	14.301	14.669	15.056	15.464	15.895	16.350	16.832	17.343	17.886	18.464									
87	19.081	19.740	20.446	21.205	22.022	22.904	23.859	24.898	26.031	27.271									
88	28.636	30.145	31.821	33.694	35.801	38.188	40.917	44.066	47.740	52.081									
89	57.290	63.657	71.615	81.847	95.489	114.589	143.237	190.984	286.478	572.957									

(PP = PROPORTIONAL PARTS*)

*Proportional parts not very reliable for angles greater than 66°.

Table 12 NATURAL LOGARITHMS

											PROPORTIONAL PARTS								
	0.00	0.01	0.02	0.03	0.04	0.05	0.06	0.07	0.08	0.09	1	2	3	4	5	6	7	8	9
1.0	0.00000	.00995	.01980	.02956	.03922	.04879	.05827	.06766	.07696	.08618	95	191	286	381	477	572	667	762	858
1.1	.09531	.10436	.11333	.12222	.13103	.13976	.14842	.15700	.16551	.17395	87	174	261	348	435	522	609	696	783
1.2	.18232	.19062	.19885	.20701	.21511	.22314	.23111	.23902	.24686	.25464	80	160	240	320	400	480	560	640	720
1.3	.26236	.27003	.27763	.28518	.29267	.30010	.30748	.31481	.32208	.32930	74	148	222	296	371	445	519	593	667
1.4	.33647	.34359	.35066	.35767	.36464	.37156	.37844	.38526	.392[illegible]4	.39878	69	138	207	276	345	414	483	552	621
1.5	.40547	.41211	.41871	.42527	.43178	.43825	.44469	.45108	.45742	.46373	65	129	194	258	323	387	452	516	581
1.6	.47000	.47623	.48243	.48858	.49470	.50078	.50682	.51282	.51879	.52473	61	121	182	242	303	364	424	485	546
1.7	.53063	.53649	.54232	.54812	.55389	.55962	.56531	.57098	.57661	.58222	57	114	171	229	286	343	400	457	514
1.8	.58779	.59333	.59884	.60432	.60977	.61519	.62058	.62594	.63127	.63658	54	108	162	216	270	324	378	433	487
1.9	.64185	.64710	.65233	.65752	.66269	.66783	.67294	.67803	.68310	.68813	51	103	154	205	256	308	359	410	462
2.0	.69315	.69813	.70310	.70804	.71295	.71784	.72271	.72755	.73237	.73716	49	98	146	195	244	293	342	390	439
2.1	.74194	.74669	.75142	.75612	.76081	.76547	.77011	.77473	.77932	.78390	47	93	140	186	233	279	326	372	419
2.2	.78846	.79299	.79751	.80200	.80648	.81093	.81536	.81978	.82418	.82855	44	89	133	178	222	267	311	356	400
2.3	.83291	.83725	.84157	.84587	.85015	.85442	.85866	.86289	.86710	.87129	43	85	128	170	213	255	298	340	383
2.4	.87547	.87963	.88377	.88789	.89200	.89609	.90016	.90422	.90826	.91228	41	82	122	163	204	245	286	327	367
2.5	.91629	.92028	.92426	.92822	.93216	.93609	.94001	.94391	.94779	.95166	39	78	118	157	196	235	275	314	353
2.6	.95551	.95935	.96317	.96698	.97078	.97456	.97833	.98208	.98582	.98954	38	75	113	151	189	226	264	302	340
2.7	.99325	.99695	1.00063	1.00430	1.00796	1.01160	1.01523	1.01885	1.02245	1.02604	36	73	109	145	182	218	255	291	327
2.8	1.02962	1.03318	1.03674	1.04028	1.04380	1.04732	1.05082	1.05431	1.05779	1.06126	35	70	105	140	175	211	246	281	316
2.9	1.06471	1.06815	1.07158	1.07500	1.07841	1.08181	1.08519	1.08856	1.09192	1.09527	34	68	102	136	170	203	237	271	305
3.0	1.09861	1.10194	1.10526	1.10856	1.11186	1.11514	1.11841	1.12168	1.12493	1.12817	33	66	98	131	164	197	230	262	295
3.1	1.13140	1.13462	1.13783	1.14103	1.14422	1.14740	1.15057	1.15373	1.15688	1.16002	32	63	95	127	159	190	222	254	286
3.2	1.16315	1.16627	1.16938	1.17248	1.17557	1.17865	1.18173	1.18479	1.18784	1.19089	31	62	92	123	154	185	215	246	277
3.3	1.19392	1.19695	1.19996	1.20297	1.20597	1.20896	1.21194	1.21491	1.21788	1.22083	30	60	90	119	149	179	209	239	269
3.4	1.22378	1.22671	1.22964	1.23256	1.23547	1.23837	1.24127	1.24415	1.24703	1.24990	29	58	87	116	145	174	203	232	261
3.5	1.25276	1.25562	1.25846	1.26130	1.26413	1.26695	1.26976	1.27257	1.27536	1.27815	28	56	85	113	141	169	197	225	254
3.6	1.28093	1.28371	1.28647	1.28923	1.29198	1.29473	1.29746	1.30019	1.30291	1.30563	27	55	82	110	137	164	192	219	247
3.7	1.30833	1.31103	1.31372	1.31641	1.31909	1.32176	1.32442	1.32708	1.32972	1.33237	27	53	80	107	133	160	187	213	240
3.8	1.33500	1.33763	1.34025	1.34286	1.34547	1.34807	1.35067	1.35325	1.35584	1.35841	26	52	78	104	130	156	182	208	234
3.9	1.36098	1.36354	1.36609	1.36864	1.37118	1.37372	1.37624	1.37877	1.38128	1.38379	25	51	76	101	127	152	177	203	228
4.0	1.38629	1.38879	1.39128	1.39377	1.39624	1.39872	1.40118	1.40364	1.40610	1.40854	25	49	74	99	123	148	173	198	222
4.1	1.41099	1.41342	1.41585	1.41828	1.42070	1.42311	1.42552	1.42792	1.43031	1.43270	24	48	72	96	120	145	169	193	217
4.2	1.43508	1.43746	1.43984	1.44220	1.44456	1.44692	1.44927	1.45161	1.45395	1.45629	24	47	71	94	118	141	165	188	212
4.3	1.45862	1.46094	1.46326	1.46557	1.46787	1.47018	1.47247	1.47476	1.47705	1.47933	23	46	69	92	115	138	161	184	207
4.4	1.48160	1.48387	1.48614	1.48840	1.49065	1.49290	1.49515	1.49739	1.49962	1.50185	22	45	67	90	112	135	157	180	202
4.5	1.50408	1.50630	1.50851	1.51072	1.51293	1.51513	1.51732	1.51951	1.52170	1.52388	22	44	66	88	110	132	154	176	198
4.6	1.52606	1.52823	1.53039	1.53256	1.53471	1.53687	1.53902	1.54116	1.54330	1.54543	22	43	65	86	108	129	151	172	194
4.7	1.54756	1.54969	1.55181	1.55393	1.55604	1.55814	1.56025	1.56235	1.56444	1.56653	21	42	63	84	105	126	147	168	189
4.8	1.56862	1.57070	1.57277	1.57485	1.57691	1.57898	1.58104	1.58309	1.58515	1.58719	21	41	62	82	103	124	144	165	186
4.9	1.58924	1.59127	1.59331	1.59534	1.59737	1.59939	1.60141	1.60342	1.60543	1.60744	20	40	61	81	101	121	141	162	182
5.0	1.60944	1.61144	1.61343	1.61542	1.61741	1.61939	1.62137	1.62334	1.62531	1.62728	20	40	59	79	99	119	139	158	178
5.1	1.62924	1.63120	1.63315	1.63511	1.63705	1.63900	1.64094	1.64287	1.64481	1.64673	19	39	58	78	97	117	136	155	175
5.2	1.64866	1.65058	1.65250	1.65441	1.65632	1.65823	1.66013	1.66203	1.66393	1.66582	19	38	57	76	95	114	133	152	171
5.3	1.66771	1.66959	1.67147	1.67335	1.67523	1.67710	1.67896	1.68083	1.68269	1.68455	19	37	56	75	93	112	131	150	168
5.4	1.68640	1.68825	1.69010	1.69194	1.69378	1.69562	1.69745	1.69928	1.70111	1.70293	18	37	55	73	92	110	128	147	165
5.5	1.70475	1.70656	1.70838	1.71019	1.71199	1.71380	1.71560	1.71740	1.71919	1.72098	18	36	54	72	90	108	126	144	162
5.6	1.72277	1.72455	1.72633	1.72811	1.72988	1.73166	1.73342	1.73519	1.73695	1.73871	18	35	53	71	88	106	124	142	159
5.7	1.74047	1.74222	1.74397	1.74572	1.74746	1.74920	1.75094	1.75267	1.75440	1.75613	17	35	52	70	87	104	122	139	157
5.8	1.75786	1.75958	1.76130	1.76302	1.76473	1.76644	1.76815	1.76985	1.77156	1.77326	17	34	51	68	85	103	120	137	154
5.9	1.77495	1.77665	1.77834	1.78002	1.78171	1.78339	1.78507	1.78675	1.78842	1.79009	17	34	50	67	84	101	118	134	151
6.0	1.79176	1.79342	1.79509	1.79675	1.79840	1.80006	1.80171	1.80336	1.80500	1.80665	17	33	50	66	83	99	116	132	149

Table 12 NATURAL LOGARITHMS (continued)

	0.00	0.01	0.02	0.03	0.04	0.05	0.06	0.07	0.08	0.09	PROPORTIONAL PARTS 1	2	3	4	5	6	7	8	9
6.1	1.80829	1.80993	1.81156	1.81319	1.81482	1.81645	1.81808	1.81970	1.82132	1.82294	16	33	49	65	81	98	114	130	146
6.2	1.82455	1.82616	1.82777	1.82938	1.83098	1.83258	1.83418	1.83578	1.83737	1.83896	16	32	48	64	80	96	112	128	144
6.3	1.84055	1.84214	1.84372	1.84530	1.84688	1.84845	1.85003	1.85160	1.85317	1.85473	16	31	47	63	79	94	110	126	142
6.4	1.85630	1.85786	1.85942	1.86097	1.86253	1.86408	1.86563	1.86718	1.86872	1.87026	16	31	47	62	78	93	109	124	140
6.5	1.87180	1.87334	1.87487	1.87641	1.87794	1.87947	1.88099	1.88251	1.88403	1.88555	15	31	46	61	76	92	107	122	137
6.6	1.88707	1.88858	1.89010	1.89160	1.89311	1.89462	1.89612	1.89762	1.89912	1.90061	15	30	45	60	75	90	105	120	135
6.7	1.90211	1.90360	1.90509	1.90658	1.90806	1.90954	1.91102	1.91250	1.91398	1.91545	15	30	44	59	74	89	104	119	133
6.8	1.91692	1.91839	1.91986	1.92132	1.92279	1.92425	1.92571	1.92716	1.92862	1.93007	15	29	44	58	73	88	102	117	131
6.9	1.93152	1.93297	1.93442	1.93586	1.93730	1.93874	1.94018	1.94162	1.94305	1.94448	14	29	43	58	72	86	101	115	129
7.0	1.94591	1.94734	1.94876	1.95019	1.95161	1.95303	1.95445	1.95586	1.95727	1.95869	14	28	43	57	71	85	99	113	128
7.1	1.96009	1.96150	1.96291	1.96431	1.96571	1.96711	1.96851	1.96991	1.97130	1.97269	14	28	42	56	70	84	98	112	126
7.2	1.97408	1.97547	1.97685	1.97824	1.97962	1.98100	1.98238	1.98376	1.98513	1.98650	14	28	41	55	69	83	97	110	124
7.3	1.98787	1.98924	1.99061	1.99198	1.99334	1.99470	1.99606	1.99742	1.99877	2.00013	14	27	41	54	68	82	95	109	122
7.4	2.00148	2.00283	2.00418	2.00553	2.00687	2.00821	2.00956	2.01089	2.01223	2.01357	13	27	40	54	67	81	94	107	121
7.5	2.01490	2.01624	2.01757	2.01890	2.02022	2.02155	2.02287	2.02419	2.02551	2.02683	13	26	40	53	66	79	93	106	119
7.6	2.02815	2.02946	2.03078	2.03209	2.03340	2.03471	2.03601	2.03732	2.03862	2.03992	13	26	39	52	65	78	92	105	118
7.7	2.04122	2.04252	2.04381	2.04511	2.04640	2.04769	2.04898	2.05027	2.05156	2.05284	13	26	39	52	65	77	90	103	116
7.8	2.05412	2.05540	2.05668	2.05796	2.05924	2.06051	2.06179	2.06306	2.06433	2.06560	13	25	38	51	64	76	89	102	115
7.9	2.06686	2.06813	2.06939	2.07065	2.07191	2.07317	2.07443	2.07568	2.07694	2.07819	13	25	38	50	63	75	88	101	113
8.0	2.07944	2.08069	2.08194	2.08318	2.08443	2.08567	2.08691	2.08815	2.08939	2.09063	12	25	37	50	62	75	87	99	112
8.1	2.09186	2.09310	2.09433	2.09556	2.09679	2.09802	2.09924	2.10047	2.10169	2.10291	12	25	37	49	61	74	86	98	110
8.2	2.10413	2.10535	2.10657	2.10779	2.10900	2.11021	2.11142	2.11263	2.11384	2.11505	12	24	36	48	61	73	85	97	109
8.3	2.11626	2.11746	2.11866	2.11986	2.12106	2.12226	2.12346	2.12465	2.12585	2.12704	12	24	36	48	60	72	84	96	108
8.4	2.12823	2.12942	2.13061	2.13180	2.13298	2.13417	2.13535	2.13653	2.13771	2.13889	12	24	36	47	59	71	83	95	107
8.5	2.14007	2.14124	2.14242	2.14359	2.14476	2.14593	2.14710	2.14827	2.14943	2.15060	12	23	35	47	58	70	82	94	105
8.6	2.15176	2.15292	2.15409	2.15524	2.15640	2.15756	2.15871	2.15987	2.16102	2.16217	12	23	35	46	58	69	81	92	104
8.7	2.16332	2.16447	2.16562	2.16677	2.16791	2.16905	2.17020	2.17134	2.17248	2.17361	11	23	34	46	57	69	80	91	103
8.8	2.17475	2.17589	2.17702	2.17816	2.17929	2.18042	2.18155	2.18267	2.18380	2.18493	11	23	34	45	56	68	79	90	102
8.9	2.18605	2.18717	2.18830	2.18942	2.19054	2.19165	2.19277	2.19389	2.19500	2.19611	11	22	34	45	56	67	78	89	101
9.0	2.19722	2.19834	2.19944	2.20055	2.20166	2.20276	2.20387	2.20497	2.20607	2.20717	11	22	33	44	55	66	77	88	99
9.1	2.20827	2.20937	2.21047	2.21157	2.21266	2.21375	2.21485	2.21594	2.21703	2.21812	11	22	33	44	55	66	77	87	98
9.2	2.21920	2.22029	2.22138	2.22246	2.22354	2.22462	2.22570	2.22678	2.22786	2.22894	11	22	32	43	54	65	76	86	97
9.3	2.23001	2.23109	2.23216	2.23324	2.23431	2.23538	2.23645	2.23751	2.23858	2.23965	11	21	32	43	53	64	75	86	96
9.4	2.24071	2.24177	2.24284	2.24390	2.24496	2.24601	2.24707	2.24813	2.24918	2.25024	11	21	32	42	53	63	74	85	95
9.5	2.25129	2.25234	2.25339	2.25444	2.25549	2.25654	2.25759	2.25863	2.25968	2.26072	10	21	31	42	52	63	73	84	94
9.6	2.26176	2.26280	2.26384	2.26488	2.26592	2.26696	2.26799	2.26903	2.27006	2.27109	10	21	31	41	52	62	73	83	93
9.7	2.27213	2.27316	2.27419	2.27521	2.27624	2.27727	2.27829	2.27932	2.28034	2.28136	10	21	31	41	51	62	72	82	92
9.8	2.28238	2.28340	2.28442	2.28544	2.28646	2.28747	2.28849	2.28950	2.29051	2.29152	10	20	30	41	51	61	71	81	91
9.9	2.29253	2.29354	2.29455	2.29556	2.29657	2.29757	2.29858	2.29958	2.30058	2.30158	10	20	30	40	50	60	70	80	90
10.0	2.30258																		

NATURAL LOGARITHMS OF 10^x and 10^{-x}

x	1	2	3	4	5	6	7	8	9	10
$\ln 10^x$	**2.30259**	**4.60517**	**6.90776**	**9.21034**	**11.51293**	**13.81551**	**16.11810**	**18.42068**	**20.72327**	**23.02585**
$\ln 10^{-x}$	$\overline{3}.69741$	$\overline{5}.39483$	$\overline{7}.09224$	$\overline{10}.78966$	$\overline{12}.48707$	$\overline{14}.18449$	$\overline{17}.88190$	$\overline{19}.57932$	$\overline{21}.27673$	$\overline{24}.97415$

Table 13 EXPONENTIAL FUNCTIONS

x	ε^x	ε^{-x}
0.00	1.000	1.000000
0.10	1.105	.904837
0.20	1.221	.818731
0.30	1.350	.740818
0.40	1.492	.670320
0.50	1.649	.606531
0.60	1.822	.548812
0.70	2.014	.496585
0.80	2.226	.449329
0.90	2.460	.406570
1.00	2.718	.367879
1.10	3.004	.332871
1.20	3.320	.301194
1.30	3.669	.272532
1.40	4.055	.246597
1.50	4.482	.223130
1.60	4.953	.201897
1.70	5.474	.182684
1.80	6.050	.165299
1.90	6.686	.149569
2.00	7.389	.135335
2.10	8.166	.122456
2.20	9.525	.110803
2.30	9.974	.100259
2.40	11.023	.090718
2.50	12.182	.082085
2.60	13.464	.074274
2.70	14.880	.067206
2.80	16.445	.060810
2.90	18.174	.055023
3.00	20.086	.049787
3.10	22.198	.045049
3.20	24.533	.040762
3.30	27.113	.036883
3.40	29.964	.033373
3.50	33.115	.030197
3.60	36.598	.027324
3.70	40.447	.024724
3.80	44.701	.022371
3.90	49.402	.020242
4.00	54.598	.018316
4.10	60.340	.016573
4.20	66.686	.014996
4.30	73.700	.013569
4.40	81.451	.012277
4.50	90.017	.011109
4.60	99.484	.010052
4.70	109.947	.009095
4.80	121.510	.008230
4.90	134.290	.007447
5.00	148.413	.006738

Table 13 EXPONENTIAL FUNCTIONS (continued)

x	ε^x	ε^{-x}
5.10	164.022	.006097
5.20	181.272	.005517
5.30	200.337	.004992
5.40	221.406	.004517
5.50	244.692	.004087
5.60	270.426	.003698
5.70	298.867	.003346
5.80	330.300	.003028
5.90	365.037	.002739
6.00	403.429	.002479
6.10	445.858	.002243
6.20	492.749	.002029
6.30	544.572	.001836
6.40	601.845	.001662
6.50	665.142	.001503
6.60	735.095	.001360
6.70	812.406	.001231
6.80	897.847	.001114
6.90	992.275	.001008
7.00	1096.633	.000912
7.10	1211.967	.000825
7.20	1339.431	.000747
7.30	1480.300	.000676
7.40	1635.984	.000611
7.50	1808.042	.000553
7.60	1998.196	.000500
7.70	2208.348	.000453
7.80	2440.602	.000410
7.90	2697.282	.000371
8.00	2980.958	.000335
8.10	3294.468	.000304
8.20	3640.950	.000275
8.30	4023.872	.000249
8.40	4447.067	.000225
8.50	4914.769	.000203
8.60	5431.660	.000184
8.70	6002.912	.000167
8.80	6634.244	.000151
8.90	7331.974	.000136
9.00	8103.084	.000123
9.10	8955.293	.000112
9.20	9897.129	.000101
9.30	10938.019	.000091
9.40	12088.381	.000083
9.50	13359.727	.000075
9.60	14764.782	.000068
9.70	16317.607	.000061
9.80	18033.745	.000055
9.90	19930.370	.000050
10.00	22026.466	.000045

Table 14 DECIMAL MULTIPLIERS

0.000 000 000 000 000 001	$= 10^{-18}$	= ten to the negative *eighteenth* power	= atto a
0.000 000 000 000 001	$= 10^{-15}$	= ten to the negative *fifteenth* power	= femto f
0.000 000 000 001	$= 10^{-12}$	= ten to the negative *twelfth* power	= pico p
0.000 000 001	$= 10^{-9}$	= ten to the negative *ninth* power	= nano n
0.000 001	$= 10^{-6}$	= ten to the negative *sixth* power	= micro μ
0.001	$= 10^{-3}$	= ten to the negative *third* power	= milli m
1	$= 10^{0}$	= ten to the *zero* power	= unit
1 000	$= 10^{3}$	= ten to the *third* power	= kilo k
1 000 000	$= 10^{6}$	= ten to the *sixth* power	= Mega M
1 000 000 000	$= 10^{9}$	= ten to the *ninth* power	= Giga G
1 000 000 000 000	$= 10^{12}$	= ten to the *twelfth* power	= Tera T
1 000 000 000 000 000	$= 10^{15}$	= ten to the *fifteenth* power	= Peta P
1 000 000 000 000 000 000	$= 10^{18}$	= ten to the *eighteenth* power	= Exa E

Table 15 ROUNDED VALUES OF PREFERRED NUMBERS*

Series	*R5*	*R10*	*R20*	*R40*	*R6*†	*R12*†	*R24*†
Approximate Ratio	1.6	1.25	1.12	1.06	1.46	1.21	1.1
	1	1	1	1	1	1	1
				1.06			1.1
			1.12	1.12		1.2	1.2
				1.18			1.3
		1.25	1.25	1.25	1.5	1.5	1.5
				1.32			1.6
			1.40	1.40		1.8	1.8
				1.50			2.0
	1.60	1.60	1.60	1.60	2.2	2.2	2.2
				1.70			2.4
			1.80	1.80		2.7	2.7
				1.90			3.0
		2.00	2.00	2.00	3.3	3.3	3.3
				2.12			3.6
			2.24	2.24		3.9	3.9
				2.36			4.3
	2.50	2.50	2.50	2.50	4.7	4.7	4.7
				2.65			5.1
			2.80	2.80		5.6	5.6
				3.00			6.2
		3.15	3.15	3.15	6.8	6.8	6.8
				3.35			7.5
			3.55	3.55		8.2	8.2
				3.75			9.1
	4.00	4.00	4.00	4.00	10	10	10
				4.25			
			4.50	4.50			
				4.75			
		5.00	5.00	5.00			
				5.30			
			5.60	5.60			
				6.00			
	6.30	6.30	6.30	6.30			
				6.70			
			7.10	7.10			
				7.50			
		8.00	8.00	8.00			
				8.50			
			9.00	9.00			
				9.50			
	10.00	10.00	10.00	10.00			

*These tables have been adapted from various international, American, and British standards.

†The *R*6, *R*12, and *R*24 tables are sometimes referred to as *E*6, *E*12, and *E*24, respectively.

Table 16 DEGREES TO RADIANS

	0.0	0.1	0.2	0.3	0.4	0.5	0.6	0.7	0.8	0.9	PROPORTIONAL PARTS 1	2	3	4	5	6	7	8	9
0	0.00000	.00175	.00349	.00524	.00698	.00873	.01047	.01222	.01396	.01571	17	35	52	70	87	105	122	140	157
1	.01745	.01920	.02094	.02269	.02443	.02618	.02793	.02967	.03142	.03316	17	35	52	70	87	105	122	140	157
2	.03491	.03665	.03840	.04014	.04189	.04363	.04538	.04712	.04887	.05061	17	35	52	70	87	105	122	140	157
3	.05236	.05411	.05585	.05760	.05934	.06109	.06283	.06458	.06632	.06807	17	35	52	70	87	105	122	140	157
4	.06981	.07156	.07330	.07505	.07679	.07854	.08029	.08203	.08378	.08552	17	35	52	70	87	105	122	140	157
5	.08727	.08901	.09076	.09250	.09425	.09599	.09774	.09948	.10123	.10297	17	35	52	70	87	105	122	140	157
6	.10472	.10647	.10821	.10996	.11170	.11345	.11519	.11694	.11868	.12043	17	35	52	70	87	105	122	140	157
7	.12217	.12392	.12566	.12741	.12915	.13090	.13265	.13439	.13614	.13788	17	35	52	70	87	105	122	140	157
8	.13963	.14137	.14312	.14486	.14661	.14835	.15010	.15184	.15359	.15533	17	35	52	70	87	105	122	140	157
9	.15708	.15882	.16057	.16232	.16406	.16581	.16755	.16930	.17104	.17279	17	35	52	70	87	105	122	140	157
10	.17453	.17628	.17802	.17977	.18151	.18326	.18500	.18675	.18850	.19024	17	35	52	70	87	105	122	140	157
11	.19199	.19373	.19548	.19722	.19897	.20071	.20246	.20420	.20595	.20769	17	35	52	70	87	105	122	140	157
12	.20944	.21118	.21293	.21468	.21642	.21817	.21991	.22166	.22340	.22515	17	35	52	70	87	105	122	140	157
13	.22689	.22864	.23038	.23213	.23387	.23562	.23736	.23911	.24086	.24260	17	35	52	70	87	105	122	140	157
14	.24435	.24609	.24784	.24958	.25133	.25307	.25482	.25656	.25831	.26005	17	35	52	70	87	105	122	140	157
15	.26180	.26354	.26529	.26704	.26878	.27053	.27227	.27402	.27576	.27751	17	35	52	70	87	105	122	140	157
16	.27925	.28100	.28274	.28449	.28623	.28798	.28972	.29147	.29322	.29496	17	35	52	70	87	105	122	140	157
17	.29671	.29845	.30020	.30194	.30369	.30543	.30718	.30892	.31067	.31241	17	35	52	70	87	105	122	140	157
18	.31416	.31590	.31765	.31940	.32114	.32289	.32463	.32638	.32812	.32987	17	35	52	70	87	105	122	140	157
19	.33161	.33336	.33510	.33685	.33859	.34034	.34208	.34383	.34558	.34732	17	35	52	70	87	105	122	140	157
20	.34907	.35081	.35256	.35430	.35605	.35779	.35954	.36128	.36303	.36477	17	35	52	70	87	105	122	140	157
21	.36652	.36826	.37001	.37176	.37350	.37525	.37699	.37874	.38048	.38223	17	35	52	70	87	105	122	140	157
22	.38397	.38572	.38746	.38921	.39095	.39270	.39444	.39619	.39794	.39968	17	35	52	70	87	105	122	140	157
23	.40143	.40317	.40492	.40666	.40841	.41015	.41190	.41364	.41539	.41713	17	35	52	70	87	105	122	140	157
24	.41888	.42062	.42237	.42412	.42586	.42761	.42935	.43110	.43284	.43459	17	35	52	70	87	105	122	140	157
25	.43633	.43808	.43982	.44157	.44331	.44506	.44680	.44855	.45029	.45204	17	35	52	70	87	105	122	140	157
26	.45379	.45553	.45728	.45902	.46077	.46251	.46426	.46600	.46775	.46949	17	35	52	70	87	105	122	140	157
27	.47124	.47298	.47473	.47647	.47822	.47997	.48171	.48346	.48520	.48695	17	35	52	70	87	105	122	140	157
28	.48869	.49044	.49218	.49393	.49567	.49742	.49916	.50091	.50265	.50440	17	35	52	70	87	105	122	140	157
29	.50615	.50789	.50964	.51138	.51313	.51487	.51662	.51836	.52011	.52185	17	35	52	70	87	105	122	140	157
30	.52360	.52534	.52709	.52883	.53058	.53233	.53407	.53582	.53756	.53631	17	35	52	70	87	105	122	140	157
31	.54105	.54280	.54454	.54629	.54803	.54978	.55152	.55327	.55501	.55676	17	35	52	70	87	105	122	140	157
32	.55851	.56025	.56200	.56374	.56549	.56723	.56898	.57072	.57247	.57421	17	35	52	70	87	105	122	140	157
33	.57596	.57770	.57945	.58119	.58294	.58469	.58643	.58818	.58992	.59167	17	35	52	70	87	105	122	140	157
34	.59341	.59516	.59690	.59865	.60039	.60214	.60388	.60563	.60737	.60912	17	35	52	70	87	105	122	140	157
35	.61087	.61261	.61436	.61610	.61785	.61959	.62134	.62308	.62483	.62657	17	35	52	70	87	105	122	140	157
36	.62832	.63006	.63181	.63355	.63530	.63705	.63879	.64054	.64228	.64403	17	35	52	70	87	105	122	140	157
37	.64577	.64752	.64926	.65101	.65275	.65450	.65624	.65799	.65973	.66148	17	35	52	70	87	105	122	140	157
38	.66323	.66497	.66672	.66846	.67021	.67195	.67370	.67544	.67719	.67893	17	35	52	70	87	105	122	140	157
39	.68068	.68242	.68417	.68591	.68766	.68941	.69115	.69290	.69464	.69639	17	35	52	70	87	105	122	140	157
40	.69813	.69988	.70162	.70337	.70511	.70686	.70860	.71035	.71209	.71384	17	35	52	70	87	105	122	140	157
41	.71558	.71733	.71908	.72082	.72257	.72431	.72606	.72780	.72955	.73129	17	35	52	70	87	105	122	140	157
42	.73304	.73478	.73653	.73827	.74002	.74176	.74351	.74526	.74700	.74875	17	35	52	70	87	105	122	140	157
43	.75049	.75224	.75398	.75573	.75747	.75922	.76096	.76271	.76445	.76620	17	35	52	70	87	105	122	140	157
44	.76794	.76969	.77144	.77318	.77493	.77667	.77842	.78016	.78191	.78365	17	35	52	70	87	105	122	140	157

Table 16 DEGREES TO RADIANS (continued)

	0.0	0.1	0.2	0.3	0.4	0.5	0.6	0.7	0.8	0.9	PROPORTIONAL PARTS 1	2	3	4	5	6	7	8	9
45	.78540	.78714	.78889	.79063	.79238	.79412	.79587	.79762	.79936	.80111	17	35	52	70	87	105	122	140	157
46	.80285	.80460	.80634	.80809	.80983	.81158	.81332	.81507	.81681	.81856	17	35	52	70	87	105	122	140	157
47	.82030	.82205	.82380	.82554	.82729	.82903	.83078	.83252	.83427	.83601	17	35	52	70	87	105	122	140	157
48	.83776	.83950	.84125	.84299	.84474	.84648	.84823	.84998	.85172	.85347	17	35	52	70	87	105	122	140	157
49	.85521	.85696	.85870	.86045	.86219	.86394	.86568	.86743	.86917	.87092	17	35	52	70	87	105	122	140	157
50	.87266	.87441	.87616	.87790	.87965	.88139	.88314	.88488	.88663	.88837	17	35	52	70	87	105	122	140	157
51	.89012	.89186	.89361	.89535	.89710	.89884	.90059	.90234	.90408	.90583	17	35	52	70	87	105	122	140	157
52	.90757	.90932	.91106	.91281	.91455	.91630	.91804	.91979	.92153	.92328	17	35	52	70	87	105	122	140	157
53	.92502	.92677	.92852	.93026	.93201	.93375	.93550	.93724	.93899	.94073	17	35	52	70	87	105	122	140	157
54	.94248	.94422	.94597	.94771	.94946	.95120	.95295	.95470	.95644	.95819	17	35	52	70	87	105	122	140	157
55	.95993	.96168	.96342	.96517	.96691	.96866	.97040	.97215	.97389	.97564	17	35	52	70	87	105	122	140	157
56	.97738	.97913	.98088	.98262	.98437	.98611	.98786	.98960	.99135	.99309	17	35	52	70	87	105	122	140	157
57	.99484	.99658	.99833	1.00007	1.00182	1.00356	1.00531	1.00705	1.00880	1.01055	17	35	52	70	87	105	122	140	157
58	1.01229	1.01404	1.01578	1.01753	1.01927	1.02102	1.02276	1.02451	1.02625	1.02800	17	35	52	70	87	105	122	140	157
59	1.02974	1.03149	1.03323	1.03498	1.03673	1.03847	1.04022	1.04196	1.04371	1.04545	17	35	52	70	87	105	122	140	157
60	1.04720	1.04894	1.05069	1.05243	1.05418	1.05592	1.05767	1.05941	1.06116	1.06291	17	35	52	70	87	105	122	140	157
61	1.06465	1.06640	1.06814	1.06989	1.07163	1.07338	1.07512	1.07687	1.07861	1.08036	17	35	52	70	87	105	122	140	157
62	1.08210	1.08385	1.08559	1.08734	1.08909	1.09083	1.09258	1.09432	1.09607	1.09781	17	35	52	70	87	105	122	140	157
63	1.09956	1.10130	1.10305	1.10479	1.10654	1.10828	1.11003	1.11177	1.11352	1.11527	17	35	52	70	87	105	122	140	157
64	1.11701	1.11876	1.12050	1.12225	1.12399	1.12574	1.12748	1.12923	1.13097	1.13272	17	35	52	70	87	105	122	140	157
65	1.13446	1.13621	1.13795	1.13970	1.14145	1.14319	1.14494	1.14668	1.14843	1.15017	17	35	52	70	87	105	122	140	157
66	1.15192	1.15366	1.15541	1.15715	1.15890	1.16064	1.16239	1.16413	1.16588	1.16763	17	35	52	70	87	105	122	140	157
67	1.16937	1.17112	1.17286	1.17461	1.17635	1.17810	1.17984	1.18159	1.18333	1.18508	17	35	52	70	87	105	122	140	157
68	1.18682	1.18857	1.19031	1.19206	1.19381	1.19555	1.19730	1.19904	1.20079	1.20253	17	35	52	70	87	105	122	140	157
69	1.20428	1.20602	1.20777	1.20951	1.21126	1.21300	1.21475	1.21649	1.21824	1.21999	17	35	52	70	87	105	122	140	157
70	1.22173	1.22348	1.22522	1.22697	1.22871	1.23046	1.23220	1.23395	1.23569	1.23744	17	35	52	70	87	105	122	140	157
71	1.23918	1.24093	1.24267	1.24442	1.24617	1.24791	1.24966	1.25140	1.25315	1.25489	17	35	52	70	87	105	122	140	157
72	1.25664	1.25838	1.26013	1.26187	1.26362	1.26536	1.26711	1.26885	1.27060	1.27235	17	35	52	70	87	105	122	140	157
73	1.27409	1.27584	1.27758	1.27933	1.28107	1.28282	1.28456	1.28631	1.28805	1.28980	17	35	52	70	87	105	122	140	157
74	1.29154	1.29329	1.29503	1.29678	1.29852	1.30027	1.30202	1.30376	1.30551	1.30725	17	35	52	70	87	105	122	140	157
75	1.30900	1.31074	1.31249	1.31423	1.31598	1.31772	1.31947	1.32121	1.32296	1.32470	17	35	52	70	87	105	122	140	157
76	1.32645	1.32820	1.32994	1.33169	1.33343	1.33518	1.33692	1.33867	1.34041	1.34216	17	35	52	70	87	105	122	140	157
77	1.34390	1.34565	1.34739	1.34914	1.35088	1.35263	1.35438	1.35612	1.35787	1.35961	17	35	52	70	87	105	122	140	157
78	1.36136	1.36310	1.36485	1.36659	1.36834	1.37008	1.37183	1.37357	1.37532	1.37706	17	35	52	70	87	105	122	140	157
79	1.37881	1.38056	1.38230	1.38405	1.38579	1.38754	1.38928	1.39103	1.39277	1.39452	17	35	52	70	87	105	122	140	157
80	1.39626	1.39801	1.39975	1.40150	1.40324	1.40499	1.40674	1.40848	1.41023	1.41197	17	35	52	70	87	105	122	140	157
81	1.41372	1.41546	1.41721	1.41895	1.42070	1.42244	1.42419	1.42593	1.42768	1.42942	17	35	52	70	87	105	122	140	157
82	1.43117	1.43292	1.43466	1.43641	1.43815	1.43990	1.44164	1.44339	1.44513	1.44688	17	35	52	70	87	105	122	140	157
83	1.44862	1.45037	1.45211	1.45386	1.45560	1.45735	1.45910	1.46084	1.46259	1.46433	17	35	52	70	87	105	122	140	157
84	1.46608	1.46782	1.46957	1.4+131	1.47306	1.47480	1.47655	1.47829	1.48004	1.48178	17	35	52	70	87	105	122	140	157
85	1.48353	1.48528	1.48702	1.48877	1.49051	1.49226	1.49400	1.49575	1.49749	1.49924	17	35	52	70	87	105	122	140	157
86	1.50098	1.50273	1.50447	1.50622	1.50796	1.50971	1.51146	1.51320	1.51495	1.51669	17	35	52	70	87	105	122	140	157
87	1.51844	1.52018	1.52193	1.52367	1.52542	1.52716	1.52891	1.53065	1.53240	1.53414	17	35	52	70	87	105	122	140	157
88	1.53589	1.53764	1.53938	1.54113	1.54287	1.54462	1.54636	1.54811	1.54985	1.55160	17	35	52	70	87	105	122	140	157
89	1.55334	1.55509	1.55683	1.55858	1.56032	1.56207	1.56382	1.56556	1.56731	1.56905	17	35	52	70	87	105	122	140	157
90	1.57080																		

Table 17 RADIANS TO DEGREES

RADIANS	DEGREES		SINE	COSINE	TANGENT
.01	0.5730	0° 34'	.01000	.99995	0.01000
.02	1.1459	1° 8'	.02000	.99980	0.02000
.03	1.7189	1° 43'	.03000	.99955	0.03001
.04	2.2918	2° 17'	.03999	.99920	0.04002
.05	2.8648	2° 51'	.04998	.99875	0.05004
.06	3.4377	3° 26'	.05996	.99820	0.06007
.07	4.0107	4° 0'	.06994	.99755	0.07011
.08	4.5837	4° 35'	.07991	.99680	0.08017
.09	5.1566	5° 9'	.08988	.99595	0.09024
.10	5.7296	5° 43'	.09983	.99500	0.10033
.11	6.3025	6° 18'	.10978	.99396	0.11045
.12	6.8755	6° 52'	.11971	.99281	0.12058
.13	7.4485	7° 26'	.12963	.99156	0.13074
.14	8.0214	8° 1'	.13954	.99022	0.14092
.15	8.5944	8° 35'	.14944	.98877	0.15114
.16	9.1673	9° 10'	.15932	.98723	0.16138
.17	9.7403	9° 44'	.16918	.98558	0.17166
.18	10.3132	10° 18'	.17903	.98384	0.18197
.19	10.8862	10° 53'	.18886	.98200	0.19232
.20	11.4592	11° 27'	.19867	.98007	0.20271
.22	12.6051	12° 36'	.21823	.97590	0.22362
.24	13.7510	13° 45'	.23770	.97134	0.24472
.26	14.8969	14° 53'	.25758	.96639	0.26602
.28	16.0428	16° 2'	.27636	.96106	0.28755
.30	17.1887	17° 11'	.29552	.95534	0.30934
.32	18.3346	18° 20'	.31457	.94924	0.33139
.34	19.4806	19° 28'	.33349	.94275	0.35374
.36	20.6265	20° 37'	.35227	.93590	0.37640
.38	21.7724	21° 46'	.37092	.92866	0.39941
.40	22.9183	22° 55'	.38942	.92106	0.42279
.42	24.0642	24° 3'	.40776	.91309	0.44657
.44	25.2101	25° 12'	.42594	.90475	0.47078
.46	26.3561	26° 21'	.44395	.89605	0.49545
.48	27.5020	27° 30'	.46178	.88699	0.52061
.50	28.6479	28° 38'	.47943	.87758	0.54630
.52	29.7938	29° 47'	.49688	.86782	0.57256
.54	30.9397	30° 56'	.51414	.85771	0.59943
.56	32.0856	32° 5'	.53119	.84726	0.62695
.58	33.2316	33° 13'	.54802	.83646	0.65517
.60	34.3775	34° 22'	.56464	.82534	0.68414
.62	35.5234	35° 31'	.58104	.81388	0.71391
.64	36.6693	36° 40'	.59720	.80210	0.74454
.66	37.8152	37° 48'	.61312	.78999	0.77610
.68	38.9611	38° 57'	.62879	.77757	0.80866
.70	40.1070	40° 6'	.64422	.76484	0.84229
.72	41.2530	41° 15'	.65938	.75181	0.87707
.74	42.3989	42° 23'	.67429	.73847	0.91309
.76	43.5448	43° 32'	.68892	.72484	0.95045
.78	44.6907	44° 41'	.70328	.71091	0.98926
.80	45.8366	45° 50'	.71736	.69671	1.02964
.82	46.9825	46° 58'	.73115	.68222	1.07171
.84	48.1285	48° 7'	.74464	.66746	1.11563
.86	49.2744	49° 16'	.75784	.65244	1.16156
.88	50.4203	50° 25'	.77074	.63715	1.20966
.90	51.5662	51° 33'	.78333	.62161	1.26016
.92	52.7121	52° 42'	.79560	.60582	1.31326
.94	53.8580	53° 51'	.80756	.58979	1.36923
.96	55.0039	55° 0'	.81919	.57352	1.42836
.98	56.1499	56° 8'	.83050	.55702	1.49096
1.00	57.2958	57° 17'	.84147	.54030	1.55741

Table 17 RADIANS TO DEGREES (continued)

RADIANS	DEGREES		SINE	COSINE	TANGENT
1.02	58.4417	58° 26'	.85211	.52337	1.62813
1.04	59.5876	59° 35'	.86240	.50622	1.70361
1.06	60.7335	60° 44'	.87236	.48887	1.78442
1.08	61.8794	61° 52'	.88196	.47133	1.87122
1.10	63.0254	63° 1'	.89121	.45360	1.96476
1.12	64.1713	64° 10'	.90010	.43568	2.06596
1.14	65.3172	65° 19'	.90863	.41759	2.17588
1.16	66.4631	66° 27'	.91680	.39934	2.29580
1.18	67.6090	67° 36'	.92461	.38092	2.42727
1.20	68.7549	68° 45'	.93204	.36236	2.57215
1.22	69.9009	69° 54'	.93910	.34365	2.73275
1.24	71.0468	71° 2'	.94578	.32480	2.91193
1.26	72.1927	72° 11'	.95209	.30582	3.11327
1.28	73.3386	73° 20'	.95802	.28672	3.34135
1.30	74.4845	74° 29'	.96356	.26750	3.60210
1.32	75.6304	75° 37'	.96872	.24818	3.90335
1.34	76.7763	76° 46'	.97378	.22875	4.25562
1.36	77.9223	77° 55'	.97786	.20924	4.67344
1.38	79.0682	79° 4'	.98185	.18964	5.17744
1.40	80.2141	80° 12'	.98545	.16997	5.79788
1.42	81.3600	81° 21'	.98865	.15023	6.58112
1.44	82.5059	82° 30'	.99146	.13042	7.60183
1.46	83.6518	83° 39'	.99387	.11057	8.98861
1.48	84.7978	84° 47'	.99588	.09067	10.98338
1.50	85.9437	85° 56'	.99749	.07074	14.10142
1.52	87.0896	87° 5'	.99871	.05077	19.66953
1.54	88.2355	88° 14'	.99953	.03079	32.46114
1.56	89.3814	89° 22'	.99994	.01080	92.62050

answers to odd-numbered problems

note The accuracy of answers to numerical computations is, in general, that obtainable with a ten-inch slide rule.

PROBLEMS 2-1

1. (*a*) 25 times R
(*b*) 6 times r
(*c*) 0.25 times I

3. (*a*) \$396.00
(*b*) \$2.75n

5. 12.5I A

7. $\frac{2}{3}C$ pF, $4C$ pF, $48C$ pF

9. (*a*) $16 + R$ Ω
(*b*) $e + 220$ V
(*c*) $i - I$ A

11. $L_2 = L_1 - 125$ mH

13. (*a*) 44 A
(*b*) 0.25 A

15. (*a*) 2.99 s
(*b*) 0.685 s

17. 0.293 m

PROBLEMS 2-2

1. (*a*) 72
(*b*) 276
(*c*) 1296
(*d*) 72
(*e*) 36
(*f*) 207

3. (*a*) Monomial
(*b*) Monomial
(*c*) Monomial
(*d*) Binomial
(*e*) Trinomial
(*f*) Binomial
(*g*) Trinomial
(*h*) Trinomial
(*i*) Monomial
(*j*) Trinomial

5. (*a*) $I = \dfrac{E}{R}$
(*b*) $E = IR$
(*c*) $P = RI^2$
(*d*) $R_1 = R_2 + R_3$
(*e*) $K = \dfrac{M}{\sqrt{L_1 L_2}}$
(*f*) $R_p = \dfrac{R_1 R_2}{R_1 + R_2}$
(*g*) $N = \dfrac{R_m}{R_s} + 1$

7. 10 197 μH Note that, all other factors remaining equal, if the number of turns is tripled, the inductance is multiplied by a factor of nine (3^2).

9. (*a*) Increased by a factor of 4
(*b*) Increased by a factor of 9
(*c*) Reduced to a value one-fourth the original

PROBLEMS 3-1

1. 71
3. -46
5. 28
7. -1081
9. 208.56
11. 4
13. $-10\frac{7}{32}$
15. $\frac{2}{15}$

PROBLEMS 3-2

1. 61
3. 213
5. 994
7. 3.84
9. $-10\frac{1}{16}$
11. (a) 67°
    (b) 26°
    (c) 159°
13. $364.80
15. 242 V

PROBLEMS 3-3

1. $11i$
3. $112IZ$
5. $4I - 5i$
7. $3IR + 13E$
9. $128\theta - 110\phi$
11. $27i^2r + 10W - 3ei + 49w$
13. $1.46eI + 3.82W + 0.75I^2r$
15. $-\frac{11}{48}\pi ft - 2\frac{1}{8}\pi Z$
17. $10\phi + 10\theta$
19. $47.6\frac{E^2}{R} - 16.4EI + 5.8I^2R$
21. $3.90IZ - 1.31IR - 0.41IX$
23. $6.64\psi - 7.1\lambda$

PROBLEMS 3-4

1. $3 - 7y$
3. $10R - 3X + 3$
5. $10\frac{E^2}{R} - 3EI$
7. $\alpha + 2\beta$
9. $17a - 10b + 6c$

PROBLEMS 3-5

1. (a) $3X + (X_C - X_L + Z)$
   (b) $\alpha + (6\beta - 3\phi + \lambda)$
   (c) $5W + (6I^2R - 3EI + 7I^2Z)$
   (d) $\frac{E^2}{R} + (-3I^2R + 7I^2Z - 4EI)$
   (e) $8\lambda + 3\mu + (-7\theta - 3\phi + 6\alpha)$
3. $X^2 + R^2 - N$
5. 16.8 + eV

7. $Z - \sqrt{r^2 + x^2}$

9. $P - I^2R - \dfrac{E^2}{R}$

11. $X_C - \dfrac{1}{2\pi f C_1}$

PROBLEMS 4-1

1. 12
3. 13.6
5. $-\frac{15}{256}$
7. 0.000 000 938
9. eit
11. $2\pi f L_1 L_2$
13. $\dfrac{1}{2\pi f C_p}$
15. $-\dfrac{\psi\mu}{\theta\phi}$

PROBLEMS 4-2

1. x^5
3. $-e^{10}$
5. $6m^4$
7. $-60m^4x^3$
9. abm^{n+p}
11. $8p^3$
13. $6a^3b^4c^3d^7$
15. $-\dfrac{\pi M X_L}{4}$
17. $-0.075e^3i^4rw$
19. a^6

PROBLEMS 4-3

1. $18a + 30b$
3. $4I^2R_1 + 8I^2R_2$
5. $4.7\lambda^2\phi + 9.4\theta\phi - 14.1\mu\phi$
7. $2\alpha^4\beta^2 + 1.5\alpha^3\beta^3 - 2.5\alpha^2\beta^4$
9. $15a^3r_1r_2 + 6a^2r_1^{\,2}r_2 - 18ar_1^{\,3}r_2$
11. $\dfrac{iI^3RZ}{3} - \dfrac{iI^3R^2Z}{6} - \dfrac{2i^2IZ^2}{9}$
13. $3I^3PR - 6Ii^2Pr + 2IP^2$
15. $0.157E^3IZ^3 + 0.314EIZ^5 - 10.5IZ^6$
17. $15\phi - 21\theta$
19. $\theta^3 - \phi^3$
21. $0.9\pi\omega + 3\eta\pi^2 + 2.5\eta\omega^2 - 6.5\pi\omega^3$
23. $\dfrac{8\lambda E^2}{3} + 4\lambda Ee - \dfrac{\lambda e^2}{2}$
25. $0.125IR - 0.025IR_1 - 0.4125IR_2$
27. 0
29. $7s$

PROBLEMS 4-4

1. $\alpha^2 + 2\alpha + 1$
3. $\alpha^2 - 2\alpha + 1$
5. $\beta^2 - 9$
7. $p^2 + 8p + 15$
9. $r^2 - 8r - 33$
11. $m^2 + 6m + 8$

13. $3\alpha^2 + 15\alpha\beta - 42\beta^2$
15. $6\theta^2 - 7\theta\lambda - 5\lambda^2$
17. $6m^2 - 5mn - 6n^2$
19. $5R^2 - 17RZ + 6Z^2$
21. $6a^3 + 17a^2 + 2a - 1$
23. $2R^3 - 2R^2r - 2Rr^2 + 2r^3$
25. $a^3 - a^2b - ab^2 + b^3$
27. $\theta^3 - \theta^2\phi - \theta\phi^2 + \phi^3$
29. $a^3 + 3a^2b + 3ab^2 + b^3$
31. $x^2 + 2xy + y^2$
33. $M^2 - 2MN + N^2$
35. $8\alpha^3 + 24\alpha^2w + 24\alpha w^2 + 8w^3$
37. $16I^4R^2 - 38I^2R + 14$
39. $10a^3 + 4a^2b - 16a^2 - 5ab^2 - 14ab + a$

PROBLEMS 4-5

1. 5
3. 5
5. $-\frac{4}{3}$
7. $-2\pi fC$
9. $\dfrac{E \times 10^8}{L_v}$
11. -6
13. -225
15. $-\frac{5}{2}$

PROBLEMS 4-6

1. $4x^2y^4$
3. $-2\theta\phi^2\psi^3$
5. $-\dfrac{4X_cZ^2}{3}$
7. $3\eta^4\lambda^3\pi$
9. $\dfrac{3mn^2p}{4}$
11. $-9c$
13. $-4\lambda^3\psi^4$
15. $\dfrac{b^7d^4}{3ac^6}$
17. $-\dfrac{\phi^6}{4\theta^{12}\psi\Omega^3}$
19. $\dfrac{90\,000\alpha^2\beta^5}{\gamma^2}$

PROBLEMS 4-7

1. $4x + 5y$
3. $12\alpha^2 - 9\beta^2$
5. $3R_1 + 6R_2 - 4R_3$
7. $\dfrac{0.005\mu^3}{\pi} + 10\mu\pi$
9. $\dfrac{3m^3}{10} - \dfrac{7m}{5} - \dfrac{6}{5m}$
11. $6 + 10xz - 5x^2z^2 - 3x^4y^4$
13. $2(\theta + \phi) - 4(\theta + \phi)^3 + 3(\theta + \phi)^5$
15. $\dfrac{(EI + P)^2}{2} - 2 + \dfrac{6}{EI + P}$

17. $\dfrac{1}{I\left(\omega L - \dfrac{1}{\omega C}\right)} - 2I\left(\omega L - \dfrac{1}{\omega C}\right) - 5I^3\left(\omega L - \dfrac{1}{\omega C}\right)^3$

19. $3(\theta - \phi)^2 - 6(\theta + \phi)(\theta - \phi)^3 - \dfrac{9(\theta - \phi)}{\theta + \phi}$

PROBLEMS 4-8

1. $x + 1$
3. $\theta + 3$
5. $2E - 6$
7. $3R^2 - 4Z - 7$
9. $K^2 + 7K + 14 + \dfrac{6}{K - 1}$
11. $E + e$
13. $E^3 + E^2e + Ee^2 + e^3$
15. $E^2 + I^2R^2$
17. $X^5 - X^4Y + X^3Y^2 - X^2Y^3 + XY^4 - Y^5$
19. $\theta^2 + 2\theta\phi + \phi^2$
21. $2R_2 - 3$
23. $10E^2 - 3E - 12 + \dfrac{7E - 45}{3E^2 + 2E - 4}$
25. $3R + \frac{1}{3}$
27. $6x - \dfrac{y}{3} - \dfrac{1}{2}$
29. $\dfrac{3L_1^2}{8} - \dfrac{L_1}{4} - \dfrac{2}{3}$

PROBLEMS 5-1

1. $x = 4$
3. $k = -5$
5. $p = 6$
7. $\pi = 5$
9. $IR = 4$
11. $\alpha = -10$
13. $E = -5$
15. $Q = -2$
17. $I = -1.4$
19. $\beta = 1$

PROBLEMS 5-2

1. $E - 75$ V
3. $d = rt$ km
5. $y = t$ yr
7. $\dfrac{Z}{t}$ kilometers per minute (km/min)
9. 110 V
11. $I = \dfrac{E}{R}$
13. 12 m by 6 m
15. 4.25, 10.75, and 8.5 m
17. $h^2 = a^2 + b^2$
19. 63, 64, 65

PROBLEMS 5-3

1. $C = \frac{Q}{V}, \quad V = \frac{Q}{C}$

3. $Z^2 = R^2 + X^2, \quad X^2 = Z^2 - R^2$

5. $R = \frac{KL}{m}, \quad K = \frac{Rm}{L}, \quad m = \frac{KL}{R}$

7. $\lambda = \frac{v}{f}, \quad v = f\lambda$

9. $L = \frac{RQ}{\omega}, \quad Q = \frac{\omega L}{R}, \quad \omega = \frac{RQ}{L}$

11. $X_C = \frac{1}{2\pi fC}, \quad f = \frac{1}{2\pi CX_C}$

13. $\phi = HA, \quad A = \frac{\phi}{H}$

15. $E = \frac{BLv}{10^8}, \quad L = \frac{E \times 10^8}{Bv}, \quad v = \frac{E \times 10^8}{BL}$

17. $E_s = \frac{E_p I_p}{I_s}$

19. $I = \frac{E - e}{R}, \quad E = IR + e, \quad e = E - IR$

21. $\theta = \omega t, \quad \omega = \frac{\theta}{t}$

23. $V = \frac{V_o + V_t}{2}, \quad V_t = 2V - V_o$

25. $r^3 = \frac{3A}{4\pi}$

27. $Z_t = \frac{F(R - r)}{C}, \; F = \frac{CZ_t}{R - r}, \; R = \frac{CZ_t + Fr}{F},$
$r = \frac{FR - CZ_t}{F}$

29. $E_b = iR_L + e_b, \quad e_b = E_b - iR_L,$
$i = \frac{E_b - e_b}{R_L}$

31. $l = \frac{Rd^2}{\rho}, \; \rho = \frac{Rd^2}{l}, \; d^2 = \frac{\rho l}{R}$

33. $A = \frac{Cd}{0.0884K(n - 1)},$
$n = \frac{Cd + 0.0884KA}{0.0884KA}$

35. $L = CRZ_r, \quad C = \frac{L}{RZ_r}, \quad R = \frac{L}{CZ_r}$

37. $\beta = \frac{\gamma\omega\alpha}{\eta}$

39. $Q = \frac{\rho hv}{e}$

41. $C_2 = \frac{V_3 - V_2}{\omega^2 L V_3}$

43. $I_n = \frac{Q - I_p p}{n}$

45. $R_1 = \frac{1}{\omega_{01} C} - R_2$

47. 4000 Ω

49. 5 m

PROBLEMS 5-4

1. $\frac{1}{4}$ **3.** $\frac{5}{4}$ **5.** $\frac{4}{6}$, $\frac{6}{9}$, $\frac{12}{18}$, etc. **7.** $\frac{1}{3}$ **9.** 32:1

PROBLEMS 5-5

1. 10 **3.** 200 **5.** $X = 15$ **7.** $IR = 8$ **9.** $Q = 0.0014$

PROBLEMS 5-6

1. $D \propto R,\quad D = kR$ **3.** $C \propto A,\quad C = kA$ **5.** $X_C \propto \frac{1}{C},\quad X_C = \frac{k}{C}$

7. $T \propto \sqrt{L},\quad T = k\sqrt{L}$ **9.** $V \propto \frac{1}{P},\quad V = \frac{k}{P}$ **11.** $L \propto \frac{1}{d^2},\quad L = \frac{k}{d^2}$

13. 1.74 kV **15.** 5400 kg

PROBLEMS 6-1

1. 6 **3.** 6 **5.** 3 **7.** 1 **9.** 4

PROBLEMS 6-2

1. 6.43×10^5 **3.** 6.53×10^3 **5.** 9.44×10^{-9} **7.** 3.67×10^{-1}

9. 2.50×10^{-1} **11.** 3.99×10^4 **13.** 2.59×10^{-2} **15.** 2.76×10^5

17. 1.08×10^{-7} **19.** 3.00×10^6

PROBLEMS 6-3

1. 1.00×10^{-2} **3.** 3.92×10^{-1} **5.** 7.14×10^{-12} **7.** 3.11×10^{11}

9. 3.20×10 **11.** 5.65×10^0 **13.** $9.42 \times 10^4\ \Omega$ **15.** $2.20 \times 10^{-1}\ \Omega$

PROBLEMS 6-4

1. 5.00×10^{-7} **3.** 1.05×10^{-4} **5.** 1.00×10^{-13} **7.** 2.87×10^{10}

9. 2.55×10^2 **11.** 1.63×10^{-3} **13.** $6.62 \times 10^2\ \Omega$ **15.** $1.26 \times 10^{-6}\ \Omega$

PROBLEMS 6-5

1. 10^{12} **3.** 10^{20} **5.** 6.25×10^{14} **7.** 2.56

9. 5×10^{-3} **11.** 30 **13.** 1.01×10^4 **15.** 1.50×10^6 Hz

17. 7.50×10^6 Hz **19.** 1.20×10^6 Hz

PROBLEMS 6-6

1. (*a*) 3100
   (*b*) 3.10×10^3

3. (*a*) 4 190 000 000 000
   (*b*) 4.19×10^{12}

5. (*a*) 6 279 999.841
   (*b*) 6.28×10^6

PROBLEMS 7-1

1. (*a*) 4.30×10^6 mV
   (*b*) 4.30×10^9 μV
   (*c*) 4.30 kV

3. (*a*) 1.35×10^{-3} kV
   (*b*) 1.35×10^6 μV
   (*c*) 1.35×10^3 mV

5. (*a*) 3.30 kΩ
   (*b*) 3.30×10^{-3} MΩ
   (*c*) 3.03×10^{-4} S

7. (*a*) 2.00×10^{-8} F
   (*b*) 2.00×10^{-2} μF

9. (*a*) 3.47×10^{-1} kW
   (*b*) 3.47×10^5 mW
   (*c*) 3.47×10^8 μW

11. (*a*) 1.32×10^0 MHz
    (*b*) 1.32×10^6 Hz

13. (*a*) 4.00×10^{-1} W
    (*b*) 4.00×10^{-4} kW

15. (*a*) 1.50×10^{-2} MHz
    (*b*) 1.50×10^4 Hz

17. (*a*) 5.50×10^4 μA
    (*b*) 5.50×10^1 mA

19. (*a*) 2.70×10^6 Ω
    (*b*) 2.70×10^3 kΩ

21. (*a*) 3.35×10^6 μH
    (*b*) 3.35 H

23. (*a*) 5.00×10^2 pF
    (*b*) 5.00×10^{-10} F

25. (*a*) 2.50×10^6 μS
    (*b*) 4.00×10^{-1} Ω

27. (*a*) 2.35×10^0 mA
    (*b*) 2.35×10^{-3} A

29. (*a*) 1.50×10^8 W
    (*b*) 1.50×10^5 kW

PROBLEMS 7-2

1. (*a*) 108 in
   (*b*) 274 cm
   (*c*) 2.74×10^3 mm

3. (*a*) 80.7 in
   (*b*) 205 cm
   (*c*) 2.24 yd

5. (*a*) 2.88 mi
   (*b*) 4.63×10^3 m
   (*c*) 4.63 km

7. 1.63 mm

9. 3.74 mH/km

11. 6×10^{-2} dB/100 m

13. 3.22×10^{-3} Ω/cm

15. 4.72 in/min

PROBLEMS 7-3

3. $X_L = 2\pi f L \ \Omega$

5. $f = \dfrac{159}{\sqrt{LC}}$ MHz

7. $\delta = \dfrac{6.62}{\sqrt{f}}$ cm

9. $R_{ac} = 83.2 \times 10^{-9} \dfrac{\sqrt{f}}{d}$ Ω/cm

13. 435 cm

15. 66.2 cm

17. (*a*) 72.4 cm
    (*b*) 75.9 cm
    (*c*) 30.7 cm

19. (*a*) 72.4 cm
    (*b*) 75.9 cm
    (*c*) 68.6 cm
    (*d*) 30.7 cm
    (*e*) 15.4 cm

PROBLEMS 8-1

1. 4.40 A

3. 6.20 A

5. 0.080 μA

7. 3.75 A

9. (*a*) 0.571 A
(*b*) 0.635 A

PROBLEMS 8-2

1. (*a*) 5.6×10^3 W
(*b*) 5.6 kW

3. 0.833 A

5. 15.0 hp

7. 1119 kW

9. 0.108 W

11. (*a*) 0.579 W
(*b*) 35.1 mA

13. (*a*) 20.8×10^{-6} μW
(*b*) 0.231 μA

15. (*a*) 90.5%
(*b*) $24.72

17. 18.4 hp

19. 2.4 kW

PROBLEMS 8-3

1. (*a*) 69.6 mA
(*b*) 47.3 V
(*c*) 1.60 W

3. (*a*) 121 Ω
(*b*) 100 W
(*c*) 80.3 W

5. 179 Ω

7. 1.43 Ω

9. (*a*) 1.8 Ω
(*b*) 1.5 Ω
(*c*) 3.48 kW

PROBLEMS 8-4

1. (*a*) 954 Ω
(*b*) 310.5 V

3. (*a*) 210 Ω
(*b*) 525 mW
(*c*) 310.5 V
(*d*) 15.5 W

5. (*a*) 188 Ω
(*b*) 33.2 mW
(*c*) 44.6 kΩ
(*d*) 350 mW
(*e*) 252.5 V
(*f*) 3.36 W

7. (*a*) 60 kΩ
(*b*) 600 μW

9. 2.5 kΩ

PROBLEMS 9-1

1. (*a*) 0.166 Ω
(*b*) 0.103 Ω

3. 789 Ω

5. 1.48 Ω

7. 0.513 Ω

9. 1.31 m

PROBLEMS 9-2

1. 13 Ω

3. 73.3 Ω

5. 0.0382 $\mu\Omega \cdot$ m

7. 2.36 km

9. 444 m

PROBLEMS 9-3

1. 4.60 Ω

3. 15.8 Ω

5. No

PROBLEMS 9-4

1. (a) 0.639 Ω
   (b) 1.499 kg
3. (a) 4.08 km
   (b) 21.3 Ω
5. 2112 Ω
7. No. 1 wire
9. (a) No. 5 wire
   (b) 96.3%

PROBLEMS 10-1

1. x^2y^2
3. $e^3i^6Z^3$
5. $16\pi^2\phi^2$
7. $-8I^3R^3$
9. $2\pi X_L{}^2$
11. $-\dfrac{1}{4\pi^2f^2C^2}$
13. $-\dfrac{125P^6}{E^3I^3}$
15. $-\dfrac{V^6}{8g^3}$
17. $\dfrac{B^6A^3l^3}{512\omega^3}$
19. $-\frac{16}{9}\pi^2R^6$
21. $\dfrac{x^{12}y^{18}}{p^{15}}$

PROBLEMS 10-2

1. $\pm a$
3. $\pm 3i$
5. $-\omega$
7. $\pm 5\lambda^2\Omega^3$
9. $3x^2$
11. 4
13. $\pm 13m^2np^3$
15. $3\theta^2\phi^4\omega$
17. $\pm\dfrac{16\pi rx^2}{17z^3\phi^2}$
19. $\pm\dfrac{25r^3s^2t^4}{4x^3z^5}$
21. $\dfrac{4a\omega^2}{5x^2z^4}$
23. $\pm\dfrac{5vt}{16a^4bx}$

PROBLEMS 10-3

1. $2(a + 3)$
3. $\theta(3 + \phi + 4\omega)$
5. $10i(2r - z)$
7. $\dfrac{ay}{36}(4ay + 12a^2 - 3y^2)$
9. $2a^2bc(ab + 4c^2 + 6bc)$
11. $36\alpha^2\beta^2\omega^2(\alpha^2\beta - 2\omega^3 + 5\beta^3)$
13. $\dfrac{1}{3648}Ii^2(57Ii + 48i^2 - 76I^2)$
15. $120\eta\theta^2\phi\omega(6\eta^3\phi^2 + 9\eta\theta^2\omega + 5\eta^2\theta\omega - 4\theta^4)$

PROBLEMS 10-4

1. $\theta^2 + 6\theta + 9$
3. $m^2 - 2mR + R^2$
5. $\alpha^2 + 32\alpha + 256$
7. $9X^2 - 6XR + R^2$
9. $F^2 - 2Ff + f^2$
11. $25\theta^2 + 40\theta\phi + 16\phi^2$
13. $81r_1{}^2 - 54r_1r_2 + 9r_2{}^2$
15. $1 + 2X_L{}^2 + X_L{}^4$
17. $36v^4 - 24v^2t^3 + 4t^6$
19. $900 - 180 + 9 = 729$
21. $36\pi^2R^4 - 24\pi^2R^2r^2 + 4\pi^2r^4$
23. $2.25\theta^4 - 1.5\theta^2\alpha + 0.25\alpha^2$

25. $\frac{9}{16}X^4 - \frac{3}{4}X^2Z^2 + \frac{1}{4}Z^4$ 27. $36\phi^4\omega^2 - 3\phi^2\omega\lambda^2 + \frac{1}{16}\lambda^4$ 29. $x^2 + x + \frac{1}{4}$

31. $\frac{1}{4} - E + E^2$ 33. $1 + 2e^3 + e^6$ 35. $L^4 - \frac{7}{4}L^2P + \frac{49}{64}P^2$

37. $\frac{b^2}{9} + \frac{bm}{3} + \frac{m^2}{4}$ 39. $R_1^2 - \frac{5}{4}R_1R_2 + \frac{25}{64}R_2^2$

PROBLEMS 10-5

1. $6e$ 3. 4λ 5. $10xy$ 7. $14\omega\pi$

9. $12mp$ 11. I^2 13. $16p^2$ 15. $\frac{1}{3}\theta\phi\omega$

17. $\frac{1}{9}\pi^2$ 19. $\pm(M + 1)$ 21. $\pm(4q_1 + q_2)$ 23. $\pm(3\alpha^2\beta + 9\gamma)$

25. $\pm(\frac{3}{5}\pi R^2 + \frac{2}{3})$ 27. $\pm(\frac{5}{6}\phi + \frac{2}{7}\lambda)$

PROBLEMS 10-6

1. $3c(a + 2b)$ 3. $2\lambda(\theta + \phi)^2$ 5. $6\alpha^2(2\alpha + 5\beta)^2$ 7. $\frac{20f_o}{\omega}(\omega_1 - \omega_2)^2$ 9. $\frac{5r}{16e}(\lambda - 4f^2)^2$

PROBLEMS 10-7

1. $\theta^2 - 4$ 3. $I^2 - i^2$ 5. $9Q^2 - 4L^2$ 7. $\frac{4}{9}E^2I^2 - P^2$ 9. $\frac{4E^4}{R^2} - \frac{9I^4R^2}{P^2}$

PROBLEMS 10-8

1. $(a + b)(a - b)$ 3. $(2\theta + 4\phi)(2\theta - 4\phi)$ 5. $(\frac{1}{2} + \theta)(\frac{1}{2} - \theta)$

7. $(1 + 15\omega)(1 - 15\omega)$ 9. $(9\theta\mu + 1)(9\theta\mu - 1)$

13. $9(ab - 2m - 3pq)(ab - 2m + 3pq)$ 15. $(5a + 10cl + 12l)(5a + 10cl - 12l)$

PROBLEMS 10-9

1. $\theta^2 + 7\theta + 12$ 3. $R^2 - R - 2$ 5. $\theta^2 + 9\theta + 18$ 7. $9\theta^2 - 3\theta - 2$

9. $I^2 - 7I + 12$ 11. $\alpha^2 - \frac{5\alpha}{4} + \frac{1}{4}$ 13. $I^2R^2 + \frac{IR}{6} - \frac{1}{6}$ 15. $\alpha^2 + \alpha + \frac{2}{9}$

17. $\frac{1}{LC} - \frac{4f}{\sqrt{LC}} + 3f^2$ 19. $\alpha^2\beta^4 + \frac{3\alpha\beta^2}{10} + \frac{1}{50}$

PROBLEMS 10-10

1. $(\alpha + 1)(\alpha + 2)$ **3.** $(R + 6)(R + 2)$ **5.** $(\beta + 6)(\beta - 4)$ **7.** $(\theta + 4)(\theta + 6)$

9. $(t + 11)(t - 2)$ **11.** $(Z^2 - 2)(Z^2 + 10)$ **13.** $(\pi + 8)(\pi - 7)$ **15.** $(\omega + 2f)(\omega - 3f)$

17. $(\theta - \frac{1}{2})(\theta - \frac{1}{3})$ **19.** $(\phi^2 + \frac{1}{5})(\phi^2 - \frac{1}{10})$

PROBLEMS 10-11

1. $x^2 - 3x - 10$ **3.** $6\phi^2 + 11\phi + 3$ **5.** $12j^2 - 2j - 4$

7. $6\omega^2 + 13\omega - 5$ **9.** $\dfrac{\omega^2}{4} + 2\omega - 32$ **11.** $6Z^2 + 13IRZ + 5I^2R^2$

13. $15X^2 - 94X - 40$ **15.** $15\theta^2 - 77\theta + 10$ **17.** $35 - 31\pi + 6\pi^2$

19. $6\alpha^2 + 31\alpha\beta + 35\beta^2$ **21.** $4a^2 - 24at + 35t^2$ **23.** $\omega^2 + 0.5\omega f - 0.14f^2$

25. $\dfrac{x^2}{4} - \dfrac{3x\lambda}{2} - 4\lambda^2$ **27.** $24Z^2 + \dfrac{4Z}{IR} + \dfrac{1}{6I^2R^2}$ **29.** $0.16p^2 - 0.62pq + 0.21q^2$

PROBLEMS 10-12

1. $(\omega + 2)(\omega - 5)$ **3.** $(2m - 3)(4m + 5)$ **5.** $(2x + 5)(3x - 2)$

7. $(3\phi + 4)(3\phi + 2)$ **9.** $(\alpha + 3\beta)(2\alpha - 7\beta)$ **11.** $(10m - 7)(4m + 3)$

13. $(8l + 3w)(10l - 2w)$ **15.** $(12\beta^2 - 9\gamma)(2\beta^2 - \gamma)$ **17.** $(9lm - w)(3lm + 2w)$

19. $6(\psi + 2\Omega)(\psi - 2\Omega)$ **21.** $(5x + \Delta)(3x - 2\Delta)$ **23.** $(8\theta + \frac{1}{2})(6\theta + \frac{1}{4})$

25. $(0.6\theta + 2)(0.3\theta - 1)$

PROBLEMS 10-13

1. $16\omega^2L^2$ **3.** $\dfrac{a^{12}b^{12}c^4d^8}{p^8q^{12}r^4}$ **5.** $\pm\dfrac{12IR}{13FX_C^2}$

7. $-\dfrac{5lm^2}{3x^4y^5z}$ **9.** $6\theta\phi^2\omega^2$ **11.** $I(R + r)(R - r)$

13. $\dfrac{e^2}{8}\left(\dfrac{3}{r_1} + \dfrac{5}{r_2} - \dfrac{7}{r_3}\right)$ **15.** $\dfrac{x}{16}(7k - 3l - 9m)$

17. $R^2 + 24R + 144$ **19.** $144I^4 + 16I^2 + \frac{4}{9}$

21. $\frac{25\beta^2}{81} - \frac{10\beta\lambda}{3} + 9\lambda^2$

23. $6r$

25. $49Q^2$

27. $\frac{\lambda^2}{16}$

29. $\pm(m + 5)$

31. $\pm(4\alpha + 10\beta)$

33. $\pm\left(\frac{\phi}{6} - \frac{\lambda}{2}\right)$

35. $6R(4i + 7I)$

37. $3i(r + 3)^2$

39. $12\omega(8\theta - \phi)^2$

41. $\alpha^2 - 4\beta^2$

43. $Z^2 - 144$

45. $\frac{576E^2}{I^2R^2} - 4P^2$

47. $(Q + 1)(Q - 1)$

49. $\left(2\omega L + \frac{1}{4\omega C}\right)\left(2\omega L - \frac{1}{4\omega C}\right)$

51. $(0.05\psi + 0.6\mu)(0.05\psi - 0.6\mu)$

53. $\lambda - 2$

55. $\frac{1}{3}\alpha + \frac{2}{7}\beta$

57. $\frac{3}{5}e - \frac{4}{9}ir$

59. $\kappa^2 - 2\kappa - 8$

61. $0.2X_C^2 - 2.9X_C - 1.5$

63. $A^2 - \frac{2A}{15} - \frac{1}{15}$

65. $24\mu^2 + 2\mu g_m - 12g_m^2$

67. $0.6R^2 + 0.1Rr - 0.2r^2$

69. $24\phi^2 + 2\theta\phi - \frac{\theta^2}{3}$

71. $(3z + 1)(2z + 3)$

73. $(\lambda - 5)(\lambda - 3)$

75. $(x - 2)(x - 0.6)$

77. $(2R + 3X)(6R - 5X)$

79. $(2E - 0.5IR)(E + 0.3IR)$

81. $\left(\frac{X_C}{3} + Z\right)^2$

83. $(2\pi + f)(8\pi - 5f)$

85. $3(x + 2)(x - 2)$

87. $\frac{3}{2i}(E - 6e)^2$

89. $\frac{c}{144d}(8a - 9b)(9a - 8b)$

PROBLEMS 11-1

1. 8
3. $4\theta\phi$
5. $0.5a^2bc$
7. $3IR$
9. $X_L + X_C$
11. $E - 1$
13. $\sqrt{L_1L_2} + M$
15. $5\left(2I + 3\frac{E}{R}\right)$

PROBLEMS 11-2

1. 420
3. 360
5. $\theta^4\phi^3\lambda^3\mu\omega$
7. $180\,m^3n^2p^4$
9. $t^2 - 5t + 6$
11. $\mu(\mu + 3)(\mu + 5)$
13. $11(3\theta - 1)(2\theta + 1)(2\theta + 3)$
15. $\left(Q + \frac{\omega L}{R}\right)\left(Q - \frac{\omega L}{R}\right)\left(4Q - \frac{5\omega L}{R}\right)\left(2Q - \frac{7\omega L}{R}\right)$

PROBLEMS 11-3

1. 18
3. xy
5. $9abd$
7. $t^2 - 2t + 1$
9. $6i + 6\alpha$
11. $\frac{12}{64}$
13. $\dfrac{\omega LR^2 - \omega LX^2}{R^3 - RX^2}$
15. $\dfrac{2ECQ - 3Q}{2E^2C^2 - EC - 3}$

PROBLEMS 11-4

1. $\frac{3}{4}$
3. $\frac{1}{13}$
5. $\dfrac{1}{ab^3}$
7. $\dfrac{5I}{R}$
9. $\dfrac{x}{x^2 + y^2}$
11. $\dfrac{a + b}{a - b}$
13. $\dfrac{x + y}{3}$
15. $\dfrac{\omega(\pi + 3\lambda)}{3\pi + \lambda}$

PROBLEMS 11-5

1. $\dfrac{a}{x}$
3. $\dfrac{2\pi fL}{X_C - X_L}$
5. $\dfrac{\omega L}{R_2 - R_1}$
7. $\dfrac{-IR}{E + e}$
9. $\dfrac{\pi R^2}{A_2 - A_1}$
11. -1
13. $-\dfrac{1}{\phi + \theta}$
15. $\dfrac{4 - \pi}{5 + \pi}$

PROBLEMS 11-6

1. $\frac{17}{8}$
3. $\dfrac{ac + b}{c}$
5. $\dfrac{4F - 5}{F}$
7. $\dfrac{4\pi + 6}{\pi + 1}$
9. $\dfrac{IR - I - E}{I}$
11. $\dfrac{(9 + 2x)(1 - x)}{x^2}$

13. $\dfrac{(R-1)(R+7)}{R^2}$

15. $\dfrac{9\lambda^2 - 4\lambda - 2}{(3\lambda+1)(3\lambda-1)}$

17. $\dfrac{-\theta^3 + 13\theta^2 + 31\theta - 45}{\theta^2(\theta-1)}$

19. $\dfrac{2\alpha^4 - 7\alpha^2 - 1}{\alpha^2 - 3}$

21. $5\frac{3}{16}$

23. $1 + \dfrac{y^2}{x^2}$

25. $R^2 + 7R + 14 + \dfrac{6}{R-1}$

27. $E^3 - E^2e + Ee^2 - e^3 - \dfrac{1}{E+e}$

29. $2x + 2 - \dfrac{x}{x^2+1}$

PROBLEMS 11-7

1. $\frac{35}{70}$, $\frac{30}{70}$, $\frac{28}{70}$

3. $\frac{36}{48}$, $\frac{21}{48}$, $\frac{20}{48}$

5. $\dfrac{\theta\omega}{\phi\omega}$, $\dfrac{\lambda\phi}{\phi\omega}$

7. $\dfrac{ei}{ir}$, $\dfrac{1}{ir}$, $\dfrac{ei^2r}{ir}$

9. $\dfrac{a+b}{a^2-b^2}$, $\dfrac{a-b}{a^2-b^2}$

11. $\dfrac{3\phi+3\pi}{\phi^2-\pi^2}$, $\dfrac{4\phi-4\pi}{\phi^2-\pi^2}$

13. $\dfrac{ac-ad}{c^2-d^2}$, $\dfrac{bc+bd}{c^2-d^2}$, $\dfrac{bc+bd-ac-ad}{c^2-d^2}$

15. $\dfrac{\pi^2-\phi^2}{\pi\phi}$, $-\dfrac{\phi^2}{\pi\phi}$

PROBLEMS 11-8

1. $\frac{33}{70}$

3. $-\frac{5}{48}$

5. $\dfrac{65IR}{48}$

7. $\dfrac{\alpha\delta - \beta\gamma}{\beta\delta}$

9. $\dfrac{ayz - bxz - cxy}{xyz}$

11. $\dfrac{10R - 3I^2 + 4}{I^2R}$

13. $\dfrac{19I + i}{6}$

15. $\dfrac{3\alpha+\beta}{\alpha^2-\beta^2}$

17. $\dfrac{3L_1 + 34}{L_1^2 + 4L_1 - 12}$

19. $\dfrac{7+25\theta}{6(1-\theta^2)}$

21. $\dfrac{23I + 133}{I(I+7)(I-7)}$

23. $\dfrac{2\pi+7}{3+\pi}$

25. $\dfrac{8\theta\phi}{\theta^2-\phi^2}$

27. $\dfrac{10-4E}{(E-4)(E-5)(E-6)}$

29. $\dfrac{12\omega^2}{\omega^3+27}$

PROBLEMS 11-9

1. $\frac{1}{4}$

3. $-\frac{1}{50}$

5. 4

7. $4xy^3$

9. $\dfrac{4}{\theta^3\omega^2}$

11. $\dfrac{\omega}{2\pi fR}$

13. $\dfrac{4x+4y}{(x-y)^2}$

15. $\dfrac{5x+y}{3x+2}$

17. $\frac{1}{\phi - 2}$

19. ϕ^3

21. $14\alpha^2$

23. $4c$

25. ϕ^2

27. $\frac{1}{3\omega L + R}$

29. $2m$

PROBLEMS 11-10

1. $-\frac{7}{8}$

3. $-\frac{7}{40}$

5. $\frac{Q\omega L_1 L_2}{L_1 + L_2}$

7. $\frac{Ir}{Ir - E}$

9. $\frac{2E(E - e)}{e(E + e)}$

11. $\frac{l - w}{l + w}$

13. $\frac{b}{\alpha}$

15. $-\frac{I^2 + i^2}{2Ii}$

PROBLEMS 12-1

1. $\phi = 8$

3. $\alpha = 8$

5. $\omega = \frac{7}{16}$

7. $\phi = 5$

9. $\lambda = 3$

11. $\omega = -12$

13. $\theta = -2$

15. $m = 12\frac{19}{24}$

PROBLEMS 12-2

1. $Q = 40$

3. $\theta = 4$

5. $r = 30$

7. $R = 2.5$

9. $b = \frac{1}{17}$

11. $\alpha = -3$

13. $\lambda = 8$

15. $\alpha = 3$

PROBLEMS 12-3

1. $I = 2$

3. $q = 3$

5. $\theta = \frac{1}{4}$

7. $\omega = 2$

9. $\pi = 5$

11. $e_o = 5$

13. $x = 3$

15. $\alpha = 13$

17. $\omega = 5$

19. $\alpha = 3$

23. 42 min

25. $x = \frac{abc}{ab + ac + bc}$ days

27. 5 h

31. 90 kg

33. \$43.75

35. 25, 600

37. $\frac{1}{2}$, $1\frac{1}{2}$, $2\frac{1}{2}$

39. 8×2 m

PROBLEMS 12-4

1. $V_o = \frac{LbV_d}{2\alpha Y_d}$

$V_d = \frac{2\alpha V_o Y_d}{Lb}$

3. $E_b = IR + e$

$e = E_b - IR$

5. $V_2 = V_3(1 - \omega^2 LC_2)$

$L = \frac{V_3 - V_2}{\omega^2 C_2 V_3}$

7. $R_t = R_o(1 + \alpha t)$

$$t = \frac{R_t - R_o}{\alpha R_o}$$

9. $r = \dfrac{eR}{E - e}$

$$R = \frac{r(E - e)}{e}$$

11. $R_1 = \dfrac{1}{\omega^2 C_1 C_2 R_3} - R_2$

13. $\beta = \dfrac{V_o - I_o R_o}{\mu V_o}$

15. $\alpha = \dfrac{b(1 + C_o)}{1 - C_o}$

$$b = \frac{\alpha(1 - C_o)}{1 + C_o}$$

17. $Z_1 = \dfrac{Z_3(E - IZ_2)}{I(Z_2 + Z_3)}$

$$Z_2 = \frac{Z_3(E - IZ_1)}{I(Z_1 + Z_3)}$$

$$Z_3 = \frac{IZ_1 Z_2}{E - I(Z_1 + Z_2)}$$

19. $R_x = \dfrac{R_y(AV_1 - V)}{V(A + 1)}$

$$R_y = \frac{VR_x(A + 1)}{AV_1 - V}$$

$$A = \frac{V(R_y + R_x)}{R_y V_1 - R_x V}$$

21. $X_p = \dfrac{X_s^2 R + Z_{ab}^2 X_s + Z_{ab}^2 R}{Z_{ab}^2}$

$$R = \frac{Z_{ab}^2(X_p - X_s)}{X_s^2 + Z_{ab}^2}$$

23. $V_n = \dfrac{I_2 R(R_1 + R_2)}{2R_1 + R_2}$

$$R_1 = \frac{R_2(V_n - I_2 R)}{I_2 R - 2V_n}$$

25. $C = \frac{5}{9}(F - 32)$

27. $R = \dfrac{R_o(BI_o - V_1)}{V_1}$

$$R_o = \frac{V_1 R}{BI_o - V_1}$$

29. $R_H = \dfrac{\mu m N}{Kg} - r$

$$r = \frac{\mu m N}{Kg} - R_H$$

31. $R = (1 - \alpha)Z_2 + (1 - \alpha + k\alpha)Z_1$

$$Z_1 = \frac{R - Z_2(1 - \alpha)}{1 - \alpha + k\alpha}$$

33. $F_2 = \dfrac{2fF_s}{\alpha F_{12} - 2f}$

$$f = \frac{\alpha F_{12} F_s}{2(F_2 + F_s)}$$

35. $\alpha = \frac{1}{H_2R_1} - S$

$$S = \frac{1 - \alpha H_2R_1}{H_2R_1}$$

37. $Z_1 = \frac{R_p(\mu e_g - e_1)}{e_1}$

$$R_p = \frac{e_1Z_1}{\mu e_g - e_1}$$

39. $C_v = \frac{C_o(f_c - fX)}{f_c}$

$$C_o = \frac{C_vf_c}{f_c - fX}$$

41. $v'_o = \frac{\omega s v'_m}{\omega s + 2\mu v'_m}$

$$v'_m = \frac{\omega s v_o}{\omega s - 2\mu v'_o}$$

43. $v_o = \frac{Vi_1R_1}{R_2(i_1 + i_2) + i_1R_1}$

$$i_1 = \frac{Rv_oi_2}{VR_1 - v_o(1 + R_2)}$$

45. $G = \frac{\mu_1\mu_2}{(1 - \mu_1\beta_1)(1 - \mu_2\beta_2)}$

$$\beta_1 = \frac{\mu_1\mu_2 + G(\mu_2\beta_2 - 1)}{G\mu_1(\mu_2\beta_2 - 1)}$$

47. $I_1 = \frac{\pi\lambda^2\gamma_1(2I_f + 1)}{\sigma_o(\gamma_1 + \gamma_f)} - \frac{1}{2}$

$$I_f = \frac{\sigma_o(2L_i + 1)(\gamma_1 + \gamma_f)}{4\pi\lambda^2\gamma_1} - \frac{1}{2}$$

49. $\lambda = \frac{\pi n' d_o(d_1 - d_o)}{1 - d_1}$

$$d_1 = \frac{\lambda + \pi n' d_o^2}{\lambda + \pi n' d_o}$$

51. $R_2 = \frac{-Z_2(Z - Z\alpha + Rk\alpha)}{Z_1 + Z_2}$

$$Z = \frac{RkZ_2\alpha + R_2(Z_1 + Z_2)}{Z_2(\alpha - 1)}$$

$$\alpha = \frac{Z_1R_2 + Z_2R_2 + ZZ_2}{ZZ_2 - RkZ_2}$$

53. $r_1 = \frac{r_2r_3}{r_4}$

$$r_3 = \frac{r_1r_4}{r_2}$$

$$r_4 = \frac{r_2r_3}{r_1}$$

55. $G = \frac{V_{out}C_fC_{fg}}{C_{fg}Q - V_{out}C_f(C_d + C_{fg})}$

57. $p_2 = \frac{CNP_Lp_1}{p\omega\varepsilon_2(\tan\delta) - CNP_L}$

59. $R_p = \frac{R_1R_2}{R_1 + R_2}$

$$R_1 = \frac{R_2R_p}{R_2 - R_p}$$

$$R_2 = \frac{R_1R_p}{R_1 - R_p}$$

61. $R_3 = \frac{E_oR_a}{\mu E - E_o(\mu + 1)}$

$$R_a = \frac{R_3}{E_o}[\mu E - E_o(\mu + 1)]$$

$$\mu = \frac{E_o}{R_3}\left(\frac{R_a + R_3}{E - E_o}\right)$$

63. $\pi = \dfrac{MNk}{4(kH_o - M)}$

$k = \dfrac{4\pi M}{4\pi H_o - MN}$

65. $b = \dfrac{d(X^2 + X'^2)}{(X + X')^2}$

67. $R_a = \dfrac{\mu R_1 R_3(E - E_o) - E_o R_3(R_s + R_1)}{E_o(R_1 + R_3 + R_s) - ER_3}$

$R_s = \dfrac{\mu R_1 R_3(E - E_o) - E_o(R_1R_3 + R_aR_3 + R_aR_1) + ER_aR_3}{E_o(R_a + R_3)}$

$\mu = \dfrac{R_aR_3(E - E_o) - E_o(R_aR_s + R_aR_1 + R_sR_3 + R_1R_3)}{R_1R_3(E_o - E)}$

69. $R_a = \dfrac{\mu R_o R_1(R_s + R_1 + R_2)}{R_2(R_s + R_1) - R_o(R_s + R_1 + R_2)}$

$R_2 = \dfrac{R_o(R_s + R_1)(R_a + \mu R_1)}{R_a(R_s + R_1) - R_o(R_a + \mu R_1)}$

$\mu = \dfrac{R_aR_2(R_s + R_1) - R_o(R_s + R_1 + R_2)}{R_oR_1(R_s + R_1 + R_2)}$

71. $R_a = \dfrac{R_1R_2R_3(\mu R_1 - R_i)}{(R_i - R_1)(R_1R_2 + R_2R_3 + R_1R_3)}$

73. $\pi = \dfrac{\alpha^2(\beta - \alpha)}{\alpha + 2\beta}$

$\beta = \dfrac{\alpha(\alpha^2 + \pi)}{\alpha^2 - 2\pi}$

75. $A = 5.89 \times 10^{-14}\ \text{m}^2$

77. $Z_2 = 6\ \Omega$

79. $R_2 = 100\ \Omega$

81. $F = C$ at $-40°$

83. $R_o = 32\ \Omega$

85. $C_1 = 3$ pF

87. $p = 133.3$

89. (*a*) Increased by a factor of 4
(*b*) Halved

91. $R = 0.9\ \Omega$

93. $R = \dfrac{n(E - Ir)}{I}\ \Omega$

$n = \dfrac{IR}{E - Ir}$ cells

95. $\mu = \dfrac{i_p(r_p + r_b)}{e_g}$

$r_p = \dfrac{\mu e_g - i_p r_b}{i_b}\ \Omega$

99. $V_o = \dfrac{S}{t} - \frac{1}{2}gt$

101. $V_o = 50.95$ m/s

103. $\boldsymbol{E}_{max} = R_b(\boldsymbol{I}_{max} - \boldsymbol{I}_{min}) + \boldsymbol{E}_{min}$

$\boldsymbol{I}_{min} = \boldsymbol{I}_{max} - \dfrac{\boldsymbol{E}_{max} - \boldsymbol{E}_{min}}{R_b}$

105. $E_p = 250$ V

107. $E_1 - E_2 = 98$ V

109. $\alpha = \dfrac{\beta}{1 + \beta}$

PROBLEMS 13-1

1. 165 Ω

3. 37.2 kΩ

5. (*a*) 50 Ω
(*b*) 340 kΩ
(*c*) 1.95 kΩ

7. 440 V

9. 112 kΩ

11. 4.47 W

13. 1.47 W

15. 1 kV

PROBLEMS 13-2

1. 5 Ω

3. 4.8 Ω

5. 6.11 Ω

7. (*a*) 2.1 kΩ
(*b*) 17 kΩ

9. $R_p = \frac{R}{n}$ Ω

11. 1.5 kW

13. (*a*) $R_3 = 22$ Ω
(*b*) $P_t = 1.84$ kW

15. 10 kΩ

PROBLEMS 13-3

1. 332 mA

3. 176 W

5. (*a*) $E_G = 230$ V
(*b*) $R_3 = 20$ kΩ
(*c*) $R_t = 70.6$ kΩ
(*d*) $I_2 = 1.86$ mA
(*e*) $I_3 = 1.4$ mA

7. (*a*) $V_1 = 702$ V
(*b*) $V_2 = 298$ V
(*c*) $R_2 = 2.2$ kΩ
(*d*) $R_3 = 6.7$ kΩ
(*e*) $I_t = 180$ mA
(*f*) $I_3 = 44.6$ mA
(*g*) $P_t = 180$ W

9. 354 Ω

11. 600 Ω

13. (*a*) 4.1 kΩ
(*b*) 10 kΩ
(*c*) 48 W

15. 730 W

17. 2.47 A

PROBLEMS 14-1

1. 1.08 Ω

3. 10 ft $3\frac{1}{2}$ in

5. (*a*) 0–10 mA: 6.11 Ω
(*b*) 0–100 mA: 0.556 Ω
(*c*) 0–1 A: 0.0551 Ω
(*d*) 0–10 A: 0.005 51 Ω

7. $R_1 = 150$ Ω
$R_2 = 15$ Ω
$R_3 = 1.5$ Ω
$R_4 = 0.167$ Ω

PROBLEMS 14-2

1. (*a*) 37.5 V
(*b*) 25 V

3. $R_1 = 9.6$ kΩ $\quad R_3 = 999.6$ kΩ
$R_2 = 99.6$ kΩ $\quad R_4 \cong 10$ MΩ

PROBLEMS 15-1

1. 1 = 60 V
   2 = 6 V
   3 = 0.6 V
   4 = 0.06 V

3. 27 kΩ: 0.114 W
   68 kΩ: 0.288 W
   75 kΩ: 0.318 W

5. $P_t = 14$ W
   $P_1 = 6.4$ W
   $P_2 = 2.4$ W
   $P_3 = 5.2$ W

7. $R_1 = 13.2$ W (use 20 W)
   $R_2 = 7.13$ W (use 10 W)
   $R_3 = 3.37$ W (use 5 W)

9. $R_1 = 3$ kΩ
   $R_2 = 3$ kΩ
   $R_3 = 10$ kΩ
   $R_4 = 500$ Ω

11. 42 Ω

PROBLEMS 15-2

1. (a) $I_1 = 0.25$ A
   (b) $I_2 = 0.0833$ A

3. (a) $I_1 = 103.2$ A
   (b) $I_2 = 46.8$ A

5. $I_2 = 4.94$ A

PROBLEMS 15-3

1. 0.0524

3. 0.0226

5. 35.9 km

7. $X = \dfrac{R_2L - R_1R_3}{R_1 + R_2}$

PROBLEMS 16-1

1. Current varies directly as the applied voltage. (Graph of current is a straight line.)

3. With velocity constant, distance varies directly as time. (Graph of distance is a straight line.)

5. (a) 2 P.M.
   (b) 480 km
   (c) 80 km

7. Third, sixth, ninth, and fifteenth

PROBLEMS 16-2

1. Latitude

PROBLEMS 16-5

1. $y = \frac{2}{5}x - 2$

3. $y = 0.113x + 1.2$

5. $R = -0.000\,667T + 0.4$

7. (a) 0.021 25:1
   (b) 47:1
   (c) 47 Ω

9. (a) 200 V to 265 V
   (c) 6667 Ω
   (d) Approximately 6700 Ω

PROBLEMS 17-1

1. $x = 6, \quad y = 2$

3. $x = 5, \quad y = 3$

5. $E = 6, \quad I = -10$

7. $\alpha = -2, \quad \beta = -3$

9. $I_1 = 5, \quad i = 5$

PROBLEMS 17-2

1. $a = 2.5, \quad b = 4$
3. $R = 3, \quad Z = 2$
5. $R_1 = 1, \quad R_2 = 3$
7. $s = -2, \quad t = 2$
9. $L = 1, \quad M = 2$
11. $I_1 = 3, \quad I_2 = -2$
13. $E = 2, \quad e = 3$
15. $\lambda = 4, \quad \pi = -1$
17. $E = -11, \quad e = 12$
19. $I = 12, \quad i = 9$

PROBLEMS 17-3

1. $E = 3, \quad I = 2$
3. $I = -2, \quad i = 3$
5. $\alpha = 8, \quad \beta = 5$
7. $E = \frac{1}{2}, \quad e = \frac{1}{3}$
9. $\theta = 16, \quad \phi = -10$
11. $F = -3, \quad f = 2$
13. $\gamma = 14, \quad \delta = -4$
15. $\varepsilon = 2, \quad \psi = 3.5$
17. $\theta = 3, \quad \phi = 1$
19. $a = 1.5, \quad b = 0.4$

PROBLEMS 17-4

1. $I = i = 1$
3. $\lambda = 6, \quad \pi = -4$
5. $x = 3, \quad y = 4$
7. $\alpha = 6, \quad \beta = 1$
9. $p = 3, \quad q = -7$

PROBLEMS 17-5

1. $a = 6, \quad b = 2$
3. $\theta = 11, \quad \phi = -5$
5. $\varepsilon = 40, \quad \eta = 5$
7. $X_C = 2, \quad X_L = 3$
9. $\theta = \frac{3}{4}, \quad \lambda = \frac{1}{2}$

PROBLEMS 17-6

1. $R = 3, \quad Z = 4$
3. $X_L = 5, \quad X_C = 11$
5. $\theta = 16, \quad \phi = 5$
7. $G = 60, \quad Y = 33$
9. $L_1 = -\frac{15}{7}, \quad M = \frac{15}{11}$

PROBLEMS 17-7

1. $\alpha = \dfrac{P + Q}{6}, \quad \beta = \dfrac{2Q - P}{3}$
3. $E = \dfrac{7a - b}{4}, \quad IR = \dfrac{b - 3a}{4}$
5. $\theta = \dfrac{3\alpha - 5\beta}{38}, \quad \phi = \dfrac{2\alpha + 3\beta}{19}$
7. $X_C = 50(Z_1 - Z_2), \quad X_L = \dfrac{20Z_2 - 10Z_1}{3}$
9. $R_1 = \dfrac{R_p R_t}{R_p - 2R_t}, \quad R_2 = \dfrac{R_p R_t}{3R_t - R_p}$

PROBLEMS 17-8

1. $\theta = -2, \quad \phi = 4, \quad \pi = 1$

3. $R_1 = 9, \quad R_2 = 2, \quad R_3 = -4$

5. $R_L = 3, \quad R_p = 5, \quad R_1 = 8$

7. $r = 5, \quad R = 6, \quad R_L = 7$

9. $s = 12, \quad t = 4, \quad v = 8$

PROBLEMS 17-9

1. $\dfrac{I_t + I_d}{2}, \quad \dfrac{I_t - I_d}{2}$ A

3. $\frac{2}{3}$

5. $\dfrac{90 + \alpha^\circ}{2}, \quad \dfrac{90 - \alpha^\circ}{2}$

7. Resistors, 10¢ each, capacitors, 20¢ each

9. $L = 60$ km/h, $\quad Q = 70$ km/h

11. $W = \dfrac{Q^2}{2C}$

13. $s = ut + \frac{1}{2}at^2$

15. $\mu = g_m r_p$

17. $R = \dfrac{L}{Cr}$

19. $Q = CE$

21. $H = 7.20$

23. $R = R_p \dfrac{E_p}{\mu E_g - E_p}$

25. $R_X = \dfrac{R_A}{R_A + R_B}(V_3 - V_2)$

$R_Y = \dfrac{R_A}{R_A + R_B}(V_2 - V_1)$

$R_T = \dfrac{R_A}{R_A + R_B}(V_3 - V_1)$

27. $R_1 = \dfrac{R_a R_b + R_b R_c + R_a R_c}{R_c}$

$R_2 = \dfrac{R_a R_b + R_b R_c + R_a R_c}{R_a}$

$R_3 = \dfrac{R_a R_b + R_b R_c + R_a R_c}{R_b}$

PROBLEMS 18-1

1. 2 **3.** 34 **5.** 0 **7.** -114

9. -0.02 **11.** 0 **13.** $bx - ay$ **15.** $bx - ay$

PROBLEMS 18-2

1. $a = 4, \quad b = 2$

3. $\theta = 10, \quad \pi = 2$

5. $I = 3, \quad i = -2$

7. $r_p = \frac{1}{2}, \quad r_L = \frac{1}{3}$

9. $R_1 = 150, \quad R_2 = 1200$

PROBLEMS 18-3

1. 21
3. -245
5. -2016
7. $x = 5, \quad y = 7, \quad z = 3$
9. $\alpha = 3, \quad \beta = 4, \quad \gamma = 7$
11. $E = \frac{1}{2}, \quad e = \frac{1}{3}, \quad IR = \frac{1}{4}$

PROBLEMS 18-4

1. 10
3. -25
5. 23
7. 0

PROBLEMS 18-5

1. 297
3. -56
5. 22.1
7. 220
9. $I_1 = 1, \quad I_2 = 3, \quad I_3 = 5$
11. $\alpha = -2, \quad \beta = 4, \quad \gamma = 1$
13. $R_1 = 5.5, \quad R_2 = 3.6, \quad R_3 = 1.3$
15. $\varepsilon = 1, \quad \eta = 2, \quad \kappa = 3, \quad \lambda = 4$

PROBLEMS 19-1

1. 1 Ω
3. 0.4 Ω
5. (*a*) 5.6 W
   (*b*) 93.3%

PROBLEMS 19-2

1. (*a*) 1.3 V
   (*b*) $1\frac{1}{3}$ A
   (*c*) 0.975 Ω
3. (*a*) 0.48 Ω
   (*b*) 30 mW
   (*c*) 5.92 Ω
   (*d*) 370 mW
   (*e*) 92.6%
5. 11.1 A
7. (*a*) 8.4 Ω
   (*b*) 600 mW
   (*c*) 15 A
9. (*a*) 385 mA
   (*b*) 12.7 V
   (*c*) 4.88 W
11. (*a*) 4.08 V
    (*b*) 1.6 V
    (*c*) 288 mW
13. (*a*) 0.0733 Ω
    (*b*) 200 A
17. $E = 1.4$ V, $\quad r = -.2$ Ω
19. $E = 2.1$ V, $\quad R = 0.665$ Ω

PROBLEMS 20-1

1. a^7
3. x^3
5. p^{q+r}
7. $l^{\alpha+\beta}$
9. x^3
11. x^{5y-2}
13. $\theta^{2\beta}$
15. I^9
17. x^6y^9
19. a^{4x}

21. $x^{4l}y^{4m}z^{4p}$

23. $\frac{E^2}{R^2}$

25. $\frac{\omega^{18}}{64\pi^6 f^{12}}$

27. $\frac{-X_C{}^6}{X_L{}^3}$

29. α^{4x-4}

31. $\frac{I^2}{R}$

33. $\frac{z^{3\lambda}}{y^{\pi}}$

35. $\frac{\theta^4}{\phi^3\lambda^{2x}}$

37. $\frac{bc}{a^3}$

39. $\frac{Ir^3}{4R^2}$

PROBLEMS 20-2

1. ± 4

3. ± 2

5. $-4a^2bc^4$

7. $\pm I^6R^3$

9. $\frac{9\lambda^6}{\omega^8}$

11. $\sqrt[2]{9}$

13. $\sqrt[3]{8\alpha}$ or $2\sqrt[3]{\alpha}$

15. $\sqrt[4]{(\theta\lambda)^3}$

17. $a^{\frac{3}{2}}$

19. $2^{\frac{4}{3}}E^{\frac{1}{3}}$

21. $\theta^{\frac{2}{3}}\omega^{\frac{4}{3}}$

23. $(\alpha\beta)^{\frac{2}{5}}$

25. $4\pi f2^{\frac{1}{3}}$

PROBLEMS 20-3

1. $\pm 2\sqrt{2}$

3. $\pm 3\sqrt{2}$

5. $\pm 5\sqrt{2}$

7. $\pm 4\sqrt{5}$

9. $\pm 12\sqrt{5}$

11. $\pm 2\theta\phi^2\sqrt{3}$

13. $\pm 20I\sqrt{6R}$

15. $\pm 18\omega f^2FT^2\sqrt{7FT}$

17. $\pm 33\alpha^4\beta^3\gamma^4\sqrt{2\alpha\beta}$

19. $\pm 28\pi^2L^2X_Lr^3\sqrt{3}$

PROBLEMS 20-4

1. $\frac{\sqrt{3}}{3}$

3. $\frac{\sqrt{10}}{5}$

5. $\pm\frac{\sqrt{3}}{2}$

7. $4\sqrt{2}$

9. $\frac{\sqrt{\lambda}}{\lambda}$

11. $\pm\frac{3\sqrt{\theta}}{4\theta}$

13. $\sqrt{\theta\lambda}$

15. $\pm\frac{\alpha\sqrt{\gamma}}{\gamma}$

17. $\frac{R^2\sqrt{\pi A}}{A}$

19. $\pm\frac{X_L\sqrt{15}}{4}$

21. $\pm\frac{4Q^2\sqrt{5}}{9}$

PROBLEMS 20-5

1. $3\sqrt{3}$

3. $\sqrt{5}$

5. $(m - p + q)\sqrt{3}$

7. $30\sqrt{3}$

9. 0

11. $6\sqrt{2}$

13. $\frac{2\sqrt{5} - \sqrt{15}}{5}$

15. $\frac{\sqrt{2\pi}}{8}$

PROBLEMS 20-6

1. $\pm\sqrt{6}$ **3.** $\pm 4\sqrt{5}$ **5.** $\pm 12\sqrt{10}$ **7.** 2 **9.** $A - D$

11. $2\alpha - 7 - 2\sqrt{\alpha^2 - 7\alpha}$ **13.** $\theta - \phi$ **15.** $\pm 6\alpha\pi\sqrt{2}$ **17.** 6 **19.** 9

PROBLEMS 20-7

1. $\pm\sqrt{5}$ **3.** $4(3 - \sqrt{7})$ **5.** $-\dfrac{3(1 + \sqrt{3})}{2}$

7. $\dfrac{x^2 - 2x\sqrt{y} + y}{x^2 - y}$ **9.** $\dfrac{3 + \sqrt{3}}{4}$ **11.** $-\sqrt{6} - 3\sqrt{3} + 2\sqrt{2} + 6$

PROBLEMS 20-8

1. j6 **3.** j12 **5.** $-\mathrm{j}Z$ **7.** $\mathrm{j}I^2X$

9. $-$j35 **11.** $\mathrm{j}\frac{4}{11}$ **13.** $\mathrm{j}\dfrac{4\sqrt{6}}{15}$ **15.** $-\mathrm{j}\dfrac{E\sqrt{P}}{P}$

PROBLEMS 20-9

1. 5 + j20 **3.** 41 − j2 **5.** 172 + j5 **7.** 20 − j2

9. 1 + j4 **11.** 9 + j18 **13.** −78 − j11 **15.** 20 + j8

PROBLEMS 20-10

1. 3 − j9 **3.** 75 − j30 **5.** $\theta^2 - \phi^2 + \mathrm{j}2\theta\phi$ **7.** $\dfrac{1 - \mathrm{j}1}{2}$

9. j1 **11.** $\dfrac{1 - \mathrm{j}1}{2}$ **13.** $\dfrac{6(6 + \mathrm{j}X)}{36 + X^2}$ **15.** $\dfrac{R^2 + \mathrm{j}2R\omega X - \omega^2X^2}{R^2 + \omega^2X^2}$

17. $\dfrac{-\phi^2 + \mathrm{j}\theta\phi}{\theta^2 + \phi^2}$ **19.** $\dfrac{R^2 - \mathrm{j}R\left(\omega L - \dfrac{1}{\omega C}\right)}{R^2 + \left(\omega L - \dfrac{1}{\omega C}\right)^2}$

PROBLEMS 20-11

1. $x = 4$ **3.** $y = 9$ **5.** $Z = 625$

7. $M = 67$ **9.** $\lambda = 1$ **11.** $\phi = 25$

13. $P_r = \dfrac{i_s^2}{2\rho^2 P_s}$

15. $\eta = \dfrac{S^2}{\alpha^2 N^2 \tau}$

17. $W = \dfrac{V^2\alpha(1 - \varepsilon_1)}{V^2 - C^2}$

19. $Q_2 = \dfrac{n^2(Y_n^2 - G)}{G^2(n^2 - 1)^2}$

21. $g_m^2 = \dfrac{G_L(G_1 - G_a^2)}{R_{eq}(G_a^2 - G_1) - G_1}$

23. $C = 250$ pF

25. $C_a = \dfrac{C_b}{(2\pi f)^2 L C_b - 1}$

PROBLEMS 21-1

1. $E = \pm 5$
3. $i = \pm\sqrt{189}$
5. $\omega = \pm 6$
7. $\lambda = \pm\frac{3}{11}$
9. $\mu = \pm\frac{4}{5}$
11. $m = \pm\sqrt{2}$
13. $\lambda = \pm 6$
15. $X_C = \pm\dfrac{\sqrt{95}}{5}$

PROBLEMS 21-2

1. $\alpha = -1$ or -4
3. $R = 2$ or 7
5. $\lambda = 1$ or -2
7. $E = 2$ or 20
9. $Q = 2$ or 11
11. $\alpha = -2$ or -25
13. $Z = 3$ or 6
15. $i = 3$ or $-\frac{7}{4}$

PROBLEMS 21-3

1. $x = 2$ or 6
3. $E = 6$ or 9
5. $i = 2$ or 25
7. $\theta = 1$ or 2
9. $M = -2$ or 24
11. $\theta = 3$ or -2
13. $\phi = 10$ or -6
15. $R = 5$ or $5\frac{1}{3}$

PROBLEMS 21-4

1. $\theta = 1$ or -4
3. $I = 7$ or -5
5. $q = \frac{3}{4}$ or $-\frac{5}{2}$
7. $Z = \dfrac{3 \pm \sqrt{129}}{12}$
9. $m = \frac{1}{4}$ or $-\frac{1}{6}$
11. $R_1 = -5$ or 0
13. $\beta = 5$ or $5\frac{1}{3}$
15. $i = 2$ or $-\frac{2}{15}$

PROBLEMS 21-7

1. (a) 16, roots real and unequal
   (b) 0, roots are equal
   (c) −80, roots are imaginary
3. 21 and 23
5. 140×160 m
7. 220 and 240

9. (a) $E = \pm\sqrt{\dfrac{PnR}{k}}$

(b) No change in E

11. $r = \dfrac{-PXx \pm x\sqrt{P^2X^2 + 4R^2(P-1)}}{2R(P-1)}$

$x = \dfrac{PXr \pm r\sqrt{P^2X^2 + 4R^2(P-1)}}{2R}$

13. $v = 1.0 \times 10^3$ m/s

15. $v = \sqrt{2gs}$ m/s

17. $h = \dfrac{v^2}{64}$ ft

19. $R = 50\ \Omega$

25. (a) 2 A
(b) 120 V
(c) $R_1 = 10\ \Omega$
$R_2 = 20\ \Omega$
$R_3 = 30\ \Omega$

27. 20 V and 15 A or 60 V and 5 A

PROBLEMS 22-1

1. 1.03 mA

3. 54 V

5. 244 V

7. (a) 0.5 A
(b) 12.3 V

9. 2.22 Ω

PROBLEMS 22-2

1. 39.9 V

3. 32 V

5. (a) 1.19 A
(b) 53.2 mW

7. (a) 1.0 A
(b) From a to b

PROBLEMS 22-3

1. (a) 1.27 A
(b) 14.6 W

3. (a) 1.64 A
(b) 2.86 A from A to b

5. (a) 95.5 W
(b) 3.18 V

7. (a) 86.3 W
(b) 16 V

9. (a) 5.19 A
(b) 167 W

11. (a) 64.8 V
(b) 2.1 kW

13. (a) 220 V
(b) 313 W

PROBLEMS 22-4

1. $R_a = 4.8\ \Omega$, $R_b = 4\ \Omega$, $R_c = 6\ \Omega$

3. $R_a = R_b = R_c = 167\ \Omega$

5. $R_1 = 16.6$ kΩ, $R_2 = 6.36$ kΩ, $R_3 = 9.06$ kΩ

7. 72.7 rA

9. 6.52 mA

11. Zero A

13. $I = 5$ A

15. 3.1 A

17. 14.7 A

PROBLEMS 22-5

1. (*a*) Constant 120-V source in series with 0.8 Ω
   (*b*) Constant 150-A source in parallel with 0.8 Ω

3. 0.187 V in series with 18.75 Ω; $I_2 = 6.52$ mA

5. 65.6 mA in parallel with 12.6 Ω; $I_5 = 30$ mA

PROBLEMS 23-1

1. (*a*) 22°, (*b*) 67°, (*c*) 49°, (*d*) −80°, (*e*) −165°, (*f*) 100°

7. 26

9. 21 600°/s

PROBLEMS 23-2

1. (*a*) $\frac{\pi^r}{3}$, 1.05^r
   (*b*) $\frac{2\pi^r}{3}$, 2.09^r
   (*c*) $\frac{11\pi^r}{12}$, 2.88^r
   (*d*) $\frac{5\pi^r}{4}$, 3.93^r
   (*e*) $\frac{19\pi^r}{12}$, 4.97^r
   (*f*) $\frac{\pi^r}{36}$, 0.0873^r

3. $330\pi^r$

5. $\frac{5\pi^r}{2}$

7. $\frac{40\pi^r}{3}$/s

9. 30 rev/min

PROBLEMS 23-3

1. (*a*) 50^g
   (*b*) 33.3^g
   (*c*) 66.7^g
   (*d*) 1.33^g
   (*e*) 250^g
   (*f*) 350^g

3. (*a*) 0.7854^r
   (*b*) 0.3142^r
   (*c*) 1.178^r
   (*d*) 2.356^r
   (*e*) 3.142^r
   (*f*) 6.283^r

PROBLEMS 23-4

1. 4.5 m and 6 m

3. $a = 4$, $c = 5$, $B = 36.9°$

5. $a = 11.4$, $c = 11.4$, $B = 20°$

7. $b = 17.7$, $c = 13$, $A = 33.8°$ **9.** $c = 10$, $A = 49.1°$, $B = 101.6°$

PROBLEMS 23-5

1. $c = 58$, $B = 15°$ **3.** $a = 18$, $A = 13°$ **5.** 12 m **7.** 19.9 m **9.** 150 m

PROBLEMS 24-1

1. $\sin\theta = \dfrac{a}{c}$ $\sin\phi = \dfrac{b}{c}$

$\cos\theta = \dfrac{b}{c}$ $\cos\phi = \dfrac{a}{c}$

$\tan\theta = \dfrac{a}{b}$ $\tan\phi = \dfrac{b}{a}$

$\cot\theta = \dfrac{b}{a}$ $\cot\phi = \dfrac{a}{b}$

$\sec\theta = \dfrac{c}{b}$ $\sec\phi = \dfrac{c}{a}$

$\csc\theta = \dfrac{c}{a}$ $\csc\phi = \dfrac{c}{b}$

3. (a) $\dfrac{OP}{OR} = \tan\beta$

(b) $\dfrac{PR}{PO} = \sec\alpha$

(c) $\dfrac{OR}{PR} = \cos\beta$

(d) $\dfrac{OP}{RP} = \sin\beta$

(e) $\dfrac{PR}{RO} = \csc\alpha$

5. $\sin\theta = 0.707$
$\cos\theta = 0.707$
$\tan\theta = 1.00$
$\cot\theta = 1.00$
$\sec\theta = 1.41$
$\csc\theta = 1.41$

7. $\sin\phi = 0.447$
$\cos\phi = 0.894$
$\tan\phi = 0.500$

9. $\sin x = \frac{8}{10}$
$\cos x = \frac{6}{10}$
$\tan x = \frac{8}{6}$
$\sec x = \frac{10}{6}$
$\csc x = \frac{10}{8}$
$\cot x = \frac{6}{8}$

11. $\sin B = \frac{4}{5}$
$\cos B = \frac{3}{5}$
$\tan B = \frac{4}{3}$
$\cot B = \frac{3}{4}$
$\sec B = \frac{5}{3}$
$\csc B = \frac{5}{4}$

PROBLEMS 24-2

1. I or II **3.** III or IV **5.** II or III **7.** I

9. IV **11.** I or III **13.** No

Q	sin	cos	tan
15.	+	+	+
17.	+	−	−
19.	−	−	+
21.	−	−	+
23.	+	+	+

Q	sin	cos	tan	sec	csc	cot
27.	$\frac{5}{13}$	$\frac{12}{13}$	$\frac{5}{12}$	$\frac{13}{12}$	$\frac{13}{5}$	$\frac{12}{5}$
29.	$\frac{-5\sqrt{41}}{41}$	$\frac{-4\sqrt{41}}{41}$	$\frac{5}{4}$	$\frac{-\sqrt{41}}{4}$	$\frac{-\sqrt{41}}{5}$	$\frac{4}{5}$
31.	$-\frac{3}{5}$	$\frac{4}{5}$	$-\frac{3}{4}$	$\frac{5}{4}$	$-\frac{5}{3}$	$-\frac{4}{3}$
33.	$\frac{-3\sqrt{34}}{34}$	$\frac{-5\sqrt{34}}{34}$	$\frac{3}{5}$	$\frac{-\sqrt{34}}{5}$	$\frac{-\sqrt{34}}{3}$	$\frac{5}{3}$

PROBLEMS 24-3

1. 0

3. ∞

5. No

7. (*a*) 1
(*b*) -1
(*c*) -1
(*d*) 1

PROBLEMS 25-1

Q		sin	cos	tan
1.	(*a*)	0.309 02	0.951 06	0.324 92
	(*b*)	0.927 18	0.374 61	2.475 09
	(*c*)	0.161 60	0.986 86	0.163 76
	(*d*)	0.793 35	0.608 76	1.303 23
	(*e*)	0.045 36	0.998 97	0.045 41
3.	(*a*)	0.033 86	0.999 43	0.033 87
	(*b*)	0.842 07	0.539 36	1.561 29
	(*c*)	0.628 10	0.778 13	0.807 18
	(*d*)	0.646 52	0.762 89	0.847 48
	(*e*)	0.822 83	0.568 27	1.447 99

PROBLEMS 25-2

1. (*a*) 27°
(*b*) 6.7°
(*c*) 61.5°
(*d*) 40.1°
(*e*) 2.14°

3. (*a*) 85.4°
(*b*) 0.5°
(*c*) 40.1°
(*d*) 58.8°
(*e*) 25.75°

5. (*a*) 13.16°
(*b*) 0.53°
(*c*) 74.38°
(*d*) 41.77°
(*e*) 47.11°

PROBLEMS 25-3

Q		sin	cos	tan
1.	(*a*)	0.956 30	−0.292 37	−3.270 85
	(*b*)	0.342 02	−0.939 69	−0.363 97
	(*c*)	0.764 92	−0.644 12	−1.187 54
	(*d*)	0.537 30	−0.843 39	−0.637 07
	(*e*)	0.066 27	−0.997 80	−0.066 42
3.	(*a*)	−0.984 81	0.173 65	−5.671 28
	(*b*)	−0.669 06	0.743 14	−0.900 40
	(*c*)	−0.175 37	0.984 50	−0.178 13
	(*d*)	−0.865 15	0.501 51	−1.725 09
	(*e*)	−0.008 73	0.999 96	−0.008 73

Q	sin	cos	tan
5. (*a*)	−0.087 16	−0.996 19	0.087 49
(*b*)	−0.294 04	0.955 79	−0.307 64
(*c*)	−0.652 10	−0.758 13	0.860 14
(*d*)	−0.045 36	0.998 97	−0.045 41
(*e*)	−0.003 49	−0.999 99	0.003 49

7. (*a*) $\phi = -47.1°$
(*b*) $\phi = 91.6°$
(*c*) $\phi = 51.3°$
(*d*) $\phi = 167.5°$
(*e*) $\phi = -69.9°$

9. 1.42 m

11. 90° **13.** 175 lux **15.** No **17.** 40.7°

PROBLEMS 26-1

1. $Z = 26.8$, $X = 15.2$, $\phi = 55.3°$
3. $Z = 600$, $R = 424$, $\theta = 45°$
5. $Z = 70.0$, $X = 29.5$, $\phi = 65.1°$
7. $Z = 1 \times 10^6$, $X = 4.65 \times 10^5$, $\phi = 62.3°$
9. $Z = 1030$, $R = 557$, $\phi = 32.7°$
11. $Z = 159$, $R = 100$, $\phi = 38.9°$
13. $Z = 0.239$, $X = 0.214$, $\phi = 26.1°$
15. $Z = 0.378$, $R = 0.0500$, $\phi = 7.5°$

PROBLEMS 26-2

1. $R = 73.6$, $X = 19.7$, $\theta = 15°$
3. $R = 17.0$, $X = 44.5$, $\phi = 20.9°$
5. $R = 7.84 \times 10^3$, $X = 6.21 \times 10^3$, $\theta = 38.4°$
7. $R = 0.932$, $X = 0.171$, $\theta = 10.4°$
9. $R = 3.12$, $X = 4.04$, $\phi = 37.7°$

PROBLEMS 26-3

1. $\theta = 60.8°$, $\phi = 29.2°$, $R = 112$
3. $\theta = 69.1°$, $\phi = 20.9°$, $X = 44.5$
5. $\theta = 8.4°$, $\phi = 81.4°$, $X = 0.109$
7. $\theta = 38.4°$, $\phi = 51.6°$, $R = 7.84 \times 10^3$
9. $\theta = 51.9°$, $\phi = 38.1°$, $X = 0.849$

PROBLEMS 26-4

1. $\theta = 9.9°$, $\phi = 80.1°$, $Z = 36.0$
3. $\theta = 47.9°$, $\phi = 42.1°$, $Z = 7.14$
5. $\theta = 2.7°$, $\phi = 87.3°$, $Z = 431$
7. $\theta = 83.6°$, $\phi = 6.4°$, $Z = 48.7$
9. $\theta = 46°$, $\phi = 44°$, $Z = 0.403$

PROBLEMS 26-5

1. 33.7° **3.** 4.21° **5.** 62.7° **7.** 9.84 m **9.** 20.2 m **11.** 97.7 m

PROBLEMS 26-6

1. (a) $\phi = 68°$
(b) $278\ m^2$
(c) $37.1\ m$
(d) $278\ m^2$

3. $4.14 \times 10^3\ mm^2$

PROBLEMS 27-2

1. $b = 7.65, \quad c = 9.01, \quad \gamma = 70°$

3. $a = 12.9, \quad c = 18, \quad \gamma = 75°$

5. $a = 33, \quad c = 91.7, \quad \gamma = 108°$

7. $a = 1.14, \quad b = 7.1, \quad \alpha = 8°$

9. $a = 11.3, \quad c = 63.6, \quad \beta = 55.5°$

11. 2.53 km

PROBLEMS 27-3

1. $a = 7.3, \quad \beta = 39.4°, \quad \gamma = 77.6°$

3. $c = 0.908, \quad \alpha = 8°, \quad \beta = 40°$

5. $c = 4691, \quad \alpha = 10.5°, \quad \beta = 21.8°$

7. $\alpha = 21.8°, \quad \beta = 38.2°, \quad \gamma = 120°$

9. $\alpha = 17.5°, \quad \beta = 50°, \quad \gamma = 112.5°$

11. 55.7 by 146.8 mm

PROBLEMS 27-4

1. $0.366 \sin\theta + 1.366 \cos\theta$

3. $0.366(\cos\theta - \sin\theta)$

5. $\frac{33}{65}$

PROBLEMS 28-1

1. 182.5 at 28.2°

3. 238 at 244.7°

PROBLEMS 28-2

1. $x = 12.4, \quad y = 27.3$

3. $x = 0.0423, \quad y = 0.864$

5. $x = -46.3, \quad y = 0$

7. $x = -56.6, \quad y = -177$

9. $x = -28.4, \quad y = 11.9$

11. 728 N, 234 N

13. 849 m

15. 131 N

PROBLEMS 28-3

1. $420\angle 81.2°$

3. $1.92\angle 39.9°$

5. $364\angle 15.1°$

7. $183\angle 0°$

9. $125\angle 270°$

11. $7.65\angle 252.2°$

13. $25.9\angle 160.8°$

15. $24.4\angle 216.5°$

PROBLEMS 28-4

1. $321\underline{/55.9^\circ}$
3. $111\underline{/19.4^\circ}$
5. $31.2\underline{/167.4^\circ}$

PROBLEMS 29-2

1. (a) $\dfrac{\pi^r}{21\,600}$ /s
   (b) $\dfrac{\pi^r}{1800}$ /s
   (c) $\dfrac{\pi^r}{30}$ /s
3. (a) 4.5°/min
   (b) $\dfrac{\pi^r}{2400}$ /s
5. (a) $72\pi^r$
   (b) $7.2\pi^r$
   (c) $3.6\pi^r$

PROBLEMS 29-3

Q	(a)	(b)	(c)	(d)	(e)
1.	100	2π	1	1	40° lead
3.	0.750	628	100	0.01	3° lead
5.	E_{max}	157	25	0.04	17° lag

13. (b) $y = 60 \sin 40\pi t$ cm
    (c) −35.3 cm
    (d) 60 cm
    (e) $10\pi^r$

PROBLEMS 30-1

1. (a) 51 A
   (b) 152 A
   (c) 115 A
   (d) −146 A
   (e) −92.3 A
3. 440 V
5. −91.7 V
7. −1.11 A
9. 210° and 330°

PROBLEMS 30-2

1. (a) 400 Hz
   (b) 2.5 ms
   (c) $e = 314 \sin 800\pi t$ V
3. (a) 40 poles
   (b) $e = 250 \sin 800\pi t$ V
   (c) −238 V
5. 600 rev/min
7. 500 MHz
9. $i = (3 \times 10^{-5}) \sin (1000\pi \times 10^6)t$ A

PROBLEMS 30-3

1. 49 V
3. 16.5 V
5. 127 V
7. 21.2 A
9. 232 mA

PROBLEMS 30-4

1. (*a*) $i = 6.5 \sin (377t + 36^\circ)$ A
(*b*) 6.46 A

3. -5.39 A

5. (*a*) $i = 283 \sin (314t - 25^\circ)$ A
(*b*) 63.7 A

7. 9.2° lag

9. 49° lead or lag

PROBLEMS 31-1

1. $20.7 + j11.3 = 23.6\angle 28.6^\circ$

3. $1100 + j400 = 1170\angle 20^\circ$

5. $-442 + j741 = 863\angle 120.8^\circ$

7. $11.2 - j36.4 = 38.1\angle -72.9^\circ$

9. $2500 - j400 = 2532\angle 9.09^\circ$

11. $10 + j9 = 13.5\angle 42^\circ$

PROBLEMS 31-2

1. $34 - j22 = 40.5\angle -32.9^\circ$

3. $20.9 + j25.13 = 32.7\angle 50.2^\circ$

5. $4.96 - j74.7 = 74.9\angle -86.2^\circ$

7. $0.0588 - j0.765 = 0.767\angle -85.6^\circ$

9. $-0.589 - j4.35 = 4.39\angle 262.3^\circ$

PROBLEMS 31-3

1. $20.7 + j11.3 = 23.6\angle 28.7^\circ$

3. $1104 + j400 = 1174\angle 19.9^\circ$

5. $-445 + j741 = 864\angle 121^\circ$

7. $11.2 - j36.5 = 38.2\angle -72.9^\circ$

9. $2478 + j798 = 2.6 \times 10^3\angle 17.9^\circ$

11. $-300 + j400 = 500\angle 126.9^\circ$

PROBLEMS 31-4

1. $33 - j5.99 = 33.5\angle -10.3^\circ$

3. $-77 + j40.8 = 87.1\angle 152.1^\circ$

5. $4.43 - j74.6 = 74.7\angle -86.6^\circ$

7. $1.92 + j0.565 = 2\angle 16.4^\circ$

9. $-0.239 - j0.146 = 0.278\angle -148.6^\circ$

11. $13.9 - j2.69 = 14.1\angle -11^\circ$

13. $\pm 12\angle 15^\circ$

15. $2.89\angle 44^\circ$

17. $4\angle 90^\circ$

19. $27\angle 33^\circ$

PROBLEMS 32-1

1. (a) 368 mA
   (b) $e = 124 \sin 800\pi t$ V
   (c) $i = 0.521 \sin 800\pi t$ A
   (d) 24.7 V
   (e) 2.98 W
   (f) 109 mA

3. 22.9 V

5. $i = 51.6 \sin (2 \times 10^4 \pi t)$ mA

PROBLEMS 32-2

1. 5.65 Ω

3. 94.2 kΩ

5. 2.54 kΩ

7. 297 mA

9. 200 MHz

11. −424 mA

13. (a) X_L is doubled.
    (b) X_L is tripled.
    (c) X_L is halved.

PROBLEMS 32-3

1. 18.1 Ω

3. 72.3 mΩ

5. 138 Ω

7. 18.8 μA

9. 4 μF

11. $i = 0.639 \sin (377t + 90°)$ A

13. 153 pF

15. 137 μA

17. X_C varies inversely as C.

PROBLEMS 32-4

1. (a) 565 Ω
   (b) 567 Ω
   (c) 388 mA
   (d) $i = 549 \sin (377t - 86.5°)$ mA
   (e) 13.6 V
   (f) 219.6 V

3. (b) 12.7 kΩ, 3.9 MHz, 75° lag

5. (a) 121 Ω
   (b) 351 Ω
   (c) 342 mA
   (d) 113 V
   (e) 41.2 V

7. 358 Ω

9. (a) 1.03 kΩ
   (b) 582 mA
   (c) 582 V
   (d) 144 V

PROBLEMS 32-5

Q	Z	I	i	PF	P
1.	$528\angle 67.8°$ Ω	416 mA	$i = 589 \sin (377t - 67.7°)$ mA	38.0%	34.7 W
3.	$2020\angle 8.88°$ Ω	54.3 mA	$i = 76.9 \sin (314t - 8.88°)$ mA	99.0%	5.92 W
5.	$558\angle 66.8°$ Ω	2.15 A	$i = 3.04 \sin [(3.14 \times 10^7)t - 66.8°]$A	39.4%	1.02 kW
7.	$15\angle 2.4°$ Ω	7.79 A	$i = 11 \sin (377t - 2.4°)$A	99.9%	9.1 W
9.	$515\angle 13.7°$ Ω	3.44 A	$i = 4.84 \sin [(15.7 \times 10^6)t - 13.7°]$A	97.0%	5.88 kW

11. (*a*) 200 Ω
(*b*) 282 Ω
(*c*) 0.75 H

13. (*a*) $55.8\underline{/-32.6°}$ Ω
(*b*) 505 μF

15. (*a*) 23.5 A
(*b*) 8.95 kW

PROBLEMS 32-6

1. (*a*) 4.53 mA
(*b*) 2.56 mW
(*c*) 328 V across the capacitor, 228 V across the coil

3. 12 kHz

5. (*a*) 0.239 mH
(*b*) 2.3 MHz

PROBLEMS 33-1

1. 720 pF

3. (*a*) $e = 311 \sin 377t$ V
(*b*) $i = 138 \sin (377t + 90°)$ mA
(*c*) 104 V
(*d*) 1.82 μF
(*e*) 17.6 mA

5. (*a*) 0.47 H
(*b*) 1.41 H

PROBLEMS 33-2

1. $40.2\underline{/41.5°}$ Ω

3. $39.2\underline{/25°}$ Ω

5. $38.5\underline{/12.5°}$ Ω

7. $68\underline{/74.2°}$ Ω

9. $114\underline{/188°}$ Ω

11. $144\underline{/1.52°}$ Ω

13. $2.65\underline{/-73.1°}$ A

PROBLEMS 33-3

1. (*a*) 5.623 MHz
(*b*) 5.627 MHz
(*c*) 25.7

3. (*a*) 113 kHz
(*b*) 5.66 MΩ
(*c*) 8.84 Ω

5. 500 pF

7. 7.08 mW

9. 0.6% leading

11. 32.2 pF

PROBLEMS 33-4

1. $Z_a = 11\underline{/-5.2°}$ Ω, $Z_b = 19.7\underline{/76.7°}$ Ω, $Z_c = 17.2\underline{/2°}$ Ω

3. $Z_1 = 74.9\underline{/39.9°}$ Ω, $Z_2 = 91\underline{/-73.1°}$ Ω, $Z_3 = 96.4\underline{/47.9°}$ Ω

5. $Z_{ab} = 64.0\underline{/2.3°}$ Ω

7. $Z_{ab} = 187\underline{/27.1°}$ Ω

9. 21.6 W

11. $Z_{ab} = 89\underline{/-25°}$ Ω

13. 84.9 W

15. $Z_{ab} = 218\underline{/-47.4°}$ Ω

17. $Z_{ab} = 81.4\underline{/67.2°}$ Ω

PROBLEMS 34-1

1. $2 = \log_{10} 100$

3. $2 = \log_7 49$

5. $0.5 = \log_4 2$

7. $1 = \log_a a$

9. $0 = \log_a 1$

11. $10^3 = 1000$

13. $5^2 = 25$

15. $6^0 = 1$

17. $5^4 = 625$ 19. $r^s = t$ 21. $x = 2$ 23. $x = 6$

25. $x = 0.5$ 27. $5 - 3 = 2$ 29. 1, 2, 3, 4, 5, 6, 7, 8, 9

PROBLEMS 34-2

1. 1 3. 2 5. 0 7. 2 9. $\overline{4}$ or $6 - 10$ 11. 1

13. 3 15. $\overline{4}$ or $6 - 10$ 17. $\overline{3}$ or $7 - 10$ 19. 3 21. $\overline{1}$ or $9 - 10$ 23. 1

25. -1.5 27. $\log 6792 + \log 20.9 - \log 176$

29. $\frac{1}{2}[\log 512 + \log 0.36 - (\log 2 + \log \pi + \log 177)]$ 31. $3(\log 159 + \log 0.837 - \log 82.2)$

33. 0.437 12 35. $\overline{2}$.437 12 or 8.437 12 − 10

37. 4.437 12 39. 7.437 12

41. $\overline{11}$.437 12 or 9.437 12 − 20 43. 7570 or 7.57×10^3

45. 7.57×10^6 47. 7.57×10^{-7}

49. 7.57×10^{10}

PROBLEMS 34-3

1. 0.845 10 3. 1.845 10 5. 2.857 94

7. 2.012 84 9. 5.582 06 11. 3.966 80

13. 9.041 79 − 10 or −0.958 21 15. 2.402 54 17. 4.497 21 − 10

19. 0.797 96 21. 0.434 30 23. 6.673 02 − 10

25. 4.845 04 27. 6.301 68 29. 8.698 97 − 20

PROBLEMS 34-4

1. 3 3. 30 5. 642 7. 101

9. 2.42×10^5 11. 8.425×10^3 13. 9.792×10^{-1} 15. 3.4251×10^2

17. 1.493×10^{-5} 19. 6.28 21. 2.718 23. 9.82×10^{-4}

25. 7.9984×10^4 27. 1728 29. 8×10^{-12}

PROBLEMS 34-5

1. 5.951 49 **3.** $\bar{2}.403\ 19$ **5.** $\bar{4}.273\ 66$ **7.** 2.559 78

9. 2.347 87 **11.** 3.855 91

PROBLEMS 34-6

1. 2.56×10^2 **3.** 2.5×10^2 **5.** 3.78×10 **7.** -2.47×10^2

9. 2.37×10^3 **11.** 1×10^4 **13.** -9.81×10^{-1} **15.** 7.59×10^4

17. -1.35×10^3 **19.** 9.4

PROBLEMS 34-7

1. 3 **3.** 20 **5.** -700 **7.** -578 **9.** 4.88×10^6

PROBLEMS 34-8

1. 4.94 **3.** 2.42 **5.** -1.25×10^{-1} **7.** 2.61 **9.** 5.95×10^{-2}

PROBLEMS 34-9

1. 164 **3.** 9.59×10^{-8} **5.** 5.65 **7.** 5.93 **9.** 9.28×10^{-1}

11. 25.3 **13.** 169 **15.** 1.16 **17.** 4.85×10^{-1} **19.** 8.76×10^{-1}

PROBLEMS 34-10

1. 2.563 09 **3.** $-0.186\ 97$ **5.** 1.95 **7.** 7.77×10^{-1} **9.** 4.7005

PROBLEMS 34-11

1. $x = 5.42$ **3.** $x = 31.6$

5. $x = 115$ **7.** $P_1 = 1.01 \times 10^4$

9. $x = 5.62 \times 10^6$ **11.** $x = 3.69$

13. $m = -1.84 \times 10^{-1}$ **15.** $x = 5 \times 10^{-1}$

17. $x = 4$ **19.** $L_1 = \sqrt[3]{L_2^{\,2}}$

21. $I_0 = I_g \cdot 10^{5044 \frac{V_g}{T}}$ **23.** (a) $E = i_c \varepsilon^{\frac{t}{RC}}$

(b) $C = \dfrac{0.4343t}{R(\log E - \log i_c R)}$

(c) $t = 2.3026RC(\log E - \log i_c R)$

25. (a) $E = \dfrac{i_L R}{1 - \varepsilon^{\frac{-Rt}{L}}}$

(b) $L = \dfrac{0.4343Rt}{\log E - \log (E - i_L R)}$

(c) $t = \dfrac{2.3026L}{R}[\log E - \log (E - i_L R)]$

27. (a) $E = \left(\dfrac{I_p + I_g}{K}\right)^{\frac{2}{3}} - \dfrac{E_p}{\mu}$

(b) $E_p = \mu\left[-E + \left(\dfrac{I_p + I_g}{K}\right)^{\frac{2}{3}}\right]$

(c) $\mu = \dfrac{E_p}{\left(\dfrac{I_p + I_g}{K}\right)^{\frac{2}{3}} - E}$

29. $i = 1.01$ A

31. 34.7 ms

33. 208 ms

35. 265 words/min

39. (a) 1.66×10^{-4} C
(b) 8 V

41. 296 mA

43. 1.26

PROBLEMS 35-2

1. $R12$ series of preferred values:
1.0, 1.2, 1.5, 1.8, 2.2, 2.7, 3.3, 3.9, 4.7, 5.6, 6.8, 8.2, 10.
Maximum % error: ±11.1%

3. $R10$ series of preferred values:
1.25, 1.6, 2.0, 2.5, 3.2, 4.0, 5.0, 6.4, 8.0, 10.
Probable published tolerance: ±15%

PROBLEMS 35-3

1. (a) 13 dB
(b) 14 dB
(c) 18 dB
(d) −22.5 dB

3. 0.735 V

5. 775 mV

7. 1.94 V, 6.44 mA

9. (a) 12 mW, 2.68 V
(b) 60 mW, 6 V
(c) 0.6 mW, 0.6 V
(d) 6×10^{-8} mW, 6 mV

11. 10^9

13. 10^3 W

15. 1×10^{-7} W

17. 10^{-8}

19. 10^3

21. 100 dB

23. 2.92×10^6

25. 54 dB

27. 1.05 V

29. (a) 223
(b) 37 dB
(c) 6.02 mW

31. 3.92 W

33. 1.11 μV

35. 0.54 dB/mi

37. 400

39. 109 kW

41. 25.8 dB

PROBLEMS 35-4

1. 270 mH
3. (*a*) 7.14 mH
(*b*) 22.5 nF
5. 356 mm
7. No. 2
9. 171.2 nF/km
11. 115 km
13. (*a*) 19 nH/cm
(*b*) 0.0584 pF/cm
(*c*) 0.295 mH
(*d*) 906 pF

PROBLEMS 35-5

1. 570 Ω
3. −1.2%
5. No
7. No. 6
9. 138 Ω
13. (*a*) 0.3 dB
(*b*) 93.5%
15. (*a*) 0.7 dB
(*b*) 85.1%
17. 604 pF
19. 5.6%

PROBLEMS 36-1

1. 5
3. 1
5. 7
7. 42
9. 35

PROBLEMS 36-2

1. 110
3. 10010
5. 11111
7. 1100001
9. 10110001

PROBLEMS 36-3

1. 2
3. 51
5. 63
7. 596
9. 3256

PROBLEMS 36-4

1. 31
3. 124
5. 245
7. 1467
9. 7530

PROBLEMS 36-5

1. 100_3
3. 102_5
5. 10000_4
7. 1000_{12}
9. 40240_6
11. 64
13. 59
15. 3
17. 70
19. 6842

PROBLEMS 36-6

1. 011 110 001
3. 101 011 010
5. 001 000 110
7. 101 010 110 110
9. 111 111 111 111
11. 5
13. 35
15. 675
17. 632
19. 252

PROBLEMS 36-7

1. 111 001 **3.** 1 110 000 **5.** 111 100 **7.** 1 000 001 **9.** 110 100

PROBLEMS 36-8

1. 001001 **3.** 001001 **5.** 100111 **7.** 010111 **9.** 000101

PROBLEMS 36-9

1. 001011 **3.** 001100 **5.** 001100 **7.** 001011 **9.** 100101

PROBLEMS 36-10

1. 1011110 **3.** 10111001 **5.** 100011111 **7.** 101100000001

9. 101110110101

PROBLEMS 36-11

1. 011 **3.** 1001 **5.** 110 **7.** 1100 **9.** 10100

PROBLEMS 37-1

1. sl **3.** $s + l$ **5.** sl **7.** $\overline{s} + l$ **9.** $sl + \overline{s} \cdot \overline{l}$

PROBLEMS 37-2

2.

a	b	c	$a + b$	$c(a + b)$
1	1	1	1	1
1	1	0	1	0
1	0	1	1	1
1	0	0	1	0
0	1	1	1	1
0	1	0	1	0
0	0	1	0	0
0	0	0	0	0

4.

a	b	c	$\overline{a}$	$\overline{b}$	$\overline{c}$	$\overline{a} \cdot \overline{b} \cdot \overline{c}$
1	1	1	0	0	0	0
1	1	0	0	0	1	0
1	0	1	0	1	0	0
1	0	0	0	1	1	0
0	1	1	1	0	0	0
0	1	0	1	0	1	0
0	0	1	1	1	0	0
0	0	0	1	1	1	1

PROBLEMS 37-4

1. (a) $Z_{xy} = \bar{a} + \bar{b} \cdot \bar{c}$
   (b) $Y_{xy} = a(b + c)$

3. (a) $Z_{LM} = (\bar{A} + B)(A + \bar{B})$
   (b) $Y_{LM} = A\bar{B} + \bar{A}B$ or $(A + B)(\overline{AB})$

5. (a) $Z_{pq} = \bar{a} + \bar{b}(ab\bar{c} + \bar{a}) + \bar{c}$ or $\bar{a} + ab + c$
   (b) $Y_{pq} = ab + a(\bar{a} + \bar{b} + c)c$ or ac

7. 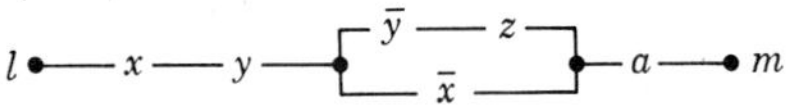

9. 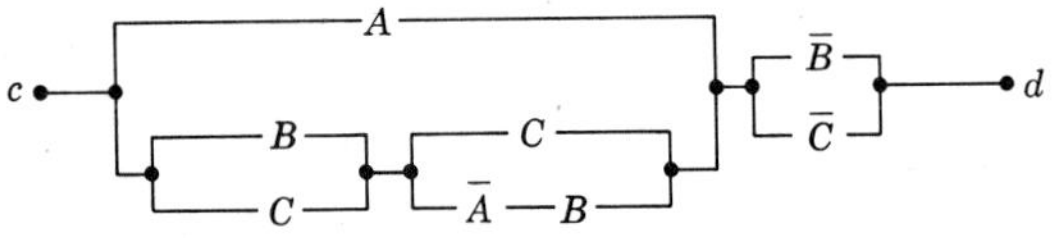

PROBLEMS 37-7

1. $\overline{\overline{a + b}} = a + b$: OR gate

3. $\overline{\overline{\bar{a} + \bar{b}}} = \bar{a} + \bar{b} = \overline{ab}$: NAND gate

index

Abbreviations, 3
  list of, 539–540
Absolute value, 15
Accuracy:
  of circuit components, 55
  of interpolation, 341
  of slide rules, 55
Addition:
  of algebraic fractions, 146
  of binary numbers, 520
  of complex numbers, 265
  of like terms, 18
  of logarithms, 472
  of polynomials, 19
  of positive and negative numbers, 15
  of powers of ten, 66
  of radicals, 259
  of unlike terms, 18
  of vectors (phasors), 370, 405, 407
Algebraic expressions:
  defined, 7
  literal, 7
Algebraic fractions, 135
  addition of, 146
  common errors, 142
  conversion to LCD, 145
  division of, 149
  lowest terms, 139
  mixed expressions, 143
  multiplication of, 148
  signs, 141
  subtraction of, 146
Alphabet, Greek, 541
Alternating current:
  average value of, 398
  defined, 392
  effective value of, 399
  equation of, 397
  instantaneous value of, 394
  maximum value of, 393
  phase angle, defined, 400
  RMS value of, 399
Alternating-current circuits:
  C, 419
  L, 416
  L and C: in parallel, 440
    in series, 429
Alternating-current circuits (*Cont.*):
  parallel resonance, 451
  R: in parallel, 439
    in series, 413
  R and C: in parallel, 444
    in series, 426
  R and L: in parallel, 443
    in series, 424
  R, L, and C: in parallel, 445
    in series, 429
  resistance in series, 413
  resistances in parallel, 439
  series-parallel, 449
  series resonance, 436
Alternating electromotive force (EMF):
  average value of, 398
  defined, 393
  effective value of, 399
  equation of, 397
  generation of, 392
  instantaneous value of, 394
  maximum value of, 393
  representation of, 394
  vector (phasor) representation for, 394
Ampere:
  alternating current, 392–393
  defined, 69
Amplitude, 387
Amplitude factor, 387
Angle(s):
  acute, 312
  complementary, 313
  defined, 312
  of depression, 357
  difference laws, 367
  of elevation, 357
  generation of, 313
  gons, 318
  grads, 318
  measurement, 314
  natural system, 316
  negative, 313
  notation for, 312
  obtuse, 313
  phase, 389, 400
    lag, 389
    lead, 389
Angle(s) (*Cont.*):
  π measure, 316
  positive, 313
  radian measure, 69, 316
  right, 312
  sexagesimal system, 314
  standard position, 313
  supplementary, 313
  sum laws, 366
Angular motion, 383
Angular velocity, 384
Antenna gain, 504
Antilogarithms, 471
Apparent power, 430
Approximations, 61
Area of triangle, 359
Atto, 72
Average value:
  of alternating current, 398–399
  of alternating EMF, 398–399
Axioms, 40
Ayrton shunt, 184

B

Base:
  change of, 459, 481
  defined, 459
Base units of SI metric system, 68
Battery cells:
  in parallel, 249
  in series, 249
Bel, 498
Bias resistor, 101
Binary numbers:
  addition of, 520
  conversion to octal numbers, 517
  defined, 515
  division of, 523
  multiplication of, 523
  subtraction of, 521, 522
Binomial, defined, 11
Boole, George, 525
Boolean algebra, 525
  addition, 528
  conjunction, 526
  disjunction, 526

Boolean algebra (*Cont.*):
multiplication, 528
tautologies, 528
Braces, 20
Brackets, 20
Bridge:
Murray loop, 199
Wheatstone, 199
Bridge circuits:
-T transform, 455
-Y transform, 455

C

Calculators, 54
Candela, 69
Capacitance, 70
of capacitors: in parallel, 440
in series, 421
defined, 42
effects of, in ac circuits, 412
of equal capacitors in series, 421
of transmission lines, 508
Capacitive reactance, defined, 420
Capacitors:
in parallel, 440
in series, 421
Cartesian coordinates, 207
Cathode biasing, 101
Cells:
in parallel, 249
in series, 249
Characteristic of logarithm, 465
Characteristic impedances of transmission lines, 510
Charge, 69
Coefficient:
defined, 7
fractional, 153
temperature, 111
Cofactor, defined, 241
Common factor, 120, 135
Complex fractions, 151
Complex numbers:
addition of, 265
defined, 265
division of, 266
multiplication of, 266
polar form, 404
rectangular form, 404
Complex numbers (*Cont.*):
subtraction of, 265
Components of a vector (phasor), 373
Compound circuits, 177
Compound fractions, 151
Conditional equation, defined, 39
Conductance, 70
Conductor, resistance of, 107
Conjugate, defined, 260
Conversion factors, 74
Coordinate notation, 206
Coordinates, rectangular, defined, 207
Cosecant, defined, 326
Cosine, defined, 325
Cosine curve, graph of, 382
Cosine law, 363
Cotangent, defined, 326
Coulomb, defined, 69
Cube of a monomial, 117
Cube root:
defined, 119
of a monomial, 119
Current:
defined, 69
direction of flow, 87, 292
Current decay in a capactive circuit, 488
Current dividers, 198
Current growth in an inductive circuit, 488
Current ratios, decibels, 502
Cycle:
alternating-current, 396
defined, 396
periodic function, 387–388

D

Decibel(s):
current ratios, 502
defined, 498
power ratios, 498
reference levels, 500
voltage ratios, 502
Decimal multipliers:
prefixes, 73
table of, 72
Degree:
of angle, defined, 313
Degree (*Cont.*):
Celsius, 71
of a monomial, 135
of a polynomial, 135
Delta circuit, 303, 454
Denominator:
defined, 138
lowest common, defined, 145
Dependent variable, 209
Determinants:
cofactors, 241
evaluation of, 235
minors, 240
properties of, 243
second-order, 234
third-order, 238
Difference, 18
Difference identities, 366
Direction of current flow, 87, 292
Discriminant, 284
Distance:
horizontal, 357
vertical, 357
Dividend, defined, 32
Divider circuits, 192
Division:
of algebraic fractions, 149
of binary numbers, 523
of complex numbers, 266, 406, 408
defined, 32
with exponents, 33
of fractions, 149
of monomials, 34
of polynomials, 35, 36
of positive and negative numbers, 32
of powers of ten, 60
with radicals, 260
using logarithms, 475
of vectors (phasors), 406
polar form, 408–409
by zero, 138
Divisor, defined, 32

E

Effective value:
of alternating current, 399
of alternating voltage, 399
Efficiency, defined, 91
Electrical units, 69–71

EMF (electromotive force):
alternating (*see* Alternating electromotive force)
of a battery, 247
defined, 87
of a generator, 392–393
Emitter biasing, 103
Energy, 70
English FPS (foot-pound-second) system, 67, 68
Equality sign, defined, 39
Equation(s):
of alternating current, 397
of alternating EMF, 397
axioms, 40
cancelling terms, 42
changing signs, 42
complex numbers, 431
conditional, 39
containing decimals, 155
containing radicals, 267
defined, 39
derived from graph, 215
exponential, 485
formula, 46
fractional: defined, 153
solution of, 156
with fractional coefficients, 153
identical, 39
inconsistent, defined, 237
indeterminate, 210
linear: defined, 210
methods of plotting, 211
literal, 46, 161
logarithmic, 484
of periodic functions, 380
for phase relations, 389
quadratic: affected, 271
complete, solution of, 272
defined, 271
discriminant of, 284
pure, solution of, 271
solution of: by completing the square, 274
by factoring, 272
by formula, 276
by graphical means, 283
pure, 271
testing solutions, 277
standard form, 275
radical, 267
Equation(s) (*Cont.*):
second-degree, 271
simple, solution of, 41
simultaneous: fractional form, 225
linear, defined, 220
literal, 161
solution of: by addition and subtraction, 221
by comparison, 224
by determinants, 236
by graphs, 220
by substitution, 223
three unknowns, 229
trigonometric, 360
Equivalent circuits:
Norton, 307
star and delta, 303
Thevenin, 307
Evaluate, 8
Exaunits, 72
Exponent(s):
defined, 10
fractional, 255
fundamental laws of, 26, 33, 254
negative, 33, 254
zero, 33, 254
Exponential equations, 485
Expression:
algebraic, 7
mixed, 143

F

Factor(s):
amplitude, 387
defined, 7
highest common, 135
prime, defined, 123
Factoring:
difference of two squares, 125
radicands, 256
trinomials, 126, 129
Fall, 357
Farad, defined, 70
Femtounits, 72
Force, 71
Formula, defined, 46
FPS (foot-pound-second) system, 67, 68
Fractional equation, defined, 153
Fractional exponents, 255
Fractions:
addition of, algebraic, 146
changing to mixed expressions, 143
common errors, 142
complex, 151
compound, 151
defined, 138
division of, algebraic, 149
equivalent, 139
multiplication of, algebraic, 148
power of, 64
reduction of: to LCD, 145
to lowest terms, 139
to mixed expressions, 143
signs of, 141
subtraction of, algebraic, 146
Frequency, 70
defined, 72, 387, 396
resonant: parallel ac circuits, 451
series ac circuits, 437
Functions:
defined, 322
periodic, defined, 383
trigonometric (*see* Trigonometric functions of angles)

G

Gain of antennas, 504
General number, defined, 6
Gigaunits, 72
Gons, 318
Grade, 357
Grads, 318
Graphs:
coordinate notation, 206
of cosine curve, 382
defined, 202
derivation of linear equations, 215
of linear equations, 208
of logarithmic equations, 484
plotting, 202, 211
of quadratic equations, 278
of simultaneous linear equations, 220
of sine curve, 380

Graphs (*Cont.*):
  solving problems by, 204
  of tangent curve, 382
Greek alphabet, 541

H

Henry, defined, 70
Hertz, defined, 70
Highest common factor, 135
Horizontal distance, 357
Hypotenuse, 320

I

Identical equation, defined, 39
Identity:
  defined, 360
  difference, 366
  Pythagorean, 360
  simple, 360
  sum, 366
Imaginary numbers:
  defined, 264
  representation of, 265
Impedance(s):
  defined, 425
  in delta, 454
  in parallel, 447
  in π, 454
  polar form, 432
  rectangular form, 432
  of resonant parallel circuits, 452
  of resonant series circuits, 437
  in series-parallel, 449
  in T, 454
  in Y, 454
Independent variable, 209
Index of a root, 10
Inductance, 70
  and capacitance in parallel, 440
  defined, 412
  effect of, in ac circuits, 412
  of transmission lines, 508
Inductive reactance, defined, 417
Instantaneous value:
  alternating current, 395
  alternating EMF, 394
Interpolation, accuracy, 337, 341
Inverse logarithmic functions, 471
Inverse trigonometric functions, 339

J

j, operator, 261
Joule, 70

K

Kelvin, 69
Kilogram, 69
Kilounits, 72
Kilowatthours, defined, 91
Kirchhoff, G. R., 292
Kirchhoff's laws:
  bridge circuits, 303
  outline for solving networks, 300
  parallel circuits, 296, 298
  series circuits, 293
  statement of, 292

L

Lag, 389
Lead, 389
Length, 68
Levels, reference, decibels, 500
Like terms:
  addition, 18
  defined, 10
  subtraction, 18
Linear equations, graphs, 208
Linear velocity, 383
Literal equations, 46, 161
Literal numbers, defined, 6
Logarithm(s):
  in calculations, 479
  change of base, 481
  characteristics, 465
    rules for, 466
  common system, 462
  defined, 459
  division by, 475
  extracting roots by, 478
  fractional exponents, 479
  mantissas of, 465, 466
  multiplication by, 474
  natural system, 463, 482
Logarithm(s) (*Cont.*):
  notation, 460
  of a power, 462
  of a product, 461
  of a quotient, 461
  raising to a power by, 477
  of a root, 462
  tables of, 468
    common, 548
    natural, 558
  use of, with negative numbers, 474
Logarithmic equations, 484
Losses, 91
Lowest common denominator, defined, 145
Lowest common multiple, 137
Lowest terms, 139
Lumen, 70
Lux, 70

M

Magnetic units, 70
Mantissas of logarithms, 465, 466
Mass, 69
Mathematical symbols, 538
Maxima and minima, 284
Maximum power, 285
Measurement systems, 67
  conversions, 71, 75
  English, 67
  SI metric, 67, 71
Megaunits, 72
Meter:
  current: multirange, 182
    sensitivity, 181
    shunting methods, 184
  length, 68
  movement, 181
  resistance, 188
  volt, 186
Metric system, 4, 67, 68, 71
Microhm-meter, 109
Microunits, 72
Mil, 109
Milliunit, 72
Minor, determinants, 240
Minuend, defined, 18
MKS (meter-kilogram-second) system, 67

Mole, 69
Monomial:
  defined, 11
  degree of, 135
Multimeters, 190
Multinomial, defined, 11
Multiple:
  defined, 137
  lowest common, 137
Multiplication:
  with algebraic fractions, 148
  with binary numbers, 523
  with complex numbers, 266, 406, 408
  defined, 24
  exponents, 26
  with fractions, 148
  graphical representation of, 29
  of monomials, 28
  of polynomials, 28, 30
  of positive and negative numbers, 24
  with powers of ten, 59
  with radicals, 260
  on slide rule, 493
  using logarithms, 474
  with vectors (phasors), 406, 408
Multiplication sign, defined, 7
Murray loop, 199

N

Nanounits, 72
Napier, John, 459
Natural logarithms, 482
Negative exponent, 33, 254
Negative numbers, defined, 13
Negative power, 424
Newton, 71
Norton's theorem, 308
Notation:
  for angles, 312
  for coordinates, 206
Numbers:
  complex: addition, 265, 405, 407
    defined, 265
    division, 266, 406, 408
    multiplication, 266, 406, 408
    subtraction, 265, 405, 407
Numbers (*Cont.*):
  general, 6
  imaginary, defined, 264
  literal, defined, 7
  negative: defined, 13
    need for, 13, 14
    square root of, 264
  positive and negative, representation of, 26
  prime, defined, 117
  real, 264
  rounded, 57
  systems: binary, 515
    octal, 517
  vector (phasor), 431
Numerator, defined, 138

O

Ohm, defined, 70
Ohmmeter, 188
Ohm's law:
  for compound circuit, 177
  for dc parallel circuits, 171
  for dc series circuits, 87
Operator j, defined, 261
Origin, 206

P

Parabola, 278
Parallel resonance, 451
Paralleled resistances, 171, 439
Parentheses, 20
Pascal, 71
Perfect square, trinomial, 122
Period, 388
Periodic function, defined, 380
Periodicity, 383
Petaunits, 72
Phase, 388, 400, 404
Phase relations, equations for, 389
Phasor algebra, 404
Phasors [*see* Vectors (phasors)]
π circuits, 303
  T equivalents, 303
Picounits, 72
Plotting methods, 202, 211
Points on graph, 202
Polar form, 404
Poles, 396
Polynomial(s):
  addition of, 19
  with common monomial factor, 120
  defined, 11
  degree of, 135
  division of, 36
  multiplication of, 28, 30
  subtraction of, 19
Positive power, 58, 424
Potential, 69
Power, 71
  in ac capacitive circuits, 423
  in ac circuits: average, 414
    instantaneous, 414
  in ac inductive circuits, 423
  in ac resistive circuits, 414
  actual, 430
  apparent, 430
  defined, 90
  maximum, delivered to load, 285
  negative, 424
  positive, 58, 424
  true, 430
Power factor, 430
Power ratios, decibels, 498
Powers of phasors, 410
Powers of ten:
  addition with, 66
  defined, 57
  division with, 60
  multiplication with, 59
  power of a fraction, 64
  power of a power, 63
  power of a product, 64
  reciprocals, 61
  root of a power, 64
  subtraction with, 66
Preferred values, 497
Prime factors of an expression, 123
Prime numbers, 117
Primes, 7
Product:
  of any two binomials, 127
  of binomials having common term, 125
  complex numbers: polar form, 408
    rectangular form, 406
  defined, 7

Product (*Cont.*):
sum and difference of two terms, 124
Proportion, 50
Proportionality, 51
Protractor, 314
Pythagoras' theorem, 360
Pythagorean identities, 360

Q

Q, defined, 437
Quadratic equations (*see* Equations, quadratic)
Quadratic formula, 276
Quadratic surd, defined, 260
Quotient, defined, 32

R

Radian, 69
defined, 316
Radical(s):
addition and subtraction of, 259
containing fractions, 257
division of, 260
in equations, 267
multiplication of, 260
removing factors from, 256
similar, 259
simplification of, 256
Radical sign, defined, 10
Radicand, defined, 256
Radix, 518
Ratio, defined, 49
Rationalizing:
complex number, 266
denominator containing radical, 260
Ratios:
current, decibels, 502
power, decibels, 498
trigonometric, 322
voltage, decibels, 502
Reactance:
capacitive, defined, 420
inductive, defined, 417
Real numbers, defined, 264
Reciprocal, defined, 62
Rectangular coordinates, 206
Rectangular form, 404
Reference levels, decibels, 500
Remainder, defined, 18
Resistance, 70
and capacitance: in parallel, 444
in series, 426
of conductor, 87, 107
and inductance: in parallel, 443
in series, 424
specific, 109
to temperature effects, 111
Resistances:
in compound circuits, 177
in Δ, 303
in parallel, 171, 439
in π, 303
in series, 96, 413
in T, 303
in Y, 303
Resistors, bias, 101
Resonance:
parallel, 451
series, 436
Resonant frequency:
parallel circuits, 451
series circuits, 437
Right triangle:
defined, 320
facts concerning, 320
procedures for solution of, 350
Rise, 357
Root:
cube, defined, 119
of an equation, 271
of a phasor, 410
of a power, 64
square, defined, 118
Root-mean-square (rms) value, 399
Rounded numbers, 57
Run, 357

S

Scalar, defined, 369
Secant, defined, 326
Second, defined, 69
Series resonance, 436
Sexagesimal system, 314
Shunt:
Ayrton, 184
Shunt (*Cont.*):
connection methods, 184
universal, 184
SI (Système International des Unités) metric system, 67, 68
base units, 68
Siemens, 70
Significant figures, 56, 80
Signs:
of fractions, 141
of grouping, 20
insertion of: preceded by minus sign, 22
preceded by plus sign, 22
of operation, order of, 7
Similar terms, 10
Similar triangles, defined, 319
Simultaneous equations, 220
addition and subtraction of, 221
comparison of, 224
determinants, 236
fractional form, 225
graphical solution of, 220
substitution of, 223
three unknowns, 229
Sine(s):
defined, 325
law of, 362
Sine curve, graph of, 380
Sinusoid, defined, 386
Slide rules, 492
accuracy of, 55
division with, 493
multiplication with, 493
squares and roots with, 495
types of, 53
vector (phăsor) calculations, 431
Slope, 357
Specific resistance, 109
Square:
of a binomial, 121
difference of two terms, 121
of a monomial, 117
sum of two terms, 121
Square root:
defined, 118
of a monomial, 118
of negative numbers, 264
of a trinomial, 122
of a trinomial square, 123

Star circuits, 303
Steradian, 69
Subscripts, 7
Subtraction:
of algebraic fractions, 146
of binary numbers, 521
of complex numbers, 265, 405, 407
defined, 18
of fractions, 146
of like terms, 18
of logarithms, 472
of polynomials, 19
of positive and negative numbers, 17
of powers of ten, 66
of radicals, 259
of unlike terms, 18
of vectors (phasors), 405, 407
Subtrahend, defined, 18
Sum identities, 366
Surd:
binomial quadratic, 260
defined, 260
division with, 260
quadratic, 260
rationalization, 261
Superscripts, 7
Symbols, mathematical, 538

T

T circuits, π equivalent, 303
Tangent, defined, 325
Tangent curve, graph of, 382
Tautologies, 528
Temperature, 69, 71
Temperature effects, resistance to, 111
Teraunits, 72
Term(s):
defined, 10
like, 10, 18
similar, 10
transposition of, 42
unlike, 10, 18
Tesla, defined, 70
Thevenin's theorem, 307
Time, 69
Time constant:
capacitive circuit, 488
inductive circuit, 487
Transmission lines:
capacitance of, 508
characteristic impedance of, 510
defined, 507
inductance of, 508
surge impedance of, 510
types of, 507
Triangles:
area of, 359
right: defined, 320
solutions of, 350
similar, defined, 319
Trigonometric formulas, 367
Trigonometric functions of angles:
of any angle, 329
of complementary angles, 327
computation of, 332
defined, 322, 324, 325
in first quadrant, 333, 337
in fourth quadrant, 334, 343
greater than 90°, 329, 342
greater than 360°, 344
inverse, 339, 346
of negative angles, 344
of 0°, 333
ranges of, 333
representation of, 334
in second quadrant, 334, 342
signs of, 330
table of, 552
in third quadrant, 334, 343
Trigonometric identities, 360
Trigonometric ratios, 322
Trinomial, defined, 11
True power, 430
Truth tables, 529

U

Uniform circular motion, 385
Units in calculations, 80
Universal shunt, 184
Unlike terms, 10, 18

V

Variable:
defined, 209
dependent, 209
Variable (*Cont.*):
independent, 209
Variation, 51
Varley loop, 201
Vector notation, 370
Vector representation, 369
Vectors (phasors):
addition of, 370, 376, 405, 407
components of, 373
nonrectangular, 378
rectangular, 376
defined, 369
division with, 406, 408
exponential form, 409
multiplication with, 406, 408
powers of, 410
roots of, 410
subtraction of, 405, 407
Velocity:
angular, 384
linear, 383
Vertical distance, 357
Vinculum, 20
Volt, defined, 69
Voltage dividers, 192
with loads, 193
Voltage ratios, decibels, 502
Voltmeters, 186
loading effects of, 188
multirange, 188
sensitivity of, 187

W

Watt, defined, 71, 90
Watthour, defined, 71, 91
Weber, defined, 70
Wheatstone bridge, 199
Wire measure, 113
Wire table, 542

X

x axis, 206

Y

y axis, 206
Y circuits, equivalent, 303, 454

Z

Zero of an equation, 284
Zero exponent, 33, 254

laws of logarithms

$a^x = N \qquad \log_a N = x \qquad N = \text{antilog}_a\, x$

$\log_a a^b = b$

$\log_a (M \cdot N) = \log_a M + \log_a N$

$\log_a \dfrac{M}{N} = \log_a M - \log_a N$

$\log_a M^n = n \log_a M$

$\log_a M^{1/n} = \dfrac{\log_a M}{n}$

$\text{colog}_a N = \log_a \dfrac{1}{N}$

$\log_b a = \dfrac{1}{\log_a b}$

$\log_b N = \log_a N \cdot \log_b a = \dfrac{\log_a N}{\log_a b}$

laws of exponents

$a^m \cdot a^n = a^{m+n}$

$a^m \div a^n = a^{m-n}$

$(a^m)^n = a^{mn}$

$a^m = \dfrac{1}{a^{-m}}$

$(ab)^m = a^m b^m$

$\left(\dfrac{a}{b}\right)^m = \dfrac{a^m}{b^m}\ (b \neq 0)$

$a^0 = 1$

$a^{m/n} = \sqrt[n]{a^m} = (\sqrt[n]{a}\ )^m$